BIOLOGY

BIOLOGY

Donald D. Ritchie BARNARD COLLEGE OF COLUMBIA UNIVERSITY

Robert Carola

ADDISON-WESLEY PUBLISHING COMPANY

Reading, Massachusetts · Menlo Park, California · London · Amsterdam · Don Mills, Ontario · Sydney

This book is in the Addison-Wesley Series in LIFE SCIENCE

SPONSORING EDITOR: James H. Funston

TEXT AND COVER DESIGNER: Catherine Dorin

ILLUSTRATORS: Oxford Illustrators
Dick Morton
Bob Trevor

PHOTO RESEARCHERS: Carol Palmer
Kimberly Colucci

Cover design: French or Flemish, late XV Century.
The Hunt of the Unicorn III, The Unicorn Tries to Escape—Detail.
Wool and silk with metal threads from the Chateau of Verteuil.

Metropolitan Museum, New York
From the Cloisters Collection—gift of John D. Rockefeller, Jr., 1937.

Second printing, May 1979

Preface

WE HAVE WRITTEN THIS INTRODUCTORY biology book for the biology student and the course instructor. Both share a common goal, but both approach the course from different perspectives. To be useful to a *student*, the book must contain the essential facts and provide an integrated treatment relating these facts to one another and to the student's world of experience. The book should have numerous illustrations and explain biological phenomena clearly. The book should (must) support and aid the student in truly *understanding* phenomena by giving sufficient explanation so that understanding becomes easy and meaningful. The biology *instructor* needs a book that has enough information to satisfy a diverse group of students likely to enroll in an introductory biology course. There should be enough information to satisfy the ambitious students, but not so much complexity or detail that the others are discouraged. The biology text should support instructors' innovation but not limit variations in emphasis and approach. The textbook should be a source of visual pleasure and intellectual adventure for both the student and the instructor.

In examining this textbook for possible adoption, the instructor can expect to find longer, more complete explanations for phenomena. This attempt to aid understanding by supplying adequate information is an important feature of our text and one that we believe will support better learning and better teaching. We have tried to avoid abbreviated and condensed explanations that force the student to simply memorize a set of technical terms. An introductory student faced with a condensed explanation usually has difficulty understanding a phenomenon. We believe the explanations we offer are clear and complete without being technical and dull. Toward this end, even the captions are self-contained mini-essays that attempt to present concepts fully and without repeating the text.

The various features of our text, including illustrations, essays, and portfolios, exist to increase student awareness of biological phenomena and to permit the student to see important relationships among diverse phenomena.

Within the limitations of a textbook, we believe that this book will *encourage* and support the student's awakening interest in biological reality. We hope that by becoming aware, students will *ask* more and better questions, and that by asking better questions, students will become more aware.

Because the science of biology is a lively, vigorous, developing discipline, we have referred repeatedly to the gaps in biological knowledge. We have tried to show what discoveries are waiting to be made, and to encourage some new uncommitted minds to participate in the adventure. We have tried to avoid dogmatic statements as revealed truth, and we have frequently discussed the observations or experiments that are the foundations of our knowledge. All introductory students must learn that scientific information rests on the *interpretation of evidence.* The most important question a scientist asks is always "What is the evidence?"

Students are interested in themselves. For this reason we have included discussion of human physiology and dysfunction in Part Six of the text. This discussion emphasizes the human organism but not to the exclusion of other animal forms. At no point have we made this an exclusively human-directed book. The human species is only one of several million biological species, and one of the most important discoveries a student can make is the interrelationship that exists between the extraordinary diversity of form and structure among representative samples of these different species.

The text contains sufficient information to permit selection of chapters for one-term or one-semester courses. The arrangement and grouping of these chapters permits easy variation in sequential use.

The authors would like to thank the many teachers who reviewed the manuscript at various stages of writing and revision: Robert Shapley (Rockefeller University), Craig L. Himes (Bloomsburg State College), Jerry Button (Portland Community College), R. H. Barth, Jr. (University of Texas at Austin), Fred Landa (Virginia Commonwealth University), John R. Jungck (Clarkson University), Sol A. Karlin (Los Angeles Pierce College), Benjamin Bowman (Macomb County Community College), Alice Holtz (Boston University), Robert B. Mitchell (The Pennsylvania State University), Frank L. Bonham (San Diego Mesa College), Bertwell K. Whitten (Michigan Technological University), Gail Patt (Boston University), Don Patt (Boston University), Susan D. Waaland (University of Washington), John D. Cunningham (Keene State College), Norman S. Kerr (University of Minnesota), Theodore O'Tanyi (Widener College), Georgia E. Lesh-Laurie (The Cleveland State University), Richard K. Boohar (The University of Nebraska—Lincoln), and Alice Steinmuller. The authors assume responsibility for any errors that persist.

Besides the academicians listed above, two members of the staff of Addison-Wesley deserve a special note of gratitude. If Catherine Dorin had merely designed the book, she would have earned our praise and deep appreciation, but she did much more. Page by page, even line by line, she designed and redesigned the text, cover, and illustrations until even she was satisfied. And to James Funston, our editor and colleague, we give public thanks for the private job of finding new ways to write a biology book that will satisfy and help both teacher and student. We applaud his unceasing desire to keep the standards high.

A final word to teachers and students. Please let us know by writing to us via Addison-Wesley if you have any suggestions for improving future editions. This book was written for you, and we would appreciate your help in making it as useful as it can be.

New York   D. D. R.
December 1978   R. C.

Abridged Contents

Contents

28 Animal Behavior I: Perception and Communication 552

29 Animal Behavior II: Animal Societies 572

30 The Individual and the Environment 589

31 Ecosystems, Communities, and Biomes 602

Epilogue: Some Biological Problems for the Future 626

Prologue: What Is Life?

1

Seedling of arctic lupine grown from seed that had been frozen in a ground squirrel burrow in Alaskan tundra for some 10,000 years.

The people of Utopia think that exploring the secrets of nature with the help of science is not only an admirable pleasure but a means of getting the most credit from the Creator. They think that He, like any artist, displayed the notable machine of Earth so that it could be studied by people, the only beings He made able to do such a great thing. It seems to them that He prefers someone who is curious and excited about His work to one who, like a mindless brute, is stupid and unmoved by such a marvelous spectacle.

Thomas More: *Utopia*, 1516.   Trans. D.D.R.

A DEFINITION OF BIOLOGY COMES EASILY: IT IS the study of life. But humans still have no satisfactory definition of *life*. We think we know intuitively whether a thing is alive or not, but intuition can be notoriously misleading. The most obvious way to find if an unknown object is alive is to poke it and see if it moves, that is, to find if it is responsive. But many things that we would never think of as living are responsive. An automobile responds if we move the proper controls, and a violin responds if a bow is drawn across the strings. Indeed, musicians commonly refer to instruments as "responsive." And movement is not restricted to animate beings. Clouds move and waves move and the heavenly bodies shift position, so that primitive peoples have ascribed a kind of life to them, but in our society we regard clouds and waves and heavenly bodies as nonliving.

Another test for life is to observe breathing, but plants do not breathe in the way that higher animals do, and an almost-closed glass jar does "breathe" with changes in temperature. After the chemical basis of breathing was discovered, the term *cellular respiration* was substituted for breathing, and indeed, cellular respiration is one of the more dependable criteria of life. But when a lupine seed sprouts after being frozen for 10,000 years in arctic ice, it is obvious that precious little respiration was going on during that time (Figure 1). Yet the seed is alive. Conversely, all

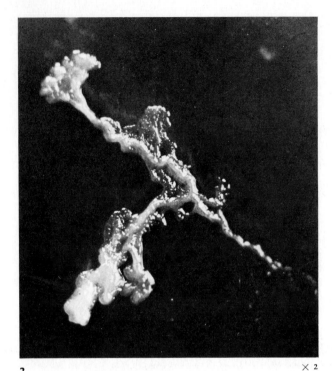

2

An illustration of the difficulty of using only one criterion for the determination of the presence of life. A plasmodium of a slime mold, having unrestricted form and size, lacks one of the standards of life, but it moves, feeds, respires, and reproduces, and so it is definitely living.

×2

the enzymatic reactions of respiration can be duplicated in test tubes, but we would never say that the contents of the test tube are alive. And viruses, which are a special problem for those who try to define life, have no respiratory activity of their own.

Growth is a common criterion of life, especially when by "growth" one means increase in size or complexity. Still, river deltas and volcanoes and crystals, all plainly inanimate, grow too.

Definite size and form are characteristic of most living things, but exceptions can be found. Some of the multinucleate protozoa and the plasmodial slime molds (Figure 2) are infinitely variable in size and form, and they are clearly alive, but a quartz crystal, for example, is more precisely formed than any living cell and is just as clearly not alive (Figure 3).

One of the surest criteria of life is reproductive ability. As some wag has said, you need not fear that some morning you will find your desk crawling with little typewriters. But it is theoretically possible—indeed, it would be realistically possible if anyone wished to spend the money on such a project—to construct a machine that would make a machine like itself, and so on. Such a self-reproducing machine would not be alive any more than a bottle-making machine is.

We are thus forced to the conclusion that life is characterized by a combination of features, including movement, responsiveness, a special kind of growth, respiratory activity, regularity of form and size, and reproductive ability, but that not all of these features are demonstrable all the time in any one organism. We are also forced to recognize that there are vague borders between life and nonlife. A virus particle floating in the air shows none of the usual characteristics of a living thing, but if it makes contact with a suitable host cell, it can inject its nuclear material into that cell (movement), and the host cell may then make copies of the virus (reproduction?). Is a virus alive? The answer depends on where the human observer subjectively decides to draw his defining line.

We should be aware of some of these uncertainties in biology, recognizing that we cannot in our present state of ignorance make clean separations in our classifications, especially since nature does not seem to have done so. But uncertainties and puzzles need not keep us from learning what we can about what we call living things. There are legal, medical, and philosophical problems involved in the question: Is a freshly decapitated man alive? A freshly decapitated man hasn't a chance of ever returning to normal activity, but that is exactly why the question is inter-

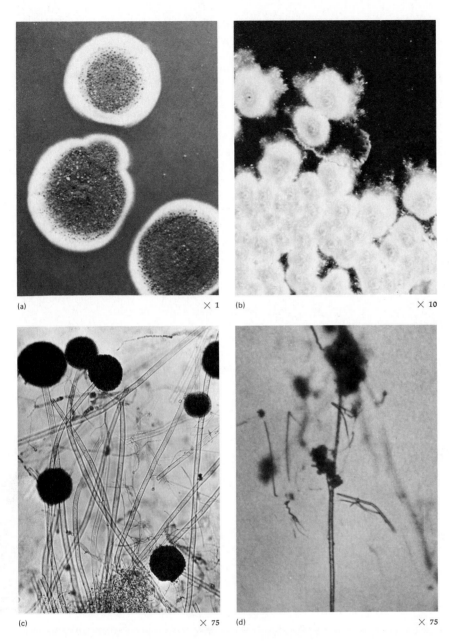

(a) × 1

(b) × 10

(c) × 75

(d) × 75

3
Deceptiveness of appearance in determining what is living and what is nonliving. (a) Colonies of the mold Penicillium growing on a culture medium. (b) Clusters of crystals "growing" on a piece of paper in a solution containing inorganic materials. (c) Microscopic view of threads and spores of the common mold Aspergillus. (d) Similarly magnified view of the crystals shown in (b). Regularity of form and definite size are apparent in both the living and nonliving examples, but only the molds meet enough of the criteria to be acceptable as living. The crystals were so moldlike in appearance and in growth rate and in microscopic structure that the paper manufacturers called in a mold specialist to help "kill" the "infection."

esting. His liver doesn't yet know that the head is gone, and for a few minutes it will keep on doing its old job. Even his brain cells work for a bit. There is an old story that during the terror of the French Revolution, some of the aristocrats waiting for the guillotine kept up their bravado by telling those next in line to watch and see how long the severed head would keep blinking its eyes—the answer is said to be about 5 to 10 seconds.

The question of the life status of a freshly decapitated person is not answerable at present. Neverthe-

less, we can recognize that a cat slithering down an alley and a buttercup blooming in a meadow are definitely alive, and we can set ourselves to studying their structures and activities and relations to the world without letting the borderline cases divert us from the task.

The definition of life as a predicament preceding death is facile but unproductive, and most biologists choose to ignore it. They prefer to stick to the job of learning what they can, regardless of the fact that it is easier to study a thing than to define it.

THE MECHANICS OF LIFE

PART ONE

1

The Chemistry of Life

Scanning electron micrograph of ten-thousand-year-old sediments from the Lake Champlain region of New York. These are the glassy skeletons of one-celled algae.

THIS CHAPTER WILL INTRODUCE YOU TO SOME of the characteristics and habits of atoms and molecules. It will explain how single atoms are structured and describe how atoms combine with other atoms to form molecules. Much of the fascination, excitement, and drama of biology in the twentieth century has surrounded the discoveries of chemistry and the application of this knowledge to the molecules that make up living systems. In some measure we want you to participate in and appreciate this excitement and drama. That is why you are asked to read this chapter and understand a few basic chemical principles. This chapter does not try to make chemists out of biology students. However, it does try to show the importance of chemical events that lie behind the familiar phenomena that characterize the living world to our human senses.

An individual's consciousness of the living world is observation. We see the leaves turn colors in the autumn and fall to the ground. We watch a baby grow and develop. If these observations are followed by careful descriptions of the phenomena, we have the beginning of science. But such descriptions do not explain the phenomena. A good description is a basis for asking a question. What causes leaves to turn color? Why do leaves fall? What causes a baby to grow and change? These questions spring quite naturally out of observations and descriptions. To answer them, biologists are increasingly drawn to chemical explanations.

In writing this chapter we have tried to point out the relevance of many of the chemical principles to phenomena discussed later in the book. However, you will want to return to portions of this chapter as you read others, and this is one reason for placing it first in the book.

There are many details in this chapter that will be relevant to discussions in later chapters. However, there are seven major points:

1. The relation between atomic structure and the capacity of atoms to form bonds with other atoms.

2. The three basic types of bonds and their chemical characteristics.

3. The nature of the chemical reaction and changes in the bonding.

4. The major classes of molecules found in living organisms.

5. The types of chemical bonds and their contribution to the structure and properties of these molecules.

6. The properties of water and how they relate to the roles water plays in living organisms.

7. The nature and role of enzymes in living cells.

Primitive humans looked at things about their own size, downward to small insects and upward to hills and clouds. Microbes and molecules were hopelessly beyond their experience, but because heavenly bodies could be seen, they were thought to be almost within reach. Only with the coming of a scientific attitude, with methods of verifying facts, did humans begin to find rational explanations for the observed phenomena of nature, and to increase their understanding of how living things work.

One of the lessons learned from the scientific study of organisms is that there are levels of organization, from subatomic particles through atoms and molecules to cells, tissues, organs, whole organisms, societies, ecosystems, solar systems, and galaxies. Each level of organization must be studied and related to the other levels of organization.

One point must be kept in mind throughout the consideration of all biological study: The whole is greater than the sum of its parts. The necessary materials for a molecule of the human blood pigment, hemoglobin, are present in a rusty nail and a breath of air, but atoms of iron, hydrogen, oxygen, carbon, and nitrogen cannot function as hemoglobin does unless they are precisely organized into the proper molecule.

In this last quarter of the twentieth century, the study of biology has reached a point at which we cannot understand much of what plants and animals do unless we know something of their chemistry. In the present chapter, we give the "basic vocabulary" of biological chemistry, which is analogous to the basic vocabulary needed to carry on a simple conversation in a foreign language.

CHEMICAL ELEMENTS AND THE STRUCTURE OF ATOMS

Matter, including biological matter, is composed of _elements_, substances which consist of only one kind of atom. Of about 103 known elements, living organisms use fewer than three dozen, and of these, only four are used in quantity. Oxygen accounts for about 65 percent, carbon about 18 percent, hydrogen about 10 percent, and nitrogen about 3 percent. The remaining 4 percent accounts for all the other elements incorporated in living material.

All matter consists of atoms, and we shall see shortly that atoms consist mostly of space, even the atoms of the hardest surface you can think of. The word "atom" was proposed at a time when atoms were thought to be the final indivisible particles of matter. The word is Greek and can be roughly translated as "uncuttable." Since the beginning of the twentieth century, however, we have known that atoms are made up of a number of particles. The best known is the atomic _nucleus_ (meaning "kernel"), with its electrically charged _protons_ and uncharged _neutrons_. Atoms also contain _electrons_, which orbit around the nucleus (Figure 1.1). Atoms are frequently compared to small solar systems, with the nucleus like the sun at the center and electrons orbiting like the planets. The nucleus contains most of the mass of an atom, since a single proton has about 1800 times the mass of an electron. The comparison between an atom and the solar system is imperfect because, unlike planets, electrons can shift orbits.

The chemical and physical characteristics (properties) of an atom are determined by the number of protons and neutrons in its nucleus and by the number of electrons surrounding the nucleus. For example, the smallest and lightest element is hydrogen, with one proton, no neutrons, and a single electron in orbit around its nucleus. The heaviest is uranium, with 92 protons, 146 neutrons, and 92 electrons. The biologically useful element oxygen has eight protons,

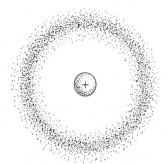

1.1

Schematic representation of an atom of hydrogen, which has one proton in its nucleus. The fuzzy halo is a way of showing that the electron orbiting the nucleus cannot be accurately drawn at any one place at any one time.

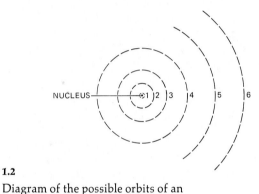

1.2

Diagram of the possible orbits of an electron around the nucleus of a hydrogen atom. In its resting state, the electron remains at the first level, but if sufficient energy is pumped into the atom, the electron may be raised to higher energy levels, that is, pushed into orbits at greater distances from the nucleus. It requires ever-greater amounts of energy to move the electron to more distant orbits, known as "shells." An electron may jump from one energy level to another, but it does not establish an orbit at distances between the possible orbits. Just as energy is required to raise an electron to a more distant orbit, energy is emitted if an electron falls to a lower level.

eight neutrons, and eight electrons. Carbon has six of each of those particles.

Atoms may be described by the numbers given to them. First, every element has an *atomic number* that tells the number of protons in its nucleus. The atomic number of carbon is 6. An atom also has an *atomic weight,* which is an expression of the relative weights of the elements, compared specifically with C^{12}. The atomic weight of carbon is about 12. (See the essay on isotopes.) The number of protons and electrons plays important roles in determining the interactions that can take place between atoms. The discussion that follows will show how the specific distribution of electrons determines the atom's capacity to combine with other atoms.

ENERGY-LEVEL SHELLS

The number of electrons and protons in any given atom determines its chemical activity, but the way in which the electrons are distributed is also important. Electrons go around an atomic nucleus somewhat as the moon or a satellite goes around the earth. Such simple examples, however, describe only such a small atom as hydrogen. In heavier elements, with more nuclear particles and more electrons, the electrons are distributed as though they were in layers, known as *energy-level shells* (Figure 1.2). There are no more than two electrons in the shell nearest the nucleus, as many as eight in the second, and larger numbers in still more distant shells. As many as seven shells have been identified in the largest atoms.

For biologists, there are two critical points about energy-level shells. First, the shell an electron occupies is a function of the electron's energy. Raising an electron to a shell more distant from the nucleus requires energy, and dropping one to a lower shell releases energy. Second, the number of electrons in the outermost shell determines how one atom will combine with another atom. We shall meet a striking example of the first point in the excitation of a chlorophyll molecule in light (Chapter 4). The second point requires additional explanation.

An atom is most stable and will not combine with other atoms when its outermost energy-level shell is full. Atoms seek the most stable form. If an atom has only one electron in its outermost shell, it can lose that electron or gain an electron. If the electron is lost, the atom will then have more positively charged protons in the nucleus than electrons in orbit. In contrast, if an atom has, say, seven electrons in its outermost shell, it becomes more stable by the capture of one

ISOTOPES AS AIDS TO BIOLOGISTS

The atomic weight of carbon might seem to be exactly 12 if a carbon atom has six protons and six neutrons. But not *all* carbon atoms have six neutrons. A few, with eight neutrons, have a weight of 14, and those heavier atoms are averaged in when atomic weights of elements are calculated and published. Since most carbon atoms on earth are $_6C^{12}$, the accepted atomic weight of carbon is close to 12, that is, 12.01115.

Those carbon atoms that have six protons but *not six neutrons* are <u>isotopes</u>. Isotopes of any element are atoms with the same number of protons and electrons but different numbers of neutrons. Although the commonest carbon atom has six protons and six neutrons in its nucleus, other isotopes of carbon have seven or eight neutrons, giving them atomic weights of 13 and 14 instead of the usual 12. Since all forms of carbon have six electrons and six protons, they all have the *atomic number* 6.

Some isotopes are naturally *radioactive*; others are made so by laboratory procedures. Radioactive isotopes are unstable because their nuclei are undergoing changes called decay. *Radioisotopes* emit radiation that can be detected. For example, radiocarbon, C^{14}, emits beta particles from the carbon nucleus. Radiation can be detected and localized by several physical or photographic means, and thus an investigator can know where a compound containing a radioactive isotope has gone. Since radiocarbon, C^{14}, and ordinary carbon, C^{12}, act the same chemically, an experimental compound containing C^{14} can be fed or injected into an organism, and the path of the compound can be followed by "watching" where the radioactivity goes. Common radioactive isotopes used in biological experiments are sulfur (S^{35}), phosphorus (P^{32}), carbon (C^{14}), and tritium (H^3).

Isotopes of oxygen and carbon were essential in discovering the details of photosynthesis. Heavy oxygen was used to trace the fate of water, H_2O^{18}. Radiocarbon was used to study the buildup of carbohydrate from carbon dioxide, $C^{14}O_2$ (Chapter 4). Isotopes of nitrogen helped explain the replication of DNA (Chapter 6).

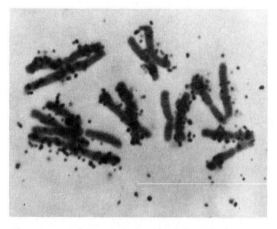

The use of radioactive tracers in cells. The gray rods are chromosome pairs, and the dark spots are places in the photographic film that were affected by tritium. Note that where one member of a chromosome pair is heavily labeled (that is, sprinkled with grains), its opposite member is not. From such observations, conclusions can be drawn concerning the methods of chromosome duplication (Chapter 8).

The presence of radioisotopes can also be used to determine the age of samples. Cosmic radiation from space generates a certain amount of C^{14}, which gets into the air as radioactive carbon dioxide, $C^{14}O_2$. The radioactive carbon dioxide is incorporated into a plant during photosynthesis (Chapter 4). When a plant dies, it ceases to accumulate C^{14}, but the radioactive carbon continues to emit beta particles as the C^{14} decays. The rate at which C^{14} decays is such that after 5730 years, half of the original radioactivity is lost. Such a time span is called the *half-life* of the isotope. After another 5730 years, half the remaining radioactivity is lost. The amount of radiocarbon detectable in samples of unknown age can be determined, and the age of the sample can be estimated. Radiocarbon dating is a common method of determining the ages of biological material, especially wood. It is useful up to about 10,000 years and has been reasonably well correlated with dates obtainable from historically dated artifacts, but the accuracy of the method lessens with older samples.

more electron. In picking up an additional electron, the atom gains a negative charge. The combining of atoms with other atoms is the result of the tendency to attain the most stable number of electrons in the outer shell.

The need to understand the action of electrons and electron shells can be shown by some reactions of the element carbon. Carbon is an especially important element because it is a part of all biological molecules and participates in all biological reactions. Because carbon has four electrons in its outermost shell, it must *gain* four electrons to fill its outer shell to capacity (Figure 1.3). In order to complete its outer shell, a carbon atom may "gain four electrons" by sharing electrons with other atoms. This sharing process results in a bond between the atoms. By sharing electrons and forming bonds, carbon atoms combine to form intricate chains and ring structures. These structures appear complicated, but they are determined by the relatively simple "combining capacities" of the constituent atoms, determined in turn by the distribution of electrons.

Because it can enter into so many chemical combinations, carbon is the basis for some of the most complex molecules on earth. Indeed, an accepted definition of *organic*, which used to be defined as "derived from living organisms," is "a chemical compound, usually complex, containing carbon." Compounds not containing carbon are therefore referred to as *inorganic*. Carbon dioxide (CO_2) is one of the few carbon-containing inorganic compounds.

The gaining, losing, and sharing of electrons in their outer shells gives atoms their capacity to combine with other atoms (Table 1.1). Combinations of atoms result in large molecules consisting of thousands of atoms. It is important to realize that large molecules have specific three-dimensional shapes because the atoms have specific relationships to each other. Large molecules made up of the same number of atoms combined in the same way will have identical three-dimensional shapes. A molecule that contains different kinds of atoms or the same kinds combined in different ways will have a different three-dimensional shape. It is important to realize at the

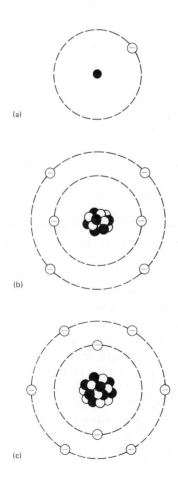

(a)

(b)

(c)

1.3

Diagrams of the structure of atoms. (a) A hydrogen atom, with one proton in its nucleus and a single electron in the first orbit. (b) A carbon atom, with six protons (black balls) and six neutrons (white balls) in its nucleus, two electrons in its first orbit, and four electrons in the second orbit. (c) An oxygen atom, with eight protons and eight neutrons in its nucleus, two electrons in its first orbit, and six electrons in its second orbit. The diagrams are drawn without regard to actual proportions because it is impossible to give a realistic picture of an atom on any piece of paper that can be manufactured. If the picture of an oxygen atom, for example, were drawn with a nucleus one centimeter in diameter, the first orbit would be about 150 meters away, the second would be about 600 meters away, and the electron would be too small to print. An atom, even of the heaviest element, is mostly "space."

TABLE 1.1

Combining Capacities for Atoms Forming Covalent Bonds

ATOM AND SYMBOL		COMBINING CAPACITY	UNSTABLE ELECTRON DISTRIBUTION IN OUTER SHELL	HOW MANY ELECTRONS CAN ATOM ACCEPT TO FORM COVALENT BONDS?	EXAMPLE (WRITTEN THREE WAYS)		
					ELECTRON DISTRIBUTION	STRUCTURAL FORMULA	EMPIRICAL FORMULA AND NAME

Hydrogen	H	1	H·	1	H:Ö:
					H
						H—O
						  |
						  H
							H_2O water

Oxygen	O	2	·Ö·	2			
Carbon	C	4	·Ċ·	4	H ·· H:C:H ·· H	H | H—C—H | H	CH_4 methane

Nitrogen	N	3	·N̈·	3	H:N̈:H
					··
					H
						H—N—H
						   |
						   H
							NH_3 ammonia

This table summarizes the principles discussed. In sharing electrons, bonds are formed. The electron distribution formula emphasizes how combining atoms gain stability in their outer shells (2 electrons for hydrogen; 8 for oxygen, carbon, and nitrogen) by sharing electrons. The *structural formula* is a simpler depiction of the molecule, where the two electrons shared are indicated by a single line connecting the symbols for the atoms. The structural formula still does not depict the three-dimensional nature of the molecule.

outset that the shapes of molecules play a great role in determining the shapes of many cells.

MOLECULES AND COMPOUNDS

Atoms may combine with each other to form a *molecule*, the smallest unit into which a substance can be divided while still retaining the chemical properties of that substance (Figure 1.4). Molecules can be made of identical atoms, such as O_2 or N_2, or of different atoms, such as H_2O or CO_2. When two or more different atoms combine, they form a *compound*, such as water (H_2O) or carbon dioxide (CO_2). A compound is a substance whose molecules are composed of more than one kind of atom.

BONDING BETWEEN ATOMS

There are several different kinds of bonds that hold atoms together. We will discuss three of the types of bonding important in living systems. Certain kinds of atoms form certain kinds of bonds when they combine with other atoms. These bonds have unique characteristics that give unique properties to the compounds so formed.

Ionic Bonds

If an atom or group of atoms loses one or more electrons, it loses part of its negative charge. It thereby becomes *positively* charged because there are now more positively charged protons in the nucleus than there are electrons surrounding that nucleus. On the other hand, if an atom or atomic group gains electrons, it becomes *negatively* charged. Such charged atoms or groups of atoms are called *ions*, and the attractive force between them results in an *ionic bond*.

To see better how the gain or loss of electrons takes place, let us consider the examples of the atoms sodium and chlorine (Figure 1.5).

A single charged atom is an ion—for example, sodium or chloride. But groups of atoms also function as ions, sometimes called complex ions. Some important examples are phosphate, $PO_4^{\equiv}$; sulphate, $SO_4^{=}$; bicarbonate, HCO_3^{-}; and acetate, $CH_3CO_2^{-}$.

Ions and the resulting ionic bonding between them have important roles to play in biological systems. The charged nature of ions is responsible for the electrical nature of the nerve impulse. Both blood and the water that bathes the cells of our bodies contain numerous ions of very great importance—Na^+, K^+, Ca^{++}. Calcium ion (Ca^{++}) is readily transported in the blood, but it is later deposited, in association with negatively charged phosphate, in a solid to form bones.

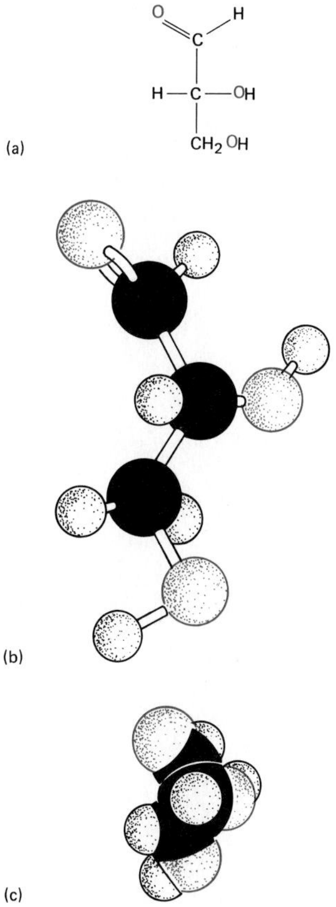

(a)

(b)

(c)

1.4

Three ways to represent the atomic structure of a molecule of the compound glyceraldehyde. (a) A structural formula, which shows with letter symbols the position and kinds and numbers of elements, flattened to fit a two-dimensional sheet of paper. (b) A ball-and-stick figure, with different sizes of balls to indicate different elements, and an attempt at showing three dimensions. (c) A "space-filling" drawing, with the outer orbits of each atom shown as parts of spheres of different sizes, shapes, and colors. Carbon is black, oxygen brown, and hydrogen white. This is the most nearly realistic rendering, but it is difficult to recognize and to remember. Students should become familiar with the conventional styles of atomic rendering.

1.5

The formation of an *ionic bond* between an atom of sodium and an atom of chlorine. An electron from the outer shell of sodium is transferred to the outer shell of the chlorine atom. The sodium atom, having lost an electron, is *oxidized* and acquires a *positive* charge. The chlorine atom, having gained an electron, is *reduced* and acquires a *negative* charge. Both atoms, neutral at the start, were not attracted to each other, but having developed opposite charges, they are now mutually attractive. Because of the electrical field generated by a positive or negative charge, a single ion will have an attraction for all other charged ions in the vicinity. (Note that the outer orbit of the sodium *atom* is no longer represented in the diagram of the sodium *ion*.)

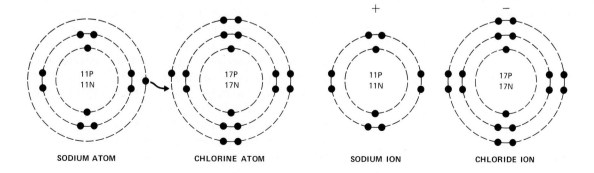

SODIUM ATOM   CHLORINE ATOM   SODIUM ION   CHLORIDE ION

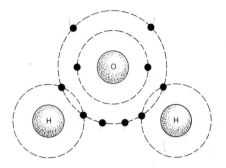

1.6

The formation of *covalent bonds* between an atom of oxygen and two atoms of hydrogen, resulting in a molecule of water. Each hydrogen atom, initially with one electron in its outer shell, shares its electron with the outer shell of oxygen, and at the same time, two electrons from the outer shell of oxygen are shared with the two hydrogen atoms. In effect, hydrogen has two electrons in its outer shell, and oxygen has eight.

1.7

Hydrogen bonding between water molecules, shown as dashed lines in color joining oxygen (large circles) of one molecule with hydrogen (small circles) of another. Weak crossbonding is thus formed within a mass of water.

Covalent Bonds

Atoms may also achieve stability by *sharing* electrons and forming a *covalent bond* (Figure 1.6). Instead of losing or gaining electrons and becoming ions, an atom may share electrons with another atom, so that the outer shells of the two different atoms become stable. For unknown reasons this sharing process holds atoms together. An example of covalent bonding is the sharing of electrons by carbon and hydrogen in a molecule of methane gas. Covalent bonding is common in the molecules making up the living cells.

Covalent bonds are strong bonds, and molecules containing covalent bonds are stable molecules. This means that it will take a great deal of energy to break them up. When we appreciate the strength of human muscles and the toughness of our skin, we are indirectly appreciating the strength of covalent bonds in molecules making up the muscle and skin. Molecules made up of "pure" covalent bonds are not electrically charged and will not be attracted to ions or to water. Fats are large molecules consisting primarily of carbon and hydrogen held together by "pure" covalent bonds. We are all familiar with how little attraction there is between oil and water. When the two are mixed together, the fat molecules of the oil remain associated with one another and not with the water.

Hydrogen Bonds

Besides ionic and covalent bondings, there is a third kind, called *hydrogen bonding* (Figure 1.7). In some ways what we call hydrogen bonding is intermediate between the "pure" ionic bond and the "pure" covalent bond, and it can be thought of as a distortion of the "pure" covalent bond. In "pure" covalent bonding, electrons are equally shared between the two atoms. An absolutely equal sharing and distribution of electrons around the atoms of a covalent bond would yield a molecule no part of which bears any positive or negative charge. However, this is not always so. Certain atoms, particularly nitrogen and oxygen, have the capacity to attract electrons toward themselves and away from atoms to which they are linked. For example, when oxygen is covalently bonded to hydrogen, the oxygen attracts electrons toward itself and away from the hydrogen atom. This unequal sharing of electrons makes the oxygen atom bear a *slight* negative charge and the hydrogen atom bear a *slight* positive charge. While maintaining its covalent bond to oxygen, the slightly charged hydrogen atom can be attracted to other negative or partly negatively charged atoms in the vicinity. When this happens, a weak electrostatic attraction is formed. It is this we call the *hydrogen bond.*

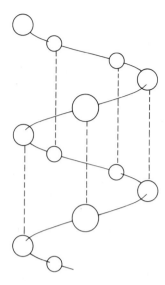

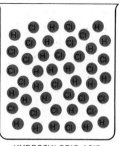 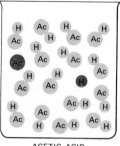

HYDROCHLORIC ACID   ACETIC ACID

1.8

Hydrogen bonding between atoms within a molecule. The circles represent amino acid subunits of a protein molecule, with the hydrogen bonds holding the subunits in place in a helix. The dashed lines in color indicate the hydrogen bonds.

Hydrogen bonds are weaker than covalent bonds, but in many molecules found in living systems there are enormous numbers of hydrogen bonds that aid in maintaining a stable three-dimensional shape. Hydrogen bonds are like sewing threads: A single stitch is not strong, but a great many stitches can hold a seam securely.

The importance of hydrogen bonds becomes apparent when one thinks of the thousands of biologically active molecules (Figure 1.8). Later in this book, we will discuss structural protein, DNA, enzymes, and hemoglobin molecules. All these very large molecules are held in proper shape by hydrogen bonds. Maintaining the shape of such molecules is necessary for them to perform their biological roles.

ACIDS AND BASES

In an everyday sense, acids are substances that taste sour and turn blue litmus paper or cabbage juice red, and bases, or alkalis, feel slimy and turn red litmus paper blue. In a chemical sense, *acids* are compounds that yield hydrogen ions (H^+), and *bases* are compounds that accept hydrogen ions. A solution that contains equal numbers of H^+ and OH^- ions is neutral. An acid solution contains more H^+ ions, and a basic solution contains more OH^- ions.

1.9

Diagram of the dissociation of two compounds: hydrochloric acid and acetic acid. Undissociated molecules are gray; those dissociated into ions are in color. All the particles in the hydrochloric acid vessel are shown as ionized into H^+ or Cl^- because hydrochloric acid is practically 100 percent dissociated in water solution. In the acetic acid vessel, only one ion pair is dissociated into H^+ and Ac^-, and even that is proportionately too many, because only about 1 percent of the acetic acid molecules are dissociated. The difference in ionization is the difference in acidity between the two compounds; that is, the greater the ionization, the greater the acidity.

DISSOCIATION AND ASSOCIATION

Although all acids are compounds that yield hydrogen ions, acids differ in their capacity to yield hydrogen ions. There are strong acids and weak acids. In water, the hydrochloric acid molecule, HCl, separates completely into positively charged hydrogen ions, H^+, and negatively charged chloride ions, Cl^-. The separation of ions in water is *dissociation*. (Figure 1.9). The recombining of the ions to form an uncharged molecule is *association*.

Hydrochloric acid dissociates entirely in water and is therefore called a strong acid. Acetic acid, the familiar ingredient in vinegar, is a weak acid because it does not completely dissociate in water.

The dissociation of HCl can be written

$$HCl \rightarrow H^+ + Cl^-.$$

The single arrow in the direction of dissociation indicates that all the HCl dissociates into H^+ and Cl^-. The dissociation of acetic acid can be written

$$Acetic\ acid \Longleftarrow H^+ + Acetate^-.$$

The shorter arrow in the direction of dissociation indicates that acetic acid is only partially dissociated. In water under most conditions, only about one molecule of acetic acid out of 100 is dissociated into the ions. The presence of two arrows indicates that two processes are going on, dissociation and association. The long arrow in the direction of acetic acid indicates that the acetate ion has a strong tendency to associate with H^+ to form a molecule of acetic acid. When this happens, acetate$^-$ is acting as a base, accepting H^+.

Water plays an important role in the process of dissociation. Without water it could not take place. Upon dissociation each ion moves away from its oppositely charged "partner" and is completely surrounded by water molecules. In this form each ion is free to move about, independent of the other ion. Later in this chapter more will be said about the fascinating properties of water and the role water plays in living systems.

In everyday life when we think about strong acids, we think about their dangerous properties. Strong acids have a capacity to inflict severe burns because they are chemically very reactive. This reactivity is primarily due to the hydrogen ion. Remember that hydrogen is the smallest of all the atoms. It is merely a nucleus consisting of one proton orbited by one electron. In the absence of the electron, the hydrogen ion is a "naked proton." It is extremely small and can move very rapidly and alter the associations of other atoms in other molecules. Hydrogen ions are important in the process of cell life. From the behavior of strong acids we can learn something about the importance, mobility, and reactivity of the hydrogen ion. We must be alert to its importance when discussing biochemical aspects of energy conversion within the cell. And we can see why it is important to keep the number of hydrogen ions "under control." How do cells do this?

Because weak acids give up fewer hydrogen ions, they are less dangerous in everyday life. Weak acids and combinations of weak acids play important roles in maintaining the concentration of hydrogen ions in the living cell. Many large molecules, such as nucleic acids and proteins, are also weak acids. By their capacity to give off and take up (by the processes of dissociation and association) hydrogen ions, they help to maintain a constant internal concentration of this highly reactive ion.

Measuring Acidity

Acidity can be expressed numerically on a scale from 0 to 14, with neutrality at 7. The smaller the number below 7, the more acidic is the solution, and the higher the number above 7, the more basic the solution. The value on the *pH scale*, as this measure is called, indicates the concentration of hydrogen ions free in water. Pure water is neutral, with a pH of 7.

The addition of acids lowers the pH in a solution numerically, and the addition of bases raises it. The numbers in the pH scale are logarithmic, each whole number representing a tenfold change in acidity, with a solution at pH 4 containing ten times the H^+ concentration of one at pH 5. The pH of 7 for pure water is due to the very slight dissociation of H_2O into OH^- and H^+.

Buffers: The Control of Acidity

The stability of the pH in a cell is critical to its survival because enzymes (see the essay on page 18) usually function only within narrow ranges of pH values, and a cell depends on the proper functioning of its enzymes for practically everything that happens.

One of the ways that pH is held reasonably steady is by means of buffers. A *buffer* is a combination of two compounds—a weak acid (for example, acetic acid) and its base (acetate)—that resists pH change even if a strong acid or base is added (Figure 1.10). A minute quantity of a strong acid, such as hydrochloric acid, added to pure water causes a large and immediate change in H^+ concentration of the water, that is, a change in pH. But if water contains a weak acid, such as acetic acid, and its base contains acetate, then the water is *buffered*, and the strong acid added will make less change in the pH of the solution than if it were not buffered. The reason is that the added H^+ associates with the acetate to form acetic acid. The acetate, in effect, soaks up the H^+.

With pH a critical factor in cell activities, it is easy to see that keeping cell contents at a fairly even level of acidity is a necessity. Cells do, in fact, have buffering compounds that protect them against sudden and drastic changes in pH, which might be destructive. Human blood, for example, is so strictly buffered that it maintains a pH within about 0.2 pH units, always close to pH 7.4.

CHEMICAL REACTIONS

Life is change. The visible, measurable changes that make life as we know it are all due to chemical changes, *chemical reactions*.

A simple chemical reaction is one in which two elements *come together* to make a new compound. Hydrogen and oxygen make water, or in the symbolism of chemistry,

$$2H_2 + O_2 \rightarrow 2H_2O.$$

This expression merely tells us that hydrogen reacts with oxygen to produce water. Energy is released, but

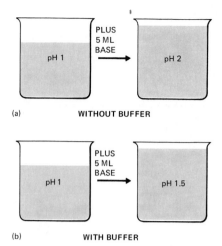

(a) WITHOUT BUFFER

(b) WITH BUFFER

1.10

The action of a buffer. (a) When 5 ml of the base sodium hydroxide is added to 20 ml of an acid solution at pH 1, there is a marked change in the acidity (hydrogen ion concentration). The pH value increases by one unit. (b) When 5 ml of the base sodium hydroxide is added to 20 ml of an acid solution that contains a buffer, there is a much smaller change in the acidity. The pH value increases by only 0.5 unit.

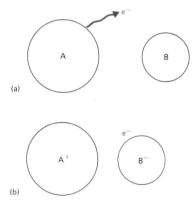

1.11

Oxidation and reduction. (a) Atom A loses an electron to atom B. (b) Atom A, having lost an electron, is said to be *oxidized*, and atom B, having gained an electron, is *reduced*. In this way entire molecules can be oxidized and reduced.

water can also be broken down to hydrogen and oxygen. That is, the reaction above is reversible. Most reactions are reversible although some are technically difficult. A slightly more complex reaction is one in which one element *replaces* another from a compound. If hydrochloric acid comes into contact with metallic zinc, the zinc atoms will replace the hydrogen, and gaseous molecular hydrogen will bubble out of the solution:

$$Zn + 2HCl \rightarrow ZnCl_2 + H_2.$$

In some reactions, there is what amounts to an exchange of atomic partners, known as a *double displacement*. Hydrochloric acid and sodium hydroxide react to yield sodium chloride and water:

$$HCl + NaOH \rightarrow NaCl + H_2O.$$

In all these examples, bonds between atoms had to be broken, so that different molecules with more stable bonds could be formed.

Oxidation-reduction reactions are especially important in biological systems. An _oxidation_ is a loss of electrons, and a _reduction_ is a gain of electrons (Figure 1.11). It must be noted that atoms do not simply yield electrons to space, but rather to some acceptor, with the result that whenever there is an oxidation (electron loss) of one atom, there must be a reduction (electron gain) by another atom. The atom that lost electrons is said to be *oxidized*, and it has *reduced* the atom that gained them. Sometimes oxidation is also accomplished by the removal of atomic hydrogen (H^+ with its electron). The relevance of oxidation-reduction reactions in living cells is that when electrons are transferred from one compound to another, energy is transferred as well, and in biology, energy transfer is of prime importance. The destruction of any organism's oxidation-reduction power means immediate death.

One kind of chemical reaction of special interest to biologists is _hydrolysis_ (Greek, "water loosening"). In a *hydrolytic reaction*, a molecule of water is added to the reacting molecule, with a resulting break in the reacting molecule. Sucrose (cane sugar) can be hydrolyzed to two simple sugars, glucose and fructose (which have the same numbers and kinds of atoms but in different arrangements):

$$\underset{\text{sucrose}}{C_{12}H_{22}O_{11}} + \underset{\text{water}}{H_2O} \rightarrow \underset{\text{glucose}}{C_6H_{12}O_6} + \underset{\text{fructose}}{C_6H_{12}O_6}.$$

Later, when we consider one of the most important molecules of living cells, adenosine triphosphate, or ATP, we will be dealing with hydrolysis whenever ATP is broken down to provide energy for cellular work. This reaction, as we will see later, yields energy:

$$\text{Adenosine}-O-\overset{\overset{\displaystyle O}{\|}}{\underset{\underset{\displaystyle OH}{|}}{P}}-O-\overset{\overset{\displaystyle O}{\|}}{\underset{\underset{\displaystyle OH}{|}}{P}}-O-\overset{\overset{\displaystyle O}{\|}}{\underset{\underset{\displaystyle OH}{|}}{P}}-OH + H_2O \rightarrow \text{Adenosine}-O-\overset{\overset{\displaystyle O}{\|}}{\underset{\underset{\displaystyle OH}{|}}{P}}-O-\overset{\overset{\displaystyle O}{\|}}{\underset{\underset{\displaystyle OH}{|}}{P}}-OH + HO-\overset{\overset{\displaystyle O}{\|}}{\underset{\underset{\displaystyle OH}{|}}{P}}-OH$$

Adenosine Triphosphate + Water → Adenosine Diphosphate + Inorganic Phosphate

THE ROLE OF ENZYMES

Louis Pasteur, the great nineteenth-century French scientist, nearly always guessed right, but he did miss once. He believed that fermentation is a strictly biological activity that can proceed only in the presence of living cells. Proof that living cells are not essential to fermentation came in 1898, when the German chemist Eduard Büchner demonstrated fermentation in a liquid from which all cells had been removed. The fermenting substance came from yeast, and the active material was called an _enzyme,_ a word that literally means "in yeast."

During the next three-quarters of a century, knowledge of enzymes increased. We now know that they are proteins of very large molecular weight . There are an estimated 10,000 different enzymes in the cells of the human. All the molecular changes that occur in these cells are controlled by enzymes.

How do enzymes function? An enzyme is a protein that _speeds up_ a chemical reaction without permanently entering into the reaction itself. Any substance that can function in this way is called a _catalyst_. Therefore, the enzyme is unchanged by the reaction it controls, and the enzyme can be used over and over again to activate the same reaction. The function of each enzyme is specific.

$$Y + X \longrightarrow Z \text{ slow,}$$

$$Y + X \xrightarrow{\text{enzyme}} Z \text{ fast.}$$

Enzymes programmed to assist one chemical reaction have no effectiveness in another reaction or with other reactants.

It is the shape—the molecular structure—of an enzyme that permits it to join both reactants together, atom to atom, and then with-

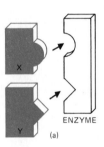 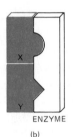 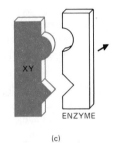

(a) Reactant X fits into one portion of the enzyme, and Reactant Y fits into the adjoining portion.

(b) Reactants X and Y are locked into position by the enzyme mold. X and Y are now side-by-side and are chemically bonded to each other.

(c) The new molecule XY is formed, and the enzyme is freed to perform the same catalytic action elsewhere.

draw after the chemical union has been completed. Reactant X will fit into one site on an enzyme and Reactant Y will fit into the adjoining site. After X and Y have been locked together by way of the enzyme "mold," the XY molecule will separate to go about its new business. Without the help of enzymes the same reaction would take place, but only at a uselessly slow pace.

It is believed that _enzyme inhibitors_, such as poisons, may prevent normal enzyme action by fitting themselves into one or both "slots" and clogging the space normally occupied by the two reactants.

If a cell is to function, it must have an adequate supply of enzymes, it must be supplied with some necessary additional substances, and it must be provided with a suitable cellular environment.

Most digestive reactions are hydrolytic, producing smaller, simpler, more soluble molecules from large, complex, insoluble ones.

Condensation reactions are essentially hydrolytic actions in reverse. Small, simple molecules are united in a condensation reaction into larger, more complex ones, and one or more molecules of water are eliminated.

SOME IMPORTANT BIOLOGICAL MOLECULES

Despite the enormous number of different compounds formed by organisms, there is a relatively small number of categories into which the compounds can be placed. The main types of compounds are water, the carbohydrates, the lipids and related compounds, amino acids and proteins, and the nucleic acids. Al-

---

What is "an adequate supply" of enzymes? Here "adequate" means first the correct enzyme for a given job. Being proteins, different enzymes can be constructed out of very different amino acids. The amino acid composition and primary structure of an enzyme give it a specific three-dimensional shape that fits it for contact with one special molecule or a group of similar molecules. The *specificity* of enzyme action is a result of the shape of the enzyme molecule, which allows it to come into close contact with the molecule or molecules, that is, the *substrates*, on which it works. Because enzymes are so large, there is usually one particular place or *active site* on an enzyme that plays the catalytic role. An enzyme succeeds by making such intimate contact with the reactant molecule that it can alter chemical bonds. Sometimes this means breaking one molecule into two, as when sucrose is converted into two simple sugars. It may also mean bringing two compounds together, as when two amino acids are combined.

What "additional substances" are necessary for enzyme function? Many enzymes cannot catalyze a reaction unless they are accompanied by extra molecules. They may be metals, like magnesium or zinc, or they may be organic compounds. Such a helping compound is a *coenzyme*, which can be experimentally removed from its enzyme. When it is removed, the enzyme is incapacitated. Many of the compounds known as vitamins are in fact coenzymes.

What is a "suitable" cellular environment? Several requirements must be met. First, the temperature is critical. If it is too cold, the general reaction rate will be too low; chemical reactions generally slow down with lowered temperature. If the temperature rises too high, the hydrogen bonds and water shells of the enzyme (a protein) will be disrupted. If such a disruption occurs, the protein is *denatured,* and its enzymatic ability is permanently destroyed. A temperature of 55°C is enough to denature many enzymes. Second, the acidity of the cell environment affects the enzyme rate. Too low or too high a pH will inhibit enzyme action. Each enzyme has an *optimal pH* at which it works fastest, and consequently pH regulation in a cell is essential to its survival. Third, the concentration of substrate must be sufficient. One might compare the enzyme-substrate relation to bricklayers building a wall. Three bricklayers (the enzymes) can build a wall out of bricks (the substrate) three times as fast as one bricklayer can, provided there are plenty of bricks. But having more bricklayers is not going to speed up the construction if each bricklayer has to spend most of his time scouting around for bricks.

An additional feature of enzymes is their *reversibility*, or ability to drive a reaction in either of two directions. Under proper circumstances, the same enzyme is capable of breaking a compound into its components or recombining them to make the original compound.

$$X + Y \xrightarrow{\text{enzyme}} Z,$$

and

$$Z \xrightarrow{\text{enzyme}} X + Y.$$

Because cellular activity depends on enzyme action, one can say that all life does too.

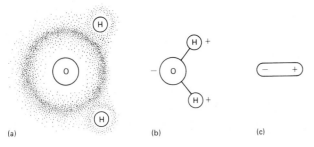

1.12

Three ways to represent a water molecule. (a) The oxygen and hydrogen atoms are shown as having distinct nuclei with hazy clouds of electrons surrounding them. Since there is at any moment a greater likelihood that there will be more electrons at the "oxygen end" of the molecule than at the "hydrogen end," the oxygen end is negative with respect to the hydrogen end. This makes the molecule *polar*. (b) The atoms are shown as skeletons connected by lines, with the charges indicated by plus and minus signs. (c) The whole molecule is drawn as a simple stick with one plus end and one minus end.

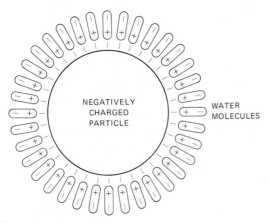

1.13

A "water shell" surrounding a small charged particle with an electrically negative-charged surface. The plus ends of the water molecules are attracted to the negative surface and form an orderly film over the surface of the particle. Such a particle, shown here as if it were sliced open, would in reality be a sphere, and the water film would cover the surface like a piece of velvet over a ball. Such water shells help keep small particles in suspension in the cytoplasm of cells.

though water is not an organic compound, it is so important that it must be considered along with the organic compounds.

Water

Although water is familiar and common and seems simple chemically, it deserves special attention. Water makes up the bulk of living things. Most animals consist of 80–90 percent water. All organisms either live in water or obtain enough water to keep cellular machinery functioning.

The cellular functions of water depend on its unique properties, which grow out of the bonding of two hydrogen atoms to one oxygen atom (Figure 1.12). Oxygen is covalently bonded to hydrogen, but these bonds are distorted because the heavier oxygen atom tends to draw electrons to itself and away from the hydrogen atoms. As a result, the oxygen bears a slight negative charge, and the two hydrogens bear slight positive charges. Because the two ends of the water molecule bear different charges, it is sometimes said to be a *polar molecule*. As a polar molecule, water can form hydrogen bonds with other water molecules and with a variety of other kinds of compounds.

Nearly all large molecules in living systems (carbohydrates, nucleic acids, proteins) have sites that are slightly charged. Water will be attracted to the charged sites and will thereby contribute to keeping such large molecules suspended in a water environment. The polar water molecule will of course be attracted to charged ions. Many surfaces within cells are charged, and water molecules will be attracted to those surfaces (Figure 1.13). The neatly packed water molecules add stability to the surfaces of cell membranes and proteins. One reason why cells are damaged by freezing and heating is that such extremes of temperature disrupt the water layers surrounding cytoplasmic particles.

The physical properties of water are also special. One important property unique to water is that in solid form it is less dense than it is as a liquid at 4°C. Consequently, ice will float in water. Among chemical molecules this is a unique property, which must be explained in terms of the structure and polarity of the water molecule. However, we are concerned here with the consequences that result from this simple physical property. If water acted like other substances, ice would not float to the top of rivers and oceans. It would sink to the bottom. This property of water was probably a requirement for the origin and evolution of life on earth. If not for this property of water, lakes, oceans, and streams would probably not exist as we know them.

Another important physical property of water is its high *heat of vaporization*. It takes a great deal of heat to turn liquid water into water vapor (the gas). When a person perspires, evaporation from the surface results in a loss of heat. Because of the high heat of vaporization, evaporation of only a little water has a large cooling effect. Such evaporation of water has a cooling effect that can be very important in maintaining the inner stability of the organism.

Water enters into the many chemical reactions of the cells directly. For example, the hydrolysis of a protein yields amino acids. Water also plays a very important indirect role. It provides the medium for the chemical interaction of all other biochemical molecules. We begin to appreciate this function when we remember that 80 percent (more or less) of most living cells is water. Much of what appears as empty space in an electron microscope picture of a cell is in fact water and the small molecules uniformly suspended in the water. The polarity, size, and shape of the water molecule make it particularly effective in suspending small molecules such as ions, sugars, and amino acids. When suspended in water, these molecules can still move, and when they collide with an enzyme of another molecule, they may undergo a chemical reaction, one that would not occur in the absence of water. For example, an absolutely dry digestive enzyme mixed with an absolutely dry food would be ineffective. This is one reason why dehydration, the removal of water, contributes to the preservation of foods and other materials.

Later in this text we will talk about ions and other molecules "in solution." By this term we refer to the capacity of water molecules to completely surround ions and other molecules and hold them in suspension but *still available for chemical reaction*.

Carbohydrates

Carbohydrates are sugars, starches, glycogen, and cellulose, all of which are composed of carbon, hydrogen, and oxygen, in the ratio of one carbon to two hydrogens to one oxygen (CH_2O). A simple sugar may have as few as three carbon atoms (a triose), six hydrogens, and three oxygens. Common sugars are five-carbon (ribose) and six-carbon sugars (such as glucose and fructose). (The suffix "-ose" generally denotes a sugar.)

Simple sugars, called *monosaccharides*, can be *condensed*, with the loss of a molecule of water and the formation of a double sugar (Figure 1.14). Condensation is the reverse of hydrolysis. For example, two molecules of glucose can be condensed to make one of maltose, which is malt sugar; or one glucose

GLUCOSE + GLUCOSE

MALTOSE + WATER

1.14

Condensation of two glucose molecules to form one molecule of a new sugar, maltose, with the elimination of a molecule of water. The hydrogen and oxygen atoms of glucose that are to be eliminated are shown in color. In reality, such a condensation involves an intermediate step, but the end result is as shown.

plus one fructose can make one sucrose, which is cane sugar. Maltose and sucrose are *disaccharides*, or double sugars. When more than three simple sugars are condensed, the result is a *polysaccharide*.

In the living world there are three very important polysaccharides, all of which are made up of glucose: starch (the storage form of glucose in plants), cellulose (the most abundant material in plant cell walls), and glycogen (the primary form of glucose storage in animal cells). Although all these polysaccharides are made up of units of glucose, they have very different chemical and physical properties because the glucose units in the three polysaccharides are linked to one another in different ways. The enzymes that control the synthesis of starch in plants join glucose in a way that is different from the way the enzymes in animal cells join glucose to form glycogen. Both starch and glycogen can be used as sources of food by animal cells. Both starch and glycogen differ considerably from cellulose. Abundant as it is, cellulose is not digestible by animals. Only a very few organisms (snails, wood-rotting fungi, and some bacteria) are able to digest cellulose to produce glucose molecules that can be used for food. Man and most other animals would starve on a diet of cellulose in spite of its glucose content. There simply is no human enzyme that can hydrolyze cellulose to produce glucose units.

Lipids

Lipids are a diverse group of biological compounds. One cannot readily characterize them except to say that they will not dissolve in water. They serve as concentrated food storage materials, as structural components of cells, and as regulatory chemicals. Familiar lipids are fats, oils, and waxes; less commonly known but equally important are the phospholipids and steroids.

Neutral fats, so called because they are not electrically charged, are mainly reserve foods. On hy-

drolysis, a neutral fat yields glycerol, also called glycerine, and usually three molecules of fatty acid (Figure 1.15). Glycerol is a three-carbon, non-intoxicating alcohol. Fatty acids are made of chains of carbon atoms, from two to dozens, usually with hydrogens attached, and ending with a *carboxyl group,* —COOH. It is the carboxyl group that makes a fatty acid an acid, because the group can yield a proton, retain an extra electron, and become —COO$^-$. Once the carboxyl group has reacted with a hydroxyl group (—OH) of glycerol, it is no longer acid, so if three fatty acids react with the three available hydroxyl groups of a glycerol molecule, the result is a neutral fat, a *triglyceride* (Figure 1.16). Oils are fats that are liquid at ordinary room temperatures. Waxes are fats compounded with some "backbone" other than glycerol.

Because there is an apparent relation between dietary fats and human circulatory diseases, emphasis is sometimes put on the kinds of fats eaten, with unsaturated fats being preferred by some dietitians. A *saturated* fat is one whose carbon atoms are all hydrogenated, with no double bonds between adjacent carbons.

1.15
Structural formula of glycerol (glycerine), the "backbone" to which fatty acids are attached in the formation of fat.

An *unsaturated* fat has double bonds between adjacent carbons instead of having hydrogen atoms attached. In double bonds four electrons are shared rather than just two.

1.16
Structural formula of a neutral fat. The three carbon atoms shown arranged vertically are from the glycerol portion of the molecule's ingredients. The long chain of carbon atoms was a molecule of a fatty acid (lauric acid). Two more identical chains are indicated by the symbol $(CH_2)_{10}$. Since every carbon atom in the fatty acid chains has two hydrogen atoms, this is a *saturated* fat. Saturated fats are thought to be more conducive to circulatory diseases than unsaturated fats are.

Phospholipids

Phospholipids are a unique group of fatty compounds that contain fatty acids, glycerol, and a negatively charged phosphate (Figure 1.17). The long carbon and hydrogen chains of the fatty acids have no attraction for water, but the charged phosphate group has a strong attraction for polar water molecules. If a single phospholipid molecule is placed in water, water will tend to be attracted to the charged phosphate group and avoid the hydrocarbon chains. If many phospholipid molecules are placed in water, their long hydrocarbon chains will tend to associate together and expose the phosphate groups to the surrounding

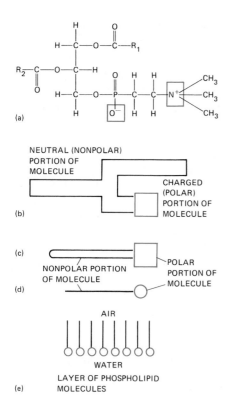

(a)

NEUTRAL (NONPOLAR) PORTION OF MOLECULE

CHARGED (POLAR) PORTION OF MOLECULE

(b)

(c)

POLAR PORTION OF MOLECULE

NONPOLAR PORTION OF MOLECULE

(d)

AIR

WATER

LAYER OF PHOSPHOLIPID MOLECULES

(e)

1.17

Phospholipids. (a) Structural formula of a phospholipid. The usual glycerol backbone of three carbon atoms is present, and two of them bear fatty acids, represented by R_1 and R_2 in the figure. The third carbon atom has attached to it a group of atoms (of the compound *choline*) containing nitrogen and phosphorus. As shown in the colored rectangles, two of the atoms may be charged, giving this portion of the entire molecule the quality of polarity. The bulk of the phospholipid molecule is nonpolar. (b) A skeleton drawing of a phospholipid molecule, showing only the rough outline of its shape, with the polar portion in color. (c) An even greater simplification of the phospholipid molecule, with the nonpolar portion represented as a stick and the polar portion as a colored square. (d) The ultimate simplification, one commonly used in model diagrams. (e) Diagram of the formation of a *monomolecular layer* of a phospholipid on a water surface. The polar portion of the phospholipid is in the water (because water, too, is polar), and the nonpolar portion of the phospholipid, which is not attracted to water, sticks up into the air. Such films are similar to those formed in cells.

water. Cell membranes are built up of phospholipids and proteins. However, it is the phospholipids that give membranes their unique sheetlike structure and many of their functional properties. In cell membranes, phospholipids are arranged so that the charged phosphate groups are on the outside and the fat portions are in contact with one another in the interior of the membrane. Later we will discuss further the structure and many roles of cell membranes. But it is important to remember that a membrane's shape and many of its properties are due to the shape, size, and properties of the phospholipid molecules out of which it is made.

Proteins

Proteins are extremely important to our modern understanding of how living things are structured and how they work. Protein molecules are large compared with such molecules as sucrose. Sucrose has a molecular weight of about 342, whereas the molecular weight of a protein may be several million. For this reason the proteins are often referred to as *macromolecules* (macro = large). Discussion of proteins will reoccur repeatedly throughout this book. A thorough familiarity with their composition, structure, and chemical properties will be helpful in understanding cellular phenomena.

A protein is made up of units, amino acids, that link together, one to another, forming a chain. Such a chain of amino acids is a *polypeptide*. Twenty different amino acids are used in the cellular synthesis of proteins. Actually, chemists can make more than 20 amino acids, but only 20 are important in living systems. All amino acids have certain chemical characteristics in common: a carboxyl group, —COOH, which gives them their acidic nature, and a nitrogen-containing amino group, —NH$_2$. Here is the general structure of an amino acid, where R could be any one of 20 different chemical groups:

$$
\begin{array}{ccc}
H & R & O \\
\backslash & | & \diagup\!\!\!\diagup \\
N\!-\!C\!-\!C & & \\
\diagup & | & \diagdown \\
H & H & OH \\
\end{array}
$$

All amino acids have this structure. The 20 amino acids in living systems differ in the particular chemical nature of the R (for radical or root) group. In the amino acid glycine, R is a single hydrogen atom. In other amino acids the R group is much larger, consisting of a chain or a ring of carbon compounds to which other acidic, basic, polar, or charged groups may be attached. Because of their different chemical compo-

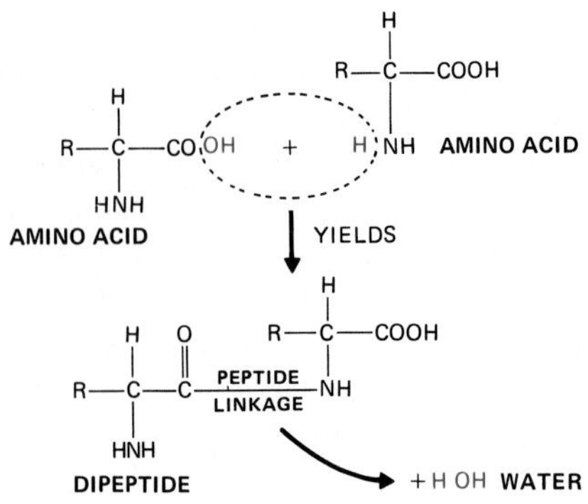

Condensation of two amino acids. The —OH (shown in color) of the carboxyl group of one amino acid and —H (also in color) of the amino group of a second amino acid are removed, forming water. The two amino acids are bound into one molecule, a *dipeptide*, through the resulting *peptide linkage*. There is still an available amino group at one end and a carboxyl group at the other end of the new molecule, and to either or both of these groups more amino acids can be attached. Thus long chains of amino acids can be built into *polypeptides*.

sition and structure these groups have very different chemical properties. It is the character of the R group that confers on the amino acid its unique properties and distinguishes one amino acid from another.

We have said that amino acids are linked, one to the other, to form proteins. How does this occur? The exact process by which living cells manufacture proteins will be discussed later in this book. But here we want mainly to emphasize the nature of the chemical bond between amino acids. Figure 1.18 shows the condensation of two amino acid molecules into a *single molecule*. As indicated in the figure, the reaction occurs between the carboxyl group of one amino acid and the amino group of the other. In the process, water (H_2O) is split out. Such a union results in the formation of a strong covalent bond between the two amino acids. In proteins this bond is referred to as the *peptide linkage.* You will note that there is still a carboxyl group at one end of the new molecule and an amino group at the other end. Both are available sites for reaction with additional amino acids. In this fashion the number of amino acids that can be linked one to another is almost endless.

The specific sequence of amino acids in a protein constitutes the protein's *primary structure*. The primary structure determines the fundamental nature of the polypeptide. For example, a polypeptide consisting of 20 identical amino acids will have certain chemical and physical properties. However, a polypeptide made up of 20 different amino acids will have very different chemical and physical properties. Furthermore, if the 20 different amino acids are arranged in a different sequence, the polypeptide will have yet another set of chemical and physical properties. We can therefore see that the composition and sequence of amino acids in a protein are very important to its structure and function. Chains of amino acids tend to twist into helices, so the protein molecule is shaped roughly like a coiled spring. The coil is maintained by hydrogen bonds between the various amino acids. This coiling is referred to as the *secondary structure* of protein. All the polypeptides discussed above show the helical structure. The different amino acid composition or sequence of proteins does not alter their capacity to form the coils characteristic of secondary structure.

In very large proteins the coils tend to fold back on themselves. This folding gives the protein its *tertiary structure*. The particular chemical nature of the different R groups of amino acids plays a very important role in determining just how the helical regions fold to produce the tertiary structure. Proteins with different amino acid compositions will fold into different shapes and have different tertiary structures. Covalent, ionic, and hydrogen bondings between R groups play important roles in maintaining a stable tertiary structure.

Still another level of protein structure is possible, the *quaternary structure*. Several folded proteins, each with its own primary, secondary, and tertiary structure, may be loosely joined into a "superprotein." The molecule hemoglobin, responsible for the transport of oxygen in the blood, is such a "superprotein" and is composed of four subunits (Figure 1.19).

Certain proteins may also join with ions, sugars, or vitamins to produce conjugated proteins. Hemoglobin is an important example of such a protein. It contains iron, which is important to the biological function of the protein. The protein component cannot work without the iron, and the iron cannot work without the protein. Both are necessary to fulfill the biological function.

TO SUM UP PROTEINS

1. Amino acids make up proteins. Different proteins have different amino acid composition.

2. Amino acids are linked, one to the other, through covalent bonds (the peptide linkage).

3. What is linked together is then coiled. The coil is held together by hydrogen bonds. Such coiling is characteristic of practically all proteins, no matter what their amino acid composition.

4. What is coiled is then folded. Proteins can be folded into an enormous number of shapes. The particular folded structure is determined by the R chains on the amino acids.

Hydrogen bonds play a major role in the forming of secondary and tertiary structure. You will recall that covalent bonds are extremely strong. Exceptionally harsh chemical methods have to be used to break such covalent bonds between amino acids, or a biological catalyst (an enzyme) must be used. However, more gentle methods, such as increase in temperature, are adequate to disrupt the hydrogen bonds responsible for secondary and tertiary structure.

The particular biological role a protein plays depends on its three-dimensional tertiary structure. If that structure is altered, the appearance, properties, and biological activity of the protein will be destroyed. Consider the familiar "egg white" of an

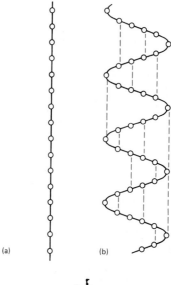

1.19

Levels of organization in proteins. (a) the *primary* structure of a protein (polypeptide) is determined by the sequence of amino acids in the polypeptide chain. Natural polypeptides contain from dozens to hundreds of amino acids. (b) The *secondary* structure of a polypeptide is brought about by the formation of hydrogen bonds between amino acids along the chain. One frequent result of such hydrogen bonding is the formation of a helix. (c) The *tertiary* structure of a protein appears when two cysteine (sulfur-containing amino acid) portions of a polypeptide chain come close together. A sulfur-to-sulfur bond (a disulfide bridge) forms, holding the polypeptide chain firmly at that point (colored rectangles). (d) The *quaternary* structure of some proteins is the result of the combination of several protein molecules into one supermolecule.

PROTEIN AND PERMANENT WAVES

When a hair shaft is produced by the cells of the human epidermis, its overall shape is determined by the shape of its protein subfibrils. The main constituent protein, *keratin*, contains a relatively large amount of the sulfur-containing amino acid cysteine. About one of every ten amino acids in keratin is cysteine. (In contrast, the protein hormone insulin has six cysteines in a total of 51 amino acids.) The sulfur of one cysteine can make an —S—S—, or disulfide bridge, with another cysteine, thereby binding the entire protein molecule into a firm shape. It follows that a high proportion of cysteine residues (component amino acids) makes a stable molecule and consequently a durable hair shape.

If the hair shaft is straight, it will remain straight because of the stability of the disulfide bridges, indicated in color in (1). Those bridges, however, can be broken by heat or chemically and restored at will. These facts are applied in giving "permanent waves." A person who has straight hair and wants it curled or has curled hair and wants it straight can have the disulfide bridges broken by heat—unpleasant and hard on the hair—or by application of proprietary solutions that accomplish the same effect (2). While the bridges are broken, the hair is bent or "set" in the desired form (3). Then it is either cooled or treated with a second solution that restores the bridges (4), and the hair will stay put. New hair pushed out from the hair follicle will have the old natural shape its possessor gave it, because the artificial treatment affects only hair that has already grown.

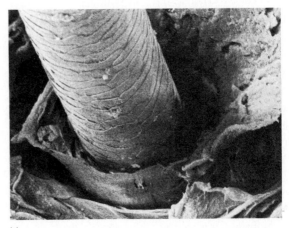

(a)

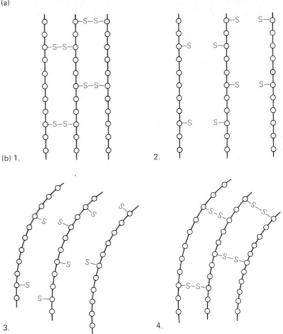

(b) 1.

2.

3.

4.

uncooked hen's egg, which is the protein albumen. When an egg is boiled or otherwise subjected to heat, its three-dimensional structure changes profoundly. A cooked egg cannot be "uncooked." The change is due to the disruption of the secondary and tertiary structure of the albumen protein by the high heat. Neither frying nor boiling breaks covalent bonds.

Nucleic Acids

The last group of compounds in our "basic vocabulary" of biological materials, the *nucleic acids*, are such special molecules that they will be considered by themselves in Chapters 6 and 7. They are mentioned here only to complete the list of important biological molecules.

SUMMARY

1. All matter is composed of *elements*, substances that consist of only one kind of atom.

2. Atoms are made up of a *nucleus*, with its electrically charged *protons* and uncharged *neutrons*, and *electrons* orbiting around the nucleus. The chemical and physical properties of an element are determined by the number of protons, neutrons, and electrons in the atoms of the element.

3. In heavier elements the electrons are distributed as though they were in layers, known as *energy-level shells*. The biologically important points about energy-level shells are that (1) the shell an electron occupies is a function of its energy, and (2) the number of electrons in the outermost shell determines how one atom will combine with another atom.

4. Carbon is an especially important element because it is a part of all biological molecules and participates in all biological reactions. Compounds containing carbon are *organic*, and compounds not containing carbon are *inorganic*.

5. A *molecule* is the smallest unit into which a substance can be divided and still retain the properties of that substance. When two or more different atoms combine, they form a *compound*, a substance whose molecules are composed of more than one kind of atom.

6. An electrically charged atom or group of atoms is an *ion*. Atoms held together by the gaining, losing, or sharing of electrons are united by chemical *bonds*. The important chemical bonds are *ionic*, *covalent*, and *hydrogen* bonds. Many large biologically active molecules are held in proper shape by hydrogen bonds.

7. *Acids* are compounds that yield hydrogen ions, and *bases* are compounds that accept hydrogen ions.

8. The separation of ions in water is *dissociation*; the recombining of the ions to form an uncharged molecule is *association*. Dissociation cannot take place without water.

9. Acidity can be expressed numerically on a *pH scale*, which is an indication of the concentration of hydrogen ions free in water. A *buffer* is a compound that resists pH change even if a strong acid or base is added.

10. A simple *chemical reaction* is one in which two elements come together to make a new compound. A slightly more complex reaction occurs when one element replaces another from a compound. Oxidation-reduction reactions are especially important in biological systems. An *oxidation* is a loss of electrons, and a *reduction* is a gain of electrons. *Hydrolysis* is a chemical reaction in which a molecule of water is added to the reacting molecule, with a resulting break in the reacting molecule.

11. Some important biological molecules are *water*, *carbohydrates*, *lipids*, *phospholipids*, *proteins*, and *nucleic acids*.

RECOMMENDED READING

Baker, Jeffrey J. W., and Garland E. Allen, *Matter, Energy, and Life*, Third Edition. Reading, Mass.: Addison-Wesley, 1974. (Paperback)

Barnett, Lincoln, *The Universe and Dr. Einstein*, Revised Edition. New York: Bantam Books, 1968. (Paperback)

Gamow, George, *Matter, Earth, and Sky*, Second Edition. Englewood Cliffs, N.J.: Prentice-Hall, 1965. (Also available in paperback)

Lambert, Robert B., "The Shapes of Organic Molecules." *Scientific American*, January 1970. (Offprint 331)

Lapp, Ralph E., *Matter*. New York: Time-Life Books, 1966.

Phillips, David C., "The Three-Dimensional Structure of an Enzyme Molecule." *Scientific American*, November 1966. (Offprint 1055)

2

Energy for Life

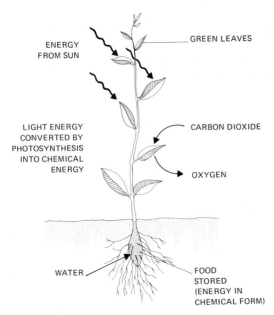

ENERGY FROM SUN

GREEN LEAVES

LIGHT ENERGY CONVERTED BY PHOTOSYNTHESIS INTO CHEMICAL ENERGY

CARBON DIOXIDE

OXYGEN

WATER

FOOD STORED (ENERGY IN CHEMICAL FORM)

2.1
Diagram of the conversion of light energy from the sun to the chemical energy of stored food. The green plant contains the necessary components to combine the elements of carbon dioxide and water to make food and give off gaseous oxygen.

ENERGY IS USUALLY DEFINED AS THE CAPACITY to do work. Biologically, we can translate this *capacity* to do work into its simplest terms: Energy is needed for the operation of all life processes. No animal moves, no plant grows, no bacterium multiplies, no cell divides without a continual input of energy.

Our most important source of energy is the sun, 155 million kilometers (93 million miles) away. The sun's energy enters the biological world by way of green plants, which use sunlight to create their food supply. This process, known as *photosynthesis*, will be discussed in detail in Chapter 4. Animals are unable to manufacture food by the use of light energy, and they depend on consumption of plants or other animals for their survival. But besides merely eating plants and animals, animals depend on the sun for their oxygen supply, which is obtained from plants as a by-product of photosynthesis.

It is well to remember that there is more to the sun's energy than its chemical use by green plants. The very heat provided by the sun in any given environment is crucial for organisms (both plant *and* animal) as they go about the business of living. Cellular use of sunlight, either directly through photosynthesis or indirectly through cellular respiration, is certainly not the only use of the sun's energy. As we shall see in this chapter and throughout the book, very few biological processes can happen *only* one way.

In photosynthesis, sunlight energizes the chemical reactions that produce oxygen and plant tissue. While they are transforming radiant energy from the sun into chemical energy used by living organisms, plants are also giving off the oxygen we breathe (Figure 2.1). In return, animals produce carbon dioxide as a waste product, and plants use this carbon dioxide to continue the entire life-giving cycle.

Plants are able to convert light energy into chemical energy. It is the ability to use chemical energy and to convert it to still other forms of energy that keeps an organism alive. Until we study photosynthesis in greater depth, we will do well to understand that organisms use food as a source of chemical energy. Chemical energy is stored in chemical compounds, and it can be applied directly to the life processes of living things. We will see later in this chapter how chemical energy can be changed to other kinds of energy.

SOME PRINCIPLES OF ENERGY CONVERSION

In general, and certainly as far as living systems are concerned, *energy can be changed from one form to another, but it cannot be created or destroyed.* This is a simple statement of the *First Law of Thermodynamics,* a "law" that biological systems adhere to with absolute strictness. If an insect eats a leaf, all the energy stored in that leaf still exists somewhere, even after the leaf has been digested and made into insect tissue. Some of it is in the chemical bonds of molecules, some in heat dissipated into the air. Any example of energy consumption and conversion would follow the same rule.

Green plants do not create energy when they synthesize food during photosynthesis, nor do animals create energy when they use food for metabolic activities. They are only converting the energy of sunlight into different forms.

At the same time, any energy conversion involves some loss of energy as far as the converting system is concerned. (Actually, the "lost" energy has been converted to heat energy that is dissipated out into the surroundings in the process. Remember, energy is neither created nor destroyed.) The insect mentioned above, having eaten the leaf, ends up with less energy in its cellular components than there was originally in the leaf. Some heat is "lost" at every step along the way as the insect digests the leaf molecules and recombines their atoms into new forms. If the insect could capture all the potential energy of its food and turn it into useful work, it would be 100 percent efficient in anything it does (Figure 2.2). *Every time an energy-yielding or energy-requiring action occurs, some energy is "lost" as heat and cannot do work.* That is the *Second Law of Thermodynamics,* and it, too, is "obeyed" absolutely by living things. It is worth noting that, as with any natural "law," if some organism could "break" it, the law would simply

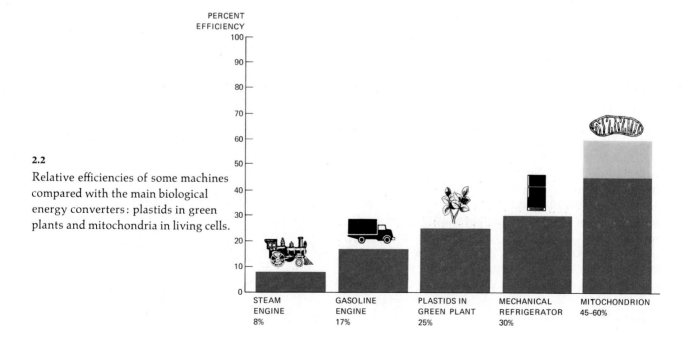

2.2

Relative efficiencies of some machines compared with the main biological energy converters: plastids in green plants and mitochondria in living cells.

PERCENT EFFICIENCY

| STEAM ENGINE 8% | GASOLINE ENGINE 17% | PLASTIDS IN GREEN PLANT 25% | MECHANICAL REFRIGERATOR 30% | MITOCHONDRION 45–60% |

cease to be, since the so-called natural laws are nothing more than our observations of how the universe operates.

The highly ordered system of cells, tissues, organisms, and populations of organisms that make up the living world is achieved only by the continued input of energy. It may seem at first glance that the living world does indeed break, or at least bypass, the Second Law of Thermodynamics, because as a plant grows, it obviously accumulates energy. However, it does not break any energy laws. The plant receives much more energy from the sun than it succeeds in storing.

HOW IS ENERGY MEASURED?

Energy is measured in several ways. Some energy units are joules for electric current, ergs for mechanical devices, BTUs (British Thermal Units) for coal and oil, and calories for food. Since these units are all interconvertible, and since it is convenient to use a common term, biologists usually use the calorie as the basic energy unit. A *calorie* (from the Latin *calor,* heat) is the amount of energy required to raise the temperature of a gram of water one degree Celsius, specifically from 14.5° to 15.5°. A *kilocalorie,* written Kcal, or more commonly Calorie, is a thousand calories.

In popular literature, as in diet instructions and on food labels, a calorie is in fact almost always a kilocalorie. Thus, if a commercial breakfast cereal is said to contain 110 calories per ounce, it means 110,-000 calories, or 110 Kcal. Biologists think and write of Calories almost exclusively, whether they are concerned with energy from the sun, the formation of carbohydrates in a green plant, or the release of energy by a respiring cell.

In practice, the calorie content of materials, especially foods, is measured by burning a weighed sample in a well-insulated container called a calorimeter, and determining how much heat is released during a chemical reaction. This is done by measuring the temperature change of a given quantity of water. The energy requirements of many biological actions, however, are measured by a combination of experimental and theoretical means that do not always yield the same answers.

Since the energy values for various compounds are usually determined experimentally, they are not known absolutely. The most recent figures are those commonly given because it is assumed that new techniques are more accurate. At present, for example, the energy content of a mole of glucose is considered to be 686 Kcal, and that of a mole of ATP to be 7.3 Kcal, but these figures, though reasonably close to correct, are subject to further refinement. Students need not be surprised to find that not all workers and all books use exactly the same numbers for the energy content of various biologically useful compounds.

THE RELEASE OF BIOLOGICAL ENERGY IS USUALLY SLOW AND STEADY

Living systems all convert energy in their own ways, and yet the basic chemical process is the same. When a burning piece of wood gives off heat, it is actually releasing energy that was contained within its chemical bonds. It would be simple if energy could be released from a cell by burning. But burning produces temperatures too high for a living cell to endure, and the burst of energy release is too short and too sudden. Organisms need a slow, steady energy source, and especially one that can be controlled and regulated, since the chemical reactions in organisms can use energy only in small packages. Cells have effective ways of regulating energy release.

Living things—from bacteria to orchids to humans—for the bulk of their energy requirements use a small organic molecule, adenosine triphosphate, customarily abbreviated as *ATP*. The ATP molecule stores energy in chemical bonds, where the energy remains locked up until it is released by rearrangement of those bonds (Figure 2.3). Although ATP stores energy as described, ATP itself is not stored and built up in ever-growing quantities. What is important is the ready availability and transferability of the energy in ATP.

The ATP molecule is built up from smaller subunits of adenine (which we will meet again when we discuss the nucleic acids in Chapter 6), a five-carbon sugar (ribose), and three phosphate groups. Most of the energy in ATP available for biological systems is contained in the energy-rich bonds between the last two phosphate groups. These energy-rich bonds are so called because they yield their energy readily. When ATP reacts with water (*hydrolysis*), the last of the three phosphate groups is separated. Such a reaction yields adenosine *di*phosphate (ADP) plus inorganic phosphate (P_i) and energy. Most of the energy released from ATP can be used for the cell's immediate needs.

Suppose that a synthesis is required—say, the joining of two small amino acid molecules into a double molecule called a dipeptide (see Chapter 1). Such a

2.3

A structural diagram of an ATP molecule.

joining requires energy, which can be obtained from the breakdown of nearby ATP.

1. ATP → ADP + P_i + Energy.
2. Amino acid + Amino acid + Energy → Dipeptide.

Two reactions such as these, starting with one set of reactants and ending with different ones, are called *coupled reactions* because they are related through the passage of a phosphate group to an intermediate compound not shown in the equations as written. The first reaction, *yielding* energy, passes a phosphate group to an intermediate, which can then react to give the final result. The second reaction *requires* energy, which is provided by ATP. Energy flow in most biological reactions is made possible by coupled reactions involving the passage of phosphate from ATP

If the generation of ATP from ADP and P_i requires an energy source, the question arises: How does the energy get into the ATP molecule? Not all reactions are completely explainable, but some partial answers are available, and some guesses can be made. We know that some reactions in cells yield energy, and when these occur, the energy made available is transferred by intermediates to the ADP + P_i reaction. The result is energy-rich ATP. The action is reversed when ATP is hydrolyzed. The energy released is transferred to whatever energy-requiring reaction the cell is carrying on. In most instances, neither the resultant ADP nor the inorganic phosphate becomes attached to the products of the reaction energized. The ATP merely provides the energy for the reaction.

Activities that require energy, such as the lifting of a weight or the cellular synthesis of a sugar molecule, are known as *endothermic* (heat in). Activities that liberate energy, such as allowing a weight to drop or breaking down a sugar molecule, are called *exothermic* (heat out). The most important exothermic biological reaction probably is

ATP → ADP + P_i + Energy.

The reverse reaction synthesizes ATP

$$\text{ADP} + P_i + \text{Energy} \xrightarrow{\text{enzyme}} \text{ATP}.$$

The synthesis of ATP is an endothermic reaction because energy from sunlight or food is required for the reaction to take place.

Cells have relatively few ways of generating ATP, and of course they must obtain energy to do so. In contrast, all biological activities require energy, and it is usually obtained from ATP. Indeed, ATP in biological jargon is called the energy "currency" of the cell, as though it were cash to pay for the energy needs of the cell.

Although ATP is a common energy source for biological activity, it is not the only one. A number of other compounds are available, some of which are used for specific functions. Guanosine triphosphate (GTP), for example, is used instead of ATP in certain steps of protein synthesis. The energy for muscle contraction is frequently supplied by another phosphorous-containing compound, creatine phosphate. Still a third compound, phosphoenolpyruvate (PEP) is used in sugar synthesis. Indeed, it would be surprising to find that any biological process used one and only one compound. Life does not proceed so.

THE PHYSICAL TRANSFER OF ENERGY REQUIRES A TRANSFER OF ELECTRONS

The two main energy-transferring processes in biology are *photosynthesis* and *respiration*. They are so important to life that they will be given special treatment in Chapters 4 and 5, but they share some features that are fundamental to all the activities of living cells. First of all, they are both essentially ATP-generating processes. However, in photosynthesis light energy is used to make ATP, which is then used to make organic compounds, whereas in respiration the energy of organic compounds is used to make

BIOLOGICAL LUMINESCENCE

One of the most striking biological examples of the conversion of energy from one form to another is the conversion of chemical energy to light energy in organisms like the firefly. Many organisms, from bacteria to fishes that dwell in salt water, are luminous also, but no luminous organisms are found among higher plants, amphibians, reptiles, birds, or mammals. This direct conversion of chemical energy to light energy is known as _biological luminescence_, or _bioluminescence_.

Aristotle and other early observers knew about bioluminescence and studied it, but it was not until 1887 that the French physiologist Raphael Dubois experimented with a luminous clam and isolated a substance he named _luciferin_, after Lucifer, the light-bearer. Luciferin emits light in hot water, and another substance, an enzyme that Dubois called _luciferase_, is active in cold water. Early in this century, E. Newton Harvey of Princeton University confirmed that bioluminescence is an enzyme-controlled process. He showed how to produce luminescence by adding water to the powder derived from a dried crustacean. Such powder, which contained both luciferin and luciferase, was later used by Japanese soldiers during World War II as a low-intensity light source when a stronger light might have revealed their field positions.

Although much information has been accumulated about the luciferin-luciferase combination, many basic questions about bioluminescence remain. The principal unanswered question is the biggest one: Exactly how do so many

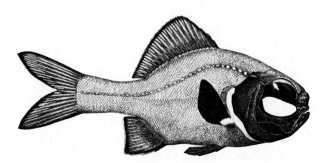

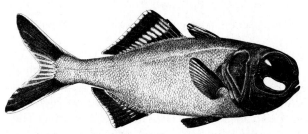

Two types of "flashlight fishes" that carry luminous bacteria under their eyes: Photoblepharon palpebratus _(top) and_ Anomalops katoptron _(bottom). Both species are shown about actual size._

diverse organisms produce their eerie and mysterious lights? In most cases it is easier to answer the question of "why," and specific examples are plentiful.

Some species use their light-producing ability as an alarm system, others use it to lure their prey, and still others, like fireflies, incorporate it into their mating rituals. One type of squid emits a radiant cloud instead of a screen of ink

ATP, which is then used for the various needs of the cell.

The essential action in these processes is the transfer of electrons from one compound to another. In photosynthesis, light energy brings about the release of electrons from chlorophyll, which pass to some electron acceptor. As the electrons pass, the energy is transferred with them. When an electron goes from a molecule, that molecule is said to be _oxidized._ The receiving molecule is said to be _reduced._

For every oxidation there is a reduction. The oxidized molecule gives up energy, and the reduced molecule receives it. In living cells, energy transfers involve oxidation-reduction reactions, and it is by means of them that energy is kept flowing.

Life on earth is a watery life, and all active cells are bathed in abundant water. The water is partially ionized, so that instead of existing as H_2O, some of the molecules dissociate into _hydroxyl ions_, OH^- (bearing an extra, unshared electron), and _hydrogen_

when it wants to elude a predator. An angler fish lures its deep-sea prey into its luminous jaws. Female marine worms near the Bermuda coast rise to the surface of the water exactly three days after a full moon and secrete a luminous circle that attracts the males, by now producing their own intermittent light. The nocturnal courtship ends with eggs and sperm being discharged into the glowing water. This relationship between mating periods, phases of the moon, and bioluminescence has been observed in many other organisms. In fact, bioluminescence can be predicted well enough to have become a scheduled tourist attraction in some Caribbean areas.

One deep-sea diver has observed that below depths of 400 meters (1300 ft), more than 95 percent of the fish are bioluminescent. Such findings suggest that some organisms in dark environments need to produce light to hunt their prey or simply to communicate. Simple organisms like fungi and bacteria can also be luminous, and several dramatic examples of symbiotic luminescence have been cited, where bacteria provide the light for their host. One Indonesian fish carries about luminous bacteria in blood-rich vessels under its eyes. When the fish does not want the illumination, it can slide a darkening lid over the bacteria without obscuring its own vision.

Although much remains to be learned about the mechanism of bioluminescence, we know from studies of fireflies that two substances are always involved: oxygen and that ever-present energy producer, ATP.

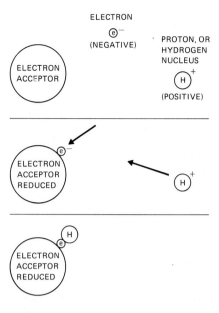

2.4

Diagram of transfer of an electron to an electron acceptor (which is reduced) and accompanying acceptance of a hydrogen ion.

Biologists frequently speak of hydrogen transfer, but what they mean is electron transfer accompanied by hydrogen pickup from dissociated water. The end result is the same regardless of the words used. The main point is that the molecule that receives the electron and the hydrogen and is therefore *reduced* is the one that now has the energy. Electron or hydrogen transfer is the main means of energy transfer in biological systems.

Energy in living systems manifests itself in several forms. *Movement* is apparent in muscle contraction, the upward push of trees growing, and the movements of chromosomes and other subcellular particles. *Heat* is released by all metabolic activities. *Light* is emitted by flashing fireflies and several luminous microorganisms. *Electrical energy* is liberated by several animals, notably electric eels. No matter what the form of energy released, ATP or some similar energy source is involved.

All these uses of energy demonstrate how the cell manages to rearrange chemical bonds and release that energy at exactly the moment it is needed. We still know little of how such a basic and vital request for energy is transmitted, and in our study of biology we will see over and over again that natural functions are so complicated that we still do not understand all of them fully. In the next chapters we will study several aspects of cells, and we will at least try to ask the right questions. Perhaps we will also be able to provide some answers to the fascinating study of life.

ions or *protons*, H^+ (lacking an electron). This is an oversimplification, but it will suffice for practical purposes. The H^+ ions are ready electron acceptors, and since they are present and free to move around, they tend to go where the electrons go. Consequently, when an electron is passed from one molecule—say, chlorophyll—to another, which then has an extra electron, an H^+ ion is likely to go along, too. The receiving molecule thus gains not only an electron but a hydrogen ion as well (Figure 2.4).

SUMMARY

1. *Energy* is usually defined as the capacity to do work. In biological terms, we can say that energy is needed for the operation of all life processes.

2. The sun is our most important source of energy. Green plants use sunlight to create their food supply in a process called *photosynthesis*. Besides producing food for plants, photosynthesis liberates oxygen.

3. The *First Law of Thermodynamics* states that energy can be changed from one form to another, but it cannot be created or destroyed. The *Second Law of Thermodynamics* states that every time an energy-yielding or energy-requiring action occurs, some energy is "lost"; that is, it is converted into heat energy and cannot do work.

4. Biologists use the *calorie* as the basic energy unit. A calorie is the amount of energy required to raise the temperature of a gram of water from $14.5°$ to $15.5°$ Celsius. A *kilocalorie*, or *Calorie*, is a thousand calories.

5. All living things, for the bulk of their energy requirements, use *ATP*. ATP stores energy in chemical bonds, where the energy remains locked up until it is released by rearrangement of those bonds. The key feature of ATP is not simply the storage of energy in the chemical bonds of the molecule, because every molecule can store energy in its bonds. Of crucial importance is the *ready availability and transferability* of the energy in the ATP molecule.

6. When ATP reacts with water (*hydrolysis*), the last of its three phosphate groups is separated. Such a reaction yields ADP, inorganic phosphate, and energy.

7. *Coupled reactions*, though not using the same beginning and final products, are closely associated through energy-carrying intermediates.

8. Activities that require energy are called *endothermic* (heat in). Activities that liberate energy are called *exothermic* (heat out).

9. The two main energy-transferring processes in biology are photosynthesis and respiration. The essential action in these processes is the transfer of electrons from one compound to another.

RECOMMENDED READING

Hinkle, Peter C., and Richard E. McCarty, "How Cells Make ATP." *Scientific American*, March 1978.

Lehninger, Albert L., *Bioenergetics*, Second Edition. Menlo Park, Calif.: W. A. Benjamin, 1971.

Lehninger, Albert L., "Energy Transformation in the Cell." *Scientific American*, May 1960. (Offprint 69)

Lehninger, Albert L., "How Cells Transform Energy." *Scientific American*, September 1961. (Offprint 91)

McCosker, John E., "Flashlight Fishes." *Scientific American*, March 1977. (Offprint 693)

McElroy, William D., and Howard H. Seliger, "Biological Luminescence." *Scientific American*, December 1962. (Offprint 141)

Miller, Julie Ann, "A Pocketful of Glow." *Science News*, February 18, 1978.

Pfeiffer, John, *The Cell*, Revised Edition. New York: Time-Life Books, 1976. (Life Science Library)

Wilson, Mitchell, *Energy*. New York: Time-Life Books, 1970. (Life Science Library)

THE LIFE OF CELLS

PART TWO

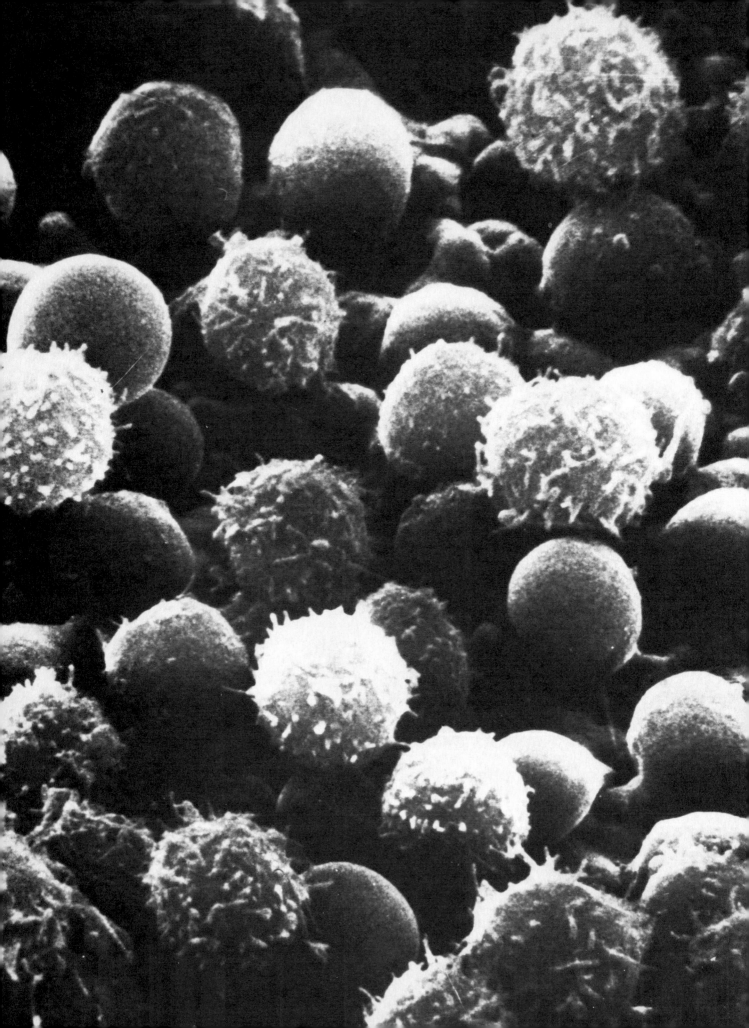

3
Cells and Cell Structure

CELLS ARE EFFICIENT CHEMICAL FACTORIES suited for the continuation of life. Compartmentalized but integrated, specialized, and precisely regulated, a living cell provides an effective environment for the chemical reactions that constitute the processes of life. In this chapter we will examine the structure and function of a model cell. No single cell contains all possible subcellular structures, and no typical cell exists. For convenience, however, and for illustrative purposes, we will build a composite picture of an imaginary cell with everything possible in it.

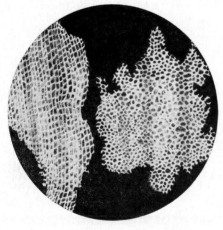

3.1
Robert Hooke saw the first biological cells ever viewed by human eyes. Because he chose to look at a thin sliver of cork, he saw only cell walls, not cell contents. His drawing of cork cells is regarded by biologists as one of the landmarks of biological history.

Scanning electron micrograph of human blood cells. The lymphocytes have rough outer surfaces with numerous microvilli.

WHAT ARE CELLS?

More than 300 years ago, in 1665, the English scientist Robert Hooke looked through one of the first microscopes and described the compartments in thin slices of cork (Figure 3.1). He called these compartments cells because they reminded him of empty chambers, and the name remained, even though Hooke also described living cells "fill'd with juices." Not until 1839, however, was it generally accepted that all plants and animals are cellular. At that time the German scientists Theodor Schwann and Matthias Schleiden, working separately, were given credit for coordinating their own and previous research to formulate the *cell theory.* The three most important

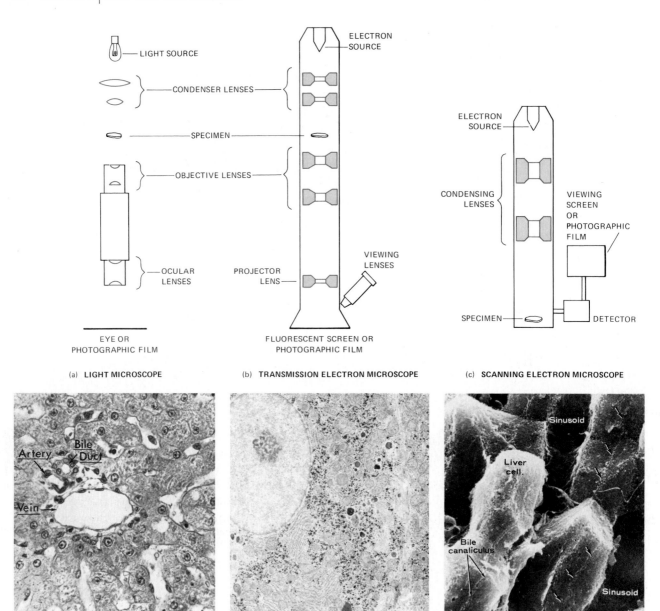

(a) **LIGHT MICROSCOPE** (b) **TRANSMISSION ELECTRON MICROSCOPE** (c) **SCANNING ELECTRON MICROSCOPE**

3.2

Three kinds of microscopy and the results they produce. (a) A light microscope, drawn inverted for comparison with the other two instruments, and a photograph of liver cells as seen through such a microscope. × 300. (b) A transmission electron microscope, which uses electrons from a hot filament instead of light. The electrons, guided by magnets called "lenses," pass *through* the specimen (transmission), so that the image on a screen or photographic plate is a shadow of the specimen, with varying densities caused by differences in the electron absorption or scattering in the specimen. The electron micrograph is also of liver cells, but at a magnification of about × 6000.

(c) A scanning electron microscope, which also uses electrons to form an image. However, instead of passing through the specimen, electrons are beamed *onto a surface* of a specimen. An image of the surface is picked up by a detector like a television receiver and made visible on a tube like a television screen. The image gives an illusion of three-dimensional depth, as shown in the scanning electron micrograph of the outer surfaces of liver cells. × 600. Electron microscopes are usually housed in special rooms with braced floors to support their massive weight. Temperature, humidity and voltage are controlled.

TABLE 3.1

Comparison of Units of Measurement in Cell Biology

MILLIMETER	MICRON (= MICROMETER)	MILLIMICRON (= NANOMETER)	ANGSTROM (= DECINANOMETER)
1	1,000	1,000,000	10,000,000
0.001 (1×10^{-3})	1	1,000	10,000
0.000001 (1×10^{-6})	0.001 (1×10^{-3})	1	10
0.0000001 (1×10^{-7})	0.0001 (1×10^{-4})	0.1	1

From *Biology of the Cell* by Stephen L. Wolfe. © 1972 by Wadsworth Publishing Company, Inc., Belmont, California 94002. Reprinted by permission of the publisher.

factors of the cell theory are that (1) the cell is the unit of structure and function of all living things; (2) all cells come from former cells of the same kind, as a result of cellular reproduction; and (3) all living things are composed of a cell (or cells) and cell products.

Originally the emphasis was on the walls of cells, but there is now general recognition that the functional part of a cell is its contents. Microscopes have been refined continually since Hooke's early version, and by the 1950s scientists were able to use electron microscopes, with sharp and clear magnifications up to 50,000 times, to obtain unique information about the inner structure and workings of cells (Figure 3.2).

In the following description of cell components, each entity must be treated more or less separately and its form and activity described as if the information had been obtained from a series of still pictures selected from a strip of movie film. Indeed, much of what is known about cells has been gained in just that way. But one must remember that cells are in fact as busy as complex factories in full production, with many things going on at once. At a given moment, a fully functional cell may be replicating its DNA, digesting complex food molecules, building up chains of polypeptides by joining up residues at the rate of several hundred a minute, pouring out energy-rich ATP, engulfing external particles, packaging and expelling internal particles, creating new cell parts. All this is going on while the cytoplasm is surging about with a speed, relative to the size of the suspended particles, like that of a river at full flood. If a person could shrink down to molecular size, enter a live cell, and watch what happens, he would find a bewildering array of movements, syntheses, creations, and destructions.

Cells are the smallest independent units of life, and all life as we know it depends on the chemical activities of cells. In life, cells vary enormously. They can range in size from Rickettsias, bacteriumlike organisms less than a micron (1 micron = 1 micrometer

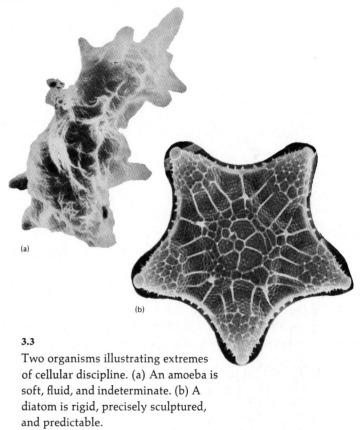

(a)

(b)

3.3

Two organisms illustrating extremes of cellular discipline. (a) An amoeba is soft, fluid, and indeterminate. (b) A diatom is rigid, precisely sculptured, and predictable.

or 1 millionth of a meter; see Table 3.1), to nerve fibers more than a meter long (but too thin to be seen by the naked eye) or the biggest of all, ostrich eggs, which are usually the size of a small grapefruit. Most cells probably range from 10 to 100 micrometers. As an example of cells more typical in size than the extremes cited above, consider your own liver cells. A quarter of a million liver cells could fit into a cube one millimeter high, a cube smaller than the small letters on this page.

In shape, cells may have the formless instability of an amoeba or the ultramicroscopic, precise delicacy of a diatom shell (Figure 3.3). Color possibilities are

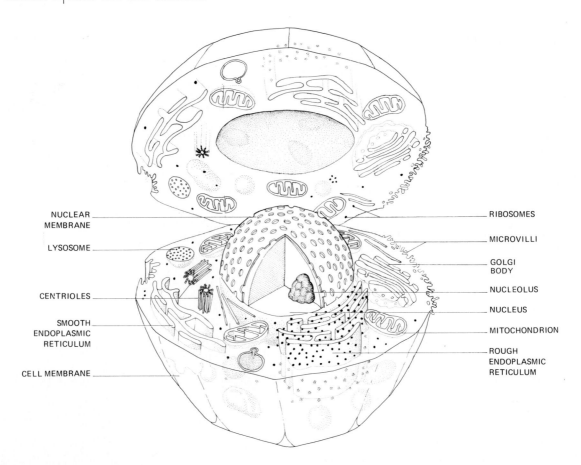

NUCLEAR MEMBRANE

LYSOSOME

CENTRIOLES

SMOOTH ENDOPLASMIC RETICULUM

CELL MEMBRANE

RIBOSOMES

MICROVILLI

GOLGI BODY

NUCLEOLUS

NUCLEUS

MITOCHONDRION

ROUGH ENDOPLASMIC RETICULUM

3.4

Diagrammatic view of an imaginary animal cell, sliced open in such a way as to show some of the cytoplasmic inclusions. Such figures are made up from a number of sources, especially from thin sections and electron micrographs, and they represent what has been called a "pedagogical" cell, to indicate that no such cell exists. Composite pictures are useful in showing proportions and distributions of parts but should not be thought of as representing any actual example. Details of the parts shown, as well as of those parts not shown in this figure, will be described and illustrated later in this chapter. No one figure can illustrate any cell in all its complexity; one might as well try to present *Macbeth* entire in a single picture.

endless, with pigments in all possible cell parts: in cell walls (ebony wood), cytoplasm (red blood cells), cytoplasmic particles (green leaves), vacuoles (red roses) or even nuclei (Indian pipe leaves). Whiteness, however, is due to reflection of light from air spaces *between* cells, or to complete reflection from tissues in aquatic organisms.

The living stuff of cells is protoplasm, and the protoplasm of a single cell is a protoplast, consisting of a nucleus and the nonnuclear portion, the cytoplasm. Protoplasm is colorless unless a pigment is present. It is slimy and somewhat liquid, but it varies from almost watery to almost solid and has the ability to change its viscosity (resistance to flow) greatly and quickly. Protoplasm is usually slightly heavier than water unless fat droplets, which are lighter than water, are incorporated into it to lighten it. It has the elasticity of a badly worn rubber band.

If one considers how plants and animals actually look and feel, this description of protoplasm seems incongruous, but the fact is that protoplasm without some extraneous coating is rarely encountered. Occasionally, if you know where to look, you can find a mass of naked protoplasm creeping about on a moist decaying log. If you do see protoplasm, it is the feeding stage of a slime mold, and it may be no bigger than a bean or as wide as your hand.

Living cells are composed of many discrete parts, and in the rest of this chapter we will discuss the structure and function of the most important subcellular units, starting with plant cell walls.

PLANT CELL WALLS AND ANIMAL CELL COATS

Because protoplasm is wet and runny, an organism that maintains any dependability of form must have some stabilizing structure. Most cells do have extracellular coats: soft envelopes in animals or stiff cell walls in plants.

Plant Cell Walls

Plants owe most of their rigidity to their walls. Walls of woody plant parts are thick and strong, and even those of soft parts, such as tender young leaves, mushrooms, and seaweeds, are firm enough to keep the plant in shape. In land plants, at least, the commonest component of cell walls is cellulose, probably the commonest organic compound in the world. Cellulose, which is built into submicroscopic fibrils in a meshwork around the outside of cells, is laid in place by the activity of the protoplasm inside (Figure 3.5). How the protoplasm inside the cell can synthesize cellulose and deposit it outside the cell is not known. It is as though a carpenter were to shingle the outside of a house while remaining inside with the doors and windows shut.

Each cell of a plant has its own cell wall. The walls of adjacent cells are held together (or held apart?) by a layer of jellylike pectin, the familiar material that makes jelly gel. The pectic layer, the middle lamella, is responsible for the structural unity of plant parts. A crisp pickle has intact middle lamellas; a mushy pickle is one whose middle lamellas have been improperly salted or have been attacked by the enzymes of some invading mold.

Molds and mushrooms (as well as many animals, such as shrimp and insects) have cell coverings containing *chitin*, a substance resembling cellulose, except that instead of being composed of simple glucose residues, it yields on hydrolysis a nitrogen-containing sugar. Bacterial cell envelopes are different from all others, being made of residues of special organic acids (Figure 3.6).

× 20,000

3.5

Electron micrograph of criss-crossed cellulose fibrils in the wall of a green alga. The orderly meshing of the fibrils, so small that it is visible only in electron micrographs, gives plant cell walls their strength, elasticity, and flexibility.

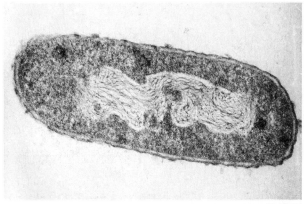

× 38,000

3.6

An electron micrograph of a thin section through a bacterial cell. The pale outer margin is the cell envelope. The dark cytoplasm contains so many ribosomes that they cannot be distinguished. The mid-portion of the cell is filled with streaks of chromosomal material, although the cell does not have a defined nucleus. The example is the colon bacillus, *Escherichia coli.*

Animal Cell Coats

In contrast to the firm cellulose walls of plants, the cellular envelope of animals is flexible and elastic. For a simple comparison of the two, pat a tree and then pat your sweetheart. The animal cell coat is made of a conjugated protein, that is, a complex of protein-plus-carbohydrate, which not only gives the cell toughness and resilience but establishes its reactivity toward other cells.

Cells have sensitive outer surfaces that can distinguish subtle chemical differences in the extracellular environment. This sensitivity can be demonstrated, and experimentally altered, by using cells in tissue culture. If a few cells of kidney or liver are collected under germ-free conditions and grown in glass vessels with a suitable nourishing fluid, the cultured cells ooze about in a thin layer on the glass surface. If kidney cells and liver cells are mixed, each will find its own kind, so that there will be clumps of kidney "tissue" and clumps of liver "tissue." This segregation occurs because the two kinds of cell coats are different enough to be recognizable to the two kinds of cells. In fact, liver cells from human and mouse are more alike than cells from human liver and human muscle.

CYTOPLASM

The nonnuclear portion of a cell is the _cytoplasm._ Nuclei and cytoplasm have been traditionally treated separately because it is easier to work on and to talk about small, simpler parts than whole, complex cells, but nuclei and cytoplasm are in fact so closely associated that they are absolutely dependent on each other. In the following discussion, cytoplasm will be described first, including the various membrane-bound particles known as _organelles._ It must be kept in mind, however, that cytoplasm without nuclei is doomed to a short, restricted life, and nuclei without cytoplasm are useless.

Cell Membranes

The actual outermost protoplasmic layer of a cell is a membrane so thin that it cannot be seen. It can be photographed in electron microscopes, but because it is only about 100 angstrom units thick, it is below the limits of resolution of a light microscope (Figure 3.7). Since an angstrom unit (usually abbreviated Å) is one ten-billionth of a meter, it would require a stack of about a million cell membranes for them to become barely visible to an unaided human eye.

× 360,000

3.7

Electron micrograph of a section cut through the cell membrane, showing the usual double lines that appear in such preparations. The commonest interpretation is that the two dark lines represent the protein layers making the "bread of the sandwich" and the paler stripe between is the lipid layer, or the "meat of the sandwich."

The _cell membrane_, also known as the _plasma membrane_ or the _cytoplasmic membrane_, has been known longer and studied more intensively than the internal cell membranes. Cells have many membranes besides the outer one, and the more that is found out about cells, the more it becomes apparent that membranes are essential to most normal cell functions.

Most subcellular particles are surrounded by membranes. These membranes guarantee the separateness of the particles so that they do not interfere with each other. Membranes also control what substances pass in and out.

Cell membranes are not only _selective_, they are _active_. In other words, they allow some substances to pass through while refusing passage to others, and they can initiate and control the _active transport_ of substances in and out of the cell.

Cell membrane structure   It is now thought that the membrane is composed of a double fatty layer that is studded with protein molecules. Some protein molecules may be on the outside of the membrane, some may be on the interior side, and others may extend through both fatty layers. If this description (or _model_) is correct, it is likely that the transport of most substances through the cell membrane takes place by way of the protein "passages."

The fluid mosaic model   Cell membranes can be measured, and they can be separated from the rest of the protoplasm and analyzed chemically. The membranes consist mainly of phospholipids and proteins. Know-

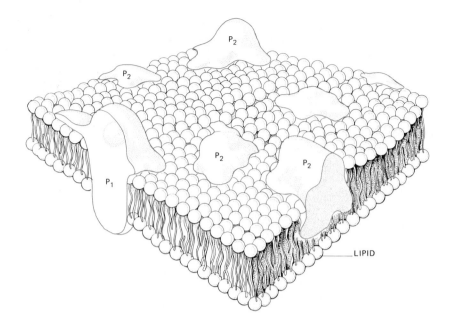

LIPID

3.8

The structure of a plasma membrane as envisioned by Singer and Nicholson. The small spheres with dangling tails represent the double phospholipid layers, one facing the cytoplasm and one on the outside. The sphere is the charged "phospho-" portion of a molecule, and the tail is the uncharged portion. The protein portions are shown as extending completely through the membrane in one instance (P_1) and simply being embedded in the membrane in others (P_2). The entire membrane is thought of as being in a constant state of flux, not a firm, immobile sheet. $\times$ 1.5 million.

ing the thickness of a membrane and the size of phospholipid molecules, one can guess that the phospholipids are arranged in a double layer (Figure 3.8). This makes sense because phospholipids have one charged end and one uncharged end, and the charged ends can stick into the watery protoplasm inside and the similarly watery external environment while the uncharged ends face each other. Furthermore, electron micrographs of sections cut properly through the membranes show pairs of lines. The problem lies in interpreting these pairs of thin parallel lines.

Any imaginary model of a membrane must take into account several demonstrable facts, especially facts of *permeability,* that is, the ability to allow passage of substances. Some substances—as for example, water and fat solvents—pass readily through membranes. Some, such as uncharged organic molecules, pass through with varying degrees of difficulty, depending in part on their size. Others, such as K^+ or Ca^{++} ions, pass only with considerable difficulty. The molecular architecture of membranes remains a puzzle, but the currently popular model, shown in Figure 3.8, includes the additional feature that the entire membrane is capable of constant rearrangement. This *fluid mosaic model* is substantiated by a number of theoretical and experimental studies, but like other membrane models, it is not universally accepted.

One recent observation that supports at least the fluid part of the fluid mosaic model comes from a study of the one-celled animals inhabiting the intestinal tracts of termites (Figure 3.9). These microscopic creatures swimming in the termite gut have a

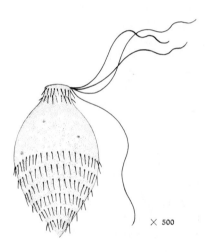

$\times$ 500

3.9

A protozoan, Hyperdevescovina, from a termite gut, showing the four flagella at the anterior end, as well as scattered bacteria attached to the "head" and "body," but absent from the "neck." With the head fixed against an object, the protozoan can slowly spin its body at a rate of one turn every one or two seconds. An observer can see that the "head" bacteria remain stationary while the "body" bacteria are going around. The "neck" is capable of accommodating to the rotation. This is strong direct evidence of the fluidity of cell membranes.

narrow neck near the front end, so that they seem to have a "head" and a "body." The body bristles with attached bacteria, and the head has four thread-like appendages, or *flagella* (see p. 56). Such an animal can stick its head against a firm object, hold it still, and rotate its body. One can watch the bacteria going around while the flagella remain fixed. The neck is a protoplasmic bridge surrounded by a membrane. It is difficult to imagine how such a spinning action could occur unless the membrane was indeed capable of unlimited molecular rearrangement, that is to say, was fluid.

3.10

The phenomenon of *diffusion*. If a quantity of a volatile substance is placed in concentrated mass in a small space, it will move from the region of greatest concentration into regions of lesser concentration. At the beginning (a), the substance is concentrated in one small place, later (b) it spreads into the regions of lesser concentration, and finally (c) it reaches a stable state when the substance is uniformly dispersed throughout the available space.

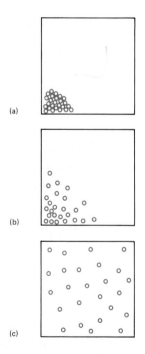

(a)

(b)

(c)

Membrane permeability and osmosis If it is to survive, a cell must maintain a proper balance between its internal and external environment. What is meant by "proper" depends on the cell in question, whether it is in a desert cactus, a freshwater worm, or a marine fish. The balance involves, among other things, regulation of the concentration of dissolved materials in water. In an enclosed space, as in a cell, the concentration of dissolved material, the *solute*, increases as the *solvent* (in cells, water) decreases, and conversely. One can think of a high concentration of solute or a high concentration of solvent. A cell can raise the concentration of solute either by bringing in more solute or by expelling solvent. Or it can lower solute concentration by expelling solute or bringing more solvent in. Still another way of changing solute concentration is by changing undissolved particles such as starch into soluble sugar and back again. All these methods are used by cells to keep their water balance.

Unless a substance is at absolute zero, −273°C, its molecules are in constant random motion. As a result of this motion, molecules left to themselves will tend to spread from a region where they are crowded to become more evenly distributed. This phenomenon of *diffusion* (Figure 3.10) is usually stated as the tendency of molecules to move from a region of greater concentration toward a region of lesser concentration. Diffusion is one of the factors affecting water balance in cells. If a sugar lump is placed gently in a vessel of water, the sugar molecules diffuse into the water while the water molecules are diffusing into the sugar, until there is an even mixing of the two. Such mixing without mechanical stirring will require weeks, but it will eventually become complete.

The rate of diffusion depends on the state of matter of the diffusing molecules. Gases diffuse rapidly; that is why perfume works. Liquids diffuse more slowly, at a rate measurable in millimeters per week. Solids are the slowest of all, their rate being measurable in micrometers per year, but if a flat piece of gold is pressed against a flat piece of silver, each will move into the other. The movement is not detectable by simple inspection, but it happens. The diffusion rate is also affected by temperature: the higher the temperature, the faster the diffusion. Molecular size, too, affects the diffusion rate, with a small glycerol molecule moving faster than a large fatty acid.

Cells must work constantly to keep themselves in order, and the word "work" here means that there must be a continuous expenditure of energy. A cell works at keeping its membranes intact, since the membranes act as gatekeepers in determining what

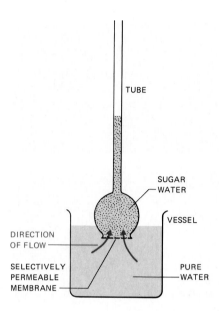

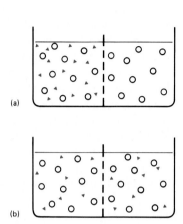

3.11

Movement of a solute through a selectively permeable membrane. In (a) the solvent (large circles) is on both sides of the separating partition but cannot pass through it. The solute (small triangles) is initially on one side of the partition but *can* pass through it. After a time (b) the solute has diffused through the partition and is uniformly distributed throughout the system. The partition, permitting the passage of one substance but not another, represents a *selectively permeable membrane.* Selective permeability is one characteristic of living cell membranes.

3.12

An osmometer, a device for measuring the amount of actual pressure that a solution can develop when separated from a pure solute by a selectively permeable membrane. The solution in the tube, a mixture of sugar and water, is separated from pure water in the vessel outside the tube. The water diffuses through the membrane, but since the sugar cannot pass the membrane, the total flow is from the vessel into the tube, and it builds up a pressure inside the tube. The fluid level rises in the tube. A mixture of sugar and water by itself exerts no *pressure,* but it does have an osmotic *value* that can be determined by means of an osmometer. Cells are difficult to use directly for *pressure measurements,* but osmotic *values* for cells can be obtained by immersing them in solutions whose osmotic values have been determined with an osmometer.

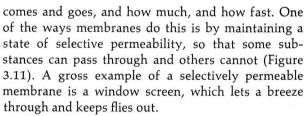

comes and goes, and how much, and how fast. One of the ways membranes do this is by maintaining a state of selective permeability, so that some substances can pass through and others cannot (Figure 3.11). A gross example of a selectively permeable membrane is a window screen, which lets a breeze through and keeps flies out.

If a cell with some solute in its sap is immersed in pure water, there will be a tendency for the solute to diffuse out, but it cannot, because the cell membrane keeps it in. There is also a tendency for water to diffuse in because the pure water outside is more concentrated, in terms of water molecules, than it is inside, where it is diluted by the solute. The total movement, therefore, is inward, and the inward

movement will continue until either the cell bursts or the back pressure exerted by the cell wall in plants or the cell envelope in animals equals the inward diffusion pressure. Tough cells, like mature plant cells, swell and get plump, or *turgid,* like an inflated tire. Delicate cells, such as red blood cells lacking a strong cover, burst when put in water. Before tender plant parts grow a stiffening tissue, they depend to some degree on being pumped full of water; otherwise they droop. Wilted leaves are those suffering excess water loss and consequent loss of turgor.

By means of an *osmometer,* one can measure the actual pressure developed when two substances, both potentially diffusible, are separated by a partition that allows the passage of only one (Figure 3.12). Since

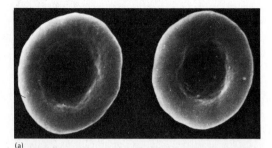

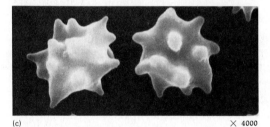

(a)

(b)

(c)   × 4000

3.13

Human red blood cells in solutions of various concentrations. (a) A cell in plasma, in which the concentrations inside and outside the cell are equal. The cell retains its shape. (b) A cell in a solution that is less concentrated than the internal solution of the cell. Water moves *into* the cell until it bursts. The cell contents are lost and the empty plasma membranes are left as "ghosts." (c) A cell in a solution that is more concentrated than the internal solution of the cell. Water moves *out of* the cell, leaving it shrunken.

such pressure can be great enough to burst cells, they must either have strong walls or keep their internal and external environments in balance. Red blood cells, for example, exist only in the well-regulated blood plasma (Figure 3.13a). If the solute concentration of the plasma were to go too low, the blood cells would explode (Figure 3.13b). If it went too high, water would diffuse out of the cells (there being relatively more water inside than out), and the cells would shrivel like a raisin (Figure 3.13c).

Osmosis, the simple movement of a solvent through a selectively permeable membrane, is frequently invoked to explain water movement into cells. The rate at which water actually passes, however, is too great to be explained merely by diffusion. Rather there must be massive movement of water, generally called *bulk flow.*

Transport across membranes   In the examples given above of cells in water, there was always some solute, such as sugars or salts, in the cell sap. But if the membranes are impermeable to the solutes, how did they get in to begin with? The answer lies in the *selectivity* of the membranes, their ability to change that selectivity with the passage of time, and their ability to move molecules actively. Some molecules do cross membranes purely by diffusion, with the membrane having nothing to do with it. Such movement is *passive transport.*

Passive transport requires nothing from a cell. A second possibility, *facilitated transport* (Figure 3.14a), does not require energy but does require a postulated carrier of some sort to "escort" a molecule across a membrane. The entry of glucose into red blood cells is the most intensively studied system of this sort.

Cell membranes, however, are usually far from passive. They can carry out a number of complex activities, which can be described better than they can be explained. Transport that requires energy is *active transport* (Figure 3.14b). Indeed, no one has explained satisfactorily how a membrane can perform such feats as the following: (1) A white blood cell can approach a bacterium in the blood plasma, send out an encircling crater around it, cover it with the crested lips of the crater, pull crater, bacterium, and a bit of plasma down into the protoplast, and there pinch off a vesicle containing the whole thing. (2) A bacterium can distinguish between two molecules that differ in their atomic configurations, bring one into the protoplast, and ignore the other. (3) A ma-

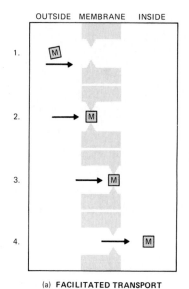

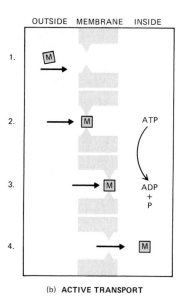

(a) FACILITATED TRANSPORT (b) ACTIVE TRANSPORT

3.14
Diagrams illustrating methods of transport of materials across cell membranes. Simple transport by means of ordinary diffusion was shown in Figure 3.11. (a) In *facilitated transport* a molecule, M, approaches the membrane (1), is bound by a carrier (2), carried through (3), and released inside the cell (4). This is not an energy-requiring activity. (b) In *active transport* the procedure is thought to be similar except for the important difference of the energy input by ATP. Active transport can be stopped by anything that interferes with a cell's ATP supply, such as cutting off its oxygen or adding a poison.

rine algal cell can bring potassium into its protoplast and keep bringing it in until there is a greater concentration of potassium inside than outside; this process is known as *building up against a gradient*.

Particle transport A white blood cell engulfing a bacterium or an amoeba engulfing a food particle is an example of what is called *phagocytosis* (Figure 3.15), which means "cell eating." A similar phenomenon is *pinocytosis*, or the taking in of fluid by the same kind of membrane action. By a reversal of such action, a particle in a cell—for example, a droplet containing a digestive enzyme—may be surrounded by an internal membrane in the protoplasm, carried to the cell boundary, and popped out through the cell membrane to the outside of the cell. All these events show that cell membranes can break and re-form quickly, and regulate the passage of substances into and out of cells. Our knowledge of cell membranes is probably naive, but enough is known to make it clear that membranes are of prime importance to cells, and that if we are ever to understand the life of cells, we will have to learn how and why membranes work the way they do.

Endoplasmic Reticulum
The cell membrane is only the outermost of protoplasmic membranes. Inward from the cell membrane, and physically connected with it in places, there is

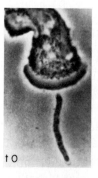
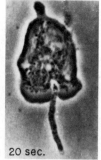

t 0   20 sec.

40 sec.   60 sec.   70 sec.

× 1500

3.15
A human white blood cell engulfing a bacterium, *Bacillus megaterium*. From the beginning, when the bacterium touches the blood cell, until the bacterium has been taken completely in, the action takes 70 seconds.

3.16

The *endoplasmic reticulum*, ER. (a) A diagrammatic cell, cut open to show position and distribution of ER. (b) Electron micrograph of a thin section through cytoplasm, showing the rough ER studded with ribosomes on the cytoplasmic sides of the membranes. (c) Drawing showing the three-dimensional structure of ER. The sheets of membrane that make up the ER branch and recombine and have holes in them. The ER membranes occur in double layers, with space between membrane pairs. The side of a membrane that touches the cytoplasm has ribosomes, but the side toward the cisterna does not. Endoplasmic reticulum with ribosomes is known as "rough," and that without ribosomes is "smooth."

(a)

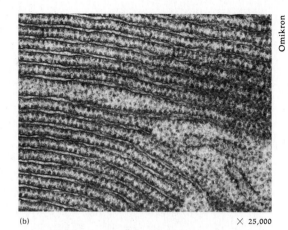

(b)   × 25,000

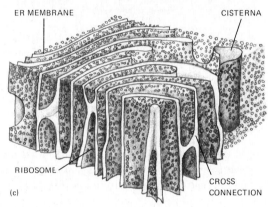

ER MEMBRANE   CISTERNA

RIBOSOME

(c)   CROSS CONNECTION

the *endoplasmic reticulum* (Figure 3.16), frequently called simply the *ER*. It is a labyrinthine complex of double membranes branching and spreading throughout the cytoplasm. The individual portions of the ER are too small to be seen with a light microscope. They can be photographed in thin sections of cells by electron microscopy, but unfortunately that technique shows only two dimensions. In section, the membranes appear as branching pairs of lines, but it must be pointed out that the ER forms a three-dimensional compartment separated from the cytoplasm by the membranes.

The ER seems to act as a system of internal channels through which various materials move. It also serves as a point of attachment for ribosomes, where proteins are assembled (Chapter 7). In electron micrographs, the ER can be seen studded with ribosomes. The terms "rough ER" and "smooth ER" refer to endoplasmic reticulum that does or does not have attached ribosomes. ("Rough ER" *does* have attached ribosomes.)

In plant cells, there is a tendency for the watery spaces, known as cisternae, between a pair of ER membranes to swell. One or more of the swellings enlarge until in old cells they may practically fill the cell, leaving only a thin layer of cytoplasm around the periphery of the cell. The central vesicle is the conspicuous *vacuole,* a poor name because it is not empty (Figure 3.17). It contains mostly water, but in the water are salts, proteins, crystals, and pigments. Many plant colors are due to pigments dissolved in vacuoles, especially the blues, reds, and purples. As usual, however, it is better not to generalize quickly, for although beets and apples and roses owe color to vacuole pigments, tomatoes, peppers, carrots, and dandelions owe theirs to cytoplasmic plastids. Vacuoles function as water reservoirs in maintaining turgor and apparently sometimes as a dumping ground for wastes.

Ribosomes

It is established that *ribosomes* are necessary for the synthesis of protein from amino acids. Like other submicroscopic organelles, ribosomes were not discovered until electron micrographs of thin-sectioned cells became available. Then they appeared everywhere

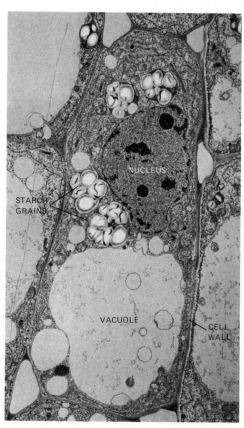

× 8,500

3.17
Electron micrograph of a thin section through a plant cell, showing the water-filled *vacuole*, produced by swelling of the cisternae between pairs of ER membranes.

3.18
(a) Electron micrograph and (b) schematic drawing of *ribosomes* from rabbit cells. Five ribosomes, connected by a strand of a special nucleic acid, messenger ribonucleic acid (mRNA), constitute a *polysome*. The subunits of the ribosomes, shown in the drawing, are not apparent in this micrograph, but they are visible in some preparations and can be separated by physical means.

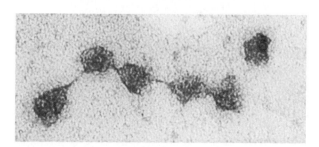

(a)

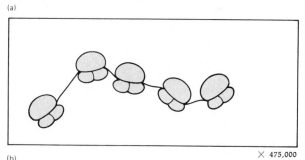

(b)   × 475,000

any protein synthesis occurs: in bacteria, on or near the ER of higher plant and animal cytoplasm, in plastids and mitochondria. Ribosomes appear in electron microscope photographs as slightly flattened spheres with a slight constriction around the middle, showing that a ribosome is a double structure (Figure 3.18). During times of active protein synthesis, a string of five or more ribosomes may be linked together, presumably by a strand of messenger RNA, to form a polyribosome, or simply a *polysome* (Chapter 7).

Final assembly of ribosomes occurs in the cytoplasm, but the initial steps take place in nucleoli. (See the section on nucleoli later in this chapter.) There ribosomal protein and nucleic acids are joined, and the incomplete ribosomes are passed to the cytoplasm. A complete ribosome of a higher animal cell is too large to be a single molecule but too small to be treated as an ordinary cell organelle.

Golgi Bodies and Lysosomes

Golgi bodies (named for their discoverer, Camillo Golgi) are collections of membranes associated with the ER. In plants, Golgi bodies are also called *dictyo-*

3.19

Golgi body. (a) Schematic view of a cell, cut open to show position and proportions of the bodies. (b) Electron micrograph of two sets of Golgi bodies in the cytoplasm of a corn root. (c) Reconstruction of Golgi bodies to show three-dimensional structure of the membranes, internal cisternae, and droplets being formed around the edges.

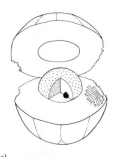

(a)

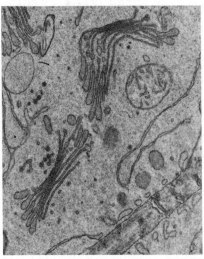

(b)   × 18,330

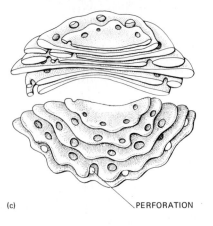

(c)   PERFORATION

somes. Although they can be seen plainly in electron micrographs of sectioned cells, their three-dimensional structure is difficult to imagine, and they cannot be observed adequately with a light microscope. Golgi bodies have been described as stacks of saucers, but if the parts of a Golgi body are like saucers, they are saucers with irregular, frilly edges (Figure 3.19). Apparently, Golgi bodies are formed by the budding off of portions of the ER, and they in turn bud portions off themselves. Substances synthesized in the cytoplasm are delivered into the compartment between ER membranes, passed on to the compartment defined by the membranes of the Golgi body, and then to their final destination, usually outside the cell.

Golgi bodies are most apparent and abundant in cells with secretory activity: pancreas cells secreting digestive enzymes, nerve cells secreting transmitter substances, plant cells laying down cell plates between newly divided cells. The Golgi bodies concentrate the substance to be secreted and deliver it to its destination. In some instances they actually engage in some synthetic work along the way.

Lysosomes can be seen easily in any microscope as droplets, but they are so characterless that they cannot be identified. They were discovered when homogenized cells had their components separated, and these spherical bodies, covered by a single membrane, proved to be extraordinarily rich in hydrolytic enzymes (Chapter 1). Lysosomes are widely distributed in all but bacterial cells, but they are most abundant in tissues that experience rapid changes such as in lungs. Since they are loaded with digestive enzymes, lysosomes can act as scavengers, ingesting and digesting "worn out" cell parts. Or they can burst, releasing the concentrated enzymes, which can then digest the cell they are in, as happens to the cells in a tadpole's tail. When a tadpole is changing into an adult frog, the lysosomes in its tail cells cause the breakdown of structure, and the tail is absorbed into the frog's body. Lysosomes can also contribute their enzymes to digest invading bacteria. Lysosomes were called "suicide bags" by their discoverer, Nobel laureate Christian DeDuve, but other people have called them "garbage disposal units."

Plastids

Responsible for the conversion of light energy to chemical energy, plastids occupy such a special place in the economy of life that they will be treated separately in Chapter 4.

Mitochondria

The main activity of the *mitochondria* is the production of ATP for use in cellular work. The conversion of organic molecules to usable energy occurs in the mitochondria, and it is for this reason that a mitochondrion is sometimes called the "powerhouse of the cell." Live mitochondria are vigorously active. They squirm, roll up, stretch, grow lumps, divide, flow together, swell, shrink, and even seem to swim against cytoplasmic currents (Figure 3.20a). Detailed understanding of mitochondria began with the advent of electron microscopes and the techniques of homogenization and centrifugation. Electron micrographs show that mitochondria, various as they may be in size and shape, have a fairly uniform internal structure (Figure 3.20b, c). "Fairly" here means that there is, as expected, some deviation from the standard. Most cells of higher organisms contain several hundred mitochondria, with other quantities ranging from one or two in some protozoa to as many as 150,000 in the egg cells of amphibians.

An outer membrane, sometimes continuous with the ER, delimits a mitochondrion. Inside, another membranous structure is folded and doubled on itself so as to form incomplete partitions, the *cristae*, or crests. High-resolution micrographs sometimes show the inner membrane to be covered with pebbly bumps that seem to be stalked, while the outside of the outer membrane has stalkless bumps (Figure 3.20d, e). The inner stalked bumps are suspected by some investigators to be the functional units of mitochondrial work, ATP production, but such an idea is not fully substantiated.

Besides being repositories for the enzymes of cellular respiration (Chapter 5), mitochondria have many other uses in cells, being concerned with calcium metabolism, rubber synthesis, antibody formation, protein storage, and flagellar movement. New functions of mitochondria continue to be found, and it has even been suggested that cancerous growth of cells is a result of mitochondrial dysfunction.

Plastids and mitochondria as symbiotic organelles
Because of their size and appearance, it was suggested long ago that mitochondria are something like cellular slaves that live in the cytoplasm of cells but are in fact bacteria. That was only an unsubstantiated idea, however, until the electron microscope revealed the structure of bacteria and subcellular particles. As more is learned about all these entities, the resemblances between bacteria, mitochondria, and plastids increase, until it is becoming accepted that mitochondria and plastids may indeed be separate organisms, which have inhabited other organisms so long that now neither can live without the other. The arguments are convincing. Besides the similarities in size and shape, bacteria and plastids and mitochondria are

3.20

Mitochondria. (a) Living mitochondria in a filament of a fungus as seen in a light microscope. (b) Electron micrograph of a sectioned mitochondrion, showing covering membrane and folds of the inner membrane (cristae). (c) Schematic drawing of a reconstructed mitochondrion to show its three-dimensional structure. (d) Electron micrograph of the inner membrane at high magnification; arrows point to the elementary particles ("lollipops"). (e) Sketch interpreting the electron micrograph of the elementary particles.

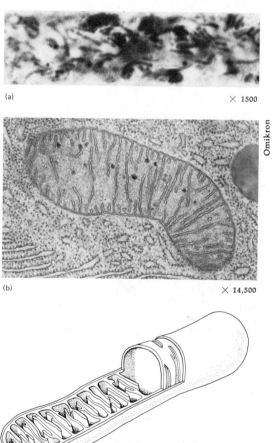

(a) × 1500

(b) × 14,500

Omikron

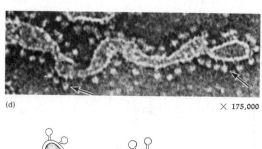

(c)

(d) × 175,000

(e)

TAKING CELLS APART
TO PUT THEM
BACK TOGETHER

In order to see most organelles in position in cells, one must slice the cells, perhaps stain them, and then look at them with a light microscope or photograph them with an electron microscope. Most such treatments are destructive, and the cells are dead practically from the start of the work. How can an experimenter know what an organelle does simply from inspecting a dead one?

The answer is that no single procedure can give a satisfactory picture. The experimenter usually must see the organelle but must also get it out of its cell, accumulate a sufficient quantity to experiment on, and get rid of foreign particles that might confuse the results. First, live cells must be broken up. The easiest and commonest way to do that is to drop a piece of tissue in a kitchen blender of the type that was originally made to stir drinks. The steel blades, whirling at 3000–4000 rpm (revolutions per minute), can disrupt cells without destroying their internal organelles, especially if everything is kept cold and a liquid medium is provided that does not upset the water balance. Common cane sugar in water is frequently used as the liquid medium. The result is a slurry mixture containing nuclei, mitochondria, and all the other subcellular particles.

The next step is to separate out the different pieces of the cells. This can be done by taking advantage of the different densities of the various organelles. Cell walls are heavier than nuclei; nuclei are heavier than mitochondria; mitochondria are heavier than ribosomes; and so on. If the complex mix is spun in a centrifuge at several thousand rpm, the effect is comparable to increasing the force of gravity from several hundred times to 100,000 or more. Particles that are only slightly heavier than the fluid medium will settle as a pellet at the bottom of the tube. By controlling the speed (and hence the force) and the time from a few minutes to several hours, one can harvest a series of different organelles selectively, starting with the heaviest. Once a pellet of mitochondria, for example, is collected, it can be tested for enzyme reactions or analyzed for chemical residues or checked to find where radioactive tracers have gone. It can even be made to carry out a particle's original functions.

Once a number of such investigations have been made, the problem of integrating the information begins. Some syntheses have been accomplished, but so far there is no machine that can function as an antiblender and put together what man has put asunder.

all bound by membranes, they all have their own DNA, which, unlike the DNA of higher organism chromosomes, is not associated with protein, and they have ribosomes that are like one another and smaller than the ribosomes of higher cells. Plastids and mitochondria have a partially independent genetic life, typical of organelles with their own DNA. Further, some animals routinely incorporate algal cells into their own cells, where the algae grow and carry on photosynthesis. One biologist has asked, "Are we really powered by bacteria?" Perhaps we are.

Microtubules

Microtubules may be up to several micrometers long, but being only about 250 Å thick, they are far too slender to be seen with a light microscope. In micrographs, they are so straight that they seem to have been drawn with a ruler (Figure 3.21). When separated from a number of different cell types, they prove to be chemically identical, made of a protein appropriately called tubulin. The tubulin subunits can assemble into microtubules or disassemble and disperse in the cytoplasm, influenced by the cellular environment. They can even assemble themselves in test-tube preparations.

Microtubules have been thought of as a kind of cytoplasmic skeleton, but they are in fact more commonly associated with movement: in cilia and flagella, in pseudopods of amoebae, in cellulose deposition, and in chromosome movement during nuclear division (Chapter 8).

3.21

(a) Electron micrograph of sectioned guinea pig marrow cell showing the long, straight, slender *microtubules* in the cytoplasm. (b) Drawing of a portion of a microtubule to show the arrangement of protein units that make up the tubule. Thirteen units make one turn of a low-pitched helix. × 1.2 million.

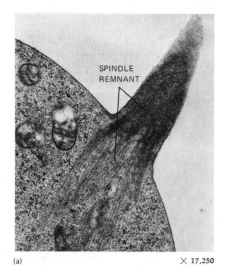

SPINDLE REMNANT

(a)   × 17,250

(b)

Centrioles

Although centrioles are barely visible under a light microscope, it has been known for nearly a hundred years that they are associated with nuclear division in animals and some lower plants. Flowering plants have no centrioles. Electron micrographs show that a centriole is a cylindrical organelle, about 0.2 micrometers thick and 0.5 micrometers long. When associated with nuclei, they occur in pairs and act as focal points at opposite ends of a cell during nuclear division. Centrioles from many organisms, from algae through mammals, show a standard structural pattern of microtubules (Figure 3.22). They are also usually associated with other microtubules, such as the mitotic asters (Chapter 8) or cilia (see p. 56), and are connected in some unknown way with cellular motility.

(a)

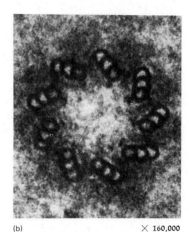

(b)   × 160,000

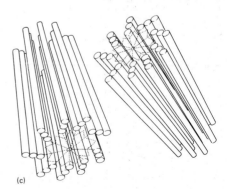

(c)

THE NUCLEUS AND NUCLEAR MEMBRANES

The word *nucleus* is derived from the Latin word for kernel, that is, the core or essential part. Indeed, the nucleus is the control center of the cell, and it contains the necessary information to direct the reproduction and heredity of new cells.

In all but the simplest organisms (bacteria and related forms), the nucleus is a well-defined region clearly delimited from the general surrounding cytoplasm by a membrane. It is roundish or flattened, usually colorless, and easily visible through a light microscope. The nucleus was first recorded in 1833 by the English scientist Robert Brown, who found it in the epidermal cells of orchid leaves and then began finding similar bodies in all the cells he investigated.

3.22

Centrioles. (a) Position of centrioles near the cell nucleus. (b) Electron micrograph of a cross section through a centriole, showing the arrangement of the tubules and the central "spokes." (c) Schematic drawing of a pair of centrioles, showing their relation to each other, one at right angles to the other, and the nine triplets of tubules that form a roughly cylindrical body of the centriole.

3.23

(a) A *nucleus,* as shown in a reconstruction from light and electron micrographs. The double nuclear membrane has many pores. Floating in the nuclear sap is a nucleolus, of which there may be one or several. Nuclei and nucleoli are variable in shape. (b) Electron micrograph of a section through a nucleus. The grainy appearance of the nuclear sap and the irregular, heterogeneous nature of the nucleolus are characteristic of such photographs.

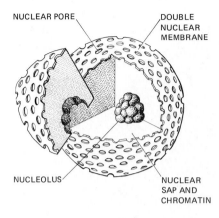

NUCLEAR PORE

DOUBLE NUCLEAR MEMBRANE

NUCLEOLUS

NUCLEAR SAP AND CHROMATIN

(a)

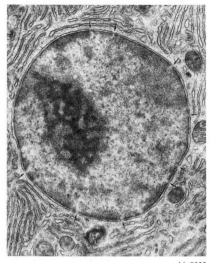

× 7000

(b)

3.24

(a) The first known figures of chromosomes were drawn by Wilhelm Hofmeister, who saw them in the spiderwort Tradescantia, without knowing the significance of what he was doing. His figures illustrate the development of pollen grains, and his chromosomes are readily identifiable. × 400. (b) Tradescantia chromosomes as seen through a modern light microscope in stained material.

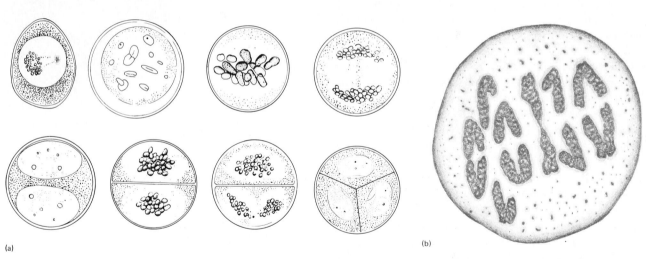

(a)

(b)

Looked at with a light microscope, the nuclei seemed not to have a distinct casing, but as we know now, the boundary was too thin to see. Present investigations, however, show the nuclear boundary to be something like the cell membrane, but with some important differences. It is not a single membrane, but two membranes, with a discernible "space" (compartment) between them. Nor is the nuclear membrane continuous. It has numerous pores in it, so that it is something like the plastic balls children call "whiffle balls" (Figure 3.23). Furthermore, the outer nuclear membrane differs chemically from the inner one. Thus the nuclear membrane separates the cytoplasm from the nuclear sap, but it does not do so completely. There must be two-way traffic between the nucleus and the cytoplasm, and all theoretical and experimental evidence says this is so. During nuclear division, the nuclear membrane usually breaks up and remains "lost" (undetectable) in the cytoplasm until the end of the division period. After two new nuclei have been formed, the ER spreads new nuclear membranes over the chromosome masses.

Chromosomes

Chromosomes are the DNA-containing structures that can be seen temporarily during cell division. Between divisions the chromosomes are not readily visible; it is then that they are busy with their two main functions, replicating themselves and regulating the synthesis of proteins.

In 1848, a now half-forgotten genius, Wilhelm Hofmeister, drew figures of chromosomes in the developing pollen grains of the spiderwort, Tradescantia (Figure 3.24). Even now, after all the years and all the studies, it is difficult to find a species or a tissue better suited for such an observation. It causes one to wonder what led him to choose that material, out of the bewildering array of possibilities. Since the time of Hofmeister, thousands of species have been scrutinized and thousands of investigations made on chromosomes, and the results have led to some of the fundamental discoveries of biology.

Chromosomes (Greek, "color bodies") were so named because of their stainability with the textile dyes available to the early biologists. By the end of the 1860s, the general behavior of chromosomes during the division periods of cells had been described. The main features of cell division (Chapter 8) were known. It seems ironic now that Gregor Mendel, who was working at the same time, formed the basic theories of inheritance without being aware of the significance of the discoveries of the cell biologists. Nor were they aware of him. It would be another half-century before the still unborn science of genetics and the infant science of cell biology would get together.

Nuclei of cells in all higher organisms have chromosomes, which vary in number from one per nucleus in a variety of intestinal worm to hundreds in some ferns and shrimp, and in size from less than a micrometer in some fungi to more than 100 micrometers in fly larvae. No membrane covers a chromosome; it lies uncovered in the nuclear sap. One or more strands of DNA run the length of a chromosome, coiled to varying degrees. During the time when a cell is metabolically active and not in the process of dividing, the coiling is minimal and the chromosome is so long and thin that it cannot be seen with a light microscope. Before and during nuclear division, the DNA coils more and more tightly. Coil upon coil is formed until a chromosome becomes a shorter, stubbier, recognizable body. The bulk of material in a chromo-

some is protein, especially histone, which is basic (in the chemical sense) because of the presence of the basic amino acids lysine, arginine, and histidine. Acidic protein is also present. In spite of continuing attempts to find the use of chromosomal protein and the way it is physically built into the chromosome body, final answers are still lacking.

Most chromosomes have a special region somewhere along the DNA where microtubules become attached. This _centromere_ (Figure 3.25) seems to be the hitching point for the connection of spindle fibers (microtubules), and it guarantees that a chromosome

3.25

A chromosome as it appears during nuclear division, reconstructed from electron micrographs. The tangled chromatin threads and the centromere, the attachment point for the spindle fibers (microtubules), are conspicuous. Many chromosomes show a constriction at some point other than the centromere.

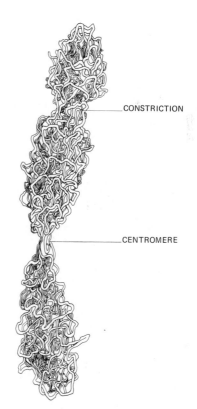

CONSTRICTION

CENTROMERE

CILIA AND FLAGELLA: MICROSCOPIC MOVEMENT

Cilia (Latin, "eyelash") and *flagella* (Latin, "whip") provide a means of locomotion for small, free-swimming organisms or a means of forcing water past fixed cells. Cilia and flagella are alike in structure and function but differ in length. Cilia are about 25 times as long as they are thick, and flagella may be thousands of times as long as thick. The important feature of cilia and flagella is their ability to bend actively. A short cilium is something like a thin paddle, and a long flagellum can send waves of movement along its length like a snake fastened by its tail. Both effectively create water currents.

A flagellum is a membrane-bound tube full of cytoplasm and provided with a set of microtubules anchored at the basal end and running parallel through or nearly through the length of the tube. There are of course some variations from species to species, but in general the fine structure of flagella is remarkably constant throughout the living world. The arrangement

× 900

Scanning electron micrograph of cilia on the epithelium of a human trachea. In cells such as these, the beating of cilia causes fluid to move along the cell surface, but in free-floating cells, such as ciliated protozoa, the cilia make the cell move.

Ellen Roter Dirksen

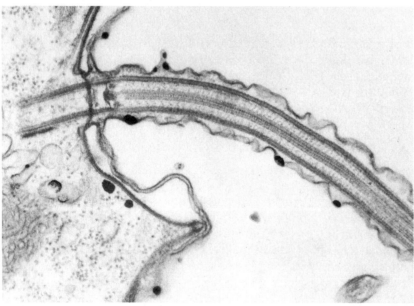

Omikron

× 30,000

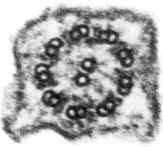

× 100,000

(a) Longitudinal section of a flagellum from a frog sperm cell, as shown in an electron micrograph. The microtubules of the flagellar shaft are anchored in the basal body in the cytoplasm. (b) Cross section of a flagellum of a protozoan, showing the nine-plus-two arrangement of the microtubules, the "arms" protruding from the outer microtubules, and a suggestion of the long structures that run the length of the flagellum. The latter appear in cross section as dots between the central and outer tubules.

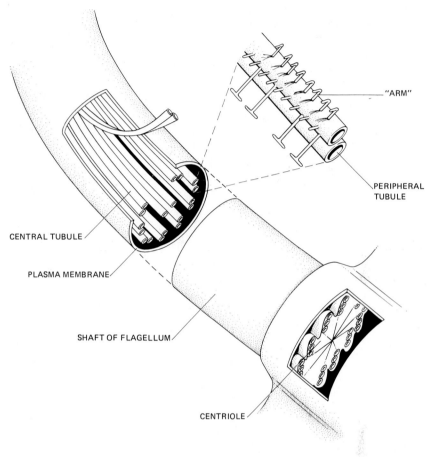

An imagined dissection of a flagellum, reconstructed from sections and broken flagella as seen in electron micrographs. The shaft of the flagellum contains the standard nine double peripheral tubules and two central ones. Rows of projections, the "arms," and T-shaped "hammerheads" are thought to act as rachets, somehow clicking past each other to make the tubules slide against one another, thus causing the shaft of the flagellum to bend. The base of the flagellum is anchored in a centriole or basal body in the cytoplasm of the cell. In the centriole, the peripheral tubules are triple instead of double, and the central tubules are lacking.

"ARM"

PERIPHERAL TUBULE

CENTRAL TUBULE

PLASMA MEMBRANE

SHAFT OF FLAGELLUM

CENTRIOLE

Diagram of power stroke and recovery of a flagellum.

DIRECTION OF DRIVING STROKE

DIRECTION OF RECOVERY STROKE

DIRECTION OF MOVEMENT OF FREE-LIVING CELL

DIRECTION OF MOVEMENT OF FLUID BY FIXED CELL

of microtubules is about the same in swimming algae, one-celled animals, mammalian sperm or fern sperm (where the special structure was first discovered), clam gills, and human mucus membranes. A section through a flagellum shows a ring of nine double microtubules around the periphery and two separate center ones. The unproved assumption is that the microtubules actively slide past one another to make the entire flagellum bend. They certainly can bend, even after being broken from cells, but only if they are provided with an energy source, such as ATP.

A ciliary beat consists of a stiff snap in one direction followed by a gentler, flexible recovery. What triggers the beat is not known, but it is evident that the impulse is controlled with great precision. In a protozoan or in a piece of ciliated epithelium, waves of movement sweep along as accurately synchronized as in a group of experienced handbell ringers.

The "invention" of flagella was probably one of the most important advances in the history of organic evolution. Surely an organism experiences an enormous advantage if it can move away from its fixed or passively floating competitors when overcrowding pollutes the environment or the food runs low.

TABLE 3.2

Presence and Functions of Cell Organelles

STRUCTURE OR ORGANELLE	IN PROKARYOTIC CELLS	IN EUKARYOTIC CELLS	MAIN FUNCTION(S)
Cell covering	Envelope containing muramic acid	Walls in plants, cell coat in animals	Protection and support
Cell membrane	Present; site of major enzymatic activity	Present	Regulation of transport
Photosynthetic chromatophores	Present in some	Absent	Photosynthesis
Plastids	Absent	Present in most plants	Photosynthesis
Lysosomes	Absent	Usually present	Contain digestive enzymes
Golgi bodies	Absent	Usually present	Concentration and passage of materials
Mitochondria	Absent	Present	Aerobic respiration
Ribosomes	Present	Present	Active in protein synthesis
Nuclear membrane	Absent	Present, with pores	Contains chromosomes and nuclear sap
Chromosomes	Continuous loop of naked DNA ("genophore")	Present, with centromeres and protein matrix	DNA replication and control of protein synthesis
Nucleoli	Absent	Present	Preparation of ribosomes
Endoplasmic reticulum	Absent	Present	Internal structure and transport
Microtubules	Present	Present	Cytoplasmic structure (?); movement
Flagella	True flagella lacking; microtubulelike structures	Present; consistent microtubule structure	Motility; water current
Centrioles	Absent	Present in all but higher plants	Mitotic activity and basal granules of cilia

will proceed to its proper destination during cell division. An abnormal chromosome without a centromere will be lost in the cytoplasm.

Nucleoli

Nuclei contain one or more roundish or lumpy *nucleoli,* visible in living cells, as well as in stained cells, where they appear as dark balls. Like chromosomes, nucleoli lack covering membranes. Nucleoli are composed of protein, presumably brought into the nucleus from the cytoplasm, and ribosomal RNA. At the onset of cell division, nucleoli disappear, to reappear after the division process is complete. Special chromosomal regions, the nucleolar organizers, generate the ribosomal RNA, and the RNA unites with protein to start the process of forming ribosomes, so that a nucleolus can be thought of as a preassembly point for ribosomes.

PROKARYOTES AND EUKARYOTES

Some organisms, such as the bacteria and blue-green algae, seem to have survived for several billion years without advancing beyond their original primitive state. These are the _prokaryotes_, meaning "before a nucleus." They differ from more advanced organisms in several structural features, the most obvious being that they lack distinct, membrane-bound nuclei. They are, indeed, characterized by their many structural deficiencies. It is noteworthy, however, that in spite of these seeming deficiencies, prokaryotes are quite effective biosynthetically. They compete successfully in a world filled with rivals.

Prokaryotes have no ER, no mitochondria, no plastids, no nucleoli, no Golgi bodies, no true flagella. A prokaryotic "chromosome" is a naked strand of DNA without the proteins that are usually associated with chromosomes. Some biologists prefer to call prokaryotic chromosomes "genophores," or "gene carriers." Such a DNA strand is usually a continuous loop, inaccurately described as "circular." Prokaryotes do have ribosomes and extensions called flagella, but the so-called flagella of motile bacteria are more like single microtubules than like the complex true flagella of higher organisms.

The _eukaryotes_, meaning "true nuclei," are the organisms whose cells have all the features described in this chapter. Apparently the evolution of these diverse organelles has made possible the range of variation that is observable in all the plants and animals, unicellular and multicellular, past and present, that make up the bulk of the living world.

SUMMARY

1. Robert Hooke first discovered and named cells in 1665. Almost 200 years later Theodor Schwann and Matthias Schleiden popularized the _cell theory_.

2. _Cells_ are the smallest independent units of life, and all life as we know it depends on the chemical activities of cells.

3. The living stuff of cells is _protoplasm_; the protoplasm of a single cell is a _protoplast_, consisting of a nucleus and the nonnuclear portion, the _cytoplasm_.

4. Plants owe most of their rigidity to their _walls_. In land plants, the commonest component of cell walls is _cellulose_. In contrast to the firm walls of plants, the cell _coat_ of animals is flexible and elastic. The animal cell coat is made of a complex of protein-plus-carbohydrate.

5. The _cell membrane_ is essential to most normal cell functions. Cell membranes are not only selective but active. They allow some substances to pass through while refusing passage to others, and they can initiate and control the active transport of substances in and out of the cell.

6. In the _fluid mosaic model_ it is postulated that the cell membrane is a double layer of phospholipids studded with protein molecules; in this model the entire membrane is capable of constant internal rearrangement.

7. Diffusion is one of the factors affecting water balance in cells. _Diffusion_ is the tendency of molecules to move from a region of greater concentration of themselves toward a region of lesser concentration.

8. _Passive transport_ occurs when some molecules cross membranes purely by diffusion, and the membrane has nothing to do with it. A cell membrane demonstrates definite, energy-requiring activity during _active transport_.

9. The _endoplasmic reticulum_ is a labyrinthine complex of double membranes. It seems to act as a system of internal channels through which various materials move. It also serves as a point of attachment for _ribosomes_, where proteins are assembled.

10. Biologists have established that _ribosomes_ are necessary for the synthesis of protein from amino acids, but the exact function of the ribosomal RNA in the process is still unknown.

11. _Golgi bodies_ are most apparent and abundant in cells with secretory activity. The Golgi bodies concentrate the substance to be secreted and deliver it to its destination, and in some instances they actually engage in some synthetic work along the way.

12. _Lysosomes_ are loaded with digestive enzymes, and they can act as scavengers, ingesting and digesting "worn out" cell parts.

13. _Plastids_ are responsible for the conversion of light energy to chemical energy.

14. *Mitochondria* are known as the powerhouses of the cell because their main work is the production of ATP in cellular respiration.

15. The *nucleus* is the control center of the cell. It contains all the necessary information to direct the reproduction and heredity of new cells. The *nuclear membrane* is a double membrane with numerous pores in it, and there is two-way traffic between the nucleus and the cytoplasm.

16. *Chromosomes* are the main DNA-carrying structures, located within the nucleus in higher organisms. They have no limiting membranes, but they do have special regions, the *centromeres*, which are attachment centers for spindle fibers during nuclear division. Chromosomes are the cell's chief self-replicating organelles; they organize *nucleoli,* and they determine the structure of cell proteins. In prokaryotes, genophores are naked loops of DNA.

17. *Nucleoli* are assembly centers in nuclei, functioning in the partial construction of ribosomes.

18. *Prokaryotic* cells lack distinct, membrane-bound nuclei, and they are characterized by their many deficiencies. *Eukaryotic* cells have all the features described in this chapter; apparently the evolution of eukaryotes has made possible the variation in plants and animals.

RECOMMENDED READING

Capaldi, Roderick A., "A Dynamic Model of Cell Membranes." *Scientific American*, March 1974. (Offprint 1292)

Dyson, Robert D., *Cell Biology: A Molecular Approach,* Second Edition. Boston: Allyn and Bacon, 1978.

Fox, C. Fred, "The Structure of Cell Membranes." *Scientific American*, February 1972. (Offprint 1241)

Hayashi, Teru, "How Cells Move." *Scientific American*, September 1961. (Offprint 97)

Hinkle, Peter C., and Richard E. McCarty, "How Cells Make ATP." *Scientific American*, March 1978.

Holter, Heinz, "How Things Get Into Cells." *Scientific American*, September 1961. (Offprint 96)

Kennedy, Donald (ed.), *Cellular and Organismal Biology* (Readings from *Scientific American*). San Francisco: W. H. Freeman, 1974. (Paperback)

Neutra, Marian, and C. P. Leblond, "The Golgi Apparatus." *Scientific American*, February 1969. (Offprint 1134)

Nomura, Masayasu, "Ribosomes." *Scientific American*, October 1969. (Offprint 1157)

Pfeiffer, John, *The Cell*, Revised Edition. New York: Time-Life Books, 1976. (Life Science Library)

Satir, Birgit, "The Final Steps in Secretion." *Scientific American*, October 1975. (Offprint 1328)

Schopf, J. William, "The Evolution of the Earliest Cells." *Scientific American*, September 1978.

Sharon, Nathan, "Lectins." *Scientific American*, June 1977. (Offprint 1360)

Staehelin, L. Andrew, and Barbara E. Hull, "Junctions between Living Cells." *Scientific American*, May 1978.

Swanson, Carl P., *The Cell*, Third Edition. Englewood Cliffs, N.J.: Prentice-Hall, 1969.

Wessells, Norman K., "How Living Cells Change Shape." *Scientific American*, October 1971. (Offprint 1233)

Wolfe, Stephen L., *Biology of the Cell*. Belmont, Calif.: Wadsworth, 1972.

4

Photosynthesis: Trapping and Storing Energy

LIFE RUNS ON SUNLIGHT. PLANTS CAN CONVERT the light energy of the sun into chemical energy, and this process is called *photosynthesis,* one of the most important chemical processes in nature. Plants store enough food with their light-trapping abilities to feed not only themselves but animals as well. Without photosynthesis there would be almost no usable energy for the cells of all living organisms. Not only does photosynthesis provide the food and energy to sustain plant life; it indirectly supplies the basic fuel for animals as well. Because animals cannot manufacture their own food, they must consume plants and other animals. But besides satisfying their nutritional needs, animals receive their oxygen supply as a by-product of photosynthesis. Without photosynthesis, therefore, life as we know it could not exist.

We still do not fully understand how plants convert the sun's energy into food, but since the discovery of photosynthesis by Joseph Priestley about 200 years ago, our knowledge has grown a great deal. In fact, so much has been learned about photosynthesis since the 1930s that it now appears that only details need to be cleared up.

WHAT IS PHOTOSYNTHESIS?

The essential facts about photosynthesis are these: Light energy is converted by means of the chlorophylls and accessory pigments into chemical energy and stored as carbohydrates; the simple compounds used are water and carbon dioxide; and the final products are organic molecules (sugars, starches, fats, etc.), and oxygen. The overall reaction is usually written

$$6CO_2 + 6H_2O \xrightarrow[\text{light}]{\text{chlorophyll}} C_6H_{12}O_6 + 6O_2.$$

carbon dioxide   water   glucose   oxygen

This process becomes more understandable if we analyze the word "photosynthesis." *Photo* means light, and *synthesis* means the process of building up a complex substance through the union of simpler substances. But the seemingly simple process of photosynthesis is aided by complex support mechanisms, as we shall see.

PRODUCERS AND CONSUMERS: AUTOTROPHS AND HETEROTROPHS

No life can exist without a constant source of energy. There are many variations in the methods used to extract and use energy, but the two major types of energy-requiring cells are called *autotrophic* and *heterotrophic*. When scientists first discovered that plants are able to manufacture their own food, organisms were reclassified as either *autotrophs*, organisms that can synthesize their food from inorganic sources, or *heterotrophs*, organisms that must acquire their food because they cannot manufacture it themselves.

The chief autotrophs are green plants, which acquire their energy by transforming the light energy

THE ROLE OF PHOTOSYNTHESIS IN THE HABITATION OF THE LAND

Photosynthesis is such a fundamental and important process that it may be natural to think that it has existed since the earth originated about 4.6 billion years ago. In fact, photosynthesis probably started about three billion years ago, and then only with simple bacteria and blue-green algae.

Life began in the water about 3.5 billion years ago; two billion years later it managed to emerge onto the land. Without photosynthesis, however, life could not have evolved even long enough for it to begin the conquest of dry land.

In a sense, some cells "invented" photosynthesis and thereby opened a new energy source for exploitation. Fermentation (energy release without oxygen intake) came first, and although it produced the minimum energy necessary for life, it was a limited and wasteful process. According to present opinion, the atmosphere of the earth was essentially devoid of free oxygen before the onset of photosynthesis. Fermenting cells did not use free oxygen, but the carbohydrates that were used to produce energy were being consumed without being replaced. However, before fermentation ran out of carbohydrate energy sources, photosynthesis evolved, first in simple organisms and

then in higher green plants. Of course, those cells that developed chlorophyll and were able to convert sunlight into chemical energy gained a tremendous advantage over cells that remained dependent on existing sources of nourishment.

Once chlorophyll evolved, it used carbon dioxide and sunlight to produce ATP and oxygen through photosynthesis. As oxygen in the atmosphere increased, so did ozone (O_3), which could absorb much of the sun's lethal ultraviolet radiation. About 350 million years ago, life on the land and in the air became possible because of the protective shield of ozone and the abundance of free oxygen. (Actually, organisms had to adapt to the free oxygen, since it was a totally new factor. The great importance of free oxygen was that it made cellular respiration possible later on.)

Why did some sea life move to the land? Why not remain in the ocean without the evolutionary migration toward the shore and onto the land? Probably because the shore provided a new habitat where the bounty was great. Sunlight, land, water, oxygen, carbon dioxide, minerals, and virgin territory with minimal competition were plentiful.

of the sun into carbohydrate plus oxygen. Heterotrophic cells, such as those found in the human body and in other animals, in fungi, and in plants that do not contain chlorophyll, obtain their energy through the breakdown of food, such as sugar, in a process called *respiration*. Note that photosynthesis builds up compounds to convert and store energy, whereas respiration breaks down compounds to release energy. Cellular respiration will be discussed in detail in the next chapter.

WHAT ARE THE REQUIREMENTS FOR PHOTOSYNTHESIS?

The principal requirements for photosynthesis are light, chlorophyll, chloroplasts, carbon dioxide, and water. Oxygen and some critical minerals must also be present.

Light

The electromagnetic spectrum is composed of radiations of wavelengths measured in nanometers, or nm (a nanometer is one-billionth of a meter), ranging from very short wavelengths (gamma rays, X-rays) to very long ones (infrared radiation, radio waves). The longer the wavelength, the less energy it possesses; the shorter the wavelength, the greater energy it possesses. That portion of the electromagnetic spectrum usable by plants for photosynthesis ranges from about 400 to 750 nm. This is approximately the same range that the human eye can detect, that is, the "visible" portion of the spectrum.

The energy in light is inversely proportional to the wavelength, so the short (violet) end of the visible spectrum contains about 1.4 times more energy than the long (red) end (Figure 4.1). The chlorophyll pigments absorb some light along the entire visible spectrum, but they absorb more at the red end. Indeed, plants appear green because of their great absorption of red light and the reflection of the green portion of the spectrum.

The *intensity* of light, which is the amount of energy delivered on a given area per unit of time, is as important as the wavelength (or "quality") of the light. If the light intensity is too low, photosynthesis ceases. If the light intensity is such that the metabolism of the plant uses up the photosynthetic products as fast as they are produced, that light intensity is called the <u>compensation point</u> (Figure 4.2). At the other extreme, light can reach a point of intensity so high that the plant can use no further light, and the plant becomes *saturated*. Many plants, especially

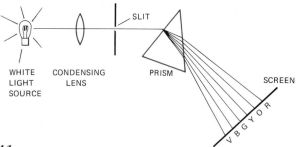

4.1

Breaking white light into its components. A source of white light, a mixture of all wavelengths and frequencies, has a beam focused through a slit and then through a prism. The glass of the prism changes the direction of the beam differently for different wavelengths, so that the shorter ones (violet, about 380 nm) are bent more than the longer ones (red, 700–750 nm). The other wavelengths are distributed between, with green about 550 nm and yellow about 600 nm. The resulting spectrum can be projected on a screen, producing a visible "rainbow."

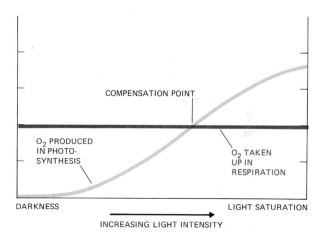

4.2

The compensation point in photosynthesis. In the dark, a green plant uses food and takes up oxygen; in light (up to a point) it accumulates food and releases oxygen. At some light intensity between darkness and saturation, the photosynthetic oxygen output equals the respiratory oxygen intake, and there is no net oxygen change. This is the compensation point. (This concept is based on the not altogether correct supposition that respiration proceeds at the same rate in light and darkness.) A compensation point is reached each day as the sun rises in the morning and again when it sets in the evening. It is also reached by aquatic plants, which may sink in water until the light is reduced.

those adapted to low light intensities, as in deep water or in tropical rain forests, are actually harmed by too much light, and chlorophyll is destroyed.

Chlorophyll, the Photosynthetic Pigment

In general, pigments are substances that absorb light. The most conspicuous photosynthetic pigments are the *chlorophylls,* the pigments in green plants that, along with accessory pigments, absorb light at the outset of photosynthesis. The two common chlorophylls of higher plants are chlorophyll *a* and chlorophyll *b*, which are chemically similar. Both are complex molecules composed of carbon, hydrogen, oxygen, nitrogen, and magnesium (Figure 4.3).

The special feature of chlorophyll is its ability to be "excited" by certain wavelengths of light. During such excitement, which lasts about one-billionth of a second, pairs of electrons are pushed farther away from their nucleus and are thus raised to a higher, unstable energy level (Figure 4.4). In some cases the excited electrons of chlorophyll *a* may merely drop back to their original energy level and emit heat and/or light as *fluorescence*, accomplishing nothing of chemical value (Figure 4.5). If white light falls on chlorophyll in solution in acetone, some of the green light not being absorbed will pass through the solution. A chlorophyll solution is transparent to green light. Some of the light will be trapped by the chlorophyll. But if there are no electron acceptors present, the excited chlorophyll electrons will merely drop back to their previous condition, giving off light as they do so. During this exchange of energy, some energy is lost (the second law of thermodynamics again) and the light released will be less energetic, that is, of longer wavelength than the stimulating light. It will be red. A concentrated chlorophyll solution looks blood red when seen in reflected light. (Light from fluorescent and television tubes is produced in similar fashion. Electrons are raised to higher energy levels from which they fall, emitting light in the process.) However, if a suitable electron-accepting molecule is available, the electrons will be transferred to the acceptor, and photosynthesis will have begun.

Chlorophyll *a* seems to be most directly involved in photosynthesis. Chlorophyll *b* and the other related pigments are able to absorb energy, but instead of using it directly, they apparently pass it along to chlorophyll *a*.

4.3
Molecular structure of a molecule of chlorophyll *a*. Chlorophyll *b* is the same, except that it has an aldehyde group, CHO, in place of the encircled methyl group, CH_3. The long chain of carbon atoms, which is uncharged and hence not water-soluble, is thought to extend into the lipid portion of the chloroplast membrane. The porphyrin ring, which resembles a ring in the red blood pigment, hemoglobin, contains a magnesium atom (in place of the iron in hemoglobin). The atoms in chlorophyll that can be energized by light are the nitrogen atoms.

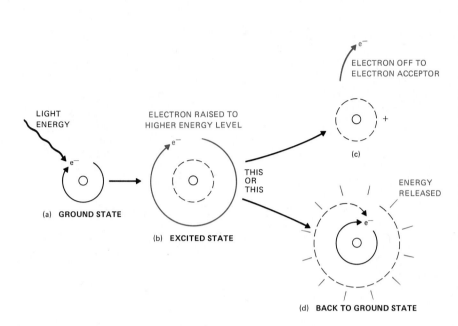

(a) **GROUND STATE**

LIGHT ENERGY

ELECTRON RAISED TO HIGHER ENERGY LEVEL

(b) **EXCITED STATE**

THIS OR THIS

ELECTRON OFF TO ELECTRON ACCEPTOR

(c)

ENERGY RELEASED

(d) **BACK TO GROUND STATE**

4.4

Schematic representation of activation of a molecule by light. When a molecule in its ground state is energized by light (a), an electron is raised to a new, higher energy level (b), leaving an unoccupied place in the electron shell, frequently called a "hole." If an electron acceptor is available, the electron can pass from the original molecule (such as chlorophyll) to the acceptor, and the energy will be passed to the acceptor (c). If there is no acceptor, the energized electron will fall back into its original position, and in so doing will re-emit energy (d). The re-emitted energy may be in the form of heat, or it may be in the form of light. If light is emitted, it will be at a longer wavelength than the original exciting light because some energy is lost in the exchange, and longer wavelengths are less energetic than short ones.

Plastid Organization

Obviously, a cell must be especially suited to the process of photosynthesis to be an effective transformer of light energy. Plant cells contain special structures for this function, and the precise agent of production is the plastid. _Plastids_ are relatively large, membrane-bound, colored organelles found only in plant cells. The special function of plastids is the synthesis and storage of food. Three types of plastids are known, and their names also indicate their colors. _Chromoplasts_ can be yellow, orange, or red. _Leucoplasts,_ containing oil or starch, are almost colorless. The photosynthetic plastids, _chloroplasts,_ contain the green chlorophyll (Figure 4.6). Photosynthesis is initiated in the chloroplast when light strikes the chlorophyll-filled organelle.

With the use of the electron microscope, we can see the complex internal structure of a chloroplast, including _grana,_ which are composed of stacks of flattened, membrane-bound vesicles, the _thylakoids._ The membranes in the grana are like other cell membranes in that they contain proteins and phospholipids, but they are special because they contain the chlorophylls, the accessory pigments, and the various electron acceptors necessary for the "light reactions" of

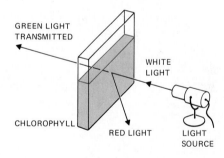

GREEN LIGHT TRANSMITTED

WHITE LIGHT

CHLOROPHYLL

RED LIGHT

LIGHT SOURCE

4.5

Fluorescence of chlorophyll. When white light strikes a solution of chlorophyll, some (especially the red and blue portion of the spectrum) is absorbed, and what is not absorbed (green) passes through the solution, that is, is transmitted. Some of the energy of the absorbed light is re-emitted as longer wavelengths in the red end of the visible spectrum. A chlorophyll solution, viewed as shown in the diagram, appears red on the side toward the light and green on the other side.

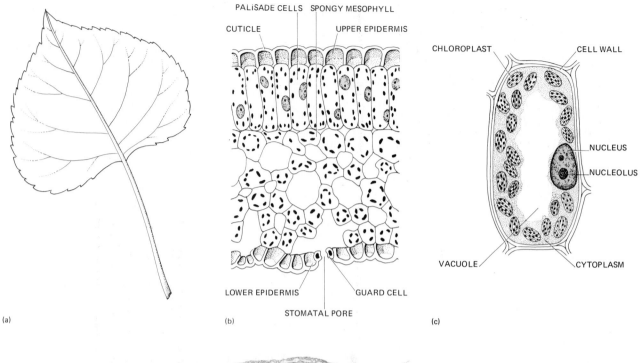

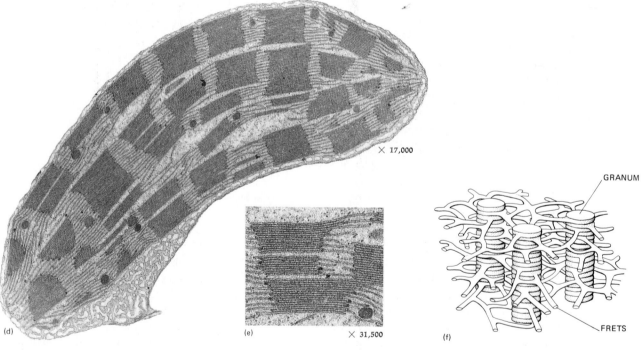

4.6

The photosynthetic machinery. (a) An intact leaf, natural size. (b) A section through a leaf. × 100. (c) A single chloroplast-bearing cell from the spongy mesophyll of a leaf. × 300. (d) Electron micrograph of a section through a single chloroplast, showing the covering membrane and *thylakoids* stacked into *grana*. (e) Electron micrograph of a section through four grana, showing neatly arrayed thylakoids making up the grana and connecting *frets* extending through the *stroma* from one granum to the next. (f) A reconstruction of several grana, shown in three dimensions. The drawing is a composite of electron micrographs made from sections cut at different angles. The grana are seen as stacks of thylakoids, connected by membranous frets between the grana. × 50,000.

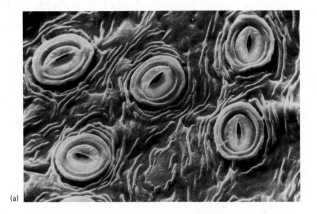

(a)

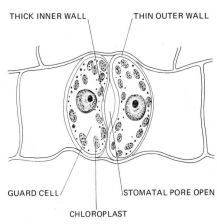

THICK INNER WALL   THIN OUTER WALL

GUARD CELL   STOMATAL PORE OPEN

CHLOROPLAST

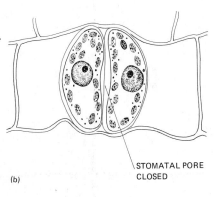

(b)

STOMATAL PORE CLOSED

4.7

(a) Scanning electron micrograph of the lower surface of a leaf, showing five open *stomata*. Stomata are distributed on leaf surfaces in such a way that they allow adequate entrance of carbon dioxide and the liberation of oxygen during photosynthesis. Because the stomata cannot be selective in choosing which gases can pass, they must also allow water vapor to escape. Water will necessarily escape because the internal surfaces of the leaf are wet with water brought up from the roots, and the air outside is usually drier than the air in the spaces of the leaf. Excess water loss can be fatal. As a compromise, leaves allow carbon dioxide, oxygen, and water vapor to move through the stomatal openings when the leaves are illuminated, but they close the openings in darkness. We can make stomata open or close and can watch them doing so, yet the exact mechanism is not known. The best that can be said at present is that light, water, and potassium movement are involved. (b) Movement of guard cells. When water enters guard cells and makes them firm, the pressure against the thin outer walls causes them to bulge outward, pulling the stomatal pore open (top). When water is lost from guard cells, they become soft, and the thin outer walls lose their bulge. The thick inner walls force the stomatal pore shut (bottom).

photosynthesis. Electron micrographs showing surface views of thylakoid membranes reveal regularly arranged bumps, each about 150 A across. These bumps, called *quantasomes*, are thought to contain a minimal array of necessary compounds and may be the actual photosynthetic units of the chloroplast. Non-grana regions of the chloroplasts, known as the *stroma*, are the sites of the so-called dark reactions of photosynthesis, to be discussed later in this chapter.

Carbon Dioxide

There are three to four molecules of carbon dioxide in every 10,000 molecules of gases in the earth's atmosphere. This is a sufficiently high concentration so that plants can pick up the carbon necessary for the buildup of carbohydrate in photosynthesis. If the concentration goes too low, say to five parts per 100,000, ordinary photosynthesis ceases, and if the concentration gets too high, carbon dioxide can become toxic. Plants could use more carbon dioxide than they normally get, up to five parts per 100 or more, but such a concentration is practically never found in nature. If it was economical to do so, crops could be "fertilized" by adding carbon dioxide to the air around leaves during daylight.

The surfaces of plants are usually covered with a waxy cuticle called *cutin* that prevents excessive water loss, infection by harmful bacteria, and damage, such as bruising. The waxy surface also shuts out carbon dioxide, and plants have had to evolve means of letting the gas in. The trick was to develop microscopic, controllable openings between cells in the leaf epidermis. These openings are *stomata* (Figure 4.7).

Water

One of the few absolute statements one can make about anything biological is that water is essential for all vital activities. It is the common cellular solvent, it helps regulate temperatures, it enters into chemical combination in countless reactions, it furnishes the essential protons for energy exchanges, it provides turgor pressure to create the stiffness of tender plant parts, it acts as a lubricant and as a medium of transport, and it even enters into the structural parts of cells. But it has a special value in photosynthesis because it furnishes one of the two raw materials from which organic molecules are synthesized. The other is of course carbon dioxide. A molecule of water, broken apart in an active chloroplast, yields two protons (H^+), two electrons, and an atom of oxygen. The protons or hydrogen ions and the electrons become incorporated into a sugar or some other energy-containing molecule, and the oxygen atom combines with another oxygen atom to form molecular oxygen, O_2, which will escape into the atmosphere. The oxygen is then available to animals for respiration.

Other Photosynthetic Requirements

Besides the direct requirements of light, raw materials, and chloroplasts, a number of indirect requirements must be met if photosynthesis is to proceed. For example, there must be some oxygen present, and mineral elements, such as iron, copper, magnesium, sulfur, nitrogen, potassium, phosphorus, and boron, which enter into the composition of the various functional compounds, are of critical importance.

UNDERSTANDING THE TWO PHASES OF PHOTOSYNTHESIS

Before discussing the "light" and "dark" reactions of photosynthesis, we should emphasize certain points to avoid confusion:

1.   The light reactions *do* utilize light energy.

2.   The dark reactions are not part of the light-trapping process. The use of the term "dark" *does not* mean that the dark reactions necessarily take place in the dark or at night. It means simply that these reactions may proceed without light as a requirement.

3.   The light and dark reactions are consecutive steps in the overall process of photosynthesis, which starts with the trapping of light energy (the light reactions) and ends with the production of sugar (the dark reactions).

4.   Photosynthesis is a continuous process. The light and dark reactions are discussed separately only to divide the light-absorbing phase from the "chemical" phase, which produces carbohydrate through complex chemical reactions. The light reactions convert light energy to chemical energy; the dark reactions use chemical energy to produce complex organic molecules.

THE LIGHT REACTIONS OF PHOTOSYNTHESIS

It has been known for more than 50 years that there are at least two steps in the photosynthetic process, one requiring light and one not requiring light. In the light reactions, two products are formed, one an energy carrier (ATP) and the other a hydrogen carrier with the long name of nicotinamide adenosine dinucleotide phosphate, usually abbreviated to NADP (Figure 4.8).

When light falls on a chloroplast, several simultaneous actions are started. There are two kinds of light-sensitive centers, designated *Pigment System I* and *Pigment System II*, which differ in their components and in the wavelengths of light they absorb. Pigment Systems I and II are alike in that they contain chlorophyll and that they can be activated by light energy.

When chlorophyll in Pigment System I is activated by light of wavelength of 700 nm, pairs of electrons are raised to a high enough energy level to be released from the chlorophyll molecule. These energy-carrying electrons are passed along a series of acceptors, one of which is an iron-containing protein, ferredoxin. The electrons ultimately are transferred to NADP, which accepts not only the electrons but also a pair of protons, $2H^+$, reducing NADP to $NADPH_2$. The compound $NADPH_2$ is able to carry its load of hydrogen (H^+ and electron) to the next photosynthetic step, the dark reactions. In the dark reactions the hydrogen will be used in the buildup of carbohydrate.

Meanwhile, the electrons have left a "gap" in the chlorophyll from which they came, and that gap must be refilled before photosynthesis can continue. That refilling is accomplished by the action of Pigment System II.

While Pigment System I was working, System II was also being activated, but by light of wavelength of 680 nm. Its chlorophyll has pairs of electrons raised to high energy levels, and they are removed from the chlorophyll, accepted by a series of electron accep-

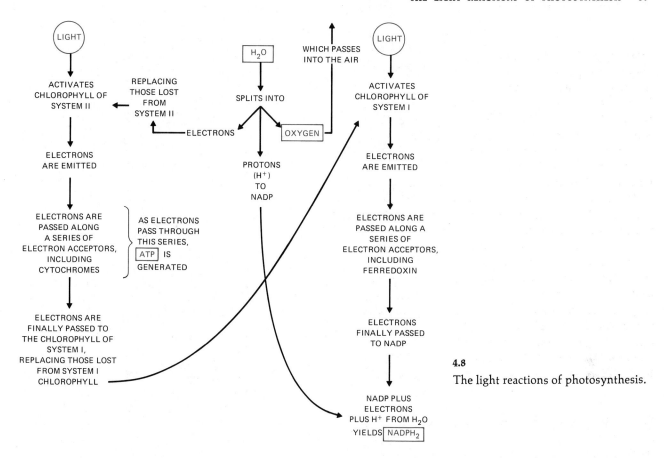

4.8
The light reactions of photosynthesis.

tors, and finally passed to the above-mentioned "gaps" in the chlorophyll of Pigment System I. This process of course leaves gaps in the chlorophyll of Pigment System II, and they in turn must be refilled.

The electrons lost from Pigment System II are replaced by electrons from new water molecules that are split into protons (H^+), oxygen, and electrons. These are the electrons that fill the gaps left in the chlorophyll of Pigment System II, and they thus keep the process operating. NADP is again reduced to $NADPH_2$ as it accepts electrons and protons. The remaining oxygen atoms form molecular oxygen, O_2, which diffuses into the atmosphere.

Another important action is going on at the same time. As the electrons in Pigment System II are passed along their chain of carriers, they become less energized at each step, but not all their energy is lost to the plant. Some of it is caught and used in the formation of ATP from ADP and inorganic phosphate. This ATP will be used as an energy supply in the dark reactions to follow.

In some primitive plants, as well as in higher plants where it is an accessory to ATP generation,

there is a process called *cyclic phosphorylation*. It is called "cyclic" because electrons are recycled in Pigment System I and "phosphorylation" because phosphorus is transferred. In this process, no new electrons are received from water, no NADP is reduced, and the only product is ATP.

The *main events in the light reactions* are: (1) the light activation of chlorophylls, with the release of energy-bearing electrons; (2) the splitting of water with the release of H^+, electrons, and oxygen; (3) the reduction of NADP to $NADPH_2$; and (4) the generation of ATP.

The Requirements of the Light Reactions	*The Products of the Light Reactions*
1. Light	1. Oxygen
2. Enzymes in the chloroplasts, structurally intact	2. ATP
3. Water	3. $NADPH_2$
4. NADP	
5. ADP and inorganic phosphate	

4.9

The Calvin cycle, the dark reactions of photosynthesis. When carbon dioxide enters the photosynthetic process, it joins the five-carbon sugar ribulose diphosphate (RuDP) to form a temporary union, which quickly breaks to two three-carbon molecules of phosphoglyceric acid (PGA). This is the first identifiable carbohydrate formed in photosynthesis. The PGA is reduced by hydrogen from NADPH, which itself was earlier reduced during the light reactions of photosynthesis. The NADP, now oxidized, is ready for another hydrogen-carrying trip, and the PGA has been reduced to phosphoglyceraldehyde (PGAL).

The three-carbon compound PGAL may be used in a number of ways. The most obvious use of PGAL is in the building of reserve food in the plant. Two molecules of PGAL can form one molecule of the six-carbon sugar fructose, which in turn can be converted to glucose. Glucose can then be condensed to starch, the usual insoluble storage form of plant food.

Some of the PGAL must be reserved for use in the regeneration of more RuDP. This can be accomplished in several ways, as shown in the schematic diagram. Two molecules of 3-C PGAL may combine to form a 6-C molecule. This 6-C molecule can unite with another 3-C to yield not a 9-C but one 5-C and one 4-C. A 4-C plus a 3-C gives a 7-C (sedoheptulose). The 7-C, joined to still another 3-C, yields not a 10-C but two 5-C sugars. The 5-C sugars thus produced receive an additional phosphate group from ATP to make the final carbon dioxide acceptor, the di-phospho-5-C sugar RuDP.

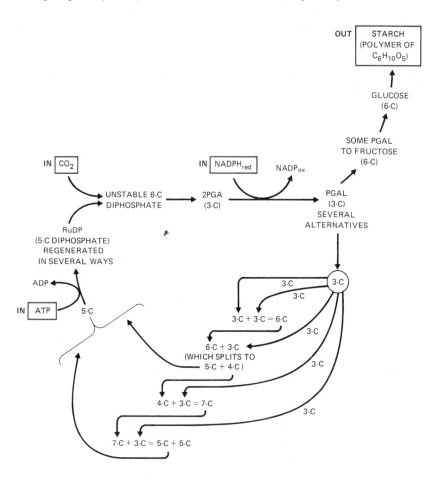

THE DARK REACTIONS OF PHOTOSYNTHESIS

The dark reactions of photosynthesis take place in the chloroplasts. During the dark reactions, complex molecules of sugar and starch, composed of carbon, hydrogen, and oxygen, are being made from the simpler molecules of carbon dioxide and the hydrogen of $NADPH_2$. The sequence of events given here was postulated by Melvin Calvin.

The first identifiable carbohydrate in most land plants is a 3-carbon compound, phosphoglyceric acid (PGA). According to the *Calvin cycle* (Figures 4.9 and 4.10), carbon dioxide from the atmosphere is enzymatically united with a 5-carbon sugar, ribulose diphosphate (RuDP), to form a 6-carbon compound, which immediately breaks down into two molecules of 3-carbon PGA.

Energy to drive the system comes from ATP, and hydrogen is brought in via $NADPH_2$, both of which have been produced in the light reactions. The energized PGA is reduced to phosphoglyceraldehyde (PGAL), a 3-carbon compound that living cells can

4.10

The Calvin cycle (simplified version).

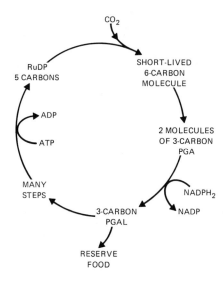

use as a starter for the synthesis of all the uncounted substances of living matter.

After PGAL has been made, it has several alternatives available. Some of the 3-carbon PGAL can be condensed to 6-carbon sugars, such as fructose and glucose. These can be further condensed to a common storage product, starch, or may be enzymatically converted to fats or (with the addition of nitrogen) to amino acids. Such conversions are begun within minutes after a chloroplast is illuminated. However, if all the PGAL were made into reserve food, there would be no new 5-carbon RuDP to pick up incoming new CO_2. Some of the PGAL, therefore, is used to generate new RuDP by means of several intermediates, including 4-carbon and 7-carbon compounds. The result is that although the bulk of PGAL synthesized during photosynthesis is used directly to make reserve food, enough is recycled to maintain an adequate supply of RuDP so that the process continues.

The Requirements of the Dark Reactions	The Immediate Products of the Dark Reactions
1. ATP	1. Carbohydrate (PGAL)
2. NADPH$_2$	2. ADP
3. Carbon dioxide	3. Inorganic phosphate (P$_i$)
4. Ribulose diphosphate (RuDP)	4. Water
5. Enzymes in the chloroplasts	

THE FATE OF WATER IN PHOTOSYNTHESIS

In general, photosynthesis can be thought of as the combination of carbon dioxide and water to yield oxygen and carbohydrate. As stated at the beginning of this chapter, the overall reaction is frequently written

$$6CO_2 + 6H_2O \xrightarrow[\text{light}]{\text{chlorophyll}} C_6H_{12}O_6 + 6O_2.$$

However, if we recall that the resulting oxygen comes from the water and not from the carbon dioxide, we can see that the equation above is not quite correct. If the molecular oxygen on the right side of the equation comes from the water (as indeed it must if the isotope experiments are correct), then how can we account for 12 oxygen atoms from only 6 molecules of water? The answer is that additional water has entered into the reaction, and new water is formed as a result. Consequently, the equation is rewritten

$$6CO_2 + 12H_2O \xrightarrow[\text{light}]{\text{chlorophyll}} C_6H_{12}O_6 + 6H_2O + 6O_2.$$

During photosynthesis, water and carbon dioxide are combined to form glucose and the important by-product oxygen. During respiration, glucose reacts with oxygen to release stored cellular energy, and carbon dioxide and water are by-products that are used again during photosynthesis. The energy released through respiration is distributed bit by bit in a controlled process that will be discussed in the next chapter.

ALTERNATIVE PHOTOSYNTHETIC SCHEMES

Much of what has been described in this chapter was determined from experiments on a green alga, *Chlorella*, or spinach, usually with an artificially high concentration of carbon dioxide. Asking themselves whether the results from such special experimental conditions could be applied confidently to plants in general, some investigators looked at photosynthetic events in a number of plants under natural conditions and found that some desert plants and some grasses, such as maize and sugar cane, do not use the Calvin cycle exclusively. In such plants, the first-formed compounds are not the expected 3-carbon PGAL, but 4-carbon acids: malic, aspartic, and oxalacetic. The Calvin scheme is consequently known as 3-carbon photosynthesis, and the *Hatch-Slack scheme*, named for its discoverers, is the 4-carbon method. (No single species can shift back and forth under varying conditions between 3-carbon and 4-carbon photosynthesis. Any given land plant always uses *either* the 3-carbon system or a combination of both systems.) The Hatch-Slack method functions when 3-carbon photosynthesis is inhibited by too high a light intensity, too high a temperature, or too low a carbon dioxide concentration. It is therefore effective in tropical or desert conditions. It has been known for a long time that a sugar cane field is one of the most productive energy traps in all nature, and part of the explanation is the use by cane of the 4-carbon method of photosynthesis, which allows the plant to take advantage of intense light and a hot climate. Thus photosynthesis, like many other biosynthetic phenomena, proves to be a process in which a given result can be achieved by more than one pathway.

SUMMARY

1. Plants can convert the light energy of the sun into chemical energy by the process called *photosynthesis,* one of the essential chemical processes in nature.

2. The essential facts about photosynthesis are that light energy, by means of the chlorophylls and accessory pigments, is converted into chemical energy and stored in carbohydrates.

3. Organisms are classified as either *autotrophs,* those that can synthesize their food from inorganic sources, or *heterotrophs,* those that must acquire their food because they cannot manufacture it themselves. The chief autotrophs are the green plants.

4. When photosynthesis developed billions of years ago, free oxygen in the atmosphere increased, and a protective layer of ozone permitted life to emerge from the ocean depths and eventually to inhabit dry land.

5. The principal requirements for photosynthesis are light, chlorophyll, chloroplasts, carbon dioxide, and water. The principal products are glucose, other organic molecules, and oxygen.

6. The *light reactions* of photosynthesis convert light energy to chemical energy. The *dark reactions* use chemical energy to produce complex organic molecules.

RECOMMENDED READING

Arnon, Daniel I., "The Role of Light in Photosynthesis." *Scientific American,* November 1960. (Offprint 75)

Björkman, Olle, and Joseph Berry, "High-Efficiency Photosynthesis." *Scientific American,* October 1973. (Offprint 1281)

Govindjee and Rajni Govindjee, "The Primary Events of Photosynthesis." *Scientific American,* December 1974. (Offprint 1310)

Lehninger, Albert L., *Bioenergetics,* Second Edition. Menlo Park, Calif.: W. A. Benjamin, 1971.

Lehninger, Albert L., "How Cells Transform Energy." *Scientific American,* September 1961. (Offprint 91)

Levine, R. P., "The Mechanism of Photosynthesis." *Scientific American,* December 1969. (Offprint 1163)

Rabinowitch, Eugene I., and Govindjee, "The Role of Chlorophyll in Photosynthesis." *Scientific American,* July 1965. (Offprint 1016)

Ray, Peter Martin, *The Living Plant,* Second Edition. New York: Holt, Rinehart and Winston, 1972.

Went, Frits W., *The Plants.* New York: Time-Life Books, 1971. (Life Nature Library)

5

Cellular Respiration: The Release of Energy

TO MOST PEOPLE THE WORD "RESPIRATION" means pumping air in and out of animal lungs. To a biologist it means the release of the free energy stored in chemical bonds of such organic compounds as sugars, starches, proteins, and fats. Respiration releases the usable chemical energy that has been converted from light energy during photosynthesis.

As we learned in the previous chapter, autotrophic organisms obtain their energy directly from sunlight, and the cells of green plants ultimately provide the energy source for all living organisms by producing glucose and oxygen during photosynthesis. Heterotrophic organisms get glucose or other organic compounds from the self-reliant autotrophs and convert it into usable chemical energy through the process of respiration. Energy must be trapped within the molecules of ATP to be usable by organisms, and respiration is a highly efficient mechanism for regulating the systematic release of trapped energy. You will see in this chapter how the overall process of respiration is made up of many separate reactions or steps, and each step is controlled by a specific enzyme.

Respiration is a strictly intracellular process. It occurs in *all* living cells to produce the moment-to-moment energy required for basic life processes. In contrast, photosynthesis takes place only in special-

TABLE 5.1

Photosynthesis and Respiration

	PHOTOSYNTHESIS	RESPIRATION
Where?	In chlorophyll-bearing cells	In all living cells
When?	In light only	All the time
Input?	CO_2 and H_2O	Reduced carbon compounds and O_2
Output?	Carbon compounds, O_2, and H_2O	CO_2 and H_2O
Energy source?	Light	Chemical bonds
Energy result?	Energy stored	Energy released
Chemical reaction?	Reduction of carbon compounds	Oxidation of carbon compounds

ized cells (Table 5.1). Respiration is precisely regulated by an organized membrane system and an array of enzymes, coenzymes, vitamins, and pigments. The general statement of respiratory events can be simply represented by the equation

$$C_6H_{12}O_6 + 6O_2 \longrightarrow 6CO_2 + 6H_2O + \text{energy (686 Kcal)}.$$
$$\underset{\text{(input)}}{\phantom{C_6H_{12}O_6 + 6O_2}} \qquad \underset{\text{(output)}}{}$$

This is essentially the reverse of the equation for photosynthesis, and it states that every molecule of glucose, when combined with six molecules of oxygen, yields six molecules each of carbon dioxide and water and liberates 686 Kcal.

Glucose is traditionally used by scientists in their equations as a starting material because much of what is known about cellular respiration was learned from studies on yeast, and yeast, like most organisms, uses glucose readily as an energy source.

THE STAGES OF RESPIRATION

Although the energy-release process in respiration is a continuous one, carried on step by step from some organic molecule, such as glucose, to the final uptake of oxygen and the release of water and carbon dioxide, it is convenient for purposes of description and understanding to divide respiration into several dis-

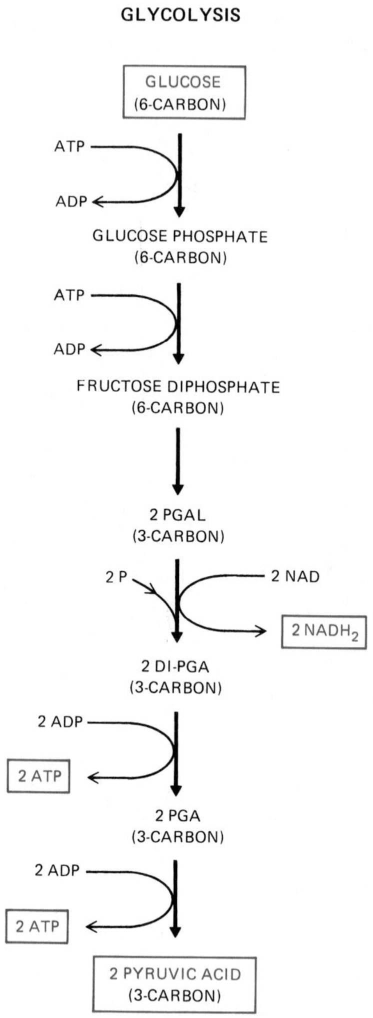

GLYCOLYSIS

5.1

Schematic representation of the glycolytic breakdown of glucose. Note especially that two molecules of ATP are used and four molecules of ATP are produced, giving a net gain of two molecules of ATP. Two molecules of pyruvic acid are formed, and NAD is reduced. No molecular oxygen is involved.

tinct parts. These different stages of the respiratory process occur in different parts of cells and under different conditions.

There are four separable aspects of respiration: (1) *Glycolysis* is the initial breakdown of glucose to an intermediate compound. Glycolysis, which means "breakdown of sugar," occurs in the cytoplasm of a cell and does not require oxygen. (2) *Fermentation* is the partial breakdown of organic molecules in the absence of free oxygen. (3) During the *Krebs cycle*, respirable molecules yield carbon dioxide and hydrogen. (4) The *electron transport system* is the main producer of ATP, and it uses molecular oxygen. The Krebs cycle and the oxidations in the electron transport system take place in the mitochondrion.

Most organisms obtain the bulk of their energy from the portion of the respiratory scheme that requires free atmospheric oxygen. Such organisms are called *aerobic* (Greek, "air" plus "life"). Even though they can achieve some energy release without free oxygen, the amount of energy so released is insufficient to sustain life for more than a short time. Consequently, aerobic organisms must have a constant and dependable oxygen supply. An air-breathing mammal, for example, can remain underwater for only a restricted time, living temporarily on stored oxygen and on the portion of the respiratory scheme that does not require free oxygen. Glycolysis and fermentation, as mentioned above, do not involve free oxygen. They are called *anaerobic* (Greek, "life without air") processes. Organisms, especially bacteria, that are able to obtain all their energy from such reactions are called anaerobes.

It must be understood that the processes listed above are inseparably connected in most living cells. They are discussed individually only for the sake of explanation.

GLYCOLYSIS: THE INITIAL BREAKDOWN OF GLUCOSE

When a molecule of glucose is to be respired, or broken down into smaller units, it is first activated in the cytoplasm of a cell. This activation is achieved by the transfer of a phosphate group from a molecule of ATP, resulting in the formation of ADP and glucose phosphate (Figure 5.1). Although the entire respiratory process is primarily a means of generating ATP from ADP and inorganic phosphate, it is necessary to expend some ATP to make the glucose accessible to the enzymes of the respiratory system. Some activation energy is required to get the action started, and it is supplied by the initial ATP. Activation energy can be compared to the energy required to start a large boulder rolling down a hillside when the boulder is blocked by a small pebble.

In fact, two ATPs are used in the activation of glucose: one to form glucose phosphate and a second one to form fructose *di*phosphate, to which the original glucose was converted. The fructose diphosphate —a 6-carbon molecule containing phosphate groups (phosphorylated) at both ends—is split into two molecules of a phosphorylated 3-carbon molecule, phosphoglyceraldehyde, or *PGAL*, which is also an intermediate in photosynthesis. Each PGAL molecule picks up an inorganic phosphate group from the surrounding cytoplasm to become *di*phosphoglyceric acid, which is oxidized to yield hydrogens to a molecule of nicotinamide adenine dinucleotide, NAD. This NAD resembles the NADP that acts as a hydrogen carrier in photosynthesis. It can exist in the *oxidized* (NAD) or *reduced* ($NADH_2$) form, and it is one of the main hydrogen carriers in cellular respiration.

At this point, there are two molecules of a 3-carbon compound, each phosphorylated at both ends. In a series of chemical reactions, these phosphate groups are transferred to four ADP molecules, generating four ATPs. Here for the first time in the scheme there is a gain in the number of high-energy phosphate bonds. Since two ATPs were required to start the initial activation of a glucose molecule and four have been formed, there is now a net gain of two ATPs. With the release of the phosphate groups from the diphosphoglyceric acid during further metabolism, two molecules of *pyruvic acid* are formed.

Pyruvic acid is considered the end of glycolysis. Note that when a cell has broken a glucose molecule down as far as pyruvic acid, it has increased its ATP supply by two molecules, and it has on hand two molecules of reduced, energy-carrying $NADH_2$ but has not released any carbon dioxide or used any oxygen. Most of the chemical-bond energy originally present in the glucose is still present in the pyruvic acid.

Once a cell has some pyruvic acid, it may do various things with it. The pyruvic acid may be used to make amino acids, or partially respired anaerobically to make such compounds as alcohol or lactic acid, or completely respired to carbon dioxide and water. Because a sequence of biochemical reactions is called a *pathway* by biochemists, pyruvic acid may

BEER, WINE, AND YOGURT

Long before they knew the biological reasons for what they were doing, people used the anaerobic activities of microbes. The curing of tobacco and tea leaves is a bacterial fermentation process, as are the preparatory steps in making vanilla flavor from the fruits of orchid vines known as "vanilla beans," or making chocolate from the fruits of cacao trees.

Humans have been making alcoholic drinks longer than they have been writing. The usual active organism is one or another variety of the yeast *Saccharomyces cereviseae.* Any sugar-containing, watery liquid can be used to make wine: plums, apples, blackberries, elderberries, oranges, dilute honey, dandelion flowers, and of course most commonly, grapes. Fermentation is allowed to continue until most or all (for dry wines) of the sugar has been broken down to carbon dioxide and ethyl alcohol, which can reach a concentration of about 12 percent. If the product is bottled before the fermentation is complete, additional carbon dioxide will be produced, resulting in bubbly wines such as champagne or sparkling burgundy.

Wine can be spoiled by the invasion of alcohol-digesting bacteria, which further ferment the alcohol to acetic acid, turning the wine into vinegar. (Vinegar literally means "sour wine.") Fortified wines, such as sherry, have their alcohol content artificially raised to about 20 percent by the addition of distilled alcohol. They do not spoil even when not refrigerated after they are opened.

Beer is similar to wine except that the source of the alcohol is not fruit or flowers but a grain such as barley or corn. Originally beers, like champagne, were bottled before fermentation was complete in order to retain the bubbles of carbon dioxide, but now commercial beers are usually artificially carbonated. The method used was invented by the same Joseph Priestley who discovered oxygen and the process of photosynthesis in the eighteenth century.

The same species of microorganism that is called brewer's yeast in a brewery is called baker's yeast in a bakery. Brewers are mainly interested in the alcohol liberated by the yeast, whereas bakers want the bubbles of carbon dioxide to swell and make the bread light. The alcohol in yeast dough is boiled off during baking.

Fresh milk ordinarily deteriorates rapidly when bacteria grow in it, as they invariably do under natural conditions. Some bacteria, however, ferment the milk sugar, *lactose,* to lactic acid, creating an environment that eliminates the undesirable bacteria. Milk that receives a starter of a proper strain of Lactobacillus can change into any of a number of cheeses, of which yogurt is a well-known type. Yogurt was originally made because it has better keeping quality than fresh milk, but now it is made in quantity because some people like the taste of it, especially when fruit and other flavors are added.

---

be thought of as standing at a sort of biological crossroad.

GLYCOLYSIS

Input		*Output*	
1.	Glucose	1.	2 molecules of pyruvic acid
2.	2 ATPs	2.	4 ATPs
3.	2 phosphate groups (inorganic)	3.	2 pairs of hydrogen atoms

FERMENTATION: ONE POSSIBLE FATE OF PYRUVIC ACID

The words "fermentation" and "anaerobic respiration" have been used by different people to mean different things, with the result that some confusion exists in the literature on the subject. Here *fermentation* is used to refer to the production of alcohol by microorganisms, as well as the formation of lactic acid in animal muscle. The critical feature is the gen-

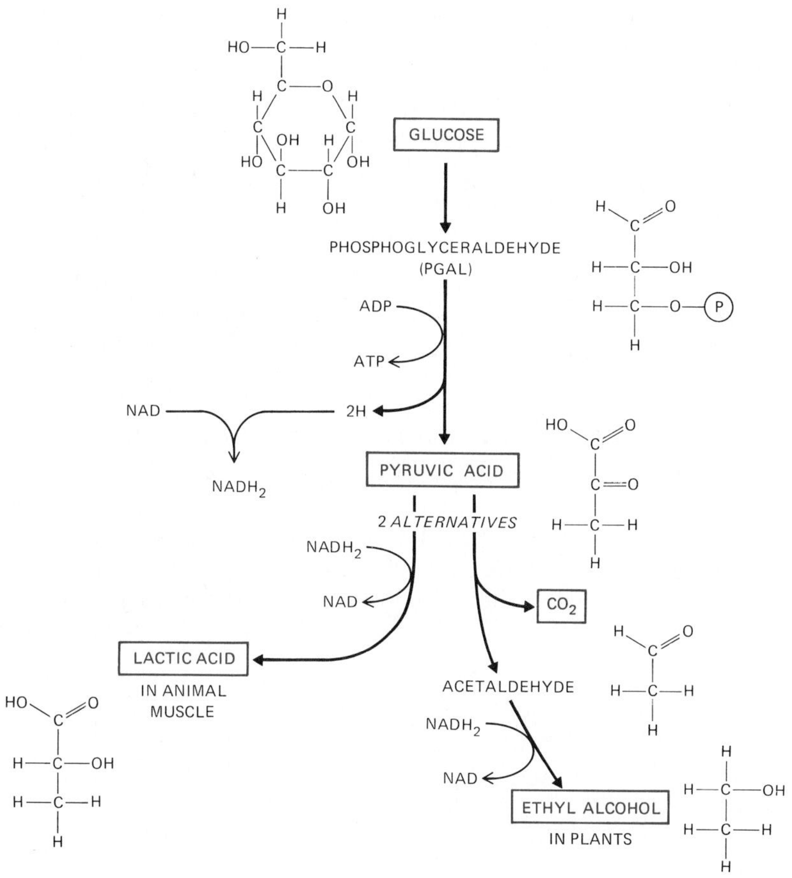

GLUCOSE

PHOSPHOGLYCERALDEHYDE
(PGAL)

ADP

ATP

NAD

2H

NADH$_2$

PYRUVIC ACID

2 *ALTERNATIVES*

NADH$_2$

NAD

CO_2

LACTIC ACID

IN ANIMAL
MUSCLE

ACETALDEHYDE

NADH$_2$

NAD

ETHYL ALCOHOL

IN PLANTS

5.2

Schematic representation of fermentation, or anaerobic respiration. Note that in animals generally fermentation yields lactic acid, and that in plants and fungi generally it yields ethyl alcohol. Note especially that CO_2 may be given out, but no O_2 is taken in, and that the total ATP yield is only the two molecules produced in glycolysis.

eration of incompletely oxidized end products without the use of atmospheric oxygen (Figures 5.2 and 5.3).

One possible product of fermentation is ethyl alcohol. In a yeast cell, for example, pyruvic acid may lose a molecule of carbon dioxide, and the resulting compound is a 2-carbon molecule of acetaldehyde. The acetaldehyde is reduced by hydrogens from NADH$_2$, which was produced during the glycolytic phase of respiration. This reduction of acetaldehyde yields ethyl alcohol, which diffuses out of the cell and accumulates in the surrounding medium. By fermentation of this sort, yeasts can change grape juice into wine, or malt suspensions into beer.

Another possible anaerobic activity of pyruvic acid yields lactic acid. In an actively working muscle, for instance, pyruvic acid can be reduced by the hydrogen from NADH$_2$ to form lactic acid. This can happen during strenuous exercise, when oxygen is not being delivered to the muscle fast enough to allow for complete oxidation of the respirable food supply. Ac-

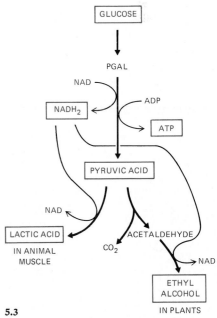

5.3

Fermentation (simplified version).

cumulation of lactic acid makes a muscle "feel tired" and can eventually stop muscle contraction until the lactate is removed and oxygen is provided in adequate amounts.

No ATP is generated in the final reactions of fermentation. The only ATP a cell gains from the glycolysis-fermentation activity is the net increase of two molecules formed during glycolysis when diphosphoglyceric acid is dephosphorylated. If a molecule of glucose contains 686 Kcal of potential energy, and the glycolysis-fermentation pathway yields only two ATPs, it can be calculated that the efficiency of the entire process is low. One ATP molecule has an energy value of about 7 Kcal, so two have a total of about 14 Kcal. Fermentation thus extracts 14/686 of the possible energy in glucose, or 2.1 percent. This is a poor showing in comparison with 25 percent for photosynthesis, or about 40 percent (as we will show later) for aerobic respiration.

FERMENTATION

Input

1. 2 molecules of pyruvic acid (from 1 molecule of glucose)

Output

1. 2 molecules of an incompletely oxidized compound (for example, lactic acid or ethyl alcohol)

2. 2 molecules of CO_2 (in alcoholic fermentation)

THE KREBS CYCLE: A SECOND POSSIBLE FATE OF PYRUVIC ACID

Most of the ATP generation and energy release in cells is made possible by a series of reactions collectively called the *Krebs cycle*, or sometimes the *tri-*

5.4

Schematic representation of the Krebs cycle. A simplified version is shown in Figure 5.5.

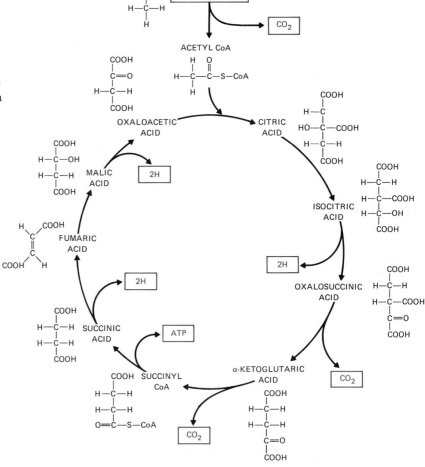

carboxylic acid or *citric acid cycle* (Figures 5.4 and 5.5).

The Krebs cycle is named for the biochemist Hans Krebs, who suggested in the 1930s the theoretical possibility of a cycle of organic acids as a means of releasing the carbon dioxide and hydrogen during respiration. Bear in mind that removal of hydrogen is *oxidation*. The several components of the cycle and their molecular structure had been known for years, but scientists did not realize what their functions might be until Krebs showed how they might work together in a so-called cycle of chemical activity. Krebs's discovery is an example of how a slow accumulation of seemingly useless information finally results in a synthesis, whereupon everything suddenly makes sense.

The reactions of the Krebs cycle depend on a precisely spaced series of enzymes located in the internal membranes of the mitochondria on the internal folds, or *cristae*. Mitochondria, like plastids, have their own membrane systems (Chapter 3).

Pyruvic acid is the compound on which the Krebs cycle depends. A molecule of pyruvic acid is enzymatically induced to give up a molecule of carbon dioxide and hydrogen ions. The carbon dioxide diffuses out of the cell, and the hydrogen is taken up by NAD to become $NADH_2$. With one carbon gone from the 3-carbon pyruvate, two remain, forming what is known as an acetyl group, so named because it resembles the 2-carbon molecule of acetic acid familiar as the sour principle in vinegar. The acetyl group is united with a molecule of Coenzyme A (abbreviated as CoA), a complex of adenine, ribose, pantothenic acid (a vitamin), and sulfur. The CoA acts as a carrier for two carbon atoms at a time and feeds them into the Krebs cycle by passing the acetyl group to a molecule of oxaloacetic acid.

Oxaloacetic acid is a 4-carbon compound. When the 2-carbon group from acetyl CoA is added to the 4-carbon oxaloacetic acid, the result is a 6-carbon molecule of citric acid. From here on around the Krebs cycle, there is an ordered sequence of reactions that serve several purposes: Some are simply rearranging reactions that prepare a molecule for the next reaction to come, some release carbon dioxide, and some release hydrogen. It is the hydrogen-releasing reactions that are the important ones for energy transfer; the rest are means to that end.

The Krebs cycle, in its bare essentials, is as follows:

1. The 2-carbon acetyl group is fed into the cycle via CoA to make the 6-carbon citric acid.

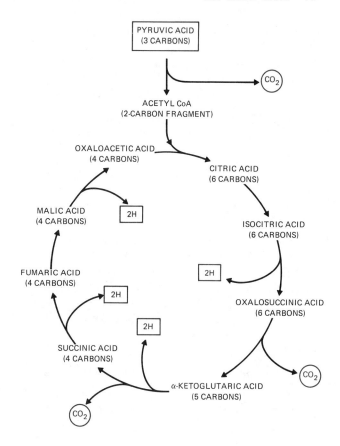

5.5

The Krebs cycle (simplified version).

2. The citric acid is rearranged to isocitric acid.

3. The isocitric acid *loses a pair of hydrogen atoms,* and the result is oxalosuccinic acid.

4. The oxalosuccinic acid *loses a molecule of carbon dioxide,* and the result is the 5-carbon ketoglutaric acid.

5. The ketoglutaric acid *loses a pair of hydrogen atoms and a molecule of carbon dioxide,* and the result is 4-carbon succinic acid.

6. The succinic acid *loses a pair of hydrogen atoms,* and the result is fumaric acid.

7. The fumaric acid is rearranged to yield malic acid, which *loses a final pair of hydrogen atoms,* yielding oxaloacetic acid, the same 4-carbon acid that accepted an acetyl group to begin with.

A review of the scheme shows the fate of the materials. First, there are in the cycle four oxidation steps during which hydrogen is given off. Since these steps are energy-releasing, they are critical to the effectiveness of the entire scheme. There are two steps during which carbon dioxide is given off. If one recalls that the original glucose molecule had six carbon atoms, that it was split into two 3-carbon PGALs during glycolysis, and that one CO_2 was released when pyruvic acid went to form acetyl CoA, all six of the carbons in glucose are accounted for in two "turns" of the Krebs cycle. Recall the general equation

$$C_6H_{12}O_6 + 6O_2 \rightarrow 6CO_2 + 6H_2O + \text{energy}.$$

Each of the reactions is catalyzed by a specific enzyme. Without these enzymes the reactions would not take place fast enough to support life as we know it. The enzymes are spatially arranged in the mitochondrial membranes so that each is in its proper place to function when its time comes.*

THE KREBS CYCLE

Input	Output
1. Acetyl CoA (2 carbons) from pyruvic acid	1. 2 molecules of CO_2
	2. 4 pairs of hydrogen atoms

5.6

The terminal oxidations. Electrons (and hydrogen ions) from the Krebs cycle are passed, via reduced NAD, to the several electron acceptors. Each acceptor then becomes a donor to the next in line until the final acceptor, oxygen, is reached and water is formed. During the passage of the electrons, ATP is generated. The cytochromes and associated electron acceptors are presumably fixed in a definite spatial arrangement in the mitochondrial membrane, but NAD, in contrast, can move. As an electron carrier, it can shuttle back and forth, carrying repeated "loads" of electrons.

---

* The Krebs cycle is called a cycle because atoms in the involved molecules may be in one compound at one time and then later in an identical molecule, and the process can be shown diagrammatically as if it were a recurring cycle. Cycles, however, are better for learning purposes than they are in reality, because no atom ever gets back to the same molecule it was once in; the entire molecular population of a cell will have changed by the time a so-called turn of the cycle has taken place.

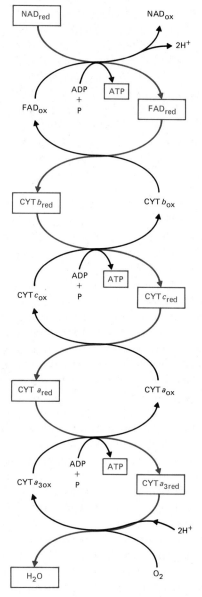

PASSAGE OF ELECTRONS
FROM KREBS CYCLE ACIDS TO O_2—
THE TERMINAL OXIDATIONS

NAD = Nicotinamide Adenine Dinucleotide
FAD = Flavin Adenine Dinucleotide
red = Reduced
ox = Oxidized

THE ELECTRON TRANSPORT SYSTEM IS THE MAIN PRODUCER OF ATP

The energy available in the chemical bonds of a sugar molecule is released as those bonds are altered. When electrons with their attendant hydrogen atoms leave a molecule to be taken on by another, the energy goes with them. In mitochondria, where most respiratory energy exchange occurs, there is an orderly arrangement of hydrogen acceptors accompanied by appropriate enzymes. The series of carriers is called the *electron transport system* (Figure 5.6). Several of the active compounds are *cytochromes*, protein-plus-pigment molecules containing iron. These cytochromes can be alternately reduced (receive electrons) and oxidized (give up electrons), with an energy loss or gain accompanying any electron transfer. Oxidation involves either the addition of oxygen or the removal of hydrogen. Most of the energy released in a cell results from the removal of hydrogen.

As a pair of hydrogen atoms is passed from cytochrome to cytochrome, some energy is necessarily lost as heat, but some of it is caught up in those still mysterious "coupled reactions" by means of which

CYANIDE, MUSHROOMS, AND OTHER FOES OF RESPIRATION

Most animals require a constant supply of oxygen, delivered all the time to all cells. The best known cause of death from lack of oxygen is probably drowning. Without free air, a person loses consciousness within seconds, and death follows in minutes. Even if a rescue is effected, oxygen deprivation for several minutes can cause permanent brain damage, because the brain is more sensitive to loss of respiration than are other tissues.

But turning off the air supply is not the only way to stop respiration. Anything that interferes with cellular respiration can effectively cause a sort of "drowning." Some poisonous mushrooms contain a variety of protein toxins that interfere with the transmission of nerve impulses, thereby preventing the contraction of the rib and diaphragm muscles that causes the physical act of breathing. When these muscles become inoperative, breathing stops and death follows shortly. Other mushroom toxins break up the oxygen-carrying red blood cells. In either instance, a poisoned person dies because of lack of cellular respiration. Antidotes to mushroom poisoning are just beginning to be developed.

Cyanide poisoning is specifically a respiratory blockage. A cyanide-treated cell cannot use oxygen because cyanide, HCN, inhibits the electron transport system and thus stops the activity of the terminal oxidations. A person dying of cyanide poisoning turns blue (cyanide means "blue" in Greek) and dies quickly. Contrary to the lore of spy stories, however, the dying is grimly painful.

Another foe of respiration is tetanus poisoning, or "lockjaw." The tetanus bacterium, *Clostridium tetani*, can exist only when oxygen is absent. It exudes a toxin 50 times as poisonous as cobra venom and 150 times stronger than another well-known poison, strychnine. Buried rusty nails are common dwelling places for such anaerobes, so that anyone stepping on a nail and driving it through the skin effects self-inoculation, stabbing the bacteria into an airless environment where they can thrive. Immunization consists of injecting a small amount of weakened toxin into the body to stimulate it to build up its own antitoxin. Anti-tetanus shots are a practical necessity for farmers, soldiers, children, and any others who have frequent skin breaks and are likely to be in contact with soil.

Ordinarily, a person will not notice a lack of oxygen as long as he can get rid of carbon dioxide. The oxygen concentration in submarines sometimes gets so low that there is not enough oxygen to allow a man to light a cigarette. But such a shortage of air causes no feeling of discomfort. In some of the great fires caused by bombing in World War II, the oxygen in entire city blocks was used so completely by the flames that people died quietly and, judging from the expressions on the corpses, quite peacefully.

ADP and inorganic phosphate are brought together in ATP. The cytochromes vary somewhat from species to species, but a representative series in order of their arrangement in a mitochondrion is cytochrome *b*, cytochrome *c*, cytochrome *a*, and cytochrome a_3.

The final hydrogen acceptor is oxygen, and when finally an oxygen atom accepts two hydrogen atoms, the result is water. This so-called water of respiration is the main source of body water in some animals, such as the larvae of clothes moths, which never drink liquid water. *The final, biologically useful product of respiration is ATP.* The carbon dioxide and the water can be considered waste.

If oxygen is present to act as the final hydrogen acceptor in the respiratory chain, each pair of hydrogen atoms released from the acids in the Krebs cycle will contribute to the generation of three ATPs. Since there are four reactions in the cycle that liberate hydrogen, there are four times three ATPs generated for each molecule of pyruvic acid (and hence acetyl

CoA) fed into the cycle. Each molecule of glucose yields two molecules of pyruvic acid, so we double the 12 ATPs to make 24. Also, from $NADH_2$ produced before the formation of citric acid and from an additional source in the cycle that was not discussed here, 12 more ATPs are generated, for a total of 36. In addition, there are the two ATPs produced during the oxygen-free action of glycolysis.

Of the final 38 ATPs, according to the usual count, all but the last-named two are the result of aerobic respiration, making aerobic respiration responsible for 36/38 or about 95 percent of the ATP energy made available from the breakdown of glucose. This figure makes it obvious why an air-breathing, warm-blooded animal like a human being, with a large, constant demand for immediate energy, dies so quickly when he or she cannot breathe and obtain oxygen.

If a molecule of ATP can yield 7 Kcal of energy, then respiration with its production of 38 molecules

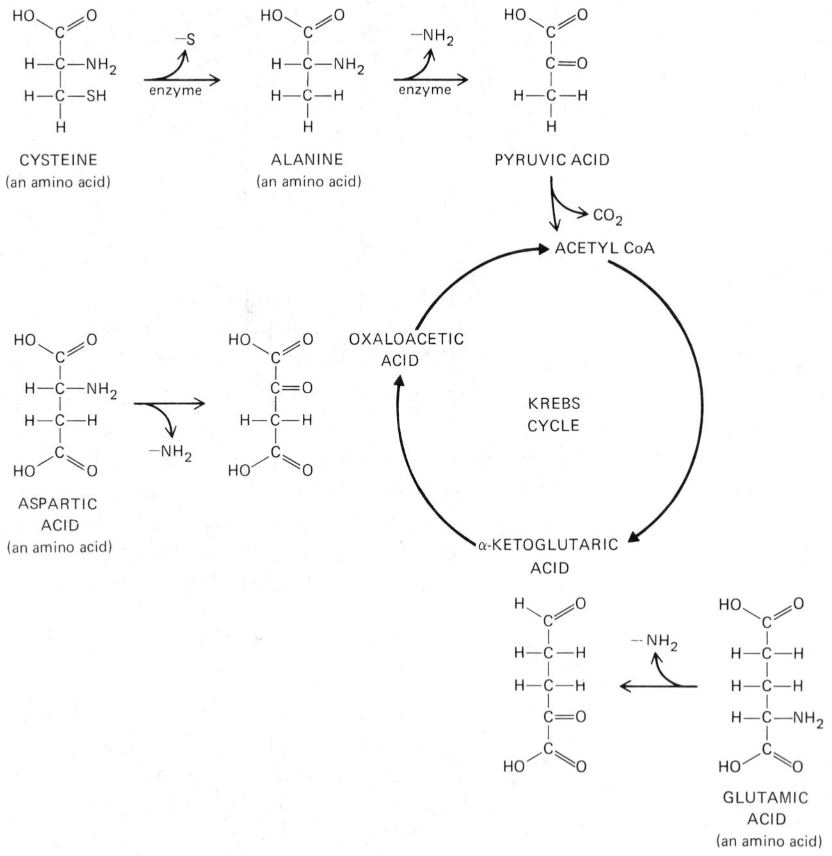

5.7

The process by which proteins are made available for the generation of ATP. After a protein has been digested to its component amino acids, such as the cysteine, alanine, aspartic acid, and glutamic acid shown here, the amino acids can be enzymatically changed to Krebs cycle acids (pyruvic, oxaloacetic, and glutaric). The Krebs cycle acids can be used to generate ATP, and the sulfur and nitrogen are either excreted from a body or recycled in plants.

5.8

The process by which fats are made available for the generation of ATP. A neutral fat can be hydrolyzed to glycerol and fatty acids, of which one example is shown here. Glycerol can be used after conversion to phosphoglyceraldehyde to generate ATP in the glycolytic pathway. A fatty acid can be broken down by the removal of two carbon atoms at a time (an acetyl group) and fed into the Krebs cycle in the same manner as is the acetyl group derived from pyruvic acid. A long-chain fatty acid can provide a number of acetyl groups and therefore many ATP molecules. Plants can use carbohydrates to make fats, and then the fats can be either used to yield energy in the form of ATP or made into carbohydrate again. Animals can work in a similar way, except that some animals (humans) seem to be unable to convert fat back to carbohydrate. Carbohydrate-to-fat is easy.

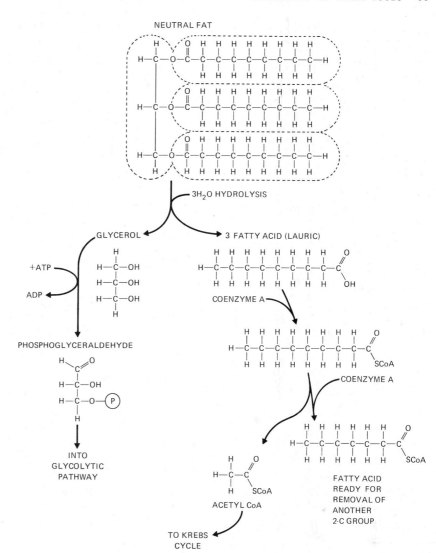

of ATP per molecule of glucose, which is "worth" 686 Kcal, is about 40 percent efficient.

> 686 Kcal = energy initially available.
>
> 7 × 38 ATPs = 266 Kcal = energy obtained.
>
> 266/686 = .388, or roughly 40 percent.

Such efficiency compares very favorably with that of human-made machines, which commonly achieve between 5 percent and 30 percent efficiency.

THE ELECTRON TRANSPORT SYSTEM
(for one molecule of acetyl CoA)

Input	Output
1. 4 pairs of hydrogen atoms	1. 4 times 3 ATPs
	2. 4 molecules of H_2O
2. 4 oxygen atoms	

RESPIRATION OF OTHER FOODS

Animals eat and plants make many kinds of food besides glucose: proteins, fats, and various starches (Chapter 22). These respirable foods can be used to generate ATP, but they must be chemically changed so that they can be brought into a form suitable for providing hydrogen to the electron transport system. *Proteins* are digested to amino acids (Figure 5.7). An amino acid can be further metabolized to yield an amino group and an organic acid. These organic acids can enter the Krebs cycle and be dehydrogenated (oxidized). *Starches* are digested to glucose, which is readily respired via glycolysis and the Krebs cycle. *Fats* are digested to glycerol (which can be taken into the glycolytic scheme) and fatty acids (Figure 5.8). Fatty acids are broken down by the removal of two carbon

METABOLISM OF FATS, PROTEINS, CARBOHYDRATES (SIMPLIFIED)

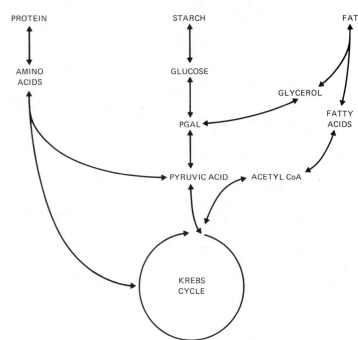

5.9

Generalized scheme of food-to-energy conversion. Protein, carbohydrate (starch), and fat can all be digested to simpler compounds (amino acids, sugar, and glycerol and fatty acids, respectively), and then the simpler compounds can be changed into molecules that can be taken up in the Krebs cycle and made to yield ATP.

5.10

Generalized scheme of cellular respiration, from a beginning sugar to the production of ATP, carbon dioxide, and water. Note especially that most of the ATP is generated by the action of the Krebs cycle in mitochondria, and that oxygen is brought into the reactions only as the final step.

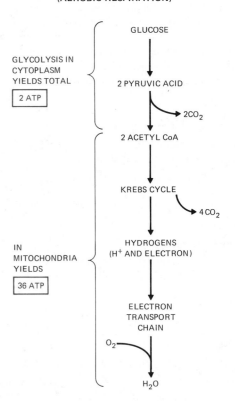

SUMMARY OF RESPIRATION
(AEROBIC RESPIRATION)

atoms at a time (acetyl groups), and these can be attached to Coenzyme A and then brought into the Krebs cycle.

Not all organisms have the necessary enzymes to make all these conversions. Most animals, for example, lack a digestive enzyme for cellulose, the most common glucose-containing compound in the world. But if the enzymes are present, practically any organic molecule can be changed into a form that provides hydrogen, thereby making it suitable for energy production.

SUMMARY

1. In a biological sense, *respiration* is the release of the free energy stored in chemical bonds of certain organic compounds. Respiration makes available to cells the chemical energy that has been converted from light energy during photosynthesis.

2. Energy must be trapped within the molecules of ATP to be usable by organisms, and respiration is a highly efficient mechanism for regulating the systematic release of trapped energy.

3. The overall process of respiration is made up of many reactions or steps, and each step is controlled by a specific enzyme.

4. Respiration occurs in all living cells to produce the energy for basic life processes. In contrast, photosynthesis takes place only in specialized cells.

5. The general events of respiration are these: Every molecule of glucose, when combined with six molecules of oxygen, yields six molecules each of carbon dioxide and water and liberates usable energy in the form of 38 ATP molecules.

6. *Glycolysis* is the initial breakdown of glucose to an intermediate compound; it occurs in the cytoplasm of a cell. *Fermentation* is the partial breakdown of organic molecules in the absence of free oxygen. During the *Krebs cycle,* respirable molecules yield carbon dioxide and hydrogen. The *electron transport system* uses molecular oxygen and is the main producer of ATP. The Krebs cycle and the electron transport system take place in the mitochondrion.

7. *Aerobic* processes require free atmospheric oxygen to carry on their normal cellular activity. *Anaerobic* processes do not require free oxygen.

8. *Glycolysis* uses one molecule of glucose, 2 ATPs, and 2 phosphate groups, and it produces 2 molecules of pyruvic acid, 4 ATPs, and 2 pairs of hydrogen atoms.

9. *Fermentation* uses 2 molecules of pyruvic acid from 1 molecule of glucose and produces 2 molecules of an incompletely oxidized compound (such as lactic acid or ethyl alcohol), and 2 molecules of carbon dioxide.

10. The *Krebs cycle* uses acetyl CoA from pyruvic acid and produces 2 molecules of carbon dioxide and 4 pairs of hydrogen atoms.

11. The *electron transport system* uses 4 pairs of hydrogen atoms and 4 oxygen atoms, and it produces 4 times 3 ATPs and 4 molecules of water.

12. Animals eat and plants make many other kinds of foods besides glucose, and all respirable foods can be used to generate ATP after they have been changed to a form that provides hydrogen to the electron transport system.

RECOMMENDED READING

Green, David E., "The Mitochondrion." *Scientific American*, January 1964.

Lehninger, Albert L., *Bioenergetics*, Second Edition. Menlo Park, Calif.: W. A. Benjamin, 1971.

Lehninger, Albert L., "How Cells Transform Energy." *Scientific American*, September 1961. (Offprint 91)

6

The Importance of DNA

6.1
Friedrich Miescher, who made the
original isolation of DNA.

MOST OF US HAVE WONDERED, AT SOME POINT
in our lives, how it was possible for us to inherit our
mother's blue eyes, our father's black hair and too-
large nose, and our Uncle Charlie's winning smile.
We may also have wondered how our cells know how
to make identical copies of themselves, or how it came
about that our sex cells have exactly half the number
of chromosomes of every other cell in our body. How
do hereditary characteristics get passed on from generation to generation, and even more to the point,
how do hereditary characteristics get passed on from
cell to cell within our own bodies?

Scientists have agreed for a long time that the
answer to these questions probably lies in the cellular
material itself. In order to find the answer, they would
first have to isolate the genetic *substance*. Then they
would have to determine its *structure*. Only then
could they hope to find out how the miracle of the
continuation of life is performed. Although our basic
knowledge of the genetic material dates back to the
middle of the nineteenth century, it took the advanced technology of our own century, plus a group
of persistent scientists, to unravel the threads of the
intricate pattern of life. Two scientists, James Watson
and Francis Crick, received the credit for putting the
pieces together in 1953. When they postulated the
structure of a molecule called DNA, they provided
enough clues so that many of the remaining questions

could be answered within a decade or two. One of the riddles of life was about to be solved.

DNA was first discovered in 1869 by Friedrich Miescher, a Swiss physician (Figure 6.1). He had not actually set out to find the substance that determined our genetic heritage, but Miescher inadvertently started one of the most dramatic searches in scientific history when he found the substance he called *nuclein.* Miescher gave it that name because it was located in the nuclei of the white blood cells he was studying. Nuclein was subsequently named *nucleic acid,* and then *DNA.* During the century following Miescher's isolation, a great deal of investigation was carried out on nuclein. Its chemical composition was determined, and many chemical properties were identified. Nuclein (DNA) was shown to be the hereditary material, but its three-dimensional structure remained obscure. The knowledge of this structure was essential to explaining how DNA could function in heredity. In 1953, almost a hundred years after Miescher's discovery, the structure of DNA was revealed by Watson and Crick in their double helix model—a discovery that has been ranked with the contributions of Newton, Darwin, and Einstein. What *is* DNA, and why is it so important?

CELLS, NUCLEIC ACIDS, AND DNA

<u>Nucleic acids</u> are extremely large molecules (macromolecules), probably the largest found in living cells. Their size results from the accumulation of small units repeated over and over again. Scientists are now certain that nucleic acids are found in all living things, and that they are the primary factors that control the genetic (hereditary) processes of life. We also know that nucleic acids are sometimes found outside the nuclei of cells, and that the nucleic acids of all living organisms, from the tiniest viruses to humans, differ in the sequence of the units. We will see later in this chapter how specific differences in the structure of nucleic acids give each organism its unique characteristics.

Nucleic acids are made up of units called <u>nucleotides</u>, and each nucleotide is made up of three components: phosphate, sugar, and nucleotide base. DNA contains four kinds of nucleotide bases: the double-ring *purines* (adenine and guanine), and the single-ring *pyrimidines* (thymine and cytosine). <u>Adenine</u>, <u>guanine</u>, <u>thymine</u>, and <u>cytosine</u> are usually referred to in a shorthand version as A, G, T, and C (Figure 6.2).

There are two main kinds of nucleic acids, each with its own sugar structure. One kind of nucleic acid

6.2

The nitrogen bases that commonly make up nucleic acids. The *purines* are *adenine* and *guanine,* and the *pyrimidines* are *thymine, cytosine* (in DNA), and *uracil* (in RNA). Note that the purines have a double-ring structure and the pyrimidines have a single-ring structure. Guanine and cytosine are able to form three hydrogen bonds each, whereas adenine, thymine, and uracil can form only two each.

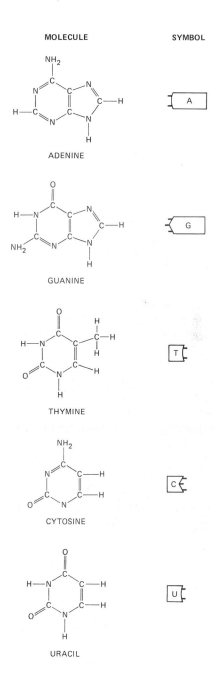

6.3

The sugars in nucleic acids. Ribose occurs in RNA and deoxyribose in DNA. The difference between the two kinds of molecules is the presence of one oxygen atom in ribose that is absent in deoxyribose. The colored atoms on the diagrams show the positions affected.

RIBOSE

DEOXYRIBOSE

contains ribose sugar made up of five atoms of carbon and five of oxygen. Nucleic acids containing ribose sugar are known as *ribonucleic acids*, or *RNA*. The other type of nucleic acid contains a sugar with one less oxygen atom, which is therefore called *deoxy*-ribose (Figure 6.3). The nucleic acid itself is designated as *deoxyribonucleic acid*, or *DNA*.

RNA molecules differ from DNA molecules in three ways besides their sugar makeup: (1) RNA contains uracil instead of thymine; (2) RNA is usually single-stranded, whereas DNA is double-stranded; and (3) RNA occurs in cytoplasm, nuclei, and several organelles in cells, but DNA is usually restricted to chromosomes or chromosomelike structures. (Small but important quantities of DNA are found in chloroplasts and mitochondria.)

In summary, nucleic acids contain sugar, phosphate groups, and nitrogen-containing bases. A combination of one of each of these materials makes up a nucleotide, or a self-contained unit within the overall molecule of nucleic acid.

PROOF OF THE GENETIC IMPORTANCE OF DNA

In 1869 Friedrich Miescher was interested in determining the content of the nuclei of cells, and he began his experiments with the white blood cells in the pus that adhered to discarded bandages. He was able to isolate a nuclear material, and from this he further isolated an unfamiliar acidic substance that was also rich in phosphorus. When Miescher announced his results in 1871, he named this new substance *nuclein*, and in later experiments with the nuclei of salmon sperm he isolated a similar substance.

Early in the twentieth century researchers discovered that chromosomes carry the hereditary information, and that chromosomes contain protein as well as DNA and some RNA. Because protein seemed more complex than the nucleic acids, scientists generally accepted the idea that only the protein was capable of carrying the hereditary information necessary to maintain life. We now know that it is DNA, not protein, that controls heredity, but it was not until the 1940s that any substantive proof became available.

In 1924 the German biochemist Robert Feulgen found that a decolorized dye, basic fuchsin, turned a bright purple in the presence of DNA. By means of this staining technique it was relatively easy to show that DNA was present in the nucleus of a cell. When it was clearly shown that there was DNA in the chromosomes of cells, scientists began to believe that DNA might be connected to heredity.

Later, Alfred Mirsky and his colleagues at the Rockefeller Institute found that the cells of any given species all contain the same amount of DNA. Sex cells contain exactly half the amount of DNA in normal body cells, but when the male and female sex cells unite in fertilization, the single resultant cell carries the same amount of DNA as other body cells.

By the time Fred Griffith began his experiments with the pneumococcus bacterium, scientists considered it possible that DNA was the substance that controlled heredity. Definite proof was still necessary, and in 1928 Griffith began the series of experiments that would ultimately bring acceptable proof in the 1940s and 1950s.

While seeking a cure for pneumonia, Griffith infected mice with two kinds of pneumococcus bacteria. Type I (or R for rough) was a harmless strain without a covering capsule, and Type III (or S for smooth) was the infective strain, with each cell covered by a protective coat (Figure 6.4). When mice were infected with the R strain, nothing happened. When they were exposed to the virulent S strain, the mice contracted pneumonia and died. These results were expected.

The unexpected happened when Griffith tried another approach. Mice were injected with the nonvirulent, uncoated R strain, and at the same time, the mice received a dose of coated S strain that had been heated and killed. To Griffith's astonishment, the mice subsequently died of pneumonia. To add to the puzzle, traces of living coated bacteria were located in the dead mice. If the uncoated bacteria were harmless, and the coated ones had been killed by heating, what caused the mice to die? And how did *live* coated bacteria get into the blood of the dead mice?

What had happened was a process called *transformation*, which allowed the harmless bacteria to acquire the genetic characteristics of the virulent bacteria. But how had this happened? What substances were involved?

In 1944, after much prior experimentation, Oswald T. Avery, Colin MacLeod, and Maclyn McCarty finally managed to isolate the factor that had caused the transformation in Griffith's experiments. First

they removed the coats from the S strain bacteria to determine if the coats were crucial to the hereditary transfer. But heated S bacteria without coats still transformed the R strain into a virulent form. The scientists then removed the protein from the S strain and left only nucleic acid in the bacteria. Again the R strain bacteria were transformed. Finally, when the nucleic acid (DNA) was separated from the S strain bacteria and only protein remained, the bacteria could no longer affect the harmless R strain.

Without a doubt, Avery had identified DNA as the effective agent of heredity. So it was DNA after all, not protein, that held the clue to life itself. That most important and thrilling discovery was considered "revolutionary" by leading scientists, who saw the consequences that would follow from it.

Another classic experiment by Alfred D. Hershey and Martha Chase in 1952 confirmed the results of Avery, MacLeod, and McCarty. Hershey and Chase were aided by the separate findings of two researchers. Thomas Anderson used electron micrographs to show that the shells of viruses become empty after

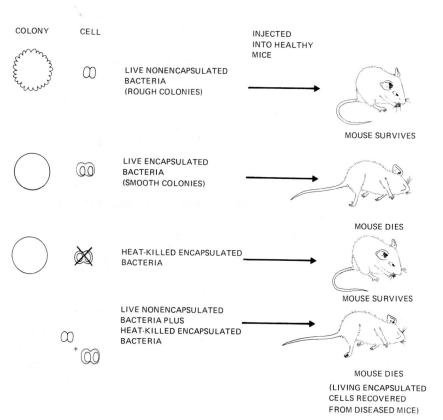

COLONY   CELL

LIVE NONENCAPSULATED BACTERIA (ROUGH COLONIES)

INJECTED INTO HEALTHY MICE

MOUSE SURVIVES

LIVE ENCAPSULATED BACTERIA (SMOOTH COLONIES)

MOUSE DIES

HEAT-KILLED ENCAPSULATED BACTERIA

MOUSE SURVIVES

LIVE NONENCAPSULATED BACTERIA PLUS HEAT-KILLED ENCAPSULATED BACTERIA

MOUSE DIES (LIVING ENCAPSULATED CELLS RECOVERED FROM DISEASED MICE)

6.4

Griffith's experiments with encapsulated and nonencapsulated *Diplococcus pneumoniae* on mice, the work that started the idea of bacterial transformation and eventually led to the demonstration of DNA as the genetic material of cells.

6.5

Viruses consist of a head of protein surrounding a single strand of nucleic acid (usually DNA but sometimes RNA). Completing the simple viral structure is a protein tail. Viruses cannot reproduce on their own. To reproduce they must infect a living cell, where they make use of the cell's enzymes and nourishment. As a result of viral infection and reproduction, the cell is usually killed. (a) When a virus infects a host bacterial cell, (b) it attaches itself to the host by its tail and (c) injects its single molecule of DNA into the host. Once inside the bacterium, the viral DNA begins to give instructions for the multiplication of new viruses. Being a parasite, the virus needs the nourishment of the cytoplasm of its host to flourish. Viruses on their own merely exist; they cannot reproduce. (d) Within 20 minutes approximately 100 new viruses, complete with protein coats, have been reproduced, and they burst from the cell ready to infect new hosts. The original host cell has been totally sacrificed to the parasitic intruders. (e) Electron micrograph of a section through a bacterial cell containing replicated virus particles. Other virus particles can be seen outside the cell. (f) A closer view of the edge of a cell with four virus particles attached to the cell surface.

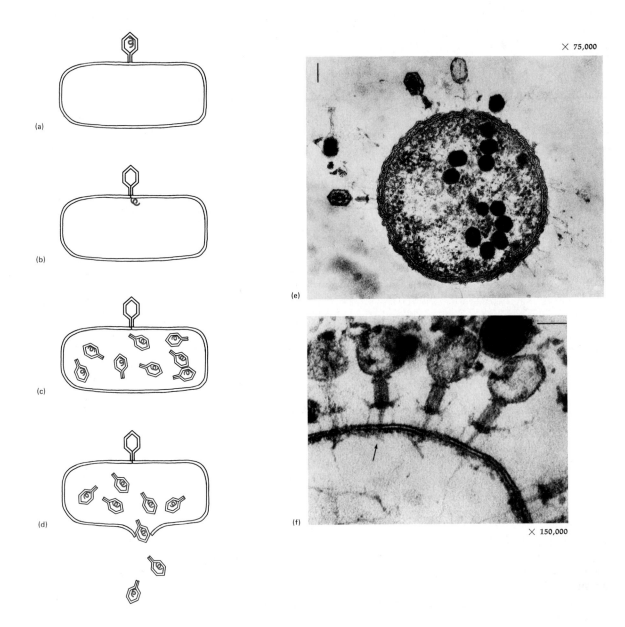

× 75,000

× 150,000

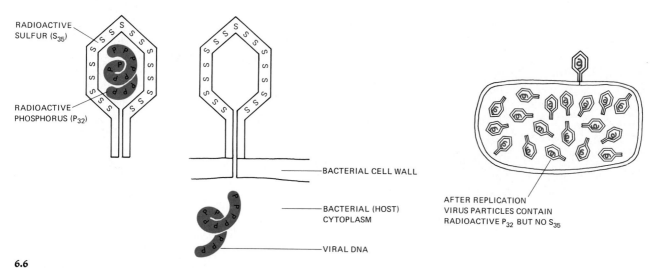

RADIOACTIVE SULFUR (S$_{35}$)

RADIOACTIVE PHOSPHORUS (P$_{32}$)

BACTERIAL CELL WALL

BACTERIAL (HOST) CYTOPLASM

VIRAL DNA

AFTER REPLICATION VIRUS PARTICLES CONTAIN RADIOACTIVE P$_{32}$ BUT NO S$_{35}$

6.6

The Hershey-Chase experiment, which demonstrated that only viral DNA, not viral protein, enters a host cell and transmits viral genetic control.

the viruses come into contact with bacteria. Roger Herriott found that when a virus attaches itself to a bacterium, the virus releases DNA into the host cell.

Bacteriophages, or *phages* for short, which are viruses that infect bacteria, have been used extensively by scientists in their search for the genetic properties of DNA (Figure 6.5). By using radioactive tracers, Hershey and Chase verified that only the DNA from the head of the virus was injected into the host bacterium; the protein of the virus shell remained outside. Subsequently, new viruses, complete with protein coats, were reproduced. These new viruses were identical to their parent virus.

Hershey and Chase reasoned that if new viruses were being created that were identical to their parent, a genetic (that is, a hereditary) force must be at work. If only viral DNA was being injected into the host cell, then it was sensible to assume that DNA was the genetic material. It was known that proteins contain sulfur but do not contain phosphorus, and that the content of DNA is the opposite, phosphorus but no sulfur. So when the viral sulfur and phosphorus were made radioactive, they could be traced to determine which substance entered the host cell and was able to direct the genetic activity. Of course, the re-

sults showed that phosphorous was on the inside and sulfur on the outside, proving that DNA alone had gone in (Figure 6.6).

This final confirmation of the work of Griffith and Avery left no further doubt that DNA was the genetic material, and it cleared the way for the frantic search for the *structure* of DNA.

The reason the *structure* of the DNA molecule was considered so important was the belief that by knowing its structure, we could explain how DNA could do the two things that any hereditary material must do: (1) make replicas of itself, and (2) regulate the activities of cells.

THE RACE FOR THE DOUBLE HELIX

After the conclusive experiments of Hershey and Chase, intensive research to discover the structure of the DNA molecule began. The chemical composition of DNA had been known for 50 years (see p. 87), but now the important questions were: How is the DNA molecule constructed? And how does DNA accomplish its work? It was these questions that would be answered in 1953 by James D. Watson, an Amer-

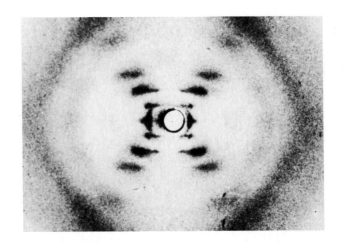

6.7
James D. Watson and Francis H. C. Crick, who published in 1953 a proposal for the structure of DNA.

6.8
An X-ray photograph of DNA. To obtain such a photograph, the crystallographer sends an X-ray beam through a small hole in a lead sheet, allowing the beam to strike a sample. If the sample is constructed in a regular pattern of repeating particles, the X-ray beam will be changed in direction in a regular way. If the resulting "diffracted beams" are caught on a photographic plate, one can develop a pattern of light and dark spots whose spacing and position can be measured. From such measurements, it is possible to calculate the spacing and position of the particles in the sample. From this particular photograph, several dimensions of spacing and arrangement of atomic groups in DNA were deduced.

ican, and Francis H. C. Crick, an Englishman (Figure 6.7).

Probably the most famous entrant in the race for the double helix was chemist Linus Pauling. Working at the California Institute of Technology, Pauling had done outstanding work in chemistry, including a description of the three-dimensional helical structure of protein molecules—the work for which he would be awarded the Nobel Prize in 1954. Confident from his success with proteins, Pauling took aim on DNA.

Most scientists agreed that DNA would be shaped like a *helix* (similar to the winding of a spiral staircase), and Crick recalled several years later that many of the most influential researchers (including Pauling) were thinking of a helical structure for DNA. "They were all building helical models," he said. "I would say, you would be eccentric, looking back, if you didn't think DNA was helical."*

Watson and Crick, searching for the structure of DNA, relied heavily on intuition and the published and unpublished research findings of others. They did no formal "research" during their attempts to build a model of DNA, and they both showed their first intuitive insights in deciding separately that the answer to the workings of DNA would be found in its three-dimensional *structure*. Watson arrived in Cambridge, England, in 1951 to work on this problem. He became acquainted with Crick, and their famous collaboration began.

Besides the assumption that DNA would be a helix and the rather complete knowledge of its chemical components, there was additional information about DNA available through X-ray diffraction photographs. Maurice H. F. Wilkins and later Rosalind Franklin were producing excellent X-ray diffraction photographs at King's College in London, and their work would be one of the most important aids to Watson and Crick (Figure 6.8).

Recall that a DNA molecule consists of sugar, phosphates, and four nucleotide bases—adenine, guanine, thymine, and cytosine. Watson and Crick knew

---

* From recorded interviews with Robert Olby in Cambridge, March 8, 1968, and August 7, 1972. Robert Olby, *The Path to the Double Helix* (Seattle: University of Washington Press).

that. They also knew from the X-ray photos of Wilkins and Franklin that DNA was most likely a coiled structure, with each twist in the coil occurring every tenth nucleotide. This was a consistent pattern, no matter how many twists there were or how long the entire strand of DNA was. What Watson and Crick still did not know was how the bases and sugar-phosphate backbones were arranged in three-dimensional space, or how the entire mechanism could duplicate itself, with all the necessary information being supplied from within the DNA.

By 1952 Linus Pauling and his co-workers at Caltech had become increasingly interested in DNA, and in December 1952 Watson and Crick learned that Pauling and R. R. Corey had built a structure for DNA. Although Pauling gave no details, Watson was convinced that Pauling had beaten them. But when Pauling and Corey published their findings in February 1953 they showed a *triple* helix structure. Most surprising for a chemist of Pauling's stature, the model contained some obvious chemical errors, and Watson calculated that he and Crick had four to six weeks before Pauling himself realized his mistakes and built the correct model.

Watson and Crick knew that in any given molecule of DNA, adenine and thymine appear in equal amounts, and guanine and cytosine also are present in equal amounts. This principle had been presented in 1949 by Erwin Chargaff, but it wasn't until 1951 that Watson and Crick were able to see the obvious one-to-one relationship between the bases. In his book *The Double Helix*, Watson tells how important the Chargaff relationships were in making the connection that would solve the puzzle:

> *Suddenly I became aware that an adenine-thymine pair held together by two hydrogen bonds was identical in shape to a guanine-cytosine pair held together by at least two hydrogen bonds. All the hydrogen bonds seemed to form naturally: no fudging was required to make the two types of base pairs identical in shape.*

Now Watson and Crick could visualize how the double-ringed adenine could combine with the single-ringed thymine, and likewise the double-ringed guanine with single-ringed cytosine without deforming the spaces between the sugar-phosphate backbones (Figure 6.9). And as Watson said, the weak hydrogen bonds fell into place between the bases naturally without any forcing or "fudging."

6.9

(a) A single nucleotide, with a phosphate group, a five-membered ring of deoxyribose, and a nitrogen base. Two such nucleotides, joined by hydrogen bonds, make a *base pair*. (b) A strand of DNA consisting of six nucleotide pairs. Note that the space can be adequately filled only when a purine, such as guanine (with its double-ring structure), is matched with a pyrimidine, such as cytosine (with its single ring). The number of atoms available for hydrogen bonding is important in determining which purine is joined with which pyrimidine. With three hydrogen bonds possible between guanine and cytosine and only two between adenine and thymine, the proper bases normally pair up accurately—guanine only with cytosine and adenine only with thymine.

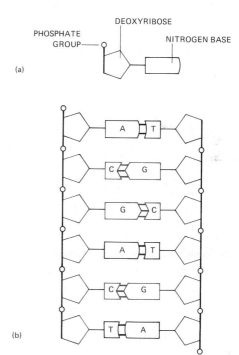

6.10

(a) The double helix form. If a ladder could be twisted, the two uprights would form a double helix. In a DNA molecule, the repeating strands of sugars and phosphate groups can be thought of as the uprights and the nitrogen bases as rungs. (b) A portion of a DNA molecule, with its double helix of sugars and phosphate and cross-connectors of bases. Every guanine base is joined to a cytosine, and every adenine to a thymine. A complete DNA molecule has not just dozens but thousands, even millions, of base pairs, and an actual molecule would be long enough to see with the unaided eye if the strand were not so thin. (c) Replication of a bacterial chromosome, which is a naked DNA strand in one continuous loop. In the first picture, one small strip has begun to separate at the upper left. In the second, almost half the chromosome has become visibly doubled, and in the third, the loop has separated except at one final sticking point.

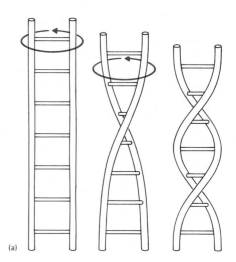

(a)

The structure was a *double* helix, twisted like a vertical ladder (Figure 6.10). Watson and Crick had won the race.

In the April 25, 1953, issue of *Nature,* Watson and Crick published their ideas about the DNA molecule in a brief article. Later in 1953 they published a follow-up article discussing how this structure could permit the reproduction of identical molecules. Their work was almost unanimously accepted, and it has since been confirmed by numerous experiments. As hoped, the structure of DNA was the key to the replicating process, and Watson and Crick modestly understated that supposition in their original article: "It has not escaped our notice that the specific pairing we have postulated immediately suggests a possible copying mechanism for the genetic material."

Pauling had been informed of the content of the original *Nature* article before it was published, and at a conference in Brussels in mid-April he declared graciously:

Although it is only two months since Professor Corey and I published our proposed structure for nucleic acid, I think we must admit that it is probably wrong. . . .

Although some refinement might still be made, I feel that it is very likely that the Watson-Crick

structure is essentially correct. In its feature of complementariness of the two chains it suggests a mechanism for duplication of a chain by a two-step process—a molecular mechanism that may well be the mechanism of hereditary transmission of characters.

I think that the formulation of their structure by Watson and Crick may turn out to be the greatest development in the field of molecular genetics in recent years.

In 1962 Watson, Crick, and Wilkins were awarded the Nobel Prize in Physiology or Medicine for their correct assessment of the structure of DNA. Rosalind Franklin had died in 1958, but her contribution to the Watson-Crick model has since been fully acknowledged.

Now that the structure of the DNA molecule had been revealed, it was possible to explain its method of replication.

HOW DNA REPRODUCES ITSELF

Let us now visualize the DNA molecule in more specific terms. If we continue to think of the molecule as a twisted vertical ladder, the side uprights consist

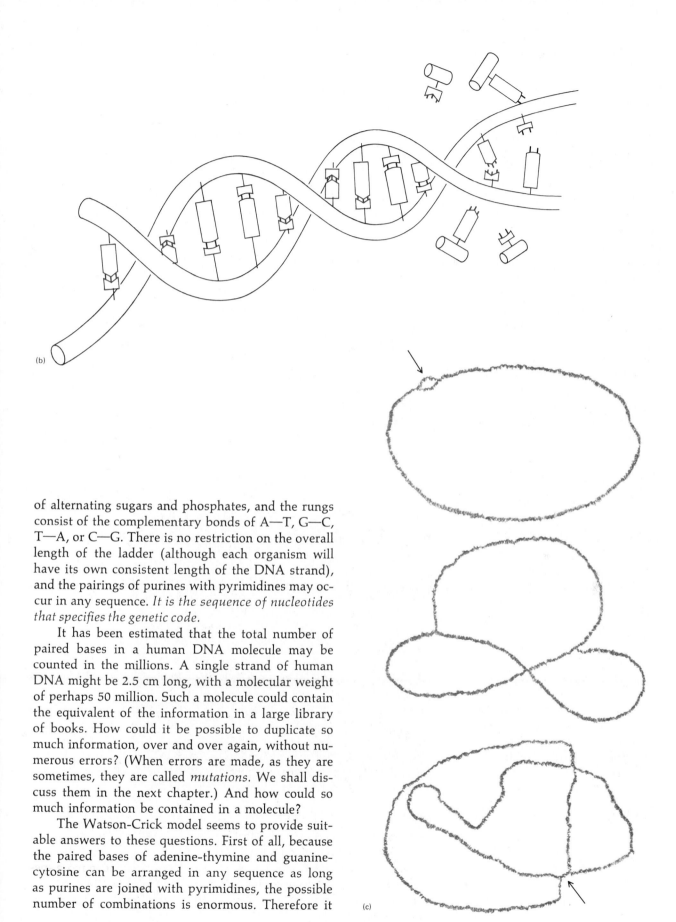

(b)

(c)

of alternating sugars and phosphates, and the rungs consist of the complementary bonds of A—T, G—C, T—A, or C—G. There is no restriction on the overall length of the ladder (although each organism will have its own consistent length of the DNA strand), and the pairings of purines with pyrimidines may occur in any sequence. *It is the sequence of nucleotides that specifies the genetic code.*

It has been estimated that the total number of paired bases in a human DNA molecule may be counted in the millions. A single strand of human DNA might be 2.5 cm long, with a molecular weight of perhaps 50 million. Such a molecule could contain the equivalent of the information in a large library of books. How could it be possible to duplicate so much information, over and over again, without numerous errors? (When errors are made, as they are sometimes, they are called *mutations*. We shall discuss them in the next chapter.) And how could so much information be contained in a molecule?

The Watson-Crick model seems to provide suitable answers to these questions. First of all, because the paired bases of adenine-thymine and guanine-cytosine can be arranged in any sequence as long as purines are joined with pyrimidines, the possible number of combinations is enormous. Therefore it

6.11

The replication of DNA. Scientists believe that the double helix starts the replication process by untwisting at one end. The weak hydrogen bonds joining the purine and pyrimidine bases are somehow released. The molecule has begun to "unzip" itself. Each adenine-thymine pair has separated, and each guanine-cytosine pair has separated. This state of separation does not exist for long. From within the cell's cytoplasm, available adenine (complete with sugar-phosphate backbone) finds its way to the newly exposed thymine, and with the help of appropriate enzymes, new hydrogen bonds are created.

On the other side of the unzipped backbone, the partnerless adenine will be joined by a free-floating thymine. Guanines will link up with stationary cytosines, and floating cytosines will form bonds with guanines still attached to the sugar-phosphate backbone. Each nitrogen base has achieved a coupling with its complementary base partner—no other couplings are normal—and the single unzipped strand of DNA has now become two exact copies of the original. It is this principle of *complementary bonding* that ensures that an exact copy will be made, and that an incredible amount of genetic information will be passed on accurately from generation to generation.

Crick has proposed that the untwisted strand begins to twist again as soon as the new bonds are formed. If so, the complicated activity of untwisting, unzipping, rezipping, and retwisting is all taking place at the same time.

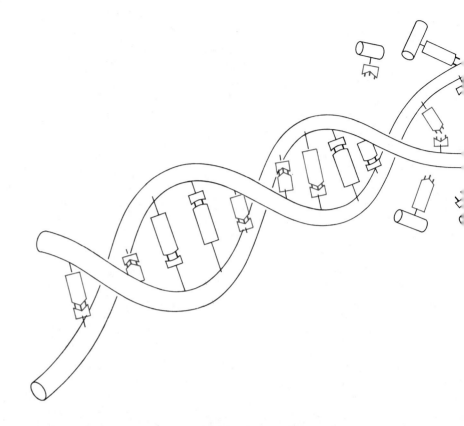

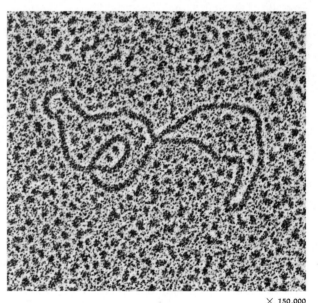

× 150,000

6.12

Electron micrograph of a gene isolated from the bacterium *E. coli*. This particular gene, estimated to be about 1.4 micrometers long, is concerned with the production of an enzyme that acts on the sugar galactose.

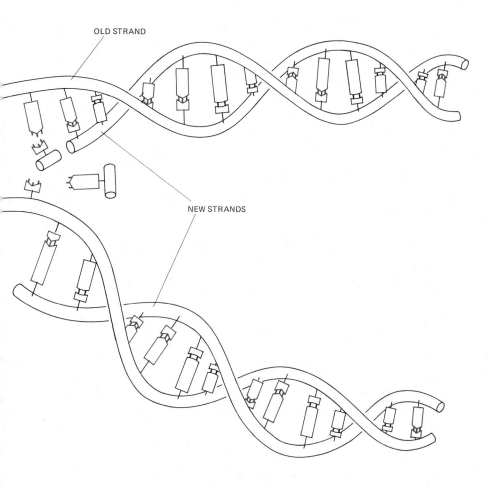

OLD STRAND

NEW STRANDS

is possible to provide enough chemical instruction to regulate the manufacture of proteins from amino acids.

In order for organisms to go on living, their cells must be able to reproduce, and because DNA is the basis of life, *it* must be able to reproduce. The Watson-Crick model permits us to understand how DNA replicates itself (Figure 6.11).

Besides its remarkable replication, DNA can also complete its remaining function, the ultimate production of protein. (About 10,000 different proteins are found in the human body.) The "central dogma" of modern biology states that *DNA makes (codes for) RNA, which makes (codes for) proteins.* Both the genetic code that holds the key to this production of life-giving protein and the function of RNA will be the concerns of the next chapter.

WHAT IS A GENE?

Before we leave this chapter, we should examine the definition of a gene, which has been modified with the advent of molecular biology. A _gene_ may be described as a length of DNA that controls a specific cellular function, either by coding for a polypeptide or by regulating the action of other genes (Figure 6.12).

SUMMARY

1. Nucleic acids are extremely large molecules (macromolecules), probably the largest found in living cells. Scientists are now certain that nucleic acids are found in all living things, and that they are the primary factors that control the hereditary processes of life.

2. DNA is made up of units called *nucleotides,* and each nucleotide is made up of three components: phosphate, sugar, and nucleotide base. DNA contains four kinds of nucleotide bases: *adenine, guanine, thymine,* and *cytosine.* Adenine and guanine are *purines,* and thymine and cytosine are *pyrimidines.*

3. There are two main kinds of nucleic acids, each with its own sugar structure. *Ribonucleic acids,* or *RNA,* contain ribose sugar. *Deoxyribonucleic acid,* or *DNA,*

contains a sugar with one less oxygen atom. RNA contains *uracil* instead of thymine. It can also be found in the cytoplasm outside the cell's nucleus.

4. *Friedrich Miescher* discovered DNA, which he called *nuclein*, in 1869, but he did not determine its function. It was not until the 1940s that any proof became available that DNA, not protein, controls heredity.

5. In 1928 *Fred Griffith* discovered the process of *transformation*, whereby DNA transferred its genetic characteristics from one organism to another. Sixteen years later, *Avery, MacLeod,* and *McCarty* confirmed that it was the DNA in Griffith's experiments that was the effective agent of heredity.

6. In 1952 *Alfred D. Hershey* and *Martha Chase* used radioactive tracers to show how viruses injected their DNA into host cells and thereby manufactured new, identical viruses. This final confirmation that DNA was the genetic material cleared the way for the search for the three-dimensional structure of DNA.

7. The three-dimensional *structure* of the DNA molecule was important to explain how DNA could do the two things that any hereditary material must do—make replicas of itself and regulate the activities of cells.

8. In 1953 *James D. Watson* and *Francis Crick,* with the assistance of X-ray diffraction photographs by *Maurice Wilkins* and *Rosalind Franklin*, identified the structure of the DNA molecule as a *double helix*. Their discovery permitted an acceptable explanation of the *replication* of DNA; it has also been shown that the *sequence of nucleotides* specifies the *genetic code*.

9. The *central dogma* of modern biology states that DNA codes for RNA, which codes for protein.

10. A *gene* may be defined as a length of DNA that controls a specific cellular function, either by coding for a polypeptide or by regulating the action of other genes.

RECOMMENDED READING

Beadle, George W., *Genetics and Modern Biology*. Philadelphia: American Philosophical Society, 1963.

Beadle, George, and Muriel Beadle, *The Language of Life: An Introduction to the Science of Genetics*. New York: Doubleday, 1966. (Also available in paperback)

Campbell, Allan M., "How Viruses Insert Their DNA into the DNA of the Host Cell." *Scientific American*, December 1976. (Offprint 1347)

Crick, Francis, *Of Molecules and Men*. Seattle: University of Washington Press, 1966. (Also available in paperback)

Crick, F. H. C., "The Structure of the Hereditary Material." *Scientific American*, October 1954. (Offprint 5)

Fiddes, John C., "The Nucleotide Sequence of a Viral DNA." *Scientific American*, December 1977. (Offprint 1374)

Kendrew, John C., *The Thread of Life*. Cambridge, Mass.: Harvard University Press, 1966.

Kornberg, Arthur, "The Synthesis of DNA." *Scientific American*, October 1968. (Offprint 1124)

Luria, Salvador E., *Life: The Unfinished Experiment*. New York: Scribner's, 1973. (Available in paperback)

Mirsky, Alfred E., "The Discovery of DNA." *Scientific American*, June 1968. (Offprint 1109)

Olby, Robert, *The Path to the Double Helix*. Seattle: University of Washington Press, 1974.

Watson, James D., *The Double Helix*. New York: Atheneum, 1968. (Also available in paperback)

Watson, James D., *Molecular Biology of the Gene*, Third Edition. New York: W. A. Benjamin, 1976.

Watson, James D., and Francis H. C. Crick, "Molecular Structure of Nucleic Acids: A Structure for Deoxyribose Nucleic Acid." *Nature*, vol. 171, April 25, 1953, pp. 737–738. (Reprinted in *Classic Papers in Genetics*, James A. Peters, ed. Englewood Cliffs, N.J.: Prentice-Hall, 1959.)

7

DNA, RNA, and the Genetic Code: Protein Synthesis

WE SAW IN CHAPTER 6 HOW DNA CAN MAKE exact copies of itself, but most organic molecules, such as fats, sugars, and proteins, are not capable of self-replication. In this chapter we will see how DNA is responsible for the synthesis of all proteins, and how proteins, in turn, may act as enzymes to control cellular reactions. These cellular reactions ultimately produce such specialized substances as fats, sugars, structural and enzymatic proteins (some proteins may be structural and enzymatic at the same time), and all the other components of a living organism. The "central dogma" of modern biology, as related in the previous chapter, states that *DNA codes for RNA, which codes for proteins*. Now we can go one step further: Some of the proteins produced are enzymes that catalyze the synthesis of small and large molecules in living systems. Directly and indirectly, enzymes regulate cellular processes. Nonenzymatic proteins of importance include some hormones (for example, insulin) and structural proteins, such as those of all cell membranes and specific proteins associated with muscles and bones.

THE ONE GENE–ONE ENZYME HYPOTHESIS

In 1909 the English physician and biochemist Sir Archibald Garrod became interested in human diseases he called "inborn errors of metabolism." Garrod investigated several diseases, including *alkaptonuria*, sometimes called the black urine disease because the affected person is unable to metabolize alkapton, which is therefore excreted with the urine and turns black when exposed to air. Garrod knew the disease was an inherited one, and he reasoned that a specific enzyme was not operative because the relevant gene was being changed, or mutated. Therefore Garrod postulated that a gene, with the aid of a specific enzyme, controlled a chemical reaction. But it took about 30 years for other scientists to see the interrelationship of genes, enzymes, and chemical reactions.

In 1941, George W. Beadle and Edward L. Tatum at the California Institute of Technology began their research with the bread mold Neurospora, an organism that normally synthesizes the vitamins and amino acids it needs. Beadle and Tatum induced random mutations by using X-rays in an attempt to "begin with known chemical reactions and then look for the genes that control them." The mutated gene could no longer produce the enzyme that controlled the necessary synthesis. They found that in almost every case they could trace a single enzyme malfunction to a single gene.

The idea that each gene controls a specific enzyme was developed, and the further idea was put forth that because enzymes are proteins, and because enzymes are involved in the making of other proteins, genes actually control protein synthesis. With this notion, now universally accepted, Beadle and Tatum originated the "one gene–one enzyme" hypothesis. The present preference is the "one gene–one polypeptide" hypothesis, because some enzymes are made up of protein *chains* called polypeptides, and each polypeptide is controlled by a separate gene.

THE IMPORTANCE OF PROTEINS

Proteins are the working molecules of cells. Besides forming the structural elements that hold cells firmly, they are responsible for the enzymatic action that makes life a dynamic and continuing phenomenon. The development and maintenance of a human body require an estimated 10,000–100,000 different enzymes, each structurally specialized to do its own particular job. Each enzymatic function is determined by the three-dimensional shape of the protein molecule, and that shape is in turn determined by the arrangement of amino acid residues in the molecule. It is obvious that the regulation of protein building is of primary importance to the survival of any organism.

A single "misplaced" amino acid can make the difference between normality and malfunction, or

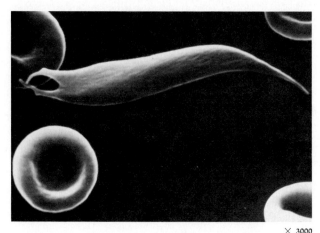

× 3000

7.1
The effect on an organism of a change in a single DNA nucleotide. The scanning electron micrograph shows normal red blood cells from a person suffering from sickle-cell anemia and one sickled cell, enormously stretched out. A substitution of one amino acid for another can mean the difference between health and sickness or even death.

7.2
A single deficiency can be totally destructive. The two seedlings of pin oak (*Quercus palustris*) were from the same parent tree. One was normally green and could grow into a mature pin oak. The other lacked one of the enzymes necessary for chlorophyll production and was white-leaved, an albino. As soon as the food supply from the acorn that produced it was used up, it starved.

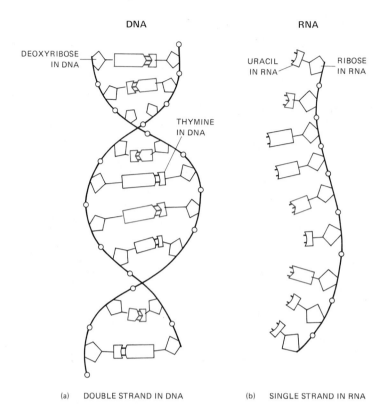

DNA

DEOXYRIBOSE
IN DNA

THYMINE
IN DNA

(a)   DOUBLE STRAND IN DNA

RNA

URACIL
IN RNA

RIBOSE
IN RNA

(b)   SINGLE STRAND IN RNA

7.3

The differences between DNA and RNA: (a) DNA is double-stranded, has deoxyribose as the sugar portion of each nucleotide, and incorporates thymine as one of its nitrogen bases; (b) RNA is single-stranded, has ribose as its sugar, and incorporates uracil instead of thymine. (RNA also contains a number of nitrogen bases besides the common ones, but their exact functions are not understood.)

even between survival and death. For example, if one of the 300 amino acids in hemoglobin* is changed, the action of the whole hemoglobin molecule is changed. A number of such changes are known, the most famous being the substitution of valine for glutamic acid in sickle-cell anemia (Figure 7.1). That seemingly small change makes an enormous difference to the life of an individual in whose blood it occurs. This example indicates the importance of the specific sequence of amino acids in a protein.

Several enzymes are involved in the stepwise synthesis of chlorophyll. The last step is the addition of two hydrogen atoms to the almost completed molecule. If the enzyme responsible for the addition of those two atoms is deficient, the chlorophyll is nongreen and useless, and the plant with such an enzymatic failure, which may appear trivial, starves (Figure 7.2).

THE TRANSCRIPTION OF RNA FROM DNA

Protein assembly starts with a process known as _transcription,_ whereby single-stranded RNA is synthe-

sized under the direction of DNA. The way in which a double-stranded DNA molecule unwinds during self-replication was described in Chapter 6. A similar separation of hydrogen bonds between the members of a pair of DNA bases opens up the molecule before transcription begins. However, when RNA instead of more DNA is to be synthesized, the process is slightly but importantly different in three ways. First, the bases in RNA differ in that RNA incorporates uracil instead of thymine. Second, the newly assembled chain of nucleotides, that is, the new RNA molecule, separates from its DNA "mold" and moves off as a single-stranded compound. Third, the RNA nucleotide contains the sugar ribose instead of the sugar deoxyribose (Figure 7.3).

During transcription, each new nucleotide must match a proper nucleotide on the opened DNA strand. Where a DNA nucleotide contains guanine, the opposite RNA nucleotide will contain cytosine; conversely, a DNA cytosine will attract an RNA guanine. A DNA thymine will attract adenine. Then the irregularity: A DNA adenine, instead of attracting an RNA thymine, will attract a uracil. (The linking of the nucleotides into a molecule of RNA is de-

---

*   Hemoglobin is a protein in blood cells responsible for carrying oxygen.

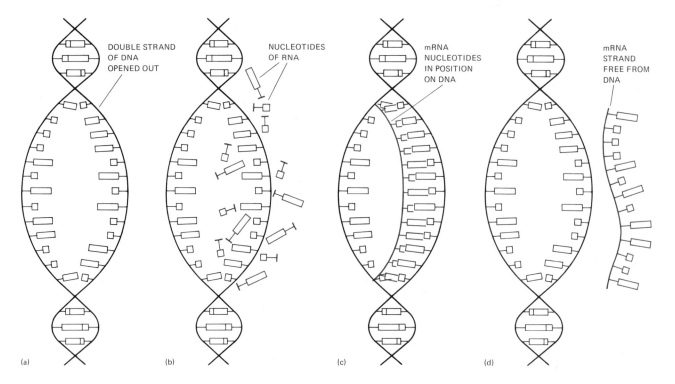

DOUBLE STRAND OF DNA OPENED OUT

NUCLEOTIDES OF RNA

mRNA NUCLEOTIDES IN POSITION ON DNA

mRNA STRAND FREE FROM DNA

(a)　　　　(b)　　　　(c)　　　　(d)

7.4

The formation of RNA from DNA. (a) The double strand of DNA in a chromosome opens, and (b) one of the strands accumulates the proper complementary bases (in color), which have come from the cytoplasm of the cell into the nucleus. (c) The complementary bases are in position along the opened DNA molecule, and (d) the string of newly assembled bases, made into an RNA strand, leaves the chromosome. The DNA is now available to produce more copies of the same RNA. Only a few bases are shown in this simplified figure, but in reality a strand of mRNA could have hundreds or thousands of bases. (e) Electron micrograph of DNA strands, caught in the act of transcribing RNA (in this instance, ribosomal RNA). The long, thin lines are DNA, and the feathery strands extending from the DNA are RNA strands whose lengths vary with the amount of RNA that had been produced at the moment of killing the cell. The short strands have just begun transcription. The naked stretches of DNA were not transcribing when the cell was killed. Amphibian oocyte.

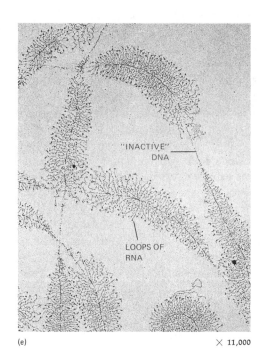

"INACTIVE" DNA

LOOPS OF RNA

(e)　　　　× 11,000

pendent on the action of specific enzymes, especially RNA polymerase.) With the exceptions noted, the new RNA is an accurately copied imprint, a kind of negative of the DNA that made it, with proper nucleotides faithfully positioned (Figure 7.4). The three kinds of RNA that are transcribed—ribosomal RNA, messenger RNA, and transfer RNA—are described in the following sections.

RIBOSOMAL RNA

Certain regions of chromosomes transcribe one of the RNAs, the *ribosomal RNA*, so called because it is destined to become part of ribosomes. As described in Chapter 3, ribosomes, which serve as the sites of protein synthesis, are cytoplasmic particles composed of protein and RNA. Although at the present writing we do not know the use of ribosomal RNA, usually called rRNA, we do know that final protein synthesis cannot occur without it.

MESSENGER RNA

A second RNA is *messenger RNA,* or mRNA, the molecules of high molecular weight that are directly responsible for the sequence of amino acids in proteins. Messenger RNA is well named, for it does indeed act as though it were carrying a message from the control center in the nucleus, DNA, to the ribosomes in the cytoplasm, telling which amino acids are to be included and where they are to be placed. After a molecule of mRNA has been transcribed on a strand of DNA, it passes to the cytoplasm and becomes attached to the ribosomes. The mRNAs are made up of chains of nucleotides, which act three at a time to specify particular amino acids, each *triplet* being said to have a "meaning" or to "code" for an amino acid.

The triplets of mRNA are consequently known as *codons.*

Even before researchers demonstrated that codons consist of triplets of nucleotides, they conceived the idea on mathematical grounds. There are only four kinds of nucleotides available (adenine, cytosine, guanine, uracil), and 20 kinds of amino acids have to be recognized. If one nucleotide meant one amino acid, there could be only four recognizable amino acids. If two nucleotides at a time were used, there could be 16 possible combinations (4^2), but that is still not enough. With three nucleotides, however, the 64 possible combinations (4^3) are more than adequate. We can say, simply, that the language of DNA-RNA transcription contains only four letters, from which 64 three-letter words may be written. These 64 words are more than enough to signify the 20 existing amino acids.

The complete set of triplet nucleotide symbols that indicate specific amino acids is called the *genetic code* (Tables 7.1 and 7.2). Experiments with mRNA

TABLE 7.1

The Genetic Code

AMINO ACID	ABBREVIATION	CODONS ON RNA
Alanine	Ala	GCU, GCC, GCA, GCG
Arginine	Arg	AGA, AGG, CGU, CGC, CGA, CGG
Asparagine	Asn	AAU, AAC
Aspartic acid	Asp	GAU, GAC
Cysteine	Cys	UGU, UGC
Glutamic acid	Glu	GAA, GAG
Glutamine	Gln	CAA, CAG
Glycine	Gly	GGU, GGC, GGA, CCC
Histidine	His	GAU, CAC
Isoleucine	Ileu	AUU, AUC, AUA
Leucine	Leu	UAA, UUG, CUU, CUC, CUA, CUG
Lysine	Lys	AAA, AAG
Methionine	Met	AUG
Phenylalanine	Phe	UUU, UUC
Proline	Pro	CCU, CCC, CCA, CCG
Serine	Ser	UCU, UCC, UCA, UCG, AGU, AGC
Threonine	Thr	ACU, ACC, ACA, ACG
Tryptophan	Tryp	UGG
Tyrosine	Tyr	UAU, UAC
Valine	Val	GUU, GUC, GUA, GUG
Initiation		(..?.. AUG)
Termination		UAA, UAG, UGA

Sixty-four possible codons ("three-letter words") can be derived from the four messenger RNA nucleotide symbols (G, C, A, U); 61 of these codons correspond to amino acids, and the remaining three codons are termination signals, any one of which will cause the protein synthesizing to cease. All signals to start protein synthesis begin with an unknown sequence of symbols ending with AUG; AUG is also the symbol for methionine. Note that some amino acids have only one codon, but others have as many as six.

TABLE 7.2

The Genetic Code (messenger RNA)

FIRST LETTER IN THE CODON	SECOND LETTER IN THE CODON				THIRD LETTER IN THE CODON
	U	C	A	G	
U	Phenylalanine	Serine	Tyrosine	Cysteine	U
	Phenylalanine	Serine	Tyrosine	Cysteine	C
	Leucine	Serine	Stop*	Stop*	A
	Leucine	Serine	Stop*	Tryptophan	G
C	Leucine	Proline	Histidine	Arginine	U
	Leucine	Proline	Histidine	Arginine	C
	Leucine	Proline	Glutamine	Arginine	A
	Leucine	Proline	Glutamine	Arginine	G
A	Isoleucine	Threonine	Asparagine	Serine	U
	Isoleucine	Threonine	Asparagine	Serine	C
	Isoleucine	Threonine	Lysine	Arginine	A
	Methionine	Threonine	Lysine	Arginine	G
G	Valine	Alanine	Aspartic acid	Glycine	U
	Valine	Alanine	Aspartic acid	Glycine	C
	Valine	Alanine	Glutamic acid	Glycine	A
	Valine	Alanine	Glutamic acid	Glycine	G

*UAA, UAG, and UGA all indicate "Stop," with no corresponding amino acid.

from a number of different organisms have shown that a given codon, say AUG, dependably specifies one particular amino acid (AUG always "means" methionine). If samples of mRNA from organisms as widely separated in evolutionary history as a yeast, a bird, a bean, and a human being all work alike, the inference is that *all* organisms have the same genetic code. The obvious differences between species are not the result of different *internal arrangements* of any triplet codon, but of the *sequence of codons*. We may speculate that no matter how different they are, all living organisms are related through the genetic code.

TRANSFER RNA

The third RNA is *transfer RNA,* or tRNA, or variously called soluble RNA, activator RNA, or adapter RNA. It is now most commonly referred to as tRNA, although the term "activator" is probably the most descriptive. This is the smallest of the RNA molecules, with a molecular weight of about 25,000 and composed of about 75 nucleotides. It is described by people with flexible imaginations as having the shape of a "clover leaf" or a "hairpin" (Figure 7.5). The size of the tRNAs is constant, although one tRNA exists for each of the 20 amino acids. The specificity resides

in the arrangement of nucleotides in one loop of the molecule, where a triplet determines how one particular tRNA will place itself on the mRNA.

A triplet, or *codon,* on the mRNA (say, adenine, uracil, adenine) will be matched by only one possible tRNA triplet (uracil, adenine, uracil). Since it is the match of the mRNA triplet, the triplet on the tRNA is an *anticodon.* The nucleotides of mRNA, that is, those of a codon, are *complementary* to those of the DNA by which it is organized, and the nucleotides of tRNA, that is, of an anticodon, are complementary to those of mRNA (Figure 7.6).

THE DISCOVERY OF THE "MEANING" OF mRNA TRIPLETS: DECIPHERING THE GENETIC CODE

To guess that a triplet, or codon, has a specific attraction for an amino acid is one thing; to show that this is actually so is something else again. The genetic code began to be deciphered when Severo Ochoa prepared a nucleotide chain in a laboratory in 1959. He made his nucleotide chain of the same nucleotide throughout. Ochoa's work was followed up in 1961 by that of Marshall Nirenberg.

The first synthetic mRNA was composed of uridines only, and so it was polyuridine, or *poly-U*. If poly-U in a vessel is supplied with all 20 tRNAs, all 20 amino acids, ribosomes, and an energy source, it will work in test tubes to form proteins without the presence of entire cells. With such a synthetic mRNA, the only amino acid selected out of the amino acid pool is phenylalanine, and the polypeptide chain becomes a chain of phenylalanines. In other words, UUU was the mRNA code symbol for the amino acid phenylalanine. The breaking of the code of mRNA

triplets had begun. If a synthetic mRNA is made only from adenine instead of uracil (polyadenine or poly-A), a polypeptide containing only the amino acid lysine is produced, and consequently AAA is said to code for lysine. Similarly, polyguanine, poly-G, was shown to code for glycine, and polycytosine, poly-C, for proline.

When mRNAs with varied nucleotide residues are considered, such straightforward methods are not

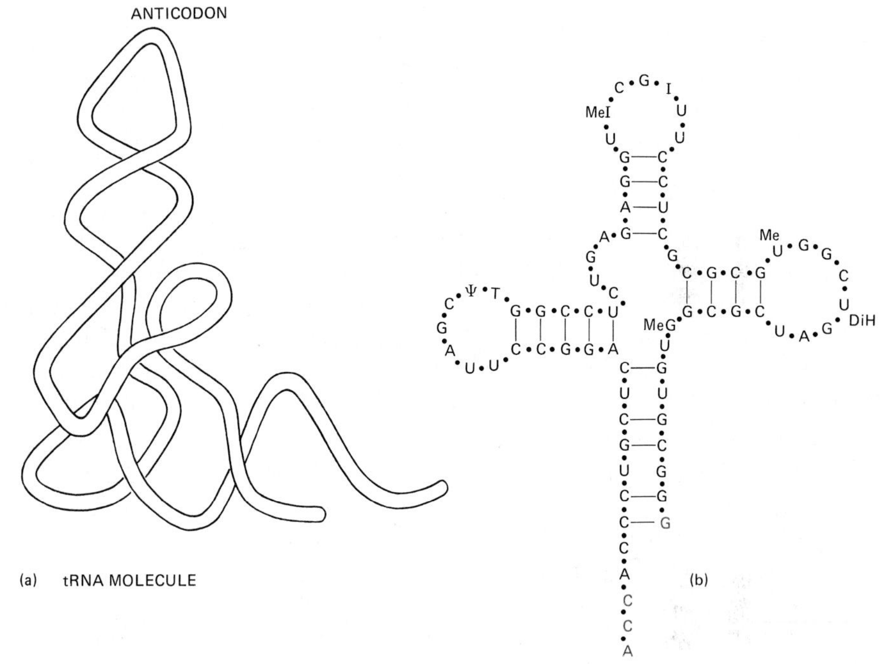

7.5

(a) A schematic view of a molecule of transfer RNA (tRNA). A *codon* in a molecule of mRNA will match the complementary bases of the *anticodon* at the end of one of the loops. (b) The sequence of bases in a tRNA molecule. Some of the base pairs are complementary, such as A-U and G-C, allowing for hydrogen bonding between them, and such bonding keeps the tRNA molecule in shape. The anticodon in this example is C-G-I. Every amino acid has its own special tRNA, yet all RNAs terminate in C-C-A at one end and in G at the other (color). The letters symbolize the usual bases: G for guanine, C for cytosine, A for adenine, and U for uracil, but tRNAs contain some less common bases, such as inosine (I), methyl inosine (MeI), dihydroxy uracil (DiHU), and methyl guanine (MeG).

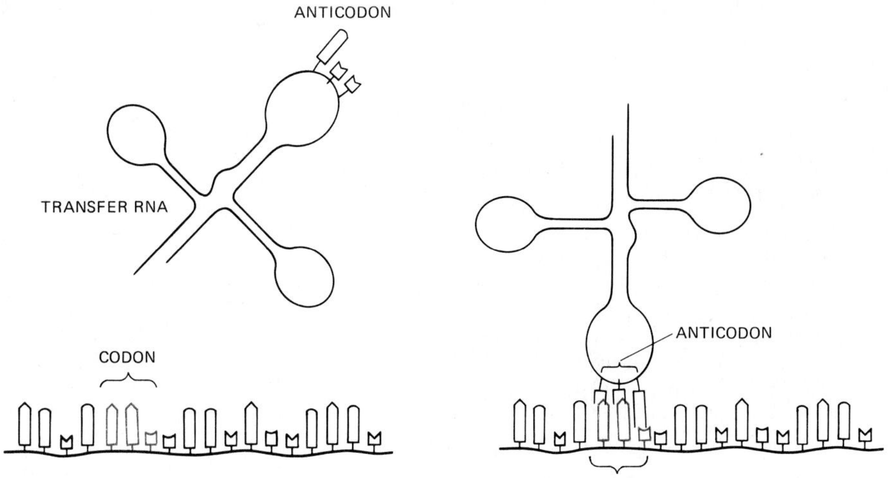

(a) tRNA MOLECULE

(b)

7.6

The positioning of a tRNA molecule in its proper place on an mRNA molecule. (a) A strand of mRNA is shown, with one codon marked to show that it has the proper sequence of nucleotides to receive the approaching tRNA molecule with its anticodon. (b) The tRNA is in place, with its anticodon nucleotides matching the complementary nucleotides of the mRNA codon.

TURNING GENES ON AND OFF

One can directly observe gene control in some cells, especially those that have large chromosomes and that are active in producing proteins. Fruit flies and midges are particularly suited to experimentation because some of the chromosomes in their secretory glands are the so-called giant chromosomes. These chromosomes are large enough to be seen with the unaided eye, and they change visibly during changes in activity.

During the larval stage (1a), before the flies become adult (1b), their secretory glands are producing large amounts of digestive enzymes and other proteins to be used in growth. Cells from such glands contain the large chromosomes (2), which are several hundred times as large as those of the usual body cells. A single chromosome, seen under a microscope, appears banded, with cross stripes definitely and recognizably placed (3). Unlike ordinary chromosomes, these are made of multiple identical strands, with each strand very tightly coiled at specific places. The thickened regions, lined up side by side, give the appearance of bands, as shown diagrammatically in (4). Such chromosomes are called *polytene* (Greek, "many threads") because of their many-stranded structure. At certain places along a polytene chromosome, enlargements appear as rather formless swellings. Sometimes called Balbiani rings (for their nineteenth-century discoverer), the enlargements are now more commonly known as *puffs* (5).

Chromosome puffs are believed to be produced in the following way. The many strands of DNA that make up the bulk of the chromosome can be thought of as a bundle of parallel fibers (6), which may uncoil, thus increasing their length. When this happens, the lengthened strands are pushed so that they bend outward (7) and, after still greater uncoiling, develop into loops (8). The hundreds of loops are uncoiled DNA strands, but the entire puff contains more than DNA, especially protein and RNA (9 and 10). Rarely, as in the fly Sciara, some new DNA is also synthesized (9).

The discovery that certain regions of chromosomes make puffs at certain times led to the supposition that a puff is an expression of genes in action. The supposition was verified in a series of experiments that were simple in their conception but highly imaginative and technically clever. (The name of one of the main experimenters is Ulrich Clever.)

When a larva was injected with a hormone that induces molting, a puff developed on a special place on a chromosome; without the hormone, the puff did not develop. This was good evidence that that particular puff was associated with molting. The nature of the puff itself could be found. A reasonable guess was that it contained RNA because RNA is required in active protein synthesis. An experiment was devised to test that guess. Cells convert uridine into uracil, and uridine is therefore a precursor of uracil. Since uracil is incorporated into RNA but not into DNA, a uracil precursor, labeled with radioactive tritium (H_3), was injected into larvae. The larvae were killed, the secretory glands were removed, and the polytene chromosomes from the glands were placed on photographic film. The developed film showed that it had been affected directly over the puffs, indicating that the radioactive tracer was concentrated in the puffs. The fact that the tracer was in a uracil precursor showed that the puffs were RNA. This was further confirmed by staining procedures. Additionally, treatment of larvae with an antibiotic that inhibits RNA transcription (Actinomycin D) prevented the buildup of radioactivity at the sites of the puffs. Puffs can be produced at will by hormone treatment or prevented by the antibiotic.

The puffing of polytene chromosomes shows that specific places on chromosomes (and therefore on DNA strands) become active when called on, and they produce RNA at such times. When those places are not called on, they relapse into inactivity. Genes can indeed be turned on and turned off, and they can be seen at work.

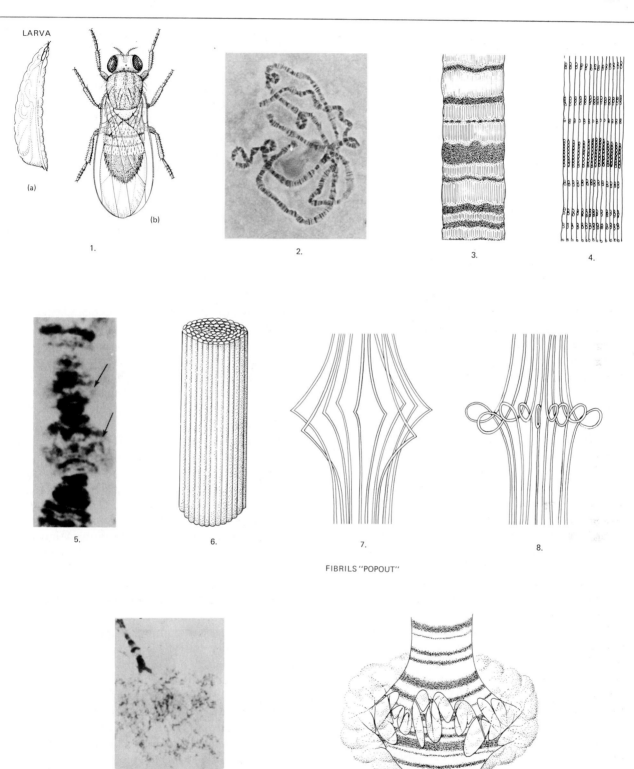

LARVA

(a)

(b)

1.

2.

3.

4.

5.

6.

7.

8.

FIBRILS "POPOUT"

9.

10.

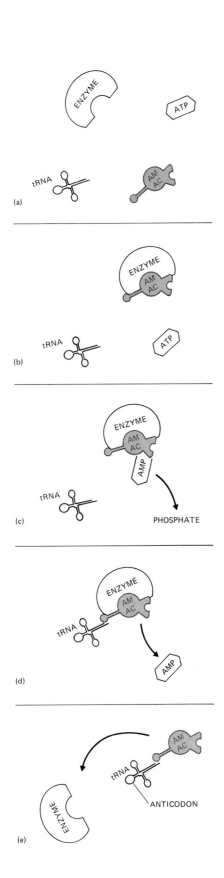

useful. The necessary indirect experiments have been performed, however, and meanings have been assigned to all possible combinations of nucleotides. For example, CUU means leucine, GCA means alanine, and so on. Not all triplets, however, code for an amino acid. For example, UAG and UGA seem not to code for anything, although there is speculation they may serve as "punctuation marks." Further, some amino acids are coded for by several triplets, as leucine is by UUA, UUG, CUU, CUC, CUA, and CUG. This repetitious ability, called the *degeneracy* of the genetic code, is thought to allow for a certain amount of error in the codons without destructive results for an organism.

THE AMINO-ACID–tRNA COMPLEX

The action of tRNA involves two main steps. One is the joining up with the particular amino acid for which a given tRNA is suited, and the other is the positioning of the amino acid in its proper place on the growing polypeptide chain. Before either of these events can happen, however, the amino acid must be *activated*. A simple amino acid floating in the cytoplasm will do nothing without both some directional guidance and a driving force. That force is provided by ATP, and the first move in starting an amino acid down its path toward protein formation is the activation of the amino acid, with the assistance of a specific activating enzyme (Figure 7.7). Two phosphate groups are liberated, and a resulting new molecule of adenylic acid monophosphate plus amino acid is formed. The amino acid, now activated, carries enough energy to become associated with its tRNA. Every amino acid has its own enzyme, which accomplishes the activation step. The next step is the forma-

7.7

A major feature of protein synthesis: activation of an amino acid and formation of an amino-acid–tRNA complex. (a) A source of activation energy (ATP), an amino acid, the amino acid's particular enzyme, and its own tRNA are free in the cytoplasm of a cell. (b) The amino acid is attached to its enzyme. (c) The enzyme–amino-acid complex is activated by ATP, which splits into inorganic phosphate and AMP. (d) The activated enzyme–amino-acid complex is joined by the tRNA and loses its AMP. (e) The enzyme is freed, and the activated amino-acid–tRNA complex is ready to move to the site on the mRNA, where a codon will accept this particular anticodon.

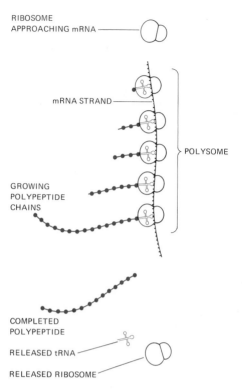

RIBOSOME
APPROACHING mRNA

mRNA STRAND

POLYSOME

GROWING
POLYPEPTIDE
CHAINS

COMPLETED
POLYPEPTIDE

RELEASED tRNA

RELEASED RIBOSOME

(a)

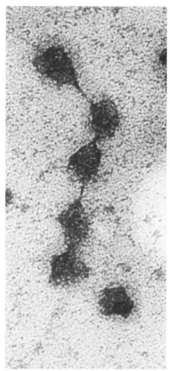

(b)

7.8

(a) Scheme of polypeptide assembly. Five ribosomes, strung along a strand of messenger RNA, constitute a polysome. One ribosome is about to attach itself to the RNA, and another has completed its turn and been released. Molecules of tRNA, each with its own amino acid, are in place, positioning the amino acids in the growing polypeptide chains. One completed polypeptide has been released. (b) Electron micrograph of a polysome of five ribosomes connected by a strand of mRNA. Note that this photograph was used (Figure 3.18) to illustrate ribosomes in Chapter 3.

tion of an amino-acid–tRNA complex. This is the final and active form of the amino acid before it is carried to the polysome, where mRNA is ready to receive it.

and delivered, all getting together a. .1e right place at the right time. Under optimal gro..ing conditions, a bacterium can make an entire new cell in as little as 20 minutes. The system is effective.

RNAs AND THE POSITIONING OF AMINO ACIDS

When the ribosomes, the mRNA, and the tRNAs with their activated amino acids are in place, it is time for the assembly of a polypeptide. The mRNA moves along a ribosome with its codons exposed, so that they can be matched with the anticodons of the tRNA. Thus a triplet of GAG in a molecule of mRNA, calling for glutamic acid, will attract only the anticodon of a tRNA that consists of CUC, and such a tRNA molecule can carry only glutamic acid. The glutamic acid–carrying tRNA will find its place on the mRNA and correctly attach its anticodon to the proper codon.

Even a small, comparatively simple bacterial cell may have an estimated 15,000 ribosomes and millions of amino-acid–tRNA complexes being formed

HOW IS A PROTEIN MOLECULE SYNTHESIZED?

The actual positioning of a tRNA on the mRNA occurs on a ribosome. The tRNA anticodon is matched with its mRNA codon by being held momentarily in place by the hydrogen bonding of the complementary bases. The tRNA carries its own specific activated amino acid. Then the next tRNA with its amino acid makes connection with the next mRNA codon, the two amino acids are joined by a peptide bond, and their bonds to the tRNA are broken. The tRNA leaves the mRNA. The bonding of the two amino acids is enzymatically controlled, and it is energized in part by GTP (guanosine triphosphate) instead of the usual ATP. After the second amino acid is hooked up, a third tRNA comes, then another and another, each amino acid bonded to the next, all

HOW TO MAKE A GENE

In August of 1976, after nine years of research, biochemist Har Gobind Khorana and his team at the Massachusetts Institute of Technology produced the first synthetic gene that is able to function within a living cell, an accomplishment that is considered as important as the discovery of the structure of DNA in 1953 by Watson

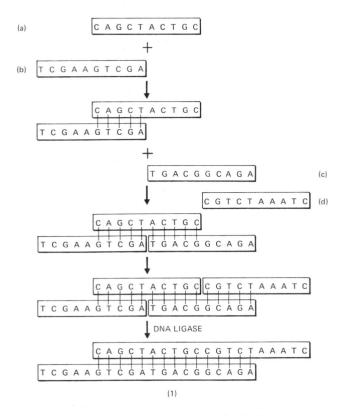

(a) | C A G C T A C T G C

+

(b) | T C G A A G T C G A

C A G C T A C T G C
T C G A A G T C G A

+

T G A C G G C A G A | (c)

C G T C T A A A T C | (d)

C A G C T A C T G C
T C G A A G T C G A T G A C G G C A G A

C A G C T A C T G C C G T C T A A A T C
T C G A A G T C G A T G A C G G C A G A

DNA LIGASE

C A G C T A C T G C C G T C T A A A T C
T C G A A G T C G A T G A C G G C A G A

(1)

The "sticky end" technique used to build the gene from single stranded pieces: The sticky end of fragment (a) left protruding after base-pairing with fragment (b) is complementary to the first few bases in fragment (c). That in turn leaves a sticky end recognizable by fragment (d). The individual single-stranded fragments are welded into continuous chains with the aid of the joining enzyme, DNA ligase.

and Crick. Part of the technique of deciphering the genetic code involved the synthesis of many different nucleotide base sequences into artificial messenger RNA. Khorana received a Nobel Prize in 1968 for his part in preparing such compounds. A strand of messenger RNA, however, is not a gene; concisely, a gene is a strand of DNA responsible for RNA assembly.

A first requirement for the test-tube manufacture of a gene is a knowledge of the base sequence of DNA, and that cannot be decided without knowing the base sequence of a specific RNA. Such a sequence has been known since 1965, when Robert Holley (corecipient of the 1968 Nobel Prize) worked it out for the transfer RNA that is associated with the amino acid alanine in yeast: alanine-tRNA. Once a transfer RNA has been mapped, the necessary bases in DNA are known, since they are complementary to the RNA bases. Furthermore, methods for incorporating genes into chromosomes have been developed (the dramatic technology of recombinant DNA, by which genetic material from one organism is inserted into another), and suitable tests have been devised for determining whether a functional gene has indeed been made.

When Khorana set out to synthesize a gene that coded for a known sequence of transfer RNA, he decided not to construct the gene by adding on one nucleotide at a time. Such a technique would have required huge quantities of material to allow for the proper chemical purification of each step and the expected loss of material at each stage of the many reactions involved. Instead, Khorana chose to construct small pieces of the molecule separately and attach them in their proper sequence later. Each piece of the chain was a single strand 10 to 15 nucleotides long. The single-stranded pieces were then joined into a double-stranded molecule, with each strand overlapping its complementary partner to provide a "sticky end" for the next strand to recognize. The separate strands were attached with the help of an enzyme called DNA ligase, which was discovered only a couple of years before Khorana used it.

The first gene to be made by such chemical methods was one that is responsible for a highly specialized RNA, a mutant transfer RNA for the amino acid tyrosine. Khorana and his collaborators chose it because it can be effectively tested in certain strains of *Escherichia coli*, the organism most favored in microbial genetics laboratories. The working part of such a transfer RNA contains only 126 base pairs, but the DNA responsible for its formation must contain about 200. Part of the reason for the extra required length is the need for two "ends," one indicating the start of the copying and the other for the end. These are the *promoter* (the first 52 nucleotide pairs) and the *terminator* (the final 21 nucleotide pairs).

The first gene to be introduced into *E. coli* cells by Khorana and his group failed to function. But another brand-new discovery guided the researchers to a bacteriophage that will not grow in *E. coli* unless a certain suppressor mutation is present. Khorana's synthetically prepared gene is such a suppressor, and after the gene was introduced into the DNA of the specific phage in *E. coli*, the phage was able to grow. Thus researchers had made a fully functioning gene, and the magnitude of the discovery was diluted only because the latest technology had made the synthesis of a gene inevitable. Nevertheless, it was a gigantic scientific achievement.

As usually happens in research, one success brings up new questions, and the questions are especially pressing in the instance of synthetic genes. What can and will be done with them? Plainly, one of the main uses of such knowledge will be the making of synthetic promoters and terminators in an effort to find out how cells succeed in controlling their own genes. Such techniques may be important in correcting genetic defects such as sickle-cell anemia. Re-

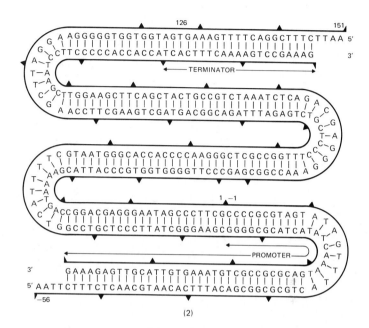

Structure of the synthetic E. coli *tyrosine transfer* RNA *gene synthesized by the Khorana group. Segments between points were linked chemically, then joined enzymatically to form the entire* DNA *double helix. (From the original drawings, Massachusetts Institute of Technology)*

searchers will also concentrate on creating "controlled mutations" in order to test their effect on both the gene and the transfer RNA. But with the development of applied aspects of such techniques come even more difficult questions. Now that genes can be made artificially and can be incorporated into whole chromosomes, will we be able to control our own genes? A human gene is 1000–3000 base pairs long. How soon will it be possible to synthesize such a complex gene? And who will decide which genes—and whose—are to be tampered with?

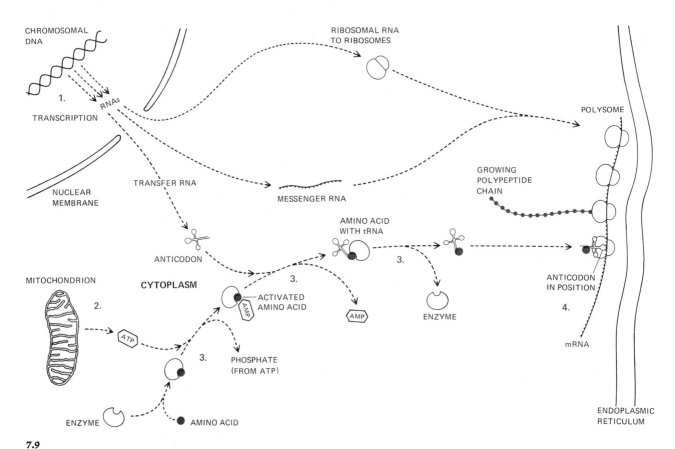

7.9

Assembling amino acids in the buildup of a polypeptide. The diagram shows (1) a DNA section of a chromosome transcribing three kinds of RNA, ribosomal, messenger, and transfer; (2) a mitochondrion producing ATP, which will be used to provide activation energy for an amino acid; (3) the sequence of events during which an activated amino-acid–tRNA complex is formed; and (4) a portion of endoplasmic reticulum with an attached polysome (mRNA and ribosomes) and a chain of amino acids being built into a polypeptide (protein).

linked by enzyme action, and all powered by GTP. The formation of mRNA from DNA was referred to as *transcription*; the formation of a polypeptide by mRNA is *translation*. Thus the central dogma can be rewritten even more accurately as

$$ \text{DNA} \xrightarrow{\text{transcription}} \text{RNA} \xrightarrow{\text{translation}} \text{protein,} $$

where the arrows indicate the direction of flow of information.

When the entire length of the mRNA has passed the ribosome, a string of amino acids will have been bonded into a polypeptide chain, with the amino end as the first-formed portion and the carboxyl as the tail. One ribosome makes one polypeptide at a time.

But ribosomes are strung together, four to six in a row, by a strand of mRNA, so as the mRNA moves along the ribosomes, it can generate as many polypeptides as there are ribosomes. Once a molecule of mRNA has passed a ribosome, that ribosome is free from the mRNA and can join another mRNA and start making another polypeptide. In similar fashion, once a tRNA has delivered its amino acid, it is liberated into the cytoplasm and can make another amino-acid–carrying trip (Figure 7.9).

The actual amount of protein synthesized can be measured with reasonable accuracy in cell-free systems in which the necessary enzymes, energy source, amino acids, and RNAs are provided. Under such artificial conditions, amino acids can be built into a polypeptide at the rate of about two a second. Estimates of the protein-building ability of an intact cell, however, indicate that under normal living conditions, polysomes can incorporate a couple of hundred amino acids per second, and they do it with much less expenditure of GTP. There is still obviously an immense amount to be learned about protein synthesis.

SUMMARY OF THE STEPS IN PROTEIN SYNTHESIS

1. The nucleus contains the DNA code message for the synthesis of proteins. The message is transcribed from one of the strands of the DNA to a single-stranded messenger RNA molecule. The DNA molecule serves as a template, or mold, so that the RNA strand is complementary to the DNA strand.

2. The messenger RNA carries the message out of the nucleus to the ribosomes in the cytoplasm. The mRNA attaches itself to the ribosomes.

3. Each of the 20 amino acids is activated by a specific enzyme. Each amino acid attaches itself to a tRNA molecule (also called adapter RNA). After an amino acid is attached to its specific tRNA molecule, the amino acid can be attached to another amino acid at a site on a ribosome.

4. Each triplet codon on the mRNA links with a complementary anticodon on a tRNA, with its attached amino acid.

5. A tRNA and its specific amino acid are attached to mRNA at a ribosome as the mRNA moves along the ribosomes. One after another, amino acids are bonded together in an ever-lengthening chain until a protein molecule is built up according to the coding order on the mRNA molecule.

6. The transfer RNA can be used again to carry a new amino acid after it has deposited its former amino acid "passenger" on the growing polypeptide chain.

MUTATIONS OCCUR WHEN A CODON IS COPIED INCORRECTLY

During the growth of organisms, cells divide many times to yield more cells that are identical or nearly identical. Before each cell division the DNA must be replicated; that is, DNA synthesis must occur, and the total amount of DNA doubles. In view of the number of times the DNA is replicated, it is not surprising that occasional mistakes are made. If a DNA molecule is not replicated correctly, every replication that follows will perpetuate the error. If this error results in a triplet that codes for an amino acid different from the amino acid originally intended, an altered gene product will result. Such a change in the DNA is a _mutation_, from the Latin word that simply means "change."*

A mutation that occurs in a body cell, that is, a _somatic_ cell, may persist for the life of the individual. If a mutation occurs early in embryonic life or in the gonads of an animal, the mutation may be passed into a sex cell and then to the next generation. Somatic mutations have practically no effect on inheritance in animals. However, they may and frequently do make a considerable difference to plants because a branch of a plant carrying a somatic mutation can produce a flower, and the mutation can pass into the cells of the flower and then on to the next generation.

Some mutations occur spontaneously for unknown reasons, but mutations can be made to happen artificially. One common way is to irradiate cells with ultraviolet light or X-rays or with gamma rays from a cobalt source. Of course, when a cell is bombarded indiscriminately with any radiation, the results are unpredictable. The experimenters who use radiation accept the death of some cells and choose whatever mutated cells (_mutants_) that suit their research purposes. Accurate control of mutations is at present impossible. As one biologist said concerning the use of radiation to produce mutations, "You might as well stand back from your Volkswagen, fling a hammer at it, and hope that it will turn into a Maserati." Another way to bring about unnatural DNA changes is by treatment with _chemical mutagens._ Experimentally produced mutants are mostly used for research purposes, but some have been used in practical programs. One such was the induction of thousands of mutants of the mold Penicillium, which were screened for their ability to produce the antibiotic penicillin. Most were useless, but eventually high-production strains were obtained, with outputs of antibiotic far beyond those of the original wild strains.

---

*   When the unmodified term _mutation_ is used, it means a DNA mutation, a change in the nucleotides. But there are also _chromosome mutations_, and the entire phrase is used when these are meant.

SUMMARY

1. The "one gene–one enzyme" hypothesis was formulated by Beadle and Tatum to indicate how each gene controls a specific enzyme. The hypothesis is currently referred to as the "one gene–one polypeptide" hypothesis, because current thinking is that the synthesis of each polypeptide is controlled by a separate gene.

2. Proteins, besides forming the structural elements that hold cells firmly, are responsible for the enzymatic action that makes life dynamic and continuing.

3. Protein assembly starts with *transcription*, in which single-stranded RNA is synthesized under the direction of DNA.

4. *Messenger RNA* carries a coded message from the nucleus to the ribosomes in the cytoplasm, telling which amino acids are to be included and where they are to be placed. The *triplets* of mRNA are known as *codons*. There are 64 possible triplet combinations, more than enough to recognize all 20 amino acids.

5. The complete set of triplet nucleotide symbols that indicate specific amino acids is called the *genetic code*.

6. *Transfer RNA* serves as an adapter, and after an amino acid is attached to its specific tRNA molecule, the tRNA can be attached to mRNA on a ribosome.

7. A protein molecule is synthesized from amino acids according to the specific coding order of the mRNA molecule.

8. *Mutations* occur when DNA is copied incorrectly or when there is a base change in the original DNA molecule. *Chromosome mutations* involve whole sections of chromosomes.

RECOMMENDED READING

Cohen, Stanley N., "The Manipulation of Genes." *Scientific American*, July 1975. (Offprint 1324)

Crick, F. H. C., "On the Genetic Code." *Science*, February 8, 1963. (Bobbs-Merrill Reprint B-49)

Crick, F. H. C., "The Genetic Code." *Scientific American*, October 1962. (Offprint 123)

Crick, F. H. C., "The Genetic Code: III." *Scientific American*, October 1966. (Offprint 1052)

Hanawalt, Philip C., and Robert H. Haynes (eds.), *The Chemical Basis of Life*, Readings from *Scientific American*. San Francisco: W. H. Freeman, 1973. (Paperback)

Holley, Robert W., "The Nucleotide Sequence of a Nucleic Acid." *Scientific American*, February 1966. (Offprint 1033)

Kornberg, Arthur, "The Synthesis of DNA." *Scientific American*, October 1968. (Offprint 1124)

Maniatis, Tom, and Mark Ptashne, "A DNA Operator-Repressor System." *Scientific American*, January 1976. (Offprint 1333)

Miller, O. L., Jr., "The Visualization of Genes in Action." *Scientific American*, March 1973. (Offprint 1267)

Nirenberg, M. W., "The Genetic Code: II." *Scientific American*, March 1963. (Offprint 153)

Rich, Alexander, and Sung Hou Kim, "The Three-Dimensional Structure of Transfer RNA." *Scientific American*, January 1978.

Temin, Howard M., "RNA-Directed DNA Synthesis." *Scientific American*, January 1972. (Offprint 1239)

CELLULAR DIFFERENTIATION

Once cells have been produced by mitosis, they usually go on to become specialized in structure (how they are built) and function (what they do). This process of specialization is called *differentiation*. Differentiated cells form the tissues and organs that carry on the many specialized life processes of all organisms.

In animals, hundreds of different kinds of tissues are formed, but most (at least in the vertebrate) can be classified as one of these: *epithelium* (including outer covering and inner linings), *connective tissue* (including cartilage, bone, and blood), *muscle tissue* (smooth, striated, and heart muscle), and *nerve tissue*.

In plants, the main kinds of tissues are *epidermal, ground tissue* (a general catch-all for tissues that are neither vascular nor epidermal), and *vascular tissues* (including conducting and strengthening tissues). We present here a collection of some of the main differentiated cells that make up tissues. Examples come from vertebrate animals and flowering plants.

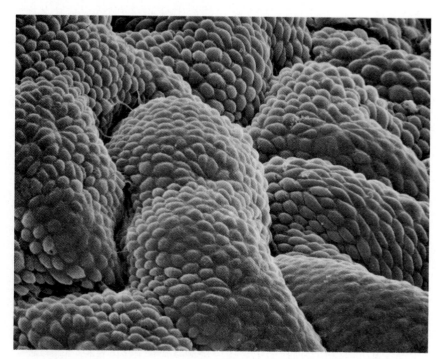

Animal epithelium. Scanning electron micrograph of the stomach lining. Each pebbly lump is a secreting cell. X 50.

Kidney tissues. Scanning electron micrograph of a portion of a kidney glomerulus. The larger, tube-shaped parts are capillaries, and the branched, rootlike processes are outgrowths from epithelial cells. X 2000.

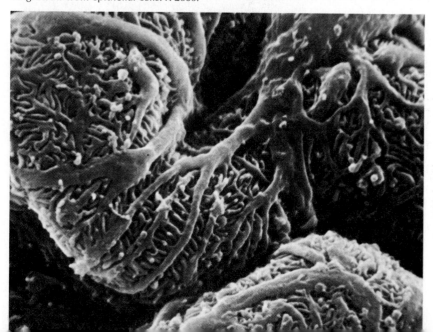

I:2

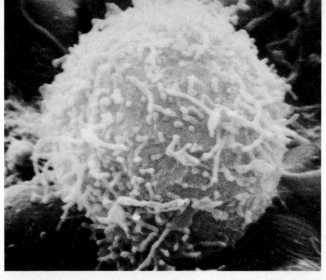

Animal connective tissue. A single lymph cell. The conspicuous surface extensions are microvilli. X 9500.

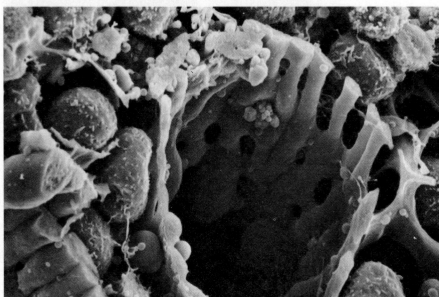

Vascular tissue in dog spleen. Scanning electron micrograph of the interior of a vein. Since the spleen is a filter that removes foreign particles and worn-out cells from blood, it has a convoluted interior, which adds to the surface area and makes filtering more efficient. X 2600.

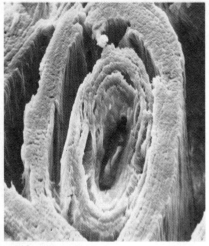

Animal connective tissue. Electron micrograph of fibers of collagen, the tough, elastic strands that help give tendons and ligaments their strength and flexibility. X 22,500.

Scanning electron micrograph of a cut surface of bone. The elliptical body is one of the units of bone: an *osteon* or *Haversian system*. The osteon has a central cavity, the Haversian canal. The main strength of bone is due to the nonliving calcium salts secreted by living cells, which are not shown here. X 250.

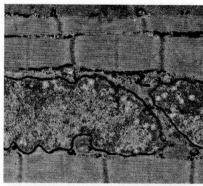

Smooth muscle. Portion of a single cell of nonstriated muscle, as is formed in arterial and intestinal walls. The large ovoid body at the top of the picture is a nucleus. Mitochondria are clumped below it. Electron micrograph of thin section, X 20,000.

Heart muscle. The fibrils are striated, but the arrangement of cells is somewhat different from that in striated skeletal muscle. Electron micrograph of thin section, X 12,000.

Striated muscle. In this electron micrograph, eight muscle fibrils are visible, stretched horizontally. The striations, which give this kind of muscle its name, are strikingly apparent. X 12,000.

Nerve tissue. Light microscope view of cells from a rat brain. The large, dark cell bodies have fibers extending from them. X 600.

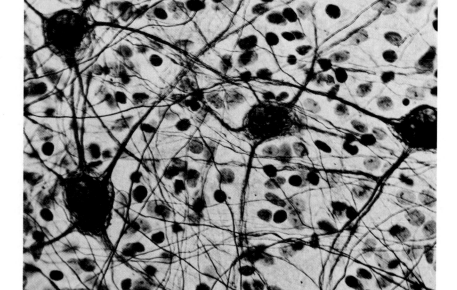

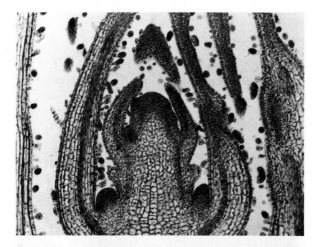

The origin of plant tissues. Section through the growing top of a stem of the white ash (*Fraxinus americana*). At the tip of the stem, cells are unspecialized and continuously dividing (growing). Some distance below the tip, differentiation begins to occur. The outer layer of cells will develop into the epidermis of the mature stem. Some of the cells under the epidermis (pale in the picture) will become ground tissue, such as pith and cortex. The round bumps on either side of the tip represent embryonic leaves beginning to form. The dark streaks, extending into the embryonic leaves, will become vascular tissue. X 25.

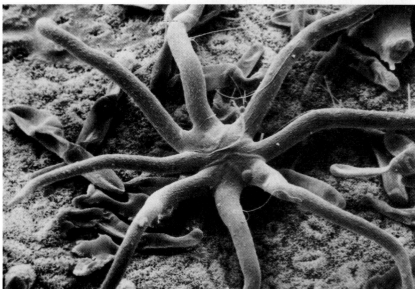

Plant epidermis. Scanning electron micrograph of the outside of an oak leaf (*Quercus gambeli*). The octopuslike object is an epidermal hair. Guard cells of stomata are visible between the "arms." X 230.

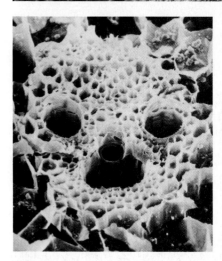

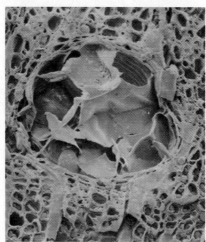

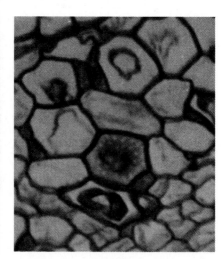

Vascular tissue. Scanning electron micrograph of a cut across a corn stem, showing the large water vessels, the heavy-walled strengthening fibers, and the large surrounding cells of the ground tissue. X 140.

Wood tissue. Scanning electron micrograph of a cut across a piece of locust wood, showing many small cells (fibers and parenchyma), and one large water vessel. X 400.

Vascular tissue. Light micrograph of a section through the phloem of squash (Cucurbita). This tissue transports food throughout the plant. The largest cells are sieve cells, one of which (just below center) was cut to show the end of the cell with its sieve plate. The small cell with a nucleus (below the sieve plate) is a companion cell. X 200.

8
Cellular Growth and Specialization

THE FINAL FORM OF AN ORGANISM DEPENDS directly on its cellular arrangement. Multicellular plants and animals start life as single cells, and they achieve maturity as a result of repeated cell divisions and changes in the form and function of individual cells. The first event in the development of complex organisms is the division of the original cell; then subsequent cells divide until a sufficient mass of cells has developed. Despite intensive efforts to find the initiating drive, no one knows what causes a cell to undergo division. Cells can be held back from division by cold and by chemical means, but experimental, predictable, controllable cell division is not yet possible. We can only provide conditions that favor growth in general and allow cell divisions to occur.

If a cell mass is to achieve the complex form of a well-developed animal, such as a human being, the cells must be directed when to divide, how often, and in what direction. The direction in which a cell divides is important to an organism. After an initial cell divides, if subsequent divisions are always in the same direction, the result will be a single strand of cells, end to end. Such organisms exist, but they are severely limited in what they can do.

Cell reproduction and placement are not enough, either, to make a functional organism. If a human egg divided the proper number of times and made the divisions in the correct directions, the result might be something that looked like a human, with arms and fingers and the rest, but if all the cells were alike, that mass of cells would be no human. It is the maturation, the differentiation, the specialization of cells that give them their final working form. Functional specialization is more striking than species specialization; a human liver cell is less like a human retinal cell than it is like a pig liver cell.

In this chapter, we shall describe some of the observable events that lead to a complete organism and give some of the current ideas on cellular control, but we cannot explain completely the forces that lead to cell division or directional growth or cellular differentiation because those forces are not known. As this book repeatedly points out, the amount that we know about life is undoubtedly much less than the amount we do not know.

THE MECHANICS OF CELL DIVISION

The story of cell division, as seen with a light microscope, has been known since the 1860s, when the German zoologist Walter Flemming, the German botanist Edward Strasburger, and the Italian zoologist Theodor Boveri, working separately, discovered and illustrated the events. Since then, details and variations have been found and facts added, especially in that period between divisions known as the *interphase*.

After a cell has divided into two, each of the new cells is about half as large as the original one (Figure 8.1). Immediately after a small new cell is produced,

it enters a period of growth known as the first "gap phase," or G_1. The G_1 lasts from a few minutes to years, or it may even be lacking in some young embryos. During that time, there is synthetic activity, mostly protein synthesis, in all cell parts except the chromosomes, and the cell increases in size until it is usually as large as its parent cell was before division.

Some time during interphase, if the cell is going to divide, it enters a new "synthetic phase," or S phase, during which the DNA of the chromosomes doubles. The important action here is the exact replication of the DNA, which is not simply a doubling of the quantity, but a doubling that provides the cell with two complete, identical sets of chromosomes. During the S phase, the chromosomes are stretched out so thin that they cannot be seen with a light microscope. That fact made it impossible to learn when duplication occurs until such new techniques as radioactive tracing were available.

MITOSIS DISTRIBUTES EQUAL DNA TO EACH NEW CELL

With the S phase past and the chromosomes doubled, the cell goes into a second gap phase, G_2, in preparation for nuclear division. Further general synthetic activity continues, but the G_2 phase seems to be less variable in time than G_1. It seems that, having once "decided" to divide, a cell goes ahead and does so. At the end of G_2, the cell begins the actual nuclear division process, generally known as *mitosis*, or the *M phase*. The word "mitosis" refers to the threadlike

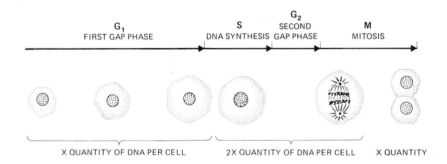

| G_1 | S | G_2 | M |
| FIRST GAP PHASE | DNA SYNTHESIS | SECOND GAP PHASE | MITOSIS |

X QUANTITY OF DNA PER CELL   2X QUANTITY OF DNA PER CELL   X QUANTITY

8.1

The life of a cell. Once produced as a result of a cell division, a new cell with its chromosomes and a standard amount of DNA grows in size during the first gap phase (G_1). Having reached full size, it goes into a phase of DNA synthesis, the S phase, during which the amount of DNA in the chromosomes is doubled. The cell then enters a second gap phase (G_2), which ends when the cell divides, parceling out its DNA equally to the new daughter cells.

NUCLEAR TRANSPLANTS

Every organism in a sexually reproducing population has its own peculiar set of genes, guaranteed to be unique by the reshuffling of genes with crossovers during meiosis and by recombinations at fertilization. The results are apparent even in such closely related groups as human families, with individual differences showing through in spite of family resemblances. Diversity is usually considered desirable in people, but uniformity is sought in such nonhuman organisms as domestic animals or crop plants. The way to achieve uniformity is to make certain that all nuclei, with their DNA sequences, are alike in all the individuals of a group.

In plants, genetic uniformity is easier to achieve than it is in animals because plants can be propagated by cuttings or by grafting, thus bypassing the entire sexual process with its endless mixing of genes. An orchard planted with Golden Delicious apples is an example of a genetically uniform population; that is to say, it is a *clone*. All the Golden Delicious apple trees in the world are asexual descendants of a single tree and constitute an enormous clone.

Theoretically, genetic uniformity is obtainable even in species that normally reproduce only by sexual means, although a number of technical and ethical problems remain. First, one must obtain a number of identical nuclei by allowing a zygote to undergo repeated mitotic divisions. By *nuclear transplantation*, a nucleus is microsurgically removed from one cell and put into another cell. If the original nucleus in an egg is destroyed and a new nucleus is brought in, the egg can grow into an embryo. Such operations have been successfully performed on what we conceited humans call the lower animals.

If a number of eggs are enucleated, that is, have their nuclei removed, and the original nuclei are replaced by nuclei all taken from the same embryo, the result is a clone. A clone of pigs or of dairy cows could be depended on to produce a steady supply of meat or of milk, subject only to the slight variations imposed by inevitable differences in environment. The beginnings of such a clone would be expensive and difficult, because expert technical skill would be required, and there would have to be some way to bring the early embryos to a condition of relative independence. They could be implanted in uteri of appropriate females, or they could even be brought through the beginning stages in an artificial nutrient medium. No mammals have yet been raised to maturity in a "test-tube" type of environment, but amphibia have, and plants are relatively easy to grow in culture. (See the essay on test-tube babies on p. 284.)

Even though effective techniques have not been perfected, people have been thinking about the possible consequences of cloning human beings. With his endless curiosity, a human wonders: What would it be like to have a thousand identical factory workers or half a million equally capable and disciplined soldiers all in the same-size uniform? Then the serious questions come. How would the cloned people *like* being unindividual? Who would rear them? (Other cloned people?) Finally the hardest question of all: Who would be empowered to decide which ones are to be cloned? The students of today may have to meet such problems, and they need to know why they are deciding what they decide.

---

appearance of chromosomes during the period (Figure 8.2).

For convenience, the M phase is conventionally divided into stages, but these stages flow smoothly from one to the next and must not be thought of as coming in jumps, as selected still pictures may seem to indicate.

Prophase

At the onset of *prophase*, the chromosomes begin to shorten and thicken, coiling upon themselves. Before the coiling begins, the total length of the DNA strands in a single human nucleus is greater than the length of the entire human body, but the strands are too thin to see. When they reach their minimum

8.2

Schematic representation (below) of *mitosis* in an animal cell. Points to be noted especially are (1) the presence of two pairs of centrioles in the interphase, (2) the beginning of nucleolar disintegration and separation of centrioles with beginning asters in early prophase, (3) the thickening of chromosomal material and establishment of poles in late prophase, (4) the completed spindle of microtubules and equatorial position of chromosomes at metaphase, (5) the splitting of the centromeres and the separation of the chromatids and their movement to

opposite poles during anaphase, and (6) the doubling of centrioles, beginning of cytoplasmic constriction, and reappearance of nucleoli in late telophase. On the next page are selected photographs and drawings from a series showing mitosis in a living cell in culture. The granules in the cytoplasm are mitochondria and other organelles. The drawings emphasize nuclear detail. The disappearance of the nucleolus, the doubleness of the chromosomes in metaphase, and the buildup of the cell plate in telophase are notably clear.

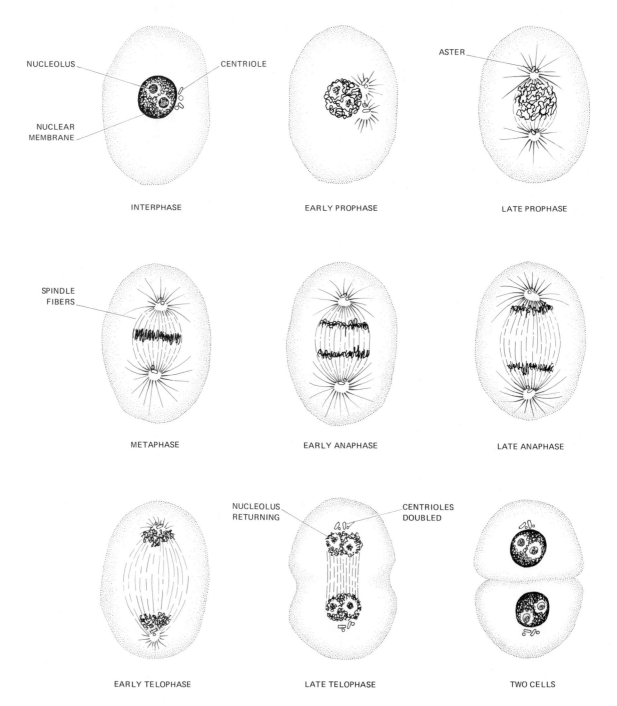

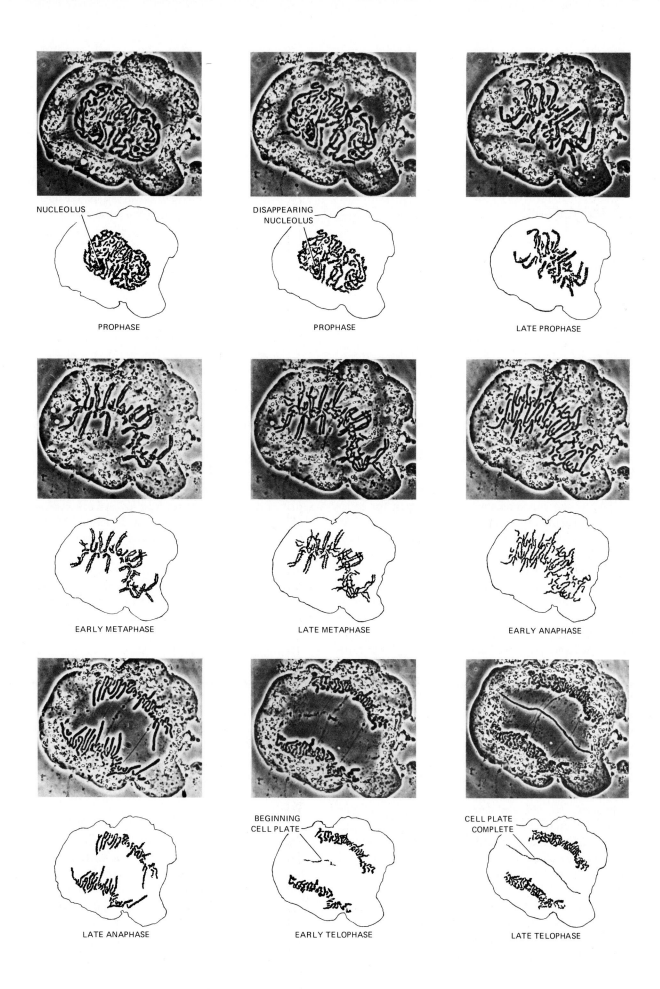

NUCLEOLUS

PROPHASE

DISAPPEARING
NUCLEOLUS

PROPHASE

LATE PROPHASE

EARLY METAPHASE

LATE METAPHASE

EARLY ANAPHASE

LATE ANAPHASE

BEGINNING
CELL PLATE

EARLY TELOPHASE

CELL PLATE
COMPLETE

LATE TELOPHASE

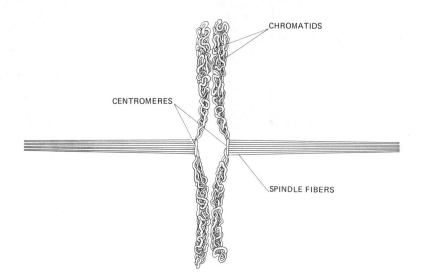

8.3
Diagram of a pair of chromosomes at metaphase, based on electron micrographs, showing the tangled threads of DNA-plus-protein and the centromeres to which the spindle fibers (microtubules) are attached.

length of only a few micrometers during late prophase, they are thick enough to be readily visible with a light microscope. By late prophase, each chromosome is double. The two halves are called *chromatids*. Each pair of chromatids has a *centromere*, to which spindle fibers are attached (Figure 8.3).

While the chromosomes are preparing for separation of their chromatids, other activities are progressing. Outside the nucleus, the two pairs of centrioles have begun to move apart, and by the end of prophase they are at opposite sides of the nucleus in what are known as the *poles* of the cell. The position of the centrioles at the poles determines the direction in which the cell will divide. Meanwhile, the nucleoli break up and disappear, as does the nuclear membrane. Between the centrioles, strands of microtubules constitute the *spindle*. In animals, a sunburst of microtubules radiates out from the centrioles, forming the *asters*. Higher plants have neither centrioles nor asters, but they do have spindles and the other mitotic components.

Metaphase
When the chromosomes have shortened to their minimum length, they move into a somewhat flattened plate across the cell and come into *metaphase.* Some spindle fibers extend from pole to pole, but others extend only from one pole and attach to the centromere of a chromosome. In early metaphase the pairs of chromatids that make up a chromosome are held together by a single centromere, but toward the end of metaphase the centromeres become double, so that each chromatid has one. It is then appropriate to call each chromatid a full-grown chromosome. Metaphase is short, lasting only a few minutes in some cells.

Anaphase
The beginning of *anaphase* is marked by the separation of the chromosomes, with one member of the pair of chromosomes starting toward one pole and the other moving in the opposite direction. Meanwhile, the poles themselves are moving farther apart, and the microtubules of the spindle are changing. The pole-to-pole microtubules lengthen while the pole-to-chromosome microtubules are shortening. The force causing anaphase movement of chromosomes is not known, although the microtubules are essential.

Telophase
Telophase is marked by the arrival of the chromosomes at the poles. Once there, they lose their distinctness, begin to lengthen, and resume their interphase condition. About this time, too, the centriole at each pole duplicates as if preparing for the next mitosis to come. The endoplasmic reticulum spreads out around the chromosome mass, making a new nuclear membrane, and the nucleolar regions of the chromosomes build new nucleoli. Nuclear division is over.

Nuclear division is over, but *cell division* is not. The new nuclei are still in the same cytoplasmic unit. Commonly, nuclear division is followed by separation of the cytoplasm into two parts. In animals this separation is accomplished by a pinching of the cell membrane, as though someone had looped a cord around the equator of the cell and pulled it tight. In plants, the cytoplasmic division is achieved by the formation of a partition, the *cell plate*, between the two portions of the cell (Figure 8.4). Cytoplasmic division is not necessarily a part of nuclear division, however. Many cells allow a normal mitosis without subsequent cell division, and the result is multinucleate cells.

Mitosis accomplishes two things: (1) the reproduction of cells and (2) the equal distribution of DNA to each. Cells do not deviate from this pattern unless they are behaving abnormally or are undergoing those special divisions called *meiosis*.

MEIOSIS REDUCES THE CHROMOSOME NUMBER

As soon as researchers discovered that chromosome counts are constant, it became apparent that there must be a time in the life of any sexually reproducing species when the chromosome number is cut in half, since nuclei of sex cells cannot go on uniting generation after generation and yet keep the count always the same. Investigation of dividing cells soon revealed the reduction divisions. In animals the task was fairly straightforward because the reductions proved to occur just before sex cell formation and they were found in the *gonads*, the gamete-producing organs. (Gametes are the sex cells.) In plants, the story is somewhat different, since the reduction divisions may occur long before gametes are made. But the reductions are nevertheless necessary, and regardless of when they occur, the general features of the process are fairly standard.

Meiosis is the term applied to the pair of divisions required for reduction of the chromosome number from the double (diploid) to the single (haploid)

8.4

Comparison between animal and plant cell division. (a) The animal cell has centrioles and asters, and the cytoplasm is divided by constriction. (b) The plant cell lacks centrioles and asters, and the cytoplasm is divided by the formation of a cell plate.

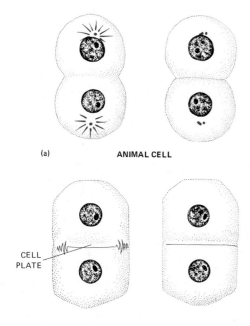

(a)   **ANIMAL CELL**

CELL PLATE

(b)   **PLANT CELL**

TABLE 8.1

Chromosome Numbers of Representative Species of Plants and Animals

ORGANISM	HAPLOID NUMBER	DIPLOID NUMBER
Ascaris	1	2
Penicillium	2	4
Fruit fly	4	8
Moss	7	14
Carrot	9	18
Corn	10	20
Frog	13	26
Hydra	16	32
Earthworm	18	36
Cat	19	38
Cobra	19	38
Human	23	46
Potato	24	48
Amoeba	25	50
Silkworm	28	56
Cow	30	60
Guinea pig	32	64
Dog	39	78
Pigeon	40	80
Horsetail	108	216

condition (Figure 8.5). There are regularly two divisions.

Superficially, meiosis resembles mitosis. The chromosomes and nuclei have the same general appearance during meiosis that they have during mitosis. But there are several fundamental differences that must be clarified if the genetic effects of sexual reproduction are to be understood. The first and most obvious is the reduction in number of chromosomes. That reduction, however, is not simply a haphazard doling out of chromosomes. Rather, it is a highly ordered, precise method of guaranteeing that every reduced cell has a full quota of the correct DNA. The second difference involves the close pairing and subsequent exchanging of parts of chromosomes. Both these points deserve special consideration and will be explained further.

In animals at least, most of the body cells are *diploid;* that is, their nuclei contain two full sets of chromosomes, one set having come from the animal's female parent and one from its male parent. The chromosomes (with some exceptions to be described later) are present in matching pairs. The two members of a matching pair are *homologous chromosomes*, or simply *homologs*. During the working life of the animal, homologs behave almost independently, ignoring each other during the mitotic cycles. But during the first prophase of meiosis, or even in the mitosis preceding the meiotic divisions, homologs approach one another, two by two. Early in Prophase I (the first prophase), when the chromosomes are long and thin, each member of a homologous pair comes to lie alongside its opposite member, normally matching point for point with great accuracy. Indeed, they become so closely associated that it is difficult to see them as double.

The picture is complicated by the fact that each homolog itself is double, consisting of two chromatids, so that what may appear as a single strand is really four strands: two homologs, each with two chromatids. Such a complex is called a *tetrad*, or a *bivalent*. The two chromatids belonging to one chromosome are *sisters*. Still during Prophase I, the chromatids undergo a series of breaks. Instead of healing themselves and returning to the way they were, however, nonsisters break at corresponding points. When healing does occur, one chromatid becomes joined up with a nonsister, and vice versa, so that there is an actual exchange of chromatid parts. Since each sister chromatid has two nonsisters with which it can make exchanges, and since up to half a dozen exchanges can be made in one tetrad, the results can become complex.

After homologs begin to separate, the points where exchanges took place cause a tangling of the chromatids, with strands crossing over one another, frequently looking like Xs when seen in a microscope. These crossover regions are consequently called *chiasmata*, from the Greek letter chi (χ). Not only are chiasmata observable, but their genetic effects can be detected in the next generation of organisms. It can be demonstrated experimentally that the Xs visible in meiotic cells do indeed represent physical exchanges of DNA, and they can be correlated with specific genetic exchanges.

At Anaphase I, the centromeres of a pair of homologs separate and the chromosomes move to opposite poles, carrying with them the newly recombined pieces of DNA that have been exchanged with nonsister chromatids. Which centromere starts toward which pole is a matter of chance, and each tetrad behaves independently of the others. Whether the original centromere came from a mother's egg or a father's sperm makes no difference; the diploid set of chromosomes is thus effectively shuffled, rearranged, and separated out. Even without chiasmata, the genetic material can be well shaken up during meiosis, but when the chiasmata are taken into account, it becomes easy to understand how among the billions of people, to take an example of only one species, no two are genetically alike. Identical twins are of course an exception, but they arise after meiosis. Humans, with their 23 pairs of homologous chromosomes could theoretically produce 2^{23} (8,388,608) kinds of gametes: If chromosome exchanges ("crossing over") between chromatids occur, the number becomes too large to think about. Meiosis is an effective method of redistributing genes.

It is important to emphasize that there is no place in the meiotic process at which one can say: This is where reduction happens. Reduction of the chromosome number and segregation of homologous chromosomes and rearrangement of parts of chromosomes are all part of the entire story of meiosis.

After Anaphase I, there is a short Telophase I, followed by a second division: Prophase II, Metaphase II, and so on. At Anaphase II the final separation of chromatids occurs, and by Telophase II there are four nuclei, each with a full set of chromosomes, with one representative of each chromosome (that is, it is haploid) but with the DNA redistributed in a way that is practically unique. It is theoretically possible that two human sperm cells could have exactly the same set of nucleotide sequences in their DNA, but the chances are about as good as your winning a national lottery every day all your betting life.

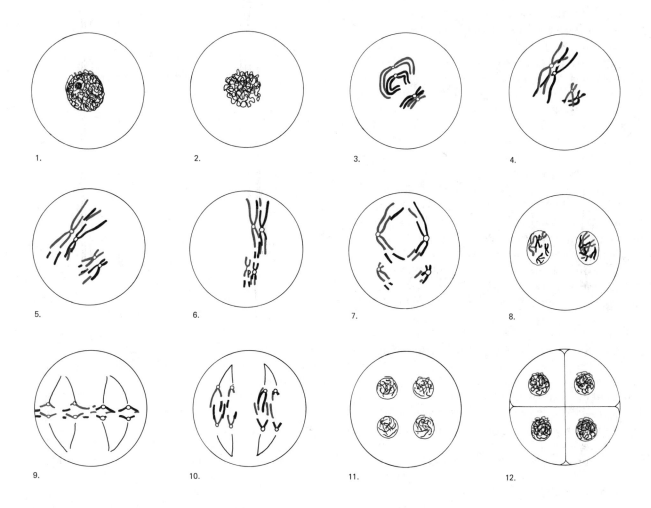

8.5

Schematic drawings of reduction division, or *meiosis*. (1) A diploid cell, about to undergo meiosis. (2) The chromatin threads become somewhat thicker, forming visible lines. (3) Chromosomes become apparent. In this scheme there are four chromosomes, one long pair with a median centromere and a shorter pair with nearly terminal centromeres. The chromosomes that were derived from the male parent (of the organism whose cells are illustrated) are black, and the chromosomes from the female parent are colored. The chromosomes line up exactly. Each chromosome is itself doubled, so that the *bivalent* (or *tetrad*) is four-stranded. (4) Strands of the bivalents (*chromatids*) come to lie across one another. (5) Parts of maternal and paternal chromatids are exchanged. All the events shown in (1) through (5) occur during the first prophase. (6) The bivalents line up on the equator of the cell, bringing the cell to Metaphase I. (7) During Anaphase I the centromeres of the bivalents separate and start toward the poles of the cell, taking the reshuffled chromosomes with them. (8) At Telophase I, which is of short duration, two nuclei are complete. (9) At Metaphase II there are two sets of spindle fibers. (10) During Anaphase II the chromatids separate, and they can now be called chromosomes. Each pole receives one long and one short chromosome, but these are different from the chromosomes the cell started with, having undergone chromatid exchanges, or crossovers. (11) At Telophase II, four nuclei, now *haploid*, are re-formed. (12) Each haploid nucleus is separated from the others. Four cells with a reduced number of chromosomes have been produced.

COMMON TYPES OF CELLULAR SPECIALIZATION

Mitosis results in two cells where formerly there was only one, but such cellular reproduction and the subsequent increase in size are not enough to make a working organism. Cellular *differentiation* must follow. (See the picture essay following p. 114.) Differentiation accounts for such diverse structures as horses' hoofs, birds' feathers, snakes' scales, human hearts and brains, oak bark, ebony wood, and orchid flowers.

A mass of cells all similarly differentiated structurally and functionally is a biological *tissue.* A knowledge of tissues is important to medical pathologists because they must know the microscopic appearance of normal tissues in order to recognize diseased ones. To an inexperienced observer, a cancerous tissue looks simply like a collection of cells, but to one who knows what differentiation (or lack of it) to look for, the difference between malignancy and normality is usually not difficult to see.

Specialization of function and specialization of structure go together. A cell with a highly specific duty, as for example, a nerve cell, a cartilage cell, or a wood fiber, may be so modified as to be scarcely recognizable as a cell. However, an unspecialized cell, such as a pith cell in a stem or a liver cell (which has so many jobs that it must not become too peculiar), looks like a textbook cell: more or less roundish and furnished with the usual organelles.

HOMEOSTASIS: THE ORGANISM REMAINS STABLE IN A CHANGING ENVIRONMENT

A dynamic, functioning organism must make an unending series of compromises between two opposing requirements. It must keep changing to eat, to breathe, to mate, to escape destruction. At the same time it must keep itself somewhat the same to maintain its strength and to retain its accumulation of experience. A balance between these two contrasting demands is attained at the cellular level and is made possible by intake and outflow of energy and material. The result is that cells seem to remain stable while working. In a functional sense, they do remain stable; some individual cells remain alive for years, but they do so by virtue of a constant turnover of cellular material. The maintenance of a measure of stability during change is called *homeostasis,* meaning "standing the same."

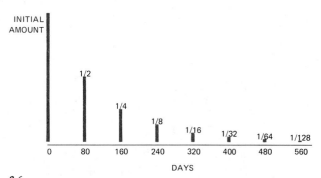

8.6
Survival of a material with a half-life of 80 days. At the end of the first 80 days, half the material is gone; after 560 days, only 1/128 remains.

By direct observation, a given cell can be seen to maintain its integrity, carrying out its normal activities for an extended length of time. It is known to have stability. But experimental treatment of cells proves that they are not stable in the same sense that a building is stable. If an animal eats amino acids with radioactive, labeled tracers, the tracers will soon show up in the structural proteins of the animal's cells. The animal has replaced some of its old proteins with new ones, and cells can make such substitutions rapidly. It is difficult to understand how a molecule can "wear out," yet some liver enzymes are taken out of circulation and replaced within a few hours. Some muscle proteins are 50 percent replaced after six months. In other words, the *half-life* of muscle protein is six months (Figure 8.6). If an average half-life of cellular protein is about 80 days, a human can expect to be rebuilt, molecule for molecule, about every year and a half. It is as though a building has a few stones removed from its structure every day and has new ones slipped into their places, so that after a time the entire building is renewed. It still has the same look and the same rooms, and it has not been out of service during the renovation.

You might ask, "If I am not the same person I was five or ten years ago, how can I remember what happened then?" Granted that giving a label is not the same as giving an explanation, a provocative answer is, "Because your cells are able to maintain homeostasis."

BIOLOGICAL FEEDBACK PROVIDES STABILITY FOR THE SYSTEM

Biological controls and indeed all self-regulating controls depend on a *negative feedback system* (Figure 8.7). A familiar example is a thermostatically controlled furnace. An action (burning of a fire in the furnace) causes a change in the system (warming of the house), which acts on a responsive sensor (the thermostat), which then causes a change in the original action (the fire is cut off) in a direction opposite to the original action. Dressing to suit the weather and driving a car are both examples of negative feedback. If you steer too much to the *right* (the action) the car goes to the *right* (the resultant change); then you *see* that the car is going to the right (the eyes are the sensors) and make a compensating adjustment to the *left*. Any dynamic system that remains stable must have an adequate control, either imposed from without (a mother wraps up a cold baby) or by negative feedback in the system itself (the mother wraps herself). The principle is in use in the chemistry of cells, the actions of organs, the behavior of whole organisms, and even in the regulation of populations.

Positive feedback, sometimes called *runaway feedback*, occurs rarely and is usually disastrous. The gambler who bets more and more as he loses until he goes broke, the alcoholic who drinks more and more until he gets so drunk he passes out, the victim of heat stroke who gets hotter and hotter as he continues to generate more internal heat until he dies, all three are suffering from positive feedback. In the example of a thermostat, if the thermostat was wired wrong, a hot room would call for heat from the furnace and would keep calling for heat until either the house burned or the fuel gave out. Positive feedback is *not* normal in biological systems.

Cellular processes are generally dependent on a feedback system. When feedback is mentioned without a modifier, it is understood to mean negative feedback since positive feedback is too uncommon to be considered. For example, when a cell is making a new DNA molecule, it must have on hand in the cytoplasm balanced amounts of adenine and thymine, and these must be synthesized enzymatically by the cell. Too much of one or the other is wasteful, and cells cannot afford to be wasteful. If there is too little of either one, however, DNA synthesis simply cannot proceed. The problem for the cell is to provide the proper amounts of both, and it does so by a feedback that shuts down adenine production if the adenine-synthesizing mechanism produces more than is being used, or that turns on the adenine producer when adenine is in short supply. The same kind of control works for the thymine producer. The result is that the cell keeps itself in order, with adequate amounts of the necessary building products. Many such examples are known, and the supposition is that cell regulation is usually if not always based on some similar arrangement.

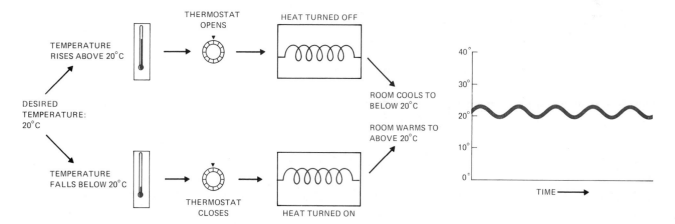

8.7

Example of a negative feedback system. When a room with a thermostat *cools* below the set point, the system calls for *heat*, and the converse also applies. Such a system does not produce an absolutely even temperature, but it does keep the temperature close to the set point. The colored line in the graph shows how a negative feedback system produces a wobbly effect. With sufficiently precise sensors, the ripple can be made so slight as to be unimportant.

GENE REGULATION:
THE JACOB-MONOD MODEL

Since every nucleus has genes for performing every function of a whole organism, and since individual cells perform only a few special functions, it follows that most of the genes in a nucleus at any one time are not doing anything. The question is one of determining how genes are regulated. What makes some "turn on" or "turn off"?

The most satisfactory current guess as to the mechanism of gene activation and inactivation comes from studies of mutants of the human colon bacillus *Escherichia coli*. The scheme is the _Jacob-Monod model_, named for the French biologists François Jacob and Jacques Monod (Figure 8.8). According to the model, several segments of DNA are involved, and these are called "genes" although they do not all act in the way described in Chapter 7. Some of the genes, called *structural genes*, do produce messenger RNA, and consequently they are instrumental in the production of polypeptides. They are the genes that function in the synthesis of enzymes. Associated with one or more structural genes is an *operator gene*, which does not contribute directly to polypeptide synthesis but acts as a switch to "turn on" the structural genes. When the operator turns on the structural genes, they begin to form mRNA, and a whole sequence of actions normally follows. Something, however, must be able to "throw the switch" of the operator. Evidence from mutated bacteria indicates that some substance, now known to be a protein, can come to lie on the operator, making it turn the structural genes off. The protein, called the *repressor substance*, keeps the structural genes turned off as long as it is supplied to the operator. The repressor protein is produced by still a third type of gene, the *regulator gene*. The regulator gene gives off repressor protein, presumably all the time. However, the repressor substance can be inactivated, and when it is inactivated, the operator turns on the structural genes.

Repressor substance can be inactivated either by some substance from outside the cell or by some product synthesized inside the cell. Either way the structural genes begin to function actively. With such a scheme, it is possible for a cell to use selectively the genes that are appropriate at a given time. For example, if a certain food molecule is not available, making an enzyme to digest it would be useless. The enzymatic machinery of the structural gene is turned off. But if that food gets into the cell, it can bind to the repressor substance, thereby interfering with its repressing ability. The operator then turns on the

8.8

Gene regulation according to the scheme of Jacob and Monod. (a) *Structural* genes are those that produce the messenger RNA responsible for the working proteins of the cell. Structural genes are made to produce or are kept from producing by a nearby *operator gene*, which acts as an on-off switch. The operator gene may be on or off, depending on whether or not a *repressor protein* is present and active. The repressor protein comes from still a third kind of gene, a *regulator*. (b) In the process of *induction*, (1) the regulator sends out an *active* repressor, which turns off the structural genes by binding with the operator gene, so no mRNA and consequently no protein are made from those genes. (2) If a usable outside material (such as lactose in *E. coli*) enters the cell, it can bind itself to the repressor protein, thereby keeping it from holding the operator in the off position. When that happens, the operator turns the structural genes on, and they start producing mRNA and then protein. (c) In *repression*, (1) an *inactive* repressor is assumed to be present, so that the operator and the structural genes are on and functioning. (2) If the structural genes are producing enzymes that are working in a cell to yield end products, those end products can bind themselves to the *inactive* repressor, thereby making it an *active* repressor, which can turn off the operator and structural genes and thus prevent excess buildup of the end product. When the inactive repressor is made active by a product within the cell, that product is called a *co-repressor*.

structural genes, which in turn start the necessary enzyme production. Such a food molecule is an *inducer*.

Some enzymes are normally present in cells, regardless of the environment. The structural genes responsible for the production of such enzymes seem to function without being switched on. Enzymes of this type are called *constitutive enzymes*. Since *E. coli* can take in glucose and metabolize it at any time, the enzymes for the utilization of glucose are constitutive. In contrast, other enzymes are produced only when there is a use for them. For example, *E. coli* cannot

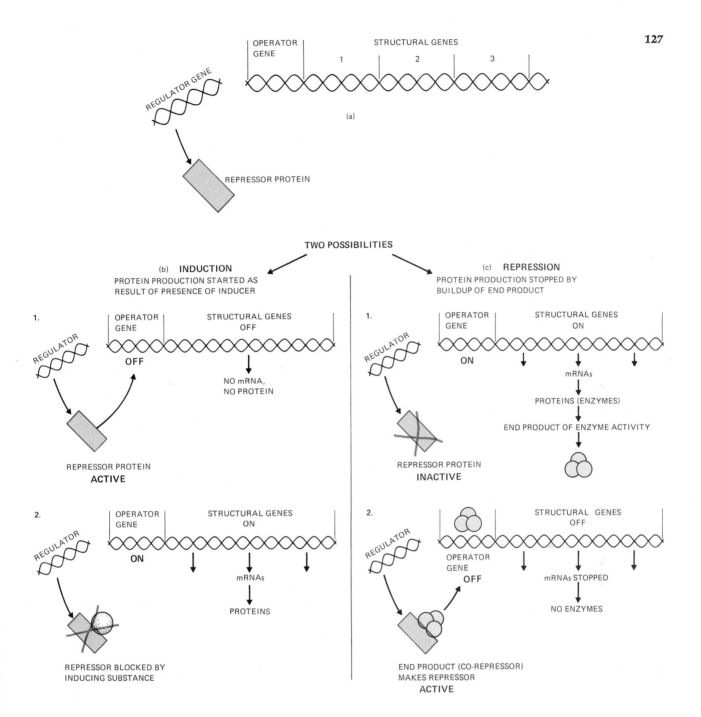

utilize the sugar lactose until the cells have remained in contact with the sugar for a while. Then the necessary enzymes begin to be synthesized, and the cells can finally take in lactose and metabolize it. The several enzymes involved, which are not produced in functional quantities without an external stimulus, are said to be *induced*. Much of what is known about gene function has come from studies on constitutive and induced enzymes.

There is the internal feedback as well. If an enzyme brings about an accumulation of some end product, then that end product can perform an action called corepression and is itself a *corepressor*. Its effect is to help the repressor substance shut off the operator switch, thereby stopping the output of end product (that is, corepressor). The actual details are more complex than this short account implies, but the principle is simple and can be reduced to two main ideas: (1) A substance from outside a cell can cause that cell to start making enzymes that will act on that substance. Thus external influences can direct the cell's specific activity. (2) An intracellular feedback

AGING AND DEATH

Aging and death have been the subjects of philosophical speculation for centuries, but only within recent years have scientific investigations been started on the physiology of aging. At present, there is little firm information, and there are no broadly accepted general principles.

Concerning aging, some facts have been obtained from studies on cell cultures. Experiments on cell cultures do not necessarily prove that what happens under artificial conditions is the same as what happens in a live animal, but they are instructive. If normal embryonic human cells are grown in culture, the culture can be maintained through about 50 generations. Cells from children survive about 30 generations, but cultures from grown people last only about 20 generations. These facts have prompted a supposition that the number of times cells can divide in a lifetime is preset and can be modified only to a minor extent by the life an individual leads. Other guesses as to the reason for aging include the hypothesis that after repeated divisions, cells come to possess more destructive mutations than they can stand, or that an increasing accumulation of toxic metabolic products leads to an eventual stoppage of activity. Experimental evidence supports all these views.

A puzzle is added by the ability of cultured cancer cells to grow and divide indefinitely. A celebrated instance is provided by the culture of HeLa cells, started in 1952 from a woman named Helen Lane. Cultures of HeLa cells have been used in so many experiments in hundreds of laboratories that their total quantity far exceeds the number of cells the original donor ever possessed. Yet HeLa cells are growing today as vigorously as ever.

If an individual's cells are "programmed" to last only so long, how does a transformation from a normal to a cancerous condition upset the program? The answer is the same as it is to most fundamental biological questions: No one knows.

If a biochemical attack on problems of aging proves to be as effective as such attacks have been on metabolic and genetic problems, then we may come to understand the process and possibly even control it. If we do, a new question poses itself, this time a question that cannot be answered by biology or chemistry or any science: How desirable would it be for a human to live indefinitely?

Average life expectancy has increased statistically because of medical and other improvements, but a rise in the average age at death merely shows that more people have escaped an early death by disease or accident and have thereby come closer to the maximum. The maximum itself has not risen. A time comes when every animal with finite size dies.

But what is death? As discussed in the introduction to this book, life eludes exact definition. It is therefore not especially instructive to define death as the absence of life, but that is what it is, an absence of the functional organization that enables a plant or animal to do things. Death has traditionally been regarded as a Thing, something positive and real, even though it cannot be measured or even satisfactorily defined in medical or legal or biological terms. Death, like cold and darkness, has terrified human beings for thousands of years, but death and cold and darkness are all negative. As physicists consider cold to be the absence of heat and darkness the absence of light, so biologists can regard death as the absence of life.

The only whole organisms that are potentially immortal are those that reproduce by fission. A single-celled alga, on dividing, ceases to be, yet the two products of the division live on, and their offspring live on, limited in succeeding generations only by the restrictions of nourishment and space. Most plants and animals, as individuals, die, but if they leave offspring, they transmit a living cell to the next generation. In a way, any organism that reproduces has a chance at a tiny bit of biological immortality.

(a)

(b)

8.9

(a) *Xenopus laevis,* the frog from South Africa that is famous among biologists because it was the organism on which animal totipotency was dem-

onstrated. This particular frog was raised in a normal way, but others grown from eggs with transplanted nuclei (b) are indistinguishable.

system can regulate the amount of any given substance that a cell manufactures. Thus a cell governs its own internal affairs.

The Jacob-Monod model applies best to bacterial cells. It is not known how the Jacob-Monod model applies to plant and animal cells, because they have much more complex systems, and they certainly have different chromosomes. Besides, many of the so-called genes of eukaryotic organisms exist not just in one copy (or two copies in diploid cells) of each gene per nucleus but in many copies, perhaps hundreds. A bacterial chromosome is like a private library in which there is one copy of each necessary book; a eukaryotic chromosome is like a large public library in which dozens of copies of the same book are stocked to meet a big demand. The discovery of all that *redundant DNA* has posed a number of questions that biologists have scarcely begun to answer. One particularly difficult question is: If there are many replicas of the same gene in a chromosome, how can a change in a single DNA codon make any difference? But it *does.*

TOTIPOTENCY: EVEN MATURE CELLS CONTAIN ALL THE GENETIC CONTENT OF THE ORGANISM

Totipotency is the ability of any nucleus in an individual to perform any of the thousands of functions it might be required to do. The term, meaning "with

all power," is believed to be applicable to nuclei in general.

For many years biologists wondered: After a cell has become fully differentiated, does it still have all its original genetic abilities, or are they lost along with the process of specialization? We now know that the first supposition is the correct one. From persuasive demonstrations in vertebrate animals and in flowering plants, we know that cells do retain all their synthetic power.

In 1952 Robert W. Briggs and Thomas J. King transplanted living cell nuclei into frog eggs and successfully produced genetically identical offspring, which were also genetically identical to their single parent donor. Shortly thereafter, J. B. Gurdon at Oxford University dissected the nucleus from an egg of the African clawed toad, Xenopus, and replaced it with a nucleus taken from a completely differentiated intestinal cell (Figure 8.9). The egg with the transplanted nucleus grew into a normal animal, complete with all the variety of cells an adult toad requires. This process, in which the nucleus from a cell of an organism is inserted into another cell, is technically known as *transplantation.*

F. C. Steward of Cornell University, using a different technique and a different organism, proved the same point Gurdon had proved. Steward and his co-workers removed a cell from the tissue called *phloem* (Chapter 16) of an ordinary commercial carrot and placed it in a germ-free medium containing an ade-

8.10

Steward's demonstration that a complete plant can be regenerated from a single differentiated cell. The ability of a phloem cell to develop into a complete plant with roots, flowers, and all other normal plant parts is taken as proof of the totipotency of plant cells.

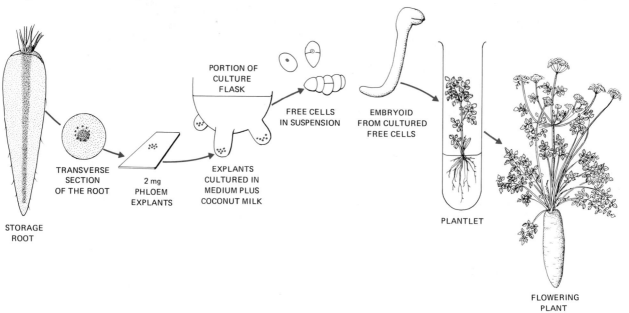

STORAGE ROOT

TRANSVERSE SECTION OF THE ROOT

2 mg PHLOEM EXPLANTS

PORTION OF CULTURE FLASK

EXPLANTS CULTURED IN MEDIUM PLUS COCONUT MILK

FREE CELLS IN SUSPENSION

EMBRYOID FROM CULTURED FREE CELLS

PLANTLET

FLOWERING PLANT

quate food supply, mineral salts, and growth substances. The cell developed into a complete carrot plant, which flowered and made seeds (Figure 8.10).

The results of the experiments of Gurdon and Steward are especially instructive because both men proved totipotency in organisms that are enormously far apart in evolutionary time. Suppose two demonstrations had been made on two closely related animals, say, a toad and a frog, or on two closely related plants, such as carrot and parsley. The original question would remain: Is totipotency limited to some special types? When two such distantly related organisms as toads (vertebrate animals) and carrots (flowering plants) can be shown to have totipotent nuclei, it seems safe to assume that the phenomenon is a general biological one, applicable to nucleated cells across a wide range of living things.

CANCER OCCURS WHEN FEEDBACK CONTROL IS MISSING

As long as cells remain under proper control, they grow normally and perform their normal functions, but if anything breaks into the feedback system, they may die or go into uncontrolled growth. Skin cells or liver cells or lung cells can start dividing and continue to do so, failing to differentiate or to stop dividing. Such unrestrained proliferation is *cancer*, meaning "crab," something that nibbles at you; it can occur in any active tissue, including that of plants.

Among the characteristics of cancer in higher animals are an abnormal, seemingly uncontrolled growth of body cells, which invade and destroy neighboring normal tissue (Figure 8.11). Cancer cells may also break off from the original mass and enter the blood or lymph channels, to be circulated throughout the body. These cells may then take hold and destroy the organ where they have settled. Such undisciplined cellular behavior is called autonomy. Besides having a somewhat independent growth, cancerous cells also appear irregular under a microscope. They may vary so much in size and structure that no recognizable formations remain.

The causes of cancer are many, probably working in different ways to produce a similar result. Similarly, an automobile may "run away" because it was left on a hill with the brakes off or because the brakes failed or because the driver went to sleep. Regardless of the reason for the lack of control, the end is the same. Cellular control can be destroyed by viruses; by certain chemical compounds; by radiation, which disrupts molecules in cells; by mechanical abrasion; and probably by still unidentified forces.

Two of the most important theories concerning the causes of cancer emphasize the genetic material and changes in that material. One hypothesis suggests that mutations occurring in the body cells may result in the uncontrollable cell growth characteristic of cancer. Another hypothesis suggests that cancer results when infectious virus DNA is inserted into the genetic material of otherwise normal cells. So far, research has not been able to choose between these two hypotheses. We *do* know that DNA extracted from cancerous cells is different in a number of ways from DNA in normal cells. Indeed, the DNA in cancerous cells is visibly changed, and it has been known for more than 65 years that a cancer in chickens is caused by a virus. Although researchers have identified no viruses that cause cancer in humans, leukemia is one form of human cancer suspected of being virally induced. Since more than 100 cancer-causing viruses have already been isolated, it is likely that investigators will some day identify a virus that causes human cancers. However, no single virus has been identified as the *sole* cause of a cancer in humans, and it is believed that neither mutations nor viral infections cause *all* cancers. (See the essay on p. 132.)

A cell without a functional feedback control for mitosis does not "know" when to stop growing. In a living body, all the requirements for growth are available: warmth, water, nutrients, vitamins. A cancerous tissue therefore goes on growing without restraint until it impairs some vital function of the organism, whereupon both organism and cancer die.

As unexplained forces may cause cells to lose their self-control, so other forces may reverse the damage. Perhaps some animal bodies develop specific cancer-controlling mechanisms, or perhaps a reversal of the original defect occurs by chance. Regardless of the cause, there are rare and occasional *spontaneous remissions*. The reason for such remissions is still as mysterious as the reasons for the start of cancer.

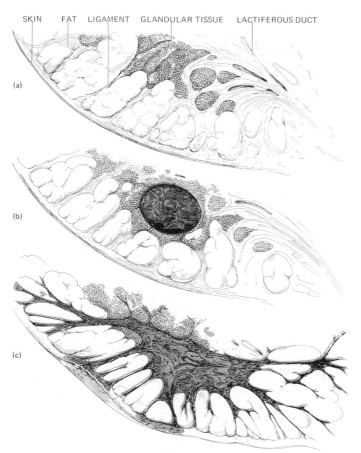

SKIN   FAT   LIGAMENT   GLANDULAR TISSUE   LACTIFEROUS DUCT

(a)

(b)

(c)

8.11

Growth pattern of cancerous tissue. The growth of a tumor in the breast ordinarily threatens life only when the tumor can spread to distant parts of the body. (a) The normal breast is organized into glandular tissue, fat, and other structures. Tumors arise almost exclusively in the glandular tissue. They are composed of cells in which the normal restraints on growth and reproduction have been removed. (b) A benign tumor can grow rapidly and become quite large, but it cannot escape the tissue in which it develops. (c) A cancerous tumor can spread throughout the glandular tissue, can often involve ligaments and skin, and can sometimes penetrate the muscle underlying the breast. In addition, cancers can in some cases migrate through the blood or the lymphatic system to establish new colonies of cells in distant, unrelated organs. This is the process known as metastasis. It is the metastatic spread of cancers that is responsible for their lethality.

CANCER:
WHAT *DO* WE KNOW ABOUT IT?

We simply do not know as much about cancer as we would like to. Most of all, we don't know how to prevent it. We *do* know that cancer is the second leading fatal disease in the United States, behind circulatory diseases, and it accounts for 20 percent of all deaths in the United States. (Every two minutes someone in the United States dies of cancer.) The current thought is that many forms of cancer are caused by environmental factors. Statistically, at least, the evidence for environmental causes is substantial. For example, Jews who emigrate to Israel from the United States or Europe contract cancers representative of the countries they leave, but their children who are born in Israel show a lower incidence of cancer in general. Lung cancers in women have only recently been appearing, whereas lung cancer in men began to be prevalent about 50 years ago. The reason for this difference may be simple: Men started smoking long before women, and cancers attributed to smoking appear to be showing up about 20 years after the first incidence of smoking. (Cigarette smokers are about four times more likely than nonsmokers to die of lung cancer.)

Cancer of the large intestine may be related to diet. Richer countries, where red meat is consumed regularly, generate more cancer deaths than countries where more cereal than meat is eaten. Studies done on women subjects indicate that deaths due to cancer of the large intestine are 40 times more prevalent among women in New Zealand than in Nigeria and 30 times more prevalent among American women than among Chilean women.

Vaccines have virtually eliminated such diseases as polio. Unfortunately, there is still no simple connection between viruses and human cancer, so that the solution of a preventive vaccine is impossible, at least for the present.

Almost half of all cancer deaths are caused by malignancies in the lung, large intestine, and breast. Cancer of the lung is the leading killer, accounting for almost 20 percent of all cancer deaths. The major areas of fatal cancer in American men are the lungs, the digestive areas, and the prostate gland. The most vulnerable areas in women are the breasts, the digestive areas, and the lungs.

 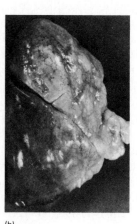

(a)   (b)

(a) A partially dissected cancerous lung, showing tumorous growths. (b) A healthy lung.

Elderly people are most susceptible to cancer, and the occurrence of cancer of the large intestine, for instance, increases about 1000 times between the ages of 20 and 80. Current thinking is that certain genes help prevent cancer until they are made ineffective by a mutation. The chances of a successful mutation are naturally increased after several unsuccessful attempts. Therefore the probability increases in direct proportion to age. Further, if it is true that several mutations are required to finally cause a cancer, then the cancer may have actually begun 10 or even 30 years before it becomes evident.

The World Health Organization has reported that approximately 85 percent of all cancer cases result from such environmental causes as smoking, overeating, overexposure to sunlight, overdrinking, and exposure to dangerous chemicals like asbestos and polyvinyl chloride, a widely used plastic. It could very well be, then, that humans are actively elevating cancer to epidemic status. We do know for certain that high doses of X-rays and similar radiation can cause leukemia and other cancers. And what could be more human-provoked than the extremely high incidence of leukemia and cancer of the breast, bowels, and brain that have occurred in Hiroshima and Nagasaki? At least one possibility for cancer prevention is clear: Humans can prevent the cancers they cause.

SUMMARY

1. The final form of an organism depends directly on its cellular arrangement. Multicellular plants and animals start life as single cells and achieve maturity as a result of repeated cell divisions and changes in the form of individual cells.

2. It is the maturation, the differentiation, the specialization of cells that gives them their final working form. Functional specialization is more striking than species specialization.

3. During the initial steps in cell division there is an exact replication of the DNA, which provides the cell with two complete, identical sets of chromosomes.

4. The actual nuclear division process is known as *mitosis*. The phases of mitosis are the *prophase*, the *metaphase*, the *anaphase*, and the *telophase*.

5. Mitosis accomplishes the reproduction of cells and the equal distribution of DNA to each.

6. *Meiosis* is the term applied to the pair of divisions required for reduction of the chromosome number from *diploid* to *haploid*.

7. Cellular *differentiation* must follow mitosis in order to achieve the proper specialization of cells.

8. *Homeostasis* is the maintenance of a measure of stability during change. Although cells are constantly being worn out and replaced, the organism maintains an overall stability.

9. Any dynamic system depends on a *negative feedback system*, which helps maintain stability by compensating for actions that may be harmful to the system. Cells are regulated by such a system of biological feedback.

10. The *Jacob-Monod* model of gene regulation postulates that a cell governs its own internal affairs through a complex system of specialized genes.

11. *Totipotency* is the principle that states that all the cells in an organism possess all the genetic capabilities of that organism, even after the cells have undergone differentiation.

12. In nuclear *transplantation* the nucleus from a cell of an organism is inserted into another cell.

13. *Cancer* occurs when the feedback systems of cells break down, causing cells to die or grow uncontrollably. The causes of cancer are many, probably working in different ways to produce a similar result.

RECOMMENDED READING

Axtell, L. M., S. J. Cutler, and M. H. Myers (eds.), *End Results in Cancer: Report No. 4*, Publication No. (NIH) 73–272. Washington: Department of Health, Education, and Welfare, 1972.

Braun, Armin C., *The Story of Cancer*. Reading, Mass.: Addison-Wesley, 1978. (Paperback)

Bryant, Peter J., Susan V. Bryant, and Vernon French, "Biological Regeneration and Pattern Formation." *Scientific American*, July 1977. (Offprint 1363)

Cairns, John, "The Cancer Problem." *Scientific American*, November 1975. (Offprint 1330)

Croce, Carlo M., and Hilary Koprowski, "The Genetics of Human Cancer." *Scientific American*, February 1978.

Gurdon, J. B., "Transplanted Nuclei and Cell Differentiation." *Scientific American*, December 1968. (Offprint 1128)

Hayflick, Leonard, "Human Cells and Aging." *Scientific American*, March 1968. (Offprint 1103)

Lane, Charles, "Rabbit Hemoglobin from Frog Eggs." *Scientific American*, August 1976. (Offprint 1343)

Mazia, Daniel, "The Cell Cycle." *Scientific American*, January 1974. (Offprint 1288)

Old, Lloyd J., "Cancer Immunology." *Scientific American*, May 1977. (Offprint 1358)

Patterson, Paul H., David D. Potter, and Edwin J. Furshpan, "The Chemical Differentiation of Nerve Cells." *Scientific American*, July 1978.

Prescott, David M., *Cancer: The Misguided Cell*. Indianapolis: Bobbs-Merrill, 1973. (Paperback)

Wessells, Norman K., and William J. Rutter, "Phases in Cell Differentiation." *Scientific American*, March 1969. (Offprint 1136)

Wolpert, Lewis, "Pattern Formation in Biological Development." *Scientific American*, October 1978.

THE DIVERSITY
OF LIFE

PART THREE

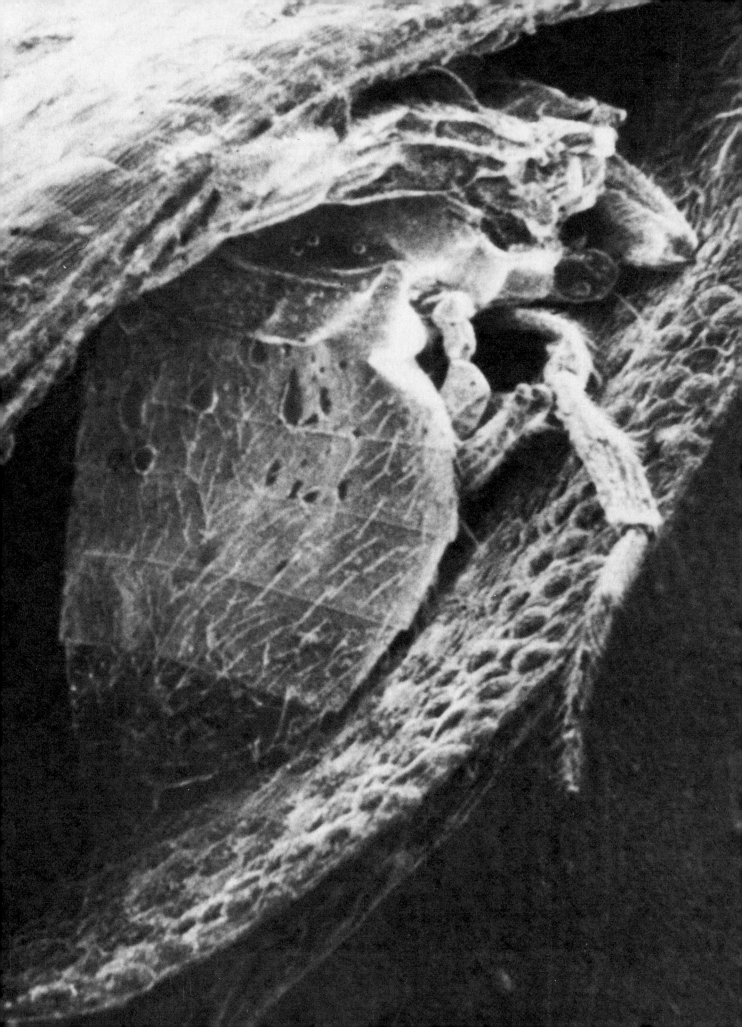

9

The Living Kingdoms: The Lower Groups and Plants

Venus flytrap leaf, cut longitudinally, which has partially eaten a fruit fly, Drosophila melanogaster.

ANYONE WHO LOOKS IN DETAIL AT THE EARTH'S organisms sees a striking contrast between constancy and diversity. Living things have many features in common, such as cellular arrangement, membrane-bound particles, energy-exchanging mechanisms, chemical controls, and on and on. Until now we have stressed general structures and functions that are common to almost all organisms, and we have done so for the very reason that they *are* common and therefore must be fundamental to the activities of life. Now, however, we will emphasize differences. In this chapter and the next we will give some idea of the major kinds of organisms alive today, with occasional mention of some that are now extinct, that lived once but are with us no more.

The sheer number of different organisms is impressive. We do not even know how many have been collected, kept, and described, because the literature on the subject is so vast and so scattered that for most groups it has never been completely catalogued. In addition, unknown numbers of organisms have never been systematically recorded. Exact counts are hard to come by for the reasons mentioned. Furthermore, in spite of a frequently used definition of a species as "a population that can produce fertile offspring," many biologists do not accept that. One biologist, a "lumper," lumps perhaps a dozen described "species" into a single broadly described one, but another authority, a "splitter," splits one "species" into a dozen narrowly described ones. Nevertheless, rough guesses can be made, and the numbers are usually greater than nonbiologists expect. Some groups of organisms are relatively small. Only about 1500 species of bacteria are recognized (although a species is harder to define in bacteria than in most groups). There are nearly 3000 species in the carrot family and about 4500 species of grass. Orchids, more numerous in the tropics but not rare in temperate zones, account for perhaps 15,000 species, not including countless cultivated hybrids. The numbers of animal species are overwhelming: 23,000 species of fishes and 8500 species of birds, for example. But the beetles are the winners, with close to 300,000 species in the world, a tenth of them living in North America. Obviously no one person can know more than a small fraction of the living species, although a good field biologist can reasonably expect to recognize most of the specimens in restricted groups. With experience, one can come to know the birds, trees, and mammals in an area. To know the common fungi, mosses, or fishes takes more time, but even an expert never learns all the crustacea or insects even in a limited region, although he may know the general group—order or family—of most specimens.

Even more striking than simply the number of kinds of plants and animals is the way in which modifications can occur. From types that are considered ancient and/or primitive, there has come a bewildering array of alterations. Some well-known examples are the tongues, jaws, and accessory mouth parts of insects, the tail feathers of birds, the leaves of ferns, the teeth of mammals, the fruits of flowering plants, and the appendages of marine crustacea. Anyone who looks intensively at a number of species or even at the individuals within a species is soon impressed by the endless variety. There is, to be sure, an underlying unity. Otherwise there could be no biological generalizations worth paying attention to. But there is, at the same time, this diversity, which is one of the main sources of biological excitement. The phrase "the diversity of life" has become almost a cliché, but like many clichés, it expresses something real and should not lose its force simply from being used too often.

In this chapter and the next, we shall describe some aspects of a few selected kinds of organisms. For every one mentioned here, thousands more exist in nature. Those included have been chosen because they are types that are in some way peculiarly in-

structive, or because they are bizarre, or because they affect human health or wealth.

As example is added to example and group after group is described, the reader will be able to find a story of ever-increasing complexity, sometimes in one direction, sometimes in another. In some groups of organisms, the usual trend toward complexity seems to be reversed as they exhibit a return to simplicity. In the groups thought to be the oldest and most primitive, structural differentiation is minimal, though even the simplest types are biochemically beyond our present understanding. You can follow a small history of life as you consider a microscopic, rather simply constructed bacterium, then an inconspicuous moss, later a finely cut fern, and finally a flowering plant, complete with its coordinated development of complicated parts. In a similar way, the development of increasing complexity in animals parallels that of plants, and you can repeat history by learning of animals so small as to be invisible and progressing through intermediate examples to the so-called pinnacles of evolution, which, to choose a few examples, might be a squid, a honeybee, a hummingbird, and a sculptor.

SCHEMES OF CLASSIFICATION

Just as there is disagreement among biologists as to species, so there is disagreement over the larger units of classification. Indeed, dozens of schemes of general classification have been proposed, and the popularity of one or the other rises or falls with the advent of new information or new opinions. The aim of the taxonomist, that is, the student of classification, is to make a system that shows the evolutionary relationships of organisms, and therein lies the difficulty. We do not know what the relationships are because most of the organisms that could clarify those relationships are dead and lost forever. The so-called family tree of living things could better be compared to a bush that is almost submerged in a murky swamp. Only the tips of some of the branches stick out above the surface of the water, and all the connecting branches are hidden. We can make guesses as to which twigs are growing on branches close to other twigs, but the guesses are speculative, regardless of the care and logic we apply to them.

The main standard of relationship is similarity. The more points of similarity that can be found between two specimens, the closer they are thought to be related. As many similarities as possible are taken into account: physical structure, embryological development, ratios of DNA base pairs, chromosome structure, reproductive methods, respiratory processes and pigments, and biochemical components. A single point of similarity may be striking but superficial, and it may lead to wrong conclusions. The tropical cecilians, for instance, look like snakes, but they are not snakes, they are more like legless frogs. The tiny plant called pearlwort, *Sagina procumbens*, growing in sidewalk cracks, can easily be taken for a moss, but it is a flowering plant, structurally and reproductively more like a carnation than like a moss plant. Organisms must be taken as whole organisms when evolutionary comparisons are made, and *all* possible characteristics must be considered.

Individuals that fit some taxonomist's definition of a species are placed in a *species*. A group of species that seem so similar that they are believed to be closely related are placed in a *genus* (plural, genera). The system of naming an organism by giving its genus and species, established by the so-called Father of Classification, Carolus Linnaeus (1707–1778), is still followed. The genus name, always starting with a capital letter, comes first, followed by the species name. Both words are italicized or underlined. The words are in the Latin form, even when the root words are Greek or some other language. However, uniform though that practice may be, it has produced some unlikely combinations, such as the tiny fungus *Nowakowskiella elegans*, named for Nowakowski, or the pecan tree *Hicoria pecan*, whose name is adapted from North American Indian words. A number of genera may be considered as belonging to a *family*. Families are grouped into *orders*, orders into *classes*, classes into *divisions* in plants or *phyla* (singular, phylum) in animals, and divisions or phyla into *kingdoms*. Any category of classification is a *taxon* (plural, taxa). A species is a taxon, as is a genus, a family, or an order.

Examples of a simplified classification of two familiar organisms, a plant (carrot) and an animal (horse):

Kingdom Plantae	Kingdom Animalia
Division Pteropsida	Phylum Chordata
Subdivision Angiospermae	Subphylum Vertebrata
Class Dicotyledonae	Class Mammalia
Order Umbellales	Order Perissodactyla
Family Umbelliferae	Family Equidae
Genus *Daucus*	Genus *Equus*
Species *carota*	Species *caballus*

THE LIVING KINGDOMS

Until recently, all organisms were forced into one of two kingdoms. Plants are green and stationary; animals eat and move about. Such a simple scheme is adequate for grass and horses, but so many living species do not fit into either a plant or an animal category that alternative systems were devised. The system used in this book is one currently favored by many biologists, but it must be remembered that many others exist, and even this one is not completely satisfactory for many kinds of organisms. In this system, the following five kingdoms are recognized:*

Monera: the prokaryotes, including bacteria, blue-green algae, actinomycetes, and viruses.

Protista: some unicellular algae, protozoa, and some intermediate forms.

Fungi: the "true" fungi (phycomycetes, ascomycetes, basidiomycetes).

Plants: red, brown, and green algae, mosses and some similar forms, and the vascular plants (ferns and their allies and seed plants).

Animals: except protozoa, all the familiar types, including sedentary sponges, many little-known phyla, and of course, human beings.

THE KINGDOM OF THE MONERA

The Monera† are the simplest, at least structurally, of all living things, and fossils resembling present-day Monera can be found in the oldest fossil-bearing rocks. They are therefore treated first in this survey of the world's organisms. The Monera are characterized more by what they lack than by what they

---

* Even five kingdoms are not enough for placing comfortably such types as viruses, slime molds, diatoms, and many one-celled (or noncellular?) organisms. At the same time, five are enough to be approaching an unwieldy number. This system is a compromise between oversimplification on the one hand and "hair-splitting" on the other, serving well enough as a working scheme until a more generally satisfactory one is produced.

† The Monera probably do not constitute a natural group, that is, a group with immediate evolutionary relationship. Chances are good that the kingdom of the Monera as treated here will not remain long in favor among biologists.

possess. They have no nuclear membrane, no well-defined nucleus with chromosomes, no protein associated with their DNA, no flagella with microtubules, no membrane-bound organelles, such as plastids or mitochondria. They all have some kind of external covering, but the materials are different from those of eukaryotic cells. There are ribosomes, but they are smaller than eukaryotic ribosomes. Perhaps present-day prokaryotes are organisms that have retained their primitive features and may therefore resemble the ancestors of eukaryotes. Being primitive does not mean being unsuccessful, however. The Monera are eminently successful in that by biological success we mean surviving in great numbers.

Viruses

Viruses were known by their activities long before actual virus particles were demonstrated. When Louis Pasteur found that diseases of silkworms and of sheep were caused by microscopically visible particles, which he called "microbes," he thought that rabies, too, should be the result of a microbial infection. However, he failed to find any particles in the cells of rabid animals, because rabies is a virus disease, and the virus particles are too small to be seen through a light microscope. Most bacterial cells are visible through a light microscope because they range from about 0.3 micrometers, close to the lower limits of the best light microscopes, up to several micrometers. Viruses, however, are usually far too small to see directly, although a few outsized ones are as large as the smallest bacteria. For a long time they were called "filterable viruses" because the minute particles could pass through filters that held back the larger bacteria. They were known to be capable of reproduction because infection, particularly of plants, could be demonstrated by the transmission of the disease known as tobacco mosaic (which causes spotted leaves) from plant to plant until only by recognizing that the causative agent had reproduced could scientists explain the spread of the disease (Figure 9.1).

A major advance in knowledge of viruses came in 1933, when Wendell Stanley concentrated enough tobacco mosaic virus to be able to purify a small quantity of it. To the surprise of the biological world, the purified virus could be crystallized. This was extraordinary. Only molecules with precise three-dimensional shape can be crystallized. When suspended in water, the crystallized virus again became infective and would produce the mosaic disease in tobacco plants. Since the ability to reproduce is one of the prime characteristics of living material, and since

(a)   × 11,000

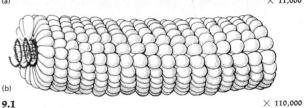

(b)   × 110,000

9.1

(a) Electron micrograph of tobacco mosaic virus, showing a silhouette of the virus particles. (b) A drawing interpreting the electron micrograph, showing the nucleic acid core of the virus particle with protein units wound in a close helix around the core.

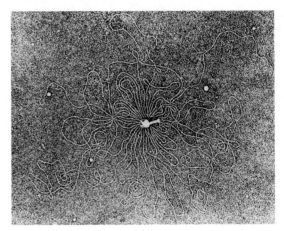

× 30,000

9.2

Virus DNA streaming from a virus particle, having been exploded by treatment with water and photographed with an electron microscope. One free end of the virus DNA can be seen at the right of the picture and the other at the left. This spectacular photograph has been reproduced thousands of times since it was published in 1962.

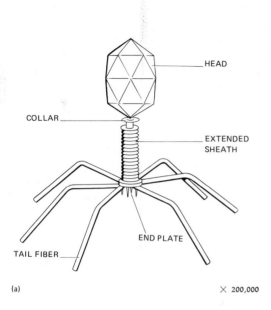

HEAD

COLLAR

EXTENDED
SHEATH

END PLATE

TAIL FIBER

(a)   × 200,000

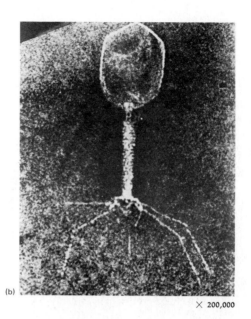

(b)

× 200,000

9.3

Bacteriophage (virus) particles. (a) A diagram of a particle, made up from a combination of sources, including electron micrographs, infection studies, and chemical experiments. The protein-coated head encloses the DNA. The protein sheath around the tail is contractile, allowing the virus to inject its DNA into a host cell. The tail fibers and end plate are involved in the firm attachment of the virus to the host cell surface. (b) Electron micrograph of a virus particle, showing the shape of the head, the tail with subunits, and the tail fibers.

crystallization is a feature of a nonliving chemical compound, viruses seem to occupy a borderline position between the living and nonliving worlds.

During the 1950s, electron microscopes were developed that had the ability to resolve particles about one-thousandth the size of those visible with light microscopes. Virus particles, called *virions*, could be photographed to show their size, shape, and fine structure. Electron micrographs reveal that virions may range from about 0.02 to 0.3 micrometers. Thus a row of about 25,000 of the small virions or about 1500 of the largest ones would stretch across a printed period on this page. Shapes, too, are various. The tobacco mosaic virus is a slender cylinder, but many viruses are roundish but with flat-sided facets like those on a cut gem. In high-resolution pictures, individual protein units of the virus coat are visible, and when a virus particle is burst open, its internal strand of DNA can be seen like a spread tangle of thread (Figure 9.2).

A virus particle, small and simply constructed as it is, has no life of its own, no respiration or known synthetic ability. It does have its nucleic acid strand—

DNA in some viruses or RNA in others—and it has a covering of protein. Several different proteins may be incorporated into the virus coat. There must be at least one enzyme that weakens the host cell wall or membrane, because all viruses grow only in some host cell. The method of attack of a cell by a virus is known best for the viruses that infect bacteria, known as _bacteriophages_ (Greek, "bacterium-eaters"), or simply as _phages_. One phage (Figure 9.3) resembles a tadpole; it has a large body and a slender tail with six fibers. These attach the tip of the tail to the outer wall of the victim and hold the phage body in place while the nucleic acid is injected through the host wall and into the cell. The protein coat of the virus remains outside. Inside the cell, the viral DNA causes the host DNA to cease normal activity and instead to carry out the genetic directions of the viral DNA. The end result of this process is the synthesis by the cell of virus nucleic acid and virus protein and their assembly into complete virus particles. In an active virus infection, as many as 200 new particles may be assembled within 20 minutes. When the new particles break out of the destroyed cell, they are

(a)   × 18

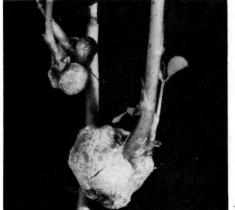

(b)   × 10

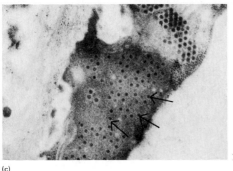

(c)   × 25,000

9.4

Virus infection of clover plants carried by insects. (a) Small insects, leaf hoppers, on a clover leaf. When the insects drive holes in the leaf epidermis, they inject the virus into the leaf. (b) The virus infection of a clover plant causes the growth of tumors on the stem. (c) An electron micrograph of an infected cell shows masses of virus particles, as well as isolated particles, indicated by arrows. Viruses as well as bacteria and fungi, when carried from plant to plant by a variety of insects, cause many plant diseases.

ready for another round of infection. Under some circumstances, a virus in a cell remains inactive and does little or no harm to the host cell. However, if the host cell undergoes a change, the virus may become destructive. For example, the virus that causes human cold sores, *Herpes simplex*, seems to be almost perpetually present, yet it fails to cause the typical symptom of ruptured cells unless the host cells are weakened, such as by another infection. If the sufferer catches a cold (another virus) or has a high temperature, the Herpes virus becomes active and ruptures cells in local regions of lips or gums, making the familiar cold sores.

The question of the origin of viruses is another of the unanswered biological questions. Viruses were once thought to be examples of an extremely primitive form of life, remaining practically unchanged from the beginning of life on earth. Their specificity of infection, however, is scarcely a primitive feature, and because viruses are as selective as they are, they are now regarded as *not* primitive. They may be "degenerate" cells that have become so completely parasitic that they have lost everything but infectivity; or they may be fragments of cells that have escaped from host cells and become established as separate but completely dependent particles.

Possibly all living cells harbor viruses, most of them so specific that they can inhabit only one kind, or few kinds, of host cells. Although many viruses seem to do little or no harm, many others are the causes of plant and animal diseases. These range from minor disturbances (as in the variegated color of Rembrandt tulips and the cold sores on human lips) to lethal disruptions (such as virus infection of elm trees and rabies in mammals). The virus of yellow fever is carried from person to person by mosquitoes; some virus-induced tumors of plants are transmitted by tiny insects called leaf hoppers (Figure 9.4). In our anthropocentric way, we think only of the human sufferer as having a disease, but the insect is as much afflicted as the larger host. Other human virus diseases are measles, common colds, the various types of influenza (named for a mysterious "influence" that affects people), poliomyelitis, mumps, and the once-dreaded but now essentially conquered smallpox. A growing suspicion that some human cancers are virus-induced is strengthened by the fact that tumors in other animals and in plants are definitely viral in origin.

Bacteria

A varied group of organisms is included in the general category of bacteria. They are considered a single taxonomic entity because they are prokaryotic, having the usual characteristics of prokaryotes.

Distribution of bacteria Bacteria have become adapted to more different living conditions than have any other organisms. They are inhabitants of soil and water, and they exist in enormous numbers on the surfaces of most, probably all, plants and animals. They can be found in Antarctic ice, in hot springs, in the upper atmosphere, in petroleum, in the digestive tracts of animals, and even inside developing seeds. Some species can withstand the high pH of alkaline springs, the crushing pressures deep in the sea, the enzymatic action of animal digestive systems, lack of oxygen, long severe drying, even in some conditions a short time in boiling water. Bacteria cannot stand long periods at high temperatures, are generally intolerant of low pH, and cannot grow actively (although they are not killed) without an adequate water supply. With such adaptability, bacteria have so completely invaded practically every bit of the earth's surface, as well as portions well above and below the surface, that the mass of bacterial cells is estimated to outweigh the mass of all the plants and animals. Bacteria are so tenacious and so widely distributed that researchers must take elaborate care to grow germ-free animals or plants or to grow uncontaminated cultures of other microorganisms. For a convincing demonstration of the ability of bacteria to get into things and grow, try keeping them out of some place where they are not wanted, such as food, wounds, or laboratory materials.

Activities of bacteria Covering the metabolic activities of bacteria is impossible in a short space, but we can consider a few functions. Bacteria generally obtain their food by bringing it into their cells through their outer membranes. The food may be in a diffusible form already, as in a sugar solution, or it may be digested by enzymes released from the bacteria. The solvent must be water, and that is why dehydrated foods do not spoil and why salted meats generally, but not invariably, do not suffer bacterial decay. Most bacteria usually require oxygen, but some can live, though inefficiently, without it. Oxygen inhibits the growth of some bacteria, which usually live deep in soil or in wounds that do not allow entrance of air. The best known is *Clostridium tetani*, the cause of tetanus, or "lockjaw," whose toxins, when liberated in a human body, can kill unless a proper antitoxin is given.

Bacteria are usually *heterotrophic*, requiring an outside food supply, but a few species are *autotro-*

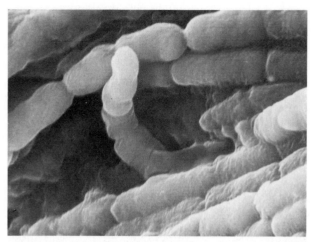

9.5 $\times$ 10,000

Scanning electron micrograph of rod-shaped bacterial cells, or bacilli.

phic, making their own food. Some of the autotrophs are *chemosynthetic*, obtaining the energy necessary for food manufacture from the oxidation of such materials as sulfur and iron compounds, nitrates, and hydrogen. A few photosynthetic species, such as the purple bacteria, obtain their energy from light, but their chlorophylls are different from those in higher plants. Further, they may use some substance other than water as a hydrogen source (Chapter 5). The photosynthetic sulfur bacteria release hydrogen from hydrogen sulfide, H_2S, and produce sulfur as a by-product, as higher plants produce oxygen from H_2O. Metabolic products, especially proteins, are exuded from bacteria, and these compounds can make important changes in the bacterial environment. Enzymes can break down materials outside the bacterial cells, or exuded toxins can affect a human or other animal body. Some exuded substances, antibiotics, inhibit the growth of other microorganisms. The antibiotics are used to regulate selectively the effects of bacteria on human and animal health, as well as in many experimental processes in biology. For example, the antibiotic *bacitracin* (named for a *bacillus* isolated from the skinned knee of a little girl named *Tracy*) is an effective germ killer in minor wounds.

Identification of bacteria All these activities are carried on with a minimal structural complexity. Bacteria are all small. The largest are only about 10 micrometers long, far below the limits of human vision. The smallest are about 0.3 micrometers, which is about at the lower limit of light microscopy (Figure 9.5). Most bacteria are about 1 to 4 microm-

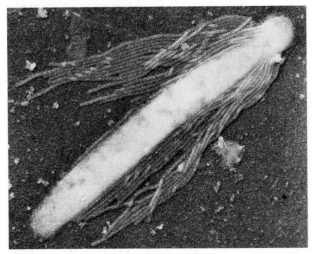

× 21,500

9.6
Electron micrograph of a bacterium
isolated from an African snail. The
flagella, clustered at one end, are remarkable in having the appearance of
twisted yarn. The preparation was
made visible in the electron microscope by a technique known as "shad-
owing," in which the specimen is
coated with a light deposit of metal.

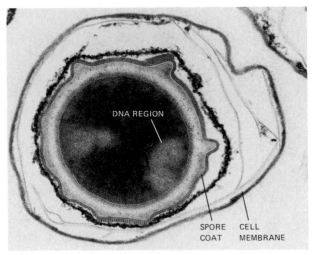

DNA REGION

SPORE COAT   CELL MEMBRANE

× 54,000

9.7
Electron micrograph of a thin section
of a bacterial cell, *Bacillus thuringien-
sis*, with a spore formed inside it. The
outer coat is the old cell envelope, and
the wavy inner layer is the spore coat.
The clear area inside the spore is the
region containing the cell's DNA.

eters across (if they are spherical, or *cocci*) or long
(if they are rod-shaped, or *bacilli*). Some of the corkscrew-shaped bacteria, called "spiral," may be seen
readily with a low-power microscope.

Besides size and shape, one identifying feature
of some bacterial cells is the presence of so-called
flagella, which are extremely slender filaments, more
like microtubules than the true flagella of eukaryotes
(Figure 9.6). The position and number of flagella is
useful in identifying bacterial types. Bacteriologists
also need to know what energy sources a bacterium
can use; whether, for example, a given species can
digest special sugars or acids or alcohols; and whether
or not it produces gases. Still another identifying feature is a type of cell wall that is peculiar to bacteria.
In contrast to the cellulose or chitinous walls of eukaryotic plants, bacterial walls are made firm by the
presence of polymers of special compounds. The kind
and amount of wall material can be determined and
used to help identify bacteria. For example, if a cell
has a relatively heavy wall, it will absorb and hold a
purple stain, even when washed with alcohol. If the
wall is thin, the color will rinse out. The procedure,
known as the *Gram stain*, can be used to determine

whether a bacterium is Gram-positive (that is, retains
the stain because it has a heavy wall) or Gram-negative. Let us say that a given species is described as a
motile (having flagella) Gram-negative rod. Then by
finding which compounds the species is capable of
using for growth, a microbiologist can identify the
organism. Other features used in identification are
the shape and texture of a mass of cells known as a
colony, the presence or absence of pigments, temperature tolerances, and the production of various gases.
Unlike most other organisms, bacteria are classified
on a basis not of how they are constructed so much
as of what they can do.

Structure of bacteria   The internal structure of a
bacterial cell is that of a usual prokaryote, without
mitochondria, plastids, endoplasmic reticulum, or nuclear membrane. Ribosomes are present—otherwise
protein synthesis would not be possible—and there
is a strand of DNA without accompanying chromosomal protein. Various granules of fatty or carbohydrate reserve materials may be scattered throughout the cytoplasm, and pigments are concentrated in
chromatophores, which are simpler than the orga-

nized plastids of eukaryotic plants. Respiratory activity is carried on in the plasma membrane, which may be infolded into complex masses, remindful of the internal cristae of mitochondria in eukaryotes. Some bacteria have *pili* (singular, pilus), which resemble flagella but are shorter and seem to function in the connection of bacterial cells to substrates or to other bacteria during mating. Further, some bacteria can build an internal, firm body known as a *spore*, which is much more resistant to heat and drying than the usual vegetative cell (Figure 9.7). Spore-forming species are harder to kill than nonspore-forming species. They may persist for years in dried soil or in clothes that have been in contact with soil.

Reproduction of bacteria   Bacterial reproduction is mostly asexual. The so-called circular chromosome duplicates itself, and then the two strands separate. This move is followed by a constriction around the cell as though a thread were drawn tight around it, and two cells result. Under optimal conditions of temperature and space and nutrients, a bacterium can divide in about 20 minutes. Sometimes bacteria mate. A protoplasmic bridge is formed between a pair of cells, and the DNA of one passes into the other. It was the discovery of this action of bacteria that led to our realization that DNA is the important genetic material.

Destructive action of bacteria   Bacteria are best known to most people because of their ability to cause disease, and indeed some of the most destructive scourges in human history have been the result of bacteria, or as they are popularly called, "germs." By far the most deadly scourge was the Black Death, or simply the *plague*, which is estimated to have killed perhaps half the population of continental Europe during the fourteenth century. In England, where as many as two-thirds of the people died of plague in the seventeenth century (Figure 9.8), there were

social, economic, and political changes whose effects are still felt, all brought about by the action of a microbe, *Pasteurella pestis*. The resulting shortage of laborers made ordinary workers really valuable for the first time in human history, and their social status rose in a way that has never been reversed. Another effective killer was *Rickettsia prowazekii*, the causal agent of typhus. Both typhus and plague are dangerous when people must live in crowded, unsanitary conditions, where rats and their accompanying fleas can carry the bacteria from victim to victim. As far as these diseases are concerned, the most effective way to prevent them is to understand how they are contracted and clean out the rats with their fleas and Rickettsias.

Other bacterial scourges of mankind have been diphtheria, caused by *Corynebacterium diphtheriae*; tuberculosis, caused mainly by *Mycobacterium tuberculosis*; and typhoid fever, caused by a Salmonella. Other Salmonellas cause a food poisoning, sometimes lethal, that is popularly but inaccurately known as ptomaine poisoning. The really dangerous food poison is that caused by the toxin of *Clostridium botulinum*. The botulism poison is so fantastically potent that a person could be killed by as little as one fifteenth-millionth of a gram, certainly not enough to see and so little that the residue on a clean-looking spoon could be lethal. Fortunately, it is uncommon, and it can be rendered harmless by heat. Therefore the boiling of suspected food for at least 15 minutes is recommended. One particularly fiendish military scheme that has been suggested is to drop into an enemy water supply enough botulism toxin to kill *half* the population. That would be enough to immobilize any society.

Plants as well as animals are susceptible to bacterial disease. One of the first identified and most serious is a destructive malady of pear trees known as fire blight. In spite of the fact that several hun-

9.8

A seventeenth-century woodcut showing London victims of the plague fleeing the city. The corpses and the skeletons brandishing spears tell of the terror that attended that most dreaded of bacterial infections, which affected rich and poor alike.

dred bacterial diseases of plants are known, however, bacteria are not the real threat to plants. As we shall see, fungi are to plants what viruses and bacteria are to animals.

Protection against bacteria Means of combating bacteria are available. Living animals have special ways of preventing or overcoming bacterial infection, using a chemical method of developing *immunity*. The subject of immunity is so important and has so many related topics associated with it that it deserves special treatment, and it will be discussed in detail in Chapter 24.

Sulfa drugs and antibiotics are used to protect animals against bacteria. These inhibit bacterial growth, although in different ways. Other ways of preventing microbial growth are used when a living body is not involved. In processing food, for instance, what is needed is some way of keeping bacteria and fungi from digesting what we would like to preserve. Common methods are freezing (keeping the temperature so low that bacterial enzymes cannot function), canning (killing bacteria by heat and then keeping live ones from the product), salting or candying (lowering the water content of the product by addition of salt or sugar), pickling (lowering the pH to a point below the tolerance limits of bacteria), or dehydration. When human consumption of a product is not expected, as in biological specimens or various structures used but not eaten by humans, preservation can be ensured by the use of a poison such as alcohol, formaldehyde, phenol, creosote, or any of a number of synthetic chemicals.

Utility of bacteria On balance, the utility of bacteria far outweighs the damage they do in causing disease and in destroying valuable materials. Without bacteria, the life of this world as we know it could not go on. The biological importance of bacteria lies in their ability to release enzymes. All biological products can be broken down by bacterial enzymes, with the result that bacteria, with assistance from fungi, maintain the cycling of materials in the living world. The enormous bulk of cellulose, lignin in wood, proteins, and fats that are produced by the higher plants could make life on earth impossible if bacterial enzymes were not available to return the oxygen, carbon, nitrogen, phosphorus, sulfur, to the air, water, and soil where they can be used over and over. Only certain manufactured substances, such as plastics and synthetic fibers, are not *biodegradable*, but there is a fair likelihood that sooner or later, given the versatility of

bacterial evolution, even such refractory substances as nylon may be attacked by some enzyme. When that happens, chances are that the enzyme will be of bacterial or fungal origin.

The decomposing abilities of bacteria may be thought of as "natural," as opposed to the specifically directed activities regulated by human control. Bacteria are used commercially in many fermentations: butter and cheese making; the curing of tea, tobacco, and cocoa; the cleaning of flax fibers for linen; the production of a long list of organic chemicals; and the commercial production of such antibiotics as bacitracin, streptomycin, and a number of other "-mycins." One feature of bacterial activity that helps in human economy as well as in the general economy of nature is the ability of bacteria in some genera to take nitrogen (N_2) from the atmosphere, reduce it to ammonia

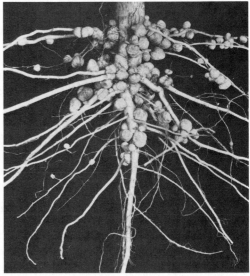

× 3

9.9

Nodules on the roots of a soybean plant. The lumps are formed by the plant after entrance of the symbiotic bacteria Rhizobium. Each nodule contains millions of bacteria that can use atmospheric nitrogen for the building of organic nitrogen compounds. The host plant profits by the process, and if the nodules are left in the ground after the plant is harvested, the soil can have its usable nitrogen content increased.

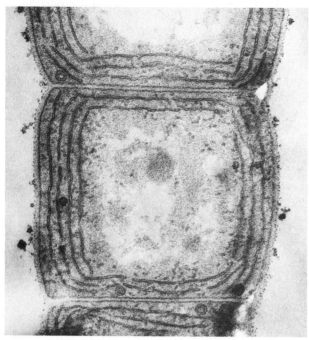

× 6800

9.10

Electron micrograph of a thin section of a cell of the blue-green alga Anabaena. There is an outermost plasma membrane. Inward from that are other membranes, which bear the photosynthetic pigments. The DNA of the cell is only faintly visible in the center. As in prokaryotic cells generally, there are no membrane-bound organelles, such as mitochondria or plastids.

(NH_3), and incorporate the ammonia into amino acids and then into protein. Nitrogen gas is not usable by most organisms, and yet nitrogen is essential to all cells. Consequently, the nitrogen "fixation" by Rhizobia in the roots of leguminous plants gives these plants an advantage (Figure 9.9). Rhizobia cannot use atmospheric nitrogen as long as they live free in soil, but they can work their way into the root hairs of susceptible plants, such as beans, clovers, alfalfa, and locust trees. Once Rhizobia are in a root, a colony develops, and the host plant grows small tumors a millimeter or two thick. In those tumors the bacteria begin a new life, trapping atmospheric nitrogen and reducing it to ammonia. Nitrogen thus fixed can be used by both the bacteria and the host plant to make any of the nitrogen-containing compounds that are essential to life.

Actinomycetes   One subgroup of bacteria, the Actinomycetes, deserves mention because the group is the source of many useful antibiotics (streptomycin, chloramphenicol, the tetracyclins) and because members of the group are so universally distributed in soil. Actinomycetes give off a volatile substance with a characteristic, familiar odor that gives soil its distinctive smell. Any gardener—indeed, any child who has "played in the dirt"—knows how fresh, moist soil smells. That smell is not soil; it is Actinomycetes.

Blue-green Algae
The blue-green algae* are prokaryotic in their lack of membrane-bound nuclei and most other cell components typical of eukaryotes, but they differ most from bacteria in their ability to carry on photosynthesis with the evolution of oxygen from water, as higher plants do (Figure 9.10). The blue-greens frequently grow into chains of cells. Each cell is physically separate from the others, not united by protoplasmic connections as typical plants are. For this reason, some biologists believe a blue-green filament is merely an aggregation of individual cells, not a multicellular organism. Nevertheless, some blue-greens produce specialized cells at specific intervals along the filament, thus showing a simple but definite power of differen-

* The word "algae" has no technical meaning, but it is commonly used by biologists and laymen to refer to small aquatic plants of many sorts, as well as to the large seaweeds. Any photosynthetic organism that is not a moss-like, fernlike, or seed-bearing plant is likely to be called an alga, just as insects are called "bugs" and whales are called "fishes" without regard to taxonomic niceties.

tiation—a characteristic of multicellular organisms (Figure 9.11). The blue-greens may be more than just a cluster of cells.

Although they are called blue-green, many members of the group are in fact red, because in addition to chlorophyll *a* and a blue pigment (phycocyanin),

they have a red pigment, phycoerythrin. Some blue-greens have pigments also in their outer coverings, giving the organisms a range of colors that can become strikingly visible. The Red Sea, for example, is so named for the occasional masses of red-pigmented "blue-green" algae that accumulate there, and the

9.11
Representative blue-green algae, about × 500. (a) Spirulina, a single coiled cell. (b) Anabaena, with enlarged, empty heterocysts. (c) Rivularia, one filament of a mass of filaments, with tapering ends and a gelatinous covering. (d) Cylindrospermum, with terminal akinetes and heterocysts. (e) Oscillatoria, the commonest of the blue-greens, capable of a swaying movement. (f) Aphanizomenon, with bundles of filaments containing empty heterocysts and dense reproductive cells, or akinetes. A nuisance in water supplies.

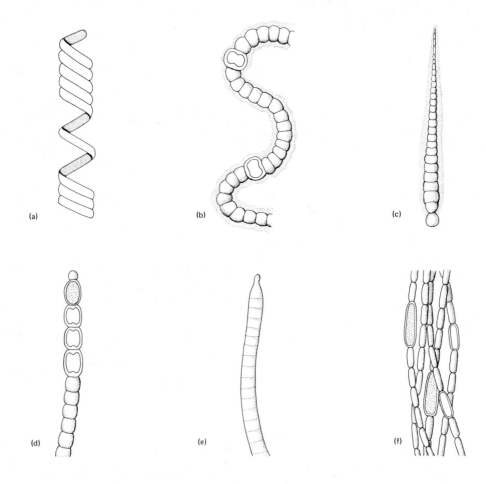

sulfur-bottomed whale has a yellow belly because of the growth of these algae.

Blue-green algae are almost as adaptable to different habitats as bacteria. They can stand temperature ranges from ice to hot springs. Much of the colored deposits around the Yellowstone hot springs is an accumulation of blue-greens. Although they are so common in soil as to be practically ubiquitous, they are typically aquatic and may grow into such massive "blooms" in public water supplies as to be severely troublesome. Besides being difficult to control or to filter out of the water, they can give drinking water a vile taste and smell, and some are toxic enough to make people and animals sick. On the positive side, blue-greens are like the Rhizobia in their ability to make atmospheric nitrogen available for amino acid and protein production. In some crops, especially rice, blue-greens provide a free source of nitrogen, in part replacing artificial fertilizer.

Coupled with their adaptability, the asexual reproductive effectiveness of the blue-greens has made them biologically successful. They can live almost anywhere, and they can proliferate at such a rate as to produce enormous masses of cells in a short time. Although some recombination of characteristics has been reported, indicating an exchange of genetic material between strains, actual mating between cells of blue-greens has not been seen. Blue-greens seem to lack true sexuality. Mating may take place, but no one has found it.

THE KINGDOM OF THE PROTISTA

The Protista are structurally more complex than the bacteria and blue-green algae. In the Kingdom of the Protista, there are advances over the cellular nature of the simpler organisms, and we can find features that show how more and more advanced organisms evolved. We also find greater diversity of form and can therefore make reasonable guesses as to how they diverged during their development from primitive ancestors.

As is true of the Monera, the aggregation of organisms included in the Protista consists of a heterogeneous assemblage. They are eukaryotic, possessing a membrane-bound nucleus and the various organelles typical of eukaryotes. As we will see, Protista exhibit a bewildering diversity of shapes, habits, and chemical components. Many are photosynthetic, others strictly heterotrophic. Some Protista form aggregations of loosely connected cells called *colonies*.

Some of these colonies show very simple specialization and integrated control of some cells and thus constitute a primitive multicellular organism. But most Protista live their lives as single-celled individuals. Some of the included phyla are groups popularly called algae. As noted earlier, the word "algae" (singular, alga) has no technical significance but is commonly applied to any plantlike organisms that are photosynthetic but structurally too different from ordinary land plants to be included with them. The various phyla grouped as Protista are probably not closely related in evolutionary history. They are thought to have diverged billions of years ago and to have developed along independent lines. Today they are far enough apart to be well established, peculiar groups. Considering all of them in this kingdom is a matter of rather arbitrary opinion. Although such a way of handling matters of classification is scarcely proper scientifically, it can be acceptable temporarily, as long as its provisional nature is made explicit. It is probably adequate for a preliminary survey whose intent is to acquaint students with some of the types of living organisms. On superficial view, the problems seem simpler than they are because the closer one looks at the details of plants and animals, the more difficult becomes the task of assigning them to rational categories. The subdivisions of this kingdom will be called phyla, as will subdivisions of the other kingdoms, in spite of the fact that in recent years botanical taxonomists have adopted the term "division" to equal the zoological term "phylum." The usefulness of having two words for comparable categories is debatable.

Phylum Protozoa

The single cell of a protozoan such as a Paramecium is a complete, functioning organism. Like most microorganisms, protozoa exist in any habitat that provides a source of nourishment and liquid water: in soil, fresh and salt water, in and on plants and animals, even floating (in desiccated form) in air. Most protozoa are microscopic, ranging from about 5 to 100 micrometers, but some large ones are several millimeters across and could pose a threat to a small beetle. Four classes of protozoa are recognized:

1. The *Flagellata*, so called because they have one or more flagella. These organelles, constructed like those of higher animals, have a membranous covering and microtubules (Chapter 3). Many members of this group are photosynthetic, and they are as much like plants as like animals (Figure 9.12). Mem-

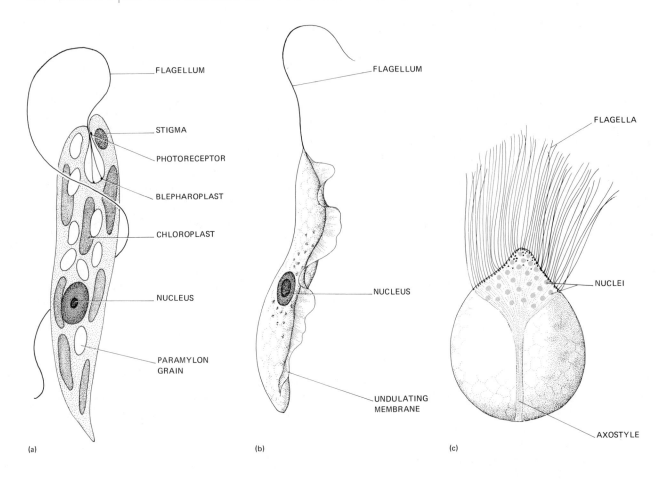

(a) (b) (c)

9.12

Flagellated protozoans. (a) Euglena, a photosynthetic, free-living cell with a red stigma, or "eye spot," close to a light-sensitive photoreceptor. The flagellum is anchored to a pair of blepharoplasts at the base of a gullet-like sac. Green chloroplasts manufacture a special kind of starch, paramylon. × 2000. (b) Trypanosoma, the cause of African sleeping sickness, transmitted by the tsetse fly. It destroys human red blood cells. A flange extends from the body from the head end to the tail end, making a soft, waving ridge, the undulating membrane. × 5000. (c) Calonympha from the gut of a termite, where, with the help of bacteria, it lives on the cellulose eaten by the termite. There are many flagella and many nuclei, as well as a central stiffening rod, the axostyle. × 700.

bers of the genera Euglena, Pandorina, and Volvox are frequently considered to be plants, but nongreen flagellates are clearly heterotrophic. One flagellate, Peranema, could be called a colorless Euglena. Most flagellates are free-living in soil or water, but some, the Trypanosomes, are parasites, causing, among other diseases, the African sleeping sickness. The parasite is passed from one person to another via the biting tsetse fly.

2.   The *Sarcodina*. The members of this class have no specific organelles for locomotion. They extend blunt or slender cytoplasmic points (*pseudopods*), and then the cell body follows the pseudopods, which can be pushed out and retracted repeatedly. Amoebas are familiar to all biology students as free-living dwellers in water. Some amoebas can cause human dysentery. Marine shelled sarcodinians through geological time have deposited billions of skeletons, which now make layers on the bottom of the sea (Figure 9.13). The "radiolarian ooze" or "Globigerina ooze" is studied by oil prospectors because the species give clues to possible petroleum deposits.

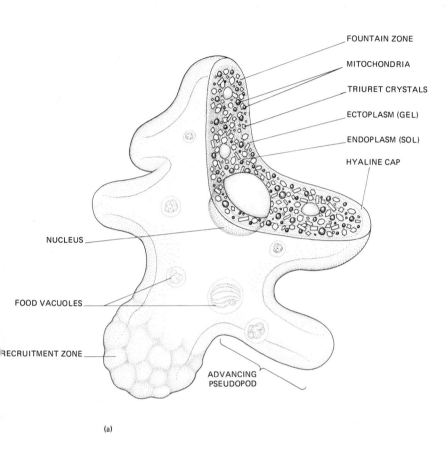

FOUNTAIN ZONE

MITOCHONDRIA

TRIURET CRYSTALS

ECTOPLASM (GEL)

ENDOPLASM (SOL)

HYALINE CAP

NUCLEUS

FOOD VACUOLES

RECRUITMENT ZONE

ADVANCING
PSEUDOPOD

(a)

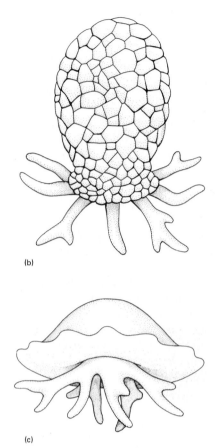

(b)

(c)

(d)   × 50

9.13
Some representative Sarcodina. (a)
Amoeba proteus, the common free-
living amoeba. × 2000. (b) Difflugia,
with a covering shell of sand particles
stuck to the outside of the animal's
cell. × 200. (c) Arcella, a sarcodinian
that secretes its own shell. × 350. (d)
Shells of various sarcodinians dredged
from deep sea sediments.

9.14

The life cycle of the malaria parasite, *Plasmodium vivax*. A mosquito, biting a victim, injects infective cells into the bloodstream (a). The infective cells enter red blood cells of the warm-blooded host (b), where they multiply (c–e) and then break out of the blood cells (f), causing the periodic fevers that are typical of malarial infection. The released parasites are gametes (g, i), which can be sucked up by a female mosquito biting the sufferer. In the mosquito's stomach, the gametes fuse (l), and the zygote (m) bores into the mosquito's gut wall (n), where it multiplies (o). When the cyst containing numerous Plasmodium cells ruptures, the parasite cells float in the mosquito's body fluid (p), some of them finding their way to the salivary glands (q), from which they can enter a new warm-blooded host when the mosquito bites it.

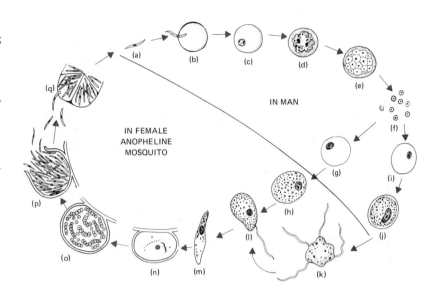

IN MAN

IN FEMALE ANOPHELINE MOSQUITO

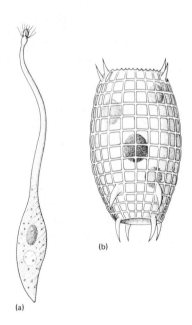

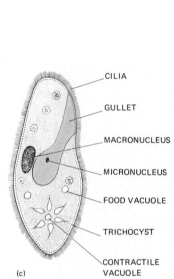

CILIA

GULLET

MACRONUCLEUS

MICRONUCLEUS

FOOD VACUOLE

TRICHOCYST

CONTRACTILE VACUOLE

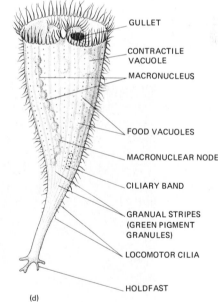

GULLET

CONTRACTILE VACUOLE

MACRONUCLEUS

FOOD VACUOLES

MACRONUCLEAR NODE

CILIARY BAND

GRANUAL STRIPES (GREEN PIGMENT GRANULES)

LOCOMOTOR CILIA

HOLDFAST

9.15

Some representative ciliated protozoans. (a) The teardrop-shaped Lachrymaria. × 85. (b) A spiny, firm-bodied Coleps, from fresh water. × 500. (c) The common "slipper animalcule," Paramecium, one of the most complex cells. Food is trapped in the gullet, from which it is taken into the cell. Small particles are circulated in the cytoplasm in food vacuoles. There is a larger macronucleus and one or more smaller micronuclei. Water content within the cell is regulated by a pair of contractile vacuoles, which alternately fill by way of star-shaped canals. One such vacuole is shown at the posterior end being filled. Under the exterior covering, the pellicle, are thousands of trichocysts, like microscopic darts that can be shot out to aid in the capture of prey. × 500. (d) *Stentor coeruleus*, a blue ciliate, × 350.

3. The *Sporozoa*. The cells of sporozoans are usually nonmotile and parasitic (Figure 9.14). The best-known members of the group are those that cause malarial fever. The word malaria, which means bad air, refers to the prevalence of the disease in swampy regions, where mosquitoes thrive. The early idea was that something evil in swamp air afflicted people. Once it was known that mosquitoes can carry the parasites and inject them into people, prevention of the disease became easier. When mosquito control is impossible, the fever can still be controlled by the use of an ancient folk remedy: quinine from the bark of Cinchonas, the Peruvian fever trees. Modern substitutes, such as atabrine, are synthetic, but the South American Indians have known for centuries about the curative power of the bark of the fever tree, having discovered it by trial and error.

4. The *Ciliata*. Protozoa with numerous short cilia, so named because of the resemblance of cilia to eyelashes (the word really means eyelid), are so different from other protozoa that some biologists classify them as a distinct subphylum. They use their accurately synchronized cilia either to move themselves through water or, if they are attached, to bring water containing food particles near or into the protist's body. Many ciliates have a groove or depression into which food can be brought, and although no mouth opening is present, solid bits can be brought inside the body by a process of engulfing (Figure 9.15). Ciliates are effective at this process, as are amoebas and even the white blood cells of mammals. Most reproduction is asexual, but mating between compatible strains of ciliates occurs, with a resulting exchange of genetic material. Mating does not include increase in numbers. Mated individuals come together, go through an elaborate set of nuclear exchanges, and then separate. However, the main genetic effect of sex is there: the possibility of recombination. Their ready availability in almost any drop of pond or gutter water, their unexpected shapes, their colors (usually rather gray, but occasionally pink or blue), their liveliness, their elaborate and controlled behavior, their unparalleled cytoplasmic complexity, and their nutritional requirements have served to make the ciliates subjects for investigation by biologists for several centuries.

Phylum Chrysophyta

Chrysophyta comes from the Greek meaning "golden plants." These organisms are named after their conspicuous yellow pigments, the carotenoids. However, they do possess two forms of chlorophyll, *a* and *c*, and can photosynthesize effectively. One group of the Chrysophyta, the _diatoms_ (Greek, "two

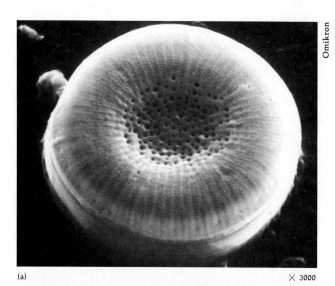

(a)   × 3000

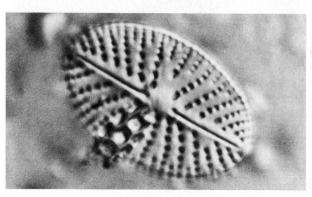

(b)   × 1000

9.16

Representative diatoms. Diatom shells are so nearly indestructible that they remain in soil and water indefinitely. The shells are the parts usually seen in collections. When alive, many of the diatoms are capable of moving along like microscopic boats. (a) *Cytotella operculata*, a circular cell. Only the shell is shown here. (b) Navicula, a common boat-shaped diatom.

cuts," from the double nature of the walls) are and have been for millions of years some of the commonest inhabitants of lakes and seas. The diatoms are unique in the structure of their shells, called valves, which fit like the halves of a Petri dish, and in their gliding manner of moving (Figure 9.16). Their cell walls, which are impregnated with glasslike silicon, are practically indestructible, and they pile up in thick layers on ocean bottoms, where they build up diatomaceous earth. In some places, (for example, California) the deposits of diatomaceous earth have been raised by geological activity. They are also commercially mined, and the earth is used as insulation, pol-

(a)

(b)   × 250

9.17

Sexuality in the yellow-green alga
Vaucheria. There are no separate, uninucleate cells, but the cytoplasm contains numerous nuclei floating free.
The hook on the right is the male
gametangium, and the dark sphere is
the female. Sperm from the male gametangium fertilize the egg in the female gametangium, entering by way
of a pore that is visible on the right of
the female gametangium just above
the hook in (b). The organism shown
in (a) is several hours younger than
the one in (b).

ishing powder, or filtering materials. The walls are so
minutely and delicately sculptured that they have
long been used as test objects in determining the quality of microscope lenses. Because they are very common and because they store reserve food in the form
of fat, diatoms are one of the main sources of food in
seas. The fat has a peculiar flavor and odor that it
retains even after it is eaten by animals. It can be accumulated in such quantity that it gives fish their
"fishy" taste. Compare the flavor of a freshwater
trout, whose ultimate food source is *not* diatoms, with
that of a marine mackerel, whose food supply *is*
largely diatomaceous.

Phylum Xanthophyta

The Xanthophyta (meaning "yellow plants") are actually greenish, but they have a bright yellow tint because of the presence of carotenoids in addition to
chlorophyll *a*. They are not rare, but there are fewer
than 500 species. The best known is an aquatic genus,
Vaucheria, which is readily recognized even without
a microscope because a mass of it, commonly found
in ponds in spring, feels so coarse and scratchy (quite
different from the slippery feel of most algae) that it
is called "brook felt" or "water felt" (Figure 9.17).

Phylum Pyrrophyta

The Pyrrophyta (meaning "flame-colored plants")
are microscopic aquatic algae, mostly with two tinsel-
type flagella attached in a characteristic manner (Figure 9.18). They are important in the ecology of the sea
because, like several other protists, they furnish food
for small animals, which are in turn eaten by larger
animals. Pyrrophyta thus serve as a base supply for
commercial fishes. Having chlorophylls *a* and *c* plus
carotenoids, they are photosynthetic. A dramatic effect of one class of Pyrrophyta, the dinoflagellates, is
produced when these organisms are exposed to air. A
churning boat propeller, the swish of an oar, or the
breaking of a wave will cause a dinoflagellate to emit
a flash of light. When millions of these creatures flash
at night, they can make the sea glitter with a luminos-
ity that sailors have long called "the burning of the
sea." (See the essay on bioluminescence on p. 32.)
Phosphorescent Bay on the south coast of Puerto Rico
was named for the spectacular displays of Noctiluca
(Latin, "night light").

Another striking activity of dinoflagellates is the
production of the so-called red tides by infestations
of red-pigmented species. Under certain still undetermined conditions, some species of dinoflagellates
(Gymnodinium or Gonyaulax, depending on the region) multiply enormously until the very water be-

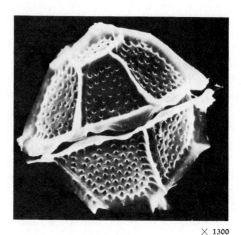

9.18

An example of the Pyrrophyta, Gony-aulax, as seen in a scanning electron micrograph. The wall plates are conspicuous, but the flagella do not show.

$\times$ 1300

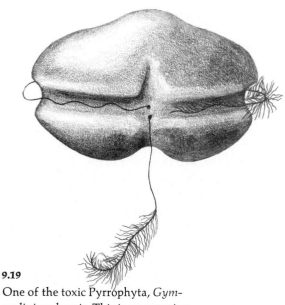

9.19

One of the toxic Pyrrophyta, *Gymnodinium brevis*. This is an organism whose poison can kill fish by the million in Florida. It is responsible for the massive blooms known as red tides. $\times$ 1675.

comes visibly colored (Figure 9.19). The trouble, from a human view, is that these minute creatures synthesize a powerful nerve poison that can kill vertebrates. With a burst of reproductive speed, the dinoflagellates become so numerous that they can furnish the main food supply for small crustaceans (shrimp and water fleas) and molluscs (clams and oysters), which feed by filtering the dinoflagellates from the water. These primary feeders are not harmed by the toxin, but people or fish can be poisoned if they eat smaller animals that have fed upon such dinoflagellates. During a severe red tide, countless dead fish may be cast up on the beach, as they sometimes are on the Florida Gulf Coast. Unsuspecting clam eaters may also be killed.

California has endured many red tides, and occasionally Atlantic infestations are carried as far north as New England. American Indians in precolonial days were aware of the danger and used to alert their contemporaries by lighting warning fires along beaches when the unseen dinoflagellates contaminated the sea.

THE KINGDOM OF THE FUNGI

Fungi look more like plants than like anything else, and they do have two definitely plantlike features: firm cell walls and reproductive bodies called *spores*.

But the cell walls are usually noncellulosic, being composed largely of *chitin*. Chitin, a polymer of an amino sugar, is more characteristic of insect bodies than of plants. Further, the spores of fungi, especially the sexually produced spores, are different from the spores of familiar higher plants in appearance and method of production. The texture of such a fungus as a mushroom is pithlike, but the cells of mushroom tissue are formed by the massing of filaments rather than by the kinds of cell division used by either plants or animals. The starchlike glycogen commonly stored by fungi as a reserve food is like the glycogen of animals, but unlike animals, fungi do not eat. Like bacteria and tapeworms, fungi obtain nourishment by absorbing predigested food from the environment. Even the chromosomes and nuclei of fungi are peculiar to the group. The chromosomes lack the proteins characteristic of other eukaryotes, and mitoses are completed within a nuclear membrane. Like animals, fungi have centrioles, but unlike animals, they do not separate new cells by constriction of the cell coating. For all these reasons, current opinion favors the classification of fungi in a separate kingdom. The fungi are considered an evolutionary group developing in a direction of their own. Fungi have been thought of as plants for so long that they are still studied mainly by botanists, but the differences between plants and fungi are so great and apparently so fundamental that now the two groups are separated. The separation constitutes a break from the old Linnean aphorism: "Rocks grow; plants grow and live; animals grow and live and feel."

The Phycomycetes

One heterogeneous class of fungi with cellulose walls and *coenocytic hyphae*, the Phycomycetes, contains several loosely related groups of fungi. *Hyphae* (singular, hypha) are the filamentous strands that fungi generally have instead of the compacted cells found in the tissues of higher organisms; *coenocytic* means that instead of having distinct cells, each with its own nucleus, the hypha is a continuous strand without cross walls. A mass of fungus hyphae is a *mycelium*. In the Phycomycetes in general, the cytoplasm flows about freely, with nuclei and other organelles suspended in the cytoplasm. Phycomycetes commonly reproduce asexually by producing cells that can swim or that are blown about, later to germinate into a new fungus. Sexual reproduction involves fusion of nuclei, but in many species, nuclei will join only if they are of proper "mating types." Since they are neither male nor female, they are called "plus" or "minus" types. A mycelium derived from a single spore is incapable of sexual reproduction. An organism that requires at least two mating types to complete a sexual cycle is called *heterothallic*; one that can mate with itself is *homothallic*. Both types occur in this group.

Several Phycomycetes have been devastating to plants, causing enormous difficulties. The downy mildew of grapes, Plasmopara, practically destroyed the French wine industry for a while in the 1880s. The blight that struck potatoes in Ireland during the late 1840s was caused by the Phycomycete *Phytopthora infestans*. It was so destructive that widespread famine resulted. The virtual elimination of the main staple food and the poverty and loss of energy that followed, coupled with poor sanitation and limited knowledge, created a double hazard of starvation and disease, much of it caused by other microorganisms, cholera and typhus. Because many were dying and those who could escape the country did, the Irish population was halved in about 10 years, and it has not fully recovered even yet. The emigrants made their presence felt all over the world, and one can trace to the potato blight the buildup of the Irish population in the United States, the rise of the Kennedy dynasty, and the election of John Kennedy to the presidency.

9.20

Fruiting bodies of ascomycetes. (a) Scanning electron micrograph of fruiting bodies of a fungus (Nectria) that causes a disease (canker) on apple leaves. Each body has a visible hole in the top through which spores are exuded. (b) Diagram of a section through a fruiting body, showing five asci (in a real one there are dozens or even hundreds), each with eight ascospores. In different species, asci contain from four to hundreds of spores.

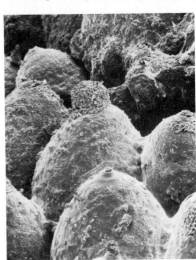

(a)   × 375

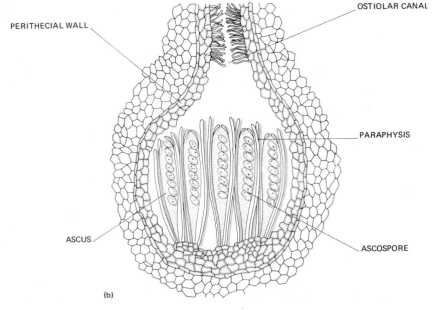

(b)

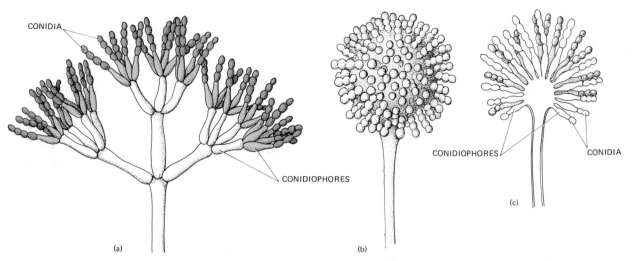

9.21

Aspergillus and Penicillium, two of the commonest imperfect fungi. Most of the blue, green, brown, and black molds that grow on fruits, paper, leather, and cloth belong to one or the other of these genera. (a) A stalk (conidiophore) of Penicillium with chains of *conidia*. Only a few conidia are shown, but a vigorously growing mold has thousands of conidia on such a "head," with hundreds in each chain. × 1000. (b) A conidial head of Aspergillus. × 200. (c) A section through a conidial head of Aspergillus, showing the manner of conidium production. × 1000. (d) Scanning electron micrograph of a few Penicillium conidia. These are young, with only a few conidia on each chain.

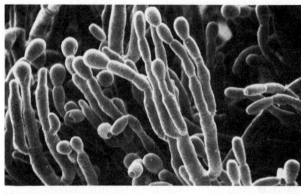

Thus world affairs can be influenced by an organism so small that when it was doing its worst, it was not even seen by the people it was indirectly killing.

The Ascomycetes

The distinctive feature of the Ascomycetes is the production of spores within a single cell, with a sexual fusion of nuclei followed by immediate meiosis and rounding up of the usually eight spores with haploid nuclei. The spore-carrying cell, known as an *ascus* (plural, asci), is the only characteristic common to all the Ascomycetes. The structure, distribution, size, and metabolic habits of Ascomycetes are so varied that no simple description can be adequate (Figure 9.20). Unlike the Phycomycetes, the Ascomycetes have cells of their hyphae compartmented, with one or two nuclei in each cell. Their walls are impregnated with chitin. Besides sexually producing spores, the *ascospores*, in asci, most Ascomycetes also form *conidia* by pinching off a piece of a hypha containing a

nucleus and enough cytoplasm to make an asexual reproductive body. Indeed, some Ascomycetes have gone so far in the use of conidia that they seem to have lost the ability to form ascospores. Since many Ascomycetes are heterothallic, failure to find ascospores in a growing fungus is no guarantee that the sexual phase is actually lacking; it may be that no biologist has found the proper mating types. But extensive searches have frequently failed to yield any ascospores, some strains of fungi under study remaining consistently asexual. Such strains are called *fungi imperfecti*, the imperfect fungi (Figure 9.21). Enough imperfect fungi have been found to produce ascospores on occasion to warrant the guess that most of them are in fact Ascomycetes. Conidia of imperfect fungi are so widely distributed that they are probably second only to bacteria in being present everywhere. There is hardly a gram of soil or square centimeter of surface, whether of plant, animal, or inanimate object, that does not carry a load of conidia, ready to

158

PARASITES, WINE, AND MONEY

The reactions among grape vines and their parasites and the vine growers furnish a complex example of the interrelations among organisms. It is the story of the French wine producers, their vineyards, the phycomycetous fungus, Plasmopara, and a root-sucking insect, the aphid called Phylloxera.

The story begins in prehistoric America, where wild grapes evolved, along with the fungus and the aphid, all surviving and establishing a balance in which the attackers invaded but did not kill. American vines, however, did not produce good wine grapes. In Europe, meanwhile, the main wine grapes from *Vitis vinifera* were untouched by Plasmopara or Phylloxera and had no resistance to either. During the 1860s, transoceanic traffic brought some of the aphids to Europe, where they attached themselves to the defenseless European grape roots and all but ruined the vines. Since destruction of the wine industry in France is almost the destruction of the French economy, the efforts to clean out the aphids can well be imagined. One successful method was suggested: Import resistant American roots, graft the high quality French vines onto them, and let the aphids do their worst.

The solution of the Phylloxera problem, however, merely brought a new threat. Along with the American vine roots came the infectious spores of the grape mildew to which the French vines were as helplessly susceptible as they were to Phylloxera. For a second time, total destruction of the wine grapes seemed certain as millions of acres of French vineyards shriveled and died.

Once more a solution was found. A professor at the University of Bordeaux, Alexis Millardet, was a student of fungi, and by the 1880s it was understood that the downy mildews were indeed parasitic fungi. He noted that the vines along a road were healthy in spite of devastating mildew away from the road. The wine growers had sprayed the roadside vines with blue vitriol (copper sulfate) to make them unattractive to grape robbers. With this information, Millardet formulated the world's first agricultural fungicide, a mix of copper sulfate and lime. Known as the Bordeaux mixture, it is still in use.

The troubles were not over. France's problems with aphids and downy mildew promised to be a bonanza for other Mediterranean countries that sought to take over the French wine economy. They hastily started enormous new vineyards, thinking that without the French competition they would have a great market. But the grafting technique took care of the aphids, the Bordeaux mixture controlled the mildew, the French wines again poured forth, and millions of drachmas and dinars and escudas were lost.

germinate into a vigorous mycelium. These are the common blue, green, brown, pink, violet, and yellow "molds" that flourish wherever moisture and a few molecules of nourishment are available.

Investigations on the Ascomycete Neurospora laid the basis for modern molecular genetics by furnishing the one gene–one enzyme hypothesis. Millions of conidia can be irradiated to make an endless supply of mutants for experimentation, and the products of meiosis, the ascospores, can be analyzed directly. Neurospora differs from other organisms used in genetics research. For example, in fruit flies, mice, maize, or peas, the products of meiosis must be allowed to mate, and then the results must be treated statistically. Because of its very peculiarity, Neurospora was almost an ideal experimental organism for the purposes of its time, the 1940s and 1950s.

Yeasts are microscopic Ascomycetes. Of the hundreds of wild yeasts that live wherever fungi can grow, including oceans and rivers, most are of no economic use. However, one species of inestimable value is *Saccharomyces cereviseae*, baker's or brewer's yeast (Figure 9.22). Mixed with flour and a little sugar, the yeast cells can respire anaerobically, yielding carbon dioxide that lightens bread. In fruit juice, the same respiratory activity produces alcohol (Chapter 5).

On the negative side, Ascomycetes are responsible for some diseases. Hardly a woman lives a whole life without ever having a vaginal infection by some

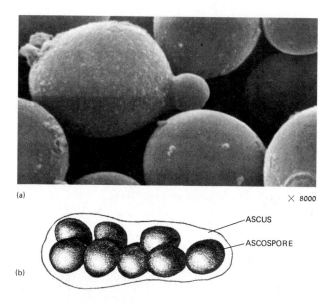

(a)   × 8000

(b)

ASCUS

ASCOSPORE

9.22

Growth and reproduction of yeasts. (a) Scanning
electron micrograph of cells of baker's yeast,
Saccharomyces cereviseae. Most yeasts grow new cells
by putting out a *bud*, which increases in size until it is as
large as the parent, and then it breaks away. (b) Spore
formation in a yeast. In this species (*Schizosaccharomyces
octosporus*) there are eight spores. The yeasts, superficially unlike most fungi, are classified with the ascomycetes because of the way they form spores. Compare this
picture with the asci and ascospores in 9.20 (b), and note
the similarity. × 1000.

× 0.33

9.23

A morel, Morchella, one of the most desirable of edible
fungi. It has never been artificially cultivated, and mushroom hunters who know where to find it jealously guard
their knowledge. The honeycombed upper part of the
fruiting body is lined with millions of asci containing
ascospores, but the growing conditions necessary for
successful development of the fungus are so special that
only about one spore from each crop of morels germinates
and grows to maturity (else the species would be increasing in numbers, and it is not).

yeastlike fungus. And various kinds of itch, athlete's
foot, and "ringworms" are due to fungi, most of them
Ascomycetes. Wild and domestic animals have their
share of fungus infections, too, but it is in plants that
these parasites cause their greatest disasters. Ascomycetes and their relatives, the imperfect fungi, produce
such an array of mildews, blights, cankers, wilts,
spots, necroses, and rots that it is a wonder there is
anything left for the insects, birds, and rats, let alone
people. The great American chestnuts, which filled
the forests of the eastern United States at the beginning of this century, are reduced to a few sickly suckers, all doomed to an early death because of the chestnut blight fungus, *Endothia parasitica*. Row after row
of the stately elms that graced the towns of New England are dead, killed by another Ascomycete, *Cera-
tostomella ulmi*. Rye is infected with ergot, *Claviceps
purpurea*, which not only damages the rye grains but
produces unfortunate effects in consumers of infected
rye flour. They undergo fantastic and horrifying de-

lusions, see their families as monsters, and may have
blood supplies to their extremities so restricted that
their ears or fingers rot off. In severe enough cases
they die. Strangely enough, the same class of fungi
that causes so much misery contains the two classic
fungus delights of gourmets, morels and truffles (Figure 9.23), and it is also the source of the workhorse of
the medical doctor's array of antibiotics, penicillin
from Penicillium.

The Basidiomycetes

The fungi included in the Basidiomycetes are so various that at first sight one would not expect them to
be even remotely related. There are mushrooms,
stinkhorns looking remarkably like vegetable phal-
luses, the messy black pustules of corn smut, the orange incrustations of blackberry rust, shelves on decaying birch trees, wads of jelly on soil or wood, the
rainbow-colored coral fungi, and the tiny golden fingers of Dacrymyces. They all have one feature in

common: the formation of *basidia* (singular, basidium), hyphal ends with horn-shaped extensions, *sterigmata*, bearing spores at their tips. These are the *basidiospores*.

The most popularly known Basidiomycetes are mushrooms, such as the commercially grown Agaricus sold in grocery stores. A widespread and economically destructive genus is Puccinia, the red rust of wheat. This parasitic fungus is like a number of rusts in that it normally requires two plant hosts: cultivated wheat and barberry bushes. During summer, the invader lives on wheat, sometimes drastically reducing the yield of grain. Spores pass from field to field in the wind, spreading the disease, which looks like red rust. Other spores live through the winter. In the following spring, they infect a different plant, or *alternate host*, the barberry. Still other spores from the barberry can infect the new wheat crop. For complexity of spore forms in such a vegetatively simple fungus, wheat rust is without equal anywhere in the living world, and it is not surprising that the rusting of wheat remained mysterious to farmers for thousands of years. Complete culture of wheat rust in the laboratory has never been accomplished. One reason that the rust has not been eliminated from wheat-growing regions is that as fast as wheat geneticists breed new resistant strains of wheat, new infective strains of rust appear. Some perhaps develop by mutation, but most come from genetic recombinations arising from sexual matings. The competition between the crop breeders and the parasite appears endless.

Basidiomycetes are synthesizers of some unique chemical compounds, some of them deadly to humans (Figure 9.24). Poisons from mushrooms may destroy red blood cells or interfere with nerve transmission. In either case the victim dies of lack of oxygen. No general rules for determining safety can be given; one simply has to know personally which species have been proved to be killers. Less lethal compounds merely confuse the consumer. The sacred mushrooms of the Mexican Indians, teonanacatl, or "God's flesh," are mostly of the Genus Psilocybe. Eating them makes people see and hear things that no one else sees or hears.

Lichens

Like most of the lower organisms, lichens puzzled early biologists. They were variously considered fungi, mosses, nonliving growths, or "queer" algae. However, the Swiss botanist S. Schwendener discovered in 1867 that a lichen is an association of fungus and alga, each so intimately associated with the other that together they make essentially a whole new or-

ganism. Later the word *symbiosis* (Greek, "living together") was coined to express the relationship, and that term has since been broadened in meaning to include any example in which organisms of more than one species live together in some special physical or nutritional relationship (Figure 9.25). Lichens have little economic value, but they are important in the ecology of some parts of the earth (Chapter 31). They also fascinate biologists because they are unique in their ability to make what looks like one organism, with the apparent attributes of a real species, yet are composed of two distinct organisms.

(a)

9.24

Representative mushrooms. (a) The fly agaric, *Amanita muscaria*, one of the moderately poisonous fungi. It is sometimes eaten by unknowing mushroom collectors because it is attractive in appearance and tastes good, but it can cause illness and hallucinations. The white gills, the ring around the stem, and the cup at the base are warning signs. (b) The parasol mushroom, *Lepiota procera*, so called because the ring is so loose around the stalk that it can usually be slid up and down. It is a large, handsome, delicious species, but its near relative, *Lepiota morgani*, with greenish instead of white gills, is poisonous (though not deadly). × 0.5.

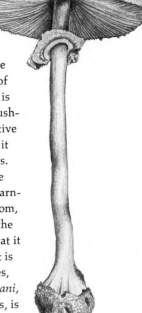

(b)

9.25

A lichen, Parmelia, on tree bark.

× 0.8

9.26

Irish moss, or carragheen, an edible red alga (*Chondrus crispus*). Millions of pounds of this alga are harvested for the sake of a gelatinous product that can be obtained from it. This *carraghenin* is used in ice cream, jellies, hand lotions, and tooth paste to give a smoothness and body. It can be boiled with milk, sugar, and flavoring to make pudding, and it has been so used for centuries in maritime countries. Soaked in whiskey and chewed, it is said to be useful as cough drops.

× 0.5

Lichens are so sensitive to air pollution that they are killed in cities, probably by the products of combustion in furnaces and autos. Strictly urban people never see lichens, but in uncontaminated regions, whole forests or whole landscapes may be gray-green with them.

THE KINGDOM OF THE PLANTS

Included here are photosynthetic organisms, mostly with cellulose walls covering the cells, usually multicellular bodies with well-developed vegetative bodies, and a life cycle in which sexual reproduction alternates with asexual. The red and brown algae are somewhat different in details from the rest of the organisms considered here, but they are in most characteristics so definitely plantlike that it seems sensible to keep them here rather than to designate more kingdoms. The green algae are like some of the Protista in certain respects, but their chemical similarity to the higher plants has convinced most biologists that there is a line of evolutionary development leading from the green algae to the more advanced groups.

Phylum Rhodophyta, the Red Algae

The Rhodophyta (meaning "rose-colored plants") are usually red because in addition to chlorophylls *a* and *d*, they have a red pigment, *phycoerythrin*. All but a few of the red algae are marine, and most are tropical, thriving especially in deep waters, where their phycoerythrin can make use of what little light energy can penetrate sea water (Figure 9.26). A red pigment is a good trap for the shorter wavelengths of light that the plants can obtain down where the energy is reduced by the filtering effect of water. Such accessory, nonchlorophyll pigments can take some radiant energy of light and transfer it to chlorophyll for photosynthetic use. Like fungi, red algae may have solid-looking bodies, but they do not have cellular tissues like those of other plants. Instead, red algae have masses of filaments compacted into well-defined leaflike or stemlike structures. Their reproductive methods are too varied and complex for treatment in a brief survey.

They are of little economic importance, although some calcareous ones are like coral animals in that they are tropical reef builders; some, such as Irish moss, *Chondrus crispus*, are edible; and some furnish

9.27 × 0.25

The brown alga Postelsia grows on the rock Pacific coast of the United States. With a definite stalk and long "leaves," it has a general resemblance to a small tree, but it has no vascular tissue.

the agar used in microbiological work. The use of agar as a solidifying agent for bacterial cultures (discovered by a Jersey City housewife) is so widespread that if the supply of agar were cut off, as it was seriously cut off during World War II, thousands of hospital, university, and research laboratories would have to change their ways of working considerably. The bulk of the world's agar supply comes from Japan, where it is produced from several red algae, mainly *Gelidium amansii*.

Phylum Phaeophyta, the Brown Algae

The Phaeophyta (meaning "dusky plants") are those seaweeds popularly known as *kelps*. Anyone who knows the beaches of North America is familiar with the rockweeds, those glistening brown masses of growth that are clearly plantlike. However, they are so different from ordinary land plants that they must be classified in a group by themselves. In spite of the similarity of their parts to real leaves and stems of higher plants, the brown algae have simpler tissue structures, and their pigments, besides the usual chlorophyll *a*, include chlorophyll *c* and the brown *fucoxanthin* that gives the plants their characteristic color. Their cell walls are supported by cellulose fibrils, but they also contain mucilaginous compounds. They store a special reserve carbohydrate, laminarin. An animal-like feature is the presence of centrioles in the cytoplasm, a feature restricted in most plantlike organisms to motile cells and entirely lacking in flowering plants. Reproduction varies but does include an alternation of sexual and asexual generations.

Some brown algae are among the most conspicuous plants in the sea. The rockweed, Fucus, covers rocks of the cooler shores of the northern hemisphere with thousands of tons of growth. The giant kelps of the American Pacific coast (Macrocystis and Nereocystis) are among the world's largest plants, growing up to 100 meters (330 ft) in length. Sargassum loosely covers more than five million square kilometers (two million square miles) of the Atlantic Ocean just north of the equator and gives that region its name, the Sargasso Sea. One brown alga, Postelsia, standing on rocky shores, splashed by breaking waves, looks like a small, sloppy palm tree (Figure 9.27). Alginates from brown algae are so useful commercially in applications requiring an inert, gelatin-like material that they are harvested, and attempts are being made to cultivate them artificially.

Phylum Chlorophyta, the Green Algae

Green algae are placed with the higher plants because they share with mosses, ferns, and seed plants three kinds of chemical compounds: (1) the photosynthetic pigments chlorophyll *a* and chlorophyll *b*, (2) the storage carbohydrate starch, and (3) cellulose as the main constituent of the cell wall. The common ability of all these plants to synthesize these substances points to a common ancestry, and it seems therefore reasonable to include in one taxonomic unit those organisms that have such striking similarity. This cannot be said of any of those algal groups that in the present book are included with the Protista.

Green algae have long been favorite objects of study both by students and by professional biologists. At first people were fascinated by the structural elegance of these microscopically delicate organisms. Then when green algae seemed to offer clues to the evolution of the vascular plants, interest heightened. Besides, they are so abundant that with a minimum of technical equipment they can be obtained alive and observed (Figure 9.28).

As a forest would seem to a giant 15 kilometers (9 mi) tall like nothing more than a thin film on the earth, so pond scums may seem to a human observer. But to a microscopist, those slimy messes become worlds of beauty, with a brilliance of color and

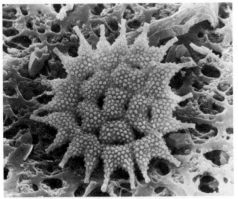

(a) × 600

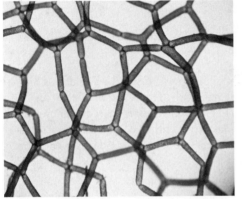

(b) × 130

(c) × 9750

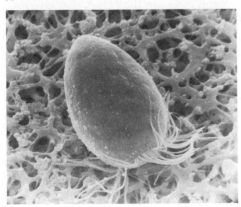

(d) × 500

a fine precision whose esthetic appeal has never been adequately captured in pictures. In the green algae, chloroplasts have evolved in many ways, so that living examples are formed into balls, stars, nets, loops, rings, and helices. Plant bodies range from single cells through branched or unbranched filaments, to sheets and even, in Fritschiella, to thickened structures that resemble parts of higher plants.

Green algae are of little direct economic importance, although they have been repeatedly suggested as food sources for people and their domestic animals because they are rapid growers and can be manipulated to form proteins and fats in high concentrations. Green algae have also been studied as possible suppliers of oxygen in submarines, but so far the scheme has met technical difficulties. Much of the research on photosynthesis has been carried out using the unicellular Genus Chlorella.

Algae are the main photosynthetic organisms in aquatic environments. Just as grasses on land serve as a prime food supply for animals, so algae in streams, lakes, and seas serve to feed small animals, fish, and eventually even the great marine mammals, such as the whales and dolphins. Green algae, along with diatoms and some of the Pyrrophyta, are so necessary to the survival of aquatic animals that they have been called the "grasses of the sea."

Phylum Bryophyta, the Moss Plants

Mosses and their relatives are typically plants of moist, shady places, but some species have enough tolerance of desiccation to allow growth on sunbaked rocks and in sidewalk cracks. Indeed, one species, *Bryum argenteum*, meaning silver moss, finds its way into urban cracks, and where it is not destroyed by constant abrasion, it can be found easily, though it is

9.28

Green algae. (a) Scanning electron micrograph of Pediastrum, notable for the geometric symmetry of the cellular arrangement of the small colony. (b) A portion of water net, Hydrodictyon, which grows in the form of delicate little sacks, readily visible to the unaided eye. Each segment, usually meeting at a three-way point, is a separate cell. (c) An electron micrograph of a unicellular green alga, Cyanidium. The alga was frozen and cracked open, and the fractured surface was photographed. The chloroplast is the grooved body in the lower part of the picture, and the nucleus is the lemon-shaped object just right of center. (d) Scanning electron micrograph of a zoospore of Oedogonium. It bears a conspicuous ring of flagella at the anterior end.

$\times$ 3

9.29

Peat moss, Sphagnum. The feathery growths are masses of leaves, each with a mix of photosynthetic and empty cells. The latter are capable of absorbing large amounts of water. The dark balls are spore capsules. Dead, partially decayed peat moss is used to soften and moisten garden soils. When sufficiently compacted, it can be burned as fuel.

$\times$ 0.67

9.30

A thick stand of moss, Catherinea. This is a mass of gametophytes, with "leaves" so small they are scarcely visible. Here "leaves" is written in quotation marks because the photosynthetic organs of mosses are not structurally comparable to the true leaves of ferns or flowering plants.

rarely noticed, in any of our cities. In contrast, the peat mosses, Sphagnum, live only in marshy places, where they are always soaking wet (Figure 9.29). In fact, all mosses must have liquid water if eggs are to be fertilized because moss sperm are motile, flagellated cells that swim to the eggs (Chapter 13).

Mosses seem to be intermediate between green algae and the higher vascular plants (Figures 9.30 and 9.31). Most live on land, but never developed the strengthening or conducting tissues which made growth of large plants possible. Mosses are not believed to be ancestral to the higher plants because fossil vascular plants can be found which are older than any fossil mosses. Consequently, the current opinion is that mosses arose from an algal ancestor, and that the vascular plants have evolved independently.

Besides the ordinary mosses and the specialized Sphagnums, the bryophytes include the liverworts and hornworts. These are either mosslike (the leafy liverworts) or ribbonlike, lying flat on soil or rocks or frequently in water. Liverworts are thought to be an evolutionary dead end, or *cul-de-sac,* having given rise to no more advanced plants. They have no economic uses, but biologists employ them as subjects of developmental problems, as well as because they enter into symbiosis with bacteria.

Bryophytes are important colonizers of bare areas (Chapter 31), but they have little direct effect on human affairs. Decayed peat moss is ground and sold for garden mulch, and compressed blocks of dead peat are cut for fuel. The "turf" of Ireland is burned extensively for home heating and even for public electrical energy. The smoke from a turf fire has a strange, sweet smell, and it has been said of the Irish in the United States that if a whiff of such smoke were wafted across the country, 10 percent of the population would burst into tears.

Phylum Tracheophyta, the Vascular Plants

The Tracheophytes are the plants that give the earth its main green covering. This group, including the ferns, the conifers, and the flowers, contains the most recent, the most complex, and the most successful plants ever to have evolved. The feature that gives the Tracheophytes their name is the system of vessels and associated cells that serve for conducting water and dissolved foods and minerals throughout the plant. The *tracheae* (singular, trachea) of plants are the water vessels, and they are so called because under a microscope they look like the tracheae or breathing tubes of animals. When plants during their evolution from aquatic to terrestrial life began to live on land, they had to have several features if they were to survive. They needed stiffening if they were to stand up and obtain sunlight, an anchoring structure to hold them firmly in the soil, a way to absorb water and minerals, a system for conducting water up to the aerial parts and transporting food down to the underground parts (which were necessarily in the dark), a waterproofing cover to prevent excess drying, and some means of allowing photosynthetic and respiratory gas exchange. These needs were met by the development of fibers and vessels of the vascular system, roots or rootlike parts, a waxy cuticle over stems and leaves, and stomata with regulatable openings. No plant can live on land and stretch its leaves far into light and air without these.

The first vascular plants were probably somewhat fernlike and still needed liquid water as a medium for transport of swimming sperm in fertilization. Eventually, however, even that need was overcome, and now the most advanced plants (the conifers and flowering plants) are free from the necessity for water when seeds are produced.

The earliest tracheophytes The earliest fossils of vascular plants occur in rocks of the Silurian Period of the Paleozoic Era, well over 400 million years ago. An instructive collection of fossils (Latin, "something dug up") comes from Rhynie, Scotland, and are named Rhynia. Those old Devonian plants were cleanly preserved, and they show such cellular details as a cuticular coat, stomata in the stem epidermis, absorptive rhizoids like root hairs, and conducting cells in the center of the stems (Figure 9.32). Rhynias were

9.31

Spore dispersal in a moss, as seen by scanning electron microscopy. (a) The tip of a moss capsule is covered by a cap until the spores are mature. (b) The cap falls off. The opening of the capsule is surrounded by a ring of *peristome teeth*, which change position with changes in moisture. They can curl down into the spore cavity or straighten, lifting spores out into the air. The spores thus freed can blow to new growing places. Shapes, sizes, and markings of capsules and peristome teeth are characteristic of each species of moss.

9.32

A reconstruction of Rhynia from 400-million-year-old fossils. It had horizontal underground stems as well as erect ones. Sporangia were borne at stem tips, but there were neither leaves nor roots. Water and mineral nutrients were taken in by hairlike rhizoids, and food was made in the photosynthetic stems. Slightly less than natural size.

(a)   × 27

(b)   × 33

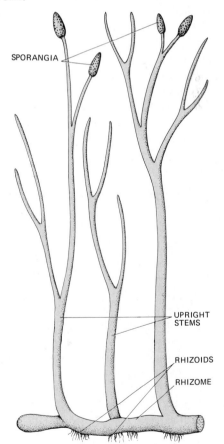

(a)   × 1

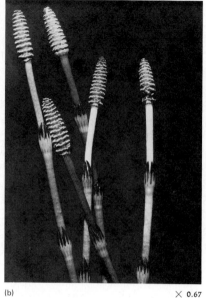

(b)   × 0.67

(c)   × 0.25

9.33

Representatives of some primitive, living vascular plants. (a) Psilotum, a living genus that resembles the long-extinct Rhynia, giving an idea of what the first land-dwelling vascular plants were like. It has green forked stems bearing the lobed sporangia visible in the photograph. The leaves are tiny scales that are photosynthetically unimportant. There are underground stems as well as the aerial ones, but there are no roots, simply rhizoids. One species of Psilotum that grows wild in Florida, where it is known as whisk fern, is indeed considered by some botanists as a true but primitive fern. (b) Horsetails. The one genus, Equisetum, is widely distributed, but it varies in size from a few centimeters to more than a meter and differs in the method of spore production. In the species shown here, the photosynthetic portion of the plant has whorls of green stems and tiny scales (photosynthetically useless) for leaves. The vegetative structures do look something like tails of horses. The reproductive portion of this species, which is brown rather than green, is unbranched. Spores are produced in cones, each consisting of a hundred or so *sporophylls*, visible in this photograph as small, pale lumps. (c) The club mosses of the genus Lycopodium, commonly called ground pine or ground cedar. They are low-growing evergreen plants that were much used for Christmas wreaths until they became scarce and were put under legal protection. The cones in this species, standing erect, bear spirally arranged, overlapping sporophylls with sporangia, in which the spores, like yellow dust, are produced by the millions.

small, half a meter or less tall, and they were about as simple structurally as a vascular plant can be. They had no leaves, no roots, and only simple two-pronged branching. At some branch tips they bore sporangia with spores. Since still earlier vascular plants are known, plants like Rhynia are not thought to be the ancestors of modern ferns and seed plants. However, they are useful in giving us an idea of how primitive land plants looked. Since a number of groups of vascular plants are still alive but not clearly related, the current opinion is that there may well have been several occasions on which aquatic plants, probably green algae, gave rise to terrestrial forms, some of which are with us yet.

Of the small subphyla of the Tracheophyta that are not commonly known, three should be mentioned even in a brief survey. They are the Psilopsids, the Lycopsids, and the Sphenopsids. The Psilopsids are represented possibly by the Florida whisk fern, *Psilotum nudum*, which resembles the fossil Rhynia but is thought more likely to be a primitive fern (Figure 9.33). The Sphenopsids now living are the horsetails of the Genus Equisetum. The Lycopsids are the club mosses, Lycopodium, and the little club mosses, Selaginella. All these are slender remnants of plant groups that were once the important plants on earth. During the Carboniferous Period some 300 million years ago, the giant Lycopods and Psilopsids were forest

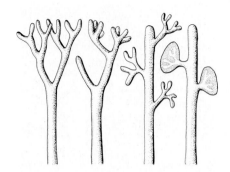

9.34
The presumed evolutionary development of leaves such as are common to most modern vascular plants. According to this reconstruction, the earliest forms were something like modern Psilotum, with negligible leaves and green forked stems. Later, stem tips grew in clusters, and the clusters flattened themselves. Finally, tissue grew to fill in the spaces between the flattened branch clusters, making a webbed blade such as now functions in ordinary leaves. The fact that fossil examples of all stages can be found is no proof that the sequence of stages as shown here actually happened, and many questions about the evolution of leaves remain unanswered.

trees, whose dead bodies yielded the coal for which the period was named. A few species have come down to the present, a reminder of what plants were before the development of flowers, but they apparently never gave rise to anything more advanced. Like the Bryophytes, they must be regarded as evolutionary dead ends.

Subphylum Pteropsida

The large group of the Pteropsida includes the great majority of the familiar plants of the world. Even though the examples range from ferns through pines and giant redwoods to grasses, lilies, roses, cacti, and sunflowers, they have two general features that convince taxonomists they are all in the same great evolutionary group. One feature is the production of spores on spore-bearing leaves (Chapter 15). The other is the possession of specially formed, veiny leaves, called *megaphylls*, meaning "large leaves." Megaphylls are indeed larger generally than the *microphylls* of some plants, such as the Lycopsids, but the critical points have to do with their veins and the way in which they are borne on branches. The veins of megaphylls run through the leaf blade either as a network or as more or less parallel strips, and the leaves themselves are presumed to have evolved as the fusion of a branch system (Figure 9.34). Further, there is a break in the vascular supply of the branch that produces a megaphyll. Instead of continuing up the branch, the vessels at the site where a leaf grows (the site is called a *node*) are interrupted for a short distance before they rejoin above the node. The break in the vessels is called a *leaf gap*, and is so characteristic of the way leaves are borne on branches in all the higher plant groups that it is taken as an indication that these plants belong in one great evolutionary category (Figure 9.35).

Class Filicineae, the ferns   In the algae it was chloroplasts with which evolution played a game of theme-

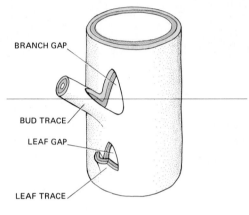

(a)

(b)

9.35
The connection of a leaf to a stem. (a) A diagram of a node, with the leaf and outer covering of the stem cut away to show the vascular supply to the leaf (the leaf trace) and the break in the vascular cylinder of the stem (the leaf gap). Above the leaf gap there was a bud (a potential branch of the stem) with its own vascular supply (bud trace) and branch gap. Such an arrangement is typical of many stems and leaves. (b) A section cut across a stem through a bud, showing the branch gap with its distinctive break in the ring of woody tissue. Sections cut above or below the node would show a continuous woody ring.

(a) × 0.08 (b) × 1 (c) × 0.1

and-variations. In the ferns it was leaves. Fern leaves range in size from tiny cliff brakes only a few centimeters tall to the lacy sheets of tree ferns three meters (10 ft) long (Figure 9.36). In compounding they may range from the simple flat frond of a hart's tongue fern through the once-divided Christmas fern and the twice-divided lady fern to the filigree of five-times divided Davallias. The stems of ferns are frequently horizontal, growing along under the soil, following the contours of the surface, branching at the tips, and spreading away from their old beginnings. After many years a fern may expand into a circle of advancing stem tips, each with a cluster of leaves coming up to make a "fairy ring." Internally, fern stems have vessels for the conduction of water and food, plus an abundance of thick-walled strengthening fibers, but different species of ferns have such different arrangements of these tissues that no general description is possible. One example shows the complexity of the arrangement (Figure 9.37). For the first time in this survey, we meet in ferns real roots with well-defined conducting tissues, not the hairlike structures of mosses or the ancient Rhynias.

Ferns produce their spores in sporangia on the leaves (Chapter 15). Sometimes the sporangia are in clusters on the lower surfaces or along margins, at tips of ordinary vegetative leaves, or on special leaves whose only function is spore formation (Figure 9.38).

People find ferns useful as ornaments because of the elegance of their leaves. On the minus side, the tough, inedible bracken is responsible for occasional infestations of pasture land. Ferns are of little importance in the human economy, although one minor

9.36

Ferns. (a) A mosaic of leaves of the bracken, Pteridium, widely distributed throughout the north temperate part of the world in fields and forests. Bracken is larger and coarser than most ferns, some of which are only a few centimeters tall and as delicate as tissue paper. (b) Instead of unfolding as leaves of flowering plants do, fern leaves unroll in characteristic fiddle heads, or *croziers*, shown in cinnamon ferns, Osmunda. (c) Ferns grow to be small trees in warm climates. These tree ferns, *Cibotium*, are in Hawaii, but some can survive as far north as Ireland if planted in protected places. A single, finely cut leaf may be as much as two meters long.

value is that well-rotted fern roots are the medium of choice for growing orchids.

Class Coniferae, the conifers   The Coniferae are variously called the conifers, the evergreens, or the softwoods, but all these names are inaccurate in part. Coniferae do not all bear cones; yews have red "fruits" that look like berries. They are not all evergreen; larches and bald cypresses are deciduous. And the wood is not necessarily soft; yellow pine is considerably harder than the so-called hardwood of basswood.

All conifers are woody plants. (How many such absolute statements can you find in this book?) Their general manner of growth by means of extension at stem and root tips and expansion of cambium under the bark is similar to that of flowering plants (Chapter 16), but the cellular structure, that is, the *histology* of the wood, in one important respect is not like that of flowering plants. The water-conducting cells are thickened and elongated, but they are not open-ended like sections of water pipe. Instead, the cells, specifically called *tracheids,* are tapered at the ends, and although they have thin pits in the walls, there are no actually open passages through which water can pass from cell to cell. In contrast, *tracheae* are open-ended water vessels, but they did not evolve until the evolution of the flowering habit.

Leaves of conifers are frequently needle-shaped, as they are in pines, firs, and spruces, but they may be scalelike, as in cypresses, arbor vitae, and white cedar, or they may be broad-bladed, as in Podocarpus. Most conifer leaves remain on the tree for more than one season, over-wintering in a green condition and thus appearing "evergreen." Any single leaf usually lasts only two summers, but some trees keep leaves for many years, with some of the ancient bristlecone pines having leaves that are 30 years old. The leaves are generally adapted to dry conditions, being well waxed, impregnated with such antidrying agents as resins, having stomata in depressed pits, and having relatively little surface exposed to the air. In pines and some others, needles are borne in bundles of two, three, five, or—rarely—one, at the ends of stubby branches that are so short as to be inconspicuous. Larches, which shed their leaves in autumn, have such definite leaf-bearing branches that the short spurs give the trees a characteristic spiny look during winter.

Reproduction in conifers. Conifers are nonflowering seed plants (Chapter 15), but they are not the only ones. They are loosely known as *gymnosperms* (Greek, "naked seeds") because the seeds are not enclosed in a covering tissue but are borne on the surfaces of spore-bearing leaves, *sporophylls*. In pine cones, to be sure, the ovules that develop into seeds are effectively tucked between the cone scales, but they are still not encased in a complete ovary, such as can be found in flowering plants. The seed in a gymnosperm is like a coin grasped in a fist: out of sight and protected, but not morphologically inside the hand in the same sense that a finger bone is inside.

As its name suggests, the conifer literally bears cones, of which pine cones are familiar examples. Those usually seen are the ovulate cones, or seed cones, which bear the female gametophytes and even-

9.37

Fern stems are usually horizontal and grow under the soil surface. They show a number of patterns in their vascular systems, but none have a cambium like that of seed-bearing trees. Instead, an apical growing point, usually with a single mitotically active apical cell, produces the tissues of the stem, or *rhizome*. In this cross section of a rhizome of the bracken fern, Pteridium, there are several large vascular bundles and a number of smaller ones, each containing a solid wood core. In the larger ones, food-conducting tissue surrounds the wood, and numerous tough fiber cells give the rhizome strength.

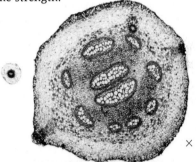

× 8

× 2

9.38

Spore production in a fern. The underside of a leaf in some genera (as here in the wood fern, Dryopteris) has clusters of sporangia, or *sori* (singular, sorus), covered with thin scales, the *indusia*. In this specimen, each sorus with its indusium looks like a raised bump. Sporangia are too small to show at this magnification, but they are in fact crowded under the indusia.

9.39

Branch and ovulate cone of Douglasfir, Pseudotsuga. The needlelike leaves and scaly cones are typical of many conifers, but the sharp bracts sticking out from between the cone scales are peculiar to this genus. Douglasfirs, among the most impressive trees in the world, furnish a large proportion of the lumber cut in the United States. They are especially useful in making plywood. About natural size.

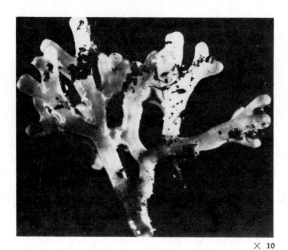

$\times$ 10

9.40

Mycorrhizal growth on roots of scrub pine, *Pinus virginiana*, showing the gnarled, distorted growth typical of such associations. Most conifers have fungi growing in or on their roots, and some will not grow without the fungus, which aids in procurement of mineral nutrients for the tree.

tually the embryos (Figure 9.39). Cones vary in size from small hemlock cones, no bigger than a fingernail, to massive cones of the sugar pine, *Pinus lambertiana*, the size of a small watermelon. They also vary in structure from thin, long, and dry, as in eastern white pine, *Pinus strobus*, to the berrylike balls of red cedar, *Juniperus virginiana*, whose cone scales are so plump they mask their really conelike nature. The *staminate cones*, which produce pollen, are much smaller, usually only a few millimeters long, and short-lived; they appear for a few weeks in spring and then dry and fall.

Distribution of conifers. Unlike many old plants that were crowded off the earth when more advanced species evolved, the conifers are thriving. After first appearing in the Permian Period of the Paleozoic Era about 400 million years ago, they multiplied and spread until by the Triassic Period they had formed large forests, as they still do throughout the northern hemisphere. Although conifers are poorly represented in lands south of the equator, they are so numerous in parts of the three northern continents that they give the landscapes their characteristic appearance. The state of Washington calls itself "The Evergreen State" because of its forests of conifers.

Among trees, conifers are record breakers. They are the tallest, with redwoods, *Sequoia sempervirens* reaching 117 meters (385 ft). (There are unconfirmed reports of taller eucalyptus trees in Australia.) They are the thickest. The Big Tree of Tule in Oaxaca, Mexico (*Taxodium mucronatum*), about 40 meters (132 ft) in diameter, was already revered by the Indians for its size when Europeans first saw it. Conifers are also the oldest, with some bristlecone pines, *Pinus aristata*, in the White Mountains of California having lived for 4600 years. These old trees are distorted and slow-growing, but in general, conifers grow fast and straight. Consequently they are valuable for timber.

Uses of conifers. The wood of the pines, spruces, firs, the various woods known as cedar, the redwoods, hemlocks, cypresses, and yews furnish the bulk of the lumber cut for general construction and other commercial purposes. In addition, conifers provide the 40 million tons of pulpwood used annually in the United States in the manufacture of paper. No other material has ever been found to have the resonant quality of spruce wood in the sound boards of pianos and the tops of violins.

Conifers and fungi. Pines—and indeed many other kinds of plants up and down the evolutionary scale—enter into symbiotic relations with fungi. In the soil,

9.41

Two representative conifers. (a) A giant sequoia, *Sequoiadendron giganteum*, in the Sierra Nevada mountains of California. For sheer bulk, giant sequoias surpass any organism that has ever lived (unless larger ones left no fossil trace). (b) A cypress swamp in Arkansas, with buttressed trees (Taxodium). These are not true cypresses of the genus Cupressus; they are deciduous conifers, shedding their needles during winter. They commonly grow in water and send up "knees" from their submerged roots, presumably as suppliers of oxygen for the cells below the surface.

(a)   (b)

a fungal mycelium penetrates a root and may grow between the cells of the host root or actually probe inside root cells. Such a fungus is called a *mycorrhiza*, from the Greek meaning "fungus root" (Figure 9.40). It obtains food and protection from the host and in return makes easier, or even possible, the entry of minerals into the root. Some pines are so dependent on their mycorrhizas that they cannot thrive without them, and emigrating peoples have been known to carry with them a little home soil to add at the bases of transplanted trees because they had learned from experience that something essential was contained in it.

Conifers have impressed people in a way that no other plants have. The "murmuring pines and the hemlocks" really do murmur, or sigh or whisper when wind passing over a billion needle leaves makes a soft vibration. The early colonists on the east coast of America were long mystified by the subtle perfume that exudes from a warm pine forest. The use of Christmas trees attests to the deference paid to a thing that can remain visibly alive during the darkest part of the year. Visitors have been moved emotionally by the great white pine forests of New England —all gone now, destroyed by the same people who admired them—by the huge Douglasfirs of the Pacific Northwest, the incomparable California redwoods, the swamp cypresses, dark and draped with gray Spanish moss, and the strange wilderness of the Jersey pine barrens (Figure 9.41).

Class Angiospermae, the flowering plants To most people, "a plant" means "a flowering plant." Since

they first evolved in the Cretaceous Period nearly 200 million years ago, flowering plants have become increasingly dominant as the main flora of the earth, until now they make the most conspicuous feature of the landscape. Except for seas, deserts, some northern forests, and a few artificial spots, such as major cities, the *look* of the world is determined by the angiosperms (Greek, "vessel seeds"). They are the basis of practically all human economy, furnishing the main requisites for survival in the form of food, clothing, and shelter. Even if we were not dependent on them for necessities, a world without flowering plants would be a poor place. Therefore the emphasis in such a book as this is on angiosperms, so much so that instead of treating them in a general survey of the plant kingdom, we give them special treatment, especially with respect to their structure and development (Chapter 16) and their physiological functions (Chapter 17). Their reproductive activities will be described in Chapter 15, and their photosynthetic ability was covered in Chapter 4.

The angiosperms have evolved in two lines, the Monocotyledon and the Dicotyledon line. These groups are named for the presence of one or two seed leaves, cotyledons, in the embryos (Chapter 15). The number of cotyledons is distinctive, but it is only one of a number of features that characterize the two plant groups. The names are commonly shortened to "monocot" and "dicot." In addition to having only one cotyledon, monocots usually have the veins in their leaves more or less parallel along the long axis. Their vascular bundles are distributed as small, isolated strands of cells running up the stems. They

9.42

Representative flowers. (a) A lily, Lilium, a monocotyledonous flower, showing the threefold nature of the flower, with six stamens, three sepals, and three petals (the sepals and petals look alike). (b) A dicotyledonous flower of the rose mallow, *Hibiscus moscheutos.* The five-parted corolla, typical of dicots, is evident. The central mass of stamens surrounds the style, which terminates in five stigmas.

(a)

(b)

usually grow little or not at all in girth, because they lack a lateral mitotic layer of cells (cambium). And their flower parts occur in threes or multiples thereof (Figure 9.42). In contrast, dicots have networks of veins in their leaves. Their vascular system consists of a central cylinder. They do have a lateral cambium, which contributes to increase in girth. And their flower parts come in fours or fives or multiples thereof.

Of the several hundred families of angiosperms, only a few are of economic importance or are well known as ornamental plants. In the monocot group, the grass family, the Gramineae, is not only the source of most of the human food in the world (wheat, rice, corn, rye, barley, millet, sorghum, sugar cane) but covers vast areas of the earth in grasslands that feed uncounted wild and domestic grazing ani-

mals. Other well-known monocots are the lilies, palms (including coconut, date, and oil palms), yams, and orchids. Among dicots are the oaks, maples, legumes (beans, peas, clovers, locusts), roses (not only the flowers, but apples, pears, peaches, almonds, plums, prunes, strawberries, blackberries), heaths (heather, blueberries, cranberries), asters in the broad sense (sunflowers, artichokes, goldenrods, dandelions), mustards, and mints. Students who live in temperate regions should not be deceived by the low number of plant families native to their home territory. The real diversity of evolution manifests itself in the tropics, where there is not only a wealth of families, but within families there are untold thousands of evolutionary "experiments," making the angiosperms the most numerous (in number of species) and most diverse plants on earth.

RECOMMENDED READING

Adler, J., "The Sensing of Chemicals by Bacteria." *Scientific American,* April 1976.

Ainsworth, G. C., *Introduction to the History of Mycology.* New York: Cambridge University Press, 1976.

Barghoorn, E. S., "The Oldest Fossils." *Scientific American,* May 1971. (Offprint 895)

Boedijn, *Plants of the World: The Lower Plants.* New York: E. P. Dutton, 1969.

Briggs, D., and S. M. Walters, *Plant Variation and Evolution.* New York: McGraw-Hill, 1969. (Paperback)

Chrispeels, M. J., and D. Sadava, *Plants, Food, and People.* San Francisco: W. H. Freeman, 1977. (Paperback)

Costerton, J. W., G. G. Geesey, and K. -J. Cheng, "How Bacteria Stick." *Scientific American,* January 1978.

Delevoryas, T., *Plant Diversification,* Second Edition. New York: Holt, Rinehart, and Winston, 1977. (Paperback)

Jaques, H. E., *Plant Families: How to Know Them,* Second Edition. Dubuque, Iowa: Wm. C. Brown, 1949. (Paperback)

Miller, Orson K., Jr., *Mushrooms of North America.* New York: E. P. Dutton, 1977.

Nester, E. W., C. E. Roberts, N. N. Pearsall, and F. J. McCarthy, *Microbiology: Molecules, Microbes, and Man,* Second Edition. New York: Holt, Rinehart, and Winston, 1978.

Raven, Peter H., Ray F. Evert, and Helena Curtis, *The Biology of Plants,* Second Edition. New York: Worth, 1976.

Tippo, Oswald, and William L. Stern, *Humanistic Botany.* New York: W. W. Norton, 1977.

Weier, T. E., C. R. Stocking, and M. G. Barbour, *Botany,* Fifth Edition. New York: Wiley, 1974.

10
The Living Kingdoms: The Animals

ANIMALS ARE TRADITIONALLY CREATURES THAT breathe. The very word "animal" refers to the air or wind that is sucked in and blown out. We call a wind gauge an *anemo*meter and a wind flower an *anem*one. A thing that does not breathe is in*animate.* If we say that an animal *respires*, we are saying the same thing with a different word, because *spirit* also refers to air. A *spirited* or *animate* organism pumps wind. Until fairly recently, only those beings specifically called animals were believed to respire, although we know now that all cells respire, and we have even adopted a useful but self-contradicting phrase, an-aerobic respiration—that is, wind without wind. Considered mechanically, many animals do not breathe. Only those with lungs or a tracheal system, as in insects, actually inhale and exhale; those with gills pick up their oxygen directly from water. Animals, then, cannot be defined in terms of respiration. Safer standards of classification include the practice of eating and the presence of a nervous system (Chapter 20), an electronically driven cellular network that makes fast communication possible throughout the body of the organism. Even the possession of a nervous system is not an absolute criterion, because some of the simplest creatures, such as sponges, do not have a nervous system and are yet accepted as animals. (Some people have claimed that plants have a kind of electrical internal communication, but no cellular counterpart to an animal nervous system has been demonstrated.) Most animals do breathe and have a nervous system, and most also eat and move, even though "most" clearly does not mean "all." The lack of a completely satisfactory textbook definition does not keep us from learning what we can about animals.

The system of classification of animals used in this book is a simplified one, which omits mention of some of the smaller and less well-known phyla. The animals to be considered here will be described according to the following outline, in which phyla, some subphyla, and classes will be named, with English names of examples given in parentheses.

Phylum Porifera (sponges)
    Class Calcarea (calcareous sponges)
    Class Hexactinellida (glass sponges)
    Class Demospongiae (bath sponges)

Phylum Coelenterata
    Class Hydrozoa (polyps)
    Class Scyphozoa (jellyfishes)
    Class Anthozoa (corals and anemones)

Phylum Platyhelminthes (flatworms)
    Class Turbellaria (planaria)
    Class Trematoda (flukes)
    Class Cestoda (tapeworms)

Phylum Aeschelminthes (roundworms)

Phylum Mollusca (molluscs)
    Class Amphineura (chitons)
    Class Scaphopoda (tooth shells)
    Class Gastropoda (snails)
    Class Pelecypoda (clams)
    Class Cephalopoda (squids)

Phylum Annelida (segmented worms)
    Class Polychaeta (clam worms)
    Class Oligochaeta (earthworms)
    Class Hirudinea (leeches)

Phylum Arthropoda (arthropods)
  Subphylum Trilobita (extinct trilobites)
  Subphylum Chelicerata
    Class Merostomata (horseshoe crabs)
    Class Arachnida (spiders)
  Subphylum Mandibulata
    Class Crustacea (shrimp)
    Class Chilopoda (centipedes)
    Class Diplopoda (millipedes)
    Class Insecta (insects)

Phylum Echinodermata
    Class Crinoidea (sea lilies)
    Class Asteroidea (starfishes)
    Class Ophiuroidea (brittle stars)

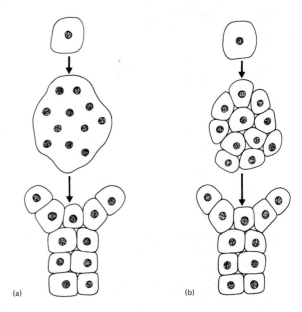

(a)   (b)

10.1
Two ways in which metazoan (multicellular) animals may have arisen. (a) If a single-celled animal underwent nuclear divisions without cytoplasmic divisions, the result would be a multinucleate organism. Then if each nucleus separated itself from the others by a surrounding membrane, a multicellular animal would be formed.
(b) A second possibility is the formation of a loosely associated mass of cells following cell divisions of a single cell. If the association became firmly established, the former colony of individuals would become a single multicellular animal. Reasonable arguments support both guesses, but no absolute answer is available.

    Class Echinoidea (sea urchins)
    Class Holothuroidea (sea cucumbers)

Phylum Hemichordata (acorn worms)

Phylum Chordata (chordates)
  Subphylum Urochordata (sea squirts)
  Subphylum Cephalochordata (lancelets)
  Subphylum Vertebrata (vertebrates)
    Class Agnatha (lamprey eels)
    Class Chondrichthyes (sharks)
    Class Osteichthyes (bony fishes)
    Class Amphibia (frogs)
    Class Reptilia (snakes)
    Class Aves (birds)
    Class Mammalia (mammals)
      Subclass Prototheria (platypus)
      Subclass Metatheria (opossum)
      Subclass Eutheria (the "beasts," human beings)

10.2

Body cavity, or lack of it, in animals. (a) In flatworms, ancient and simple animals, there is an outer ectoderm, an inner endoderm lining the digestive system, and a loosely formed mesoderm between. There is no "empty space" in the body, that is, no coelom, or body cavity. (b) In roundworms, the ectoderm, mesoderm, and endoderm are present, and in addition there is a coelom, but the coelom is not completely enclosed in mesoderm. (c) In most animals, the coelom is a "space," with mesoderm covering the endoderm-lined alimentary tract and lining the outer covering of ectoderm.

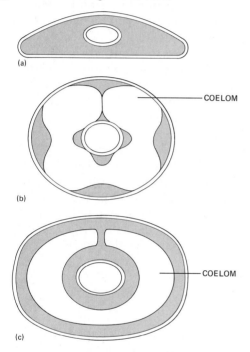

BASES OF ANIMAL CLASSIFICATION

Biologists can work directly only with individual organisms, but after they have observed thousands of individuals, patterns of similarity become apparent. The similarities serve as bases for the general classification schemes. In the arrangement of animal phyla and the lesser subdivisions, several characteristics are used as indicators of relations. These characteristics are considered basic enough to tell us which groups of animals are closely related.

1.   The first decision that must be made concerns the cellular nature of the organism. If it is one-celled (or if it is considered noncellular), it belongs with the protozoa, which in this book are classed with the Protista (Chapter 9). If it is a multicellular individual, it belongs with the metazoa, which are all the animals

to be treated in this chapter (Figure 10.1). The evolutionary origin of the metazoa is uncertain. There is little argument over the clear likelihood that they came from some protist ancestor. But here agreement ends. One possibility is that one-celled animals remained in contact after dividing, forming a mass of cells that began to function as a single entity, finally becoming an integrated, single organism. A second possibility is that a one-celled animal underwent repeated nuclear divisions without cytoplasmic division, so that a large, multinucleate protoplasmic mass began to function as a single organism and subsequently interposed cell membranes between nuclei, thus becoming a multicellular organism. Still a third suggestion is that multicellular animals developed from multicellular, colonial-type plants, having lost their photosynthetic ability. Since living intermediates can be found to support each of these possibilities, the question of the origin of the metazoa remains unanswered.

2.   A second feature of fundamental importance is the symmetry of the animal body. As metazoans evolved, they might have grown without any symmetry, simply becoming irregular masses. They might have developed into spheres, a form that has not apparently been successful; even Volvox, which certainly looks spherical, has some difference between one side of the ball and the other. A third "choice" was radial symmetry, that is, the symmetry of a wheel. However, radial symmetry has not been very successful, either. When animals became more motile, the radially symmetrical animal could not glide easily through water. Now only a few animals, such as jellyfishes, are radial—at least, in the adult stage—and they are classified in a phylum thought to deviate from the main lines of animal evolution. The final possibility was bilateral symmetry, in which the animal is two-sided and the sides are mirror images of each other, or nearly so. With bilateral (Latin, "two-sided") symmetry came differentiation along other planes than the one that delineates the two sides. Swimming animals, which the early metazoans almost certainly were, came to have a head end and a tail end, as well as a top (or dorsal) side and a bottom (or ventral) side. Once these several axes were established, there was also a right and a left. In our synopsis of the animal kingdom, everything above Phylum Coelenterata has bilateral symmetry.

3.   A third characteristic of animal bodies that is useful in delimiting large groups is the gross structure of the body itself. In the familiar higher animals, there is a body cavity called the _coelom_, which contains many of the main organs (Figure 10.2). Some of

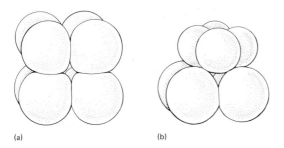

(a)   (b)

10.3

Cleavage patterns. (a) An animal zygote begins to form
an embryo, and after three cell divisions, it has eight cells.
If at this stage one cell is directly above a lower one, the
procedure is called *radial cleavage,* and it is characteristic
of the animal phyla that include insects and clams. If
the cells are separated from one another, they cannot
develop into separate, complete animals. (b) A different
method, *spiral cleavage,* results in an eight-celled embryo
in which one cell is not directly above a lower one. Spiral
cleavage is characteristic of the two animal phyla that
include starfishes and mammals. If separated from one
another before too many cell divisions have occurred,
the cells of an embryo using spiral cleavage may produce more than one complete animal. These two methods
of cleavage are so constant and so dependable that they
are thought to indicate a fundamental difference between
the kinds of animals that employ them.

the primitive animals (the flatworms, for example)
have no such cavity; they have an outer epidermal
covering, a mass of tissue beneath that, and an inner
digestive tract. In our synopsis, all animals above
Phylum Platyhelminthes have some sort of coelom.
It is necessary to say "some sort of coelom" because
some animals have a variation of the typical coelom
of, say, vertebrates.

4. A fourth mainly embryological pattern serves to
separate the coelomate animals into two great groups.
During development, the first opening into the embryo becomes the mouth in the protostomes (Greek,
"first mouth") and the anus in the deuterostomes
(Greek, "second mouth"). The deuterostomes are
Phylum Echinodermata, the starfishes and their kin,
and the Chordata, the phylum in which humans are
included. The protostomes are all the rest.

Associated with the method of mouth formation
is the way in which the cells of early embryos behave.
In protostomes, the first few cells are arranged in a
somewhat *spiral* fashion, and they are <u>*determinate*</u>;
that is, the cells at an early stage are so prearranged
that they are destined to become some specific part
of the future animal (Figure 10.3). You cannot break
up the cells of a protostome embryo and obtain
several separate and complete animals. The deuterostomes, on the other hand, generally have *radial*

10.4

Methods of alimentary tract formation in different animals. (a) After cleavage, when an animal embryo undergoes the infolding process known as *gastrulation,* it
forms a hole where the infolding occurred, the *blastopore.* (b) In some animals—for example, molluscs and
arthropods—when the alimentary tract is completed, the
mouth is formed where the old blastopore was, and the
anus breaks through at the opposite end. This is the
protostome method of development. (c) In the echino-

derms and chordates, the anus is formed where the old
blastopore was, and the mouth breaks through at the
other end of the alimentary tract. This is the *deuterostome*
method of development. Like the method of cleavage,
the method of mouth and anus formation is regarded
as a fundamental aspect of animal development. It was
established before the evolution of two main lines of
animal descent, and it separates the protostome and
deuterostome phyla.

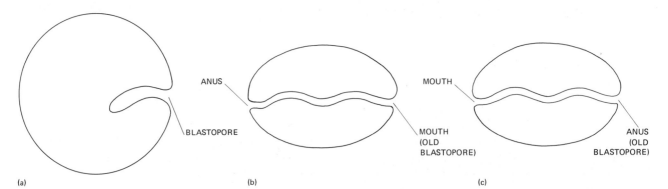

ANUS

BLASTOPORE

MOUTH
(OLD
BLASTOPORE)

MOUTH

ANUS
(OLD
BLASTOPORE)

(a)   (b)   (c)

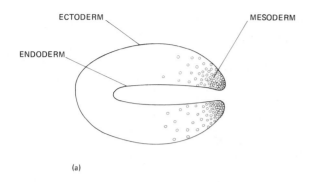

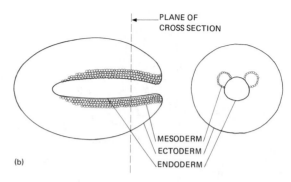

10.5

Origin of the mesoderm in protostomes and deutero-stomes. (a) In protostomes, in which the blastopore is destined to become the mouth of the adult animal, meso-derm rises as an outgrowth of cells in the region of the blastopore where ectoderm meets endoderm. The meso-derm cells divide and migrate, eventually filling the space between ectoderm and endoderm. (b) In deuterostomes,

in which the blastopore is destined to become the anus of the adult animal, mesoderm arises as an outpocketing of the endoderm. In a long section, it can be seen that mesoderm matures from the blastopore region toward the other end of the embryo. In a cross section, pockets of mesoderm can be seen growing out from the endoderm (the future gut).

cleavage, and the cells of the early embryo are _inde-terminate_; that is, there is no preordained part of the future animal that will necessarily come from any one of those cells. You can separate cells, called _blasto-meres_, of an early deuterostome embryo, and they can grow into complete, separate animals (Figure 10.4). As an aid to remembering this, think of the possibility of identical human twins. People are deu-terostomes and so are starfishes, regardless of how different they seem.

Besides the difference in the position of mouth and anus in the protostomes and deuterostomes, which gives the two groups their names, there is a difference in the way the between-layer of mesoderm grows (Figure 10.5). In the protostomes, the meso-derm begins near the blastopore, and the coelom re-sults from the fusion of cavities in the mesoderm. In the deuterostomes, the mesoderm begins as growth from the embryonic gut, and the coelom results from a split between the inner and outer layers of meso-derm. There are further differences between the pro-tostomes and the deuterostomes, especially in the larval forms of those animals that have larvae, but the differences given here suffice to show that there is a fundamental separation between the two groups. Biologists believe that such difference and such con-stancy of difference indicate an evolutionary relation of long duration, and consequently the protostomes and the deuterostomes are represented in schemes of

classification as being separate branches of the evo-lutionary tree.

Section 1:
Sponges Through Insects

PHYLUM PORIFERA, THE SPONGES

The Porifera, meaning the "pore bearers," are among the simplest of animals. They do form definite bodies, some with complex systems of water channels, and there is some cellular differentiation. They can build some fantastically intricate forms, but the elementary nature of their cellular relations is shown by the fact that an entire sponge body, after being broken up and passed through a fine sieve, can reassemble itself into a functional animal. The original demonstration, one of the classic experiments in developmental biol-ogy, was performed in 1907 by H. V. Wilson, the same biologist who once chided the senior author of this book (when he failed to arrive at the laboratory on time), "If you are going to be a nonconformist, you had also better be a genius." In obtaining food, a sponge drives water in through pores in the sides of the body wall, creating currents by the beating of

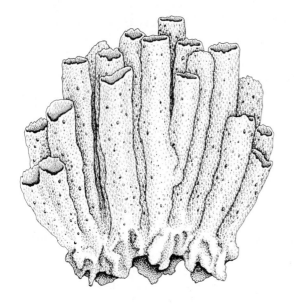

10.6
A tube sponge, *Callyspongia vaginalis*. Water containing bits of food is drawn in through the body walls by the action of flagellated cells. The usable parts are caught, and excess water and unused material are pushed out through the openings at the top. × 0.5.

flagella (Figure 10.6). The flagellated cells have rings around the bases of the flagella. Such cells are called *collar cells*, and they resemble similar cells in some protozoa. As water passes through the sponge pores, particles of food are captured and taken into the feeding cells. The rest of the water is pumped into the central cavity of the body and then out the top through an opening, the *osculum*.

There are three classes of sponges. The Calcarea have needle-shaped lime crystals scattered throughout the middle cell layers. They are marine animals with no uniformity of shape, and they vary in the arrangement of their pores. The Hexactinellida also make crystalline skeletal elements, *spicules*, but of siliceous material (hence the common name for the group, glass sponges) and usually with six-parted geometry (Figure 10.7). Some of the Hexactinellida (Greek, "six rays"), such as Euplectella, the Venus flower basket of Asian oceans, make such delicately woven glassy skeletons that when they were first brought to Europe, people did not believe they were natural objects. The Demospongiae (Greek, "sponges of the people") have a proteinaceous skeleton as well as spicules. The skeleton is the spongy part that is preserved for human use. Sponge fishermen pull the live sponges from the sea, let them die and decay on the beach, and wash away everything but the fibrous, flexible protein, spongin.

Sponges have little organization of function. Most metabolic activities are carried on by individual cells, without any visible coordinating nerve network or internal circulation. Reproduction is by asexual fragmentation or by the sexual union of eggs and sperm.

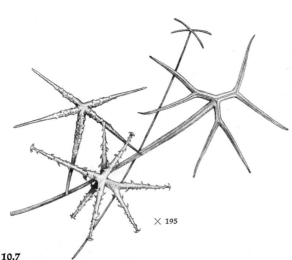

× 195

10.7
Sponge spicules, the siliceous or glassy particles that help support the sponge body. Spicules are formed inside the sponge body and are carried after completion to their permanent place in the body wall, taken there by wandering ameboid cells. Sponge spicules come in a wide variety of shapes. Those shown here are four-rayed, somewhat resembling jackstones or the iron caltrops that used to be thrown against enemy cavalry charges to cripple the horses. In addition to spicules, sponges make a network of proteinaceous *spongin*, which is the substance of commercial bath sponges.

PHYLUM COELENTERATA

The coelenterates (Greek, "hollow intestine") are the polyps, jellyfishes, sea anemones, and corals. They are characterized by their radial body symmetry, the

10.8

The common freshwater coelenterate Hydra. The body and tentacles are soft and extensible, so that the animal can shrink down to a length of three or four millimeters or stretch out to 50 millimeters. Stinging cells in the tentacles can paralyze small animals swimming by, after which the prey is brought to the mouth and ingested. Hydra can remain attached by the basal foot to a solid surface, or it can move about by turning a series of somersaults. In spite of its seeming simplicity in comparison with an insect or a mammal, Hydra has a number of different cell types, indicating that it has progressed a long way from its supposed protozoan ancestors. × 25.

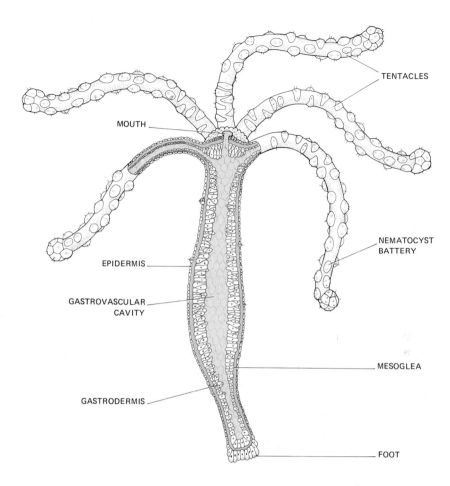

TENTACLES

MOUTH

NEMATOCYST
BATTERY

EPIDERMIS

GASTROVASCULAR
CAVITY

MESOGLEA

GASTRODERMIS

FOOT

presence of stinging cells, a simple network of nerve cells, tentacles, and a saclike _gastrovascular cavity_, in which one opening serves both for intake of food and outflow of waste. Some coelenterates exhibit a phenomenon called _dimorphism_ (Greek "two forms") or _polymorphism_ (Greek, "many forms"), with individuals differing from one another. Such differences are evident in some of the marine colonial species, which look more like plants than like animals, growing as little tufts attached to old shells, pebbles, or pilings. At the ends of some of the branches there are attached, jellyfishlike members of the colony. On other branches of the same colony, "buds" give rise to somewhat different jellyfish. The permanently attached, feeding individuals are called _hydranths_, and their function is nutritive. The budded _medusae_ (singular, medusa) are reproductive and are set free from their branches to swim away. The medusae, themselves produced asexually, bear eggs and sperm, which unite. The zygote then devel-

ops into a ciliated larva called a _planula_ ("a little wanderer"). The attached, tree-shaped stage is the polyp stage, and in some species it is the larger, more conspicuous stage, but in others the reverse is true.

Coelenterata that have a conspicuous polyp stage are in the Class Hydrozoa. One of the best known in the class, Hydra, has only a polyp stage (Figure 10.8). Hydras, common in fresh water (not a usual place for a coelenterate), are used in experiments on regeneration because they can grow new parts when dismembered or mutilated. Also, because of the primitiveness of their nervous systems, they are instructive in experiments on feeding, light and temperature reactions, and tolerance of stresses. They are favorites in biology laboratory courses for observations on their ability to capture moving prey with the stinging cells, the _nematocysts_, with which their tentacles are equipped. The Class Scyphozoa contains the free-swimming jellyfishes, marine creatures ranging from a few centimeters to a meter or more across (Figure

10.9
A free-swimming jellyfish, Chrysaora. It is essentially an inverted Hydra, with an umbrella-shaped body from which the tentacles hang. The mouth is in the center of the underside of the umbrella, shown here as a square at the top of the picture. Digestive canals radiate out from the mouth to the circumference of the body. Like Hydra, jellyfishes have stinging cells in the tentacles, some of them strong enough to be painful and even dangerous to large animals. × 1.

× 1

10.10
Skeleton of a reef-building coral, *Oculina diffusa*. Each of the flowerlike structures is an individual animal, forming part of a large colony. The living parts of the corals have been washed away, leaving only the calcareous hard skeleton.

10.9). In the Class Anthozoa are the sea anemones and corals (Figure 10.10). Corals differ from other coelenterates in secreting hard, limy supporting structures that can be built over a period of centuries into such massive piles as to form habitable islands, the coral atolls of the tropic seas. (Some calcareous algae can also build atolls.)

Other interesting coelenterates are the pink corals of the Mediterranean Sea, which are made into jewelry, and the Portuguese man-of-war. This last named "animal" (*Physalia physalis*) is not an individual but an aggregation of individuals living as a colony, floating about on the surface of the sea, buoyed up and kept moving in the wind by an inflated air float, which shimmers with iridescent colors (Figure 10.11). Drifting below are meter-long streamers provided with deadly nematocysts, which can sting, paralyze, and kill any unwary creature they touch. They can be dangerous even to as large an animal as a human being. Since Physalia is not one animal but a collection, or colony, of animals, one separate tentacle with all its little stinging individuals can remain active after being washed up on a beach. Sitting bare-legged on a man-of-war tentacle is an experience long to be remembered.

PHYLUM PLATYHELMINTHES, THE FLATWORMS

The flatworms are the simplest, in evolutionary terms, of the bilaterally symmetrical animals. The body has an outer epidermis, a digestive tract (in the free-living species) lined with endodermis, and a mass of mesoderm between (Figure 10.12). As the mesoderm is practically full of material, mostly cellular, there is no coelom. Most flatworms have contractile muscle cells and a definite nervous system, with a pair of ganglia at the end, a nerve down each side, and cross-connections like the rungs of a ladder.

The Class Turbellaria includes such free-living flatworms as the planarian, Dugesia, familiar in many biological laboratories as a subject for experimentation on feeding, regeneration, and learning. On being cut into pieces or variously mutilated, planaria can make whole new worms from small parts or heal in bizarre ways, but they are at the same time well-organized animals and can provide information on problems of polarity and cellular differentiation. They have even been trained to learn simple mazes, gliding along on their epidermal cilia, and turning right or left as they have been conditioned to do. To a person accustomed to the idea of a worm with a mouth at the head end, a planarian, which has a mouth in the middle of its underside (but no anus),

is an improbable animal, especially since it has eye spots that give it a comical cross-eyed appearance. Turbellarians live in fresh and salt water, where they feed on debris and small animals. In spite of a general simplicity of body form, turbellarians and flatworms in general have enough specialization of tissues into muscles, nerves, gonads, and digestive tract to be regarded as true organ-forming animals.

A second group of flatworms, the Class Trematoda, includes the parasitic flukes that live in the lungs, livers, and other organs of many animals of many phyla. One of them, the sheep liver fluke, shows to what complex lengths parasites can go in completing their life cycles. In requiring more than one host and in its multiplicity of forms, a liver fluke rivals the red rust of wheat (Chapter 9). Known in its general outlines since the classic research of R. Leukart and A. P. Thomas in 1883, the life cycle of the liver fluke includes, in addition to a planarianlike form that inhabits livers, a second host (an aquatic snail), and several larval stages. The human lung fluke infests still a third host, a crab.

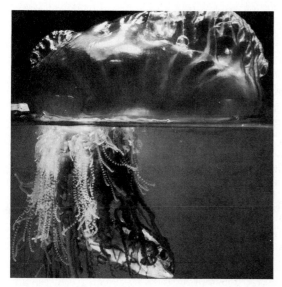

× 0.25

10.11

A coelenterate: the Portuguese man-of-war, Physalia, with a captured fish. These colonial coelenterates live in the warm waters of tropical seas, where they drift about, blown by wind on the sail-like crest.

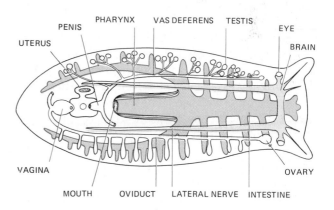

10.12

A flatworm. The dark rectangular structure in the middle of the body is the pharynx with a mouth at its end. The highly branched digestive system is visible, almost filling the body with one main branch toward the head end (right) and two less conspicuous ones toward the tail.

10.13
A tapeworm from a sheep's intestine, *Moniezia expansa*. The slender, tapered end is the head end, with hundreds of segments growing larger and larger toward the older end. The oldest segments are filled with masses of tapeworm eggs.

10.14
A roundworm that lives on the roots of plants. These worms, which inhabit soil, can be destructive to many horticultural and crop plants, causing tumors and destroying the conducting tissues of the hosts. They have no effective means of locomotion, but they whip about violently, sometimes managing to progress forward. Because they are transparent, their internal organs are readily visible when they are alive. Many roundworms (nematodes) cause animal as well as plant diseases.

× 0.5

× 130

The parasitic tapeworms are in the Class Cestoda (Figure 10.13). A human tapeworm is an example of the way in which a parasite can undergo an evolutionary simplification, or, as it is frequently called, degeneration. The body of a tapeworm consists of a head, the *scolex*, provided with suckers for clinging to the host's intestinal wall, and back of the head a string of segments that are little more than bags of eggs. There is an outer covering and a pair of nerves that extend down each side of the flattened row of sections, but there is no mouth, since the animal never eats, and no digestive tract. All metabolic requirements are met by passage of nutrients and waste products inward and outward through the tough epidermis and cuticle. Tapeworms, which infest many animals besides humans, get into their hosts by being eaten, usually while the tapeworms are in an embryonic or larval stage. They can be avoided by practicing clean personal habits, and by not eating worm-bearing meat, especially beef, or by demanding adequate cooking. A human tapeworm five meters (16 ft) long, with thousands of *proglottids* (units) can weaken a human, but such a worm is dwarfed by the broad fish tapeworm, which inhabits humans, fish, and small crustaceans, and which may carry many more thousands of proglottids in its 20-meter (65-ft) length.

PHYLUM AESCHELMINTHES, THE ROUNDWORMS

The roundworms are more complex than flatworms in body structure (Figure 10.14). Flatworms have either no mouth or a mouth located centrally on the underside of the body, but roundworms have a mouth at the anterior end, a tubular alimentary canal, and an anus at the posterior end. The flatworms have a mesoderm between ectoderm and endoderm but no definite coelom in the mesoderm; roundworms do have a body cavity, although it is not such a coelom as more advanced animals have. The roundworm is said to have a *pseudocoelom*, that is, a body cavity that is not lined inwardly and outwardly with mesoderm but has instead a body wall with an outer ectoderm over a muscular mesoderm.

The most notorious roundworms are human parasites: hookworms, which inhabit intestines; trichina worms, which infest pig muscles and are passed to human hosts via incompletely cooked pork; filaria worms, which are carried by mosquitoes and cause the monstrous deformities of elephantiasis; pinworms, which at one time or another infect almost every human being; and the guinea worm. The last is a slender, meter-long parasite that has earned its fame not so much from the damage it does as from the bizarre cure once devised for those afflicted with it. Before modern chemical methods were available, guinea worms were carefully tweezed out from under the victim's skin and slowly wound up on a stick to ease the attacker out.

The roundworms noted above are known from their destructive effects on people, but most roundworms are microscopic, free-living, harmless animals, present in and on animals, plants, soil, and water. They occur in such enormous numbers that it is impossible to avoid eating them. People who study soil or natural waters microscopically meet the transparent, whipping nematodes so constantly that unless they do not object to the idea of swallowing thousands of worms they develop a habit of washing raw food meticulously.

Some roundworms, commonly called nematodes, are destructive to plants. Root-eating nematodes can so thoroughly fill the roots of plants that nothing is left but a squirming mass of worms. With its water supply ruined, the plant suddenly wilts and dies. When infestations are severe, either the soil must be poisoned or the crop must be changed to a species that nematodes do not attack.

10.15

The shell of a clam, *Tridacna squamosa*, in the genus with the giant clams of the South Pacific, which are frequently used for fountains and garden ornaments. These animals are built essentially like the edible clams of temperate oceans, with two shells (which gives them the name *bivalves*) enclosing the soft, living body of the animal. Some Tridacnas grow to be nearly a meter wide and weigh several hundred pounds.

PHYLUM MOLLUSCA, THE MOLLUSCS

The Mollusca (Latin, *mollis*, soft) are named for their soft bodies, which in most molluscs are encased in hard shells. Although molluscs are various in size and shape, they have some common features: a mass of muscle on one side of the body known as the *foot* (easily seen in snails), a fold of tissue over the body called a *mantle*, and a rasping organ, the *radula*, or tongue, with a filelike surface. Molluscs have a well-developed set of functional systems: a digestive tract, effective sense organs coupled with an integrating nervous system, circulation of blood pumped by a heart, excretory organs (*nephridia*), respiratory organs (gills or functional lungs), and gonads. They range in size from almost microscopic snails to the Tridacna clams of the South Pacific Ocean, which grow up to a meter long and weigh 150 kilograms (330 lb) (Figure 10.15). The huge squids of the North Pacific, the largest of the invertebrates, can in 10 years reach a length of 16 meters (52 ft).

Molluscs are important to humans in many ways. We eat clams, oysters, snails, mussels, squids, octopuses, and periwinkles. Many snails are hosts for disease-producing parasites (Figure 10.16). Snails and slugs are devastating pests to crop plants in moist regions. A group of molluscs known as shipworms

About natural size

10.16

A garden snail. This snail is a common European land mollusc, similar to those served in French restaurants as *escargots*. Snails can be destructive to cultivated plants because their tongues, like microscopic files, can rasp away plant tissues. They have the ability, rare in animals, to digest cellulose. Although each individual has both ovaries and testes, one cannot fertilize itself, but when they mate, each fertilizes the other.

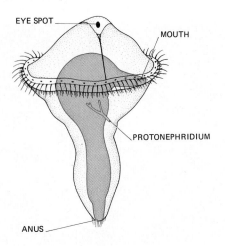

10.17

A trochophore larva, a characteristic immature form common to both molluscs and annelids. The mouth is at one side, and the anus is at the posterior end. Although molluscs and annelids are quite different during adult life, the appearance of similar larval forms, such as trochophores, is taken as an indication of evolutionary relationship of the two phyla. × 100.

bore into wood so effectively that they can scuttle a wooden ship within months or reduce massive pilings to a fragile lacy net.

With innumerable individuals and about 100,000 species now known, the Mollusca are among the most successful of animal phyla. They have been in existence for at least 500 million years and remain one of the three or four most important phyla. Their embryological development shows that they belong with the evolutionary line of the protostomes, and their cleavage is determinate and spiral. Some of the primitive molluscs show *segmentation*, and this characteristic, together with the presence of a *trochophore* larval form (Figure 10.17), shows that the molluscs have affinities with the segmented worms, or annelids, the next phylum to be described.

Classes of Mollusca

Two classes of the Mollusca that are not generally familiar are the Class Amphineura, the segmented chitons, and the Class Scaphopoda, the slender, conical tooth shells. The remaining three classes are better known, largely because they are edible or because they are the producers of most of the sea shells. The Class Pelecypoda (Greek, "hatchet foot") are the bi-

10.18

The internal structure of a clam, shown with one valve and portions of the body removed. The mass of the body is the foot, by means of which the clam can plow its way along the ocean bottom. A digestive tract leads from the mouth to a coiled intestine, which winds through the foot past a digestive gland and directly through the heart, finally ending in the anus. The whole soft body of the clam is covered by a pair of mantles, one of which has been cut away for this figure. The two flaps of the mantle partially meet at the posterior end, leaving two openings: an incurrent siphon, through which water is let into the mantle cavity, and an excurrent siphon, through which water is expelled. Currents are created by the action of beating cilia on the gills, which exchange oxygen and carbon dioxide between the water and the clam's body. A clam has no brain but does have collections of nerve cells, especially in ganglia below the esophagus, in the foot, and behind the foot. The valves may gape open, but they can be held tightly shut by two stout muscles, the anterior and posterior adductor muscles.

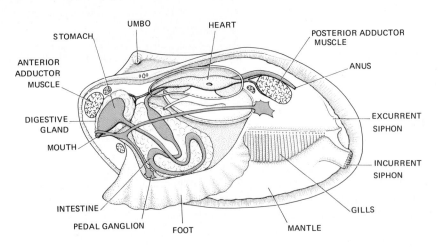

valves: clams, mussels, oysters, scallops, and thousands of similar kinds of animals. The body of a bivalve, such as a clam, is covered with two fleshy lobes called the *mantle*, which secretes the matched pair of shells and can close over the body, leaving two openings at the posterior end (Figure 10.18). If the hinge side of the shells is the dorsal side, then the mantle opening that allows water to enter the mantle cavity, the *incurrent siphon*, is ventral, and the other, which allows water out, the *excurrent siphon*, is dorsal. Water is moved over two pairs of *gills* by the action of beating cilia and made to flow in the incurrent siphon, through and past the gills, and by the *mouth*, which picks up tiny particles of food. The water then flows out the excurrent siphon, carrying with it dissolved carbon dioxide, alimentary wastes, and sometimes sex cells.

The bulk of the body is the *foot*, a mass of muscle a clam can use to plow slowly along through bottom mud. In the body are an alimentary canal; a pulsing heart, which sluggishly pumps body fluid through a system of cavities; and a nervous system, which consists of a pair of ganglia above the mouth and connecting nerves. There are sensory organs, which are associated with water testing (a mollusc can taste and smell), balance, and light perception. Scallops have a row of bright blue eyes along the edge of the mantle lobes. Clams produce eggs, which collect in the gills, where they are fertilized by sperm brought in through the incurrent siphon. The young clams, called *glochidia*, begin development in the gills of the adult, but they are released into the water, where they attach themselves temporarily to fish gills until they are large enough to drop off and start independent life.

Although many bivalves are both male and female, they usually do not produce both eggs and sperm at the same time, so that self-fertilization is avoided. The shells of bivalves are hinged by a tough ligament that tends to spring the shells apart, but a set of muscles, especially the two large anterior and posterior *adductor muscles*, can hold the shells together. The scallops sold for human food are the adductor muscles only; the rest of the scallop body, except the gonads (which are considered a special delicacy in some regions), is discarded. Some unscrupulous restaurateurs have been known to stamp out cylindrical chunks of shark meat to sell as "scallops," and it is not easy to tell the difference.

The Class Gastropoda (Greek, "stomach foot") are the whelks, various coiled sea-shell animals, snails, and slugs (shell-less gastropods) (Figure 10.19). The land-dwelling snails are the only gastropods in which lungs have evolved, but even they must

× 0.5

10.19

A whelk. Whelks are marine gastropods and as such are essentially oversized, fancy snails. When the animal is alive, its body almost fills the shell, but it can extend its massive foot and move around, or it can withdraw into its shell, closing the opening with a tough, tight-fitting lid, or *operculum*. Whelks are similar to, but smaller than, conchs (pronounced "conks"), which are also marine gastropods.

remain moist. The foot is conspicuous in gastropods, especially in the land snails, which use the foot for locomotion. A slick secretion is oozed out of the foot, and the animal travels along the bed of slime, rippling its muscles to propel itself forward. Some gastropods have a tightly fitting lid, the *operculum*, a horny plate with which they can cover the opening in the shell and protect themselves against predators and survive times of drying. The shells are either right-handed or left-handed, the direction of coil being determined by the egg regardless of the influence of the sperm. Hundreds of millions of garden snails, *Helix pomatia*, are prepared annually in French kitchens. The "sound of the sea," heard when a large conch shell is held to an ear, is only the resonance of sounds already in the air, but they are too faint to be noticed without the help of the air column inside the shell.

The Class Cephalopoda (Greek, "head foot") are the squids and octopuses. Squids have ten tentacles and octopuses have eight, as their name indicates

× 0.33

10.20

An octopus. Octopuses, squids, and cuttlefishes are almost shell-less molluscs. Only the squids have a small bony plate inside the body. Octopuses have eight arms (squids have ten) furnished with double rows of suckers that are useful in capturing and holding prey. In spite of their reputation for ferocity in folklore, these animals are usually rather small, timid creatures, and they are exceptionally sensitive and intelligent for invertebrates.

(Figure 10.20). Cephalopods are highly developed animals with sensitive perception of light and touch. The eyes of squids have a special interest for biologists because they are structurally similar to vertebrate eyes. Since squids and vertebrates are far removed from each other in evolutionary history, the fact that both of them have developed a similar set of light-sensitive organs furnishes an instructive example of convergent evolution (Chapter 26). One feature of the squid eye that could teach humans a lesson in humility is the position of the retina. In human eyes, light must pass through a layer of nerve cells before arriving at the light-sensitive retina, and it is therefore less precisely focusable than it might be. In squid eyes, the retina is in front of the retinal nerves and receives light directly from the lens, clearly a superior arrangement. Besides being sensitive, squids are probably the most readily trainable of the vertebrates. They can learn and remember, and therefore, like rats and pigeons, they have been used experimentally by students of behavior to study how learning is accomplished.

PHYLUM ANNELIDA, THE SEGMENTED WORMS

Many animals, including humans, show evidence of the serial repetition called _segmentation_, but in few is segmentation as noticeable as it is in the Annelida. Not only do they illustrate segmentation, but they also have simple, clearly identifiable organs for most animal functions. Consequently annelids, especially earthworms (angle worms, fishing worms), are much used in elementary biology courses for dissection and instruction. In vertebrates and arthropods, segmentation is demonstrable, but in annelids it is obvious. An earthworm, even externally, looks like an animated stack of coins, and the external segmentation, or _metamerism_, as it is sometimes called, is continued inside the animal's body as a series of incomplete partitions through which pass the digestive and other organs. Segmented organization interests biologists because they think it is a fundamental body plan for animals that evolved later from the first segmented types.

The Class Oligochaeta (Greek, "few hairs") is the best known class of annelids, because it contains the earthworms (_Lumbricus terrestris_). They are so common in soil almost all over the world that they are familiar even to city dwellers, who may see them squirming on sidewalks after rain. Any small patch of soil is likely to contain thousands of earthworms during moist weather, when they make little mounds of earth by casting up pellets outside their burrows. The "yard" at Harvard University is a veritable worm farm. With the arrival of hot, dry summers, earthworms disappear deep in the soil, giving rise to the myth that they turn into June bugs. With their perpetual eating through the soil, passing it through their alimentary tracts as they go, they stir it up, aerate it, turn it over, and keep it soft. In a set of classic observations, Charles Darwin, who put his mind to many a biological subject, found earthworms to be a major factor in maintaining the fertility of the earth, casting up as much as 18 tons of soil per acre per year.

Besides segmentation, the body of an earthworm shows many features of organization like those of more advanced animals. The symmetry is bilateral, with a definite head end where the nerve ganglia are (a phenomenon of "head-endedness" called _cephalization_), a true coelom lined with mesoderm, and all the functional organs that a well-developed, independent animal needs. The body wall, covered with a protective cuticle, has longitudinal muscles that can shorten the worm and circular muscles that can squeeze it out long, plus four pairs of whiskerlike _setae_ (singular, seta) per segment, which can be pushed out or pulled in. With this arrangement, the worm can extend itself with its hinder setae anchored, thus pushing forward. Then it releases its hinder setae, anchors its forward ones, and shortens its body, drawing the tail end up. Coordinated and speedy, an earthworm can elude all but the fastest predators.

The alimentary tract of an earthworm is a straight tube through the length of the body. It extends from a muscular *pharynx* back of the mouth through a slender *esophagus* to a thin-walled storage *crop,* followed by a tough, muscular grinding *gizzard,* and finally a long *intestine,* whose dorsal wall has its surface increased by a longitudinal fold, the *typhlosole.* Eating soil as it does and getting what organic material it can, an earthworm usually keeps its gut supplied with many living protozoa and roundworms. The nervous system consists of a pair of ganglia above the pharynx, connected by a nerve ring to a ventral nerve cord that enervates the whole body, with a ganglion in each segment and branch nerves from that (Figure 10.21). Without eyes, the earthworm uses its epidermis as a light-sensitive organ. It is sensitive to touch and to soil vibrations. There are testes and ovaries in each individual, but cross fertilization is accomplished by transfer of sperm from one worm to seminal receptacles of another (Figure 10.22). Later, a band around the body, the *clitellum,* is slipped along the body, as a ring is drawn off a

10.21

Internal structure of an earthworm. Segmentation, visible on the outside of the body, is striking internally because each segment is divided from the others by a partition through which nerves and other structures pass. Many of the organs are repeated in more than one segment: hearts (aortic arches), ganglia, and nephridia, for example. Most of an earthworm's functional parts are in the anterior end, with the first 40 segments containing the major digestive, nervous, sexual, and circulatory elements.

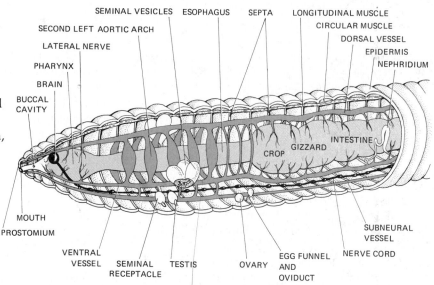

10.22

A pair of earthworms (or fishing worms or night crawlers) mating. Like some molluscs, these annelid worms have gonads of both sexes, and when they mate, they exchange sperm. Fertilization of the eggs occurs later when the eggs are laid. The segmented body form of annelids is clear even on the outside of the body. Although segmentation is less obvious in other animals, its presence can be demonstrated in many other groups, including arthropods and vertebrates.

× 1

finger, and eggs are deposited in it, to be fertilized as the band passes the seminal receptacles. The clitellar band passes over the head end of the worm to become a kind of cocoon filled with fertilized eggs. Embryological development occurs in the cocoon in the soil. The excretory system is made up of a number of *nephridia,* two in most segments. A single nephridium is a sinuous tube with a ciliated funnel at its free end, capable of picking up wastes and taking them to the outside of the animal.

Earthworms have a *closed circulatory system,* that is, a system in which blood (red in earthworms) is pumped through a set of vessels and stays in the vessels, as opposed to an *open circulatory system,* in which blood oozes about through various fluid-filled spaces. In earthworms, blood is pumped from the tail toward the head by pulsations of a vessel above the alimentary tract, then helped along by the action of five pairs of *aortic arches,* or "hearts," to a ventral vessel that carries blood toward the tail. Small branches supply the segments along the way. No special respiratory system is needed in such small animals when they live in moist environments where sufficient oxygen and carbon dioxide can be exchanged directly through the epidermis.

Other annelids, members of the Class Polychaeta, are equipped with pairs of appendages at the sides of the segments and with various head appendages, such as eyes, jaws, and tentacles. The side appendages are used for moving through water and for exchange of dissolved gases. Common polychaetes are the clam worm (Nereis) and lugworms, which live in sandy beaches (Arenicola); tube and fanworms (such as Spirographis); and the "palolo worm" of Samoa, whose annual swarming makes feasts for the islanders. Leeches are in the Class Hirudinea (Figure 10.23). Leeches are such famous bloodsuckers that the word, which originally meant physician or healer because doctors used leeches to bleed patients, has now come to be used metaphorically to refer to any person who parasitizes another. At the head end, a leech, such as *Hirudo medicinalis,* has a sucker equipped with three sharp teeth. It can attach itself firmly to skin, puncture a blood vessel of the victim, secrete an anticoagulant, and suck blood until the body of the leech is bloated. In an old romantic movie, "The African Queen," Humphrey Bogart played the archetype of the rough and tough adventurer, who carelessly braves bullets, jungles, disasters, and threats of disasters, but who is reduced to shivering horrors on climbing out of the Nile River to find himself dripping with leeches. Someone associated with that movie must have been attacked by leeches to know the revulsion that such an event brings.

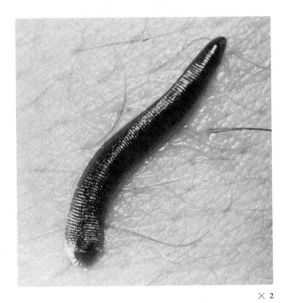

× 2

10.23

A leech, *Hirudo medicinalis.* These animals can draw up almost into a ball or stretch out to a length of many centimeters. When a problem of "too much blood" was considered a medical matter, leeches were attached to patients and allowed to suck out blood. The word "leech" originally meant a physician, but it was later applied to the animal and finally came to mean the animal only. Except for their aquatic habitat and bloodthirsty temperament, leeches are much like earthworms.

PHYLUM ARTHROPODA

Although the human species has made such an obvious impact on the earth that the present is sometimes called the Age of Man, the arthropods at the same time have such a number of species and such a number of individuals that calling this the Age of Arthropods could more easily be defended. Except for microorganisms, there are more arthropods, any way you count them, than there are members of any other phylum of animals. Insects have taken the land as their province, as crustaceans have taken the sea. Every imaginable habitat and every imaginable behavior (including habitats and behaviors that are practically unimaginable) have been tried out successfully by arthropods.

The arthropods are named for their paired, jointed appendages (Greek, "joint foot"), which are covered by a horny *exoskeleton* composed largely of *chitin,* a polymer of acetyl glucoseamine. Arthropods are protostomes, as shown by their embryonic devel-

× 1

10.24
A trilobite, showing the two longitudinal furrows that divide the animal's body into three parts, giving it the name. Although now all extinct, trilobites were once among the commonest of animals, living all over the earth and evolving thousands of species. They endured for millions of years and left abundant fossils, especially in Cambrian rocks. They had well-developed compound eyes like those of modern insects and, like arthropods generally, pairs of jointed appendages.

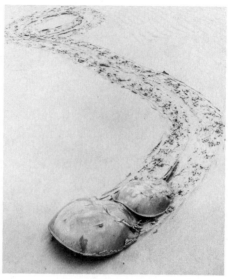

× 0.06

10.25
A pair of horseshoe crabs mating. The male, the smaller of the two, fertilizes the eggs after the female has laid them in the sand, and the female then covers the eggs with sand. In a couple of weeks they hatch, long after the couple has returned to the water. Horseshoe crabs, also called king crabs, look so similar to crustaceans that they were for a long time classified with them. Actually more nearly related to spiders, they are now usually considered to be in a class of their own.

opment, and cleavage is spiral and determinate. They have a true body cavity, lined on both sides with mesodermal tissue. Beyond these few main characteristics, the arthropods are such a diverse lot of animals that they can be considered further only by treating the subphyla and classes separately.

Subphylum Trilobita

In our present treatment, most animals known only from fossils will not be considered, but the *trilobites* (Latin, "three lobes") deserve special attention because they were so numerous (with millions of fossilized individuals still in existence), because they endured such a long time (over a period of more than 400 million years), and finally because their Precambrian ancestors, or something like them, must have given rise to the present-day Arthropods (Figure 10.24). Before they became extinct in the Permian Period about 200 millions years ago, trilobites were common. They existed in a number of forms, all with a general fundamental plan. Two longitudinal

grooves, running lengthwise from the head through a faintly segmented thorax and a clearly segmented posterior abdomen, divided the body into three regions. Trilobites had eyes, antennae, and down each side of the body a row of forked, plumy appendages. They generally ranged from about 5 to 50 centimeters (2 to 20 in.) in length. Trilobites lasted well past the time of arrival of such major arthropod classes as crustaceans and insects, and for their time they were immensely successful.

Subphylum Chelicerata

The Chelicerata (Greek, "claw horn") differ from other arthropods in having claws as the first pair of appendages, no antennae, and a body with unsegmented *cephalothorax* (head and thorax in one section) and abdomen. There are usually four pairs of legs.

One small, ancient class, the Merostomata, contains the king crabs, or horseshoe crabs (Figure 10.25). They have remained on earth almost un-

(a)   × 3

(b)   × 1.5

10.26

Arachnids. (a) A spider, Dolomedes,
sucking the blood from a small fish
it has caught. The four pairs of legs
and the two-parted body are evident.
Spiders are generally carnivorous, but
live mainly on insects. (b) A scorpion
in defensive position, with its pinchers
extended (formidable-looking but not
dangerous) and its tail raised over its
back. At the tip of the tail is a venomous sting attached to a poison bulb.
A scorpion's sting is not dangerous
but can be painful and can kill small
animals. A mother scorpion carries her
young for a while on her back. The
scorpion's shape is so characteristic
that it has been recognized for centuries as special, even giving its name
to the constellation Scorpio.

changed since the Triassic Period, 180 million years
ago or more, and somewhat similar animals have
lived since the Cambrian Period, more than 500 million years ago. Only a few species have survived, but
those are locally abundant. Resembling crustaceans,
which they were thought to be for many years, horseshoe crabs have a tough *carapace,* jointed to a smaller
abdomen, and a slender, sharp, tail-like *telson.* The
horseshoe crabs of the Atlantic coast are favored animals for furnishing body fluids to be used in immunological research. They come out of the sea to breed,
leaving their eggs in the beach sand, where they develop into trilobitelike larvae before assuming their
adult form.

Another class of Chelicerata is that of the Arachnida, the spiders, scorpions, mites, ticks, and harvest-
men, or daddy-long-legs (Figure 10.26). The garden
spider, Argiope, can serve as an example of a familiar
spider. It has a two-parted body: an anterior *cephalothorax* and a posterior *abdomen.* There are four pairs
of legs, a pair of *chelicerae* with poison fangs at the
tips, and a pair of *palps,* which are a combination of
sensory and chewing appendages. There are eight
simple eyes but no antennae. At the tip of the abdomen, the *spinnerets* are capable of extruding a silky
filament used for web building. Gas exchange is accomplished by means of thin-layered folds of tissue
called *book lungs.*

Mating in spiders is frequently preceded by elaborate species-specific courtship activities by the small
male. He goes through stereotyped movements of
waving his palps or assuming special positions, presumably to ensure recognition by the larger female,
who might otherwise take him for food. She may eat
him later anyway, but only after her eggs have been
fertilized. When she lays the eggs, usually in a cocoon
of spun web material, up to several hundred small
spiders hatch out, and some of them eat others before
breaking out of the cocoon.

In spite of their dark reputation in popular folklore, spiders are generally harmless creatures, more
our allies than our enemies because they feed largely
on insects. They rarely bite anyone, and then only
when pressed. Even the infamous black widow (*Latrodectus mactans*), with her pair of red triangles on
the underside of her abdomen, is so peaceable that
inducing her to bite for experimental reasons is difficult. Black widows will on occasion bite, but the number of instances in the United States has dropped
since the almost universal installation of plumbing in
houses. The other potentially dangerous American
spider, with the sinister name of brown recluse (*Lox-
osceles reclusa*), lives in the lower midwestern and

southeastern states. Spiders make satisfactory pets if one can overcome the prejudice of years against their hairy legs and sudden movements. Scorpions do not bite. They sting with their tails, and though the sting is painful, it is usually not lethal.

The other arachnids are either completely harmless, as the daddy-long-legs is, or do their damage by transmitting disease. The most dangerous ones are ticks, which not only carry Rocky Mountain spotted fever but can cause painful allergenic reactions in humans and domestic animals. Along with those other bloodsuckers, leeches and vampire bats, ticks inspire revulsion.

Subphylum Mandibulata

The Mandibulata (Latin, "jaws") differ from the Chelicerata in having the front appendages in the form of antennae and chitinous jaws that work from side to side (instead of up and down as vertebrate jaws do). Containing as it does the crustaceans and the insects, this subphylum has the greatest number of species and individuals of any group of present-day organisms.

Class Crustacea   The crustaceans—crabs, shrimps, lobsters, crayfish, barnacles, and a host of smaller animals—are mainly aquatic, and they occupy a position roughly comparable to that of the insects on land (Figure 10.27). A crustacean typically has two pairs of antennae; a pair of mandibles, or jaws; and two pairs of paddlelike appendages, or maxillae. Other appendages, attached to segments along the body, are variously modified for grasping, walking, swimming, or for sexual activities. The body is covered with an exoskeleton of calcium-impregnated chitin. The general body plan is segmented, but the segmentation is obscure in some parts. There is usually a head and

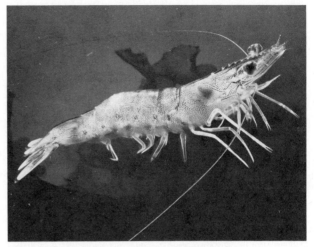

(a)   × 1

× 0.33

10.27

Representative crustaceans. (a) A shrimp, Palaemonetes, one of the common commercial shrimps or prawns. (b) An edible crab. In crabs the abdomen is reduced and tucked under the thorax so completely that it is invisible from the top side. Crabs are specially modified in shape, but are otherwise much like shrimps. (c) A cluster of barnacles. These active barnacles have extended their feathery *cirri* (singular, cirrus), with which they agitate the water and obtain food. Except during larval life, barnacles are firmly attached to some solid surface, such as rocks, driftwood, or ship bottoms. Barnacles were long thought to be molluscs, but their bodies are clearly crustacean in structure.

(c)   × 0.5

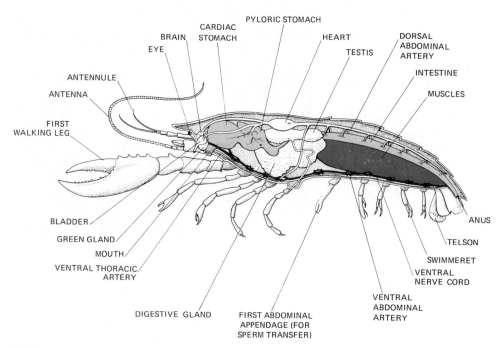

BRAIN
CARDIAC STOMACH
PYLORIC STOMACH
EYE
HEART
TESTIS
DORSAL ABDOMINAL ARTERY
ANTENNULE
INTESTINE
ANTENNA
MUSCLES
FIRST WALKING LEG
ANUS
BLADDER
TELSON
GREEN GLAND
SWIMMERET
MOUTH
VENTRAL NERVE CORD
VENTRAL THORACIC ARTERY
VENTRAL ABDOMINAL ARTERY
DIGESTIVE GLAND
FIRST ABDOMINAL APPENDAGE (FOR SPERM TRANSFER)

10.28

Internal anatomy of a lobster, a representative crustacean. Notable organs are the small heart, which pumps body fluid but does not have a closed circulatory system like that of vertebrates and is not used as a carrier of respiratory gases; a stomach with internal grinders; a nerve cord, which passes along the ventral surface, with ganglia in each body segment; and the large, powerful abdominal muscles.

thorax, although those two parts may be more or less fused into a cephalothorax, and there is usually a clearly segmented abdomen. The evolutionary source of the Crustacea is unknown, but one generally accepted guess is that early crustaceans were like trilobites.

Crayfish, which are structurally like lobsters, are frequently used as examples in descriptions of Crustacea. The body of a crayfish is divided into two recognizable parts: a cephalothorax, with a pointed *rostrum* at the anterior end; and a jointed, flexible abdomen. The cephalothorax is covered by a firm *carapace* that covers the sides of the body like overhanging eaves of a roof, protecting the feathery gills through which oxygen and carbon dioxide are exchanged. The abdomen has six segments, each with its firm exoskeleton but united to the next by a flexible membranous joint, as are the joints of the appendages. At the tail end there is a paddle-shaped extension, the *telson,* which the animal uses to escape danger. When a crayfish is startled, it can curl its abdomen, which is mostly filled with muscle, thus giving its telson a sudden pull forward, and that movement jerks the whole animal backward. There are five pairs of *walking legs,* the first of which is tipped with a strong pinching claw, all attached below the cephalothorax. Five pairs of *swimmerets,* like hairy forked paddles, grow from the underside of the abdomen.

Crayfish feed by grasping large particles with their pincers or by fanning small ones to the mouth. A pair of antennae and a smaller pair of antennules help it find food by smelling or tasting. The *mandibles* can tear food but not chew it. It is chewed in the stomach, where three horny teeth form a *gastric mill,* which grinds coarse food. Fine particles are digested by a *digestive gland* that secretes hydrolyzing enzymes; the coarse material is passed on through the intestine and out the anus. Other body wastes are collected by a *green gland,* which functions as a kidney and passes the waste out through an opening near the base of the antennae. The circulatory system is open; that is, the yellowish blood is pumped by a dorsal heart into vessels that empty into the tissues. Channels through various tissues allow the blood, with its copper-containing pigment (contrast the iron-containing pigment of red-blooded animals), to flow past the gills and eventually to make its way back through the heart. A ganglion above the esophagus, which serves as a brain, is connected by a ring of nerves to another ganglion below the esophagus, then by a ventral nerve cord along the body, with a ganglion in each segment and branching nerves from them to the rest of the body (Figure 10.28).

When crayfish mate, the male uses his first pair of swimmerets to transfer sperm from his body to *seminal receptacles* in the female's body. The sperm remain in the receptacles until the female lays eggs, several hundred at a time, and the eggs are then fertilized. The eggs are attached to the mother's swimmerets until they hatch out as miniature crayfish. When a female crayfish or lobster is carrying eggs, she is said to be "in berry."

When crustaceans or any animals with exoskeletons grow, they swell inside their exoskeletons until something must give. When hormonal conditions so direct, the outer covering splits, and the animal pulls its soft inner parts completely out of the old shell, down to the tips of claws, antennae, and even eyes, like pulling a hand out of a glove. After this process of *molting*, the body is temporarily vulnerable, but it quickly hardens a new and larger covering. "Soft-shelled crabs" are crabs caught while in an early post-molt condition. Most of the increase in size occurs just after the molt, when the crayfish swells with the intake of water. As long as it lives and grows, it must continue to molt at intervals (refer to Chapter 18).

Body form is as variable in crustacea as one would expect in so large and ancient a group. There are the familiar lobsters; Alaskan crabs with broad cephalothoraxes, little abdomens tucked under, and legs with a spread of three meters; tiny ostracods, swimming like minute bivalves; copepods, with one eye, named *Cyclops* for the mythical Homeric monster; brine shrimp, popularly sold in dry dust to hatch out in home aquaria as "sea monkeys"; barnacles with calcareous plates, long thought to be molluscs. But if body form varies, it is with the appendages that evolution has been most prodigal in its diversity. From the rather simple antennules of lobsters and the two-branched swimmerets (called *biramous*, meaning "two oars"), there have developed uncounted differences in length, shape, and ornamentation. Some of the deep-sea crustaceans have such feathery appendages, several times body length, that they seem to be all frills.

Crustaceans are important to humans in a number of ways. Aside from the obvious direct consumption of crabs, crayfish, shrimp, prawns, and lobsters as human food, there is the indirect feeding of most marine fishes, which live largely on small crustacea. The cool waters of the Antarctic and adjacent oceans are rich in small shrimplike krill, the main food source of the baleen whales and other large marine animals. Barnacles, attaching themselves to the hulls of ships, can create enough drag to reduce speed considerably, and they have to be scraped off. During the days of sailing ships, sailors routinely beached their ships, allowing them to tilt on one side to expose the underwater parts for scraping, a job called "careening." Crustacea are also indirectly responsible for disease in that they are hosts for parasites, such as the broad tapeworm, that also infect humans.

Classes Chilopoda and Diplopoda The Chilopoda and Diplopoda are small classes of Arthropoda with little biological or economic importance. The first are the *centipedes* (Latin, "hundred feet"), segmented animals with one pair of legs on most segments (Figure 10.29). Although they are frequently called "thousand-leggers," they should more accurately be called thirty-leggers. Though some tropical centipedes can

× 2

10.29

A representative chilopod: a centipede, also called a hundred-legger. Some centipedes have a slightly poisonous bite, but their relatives, the millipedes (with two pairs of legs per segment instead of one) are quite harmless.

bite painfully, they are mostly harmless. The temperate ones live on insects, but their help in killing roaches and bedbugs is not generally appreciated in houses, because they are too slithery and too much like small dragons to appeal to nonbiologists. The Diplopoda are the *millipedes* (Latin, "thousand feet"), differing from the Chilopoda most conspicuously in having two pairs rather than one pair of legs on each segment of the body. They are mostly vegetarian scavengers, living in moist, dim places, rarely moving in with humans. (There are several other minor classes of Mandibulata that are not treated here.)

Class Insecta   The insects (Latin, "cut into," in reference to the thin neck and wasp-waist) are sometimes known as the Hexapoda (Greek, "six feet"). They represent the culmination of terrestrial evolution in the arthropod line, with a bewildering variety of forms and activities. With some 900,000 described species, and with new species being found all the time, insects occur in more forms than do the members of any other biological group. (See the essay on p. 195.)

Insects affect people in every aspect of daily living. They eat our crops and flowers, destroy our stored grains, gnaw holes in clothes and carpets, tunnel through the woodwork, pulverize books, bite our skins, pester our livestock, parasitize our pets, and infest our food. They carry plant diseases, such as yellows, mosaics, witches' brooms, rots, knots, cankers, scabs, stains, mildews, blacklegs, wilts, blights, streaks, and stunting. They also transmit animal diseases, including human diseases: tapeworm, lung fluke, and guinea worm infections, malaria, sleeping sickness, Chagas's disease, kala-azar, cattle fever, dysentery, plague, tularemia, anthrax, typhoid fever, cholera, typhus, spotted fever, yellow fever, dengue, and encephalitis. Yet the same bees that may sting painfully—even to death if the victim is allergic to the stings—provide honey, the only sweetener people knew for thousands of years until the recent advent of refined sugar. Bees are responsible for the pollination of most of the fruits that people eat and most of the flowers that people enjoy, although the staple grains are wind-pollinated. Seed production of forage plants, such as clover, is dependent on the pollinating ability of insects, and hence even our meat supply depends to some extent on insects.

The general activities of nature and the economy of man depend on the insects. Insects give silk, help the fisherman catch fish, make shellac and dyes, aerate soil, help keep down undesirable weeds, feed song birds, and may even provide medicines. Butterflies, moths, and beetles are sources of artistic inspiration, and they offer pleasure to many amateur and professional *entomologists* (students of insects). Ants, grasshoppers, grubs, and caterpillars are eaten by humans in many parts of the world. In our competition with insects, we find insects themselves our allies against their own kind. Ladybird beetles eat scale insects, and mantises eat almost anything they can catch. Wasps parasitize caterpillars, and aquatic bugs eat mosquito larvae. Finally, insects are great scavengers, helping clean up the dung, feathers, and dead carcasses of animals. On balance, in spite of the trouble they can cause, insects probably do more good than harm, and certainly without them life as we know it would change enormously—and probably for the worse.

As evidence of the intimacy with which we regard insects, we have applied a wealth of names to thousands of them, many of them picturesquely illuminating. Heading the list are insects named for animals: toad bugs; lizard beetles; cuckoo wasps; hawk moths; elephant beetles; tiger, rhinoceros, and ermine beetles; leopard, owlet, and virgin tiger moths; zebra swallowtails; and orange dogs. Some insects are named for other insects: flea beetles; spider beetles (although spiders are not insects); and bee moths. Some are considered outlaws: assassin bugs; robber flies; ambush bugs; and masked hunters, which earned their title by accumulating under-the-bed dust on their heads. Less sinister are the checkered skippers, the pinching bugs, and the kissing bugs, which tend to bite people around the mouth. Some seem military, such as the army worms and army ants, the bombardier beetles and the sword-bearing crickets. Glamour comes with the handsome earwigs, bog damsels, painted beauties, and darling underwings; and the supernatural world is represented by fairy moths, ghost moths, death-watch beetles, phantom craneflies, and black witches. There are thick-headed flies and confused flour beetles. Even modern technology enters with wheel bugs, which have an obvious piece of cogwheel growing out of the thorax; railroad bugs, with green lights in front and red lights behind; and the short-circuit beetles, which chew their way into lead-covered electrical cables. These are all living insects, and the names can be found in standard entomological works.

In the some 300 millions years that insects have inhabited the earth, they have come to occupy practically every available habitat except the sea. One genus, Halobates, does ride the surface of salt water, even far out at sea, and many live in salt or brackish marshes, but almost all insects are terrestrial or live

THE AGE OF INSECTS: RIGHT NOW

There are more insects alive today than any other kind of animal. About one billion billion insects inhabit the earth, almost a billion insects for every human being. A number as large as one billion billion is almost meaningless to us, who usually deal in hundreds or thousands. So consider this instead: All the insects in the world *weigh* more than all the other animals put together. There are more than 900,000 described species of insects (some estimates range as high as three million, and it is possible that even that number is conservative), and new species are being discovered all the time. These numbers become even more impressive when we consider that many organisms we think of as insects are not really insects at all. For instance, spiders, centipedes, and beach fleas are related arthropods, but they are not insects. Of all the insects, there are more beetles than any other kind, and in fact, there are more beetles than any other kind of *organism*. Nearly 30,000 species of beetles live in the United States alone, and there are almost 300,000 species of beetles worldwide. All in all, insects make up about 75 percent of the animal kingdom.

Any organism with such impressive credentials certainly has been successful in coping with the environment. It is true that insects have been on the earth many millions of years longer than chordates, the phylum that includes human beings. Nevertheless, the quantitative difference is huge: Chordates include about 45,000 species, compared with over a million for arthropods. Besides an early start, what advantages have made insects so successful?

Insects are specialized, particularly those, like butterflies, that undergo the four distinct stages of complete metamorphosis, each of which concentrates on only one or two functions at a time, such as feeding during an early stage and reproducing at a late stage. Many insects will eat practically anything, they are adaptable to a broad range of environmental conditions, they are usually small and relatively inconspicuous, and they possess a natural suit of armor, with body parts that show a staggering array of functional variety. The first flying animals were insects. The ability to fly is undoubtedly a vital reason for the successful evolution of insects, especially when it is combined with the ability to fold their wings onto their backs when not in flight. In a sense, the insect possesses the advantages of a flying organism while retaining the crawling agility of a terrestrial one.

Insects began to fly at the right time—many of their natural predators were then developing—and for about 50 million years insects were the only flying animals. Thus they had an enormous evolutionary advantage. Whenever the environment changed, not only were insects able to adapt to the change; they usually took advantage of it. They benefited from new environments and new food sources, and some insects, like the honeybees, developed highly structured societies based on an entirely new environmental factor: flowers.

There are more insects now than ever before. They have withstood weather, famine, predators, and DDT. The present era is usually known as the "Age of Mammals." Perhaps it would be more appropriate to call it the "Age of Insects."

in fresh water. They are not usually noticed because most of them are small, but even in apparently inhospitable regions, soil is teeming with insects. They live all over animals, penetrating their skins, clinging to fur and feathers, and crawling into any available opening. They eat plant roots, hollow out stems, suck or chew leaves or tunnel along between upper and lower epidermises of leaves, and eat their way into fruits and seeds. People and their houses offer special if unwilling protection to a host of insects, the more obnoxious ones being fleas, lice, bedbugs, cockroaches, silverfish, earwigs, and houseflies. Insects may thrive in icy streams, hot springs, snowfields, oil slicks, caves, manure heaps, haystacks, and cadaver vats in medical schools. Unless the larger scavengers (vultures, opossums, crows, jackals, hyenas, eagles,

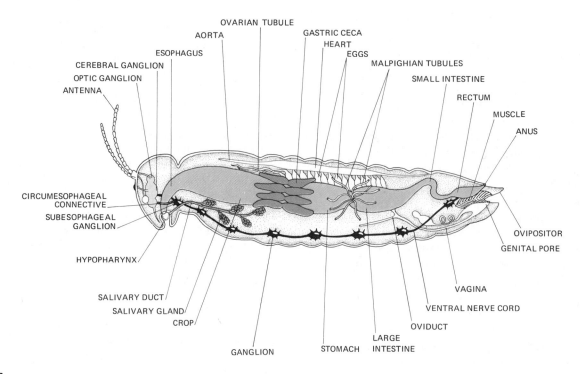

10.30

Internal structure of a grasshopper as a representative insect. Note especially the cerebral ganglion, or "brain," the ventral nerve cord with repeated ganglia, the large digestive tract with out-pockets (gastric ceca), the large crop for storage of food, and the digestive glands (salivary). The Malpighian tubules are excretory organs.

etc.) get to them first, dead animals are consumed by a succession of insects from the fur and skin eaters to the carrion beetles and fly larvae, or maggots. By far the most delicate and accurate way to prepare skeletons for study is to let the Dermestid beetles clean the bones.

The large insects attract attention by their very size, since a 12-centimeter (5-in.) beetle, a 30-centimeter (1-ft) stick insect, or a 25-centimeter (10-in.) moth may be spectacular. However, the great majority of insects are so small that they pass unnoticed by most people. The average insect is less than five or six millimeters (¼ in.) long, and some tiny ones are recognizable only with a lens and identifiable only with a microscope.

Communication among insects is as varied and complex as the other insect activities. Much communication is visual; males especially have specific shapes or colors, or they are capable of stylized movements that make them recognizable to females. Nocturnal insects can recognize and find one another in the dark. Some do so by means of flashing lights, others by means of characteristic sounds, usually produced by scraping one body part (such as a rasping leg) across another part. The most sensitive communication method is through smell; insects have much more accurate perception of smell than vertebrates do (Chapters 28 and 29).

The structure of insects.   The usual body plan of an insect consists of a head attached by a more or less flexible neck to a thick thorax, which continues as a segmented abdomen (Figure 10.30). The head bears a pair of antennae (which range from short and stubby to long and plumy, with many variations in between), a pair of compound eyes and several simple eyes, and a set of mouth parts. The compound eyes are made up of a few to several thousand individual components, the *ommatidia*, each one a slender cone with a light-sensitive receiver at the base. Such eyes are not so effective in perceiving still images as they are in perceiving motion, but they have wide angles of vision in that they point in many directions. A dragonfly can see in almost all directions at once because its bulging eyes point practically every way except inward. The mouth parts of insects are adapted to chewing, piercing, lapping, or sucking. In some insects, such as mayflies, they are so reduced that the

× 10

10.31

The face of an insect. The most conspicuous features are the long antennae, the bulging compound eyes with hundreds of simple eyes (ommatidia), and the heavy mandibles that work from side to side. This specimen is a yellow jacket, Vespula.

adult form does not eat (Figure 10.31). A primitive set of mouth parts, from which more specialized types are thought to have evolved, consists of a hinged upper *labrum* overlying a pair of side-to-side *mandibles*, or jaws; a second pair of weaker appendages, also working laterally, the *maxillae*; and a median *labium*, like a lower lip. A pair of *palps*, sensory organs, is attached to each of the last two parts. From this basic arrangement, mosquitoes have developed stilettoes for piercing skin (although mosquitoes live mostly on plants and only females bite animals), flies have a bulbous sponging set of appendages, and butterflies have a structure that can lap up nectar or coil up when not in use like a watch spring.

The insect thorax bears the locomotor organs: three pairs of walking legs and in most insects two pairs of wings. Wings, which are extensions of the thorax above, are separate from legs and not structurally comparable to legs. Wings may be leathery, horny, membranous, scaly, feathery—or lacking—but they are usually characteristic of members within a group. Separation of insects into orders is usually based on wing structure. Wings have been of enormous importance to insects because flight has enabled them to escape predators, to take advantage of otherwise inaccessible habitats, and to be carried on wind currents wherever wind blows. An occasional insect spatters against the windshield of an airplane flying at a height of six to eight kilometers (four to five miles). Walking or running is a more stable means of moving for an insect than it is for a two- or four-legged animal, because the insect always has three feet on the ground. With strong leg muscles, an insect can run, leap, and carry effectively, and it can cling to smooth surfaces with claws, hairs, and suckers without having to expend energy holding on.

The insect abdomen is without appendages except for the genitalia and, in females, egg-laying apparatus. Along the sides of the abdomen, breathing-holes, *spiracles*, provide access for oxygen and carbon dioxide exchange, leading into a system of branched tubes known as *tracheae*. Adult insects have no lungs, but they do pump air in and out of the tracheae through the spiracles, contracting and relaxing the abdominal muscles in a way that resembles vertebrate breathing. Respiratory gases are delivered directly to the tissues, in which colorless blood oozes by means of an open circulatory system. There is a heart, but movement of the blood is accomplished more by body movement than by forceful heart action.

The digestive tract in insects leads from the mouth to an enlargement of the esophagus called a crop, then to a *proventriculus*, a gizzardlike grinding sac, and finally to the hindgut. Along the way, digestive enzymes work on the food, and nutrients and water are absorbed through the gut wall until only a compressed, rather dry pellet of fecal matter is passed out. The diet of insects is so various that no brief description of the enzymes can be made. Many insects have special diets for which special enzymes are provided. Some, such as the termites, "digest" cellulose, not by their own enzymes, however, but by the enzymes of the bacteria that accompany the protozoa in the termite gut. The larvae of some flies spit out enzymes that digest the meat on which they are growing, and then all the larvae have to do is lap up the digested protein. Mosquitoes inject a droplet of saliva into the puncture of an animal skin, thus preventing coagulation of the blood and making the characteristic stinging irritation. Mayflies do not feed during the adult stage; they mate, lay eggs, live a few days, and die.

Reproduction of insects. The social life of insects is so specialized and complex that it will receive treatment on its own in Chapter 29. The sexes are separate in insects, and fertilization is internal. Beyond that,

10.32

Metamorphosis in the monarch butterfly, *Danaus plexippus.* A tiny caterpillar hatches from an egg and eats
voraciously until it increases as much
as ten thousand times in weight (a).
When it reaches its full size, it forms
a protective covering about itself and
becomes a pupa (b). The pupal coat
is formed inside the skin of the caterpillar. During the pupal stage, the
animal seems inactive, but inside the
case the larva is completely reorganizing to form a new kind of body,
that of the adult or *imago.* When the
imago is mature and external conditions are suitable, the chrysalis splits
down the back (c) and the soft, moist
butterfly crawls out (d). Blood is
pumped into the delicate wings, and
as soon as they dry and harden, the
butterfly is ready to fly (e).

(a)

(b)

few generalizations can be made. In the simplest
types, eggs are laid, and they hatch into small copies
of adults. In many orders of insect, the newly
hatched individual is somewhat like an adult but
has juvenile features. The young, called *nymphs,* go
through several stages of development, the *instars,*
becoming larger and more mature with each molt, and
finally arriving at adulthood. The series of changes is
gradual or *incomplete metamorphosis.* In the orders
that contain flies, beetles, or butterflies, there is *complete metamorphosis* (Figure 10.32). From the egg
comes a *larva* (Latin, "ghost"), which may be a maggot, grub, caterpillar, or other rather wormlike, actively feeding form. After a series of molts, the larva
covers itself with a protective coat, either of chitinous
or of spun fibrous material (silk), and goes into a period of apparent inactivity. In fact, it is during this
pupal stage that major changes occur, eventually giving rise to the *adult* or *imago.* When the pupa opens,
the imago climbs out, soft and moist, leaving the pupal shell behind. The new imago stretches and plumps
up its body and appendages, dries, and is ready for
adult life. (Further discussion will be found in Chapter 18.)

The orders of insects.   Of the 27 orders of insects
recognized by most entomologists, many are small or

not commonly known, or they consist of such minute
animals that they are seen only by specialists, but half
a dozen are so conspicuous and widely distributed
that they are known to some extent by almost everyone. One relatively primitive order is the Order Orthoptera (Greek, "straight wings"), which includes
grasshoppers, locusts, katydids, mantises, cockroaches, walking sticks, and crickets (Figure 10.33).
Metamorphosis is incomplete. The front (prothoracic)
wings are stiff and leathery when present, and the
hind wings are membranous, folded like a fan. With
their chewing mouth parts, grasshoppers and locusts
can eat leaves easily, and when a sufficient mass of
these insects settles on a growing crop, it can quickly
devastate a field, reducing it to a mass of stubble.
Plagues of locusts, that is, grasshoppers, have long
been a terror to farmers, but new techniques of fore-
casting the mass movements and diverting them
promise help for the future.

   The Order Coleoptera (Greek, "shield wings")
are the beetles and weevils, insects with usually four
wings. The stout prothoracic wings cover the membranous hind ones (Figure 10.34). The hind wings are
the active flying ones. The forewings are held out
stiff during flight and laid down during rest to meet
in a midline down the back. Metamorphosis is complete; the larvae are known as *grubs.* Beetles have

(c) (d) (e)

10.33

A representative insect: a cockroach, *Blatta orientalis,* producing an egg case. Relatively primitive insects, cockroaches have been and remain enormously successful animals in terms of survival and sheer numbers. Cockroaches can be kept under control by cleanliness, constant vigilance, and programs of poisoning, but they cannot be completely eliminated. Three species of roaches, some called "water bugs" or Croton bugs, infest houses— this one and *Blattella germanica* and *Periplaneta americana.*

10.34

A weevil on a hairy petal of an iris flower. The long snout on this weevil, which is a kind of beetle, has a pair of jaws at the tip with which it can gnaw at plant tissues. It can eat the tissues or make holes in which to lay eggs. The eggs hatch into larvae (grubs), and they, too, eat plant tissues. Weevils are well adapted to live on hard, dry foods, and they can be especially destructive to stored grains, peas, and beans.

× 1

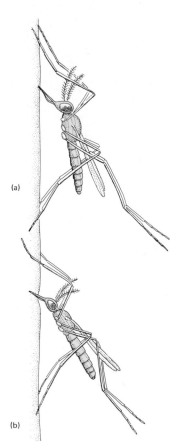

(a)

(b)

10.35

A lepidopteran: the viceroy butterfly, *Limenitis archippus*. The spreading wings and the slender filamentous antennae show that this is a butterfly, not a moth.

10.36

Mosquitoes, members of the same order of insects as flies. (a) A Culex mosquito in a biting stance, with her body parallel to the surface. (Only females bite animals, because they require blood before they can lay eggs.) These mosquitoes are annoying but do not carry human disease. (b) An Anopheles mosquito in a biting stance, with her body tilted head downward. Anopheles is the malaria carrier. × 7.

chewing mouth parts, but these may be placed at the end of a long snout, as in weevils, so that the head looks as if it had a proboscis. There are more beetles than any other group of organisms on earth, and nearly 30,000 species of them live in the United States.

In the Order Lepidoptera, (Greek, "scale wings") the butterflies, moths, and skippers have four wings, which, like much of the body, are covered with loosely set, microscopic scales. As a result they have a dusty appearance (Figure 10.35). Metamorphosis is complete: egg to larva (called a *caterpillar* in this order) to pupa (a *chrysalis* or *cocoon*) to adult. Some of the mouth parts are modified into a long proboscis, which is coiled out of the way when the insect is not sucking. The adults are generally harmless to humans, but the larvae of the coddling moth, the gypsy moth, the clothes moth, the army worm, and many others are such voracious eaters that they can cause great economic damage. On the positive side, silkworms, which spin their silken cocoons from salivary glands near the lower lip, are positively useful.

In the Order Diptera (Greek, "two wings"), the hind wings are modified into halteres, or balancers (Figure 10.36). These are the true flies, mosquitoes, gnats, and midges. They can be recognized in reality by the one pair of wings and in written words because entomologists write the names of true flies as two words, as in "house fly," but those of nonfly orders as one word, as in "dragonfly" or "stonefly." Anyone who has been bitten by a horse fly knows that some flies have piercing mouth parts. Others, however, like house flies, have a bulbous sponging enlargement at the end of a proboscis. In metamorphosis, which is complete, the larva is a wormlike maggot. Diptera are sometimes beneficial in parasitizing harmful insects,

× 2

Wingless insects of the Order Aptera. These are young and mature examples of the silverfish, *Lepisma saccharina*, which lives commonly in houses, eating the glue and paper-filler in books. Among the most primitive of living insects, they have no such drastic changes of form during metamorphosis as do beetles or butterflies. When a silverfish hatches from its egg, it is practically a small copy of an adult, but it grows and molts at intervals over a period of several years.

in damaging noxious weeds, and in acting as scavengers of dead plant and animal matter. But they are sometimes scourges of cultivated plants, and they are dreaded stingers, biters, and carriers of many human and animal diseases, including malaria, yellow fever, African sleeping sickness, typhoid fever, and dysentery. In mosquitoes, only the females bite animals, since they alone require a blood meal in order to lay eggs.

To an entomologist, the only true bugs are in the Order Hemiptera; all other "bugs," despite popular terminology, are technically not really bugs. Bed bugs, stink bugs, lace bugs, and giant water bugs are true bugs, with the typical thickened forewings crossed over the delicate hindwings to form a triangle on the top of the thorax. But common waterbugs are cockroaches of the Order Orthoptera, and lightning-bugs are beetles of the Order Coleoptera. The piercing and sucking mouth parts of bugs, sometimes almost as long as the insect, can be folded back under the thorax. Metamorphosis is incomplete, with nymphs somewhat resembling small adults. Besides causing pain and annoyance, bugs are frequently ill-smelling, and they are common parasites on crop plants.

The Order Hymenoptera (Greek, "membrane wings") are the bees, wasps, and ants (but not termites, which are in the Order Isoptera). The hindwings are loosely connected to the forewings by a row of tiny hooks. The mouth parts are mostly for chewing, but honeybees have a tubular sucking apparatus. Metamorphosis is complete, with grublike larvae. The Hymenoptera are special in their method of sex determination: Females develop from fertilized eggs, males from unfertilized, or *parthenogenetic*, eggs. As pollinators of many useful plants, as pro-

ducers of honey, and as predators on destructive insects, hymenopterans are among the insects most useful to people.

One final order of insects, the Order Aptera (Greek, "no wings"), should be mentioned because they are believed to represent the most primitive of living insects. However, their wingless condition is probably secondary; that is, they probably evolved from earlier, winged forms. They have chewing mouth parts and incomplete metamorphosis, as exemplified by those common household pests the silverfishes (Figure 10.37).

Section 2:
Echinoderms Through Mammals

PHYLUM ECHINODERMATA

The echinoderms (Greek, "hedgehog skin") are almost radially symmetrical animals with a trace of bilaterality in the adult forms and clear bilaterality in the larvae. They are therefore thought to have evolved from a bilateral ancestor, and their radial shape is a secondary development. In their embryological development, echinoderms are different from the previously described phyla, but they are like the chordates, which will be treated later. Echinoderms are deuterostomes; that is, the formation of the body during early development is characterized by the growth of an alimentary tract with a mouth breaking through at the anterior end, away from the embryonic blastopore. Also, cleavage is radial and indeterminate, as it is in the chordates. Such general and fundamen-

× 2

10.38

The underside of a starfish (Asterias). The grooves under the arms are filled with waving *tube feet*, which the animal can push out by filling them with water or retract by contracting muscles in each tube. At the end of a tube foot, there is a sucker that works like a vacuum pad. Using many of those suckers, a starfish can exert a strong and continuous pull and open a tightly shut mussel, as this one is doing.

× 0.25

10.39

The underside, or oral surface, of a starfish. The white spots on the body are calcareous ossicles that give the outer skin its hardness. Tube feet are visible along the groove in each arm. The mouth is at the center of the star, and a light-sensitive organ is at the end of each arm.

tal events led biologists to classify the echinoderms and chordates as one major branch of evolutionary development, assuming that these two groups had a common ancestor, and that they are different from the other phyla, such as the Mollusca and Arthropoda.

Echinoderms are strikingly different in their adult forms from any other animals. Aside from their apparent radial symmetry, which is unusual for animals, they have embedded in the body wall skeletal elements of calcareous plates or spiny spikes. Some echinoderms (sea urchins) have shell-like *tests*. One special feature, a *water vascular system*, can be used as a hydraulic pump to extend the soft, extensible *tube feet*, which have muscular contractility and carry terminal suckers (Figure 10.38). Echinoderms use their tube feet for movement, capturing food, respiration, and sensory perception. Another peculiarity of echinoderms is the possession in some species of *pedicellariae* (singular, pedicellaria), hard, jawlike pairs of pincers that act like tiny pliers with crossed jaws, and that are useful for cleaning the skin surface. Echinoderms are mostly sluggish, and like most radially symmetrical animals, they have a poorly developed nervous system and simple sensory receptors. They are widely distributed in practically all marine habitats, from the shallowest to the deepest seas. They are of little economic importance, although some can be destructive to shellfish. A few are eaten, for example, sea cucumbers by Pacific islanders.

In spite of millions of fossilized echinoderms, the earliest members of the phylum are missing, and the ultimate origin of the group is uncertain. They have been in existence since the Cambrian Period, more than 500 million years ago, and have been abundant ever since, but they are a peculiar side branch of animal evolution and have given rise to no more advanced types.

Classes of Echinoderms

Five classes of echinoderms are living. The Class Crinoidea contains the stalked sea lilies, with five branching arms at the top and a mouth and anus on the upper surface. Some rocks are almost solid masses of fossilized sea lilies, but living ones are mostly restricted to deep seas and are not as common as the other classes. They are the most primitive of echinoderms. The Class Asteroidea is familiar to anyone frequenting temperate or tropical shores. Its members are the starfishes (Figure 10.39). The five-rayed body is stiffened by endoskeletal plates, protecting both the internal water vascular system and the extensive digestive system. The *ambulacral groove* along the underside of each arm is lined with tube feet, with which the starfish can creep and capture prey. Starfishes have strong powers of regeneration, being able to replace lost arms or even to make two complete animals from one torn in two. The Class Ophiuroidea contains the brittle stars, which are like starfish but have slender or much-branched arms and no ambulacral grooves or pedicellariae (Figure 10.40). The Class Echinoidea is characterized by the presence of a firm globular or flattened skeleton or test, from which movable spines radiate. Some common echinoids are sea urchins, sea biscuits, and sand dollars, whose bleached tests are cast up by millions on beaches all over the world. The greatest human use of echinoids is in experimental biology, because sea urchin eggs are favorite subjects for embryological investigation. The Class Holothuroidea contains the sea cucumbers, which are soft-bodied animals like flabby, wet sacks with a ring of tentacles at the oral end. When disturbed, a sea cucumber can throw up its insides and cast away the whole mass, later regenerating a new set of organs. Some Asian people consider the sea cucumber a gastronomic delicacy, calling it "trepang."

PHYLUM HEMICHORDATA

The hemichordates are soft-bodied, usually wormlike marine animals, little known except to biologists, who regard them with interest because they are thought to occupy a unique position in the evolutionary development of the vertebrates. Their determinate cleavage and deuterostome methods of embryonic development show that they belong in the echinoderm-chordate line of evolution. They are like echinoderms in their larval forms in that they have a hydraulic system like the water vascular system of starfish and a subepidermal nerve net. They are like chordates in having gill slits (although they use them mainly for feeding) and a hollow, dorsal nerve cord (although it is restricted to a short portion of the body). They also have a rodlike structure that for many years was thought to be comparable to the notochord of vertebrate animals. The *notochord* is a cellular rod, covered with a double layer of fibers, extending almost the entire length of the animal just ventral to the dorsal nerve cord. It acts as a stiffening support in those animals that retain it throughout life and in the embryos of those animals that replace it with a backbone. Recent interpretations indicate that what was thought to be a notochord in the hemichordates is rather an outgrowth of the mouth, and it has been renamed a *stomatochord*. One of the better-known hemichordate examples is an acorn worm with the euphonious name of Balanoglossus, a secretive mud-dweller whose main claim to fame is its intermediate position in evolution between the echinoderms and the chordates (Figure 10.41). The hemichordates are sometimes called the prechordates because of the possibility that they resemble the ancestors of the chordates.

10.40

Brittle star of the Class Ophiuroidea. Brittle stars differ from ordinary starfish in having a central disk that is distinct from the slender arms. They are called brittle stars because their arms break easily. Loss of an arm or two does not seriously inconvenience a brittle star, because it can grow new ones within a few weeks.

10.41

A hemichordate, Dolichoglossus, closely related to Balanoglossus. This wormlike dweller in New England sea water bears little resemblance to the familiar vertebrate animals, but its possession of gill slits makes its inclusion in a phylum close to the chordates reasonable.

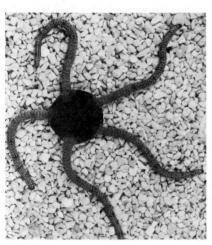

× 1

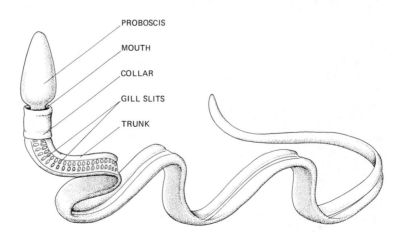

PROBOSCIS

MOUTH

COLLAR

GILL SLITS

TRUNK

PHYLUM CHORDATA

Although chordates vary from tiny, immobile sea squirts, through such unlikely phylum-mates as hagfishes, to chickens and whales and humans themselves, they are all classified in one great biological group because they have several features in common and are therefore thought to have evolved from a common ancestral group. No one knows what manner of animals those ancestors were, or even if there was one common ancestral group. The evidence, however, seems reasonable that the chordates share a beginning with the echinoderms, since both phyla have the same indeterminate cleavage of the early embryo, both develop an anus from a blastopore of the embryo and a mouth that breaks through later at the anterior end, and both form mesoderm and coelom in the same way. Both phyla, then, are deuterostomes, and current opinion has it that echinoderms and chordates represent one important, two-forked line of evolutionary development.

In addition to the characteristics they share with echinoderms, the chordates have a number of features uniquely their own. (1) Every chordate has a *notochord* at some stage of individual development, although it is a temporary embryological organ in most species. (2) A *tubular nerve cord* runs the length of the body, near the dorsal surface, and is enlarged and complicated at the head end into a brain. (3) *Gill slits* are openings in the region of the pharynx that lead from the internal alimentary canal to the outside of the animal. Gill slits function in respiration in fishes, but they are closed early in embryological develop-

ment in air-breathing animals. The slits are clearly evident in young human embryos. The bars of tissue between the slits, called *gill arches*, remain, in greatly changed form, as such structures as human tonsils. (4) A *tail*, extending beyond the anus, is always present at some stage of development, even though it may be absorbed or overgrown in such animals as frogs and humans. Other characteristic features of chordates are shared with animals of other phyla: bilateral symmetry; evidence of segmentation in the body; a well-defined head end; the usual three embryonic germ layers of ectoderm, mesoderm, and endoderm; a well-developed coelom; paired appendages in most groups; and a firm internal skeleton, or endoskeleton. A few have external skeletons, or exoskeletons.

Subphylum Urochordata

Little known except to biologists and beachcombers, the sea squirts (Figure 10.42) are so different in general appearance from most chordates that they would not be recognizable as such, except for their larval stages. Larval sea squirts, however, have the essential features of chordates: the notochord, the hollow dorsal nerve cord, and the pharyngeal gill slits. After a free-swimming stage, the larva attaches itself to a solid substrate and undergoes such a series of changes that by the time it has become an adult it is a different-looking animal. For many years, in fact, it was thought to be a mollusc. Only its larval features made possible a correct understanding of its classification. The adult is a *sessile* (attached by a broad base) animal, usually growing in colonies on marine vegetation, where it feeds by filtering particles

10.42

An Ascidian, or sea squirt. During the adult stage of a sea-squirt's life, shown here, the animal is not recognizable as a chordate or perhaps even as an animal. Sea squirts increase the number of individuals by growing buds, all of which are connected by a common water system. They are reactive and can eject streams of water when they are disturbed. Their evo-

lutionary connection to other chordates is evident only during the free-swimming larval stage, when the tiny animals have a notochord, a dorsal nerve cord, and gill slits. When they settle down to a sessile style of life, they change the entire body form, lose their original larval features, and even turn to an asexual method of reproduction.

× 5

× 4

10.43

A lancelet or Amphioxus. This tiny fishlike animal, a sand-dweller along sea beaches, serves as a prototype chordate. It has a dorsal nerve cord, and it retains its notochord and gill slits throughout its life. In this photograph, the V-shaped, segmented muscles are faintly visible, but most of the internal organs are opaque and so cannot be seen. Amphioxus may well represent the kind of animal from which the more advanced chordates, the vertebrates, evolved. However, they are still thriving. In some places, such as in the Pacific Ocean along the China coast, they are abundant enough to be caught and eaten.

out of the water and forcibly ejecting spurts of water from its excurrent siphons—hence the name "sea squirt."

The resemblance of the larva to a hypothetical chordate ancestor and the lack of resemblance of the adult to anything else have prompted the suggestion that the entire vertebrate line of evolution has been from a larva that gained the power of reproduction. Some larvae in flatworms, insects, and amphibia can indeed reproduce, so such a suggestion is not as far-fetched as it might seem at first inspection. The urochordates themselves, however—the adult stages in particular—never gave rise to anything more advanced than urochordates, and the group is interesting to biologists because of the primitive nature of the larvae, which have all the necessary chordate features, and yet the adults have retained only one: the gill slits. Adults diverge further from typical chordate behavior by surrounding themselves with plantlike cellulose.

Subphylum Cephalochordata

The most famous cephalochordate to biologists is a small fishlike creature known as the lancelet, Branchiostoma (or Amphioxus) (Figure 10.43). Lancelets burrow backwards in the sandy shores of warm seas, each only about 5–7 centimeters (2–3 in.) long, where they feed by filtering the water. They resemble larval sea squirts, but they remain permanently free-living. They have the notochord, dorsal nerve cord, gill slits, and postanal tail characteristic of chordates, and they are so like what zoologists imagine the prototype chordate to have been that they have been called "the blueprint of the phylum." They also have a seg-mented muscular body; and V-shaped (with the V lying on its side) *myonemes* or muscle segments make up the bulk of the dorsal part of the body. Such an arrangement, typical of vertebrate bodies, contrasts with the more symmetrical muscular arrangement in most worms. It thus foreshadows the body structure of higher forms.

Subphylum Vertebrata

To most nonbiologists, the vertebrates are *the* animals. In addition to the usual chordate features, vertebrates have a number of structures all their own. These include a special kind of covering, generally called skin, consisting of at least two layers, one of ectodermal and one of mesodermal origin, with capacity for forming a number of accessory parts, such as glands, hair, nails, claws, hoofs, feathers, and scales. There is a spinal column of vertebrae, from which the entire subphylum takes its name, and that column usually has two pairs of bone-supported appendages. In the closed circulatory system, the ventral heart pumps red blood throughout the body, although structural details vary from class to class of vertebrates. A pair of kidneys removes waste from the blood. The brain at the head end, with five vesicles, is encased in some kind of skull, is connected to the typical dorsal nerve cord, and is provided with 10 or 12 cranial nerves. An endocrine glandular system, distributed throughout the body, helps regulate physiology and behavior by means of hormones. The typical body has a head, trunk, and tail, usually fore and hind appendages, and sometimes a neck. Sexes are usually separate. This, then, is the vertebrate plan, and it has been one of the great evolutionary suc-

10.44

A longitudinal section through a stylized vertebrate animal. Only two features are common and lasting throughout the life of most vertebrates: the row of *vertebral bones* and the *dorsal nerve* with a brain at the anterior end. Other features, although present during early development of an individual, may disappear or at least be inconspicuous in later life. Examples are the cartilaginous *notochord* and the openings from the pharynx to the outside, the *gill slits*. Other characteristics, not all necessarily present in adult forms, are bilateral symmetry, body segmentation, a definite head and tail, a true coelom, and paired locomotor appendages.

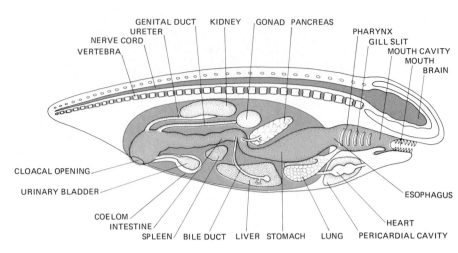

cesses, along with the arthropod plan, in the earth's history.

The taxonomic arrangement of vertebrates, especially fishes, is always in a state of change. As in most systems of classification, there is never complete agreement among zoologists as to which scheme is most satisfactory. The arrangement given here is a conservative one, simplified by the omission of some of the smaller and less well-known groups.

Class Agnatha In the Class Agnatha (Greek, "no jaws") are the lamprey eels and hagfishes. As the name indicates, these animals have no jaws but have a sucker, some with teeth, by means of which they attach themselves to living or dead fish. Lamprey eels are such serious predators that once they began to live in the Great Lakes of the United States during the 1950s, they almost destroyed the commercial fisheries there. The parasites attach themselves to a fish and suck out its juices until it dies. Attacking the larval stages, which develop in small tributaries of the lakes, has brought them under partial control. Lampreys have a bony skeleton, eyes, brains, and either male or female gonads, but living as they do on fish juice, they do not need and do not have stomachs. Hagfishes, in contrast, are not parasites, only scavengers. A special feature is that they are hermaphroditic vertebrates, although no one hagfish produces both eggs and sperm at the same time. Both types are regarded as primitive examples of vertebrate development.

Class Chondrichthyes The Class Chondrichthyes (Greek, "cartilage fishes") contains the sharks, skates, and rays (Figure 10.45). Their skeletons are of cartilage instead of bone, and this condition is considered a secondary development rather than a primitive feature. The typical shark body is streamlined into a spindle shape, but the basic shape is obscured in rays by the outgrowth of lateral "wings," which give the fishes something of a butterfly appearance. The mouth on the underside of the head is equipped with rows of teeth, which are continually replaced as they are lost. Five gill slits allow the passage of water from the pharynx over the gills to the outside. Nostrils, indentations in the underside of the head, are functional as smelling organs but do not open to the inside. A pair of anterior *pectoral fins* and a pair of posterior *pelvic fins,* supplemented by two median dorsal fins and an asymmetrical tail, provide movement. Males use part of their pelvic fins as clasping organs during mating. The skin feels rough because of the presence of tiny enamel-covered scales, formed in the same way as the teeth in the mouth. The skin scales are so sharp and so firmly anchored that shark skin can be used for sandpaper. Along each side of the body a *lateral line organ* acts as a sensitive receptor of pressure changes. With that sensory organ the fish can detect currents, prey, and mates. Sharks also have functional eyes, but they have no eyelids, so it is impossible to know whether or not they sleep. Sharks, however, do go into periods of immobility, during which they act as if they must be asleep.

Sharks are neither so ferocious as fictional stories imply nor so mild-tempered as some skin divers insist. They are generally rather innocuous creatures, but well-authenticated instances of shark attack on swimmers have been recorded, especially by great white sharks, mako sharks, and tiger sharks. Rays, too, can be dangerous because of their rasping tails, but they are more impressive than actually injurious, with their enormous winglike flaps. They can leap out of the water, wave themselves ponderously, and because of their great size seem to be flying in slow motion. Other rays have the ability to build up enough electric potential to give a stunning shock even to a human adult.

Blood circulates through a closed system, which carries it past the gills, through the body, and back to a two-chambered heart with an atrium and ventricle in tandem. The increase in complexity of the hearts and circulatory systems of more advanced animals is important to watch for in the descriptions coming next, because the different animal groups show a steady progression from the simple shark heart to the double pump of _homeothermic_ or "warm-blooded," animals. In the Chondrichthyes, sexes are separate and fertilization is internal. Some sharks are _ovoviviparous;_ that is, the females retain eggs within their bodies until young sharks are partially developed. Other species are truly _viviparous_, nourishing the young within a uterus until they are born alive. (Animals that lay eggs are _oviparous_.)

Class Osteichthyes The bony fishes of the Class Osteichthyes (Greek, "bone fishes") occupy in water the position that birds and mammals do in terrestrial environments (Figure 10.46). The bony fishes have evolved into some 18,000 modern species (as compared with only about 3000 Chondrichthyes) inhabiting the waters of the earth, from shallow ponds to the everlasting darkness of marine deeps. They have evolved into many shapes, from the pencil-like slenderness of pipefish to the squatty lumpiness of toadfish. The lungfish can survive for months out of water, and some catfish can migrate across stretches of land, struggling along with their fins on the ground.

Some mature minnows may be only a few centimeters long, whereas swordfishes may be 4 meters (12 ft) long, but the bony fishes do not rival the great sharks, with their 10-meter (35-ft) length. Most fishes are shaped to glide easily through the water, though some of the sluggish ones are variously bulbous. Their skins may be scaleless, or they may be furnished with any of several kinds of scales, some of

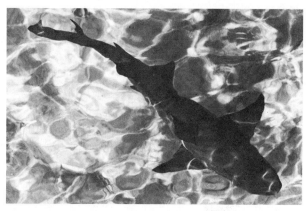

10.45
A hammerhead shark, swimming in South Atlantic waters. Sharks are special fishes in several ways: They have a cartilaginous instead of a bony skeleton, their tails are asymmetrical, their mouths open on the underside of the head, and they maintain a high concentration of urea in their blood. Like many animals with a reputation for ferocity, most sharks are not especially dangerous to human swimmers. However, they do attack people often enough to breed a healthy fear in regions where they abound.

× 0.1

10.46
A representative of the bony fishes, or Osteichthyes, named for their possession of a bony skeleton that distinguishes them from the sharks and rays, which have cartilaginous skeletons. This one is a salmon, Salmo, swimming up an Alaskan waterfall on its way to the spawning territory where it will lay (or fertilize) eggs.

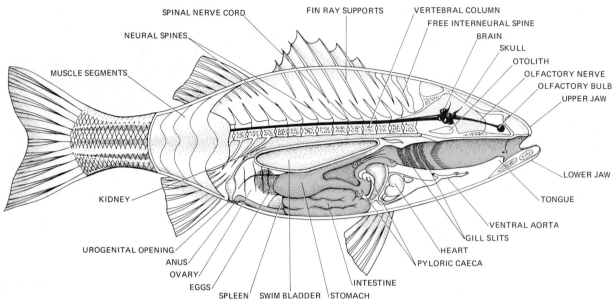

10.47

The internal structure of a fish. The head is large in proportion to the rest of the body, but the brain is small. The bulk of the body is muscle, clearly segmented, with a small space along the ventral side containing the digestive, excretory, and reproductive organs. The external nares or nostrils are pits which do not open into the pharynx, but are supplied with sensory nerves leading to the olfactory bulb. The gill slits allow water taken in by the mouth to pass over the gills and out, allowing the gill arteries to give off carbon dioxide and take in oxygen. The swim bladder can be inflated or deflated to maintain a suitable specific gravity for the fish's body.

× 0.1

10.48

A coelacanth fish, *Latimeria chalumnae*. This rare animal has been caught only a few times off the east coast of Africa. It belongs to the group known as lobe-finned fishes, which interest biologists because they may represent an evolutionary link between the usual fishes (with bony rays in their fins) and the amphibia (with legs). Added interest rises from the fact that lobe-finned fishes were known from fossils long before a living specimen was seen. They must have been fairly common some 700 million years ago, lasted for 400 million years, and then become rare. The first living one known to biologists was caught in 1938 at a depth of about 80 meters.

which, like growth rings in a tree trunk, indicate the age of the animal. The mouth is terminal, as contrasted with the under-the-head shark mouth, and the gill slits are covered by a bone-supported flap, the *operculum*. Both paired and median fins are strengthened by ribs of bone or cartilage. In bony fishes, the tail fin is typically symmetrical.

Like Chondrichthyes, Osteichthyes have a two-chambered heart, with one atrium and one ventricle, pumping red blood with nucleated red blood cells throughout the body. Oxygenation of the blood occurs when it is pumped past the gills. Some fishes also have an air chamber, or *swim bladder*, which can be filled with gas or emptied, keeping the buoyancy of the body within necessary limits (Figure 10.47). The sexes are separate and fertilization is usually external.

One rare fish deserves special mention. In 1938, commercial fishermen caught a specimen of a group of lobe-finned fishes, or coelacanths, of the Genus Latimeria (Figure 10.48). Lobe-finned fishes had been long known from fossils, but they were believed to be extinct. The discovery of this "living fossil" stirred up interest among zoologists because the paired, lobed fins resemble the fore and hind limbs of amphibia. It is thought that from such fins could have come the tetrapod (four-footed) body plan that became established as typical of land-dwelling vertebrates. Over the years, several dozen more Latimerias have been hauled up from the deep seas off the east coast of Africa, and they have been intensively studied, as is fitting for a relic that has survived almost unchanged for millions of years.

Fish have been a main source of human food, probably as long as people have been on earth. With the recent enormous increase in human population, there has been greater and greater emphasis on catching sea food, and some unrealistic optimists have thought that the oceans would be our salvation. But with ever-greater numbers of fishermen and ever-more efficient methods of capturing fish, the fish population is being depleted rapidly, and there are estimates indicating that it is being depleted faster than it can recover. Some schemes for increasing the food supply from the sea include processing "trash" fish for protein flour; developing intensive fish farms utilizing algae as fish food; and inducing the upwelling of mineral nutrients from the depths of the sea, where they are now not available for algal growth and therefore for the subsequent nourishment of organisms higher on the feeding hierarchy. Each of these methods presents technical problems, and all

must for now be regarded as possibilities rather than probabilities.

Class Amphibia   In the Class Amphibia (Greek, "double life") are the animals that came out of the water and established a successful life on land. Getting onto land offered some advantages, especially escape from competition, availability of a greater number of places to live, and a freer oxygen supply. But with the advantages there were problems. Temperature fluctuations in air are greater than they are in water. Air has great drying power, which can be dangerous to any organism. And air lacks the great buoyancy of water, so that a land animal needs stouter support than an aquatic one. Amphibians took advantage of the improved possibilities and adapted themselves to the difficulties, mainly by remaining close to water, by keeping a soft, moist skin, by developing lungs, and by evolving a sturdy, bony skeleton with a strong vertebral column and four legs.

Of the three orders of amphibians, two are familiar. In the Anura (Greek, "no tail") are frogs and toads. In the Urodela (Greek, "with a tail") are the salamanders, efts, newts, mud puppies, and Congo eels (Figure 10.49). The third order, the Apoda

(a)   × 2   (b)   × 1

10.49

Two representative amphibians. (a) A tree frog. Frogs are typically tailless, and they have unusually strong hind legs. Tree frogs have in addition a specially adapted set of suction discs on the digits of their front feet and can therefore cling effectively to twigs or even to smooth surfaces. Being amphibians, they may spend most of their lives on land (or on vegetation), but they must pass their larval or tadpole stages in water, even if that water is only a few milliliters trapped in a leaf cup. (b) A salamander. In European folklore, salamanders were thought to live in fire, probably because they lived in moist places, such as old logs, and if a log was added to a fire, the salamanders would be driven out. The short, rather weak legs and the long tail are typical. Salamanders scramble along on their bellies, wriggling their bodies and pushing with their feet. Many are brightly colored, with red, yellow, orange, and black markings.

(Greek, "no feet"), contains a few rare tropical, legless amphibia, which are mainly biological curiosities. They are the cecilians, which look more like large worms than like amphibia. All three orders share a number of characteristics. The bony endoskeleton is the main body support; the notochord is absorbed during development. Breathing is mostly by means of skin and lungs, although external gills are present in larvae. The circulatory system shows an advance over that in fishes. There is a three-chambered heart, with two atria and one ventricle, and a double circulation —that is, one set of blood vessels to the body as a whole and a separate one to the lungs. The two systems, however, are not entirely separated as they are in birds and mammals, and the amphibia are _poikilothermal_, or "cold-blooded." There are usually four limbs (none in the cecilians), with powerful hind limbs in frogs, but weak, rather ineffective legs in salamanders. Salamanders and other urodeles can scarcely be said to walk. Rather they slither along on their bellies, pushing themselves forward by a combination of wriggling and flapping their legs. The original use of lobed fins by prototype amphibia, and later of legs, was probably not so much to get _onto_ land as to get _across_ land in an effort to get back to water from a drying mud hole.

Modern amphibia still require water for the early development of embryos and larvae. Most frogs, toads, and salamanders lay eggs in ponds, where the young begin to grow, but a number of methods enable some species to bypass such obvious nurseries. Some lay eggs in small pools of water in cupped leaves, and some carry the young in folds of parental skin or even in the mouth of the father. Some amphibia attain sexual maturity and reproduce during larval life. For example, larvae of the Mexican axolotl reproduce, but if conditions are dry, they may grow into equally reproductive adults. The mud puppies of the eastern United States remain permanently larval.

Amphibia are of little economic use beyond some human consumption of frog legs for food and the aid they provide in eating insects. There is a fair industry that provides frogs for teaching purposes, because frogs are favorite subjects in biological laboratories. They are cheap, and they provide a reasonable model of vertebrate structure and function, but the demand for frogs for millions of biology students threatens to reduce the frog population severely, and efforts are now made to replace them with fetal pigs.

Class Reptilia   The Class Reptilia (Latin, "crawl"), containing the snakes, lizards, turtles, and crocodilians, is probably second only to the Arachnida (spi-

10.50

Two of the main living types of reptiles. (a) A painted turtle. Turtles are special in their chunky body form and their possession of a top and bottom shell. The terrestrial ones have clawed toes. Sea turtles have flippers and are at home only in the water. (b) A green python, one of the world's large snakes. Because they are legless, snakes have had to make some adaptations in their body structure, including hinged jaws and a streamlining of internal organs that fit in the slender tubular body. The presence of legs in the ancestors of snakes is suggested by the existence of vestigial leg bones in some pythons. Snakes have long held a prominent place in the mythology, folklore, and religion of many peoples—from the garden of Eden with Adam and Eve to Appalachian mountain country with its rattlesnake cults.

(a)   × 0.5

(b)

10.51

One indication of the relationship that exists between reptiles and birds: the presence of a single *occipital condyle*, the bearing surface of a skull on the first vertebra. In (a), which shows a reptile, a turtle, and in (b), which shows a bird, a gull, the arrows point to the single occipital condyle. In contrast, (c) shows the skull of a mammal, an opossum, with two occipital condyles.

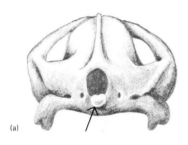

(a)

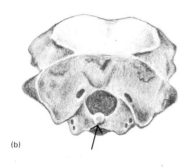

(b)

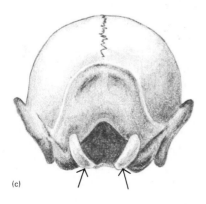

(c)

ders) in its association in popular mythology with a mysterious and sinister world (Figure 10.50). Modern reptiles are in fact a lingering remnant of a once dominant land fauna that included dinosaurs and various flying and swimming reptiles. They are no colder than most objects in nature, and indeed most maintain a warmer body temperature than the surrounding air. They are not slimy unless they have slimy algae growing on them (as some do). They are no more ferocious than any other animal that must eat to live, and they are no more secretive than any other animal that must escape its hungry predators.

Reptiles probably evolved from an amphibian ancestor nearly 300 million years ago, developing the special characteristics that make them distinctively reptilian. The endoskeleton is that of the usual vertebrate, but there is an additional exoskeleton of horny plates or scales, and the skull has a peculiar feature: one *occipital condyle* (Figure 10.51). That is the bearing surface where the skull rests on the first *cervical* vertebra, or neck vertebra. Other classes of vertebrates, except for birds, have two occipital condyles. Breathing is accomplished by means of lungs, gills having been left permanently behind in evolutionary development. In most reptiles, the heart is three-chambered, but in alligators and crocodiles it is four-chambered, so that these animals are the first (as treated in this evolutionary sequence) to have such an organ. They are still not homeothermic, however. Reptiles were the first vertebrates to become true land-dwellers. Their ability to breathe air, their possession of effective walking legs with five toes each, and the strength to stand on those legs all contributed to their terrestrial success. None of these features, however, would have been enough to free the reptiles completely from some aquatic stage without one additional development: the ability to reproduce without free water. That problem was overcome with the rise of the ability to lay *amniotic eggs*. Such eggs are water-filled and covered with a water-retaining sheath, which in reptilian eggs is leathery. They are called amniotic because they contain, among other membranes, the *amnion*, one of the special extraembryonic membranes (Chapter 14).

Four orders of reptiles are recognized. The Order Rhynchocephalia contains only one species, *Sphenodon punctatum*, the tuatara. This is another "living fossil," a lone survivor of an ancient reptilian group, now alive only in a few islands of the Cook Strait in New Zealand. One of its remarkable features is a scale-covered third eye.

The Order Testudines is the order of the turtles and tortoises. These vary from the small box tortoises of North American woods to the great sea turtles and

10.52

An American alligator, *Alligator mississippiensis*. By making water holes in the swamps of Florida and the Gulf Coast states, alligators provide living (and sometimes dying) quarters for many other animals that would otherwise not be able to inhabit the areas during the dry seasons.

the giant land tortoises of the Galapagos and Seychelles Islands. In the Testudines, the vertebrae are fused to a dorsal shell, or *carapace*, and the ribs are fused to a ventral *plastron*. Testudines have no teeth, but they have horny jaws. The sea turtles are known for their ability to find their way year after year across the oceans to specific spawning grounds, where they mate and lay eggs on sandy beaches. Like many edible marine animals, these turtles have been hunted by humans until their numbers have been dangerously reduced.

In the Order Squamata (Latin, "scaly") are the snakes and lizards. Snakes can be thought of as essentially legless lizards, although boa constrictors sometimes even yet grow a few vestigial bones where the hind limbs of ancestors used to be. Lacking chewing teeth, snakes either swallow their prey alive or crush it in coils until it ceases to breathe, or they kill it by poison. Having a lower jaw that can drop far down from the upper jaw, and having no breast bone to restrict the rib cage, a snake can swallow an object larger than itself. Snake poisons act on either the nervous system or the blood (or both) of the victim, in either way causing death from lack of oxygen. In the United States, rattlesnakes, copperheads, cottonmouth moccasins, and coral snakes are the main venomous reptiles, and hundreds of bites are reported annually. Nonetheless, snakes are in general timid animals, which will escape if they can and will bite only when alarmed. Even nonvenomous snakes can bite painfully, but their most effective defense is the voiding of foul-smelling excreta onto an attacker. Snakes eat mainly small animals, especially rodents and insects, though some of the large tropical snakes, such as the great pythons and anacondas, can handle prey as large as a pig. Snakes in turn are eaten by

other snakes and by predaceous birds and mammals.

Snakes can see, they can smell, mainly with the help of the extrudable tongue, and they are sensitive to temperature changes, but they have no ears and can perceive sound only if tremors shake the ground. Snakes can move by coiling up and lunging forward, but they progress mainly by the muscular movement of the backward-directed scales on their bellies, which they use to anchor against rough surfaces. A snake on a polished floor is practically helpless. Lizards, in contrast, have achieved true terrestrial locomotion, and many are fleet of foot. Some can rise on their hind legs and run, even skittering along over the surface of water when they are in a hurry. (See the picture essay in Chapter 21.)

The Order Crocodilia contains the broad-snouted alligators and the slender-snouted crocodiles, both inhabiting the United States (Figure 10.52). Crocodilians are well-known animals that were once numerous in Africa and the Far East, but they have been so mercilessly hunted for their skins that they are now severely depleted. In many places they are legally protected, and in some even the selling of alligator-hide products is prohibited. For all their carnivorous ways—some aggressive crocodiles are capable of capturing deer—they have a remarkable fondness for marshmallows.

Class Aves All birds are in the Class Aves. They make a rather homogeneous grouping, apparently having evolved as one branch from an earlier reptilian stock. Their scaly feet and the possession of a single occipital condyle are good indications of their relation to reptiles, but even more convincing is the evidence from fossils. The famous Archaeopteryx fossil, discovered in Bavaria in 1877, is the kind of fossil that gladdens the evolutionary taxonomist, because it has obvious features common to both the old and the new. Archaeopteryx had the teeth, the long reptilian tail, and the clawed wing digits that show its reptilian nature, but it also had the two distinguishing characteristics of birds: wings and feathers.

Many of the structural and behavioral features of birds are associated with their ability to fly. The streamlined body shape, the light, hollow bones, the air sacs in the body, and the smooth covering of feathers are all part of the pattern of a bird. (See the picture essay in Chapter 21.) Being able to fly away from danger enabled birds to survive without many of the elaborate defenses that earthbound animals developed. Even their nervous systems and behavior patterns were influenced by the ability simply to leave problems behind. Birds can learn, they are fantastically agile, and their perceptions (especially sight) are acute. Birds are homeothermic, and they have four-chambered hearts with pulmonary and body circulation completely separated. With the effective insulation of feathers, they maintain a higher body temperature than do most other animals, usually 40–42°C (104–107°F). Another avian development is the egg with a hard, calcareous shell, which must be kept warm during incubation. The requirement for parental care has led to elaborate behavior patterns of mating, nesting, care for the young after hatching, and even extensive migration (Chapter 28).

For all its advantages, flight puts some limitations on birds, and several independent bird genera have lost the ability to fly. Some birds became big enough to defend themselves; others evolved in a place where they had no dangerous enemies. In both types, flight was not necessary for survival. The ostriches of Africa, the rheas of South America, the kiwis of New Zealand, the penguins of Antarctica, and the cormorants of the Galapagos Islands are all flightless, and so were other, now-extinct species. Occasionally birds become nuisances, as when they get into airplane jet engines or accumulate, as starlings do, in flocks of millions where they are unwanted, but in general birds are regarded as friendly to humans. They are useful scavengers, and they eat unimaginable numbers of harmful insects. Some are themselves edible, and some effect pollination of flowers. They are frequent subjects of biological experimentation, and with their pleasing calls and colorful feathers, they add to our esthetic enjoyment of what would be a less attractive world without them.

Class Mammalia The Class Mammalia has been divided into three subclasses. The most primitive, Subclass Prototheria (Greek, "first beasts"), contains the egg-laying mammals of Australia, the duck-billed platypus (Ornithorhynchus) and the spiny anteater (Tachyglossus) (Figure 10.54). The name of the Mammalia is derived from the Latin word for breast,

10.53

A bird is easy to define: an animal with feathers. Feathers, developing on reptilian animals during the Jurassic Period, some 150 million years ago, were probably at first useful as an effective covering to keep the animals warm. Only later were they used to help in flying. The example here is a pelican in Florida.

× 0.12

10.54

The duck-billed platypus, *Ornithorhynchus anatinus*. The platypus survives only in Australia and Tasmania, and it has proved extremely difficult to keep alive in captivity in other parts of the world. It is a queer enough animal in shape and habit, with its ducklike snout and aquatic preference, but its more unusual features include webbed hind feet with poison glands in the claws, its habit of laying eggs, and the feeding of its young by having them lap milk from the mother's teatless underside. This cat-sized animal eats small aquatic creatures.

× 0.2

10.55

An American opossum and her family of half-grown young. When born, baby opossums are naked, ill-formed objects, scarcely more than embryos, about the size of honeybees. They have no eyes or ears, and their hind legs are poorly developed. But they have mouths and vigorous front feet, and they can find their way into their mother's pouch, where each baby finds a nipple to attach itself to. There it hangs, sucking milk until it develops the completed organs that most mammalian babies have at birth. Once it grows hair and active legs, it can come out of the pouch and will ride on its mother's back, as shown here, frequently holding on to her tail with its own tail. Opossums have not evolved in any special direction, and perhaps as a result, they can adapt to a variety of habitats. They have even extended their territories with the advancing suburbs of modern America, and they are now fairly common in northern regions where they were a rarity as recently as 25 years ago.

and the presence of breasts for feeding the young is one of the most distinctive features of the class.* Although the Prototheria do not possess breasts like those of other mammals, they do have milk glands, modified sweat glands that secrete milk into the fur along the mother's belly, where the young can lick it. The Subclass Metatheria (Greek, roughly translatable as "next beasts") are the marsupial mammals, the opossums, kangaroos, wombats, wallabies, and koala bears, which bear young in a poorly developed condition, then keep them in an abdominal pouch, the *marsupium*, where each embryonic baby attaches itself to a nipple and nurses until it achieves a measure of independence (Figure 10.55). In America, opossums are the only marsupials, but in Australia marsupials are (or were before the arrival of Europeans) the dominant animals. The third subclass, the Eutheria (Greek, "true beasts"), is made up of the placental mammals, those whose young are developed in the maternal uterus and are fed after birth by nursing.

Mammals and insects now dominate the earth. Mammals inhabit practically every available space, from the arctic ice to the forests, grasslands, and deserts of land and to the depths of the sea. They have been increasing in numbers (but not necessarily in numbers of species) and importance since the dinosaurs went into decline toward the end of the Mesozoic Era, about 70 million years ago, but fossil animals from the Jurassic Period, 150 million years ago, show a combination of reptile and mammal characteristics. Those were the *therapsids* (Greek, "look like beasts"), which may have been the ancestors of mammals, although the really important identifying marks of mammals, breasts and hair, are not preserved in those early fossils. The early mammal-like animals were small, many no larger than rats, but they stood up on their four legs, developed a greater agility than the reptiles had, and along with that agility had an effectively coordinated nervous system. Being smarter than the reptiles they were displacing, they took advantage of all their other emerging adaptive features to increase their domination of the earth. As more and more truly mammalian features evolved, the animals became warm-blooded and thus were able to enter and remain in places that were effectively closed to the cold-blooded reptiles.

---

* "Mama" is an easily articulated, repetitious sound, uttered by babies in many countries and in many languages, including Indogermanic, Greek, Russian, Welsh, Irish, and the Romance languages. It has apparently been taken to mean both the mother herself and the mother's nourishing breasts.

10.56

An insectivore, the star-nosed mole, *Condylura cristata.* Its unusual face has a ring of sensitive feelers around the snout. Like most moles, this North American mole lives out of sight, under the soil or swimming in water. Unlike most moles, however, it is sociable, at least with its own kind.

Powerful diggers, moles can practically swim through the soil, pushing ahead with their noses and throwing the soil back with their front feet. They live on insects and other small animals, and they eat almost constantly.

10.57

A rodent, the tiny European harvest mouse, *Micromys minutus.* Only about six centimeters long, harvest mice have so great a surface-to-volume ratio that they lose heat rapidly and must eat during most of their waking hours. These, clinging to stalks of wheat, are eating the grains.

Associated with increased intelligence was a series of improvements in sensory receptors: movable eyelids and directional external ears. Their feeding habits were also helpful. With teeth in both jaws, jaw joints that could work sidewise as well as up and down, and a mouth-and-nose combination that allowed breathing during eating, the evolving mammals could nourish themselves better than any of the previous vertebrates. Finally, the reproductive method of keeping the young in a uterus and feeding them milk after birth got the mammals away from the hazards of laying eggs in nests. Once established, mammals replaced the next higher group, the reptiles, relatively fast. Within a space of some 25 million years (that is "relatively fast" for such a biological development), they reached a position of major importance about 50 million years ago and have kept it ever since.

Since the human species is mammalian, more research has been done on mammals than on other kinds of animals; consequently, we know more about them, and most people are more interested in them.

In this book, major emphasis will be on the nutrition, physiology, and reproduction of mammals, especially humans, and the student is referred to those chapters which treat specific topics. Here we will briefly note the 10 major orders of mammals. Some of the smaller and less well-known orders are not included.

The orders of mammals. The Insectivora (Latin, "insect eaters") are moles and shrews, small mouse-like animals that live close to or under the ground, eating grubs, beetles, and ants (Figure 10.56). The Rodentia (Latin, "gnaw") are squirrels, rats, mice, beavers, and muskrats (but not rabbits), with their front, or *incisor*, teeth adapted to chiseling action (Figure 10.57). Tough, aggressive, and adaptable, rats compete with people for food and carry several dangerous diseases; they are probably our most formidable biological enemy. Carnivora (Latin, "meat eater") are cats and their wild kin, dogs, bears, otters, foxes, and most of the animals used for their fur. Carnivores have long, sharp *canine* (Latin, "dog") teeth, but small, weak incisors. Their teeth are thus

10.58

A carnivore, the North American timber wolf, *Canis lupus*. Although wolves have been feared and hunted for centuries, they have recently begun to find friends among ecologists, who consider them a part of nature that should be allowed to survive. Wolves are thought to serve a beneficial service in culling weak animals from caribou herds in the north. They once ranged over most of North America, but they are now generally restricted to wild areas in the northern states and Canada. Their variation in color—black, white, and many shades of gray—is thought to be due to slight genetic differences. All timber wolves are regarded as belonging to the same species.

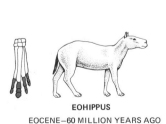

× 0.08

10.59

A member of the Artiodactyla, animals with an even number of hoofs on each foot. This is a female deer with her new fawn. The fawn will keep its spotted coat for the first year.

× 0.1

10.60

The development of the modern horse. By comparing older and older fossils, going from the obviously horse-like bones of recent types to older, smaller, and less horselike ones, paleontologists have been able to reconstruct a convincing series. One ancestor was Eohippus, a dog-sized beast with four toes. It lived in North America when the climate was warm, some 60 million years ago. It was followed by a somewhat larger descendant, Mesohippus, which had several toes but put the mass of its weight mainly on the middle toe. Mesohippus, an Oligocene creature of some 40 million years ago, gave rise about 15 million years later to Merychippus, an obviously horselike animal with a definite hoof on its middle toe, although the side toes were still present. Pliohippus, living about ten million years ago, had lost the side toes and used the middle toe exclusively. Its feet were like those of a present-day horse, but its body was small and stocky. The modern horse, Equus, has been on earth only about two million years. It disappeared from the region of its ancestors, so there were no representatives in North America when Europeans reintroduced the horse in the sixteenth century. The wild mustangs now inhabiting the western plains of the United States are escaped descendants of the horses brought to Mexico by the Spanish explorers.

EOHIPPUS
EOCENE—60 MILLION YEARS AGO

MESOHIPPUS
OLIGOCENE—40 MILLION YEARS AGO

MERYCHIPPUS
MIOCENE—25 MILLION YEARS AGO

well adapted to tearing flesh but not very good for grinding and chewing (Figure 10.58). The Artiodactyla (Greek, "even toes"), including the pigs, cattle, deer, goats, sheep, camels, hippopotamuses, and giraffes, walk on the tips of two toes on each foot, the other three toes of ancestral types having been reduced to a functionless condition (Figure 10.59). The hoofs are comparable to the claws of carnivores or the toenails of primates. Most artiodactyls are grazing animals, and they have a special digestive system (Chapter 22) that allows them to eat quickly and later "chew their cud" in safety at leisure. Domesticated since the beginning of civilization, camels and cattle

are among man's most useful animals. The Perissodactyla (Greek, "odd toes") are the animals that run and walk on the hoof of a modified middle digit: horses, zebras, the tapirs of Central America, and rhinoceroses (Figure 10.60). The Cetacea (Greek, "whale") are mammals that have returned to the sea (Figure 10.61). Besides whales, there are other cetaceans: the dolphins, porpoises, and narwhals. The last-named bear a long bony spike in the head, which used to pass for a "unicorn horn." The Chiroptera (Greek, "hand wing") are the bats, more helpful than not, in spite of the actual transmission of rabies and some vampirism of animals by tropical bats (Figure

10.61
A cetacean, the humpback whale, *Megaptera novaeangliae*. Humpbacks are relatively large whales (up to about 15 meters) with unusually long flippers that are plainly visible when the animal leaps out of the water. They are vocal animals, communicating with one another under the water by a series of complex, repeatable sounds, popularly known as the "Song of the Humpback Whale." The sounds have great carrying power in the warm waters of the Caribbean Sea, where they have been intensely studied by marine biologists.

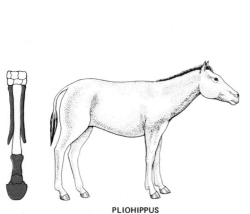

PLIOHIPPUS
PLIOCENE—10 MILLION YEARS AGO

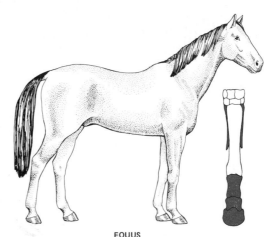

EQUUS
PLEISTOCENE AND RECENT—2 MILLION YEARS AGO

× 0.75

10.62
A bat, of the Order Chiroptera. Aside from being the only flying mammals —they were even classified with the Primates by Linnaeus—bats show a number of unusual features. They mate a couple of months before fertilization of the eggs occurs, and the female keeps the sperm alive in her uterus during the waiting period. The female has two functional teats from which the young nurse, but she also has an additional pair that does not give milk, serving only as something for the young to cling to while the mother hangs or flies. Male and female are alike except for the sexual organs, and the male has a bone in his penis. Females are impatient with the young, biting them vigorously when the babies are annoying. Contrary to popular myth, they do not entangle themselves in women's hair, and they can see quite well.

10.62). Many plants are pollinated by bats, and many insects are eaten by them. They are the only mammals that fly actively; other animals, such as "flying" squirrels, merely glide on outstretched folds of skin between limbs. The ability of bats to use echolocation in navigating and finding food is discussed in Chapter 28. The Lagomorpha (Greek, "rabbit form") are the rabbits, hares, and pikas. They are somewhat like rodents, but their teeth are different enough from those of rodents to justify placing these mammals in a different order. The Proboscidea (Greek, "front feeder") are the elephants, the only living animals in which the nose is elongated into a trunk and the incisor teeth are in the form of tusks. In the Primates (Greek, "first") are a number of familiar animals, such as the Old World monkeys, the New World monkeys, chimpanzees, gorillas, baboons, orangutans, and human beings, as well as some lesser-known, more primitive primates: tree shrews, tarsiers, and lemurs. They have grasping digits with flat nails, five to a limb.

THE EVOLUTION OF HUMAN BEINGS

We have relatively little factual material to help complete the story of human historical development, with only an occasional fragment of bone that escaped destruction, or the refuse in ancient campsites, or bits of shaped rock that look as if they might have been used as tools. Not much remains, but geologists, anthropologists, and archeologists have picked painstakingly through tons of rubble on all the continents looking for evidence of our own forebears. Alone among animals, we possess the curiosity that makes us look back into the past to ask, "Where did we come from?"

Humans as a species are "growing older" in two ways. The first is the obvious way, simply by existing, and that is a slow way by human reckoning. The second way is only a measure of our knowledge about our own history. We are finding that humanlike creatures have been inhabiting the earth for a much longer time than we had suspected. Within recent years, the estimated age of so-called modern humans has been extended. Instead of a presumed age of perhaps 10,000 years, which was a common estimate 50 years ago, new discoveries of fossils and tools have pushed the early dates back hundreds of thousands of years, and the current guess is that the living species, *H. sapiens*, may be a million years old. Humanlike species, similar enough to modern humans to be thought of as members of Genus Homo, lived in East Africa as much as five million years ago. These figures are still being revised so rapidly that final dates have surely not been determined.

Although the age of humans has been revised from a few thousand to several million, and that is certainly a great change, we are still recent arrivals. We must note that the oldest dated rocks are about 3.8 billion years old, and that the earliest fossils of any kind are also more than three billion years old. On a time scale measured in billions of years, a million or two one way or another begins to seem less overwhelming. If we let the width of this page represent a span of 3.8 billion years, then the difference between the older estimates of human duration and the present-day reevaluations are too small to measure. One common way to emphasize the shortness of human history is to compare biological time to a 24-hour day. If we assume that life began more than three billion years ago and call that time span 24 hours, then humans came on the scene a few seconds before the end of the 24-hour period.

Some 20 to 25 million years ago in Europe, Asia, and Africa, there were animals that lived in trees, ate fruit, and were somewhat apelike. From their fossils we can deduce something of their structure and habits, since the kinds of limbs and teeth an animal has can give clues to its diet and means of moving. These old beasts, now thought to be the forerunners of both present-day apes and modern humans, have been named *dryopithecines,* meaning "apes of the woods." Remains of dryopithecines are not rare, but there are no clearly intermediate fossils to tell us what kinds of development led from them to later fossils. The story is still incomplete and consequently speculative, but more old fossils are being found, and we can hope that some of the blank spaces in our knowledge may be filled in.

Informative fossils, along with indications of social activities, have been found in Africa and in Eurasia. Between one and five million years ago, several species of prehuman animals were already walking on their hind limbs. Called Australopithecus, these animals were small (under 50 kg, or 110 lb) and had brain cases only one-third to one-half that of modern humans. The presence of shaped pebbles associated with their remains is taken as an indication that they had a rudimentary idea of making tools. They are all extinct, but one (*Australopithecus boisei*) survived up to at least half a million years ago. Species of Australopithecus are not thought to be the ancestors of modern humans, although their structure was clearly humanlike.

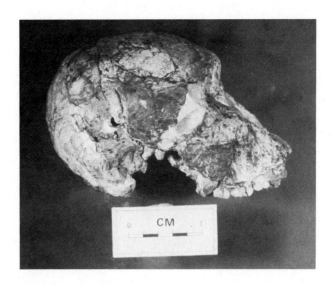

10.63
One of the most famous of the old skulls, known as ER 1470. It was found near Lake Rudolf in northern Kenya, and it is thought to have belonged to a man, that is, a creature assignable to the Genus Homo. Some 2,900,000 years old, this skull has caused students of evolution to reevaluate once more the antiquity of the human species.

Whether or not a given fossil is called Australopithecus or Homo depends on technical points of bone structure, as well as on the opinions of investigators. Fair agreement exists, however, among many seekers after early humans that a now-famous fragment of a skull known as ER 1470 belonged to a creature worthy of being called a man of Genus Homo (Figure 10.63). Modern dating techniques have shown ER 1470, found in East Africa, to be more than two million years old, probably closer to three million. The owner of this old skull was small, only about 1.3 meters (4 ft) tall. He had a brain capacity of about 800 cubic centimeters, which was large for his time, and he knew how to shape and use pieces of rock as tools. Because he was too manlike to be excluded from the Genus Homo but was at the same time not a fully evolved modern human, he has been provisionally put in a different species, *H. habilis,* which means clever or dexterous man, to distinguish him from *H. sapiens,* the wise man—a subtle but definite distinction. At the present writing, ER 1470 is the oldest identifiable Homo known.

10.64

A diagram of a possible line of descent for *Homo sapiens*, with skulls, reconstructions (except for modern man) of entire heads, and representative tools for each of several prehuman types. The suggestion is that about 12 to 14 million years ago, a primitive primate group gave rise to two divergent groups, one leading to the modern apes and one leading to Homo. Then, some one and a half to two million years ago, an unknown ancestral type diverged into several different groups, including Australopithecus, who made the crudest tools, and Pithecanthropus, who learned to shape tools completely. By the time Neanderthal man evolved, the jaw was receding visibly and the brain case was enlarging, until both those developments reached their present state in the modern human. It is problematical but possible that Neanderthal man interbred with Cro-magnon man and was essentially lost in the hybridization.

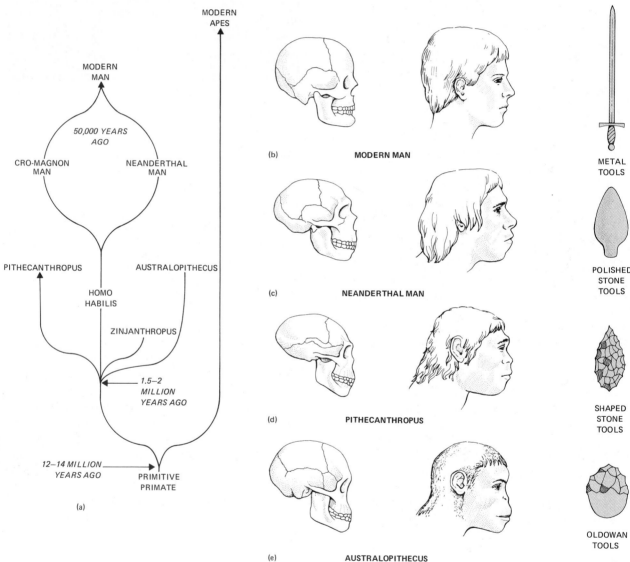

Another called *H. erectus* (although *H. habilis* and the australopithecines already walked on hind limbs) spread successfully over the Old World. Fossils of this species are associated with tools that are more advanced than the flaked tools of earlier hominids. The simple broken rocks that were in use practically unchanged for perhaps a million or two years were called Oldowan tools, named for the Olduvai Gorge, where some of them were found. *Homo erectus* improved on Oldowan-type tools by chipping them on two sides. To a modern student, that may not seem much of an improvement, but when one considers how long those dim minds took to achieve it, it becomes more impressive. *Homo erectus* was similar to a better-known type, the _Neanderthal man_. Neanderthals, named for specimens from the Neander Valley in Germany, actually lived in many places. They had sloping foreheads, relatively large brains, and rather prominent supraorbital ridges, those lumps of bone over the eyes that are characteristic of modern apes. The last of the Neanderthals lived about 40,000 years ago, but whether they died out or were absorbed by migrating waves of more advanced types is not known.

The last in the series is of course *Homo sapiens* (Figure 10.64). Existing for at least a million years, probably longer, this species spread all over the surface of the earth while land connections still existed between Asia and Australia and between North America and Asia. When those connections were later covered by the water of rising seas, the various populations were isolated. Once geographically separated, the diverging populations developed some differences, but never enough to cause loss of interfertility. All the races of present-day humans are interfertile, and mixing of genes has occurred whenever people have come back together even after thousands of years of separation.

The closest living relatives of the human species include the apes and monkeys, with the chimpanzees of equatorial Africa being the most nearly human. The ape-monkey, or anthropoid, line of evolution seems to have become separated from the hominid line (the line leading toward modern man) about 15 million years ago. Besides the easily observed similarities between humans and chimpanzees, most of the chromosomes of the two animals can be matched with considerable accuracy. There are only three chimpanzee and two human chromosomes that do not match. However, 21 human chromosomes match 21 chimpanzee chromosomes clearly, and the similarities so heavily outnumber the differences that biological relationship is almost certain (Figure 10.65).

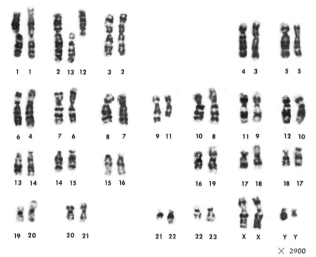

× 2900

10.65

Comparison of human and chimpanzee chromosomes. In these *idiograms* of the two animals, the human chromosomes are shown paired with those of the chimpanzee. In each pair, the human is on the left, numbered consecutively from 1 to 22, plus X and Y. The numbering system for the chimpanzee is slightly different from that of the human, and some differences of structural detail can be seen, but there is generally consistent similarity in the number, size, and banding patterns of the animals. In view of the structural, chemical, and behavioral similarities apparent at other levels, it would be surprising to find great difference at the chromosomal level.

Some of the features and abilities leading from prehuman forebears to modern human beings are biological; some are cultural. Among the structural changes are those that allowed upright locomotion or brought about a reduction in the canine teeth, the long, sharp, tearing teeth that are prominent in dogs (hence the name) and in such anthropoids as baboons. The hominid brain reached its present size of about 1200 cubic centimeters sometime between 100,000 and 500,000 years ago. The use of stones as tools may go as far back as five million years, but speech, in the sense of grammatic sound communication, is possibly only a million years old. Structural features of ancient skulls lead investigators to think that certain sounds, especially some of our vowel sounds, would have been impossible for some of the primitive types. Finally, they could not have known the subtlety, variety, and complexity of speech as we know it. Certain activities that are characteristic of today's humans are recent: the domestication of plants and animals, the use of metal, the settlement of towns, and finally, the practice of writing. These are all less than 10,000 years old, a mere flicker of time in comparison with the length of life on earth. With the invention of writing, *H. sapiens* came into the period of "history."

RECOMMENDED READING

Bakker, Robert T., "Dinosaur Renaissance." *Scientific American*, April 1975. (Offprint 916)

Blair, W. F., A. P. Blair, P. Brodkorb, F. R. Cagle, and G. A. Moore, *Vertebrates of the United States*, Second Edition. New York: McGraw-Hill, 1968.

Borror, D. J., and D. M. Delong, *An Introduction to the Study of Insects*, Third Edition. New York: Holt, Rinehart & Winston, 1971.

Buchsbaum, R. M., and L. J. Milne, *The Lower Animals: Living Invertebrates of the World.* Garden City, N.Y.: Doubleday, 1960.

Dawson, T. J., "Kangaroos." *Scientific American*, August 1977. (Offprint 1366)

Hickman, Cleveland P., Sr., Cleveland P. Hickman, Jr., and Frances M. Hickman, *Integrated Principles of Zoology*, Fifth Edition. St. Louis: C. V. Mosby, 1974.

Hickman, C. P., *Biology of the Invertebrates*, Second Edition. St. Louis: C. V. Mosby, 1973.

Hyman, L. H., *The Invertebrates*. Five volumes. New York: McGraw-Hill, 1940–1959.

Jones, Jack Colvard, "The Feeding Behavior of Mosquitoes." *Scientific American*, June 1978.

Kaston, B. J., and E. Kaston, *How to Know the Spiders.* Dubuque, Iowa: Wm. C. Brown, 1953.

Pennak, R. W., *Fresh-water Invertebrates of the United States.* New York: Ronald Press, 1953.

Peterson, R. T., *Field Guide to the Birds*, Second Edition. Boston: Houghton Mifflin, 1947.

Simons, Elwyn L., "Ramapithecus." *Scientific American*, May 1977. (Offprint 695)

Valentine, James W., "The Evolution of Multicellular Plants and Animals." *Scientific American*, September 1978.

Walker, E. P., *Mammals of the World.* Three volumes. Baltimore: Johns Hopkins University Press, 1968.

Washburn, Sherwood L., "The Evolution of Man." *Scientific American*, September 1978.

Waterman, T. H. (ed.), *The Physiology of Crustacea: Metabolism and Growth*, Vol. 1. *Sense Organs, Integration, and Behavior*, Vol. 2. New York: Academic Press, 1960–1961.

Welty, J. C., *The Life of Birds.* Philadelphia: W. B. Saunders, 1962.

GENETICS AND ANIMAL DEVELOPMENT

PART FOUR

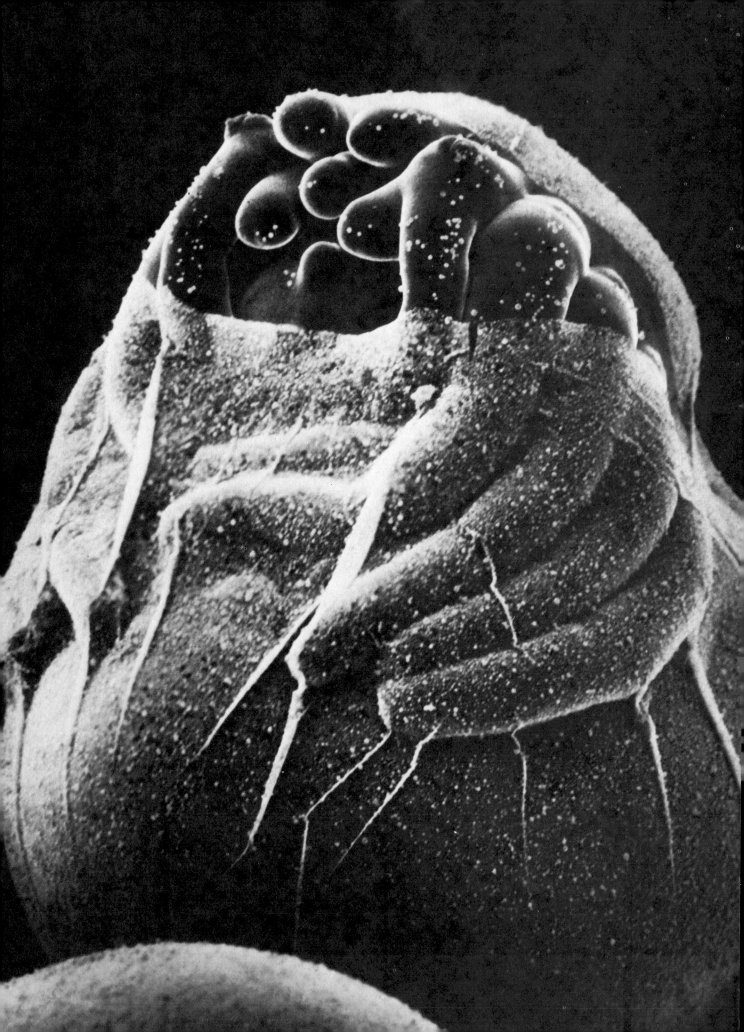

11
Mendelian Genetics

THAT PLANTS AND ANIMALS, INCLUDING HU-
mans, breed their own kind is such an obvious fact
of nature that it has been recognized as long as we
have any record. But knowing that something is so
is different from knowing *why* it is so, and only recently, in terms of human history, have we begun to
approach a detailed explanation of what has been
accepted for uncounted generations. Probably the
simplest and most generally accepted "explanation"
for the dependability of inheritance was that there
was something "in the blood" that accounted for similarities within families. Even plant sap was thought
of as "blood." The idea shows in many expressions,
such as "blood will tell" or "blue blood" or "blood
lines" or "blooded horses." Inheritance was accepted
as a mystery, but in the days before experimental
science was applied to it, there was only untestable
guesswork, and that didn't get anybody anywhere.

With the rise of an experimental biology in the
eighteenth and nineteenth centuries, many attempts
were made to work on problems of inheritance, but
they were too vaguely descriptive, and they made the
mistake of looking at too much at once. That is, early
investigations on genetics took into account whole
plants and animals, and the results were so complicated that no one could draw conclusions from them.
The very word "genetics" didn't exist until the English biologist William Bateson coined it in 1906, al-

*Scanning electron micrograph of
an embryo of a black widow
spider, Latrodectus* mactans, *about
to emerge from its egg. The legs
and mouth parts are already
visible.*

225

though the Book of Genesis ("beginnings") is several thousand years older. (The term "gene" was not introduced until 1909 by Wilhelm Johannsen, a full three years after Bateson introduced the word "genetics.")

MENDEL AND THE BEGINNING OF STATISTICAL GENETICS

Gregor Mendel was born to peasant parents in 1822 in Moravia, now part of Czechoslovakia. A sensitive child given to anxiety-induced illnesses, he was nevertheless an outstanding student throughout high school and college, and he received an excellent education. In 1841, Mendel's health was upset once again by a lack of money and reduced rations, and in his own words, he "felt compelled to step into a station of life which would free him from the bitter struggle for existence." Two years later, at the age of 21, Mendel joined the Augustinian monastery in Brunn, Austria, as a novice. Thus he was freed from the stress that had made his early life so difficult. Besides giving him freedom from financial worries, Mendel's entrance into the monastery placed him among many full-time lay teachers who provided a genuine cultural and scientific environment suitable for Mendel's own intellectual development.

In his early years as a priest, Mendel could not successfully complete his duties as an attendant to the sick, but he subsequently found contentment as a supply teacher at the local high school, and he was graciously relieved of all pastoral duties. In the summer of 1850, Mendel took an examination in an attempt to qualify as a permanent teacher, but he failed the final portion of the exam. This momentary setback, however, was to lead to one of the most important developmental stages of his life, because the following year it was recommended that he enter the University of Vienna to enrich his apparently inadequate background in the natural sciences. From 1851 to 1853 Mendel was a student at the university, where he studied botany, mathematics, and physics with some of the leading scientists of that time. It was surely during this stimulating period that Mendel learned of Schleiden's work with the cellular composition of plants and the hybridization experiments of Kolreuter and Gartner, and developed the basic statistical concepts he was to use so fruitfully in his pea experiments. Indeed, Mendel's knowledge and practical application of mathematical statistics was one of the reasons for his unique success in an experimental area where so many others had preceded him without significant scientific accomplishment.

11.1
Johann Mendel, better known under his monastic name of Gregor (1822–1884). Known during his lifetime mainly for his opposition to government taxation of monasteries, he is now recognized as the man who started the science of genetics. His paper on inheritance in peas was published in 1866 (but bore the date 1865) in a local journal, but it remained practically unnoticed until it was rediscovered 16 years after his death.

When Mendel began growing peas in 1856 in his monastery garden, he started a whole new science, one that was to change all biological thinking. The 1850s and 1860s were decades of enormous development in biological knowledge. Darwin's revolutionary ideas on evolution were born, and there was a dazzling burst of information on cell structure, the chemistry of living material, internal anatomy, and geographical distribution. And there was Mendel, at once the luckiest and unluckiest of men, obscurely breeding peas. He was lucky because he had enough imagination and insight to invent questions and to devise experiments that could provide answers to them, and he had enough training and perception to recognize his results as answers. He also had predecessors who understood the sexual nature of plants and who had provided him with agriculturally useful peas, which had been bred by trial and error until

TABLE 11.1

Mendel's Seven Pairs of Contrasting Traits in Garden Peas

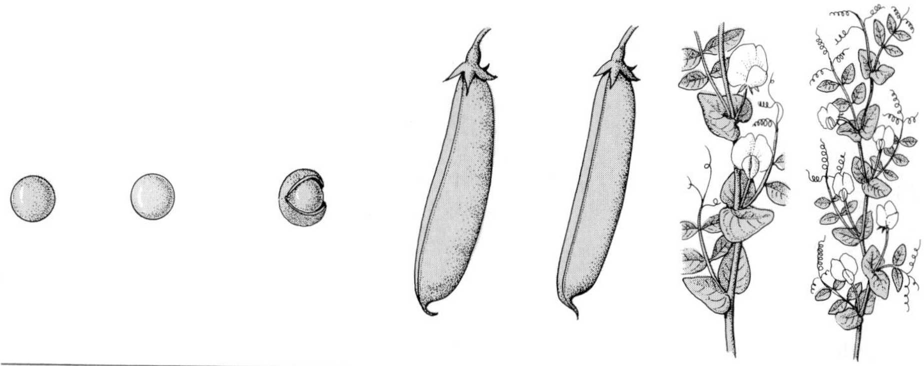

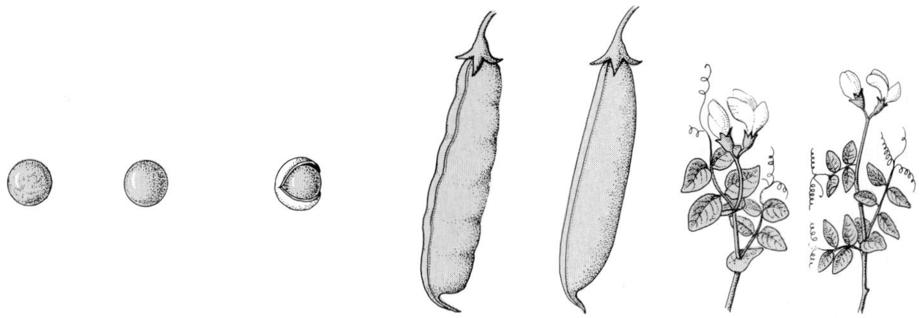

TRAIT STUDIED

	SEED SHAPE	COTYLEDON COLOR	SEED COAT COLOR	POD SHAPE	POD COLOR	FLOWER POSITION	STEM LENGTH
DOMINANT	Round	Yellow	Gray	Inflated	Green	Axial	Tall
RECESSIVE	wrinkled	green	white	constricted	yellow	terminal	dwarf

they were genetically dependable, that is, until they "bred true," or produced offspring like themselves. (Mendel would not have spoken of them in those terms.) Finally, he was lucky in that he chose to investigate seven visible characteristics in peas, and peas have seven chromosomes (Table 11.1). What is most improbable is that all his chosen characteristics had their genetic locations on different chromosomes. As will become apparent later, if it had not happened that way, he might have been hopelessly misled, although with a mind like Mendel's he might have worked his way through those difficulties as well.

He was unlucky in that after working with peas he turned to honeybees, one of the most unlikely animals for genetic studies because male honeybees are hatched from unfertilized eggs, and that fact seriously alters the possibilities of parental combinations of genes. Mendel's misfortune was compounded

when he also attempted his crosses with hawkweed. He did not know that hawkweed produces seeds asexually, and so there could be no genetic combinations of parental traits. Finally, Mendel was unlucky in that he was born at the wrong time. In 1865 he presented the results of his pea experiments to the Brunn Society of Natural Science and published his conclusions in the Proceedings of the Society the following year. Nothing happened. No one agreed, no one disagreed, no one did anything. Why? Volumes have been written trying to explain why clear, convincing explanations of fundamental, thoroughly documented work should have been passed over by the sharpest scientists of the time. Perhaps if Mendel had been an established scientist instead of an unknown "hobbyist" his experimental results would have awakened the scientific community. Perhaps Charles Darwin had completely captured the attention of world scien-

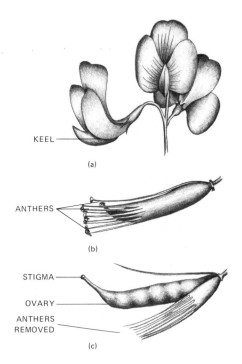

KEEL

(a)

ANTHERS

(b)

STIGMA

OVARY

ANTHERS
REMOVED

(c)

11.2

Pollination of pea flowers in hybridization experiments. (a) In entire flowers, egg-bearing ovaries and pollen-bearing anthers are concealed in the boat-shaped keel. (b) The keel is removed to show the anthers, partially connected to each other by their filaments. The ovary is in the tube formed by the filaments. In artificial pollination the keel is not actually removed; it is only pushed down to expose the anthers. (c) Before the ovary is mature, the flower is opened and the anthers are removed. Pollen from a chosen plant is touched to the virgin stigma of the plant to be pollinated, and the flower is covered, usually with a paper bag, to keep unwanted pollen out.

tists when he published his highly controversial ideas about evolution in *The Origin of Species* in 1859. But most likely, the scientists of Mendel's time were simply not ready to understand the significance of his work. By 1900, chromosomes and cell division had been described, and Mendel's observations could be directly linked to currently acceptable principles of cell biology. As is so common in the history of science, the world was now ready for the idea, and the rush was on. Almost simultaneously, the Dutch botanist Hugo de Vries, the German botanist Carl Correns, and the Austrian botanist Erich von Tschermak independently discovered Mendel's genetic laws and had little trouble establishing their importance.

Mendel made two decisions that brought him experimental success where others had failed. First, he narrowed his interest to specific details. He paid at-

tention to only one or two characteristics at a time, not allowing himself to be confused by the overwhelming complexities of an entire plant. Second, he counted his plants, keeping accurate records of everything he grew. He started the practice of statistical treatment. Once these two technical necessities were established, the way was open to all that was to follow.

THE MENDELIAN LAW OF SEGREGATION

One of Mendel's varieties of peas grew about a meter (3 ft) tall, as contrasted with another that grew as a low bush. He removed the pollen-bearing anthers from one before they ripened. This prevented self-fertilization. Then he dusted pollen from the other on the virgin flowers and covered the flower with a sack to prevent the entrance of other pollen, just as plant breeders do today (Figure 11.2). When the seeds ripened, he kept them for planting the next season. When the plants grew, Mendel found that they were all tall, whether the seeds had come from tall plants pollinated from low ones or vice versa. These new plants were the F_1, or the *first filial generation*. He allowed the F_1 plants to pollinate themselves, as peas do, and again he kept the seeds. The following season, he grew 1064 plants (the F_2 generation) and found that 787 of them were tall and 277 were dwarf. The numbers struck him. They showed a ratio of almost 3 to 1. He was certainly close: 798 to 266 would have been exactly 3 to 1 (Table 11.2).

From the numbers, Mendel deduced what had happened. First, he realized that a characteristic (tall) could appear in all individuals in the first generation, to the exclusion of another characteristic (short). However, the masked characteristic was still present in the F_1, as was shown by its reappearance in the F_2 (second filial generation). The characteristic that showed he called *dominant*, and the one that remained temporarily hidden he called *recessive*. (Mendel did not know what genes were; he called them "factors.") The recognition of dominant and recessive characteristics is his first contribution. He noted further that when one individual has two alternative possibilities, the two can *segregate* in the next generation. That was the basis for his *Law of Segregation*, one of the fundamental principles of modern genetics, which states that contrasting characteristics in an individual do not blend or contaminate one another; rather, they segregate when gametes are formed, and any single gamete receives only one characteristic (Figure 11.3). When members of a genetic pair have different characteristics, they are con-

TABLE 11.2

Results of Mendel's Monohybrid Crosses on Seven Pairs
of Characteristics in the Garden Pea

P CROSS	F$_1$ GENERATION	F$_2$ GENERATION		ACTUAL RATIO
Round × wrinkled seeds	All round	5474 Round	1850 wrinkled	2.96:1
Yellow × green cotyledons	All yellow	6022 Yellow	2001 green	3.01:1
Gray × white seed coats	All gray	705 Gray	224 white	3.15:1
Inflated × constricted pods	All inflated	802 Inflated	229 constricted	2.95:1
Green × yellow pods	All green	428 Green	152 yellow	2.82:1
Axial × terminal flowers	All axial	651 Axial	207 terminal	3.14:1
Tall × dwarf stem	All tall	787 Tall	277 dwarf	2.84:1
All characteristics combined		14,889 Dominant	5010 recessive	2.98:1

11.3

Illustration of Mendel's Law of Segregation. The parents, P, are homozygous (pure) *tall* (one parent) and homozygous (pure) *short* (the other parent). The offspring, the F$_1$ generation, are tall, and all are heterozygous (hybrid). When the F$_1$ plants are mated, the offspring, the F$_2$ generation, are about three-fourths tall and about one-fourth short. When gametes in the F$_1$ were being formed, the alleles (or genes) for tallness and shortness, present in the parent plants, were *segregated,* with the result that each gamete carried only one allele, either for tallness or for shortness, but not both. In genetics, the word "segregation" means just what it means in nontechnical English: separated according to kind.

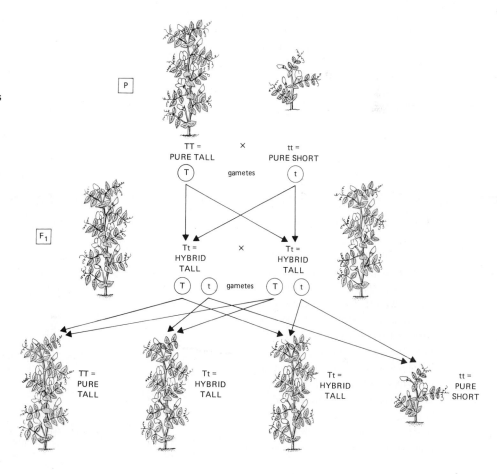

11.4

A "checkerboard," or *Punnett square,* named for the geneticist R. C. Punnett, can be used to show the predictable outcome of a genetic cross. All possible kinds of sperm from the male parent are displayed across the top, and all kinds of eggs down one side. The boxes can be filled in to indicate the kind of fertilized egg produced by each pair of gametes. Then the visible characters (phenotypes) can be determined and counted. This method gives the numbers of kinds of offspring that can be expected from any cross in which the genetic makeup (genotypes) of the parents is known. In actual crosses, in which dozens or hundreds or more of individuals are used, the numbers only approach the theoretical ratios obtained from Punnett squares or from calculations. In this example, a heterozygous pea plant produces two kinds of pollen, T and t, which are written across the top of the square. The eggs, also T and t, are written down the left side, and the resulting genotypes are written as letters (TT, Tt, or tt). The phenotypes are drawn as images, and the three tall to one short are evident.

11.5

A Punnett square showing the expected results of crossing two individuals, both heterozygous for two pairs of alleles: a dihybrid cross. The parents are tall and have yellow seeds, so the kinds of sperm that can be produced are TY, Ty, tY, and ty, and the kinds of eggs are the same. Filling the boxes appropriately shows that of the 16 squares, only one has the genotype TTYY (phenotypically tall, yellow), only one is ttyy (phenotypically short, green), nine are tall and yellow (but not genotypically identical), and so on. In a monohybrid cross, one pair of alleles gives two kinds of gametes and a matrix with four boxes. In a dihybrid cross, two pairs of alleles give four kinds of gametes and a matrix with 16 boxes. In a trihybrid cross, three pairs of alleles would give eight kinds of gametes and a matrix with 64 boxes. In an actual crossing, in which perhaps thousands of pairs of alleles are involved, Punnett squares are too cumbersome to use, and the results are calculated. The general rule is: the number of possible gametes is 2^n, where n is the number of alleles under consideration.

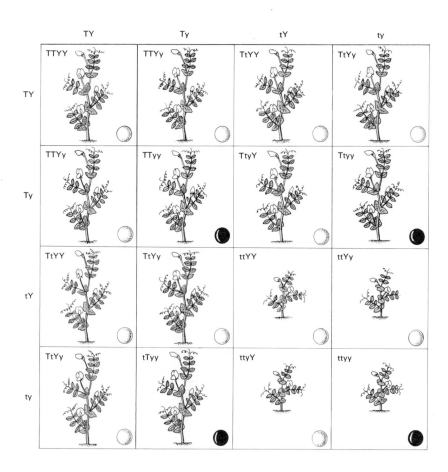

sidered to be alternative forms of the same trait and are called _alleles_. For instance, _short_ may be the alternative form of the dominant _tall_.

In that first series of experiments, Mendel had performed a _monohybrid cross_, in which the mated individuals differed in the expression of a single trait. (Mendel crossed a _tall_ plant with a _dwarf_ plant, for example.)

MENDEL'S LAW OF INDEPENDENT ASSORTMENT

After completing his monohybrid crosses, Mendel next considered the pattern of inheritance of two traits in a single mating; that is, he completed a _dihybrid cross_ (Figure 11.5). (When writing of monohybrid or dihybrid crosses, the geneticist means that attention is being paid to only one or two pairs of alleles, not that the organisms actually differ in only one or two features. In reality, it is unlikely that a pair of organisms would differ in only one pair of alleles.) Mendel crossed a plant that produced _round, green_ seeds with a plant that produced _wrinkled, yellow_ seeds, for example. Some of his peas had yellow seeds, and some had green. Further, some were round when ripe, and some were wrinkled. When the two features were considered separately, the yellow color proved to be dominant over green, and round shape was dominant over wrinkled. Therefore, if each dominant characteristic was crossed with its contrasting recessive, all offspring of the F_1 showed the dominant trait. But if the two pairs of characteristics were treated simultaneously, a new phenomenon appeared. Each pair of alleles behaved as if it were unaffected by the other pair of alleles. The number of plants with dominant seed _color_ was three times that with recessive color, and the same was true of seed _shape_. But if Mendel crossed pea plants, each of which had two pairs of dominant alleles, and inbred them (allowed

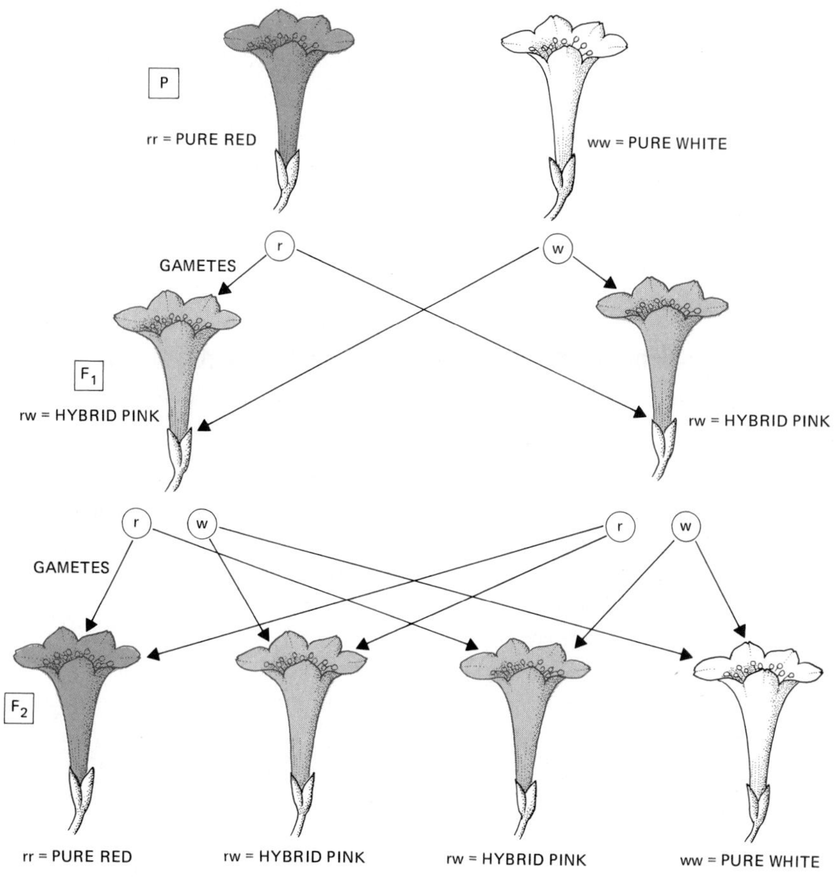

11.6

Incomplete dominance. Not all inheritance follows the dominance-recessiveness principle that Mendel found in peas. Incomplete dominance, or _intermediate inheritance_, occurs when no one trait is completely dominant or recessive. When a red four o'clock plant is crossed with a white four o'clock, the F_1 produces all pink flowers, but the F_2 produces red, pink, and white in a 1:2:1 ratio. Incomplete dominance occurs when a heterozygous organism contains one allele that produces one protein and another allele that produces a different protein. The visible trait shows the effects of both alleles.

self-pollination) in the F₁, then new combinations of alleles appeared. Although the parents, the so-called P generation, were round-and-yellow or wrinkled-and-green, some of the F₂ plants were round-and-green or wrinkled-and-yellow. *Recombination* had occurred. (You will recall that recombination indicates new gene combinations.)

In present-day genetic research, whether on humans, bacteria, or viruses, recombination is one of the bases on which progress depends because it is mainly by observing recombination, visibly expressed, that geneticists can determine what genes have done. These observations led to Mendel's second generalization, the *Law of Independent Assortment*, which states that when one pair of alleles segregates during the formation of gametes, its manner of segregation is not affected by the manner of segregation of a second, different pair of alleles. When one pair of alleles, A and a, segregates, the direction of passage of A and a are not influenced by the direction of passage of another pair of alleles, B and b. Note that the simple English phrase *independent assortment* means just what it says.

Heterozygous and Homozygous

It had long been known that children resemble their grandparents in special but unexpected ways. Mendel's experiments with dihybrid crosses helped explain genetic traits that skip a generation. A man may be carrying a recessive gene that is not expressed visibly. Geneticists would say that he is *heterozygous* for that characteristic; that is, the two alleles are not identical (Aa, for example). If the man marries a woman who is also heterozygous for the same characteristic (technically, for that *locus*, or place on a chromosome), then the couple could have a child with both recessives expressed. The child, possibly resembling a grandparent, is *homozygous*, that is, with both members of the allelic pair the same (aa, for example).

Genotypes and Phenotypes

If an individual is homozygous for a recessive characteristic, then that characteristic will be visibly expressed, and if one knows that the characteristic is recessive, one also knows that the individual must be homozygous for that characteristic. However, if the dominant characteristic is expressed, one cannot tell by inspection whether the individual is heterozygous or homozygous (Figure 11.7). An individual's genetic makeup, sometimes hidden, is that individual's *genotype*, and it may be either homozygous or heterozygous. The gene that is visibly expressed determines how the individual looks or acts and determines the *phenotype*, meaning "that which shows." For example, a large amount of dark pigment in the iris of human eyes behaves generally as dominant over lesser amounts of pigment, brown eyes being pigmented and blue eyes being less pigmented (Figure 11.8). If a

11.7

A *test cross*. If an organism shows a dominant characteristic, its genotype cannot be determined simply by inspection. In order to determine its genotype, it can be crossed with a phenotypic recessive, whose genotype is obviously that of a homozygote. In the example given here, a tall pea plant of unknown genotype, T?, is crossed with a short one, tt. In this cross, if all the offspring come out tall, then the genotype of the tall parent must be TT; but if half the offspring come out tall and half short, then the genotype of the tall parent must be Tt. Test crosses are used in plant and animal breeding programs.

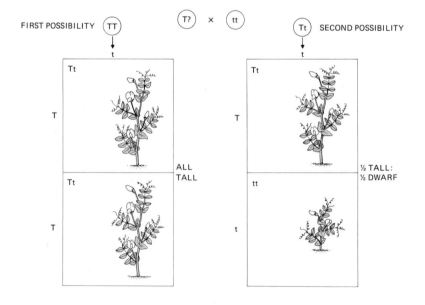

FIRST POSSIBILITY (TT)   (T?) × (tt)   (Tt) SECOND POSSIBILITY

11.8

Inheritance of human eye color. Brown (Br) is dominant over blue (bl), so a phenotypically brown eye may be either homozygous (BrBr) or heterozygous (Br bl), but a blue eye is necessarily homozygous (bl bl). In the example shown here, a heterozygous brown-eyed individual (Br bl) mates with a blue-eyed one. The Punnet square, with bl genes at the top (the only kind available from a blue-eyed person) and Br and bl down the side, shows that half the children should be brown-eyed and half should be blue-eyed. This is a reasonably satisfactory explanation, but it is made simpler than reality, since various shades of brown, gray, greenish, and blue appear in human populations. In fact, several genes are involved in determining eye color.

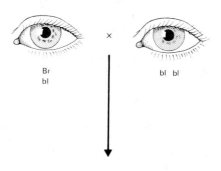

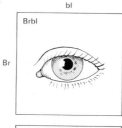

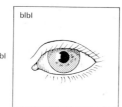

11.9

Walter S. Sutton. As a graduate student at Columbia University, Sutton saw the parallel between Mendelian inheritance and chromosome behavior during meiosis and fertilization, and he suggested that the cell parts that carry the hereditary material are specifically the chromosomes. That was in 1903. It would take another 40 years to find what portion of the chromosomes was genetically active, and 50 to describe the molecular structure of DNA.

heterozygous brown-eyed man marries a woman genetically like himself with respect to eye pigment, he could produce sperm with the "blue gene," and she could make an egg also with a "blue gene." If these gametes meet, the result is a homozygous blue-eyed child, possibly with a blue-eyed grandparent. From such an instance, it is apparent that when an individual shows a phenotypically dominant characteristic, the way to determine its genotype is to know the genotype of its parents or offspring.

Mendel did not use the words *phenotype* and *genotype*, but he obviously understood the basic concept of visible and hidden characteristics. Without using the exact words, Mendel described the nature of phenotypes and genotypes in peas: "It can be seen how rash it may be to draw from external resemblances [phenotype] conclusions as to their internal nature [genotype]."

Earlier in this chapter we discussed monohybrid crosses, which may now be related to genotypes and phenotypes in this way: The product of any monohybrid cross is 1:2:1 for the genotype and 3:1 for the phenotype (see Figure 11.4).

HOW IS MEIOSIS RELATED TO MENDELIAN PRINCIPLES?

Mendel's observations are especially remarkable when we remember that he knew less about chromosomes and cell division than you do. Mendel was able to offer an explanation of the results of his crosses without knowing the cytological mechanism that made them possible. But very soon after Mendel's work was rediscovered in 1900, his principles could be reinterpreted in terms of cytology. In 1903 Walter S. Sutton (Figure 11.9), while still a graduate student at Columbia University, demonstrated a parallel between the behavior of chromosomes during meiosis and the behavior of pairs of Mendel's "factors." Sutton called attention to "the probability that the association of paternal and maternal chromosomes in pairs and their subsequent separation during [meiosis] may constitute the physical basis of the Mendelian law of heredity."

Both the *Law of Segregation* and the *Law of Independent Assortment* could now be interpreted in terms of Sutton's *chromosome theory of inheritance,*

PROBABILITY AND CHANCE IN MENDELIAN GENETICS

Carrying the idea of independent assortment further, Mendel saw that the number of individuals with every possible combination of characteristics was close to what one would expect if the pairs of alleles were distributed by chance. For example, if you toss two pennies at a time and count the heads and tails, you will find that after a hundred or so pitches, you have about 25 percent both heads, 50 percent heads-and-tails, and 25 percent both tails. If you think of heads as being dominant in a Mendelian sense, the 25:50:25 can be expressed as 75:25, or 3:1—or what peas demonstrate in the F_2. Next, if you toss pairs of pennies and pairs of dimes simultaneously many times, you can obtain 16 combinations. Again taking into account the idea of dominance, you will find that 9 out of the 16 possibilities will show two heads, 3 will show one head, 3 the other head, and 1 all tails. The proportions are the same ones that peas exhibit when two pairs of alleles are considered at once, 9:3:3:1. Since the outcomes of pitching coins are the result of chance, and since the ratios obtained in making crosses give the same numbers that chance gives, the conclusion is that segregation and assortment of genetic features must also be a matter of chance. Chance distribution in chromosomes during meiosis and fertilization is the same as chance distribution for coins and other inanimate objects.

Two basic principles of probability should be considered in the overall discussion of probability and chance in Mendelian genetics:

1. If a coin is tossed 100 times, and it comes up heads every time, the chance of coming up heads again on the next toss still remains 50 percent, the same probability of coming up heads on the first toss, or the seventeenth, or on any

11.10

The parallel between the physical distribution of chromosomes and the phenotypic development of organisms. In these diagrams, chromosomes bearing a dominant gene for tallness in peas (T) and those bearing the dominant gene for yellow (Y) are in color, and those bearing the recessive alleles, t and y, are white. To make the distinction more readily visible, one pair of homologs (T and t) are straight, and the other (Y and y) are bent. At meiosis, the chromosomes are distributed randomly, with the restriction that each gamete must have a full set of chromosomes. The result is that one-fourth of the gametes carry the genes T and Y, one-fourth carry T and y, one-fourth carry t and Y, and one-fourth carry t and y. The distribution of chromosomes themselves can be followed by observing them with a microscope. The distribution of the genes can be followed by making a test cross, which shows that the active genetic factors act in a manner comparable to that of chromosomes. The similarity of behavior of chromosomes in meiosis and the distribution of genes in inheritance gives strong support to the idea that genes are carried by chromosomes.

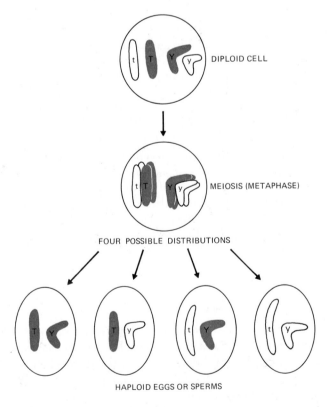

DIPLOID CELL

MEIOSIS (METAPHASE)

FOUR POSSIBLE DISTRIBUTIONS

HAPLOID EGGS OR SPERMS

given toss. *The result of one chance event does not affect the results of later occurrences of the same chance event.* In human terms, parents who already have two daughters are just as likely to have another daughter as a son on the third attempt.

2. The probability of one die coming up with any of its six numbers is $\frac{1}{6}$. Therefore, the probability of two dice each producing the same number during one toss is $\frac{1}{6} \times \frac{1}{6} = \frac{1}{36}$. *The probability that two independent events will occur together is the product of their probability of occurring separately.* Again, in human terms, the chance that a woman's first child will be a boy is $\frac{1}{2}$, and the chance that her sister's first child will be a boy is also $\frac{1}{2}$. The chance that both sisters will have first children who are boys is $\frac{1}{2} \times \frac{1}{2} = \frac{1}{4}$. In other words, the chances are 3 to 1 against the occurrence of such an event.

which proposed that genes are the determinants of heredity and that they are located on the chromosomes. This was the first specific reference to genes as the basic units of inheritance. Just as Mendel's "factors" segregated during gamete formation, so members of a pair of chromosomes segregate into different gametes during meiosis. Also, during meiosis chromosome pairs align randomly along the metaphase plate, providing a cellular parallel for the principle of independent assortment (Figure 11.10). Thus it was shown that there is a clear, understandable relation between the reproductive activities of cells and inheritance in whole organisms. The new science of genetics had come into being.

GENETICS PROBLEMS

Using the symbols T (for the "tall allele" in peas) and t (for dwarf), Y (for yellow seeds) and y (for green), complete the following exercises.

1. A phenotypically tall pea plant is pollinated by a dwarf plant, and the seeds of the first generation hybrid produce 327 tall plants and 321 dwarfs. Give the genotypes of all the plants.

2. Two tall pea plants, both heterozygous for height, were crossed, and 860 seeds were collected. What number of what phenotypes would be expected to grow from the seeds?

3. A tall, yellow-seeded plant, heterozygous for both characteristics, was pollinated by a dwarf, green-seeded one. From the 680 seeds collected, the plants were: 172 tall, yellow; 168 tall, green; 167 dwarf, yellow; 173 dwarf, green. Give the genotypes of all plants.

4. A plant of genotype TTYy was crossed with one of genotype ttYy. If the cross yielded 800 seeds, what number of what phenotypes probably resulted? What were their genotypes?

5. Two tall, yellow-seeded plants were crossed, and some dwarf, green-seeded plants resulted.
   a) What were the genotypes of the parents?
   b) What possible genotypes might there be among the tall, yellow-seeded offspring?

6. A TtYy pea plant self-pollinates, and one seed is picked at random for planting.
   a) What is the chance that the seed will produce a tall, green-seeded plant?
   b) If the plant actually turns out to be tall and yellow-seeded, what is the chance that its genotype is TTYY?

7. If you have a phenotypically tall plant, describe two ways in which you could learn its genotype.

8. In human eye color, one can assume for practical purposes that brown is dominant over blue.
   a) If a brown-eyed couple have a blue-eyed child, what are the chances that a second child will also be blue-eyed?
   b) If the couple have four blue-eyed children consecutively, how could you explain the phenomenon?

9. If a brown-eyed man with two brown-eyed parents marries a brown-eyed woman who also has two brown-eyed parents, is there any chance that the couple could have a blue-eyed child?

10. A dominant gene (S) in snapdragons produces bilaterally symmetrical ("irregular") flowers; the recessive allele (s) produces radially symmetrical flowers. Another pair of alleles controls color, yielding red (A_1A_1), pink (A_1A_2), or white (A_2A_2). If a cross is made between two snapdragons of genotype SsA_1A_2, what will be the phenotypes of the F_1? Give the proportions of each type.

ANSWERS TO GENETICS PROBLEMS

1. Parents, Tt and tt. All the tall offspring were Tt, the dwarfs tt.

2. 645 tall, 215 dwarf.

3. Parents TtYy, ttyy; the offspring, TtYy, Ttyy, ttYy, and ttyy.

4. 600 phenotypically tall and yellow, of which there were 200 TtYY, 400 TtYy; and 200 tall and green, Ttyy.

5. a) Both TtYy.
   b) TTYY, TtYY, TTYy, and TtYy.

6. a) $\frac{3}{16}$. b) $\frac{1}{9}$.

7. You might find the phenotypes of its parents, but a surer way would be to cross it with a short plant, tt—that is, make a test cross.

8. a) $\frac{1}{4}$. b) In such a small sample as four children, such a deviation from theoretical expectation would be common.

9. Yes, because one heterozygous grandparent on each side of the family might have contributed a "blue-eyed gene" to each member of the married couple, who then could produce a homozygous, blue-eyed child.

10. Irregular red, $\frac{3}{16}$; irregular pink, $\frac{6}{16}$; irregular white, $\frac{3}{16}$; regular red, $\frac{1}{16}$; regular pink, $\frac{2}{16}$; regular white, $\frac{1}{16}$.

SUMMARY

1. Pre-Mendelian thinkers about inheritance usually speculated that there was something in the blood that accounted for similarities within families. Even in the eighteenth and nineteenth centuries, the attempts to work on problems of inheritance were too vaguely descriptive, and because they tended to look at too much at once, the results were too complicated to understand.

2. Mendel made two decisions that brought him experimental success where others had failed. First, he paid attention to only one or two characteristics at a time. Second, he started the practice of statistical treatment.

3. Mendel's *Law of Segregation* states that contrasting characteristics in an individual do not blend or contaminate one another; they segregate when gametes are formed, and any single gamete receives only one characteristic.

4. The *Law of Independent Assortment* states that when one pair of alleles segregates during the formation of gametes, its direction is not influenced by the direction of another pair of alleles.

5. Mendel's hypotheses may be summarized (in current genetic terminology) as follows:
   a) Indivisible units called *genes* (Mendel called them "factors") function as pairs to determine hereditary traits. One member of the pair is contributed by each parent.
   b) Any diploid cell contains two genes for any particular trait; these genes may be alike (*homozygous*) or different (*heterozygous*).
   c) Alternative versions of genes, such as tall vs. short, are called *alleles*.
   d) If the two alleles are different, the *dominant* one will be expressed phenotypically while the *recessive* one remains hidden.
   e) One member of each pair of alleles is segregated into a gamete at the time of gamete formation.

Each member of the pair is unaffected by the other. One allele is just as likely as the other to go into a given gamete.
   f) Gametes are united randomly during fertilization, and the distribution of traits among offspring is predictable if the genotype of the parent is known. Inherited traits are therefore varied but predictable.

6. Soon after Mendel's work was rediscovered in 1900, both the Law of Segregation and the Law of Independent Assortment could be interpreted in terms of Sutton's *chromosome theory of inheritance*. Sutton showed a clear, understandable relation between the reproductive activities of cells and inheritance in whole organisms.

RECOMMENDED READING

Beadle, George, and Muriel Beadle, *The Language of Life: An Introduction to the Science of Genetics*. New York: Doubleday, 1966. (Also available in paperback)

Boyer, S. H. (ed.), *Papers on Human Genetics*. Englewood Cliffs, N.J.: Prentice-Hall, 1963.

Dunn, L. C., *A Short History of Genetics*. New York: McGraw-Hill, 1965.

Iltis, Hugo, *Life of Mendel* (translated by E. and C. Paul). New York: Hafner, 1966.

Moore, John A., *Heredity and Development*, Second Edition. New York: Oxford University Press, 1972. (Paperback)

Sturtevant, Alfred H., *A History of Genetics*. New York: Harper & Row, 1965.

Winchester, A. M., *Genetics: A Survey of the Principles of Heredity*, Fifth Edition. Boston: Houghton Mifflin, 1977.

12
Genetics and Chromosomes

ONE OF THE IRONIC TWISTS OF HISTORY IS THAT the original work of Mendel lay practically unnoticed for 35 years. When it was at last disinterred, a rush of experimenters set about proving or disproving his results. Although peas were ideal for the kind of work Mendel did, they are too large and too slow to produce the masses of statistical information geneticists need to obtain new information. In their search for a more useful organism, investigators chose fruit flies, of the genus Drosophila, those small insects that gather in swarms around decaying fruit. Since the early part of this century, and still today, fruit flies offer some special advantages. They are so small that a laboratory can raise millions of them inexpensively, they produce a generation in less than two weeks, they have only four (haploid) chromosomes, and during the larval stages their glands have relatively huge chromosomes that can be seen with the naked eye. (Still later, geneticists turned to the fungi, then the smaller and faster bacteria, and finally to the smallest and fastest of all, viruses.) Analysis of human beings has also contributed to our understanding of genetics. A great deal has been learned from family pedigrees and studies of human chromosomes from blood cells, even though breeding experiments are not possible.

12.1

Hugo de Vries, holding two Oenothera flowers. He is remembered not only for his part in the rediscovery of Mendel's work but for his enunciation of the idea of *mutations*. Either contribution alone would have been enough to establish him as one of the main founders of the modern science of genetics.

12.2

An evening primrose (Oenothera). This is the genus of plants on which de Vries based his theory of mutations. He obtained hundreds of new kinds of evening primroses in his experimental garden in Holland. It is now known that all but a few of his "mutations" were really chromosome rearrangements rather than DNA changes, but nevertheless his idea proved to be so useful for explaining an evolutionary mechanism that it is one of the fundamental concepts of modern genetics.

MUTATIONS MUST BE EXPRESSED VISIBLY

Before one tries to understand how experiments on heredity were performed, it is useful to know how differences in organisms can be employed and how their significance came to be appreciated. Differences, sometimes trivial and sometimes great, are frequently the result of inborn forces in an organism, and their visible expression is affected little or not at all by external forces. Inherited peculiarities, whether one is considering tallness in peas or animal eye colors, are *mutations*, and the story of how mutations came to be recognized by biologists starts in the Dutch garden of a botanist, Hugo de Vries (Figure 12.1).

De Vries, a supporter of Darwinian evolution, was interested in plant breeding and in biological variation, especially in the genus Oenothera, plants known in English as evening primroses. Among the results of his breeding experiments, de Vries noticed that at unexpected intervals he obtained unusual kinds of plants, one of the most famous being an unusually large type that he appropriately called *gigas*, the giant (Figure 12.2). Everyone who grows plants knows that variations are common, but de Vries

found that his unusual plants *bred true*; that is, they produced offspring identical to themselves. "Here," said de Vries, "is the way Darwin's variations come into being. They occur suddenly, as large changes, and the changes are inherited."

De Vries called these changes mutations, from a Latin verb that means "to change." He wondered if anyone else had ever described such changes, and while searching through the published literature, he came upon Mendel's forgotten paper on inheritance in peas. De Vries recognized Mendel's paper for what it was: a simple, clear, original, satisfactory explanation of the principles of inheritance. The obscure monk had been rediscovered, and as the twentieth century began, evolution and genetics met, to grow more and more intimate with the passing years.

Ironically, de Vries may be more important for having found Mendel than for having explained the basics of mutations. As it turned out, the "mutations" in evening primroses are not really mutations in the present-day sense. They are instead only chromosome rearrangements of a sort peculiar to evening primroses and a few other plants. Stable, heritable changes, however, were sought and found in many other plants and animals, until we now use mutations as the basis for practically all genetic research. The "mutations" of de Vries and the true gene mutations of practical geneticists have both been used to gain information on patterns of inheritance and interactions among genes. Now when a biologist speaks of a mutation, he means a gene mutation, which is a change in DNA. If he merely means a chromosome rearrangement, he explicitly calls it a "chromosome mutation."

The Frequency of Mutations

Mutation rates are most readily determined with experiments on bacteria and viruses, in which one mutation in a billion can be detected (Figure 12.3). These experiments show that not all genes mutate at the same frequency. If a gene is chemically stable, it will mutate less frequently than will an "unstable gene." It is not unusual for one mutation to occur in a million DNA replications, but some unstable genes may mutate once in 2000 replications. On the other hand, an

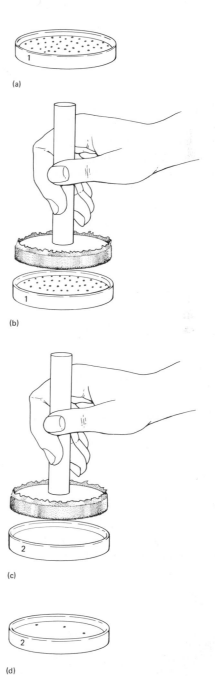

(a)

(b)

(c)

(d)

12.3

The velvet replica method of demonstrating the origin of mutants. (a) A Petri plate has numerous bacterial colonies growing on nutrient media. Since the plate was seeded with a dilute suspension of bacteria, it is assumed that each colony arose from the growth of a single cell and is therefore genetically homogeneous; that is, it is a clone. (b) A sterile velvet disk is touched to the surface of the plate. Some cells of each colony stick to the velvet, making an exact copy of the whole plate. (c) The disk with its load of bacterial spots is touched to a new plate, and it transfers some cells to the new medium. The new medium has some experimental difference from the original one; for example, it may contain penicillin. (d) Most of the bacteria thus transferred are unable to grow, but a few survive and grow until they form visible colonies. These are penicillin-resistant. Because the second plate is a duplicate of the first, it is possible to identify the colonies on the first plate that provided the resistant colonies on the second. The question that can now be answered is: Did the penicillin in the experimental plate cause the mutation to penicillin resistance, or was the mutation already present before the bacteria were placed on the experimental plate? Cells from those colonies that provided mutants, when transferred to penicillin-containing medium, grow. Therefore the mutation was already present before the cells came in contact with the antibiotic. The antibiotic in the bacterial environment did not cause the mutation.

exceptionally stable one may remain unaltered through billions of replications (Table 12.1).

Causes and Effects of Mutations

Gene mutations occur when nucleotides in a DNA molecule are added, left out, or changed. Frequency of cell divisions varies from once in less than half an hour in bacteria, through several times a day in humans, to years in some quiescent cells. Before a cell division can take place, there must be replication of DNA, transcription of RNA, and synthesis of protein. The process almost always works accurately, but occasionally something goes wrong, and the DNA nucleotide sequence is altered. That is a mutation. Sometimes the mutation is not phenotypically detectable, and no further change takes place. For example, if a TCG triplet in DNA is changed to TCA, the resultant RNA change is from AGC to AGU, and the same amino acid, serine, will be incorporated into protein as before the change (see Figure 7.9). At the other extreme, one change may result in the death of an organism. Such a mutation, regardless of its method of operation, is known as a *lethal mutation*. Ordinarily, geneticists are most concerned with mu-

tations whose effect is between the undetectable and the lethal.

We still do not know what causes natural gene mutations, but mutations may be induced artificially by any radiation that has sufficient energy to cause chemical changes of molecules in cells. The most effective agents of radiation have been X-rays, gamma rays, neutrons, and ultraviolet light. We have also learned that radiation from radioactive elements, such as that from the dust of nuclear explosions, will cause mutations that may result in a high incidence of leukemia, for example. The first person to demonstrate the effectiveness of radiation to induce genetic mutations was Hermann J. Muller, who developed a technique for proving the presence of lethal mutations in Drosophila with X-rays in 1927 (Figure 12.4). He subsequently won a Nobel Prize in 1946. By that time, the first atomic bombs had been used in Hiroshima and Nagasaki, and the possibility of mutations in human beings caused by human action began to be realized.

Mutations in germ cells may be accelerated by increasing the temperature, by introducing "chemical mutagens," such as the oxides of nitrogen from auto-

TABLE 12.1
Spontaneous Mutation Rate of Some Human Genes

MUTANT TRAIT	APPEARS ONCE IN EACH	MUTATION FREQUENCY PER MILLION
Dominant		
Pelger anomaly—abnormal white blood cells, reduces resistance to disease	6,250 births	80
Chondrodystrophic dwarfism— shortened and deformed legs and arms	11,500 births	42
Retinoblastoma—tumors on retina of eye	21,750 births	23
Aniridia—absence of iris	100,000 births	5
Epiloia—red lesions on face; later, tumors in brain, kidney, heart, etc.	41,500 births	12
Recessive Autosomal		
Albinism—melanin does not form in skin, hair, and iris	18,850 births	28
Amaurotic idiocy (infantile)—deterioration of mental ability during first months of life	45,450 births	11
Total colorblindness	17,850 births	28
Recessive X-linked		
Hemophilia	15,625 births	32
Muscular dystrophy, Duchenne type	11,625 births	43

From *Genetics: A Survey of the Principles of Heredity*, Fifth Edition, by A. M. Winchester. Copyright © 1977 by Houghton Mifflin Company. Reprinted by permission.

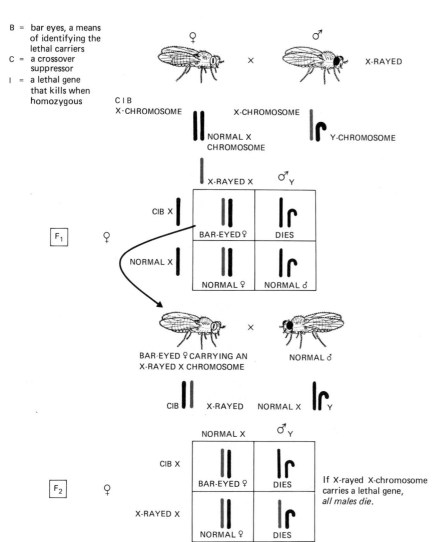

B = bar eyes, a means of identifying the lethal carriers

C = a crossover suppressor

l = a lethal gene that kills when homozygous

12.4

H. J. Muller's method of demonstrating that X-rays can cause mutations. The female fruit flies he used had three known factors: Bar (narrow) eyes that made them visibly identifiable, a "cross-over suppressor" that kept the chromosomes relatively stable, and a lethal factor that kills when it is in the homozygous condition. The males in the preliminary cross received a dose of X-rays, but not enough to kill them. The results of the first mating are shown in the first Punnett square. In the F_1 generation were some Bar-eyed females, known to be carrying one lethal gene and possibly carrying another from the X-rayed fathers. These were mated in the second cross to normal males. The results of the second cross are shown in the second Punnett square. Half the males died because they received a lethal gene from their mothers and had no compensating gene on the father's Y chromosome. If the X-rayed males in the first cross had indeed suffered a lethal mutation, then the mutant gene, inherited through the F_1 mother, would kill the F_2 males. In effect, lethals caused by X-rays lead to the absence of males in the F_2. In a number of such experiments, a disproportionate number of crosses yielded all females, showing that X-rays do cause genetic change.

mobile exhausts, mercury, DDT and other pesticides, food additives, and a wide variety of other organic compounds, such as dyes and the ingredients of tobacco smoke.

The Fate of Mutations

Most mutations are detrimental. The effect is somewhat as if you made a thoughtless change in the circuitry of your finely tuned stereo sound system. The chance of your improving its performance by such a change is slim indeed. Existing organisms must be already well adapted to their environment, as proved by the very fact of their survival. Any change is more likely than not to be a change for the worse, that is,

one that may make an organism less well suited to its environment. It is obvious, however, that over the long years occasional mutations have been advantageous, as favorable mutations are the basis for the genetic variability that makes evolution work.

Besides being usually harmful, mutations are also usually recessive. Therefore, most new mutations are not expressed phenotypically, although they can be expressed if by chance mating they happen to meet in a homozygous condition. In view of the relative rarity of mutations and the improbability of bringing together an identical pair of mutant genes, it is apparent that many mutations may exist in a population generation after generation without being expressed. Also,

12.5

Comparison of possible mutation passages in animals and plants. (a) In animals generally, a mutation (caused perhaps by radiation or a mutagenic chemical) will be passed on to the next generation only if the mutation occurs in a gonad: ovary or testis. (b) In plants, a mutation in an anther or in an ovule may be passed to the next generation in a manner somewhat similar to the animal method. But a different possibility also exists. If a mutation occurs in a vegetative bud and the bud grows into a branch, all the descendants of the mutated cell will carry the mutation. Those mutated cells may make up much, or even all, of the tissue of a flower bud, and the ovules and anthers in that bud may have the mutation, so that the egg cells and pollen grains will have it, too. Plants differ from animals in that they have perpetually embryonic growing regions; animals do not.

(a)

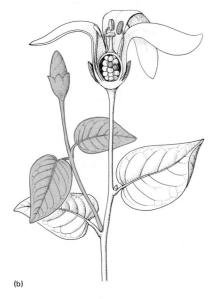

(b)

if the possessor of a gene mutation dies without having passed it on to offspring, the mutation may be eliminated without ever having been expressed at all.

When a mutation occurs in a cell, it will be passed on to the descendants of that cell, but what eventually happens to it depends on the kind of cell in which the mutation originally appeared. Only those mutations that occur in gonad cells in animals have any real chance of being passed on to later generations (Figure 12.5). A mutation in a liver cell (which is a body cell, or *somatic* cell), for example, will be lost when the animal dies, but a mutation in an ovary may be included in an egg and then in offspring.

SEX-LINKED TRAITS IN DROSOPHILA

In general, the reality of Mendelian ratios was confirmed by new experiments in the early 1900s, but soon inexplicable irregularities appeared. In 1910, Thomas Hunt Morgan and his associates at Columbia University began a series of classic experiments with the fruit fly, Drosophila. In the course of trying to find the origin of mutants, Morgan noticed among his stock of fruit flies a fly with white eyes instead of the normal red eyes (Figure 12.7). The mutant fly, a male, was mated with a normal red-eyed female, and all their F_1 offspring had normal red eyes. Morgan correctly concluded that red eyes are dominant, and these results were therefore not unusual. When red-eyed F_1 males were mated with red-eyed F_1 females, the F_2 generation contained 3470 red-eyed flies to 782 white-eyed.

It was at this point that Morgan made a crucial discovery. *All* 782 of the white-eyed flies were *males*. This was a tremendous shock, and Morgan realized it could not be a coincidence. He deduced the connection between white eyes and sex (specifically, the

12.6

Thomas Hunt Morgan (1866–1945), who, with A. H. Sturtevant, H. J. Muller, and C. B. Bridges, used fruit flies to help establish many of the principles on which modern genetics is based. Morgan and Muller both won Nobel Prizes for their contributions to biology.

12.7

Eye color in fruit flies. The usual, or *wild-type*, eye of *Drosophila melanogaster*, is red, but unpigmented white eyes occur in some mutants. A number of intermediate eye colors are known, and they are named for the intensity and hue of the pigments, such as eosin, apricot, or sepia. The white-eyed mutants are famous among biologists because of their historical position in the early research on chromosomes and sex determination.

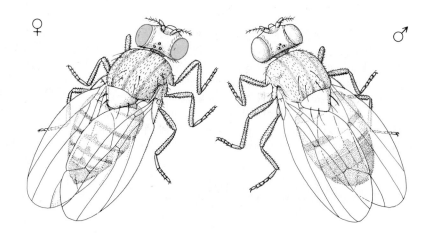

MARY LYON AND CALICO CAT

Normal female mammalian cells contain two X chromosomes. And in every female cell there is a small object that rests against the wall of the nucleus. This inert body was discovered by Murray Barr, and so it is called a *Barr body*, or more descriptively, a *sex chromatin body*, stainable and microscopically visible. Barr found the sex chromatin body in female cats (male cats do not have such a structure), but all he knew about it was that it was there. He did not connect it with chromosomes.

Later, Barr bodies were found in many mammals, including human beings, but always in females only. It was also discovered that each female cell contained one less Barr body than the number of X chromosomes present. Ordinarily this meant one Barr body for each cell, but abnormal XXX cells, for instance, have two Barr bodies. Apparently Barr bodies had something to do with sex, and since they reacted to stains that usually work for DNA, it seemed likely that they contained genes.

Appropriately, it was a woman, Mary Lyon, who discovered the significance of the sex chromatin bodies. She proposed the "single active X hypothesis," which stated that the sex chromatin body is an X chromosome that did not uncoil during mitosis. It remains as a nonfunctioning mass, clumped against the nuclear membrane. Why? Probably to balance the sex chromosome with the autosome. The inactivation of one X chromosome in a female restores the same balance as that in a normal male cell, which has only one X chromosome in both the sex chromosome and the autosome.

Lyon postulated that both X chromosomes function during an animal's early development, but that only one of them remains permanently active, the second one retiring and becoming a metabolically inert Barr body, or sex chromatin body. An instructive example of how such a development can occur is found in cats called calico or tortoise-shell, which have patchy markings of black and orange-yellow. A calico cat, typically a female, has two X chromosomes, one carrying a gene for black fur and the other carrying its allele for orange-yellow. On some parts of such a cat, the black gene goes dormant during embryonic life, making an orange-yellow patch, but on other parts, it is the colored gene that goes dormant, leaving the black to be expressed. The result is the varied coat.

Male cats normally have an XY chromosome pair. The Y chromosome carries no color gene, and the X carries either black or orange-yellow but not both. Since the black-orange-yellow locus is not the only one affecting fur color, some males *can* be mottled, but they cannot be true calicos unless they happen to have the abnormal XXY complement, and then they are sterile. Now, after centuries of guessing, we know something of why calico cats are females, but we still do not know why one X chromosome or the other can be effective to the exclusion of the opposing one.

12.8

Fruit fly chromosomes. (a) A cell from a male fly. The cell has three pairs of autosomes in matching pairs, shown in black, and one pair of unmatched sex chromosomes, shown in color. The colored *straight* sex chromosome is designated X; the *bent* one is designated Y. (b) A cell from a female fly. Again the autosomes are shown in black and the sex chromosomes in color. In females, all the chromosomes occur in matching pairs.

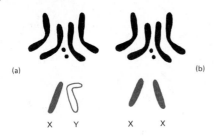

male sex) and thereby made the first identification of a trait in Drosophila that was linked to the sex of the fly. A *sex-linked trait* may be said to be a visible feature that is specifically associated with the sex of its possessor.

Chromosomal Sex Determination in Drosophila

Once an inherited bodily trait was known to be linked to sex inheritance in fruit flies, Morgan proceeded to examine Drosophila chromosomes to see if chromosomes were related to the determination of sex, and if W. S. Sutton (Chapter 11) had been correct in hypothesizing that genes are located on chromosomes. (They are.) Another proposal by Sutton could also be tested: Are several genes usually located on one chromosome? (They are.)

When Morgan examined the nuclei of somatic cells of Drosophila, he found four pairs of chromosomes, or a diploid number of eight. In females there are four matching pairs, but in males only three pairs

XO-XX TYPE OF SEX DETERMINATION

XY-XX TYPE OF SEX DETERMINATION

DIPLOID CHROMOSOME COMPLEMENT OF MALE

DIPLOID CHROMOSOME COMPLEMENT OF FEMALE

TWO TYPES OF SPERM RESULT FROM MEIOSIS

ONE TYPE OF OVUM RESULTS FROM MEIOSIS

AUTOSOMES

SEX CHROMOSOMES

♀ SEX DETERMINED BY TYPE OF SPERM ENTERING THE OVUM ♂

DIPLOID CHROMOSOME COMPLEMENT OF MALE

DIPLOID CHROMOSOME COMPLEMENT OF FEMALE

TWO TYPES OF SPERM RESULT FROM MEIOSIS

ONE TYPE OF OVUM RESULTS FROM MEIOSIS

♀ SEX DETERMINED BY TYPE OF SPERM ENTERING THE OVUM ♂

12.9

Cell diagrams of two kinds of sex determination. In the XX-XY type, there are two different "sex chromosomes." (Compare Figure 12.8.) In the XX-X0 type, there are two kinds of sperm, those carrying an X chromosome, and those that carry no sex chromosome (0), that is, carrying

autosomes only. An 0-type sperm, fertilizing an X-carrying egg, produces an X0 zygote, which develops into a male. An X-type sperm, fertilizing an egg (all eggs have an X chromosome), produces an XX zygote, which develops into a female.

RECOMBINANT DNA

"On Wednesday, June 23, 1976, the Director, National Institutes of Health, with the concurrence of the Secretary of Health, Education, and Welfare, and the Assistant Secretary for Health, issued guidelines that will govern the conduct of NIH-supported research on recombinant DNA molecules. . . . The NIH Guidelines establish carefully controlled conditions for the conduct of experiments involving the production of such molecules and their insertion into organisms such as bacteria."

So begin the official guidelines for recombinant DNA research. If the guidelines sound so ominous, what about recombinant DNA itself? What is "recombinant DNA," and why is it controversial? The technique of recombinant DNA involves the insertion of genetic material from one species into another, with results that are not completely predictable. For instance, in 1973 Stanley N. Cohen of the Stanford University School of Medicine and Herbert W. Boyer at the University of California School of Medicine at San Francisco inserted genes from an African clawed frog (*Xenopus laevis*) into the bacterium *E. coli*. This operation differed from the nuclear transplants of Xenopus completed several years earlier by J. B. Gurdon. Gurdon transplanted the nucleus from a cell of a frog into an enucleated cell of the same species. In contrast, Cohen and Boyer transferred a length of DNA from a vertebrate animal into the prokaryotic cell of a bacterium. Fusion of foreign genetic material with the chromosome of a host cell, successful about one percent of the time, results in such a recombination of DNA as to make an essentially new organism, a hybrid of the two organisms involved. Just as a child may differ from its parents in many ways, so may a hybrid show properties completely different from both host and donor organisms. Organisms created through recombinant DNA are permanent and cannot be recalled.

The basic technique of recombinant DNA separates short segments of DNA from any two cells. Ligase enzymes are then used to splice the segments into viruses or extrachromosomal particles called *plasmids*. The plasmids can then be accepted by host *E. coli* cells, and they may reproduce along with the host or actually combine with it to form a new hybrid organism capable of independent self-perpetuation.

There are four essential elements of recombinant DNA technology. (1) DNA molecules from different sources must be broken and rejoined. (2) A suitable gene carrier (plasmid) that can replicate both itself and a foreign DNA segment must be linked to the joined DNA molecule. (3) The composite DNA molecule must be introduced into a functional bacterial cell. (4) The recipient cells that have acquired the composite DNA molecule must be successfully selected from a large group of cells.

Recombinant DNA has also been called plasmid engineering, genetic engineering, and genetic manipulation. Words like "genetic manipulation" immediately raise moral and ethical questions, and indeed, the controversy that has arisen around recombinant DNA research is not likely to be settled for many years, in spite of pleas by Nobel laureates and legislative bills by United States senators.

---

match, and they have counterparts in females. One chromosome pair in males is mismatched. The mismatched pair, called the XY pair, has one chromosome (the X chromosome) like those in females, and one odd, hook-shaped one (the Y chromosome) (Figure 12.8). The XX pair in females and the XY pair in males are the _sex chromosomes_; the other three pairs, identical in both male and female, are _autosomes_.

Even before Morgan began his observations of Drosophila chromosomes, researchers had thought that sex was determined by chromosomes, and there already was some knowledge of sex chromosomes and autosomes. In fact, two types of chromosomal sex determination had been postulated: the X0-XX and XY-XX types: In some insect species the male has a single X (sex) chromosome and the female has two X chromosomes. The male is symbolized as X0, with the 0 indicating the absence of a homolog of X, and the female is XX (Figure 12.9). Both humans and Drosophila have the _XY-XX_ type of sex chromosomes, which also happens to be the most common type. The female has two chromosomes, _XX_, and the male has one chromosome identical to the female's, plus one of the shorter, hooked variety mentioned earlier. The male is designated as _XY_.

As a result of meiosis, the egg contains the haploid number of autosomes and one X chromosome. Half the sperm contain the haploid autosomes and one X chromosome, and the other half contain the haploid autosomes and one Y chromosome. If an egg is fertilized by a sperm containing the X, the resultant zygote will contain a diploid set of autosomes and two X chromosomes; this XX zygote will develop into a female. If the egg is fertilized by a Y-bearing sperm, the result is an XY male with the usual diploid autosomes.

So the ancient riddle of the determination of sex turned out to be a simple matter of Mendelian assortment. Morgan could finally say: "The old view that sex is determined by external conditions is entirely disproved. We have discovered an internal mecha-

nism by means of which the equality of the sexes is attained. We see the results are automatically reached, even if we cannot entirely understand the details of the process." For his work relating heredity to chromosomes, Morgan received the Nobel Prize in 1933.

NONDISJUNCTION IN DROSOPHILA

Calvin Bridges was one of Morgan's co-workers. He guessed that at meiosis, when homologous chromosomes normally separate, something occasionally held some of them together so that an egg might contain two Xs instead of the usual one X. It was also possible that both X chromosomes might go into the polar bodies, leaving the egg with no X chromosomes. Microscopic examination of cells from certain female flies with unusual sex ratios among their offspring verified that their eggs did indeed have either an extra X chromosome or no X chromosome at all. This fail-

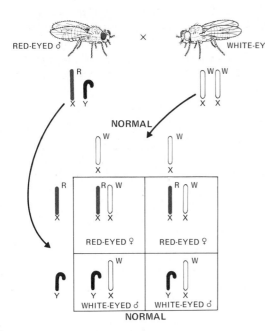
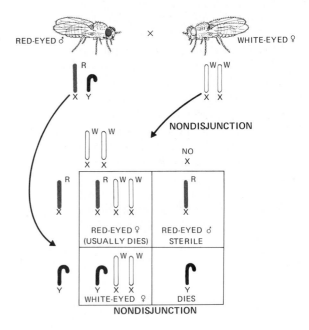

12.10

Normal and abnormal gamete production in fruit flies. In the normal distribution of sex chromosomes (and of other chromosomes as well), one member of each homologous pair should go into a gamete. Thus every egg should have a set of _autosomes_ and an X chromosome, and every sperm should have a set of autosomes and either an X or a Y chromosome. If meiosis fails to proceed properly, the two X chromosomes in the ovary of a female may stick together so tightly that both of them go into an egg, or

they may both go into a polar body, leaving an egg without any X chromosome. This is _nondisjunction_. The results of a mating involving a nondisjunction are shown in the right-hand Punnett square. If a female suffers a nondisjunction and is mated with a normal male, her daughters will have three X chromosomes, and they usually do not survive. Some of her sons will have the XXY condition, and some will have only one Y.

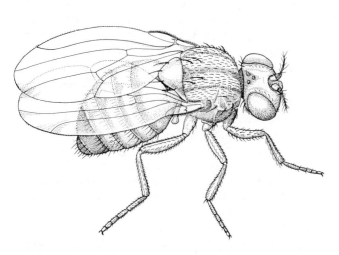

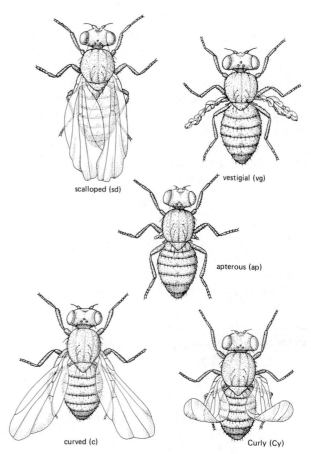

scalloped (sd)

vestigial (vg)

apterous (ap)

curved (c)

Curly (Cy)

12.11

Some mutations of wing structure in Drosophila. A normal or *wild-type* fly is shown directly above about 250 times natural size, and the deviations are to the right, each with its name and genetic abbreviation. The *apterous* mutant (Greek, "without wings") is designated ap, and the use of lower-case letters indicates that it is determined by a recessive gene. Curly, in contrast, is Cy, and the upper-case initial indicates a dominant gene. Dozens of wing mutations have been described, as have hundreds of mutations involving other fly structures and colors.

ure of chromosomes to separate during meiosis is called *nondisjunction* (Figure 12.10).

Nondisjunction and Sex Determination

If a two-X egg is fertilized by an X-carrying sperm, the XXX zygote develops into a sterile female. If the two-X egg is fertilized by a Y-carrying sperm, the XXY zygote also develops into a sterile female. This fact shows that in flies *the Y chromosome itself does not determine maleness.* Numerous crosses with a variety of nondisjunctions demonstrated that in flies the sex of an individual is determined by the ratio of X chromosomes to the number of autosomes. This is *not* true of human sex determination, as later investigations showed. In humans, the Y chromosome is a male determiner, since an XXY zygote develops into a male, and an X0 (having no Y) becomes a female.

Regardless of the difference between flies and humans, the important fact to come out of all this work is that chromosomes do indeed determine sex. An even more important conclusion was that if chromosomes could control sex, then they might be able to control everything an organism has or does. Thus it was established that of all the things that might imaginably control any biological activity, it is the chromosomes that are the chief regulators.

GENE LINKAGE AND CROSSING OVER

Morgan discovered many other irregularities in the fruit fly besides the white-eyed mutant. For example, there are mutant black bodies as opposed to the normal wild-type gray bodies, and there are mutant reduced (vestigial) wings as opposed to normal wings (Figure 12.11). Gray bodies and normal wings are both dominant. If a black-bodied, vestigial-winged fly is crossed with a gray-bodied, normal-winged one, the F_1 flies are all gray-bodied and normal-winged, as one would expect. Next, if the F_1 hybrids are inbred, one would expect $9/16$ to show both dominant characters (Gray, Normal), $3/16$ one dominant character (black, Normal), $3/16$ the other dominant character (Gray, vestigial), and $1/16$ both recessives (black, ves-

12.12

Crossing over. (a) a pair of homologous chromosomes is represented, one in color and one in outline, each with a "gene," shown as a circle. (b) Each of the chromosomes duplicates, but the two *chromatids* are held together at the *centromere*. (c) During meiosis, two chromatids, each from a different chromosome, cross over each other. At the point of contact, they both break and rejoin, but not in the original way. (d) As the chromosomes move apart during anaphase, the chromatids, which have now exchanged parts, begin to separate. (e) After meiosis is complete, each resulting reduced cell has a complete chromatid (now a chromosome), but the two that have experienced a crossover are unlike anything that went into the system before meiosis. The new chromosomes are physically and genetically different. In this example, only one crossover is shown, but in a real cell, there may be several crossovers, making complex rearrangements possible.

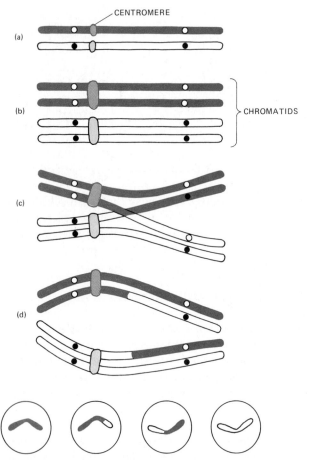

tigial). However, the expected ratios do not occur. Instead, the F_2 generation has too many individuals resembling one parent or the other and too few recombinants.

In practice, geneticists do not always inbreed the F_1 progeny as Mendel did. They use instead a *test cross* in which an experimental organism is mated with a phenotypic recessive. The phenotypic recessive is of course homozygous, since otherwise the recessive would not be apparent. A test cross is informative because the results can be assessed for genotype by direct inspection; there are no confusing dominant features to obscure the genotype. To make a test cross of those flies that do not show Mendelian assortment, the geneticist mates one of the F_1 females (Gray-bodied and Normal-winged, but heterozygous for both features) with a double-recessive male (black-vestigial). The kinds and numbers of flies among the offspring, puzzling at first, are as follows:

Gray, Normal	41.5%	} 83%
black, vestigial	41.5%	
black, Normal	8.5%	} 17%
Gray, vestigial	8.5%	

There are more black-vestigials than were expected and not enough recombinants, and it is especially notable that the majority of the flies are like one parent or the other. Instead of assorting themselves as the genes did in Mendel's peas, the genes for body color and wing shape seem to be stuck together in most, but not all, of the flies. This observation set the stage for still other important genetic discoveries.

Gene Linkage

When genes tend to be inherited together, they are said to be *linked*. Genes that are linked together compose a *linkage group*, and the specific genetic traits carried by those genes are inherited as a group, not separately. With body color and wing form in fruit flies, we saw that linkage resulted in 83 percent of the flies having the same genetic character as the parents; only 17 percent were unlike either parent.

When further crosses were made with flies and with a number of other organisms, more and more examples of linkage were discovered, until lists of mutations were built up. Some genes were closely linked, with low percentages of recombinants, and other genes were less closely linked. *Linkage lists* were made for fruit flies after hundreds of trial matings until it became apparent that four linkage groups exist in fruit flies. In corn, another favorite experi-

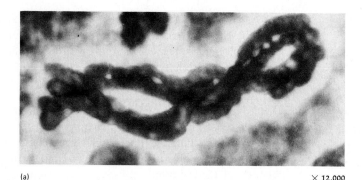

(a)   × 12,000

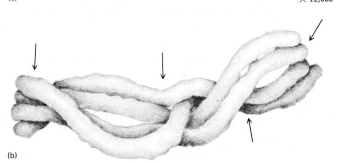

(b)

12.13
Crossing over. (a) Photomicrograph of chromosomes in grasshopper testis during meiosis. Four chromatids are twisted about one another, but they are still in contact at four places, where chromatid exchange is occurring. (b) Drawing made from the photograph, emphasizing the distinctness of the chromatids.

mental organism, ten linkage groups were established. The revealing fact about these experimental organisms is that fruit flies have *four* linkage groups and *four* chromosomes, and corn has *ten* linkage groups and *ten* chromosomes. Is there a cause-and-effect relationship between genetic traits and chromosomes, or are these apparent connections between the number of linkage groups and chromosomes merely coincidental? Of all the possible cell organelles, is it specifically the chromosomes that carry the genes?

Crossing Over

The numerical equality of linkage groups and chromosomes indicated that chromosomes are indeed the genetically important parts of cells, and other evidence in support of this notion was—or soon became—available. Segregation and independent assortment of alleles parallel the activity in chromosomes during meiosis and fertilization. (Fortunately for Mendel, he chose traits in peas that were not linked together; otherwise his demonstrations of independent assortment would not have occurred as they did. Independent assortment works only when the observed genes are located on separate chromosomes.)

If one assumes that a chromosome is a string of genes, one can explain segregation of genes at meiosis, as well as linkage of genes. But if genes are sometimes linked together on a chromosome, how do they succeed in making any recombinations at all? The supposition was that even linked genes do not always remain together as a group, and that chromosomes can break and exchange parts during meiosis, making genetic *crossovers* (Figure 12.12). These crossovers are expressed as recombinants. During a crossover, genes from a *maternally* derived chromosome become associated with genes from a *paternally* derived chromosome.

The physical appearance of chromosomes during meiosis supported the idea of crossing over, although it is usually impossible to know whether actual chromosome exchange has occurred simply by looking. But by the use of specially marked chromosomes or radioactive tracers, it has now been convincingly demonstrated that chromosomes can break apart during meiosis, and when chromosomes exchange pieces to mend the break, blocks of linked genes are exchanged (Figure 12.13). Morgan and Bridges had guessed correctly.

Once the principles of nondisjunction, gene linkage, and crossing over were basically understood by Morgan and his colleagues, there remained little doubt that genes located on chromosomes are the carriers of hereditary traits. Morgan's definitive volume, *The Mechanism of Mendelian Heredity*, published in

12.14

Evidence that chromosomes do actually trade parts. A strain of corn (*Zea mays*) that had colored, starchy grains had a mismatched pair of chromosomes. One chromosome was normal, and the other was distinguishable *at both ends:* One end had a knob, the other a visible extension. When this strain was crossed with a contrasting strain (with colored, nonstarchy grains and a normal pair of chromosomes), the offspring showed recombinations. Some were colored and starchy, some colored and nonstarchy, some colorless and starchy, and some colorless and nonstarchy. The positions of the genes on the chromosomes were known. All the offspring that carried the genes for colored grains had knobs, but they did not have the extension present in the original cross unless they were also nonstarchy. All the nonstarchy plants had the extension. The physical exchange of chromosome parts exactly parallels the genetic exchange. This classic experiment by Harriet Creighton and Barbara McClintock proved the reality, long suspected, of chromosomal crossing over.

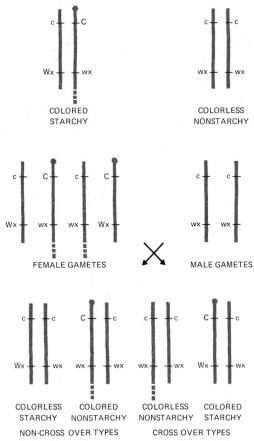

ZEA MAYS

COLORED STARCHY   COLORLESS NONSTARCHY

FEMALE GAMETES   MALE GAMETES

COLORLESS STARCHY   COLORED NONSTARCHY   COLORLESS NONSTARCHY   COLORED STARCHY

NON-CROSS OVER TYPES   CROSS OVER TYPES

1915, convinced most scientists that chromosomes contain the mechanism of heredity, a concept that many biologists rank in importance with the cell theory and Darwin's theory of evolution.

Chromosome Mapping

The idea of crossing over was brand new and still unproved when another Morgan colleague, Alfred H. Sturtevant, speculated that genes were arranged in a linear sequence. He attempted to find out if the frequency of recombinant crossovers was somehow related to the positions of the genes along a chromosome. Sturtevant supposed that the closer together the genes were, the less would be the chance for their crossing over and subsequent recombination.

Repeated matings of fruit flies showed that when two genes are linked, they produce recombinations in dependable proportions. For example, the "forked bristles" trait is linked to "fused veins" in fruit flies, with recombination in 2.8 percent of the offspring. However, the proportions differ when different pairs of linked genes are studied. The linked traits of "fused veins" and "carnation eyes" give a 3 percent recombination. Sturtevant began constructing *chromosome maps*, which located genes along a chromosome based on the percentage of crossovers (Figure 12.15). He translated the recombination percentage into distance, letting one percent equal one "map unit." In 1915, only two years after he began his research, Sturtevant was able to map more than 85 mutant genes in Drosophila. (By 1977, scientists had mapped 210 of an estimated 50,000 human genes; it is expected that by 1985 the location of 1000 human genes will be pinpointed on the human body's 46 chromosomes.)

CHROMOSOMAL ABERRATIONS SOMETIMES OCCUR

Chromosomal replication must be a dependable process. Otherwise the millions of plants and animals would not be able to maintain their stability. Chromosomal aberrations are less common than gene mutations, but nevertheless, mistakes do occur. The best known of the physical rearrangements in chromosomes are *deletions, duplications, inversions,* and *translocations* (Figure 12.16).

A *deletion* occurs when a whole chromosome or a piece of a chromosome is lost. In general, such losses are likely to be detrimental. *Duplications* are less dangerous, but they can have effects on organisms. An *inversion* occurs when a chromosome becomes looped

12.15

A chromosome map representing the sequence of gene positions along the X chromosome of the fruit fly, Drosophila. The numbers indicate the distance from one end of a chromosome, as determined by crossing experiments, and they are based on the frequency of crossing over. Since these *map distances* are numerically equal to percent crossovers between loci, they do not necessarily indicate actual physical distances on a chromosome. It is known that some parts of chromosomes permit crossing over more readily than do other parts, and consequently a genetic map is not intended to show actual distances between loci. In an analogy with human travel, it can be noted that people who live in crowded cities express distances in time rather than distances. A person who lives "thirty minutes from the office," may not even know how far that is in miles. Scientists hope that knowing the location of genes on chromosomes will facilitate research into hereditary diseases and genetic engineering.

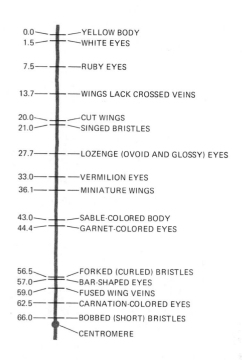

0.0	YELLOW BODY
1.5	WHITE EYES
7.5	RUBY EYES
13.7	WINGS LACK CROSSED VEINS
20.0	CUT WINGS
21.0	SINGED BRISTLES
27.7	LOZENGE (OVOID AND GLOSSY) EYES
33.0	VERMILION EYES
36.1	MINIATURE WINGS
43.0	SABLE-COLORED BODY
44.4	GARNET-COLORED EYES
56.5	FORKED (CURLED) BRISTLES
57.0	BAR-SHAPED EYES
59.0	FUSED WING VEINS
62.5	CARNATION-COLORED EYES
66.0	BOBBED (SHORT) BRISTLES
	CENTROMERE

around itself, breaks, and rejoins with a portion of the chromosome reinserted backward. The frequent result is that some of the products of meiosis are sterile. A *translocation* is the transfer of a part of one chromosome to a new place, usually to some other chromosome. Many plants live successful lives in spite of translocations. In other cases, as in some human examples described later, a translocation can so disrupt the functioning of the whole organism as to make it essentially unworkable.

MULTIPLE ALLELES AND BLOOD GROUPS

So far, alleles have been considered as if only two possibilities exist—for example, red eyes or white eyes in Drosophila. Of course, any diploid organism normally does have only two alleles, and they may be identical, making the organism homozygous for that trait, or they may be different, making the organism heterozygous. Suppose, however, that in a separate organism a mutation occurs at the "red eye" position (*locus*) on a chromosome. Then instead of two possible alleles in the population, there are three. Indeed, a whole series of alleles for eye color has been found in fruit flies, ranging from deep red through a variety of shades of pink. When a population carries more than two alleles at a specific locus, the alleles are called *multiple alleles*. Any one diploid individual can have only two alleles at a specific locus.

12.16

Chromosomal aberrations. (a) Deletion. A segment of a chromosome is lost from a midsection, and the broken ends heal together. (b) Duplication. A segment of a chromosome is repeated one or more times. (c) Inversion. A segment of a chromosome is removed and reinserted so that a portion of the gene sequence is reversed. (d) Translocation. A segment of a chromosome is detached from one chromosome and becomes attached to another one, either to its homolog (in which case the receiver would also have a duplication) or to a nonhomologous chromosome. All aberrations are potentially capable of causing phenotypic or genetic changes in the organisms possessing them, but they are not always apparent, and the organisms may seem normal.

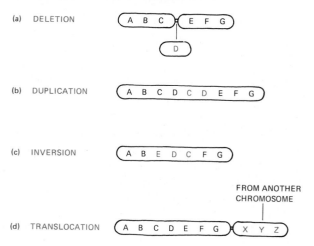

TABLE 12.2
Blood Types and Their Reactions

BLOOD TYPE	CELL ANTIGEN	PLASMA ANTIBODY	GENOTYPE	THEORETICALLY CAN GIVE BLOOD TO	CAN RECEIVE BLOOD FROM
A	A	Anti-B	AA or AO	A, AB	A, O
B	B	Anti-A	BB or BO	B, AB	B, O
AB	AB	None	AB	AB	A, B, AB, O
O	None	Anti-A and Anti-B	OO	A, B, AB, O	O

TABLE 12.3
Inheritance of Human ABO Blood Types

BLOOD TYPES OF PARENTS	CHILDREN'S BLOOD TYPE POSSIBLE	CHILDREN'S BLOOD TYPE NOT POSSIBLE
A + A	A, O	AB, B
A + B	A, B, AB, O	
A + AB	A, B, AB	O
A + O	A, O	AB, B
B + B	B, O	A, AB
B + AB	A, B, AB	O
B + O	B, O	A, AB
AB + AB	A, B, AB	O
AB + O	A, B	AB, O
O + O	O	A, B, AB

these antibodies are already present in the blood and will defend against any antigen not already present in the blood.) As an example, blood type A carries antigen A in its cells and antibody anti-B in the plasma. This means that mixing blood types A and B will cause *agglutination*, a clumping of blood cells (Figure 12.17), because the anti-A antibody in the B-type blood will affect the B-antigen, and vice versa.

Obviously, blood grouping is important in transfusions. For example, if a person with blood type A gives blood to a person with blood type B, the anti-A antibody in the recipient's blood will cause a clumping of the A-type cells in the donor's blood, and the result will probably kill the recipient. In contrast, a person with blood type AB has neither anti-A nor anti-B plasma antibody and can safely receive any blood of the ABO series. A person with blood type O has no cell antigens and can safely give blood to any of the other types. The person with blood type O is therefore known as a *universal donor*. However, if these procedures were as safe as this oversimplification makes them sound, the special precautions doctors take when blood transfusions are to be given would not be necessary.

Blood types are sometimes used in paternity suits, although blood types cannot demonstrate who *is* the father, only who is *not*. A famous paternity case involved Charlie Chaplin. His blood type was O, the baby's was B, and the mother who made the claim had type A. Look at Table 12.3 and see if Mr. Chaplin could have been the father.

From the genotypes and phenotypes in Table 12.3, and by application of simple Mendelian rules, the inheritance of ABO blood types can be determined. For example, a woman with type A blood may

Multiple Alleles and Blood Grouping

A well-known example of multiple alleles is the ABO blood-grouping system in humans. In this system, at least three alleles are available to determine the blood type of any person. Table 12.2 shows that each blood type contains one or more antigens in the red blood cells, and it also contains in the plasma those antibodies *for which it has no antigen*. (An *antigen* is usually a foreign substance that causes a body reaction that subsequently stimulates the production of an *antibody*, a protein that attempts to neutralize the antigen and its effect on the body. Normal human blood plasma, however, contains antibodies that are not produced by the presence of a foreign antigen;

be AA or AO. If she marries a man with Type B blood (BB or BO), she may have children with type A (AO), type B (BO), AB, or O. If she did indeed have four children with all the possible blood types, one would know that both she and her husband were heterozygous for the ABO series.

It is necessary to state "for the ABO series" explicitly because the ABO system is not the only one. Another blood system involves the Rh factor, so called because it was first discovered in *Rhesus* monkeys in the 1930s. Of the several genes responsible for about nine different blood antigens, one is of clinical interest because it sometimes causes Rh disease, or erythroblastosis fetalis.

Rh disease can kill a young infant if the blood of the newborn is not replaced with fresh blood. In general, the circulation of an unborn child is separate from the mother's circulation, but sometimes the infant's blood cells may leak into the mother's blood during late pregnancy or at birth. If the fetal cells carry an Rh antigen inherited from the father, and if the mother has no such cells, she will produce anti-Rh antibody (Figure 12.18). This series of events happens only when an Rh-negative woman has a child by an Rh-positive man and usually does not affect the

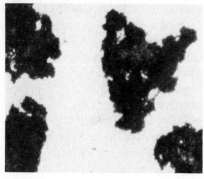

× 110

12.17
The technique of blood group determination. Blood of an unknown type is spread on a glass slide, and an antibody is added. If there is no reaction, the cells remain scattered, as shown on the left. If, for example, anti-B antibody is added, and the cells clump together, as shown on the right, the cells are sensitive to the anti-B antibody and the blood is therefore type B. This is agglutination.

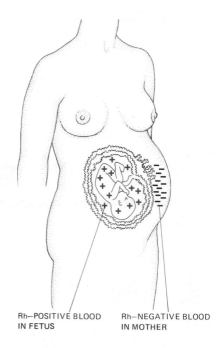
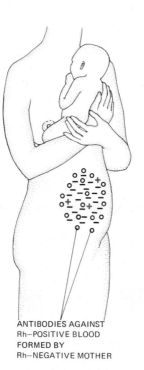
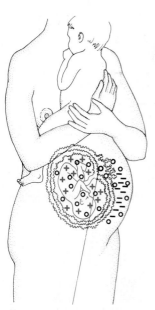

Rh—POSITIVE BLOOD IN FETUS   Rh—NEGATIVE BLOOD IN MOTHER

ANTIBODIES AGAINST Rh—POSITIVE BLOOD FORMED BY Rh—NEGATIVE MOTHER

MOTHER'S ANTIBODIES PASS THROUGH PLACENTA AND MAY DAMAGE NEXT BABY'S Rh—POSITIVE BLOOD

12.18
The development of Rh disease, or *erythroblastosis fetalis*, in an infant. The Rh-negative mother carrying an Rh-positive fetus may develop anti-Rh antibody. A first child does not have time to be affected, but a second child may receive some of its mother's antibody, which may destroy that second child's red blood cells unless appropriate countermeasures are taken.

first-born child, which is born before it has time to receive a dangerous amount of maternal antibody. Once sensitized to the Rh antigen, however, a woman carrying anti-Rh antibody may give some of that antibody to a second child. Therefore, when an Rh-negative woman has an Rh-positive child by an Rh-positive man, she should be alerted to the danger to possible later children and receive shots that prevent her from making anti-Rh antibody. If Rh denotes the dominant gene for Rh antigen production, and rh the recessive gene for lack of it, then the only combinations that could lead to difficulty would be the conception of an Rh/rh child from an rh/rh mother by an Rh/Rh or Rh/rh father. Further, there must be some exchange between maternal and fetal circulation, which may or may not occur.

Although only the ABO and the Rh antigen systems have been included here, there are in fact about 80 described blood antigen systems. When one considers the number of combinations of blood antigens possible, it appears likely that there are enough different arrangements to provide every human being with his own individual blood type. Since only a few types seem to be medically important, those few are well known, but the rest are present and functioning.

SEX-LINKED GENES IN HUMAN BEINGS

In animals that, like humans and fruit flies, have an XX-XY type of sex determination, any gene on an X chromosome can be heterozygous only in a female

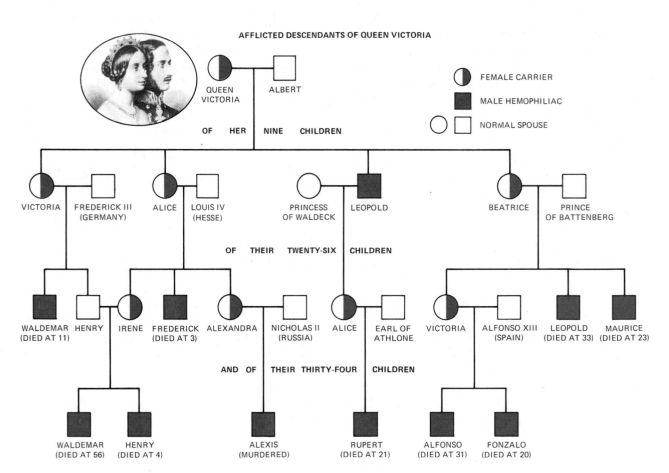

12.19

A famous example of a human mutation: hemophilia in the descendants of Queen Victoria. The chart shows the relation of the queen to several royal houses into which her children married, and the results of those marriages. Note that four of Victoria's nine children were either carriers or actually afflicted, and that of her other descendants, ten males were hemophiliacs and four females were carriers. Because it has no afflicted members, the British royal family of today is not represented on the chart.

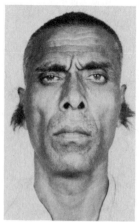
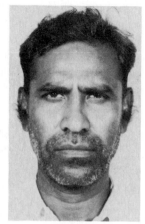

12.20
A human trait inherited on the Y chromosome: hairy ears. These three Indian brothers show the feature. A characteristic is thought to be carried on the Y chromosome if it is passed from father to sons, is inherited by all the sons of a father, and is not inherited by any daughters or children of daughters.

(with her XX pair of homologs). Since Y chromosomes do not carry alleles to match anything in the X chromosome, a male cannot be heterozygous for any sex-linked locus. He cannot be homozygous, either. His condition is called *hemizygous*. Any gene he has on his X chromosome will be expressed phenotypically, even if it is recessive, since there is no possibility that it will be masked by a dominant gene. Because of this fact, sex-linked characteristics in humans tend to be expressed mainly in males, who *do not* pass such characteristics to their sons. They may well pass it to their grandsons, however, so this kind of inheritance is sometimes called "skip-generation" inheritance.

The most famous example of sex-linked inheritance in humans is hemophilia, the bleeder's disease. Queen Victoria of England was apparently a mutant heterozygous carrier for the hemophilia gene, which she transmitted to some of her children (Figure 12.19). She herself, of course, was not a hemophiliac. The gene spread through several royal families in Europe, sometimes with serious political and personal results. Victoria's granddaughter Alexandra, a carrier, gave birth to Alexis, the son of Czar Nicholas II of Russia. Alexis was a hemophiliac. The best Russian doctors could not cure him, but the opportunist monk Rasputin claimed he could erase the disease, and the royal family fell under Rasputin's influence. Rasputin added to the country's political corruption, and the Russian Revolution, which foreshadowed the current Soviet doctrines, erupted into a bloody overthrow of the Czarist regime. Alexis would have become Czar of Russia if the 1914 Revolution had not killed him. Because he was a hemophiliac, the course of world history was dramatically affected.

Several kinds of *colorblindness* are sex-linked, with the genes, like those for hemophilia, on the X chromosome. All the many genes located on X chromosomes give inheritance patterns like those in white-eyed flies, provided the species is of the XX-XY type. The Y chromosome has few genes, but when a mutation occurs in a Y chromosome, it is passed from fathers to sons only, and daughters cannot even be carriers (Figure 12.20).

Colorblindness is more likely to occur in men than in women, and in fact there are 5 to 10 times more colorblind men than colorblind women. A recessive gene on the X chromosome carries the colorblindness trait, and there is no allele on the male Y chromosome to counteract it. In order for a man to be colorblind, he needs to receive only one X chromosome with a gene for colorblindness from his mother, whereas a woman must receive the colorblindness trait from both mother and father. Therefore, if a male receives the colorblindness gene from his mother's X chromosome, he will be colorblind. (It is as though the recessive X chromosome suddenly became dominant, and in a sense it *is* dominant to the neutral Y chromosome.)

SOME EFFECTS OF HUMAN CHROMOSOMAL ABERRATIONS

A number of human deficiencies and deformities are caused by chromosomal deletions, duplications, or translocations. They can be recognized by microscopic examination and correlated with the set of symptoms that characterize the phenotype. A *karyo-*

type, or chromosome set of a male has 22 matched pairs of chromosomes plus one unmatched pair, the XY pair (Figure 12.21). A female has 23 matched pairs, including the XX pair. Any deviation from the normal is likely to have a harmful effect on the development of the individual suffering from the deviation.

Nondisjunction in Human Chromosomes

One common misfortune is a malady known as *mongolism*. The eyelid characteristic of the Mongolian physiognomy suggested the name, but other symptoms include mental retardation and abnormal hands. It is now more frequently called *Down's syndrome*, named for one of the investigators of the disease, or *trisomy-21* (Figure 12.22). If an egg undergoes an inaccurate meiosis, two members of the same homologous pair of chromosomes may go into the final egg, and if that egg is then fertilized by a normal sperm with one homolog, the zygote will have three copies. Such a zygote is a *trisomic*, caused by nondisjunction, the failure of a pair of homologs to separate at meiosis.

If nondisjunction can result in an egg with both homologs, it can also result in one with neither. Such

an egg, fertilized by a normal sperm, will become a zygote with only one homolog, and it is therefore a *monosomic*. One partial monosomic of the fifth human chromosome causes the *cat's cry syndrome*, named for the special sound produced by infants suffering from the deficiency. Such infants have deformed heads and usually die early.

ABERRATIONS IN HUMAN SEX CHROMOSOMES

A number of examples of sex chromosome aberrations have been described. An individual with only one sex chromosome, the X0 condition, is female but has permanently infantile sex organs, fails to develop normal secondary sex characters, and has abnormal body structure (*Turner's syndrome*). An individual with *Klinefelter's syndrome* has an extra X chromosome, which produces a genotypic arrangement of XXY and a total of 47 chromosomes. Such a person is usually phenotypically male but has underdeveloped male genitals and some evidence of female breasts. The body is usually somewhat female in appearance, with long legs, sparse body hair, and some

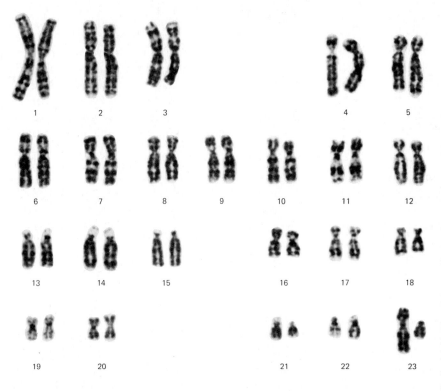

1 2 3 4 5
6 7 8 9 10 11 12
13 14 15 16 17 18
19 20 21 22 23

12.21
The chromosomes of a man. Preparation of a *karyotype* involves taking a tissue sample, usually blood, and culturing it. Mitosis is stimulated by the addition of an extract of beans (haemagglutinin), then stopped at midmetaphase by the addition of an alkaloid (colchimide), and the cells are swollen by being put in water. Cells are killed, partially digested by a protein-splitting enzyme to make the individual bands on the chromosomes appear, and finally stained. A cell with well-separated chromosomes is photographed, and each chromosome is cut out of the photograph. All the chromosomes of the cell are arranged in order from the longest, number 1, to the shortest, number 23, followed by the sex chromosomes—XX for a female or XY, as here, for a male. The X chromosome is about the size of number 10, but the Y is about like number 22.

hip development. The individual is sterile and may be mentally retarded. It was not until 1956 that the cause of Klinefelter's syndrome was discovered. Recently a woman athlete with an XXY configuration was disqualified from a track meet, and as a result of that incident, all female athletes competing in the Olympics and other international meets are required to take a chromosome test.

A person with an XYY condition is an apparently normal male, but there is some speculation that there may be an unexpectedly high proportion of individuals with XYY known to our society as "overaggressive." The hypothesis of an "XYY syndrome," which results in antisocial behavior in males, has led to a still unresolved argument among geneticists, sociologists, penologists, lawyers, and doctors. The notion that XYY males are "criminal types" is in fact losing credibility among scientists. No one really knows whether having an XYY karyotype does in fact predispose one to violence—or if it does, what should be done about it.

MULTIPLE FACTOR INHERITANCE

In some instances, a visible characteristic can be influenced by more than one pair of genes. When several pairs of genes act on a single characteristic, they are called *multiple genes,* and they are usually concerned with some quantitative aspect of inheritance, whether simply size, or perhaps amount of a pigment synthesized. Consider, for example, the overall growth of a human body, in which there is no such distinct division into tallness or dwarfness as occurs in garden peas. Most humans are neither giants nor dwarfs. A few are very short, many are of medium height, and a few are very tall. A range, with many intermediates, is common in a variety of features, such as skin color or eye color in humans, grain color in wheat, and size of cob in corn.

Not counting environmental forces, which further complicate the system, there seem to be several genes capable of adding quantitatively to many phenotypic features. In the simplest instance, one pair of alleles may be responsible for the synthesis of a red pigment in a flower. A flower with two nonproducing alleles would be white; one with one producer and one nonproducer would be pink; one with two producers would be red. The offspring from an inbred hybrid would give red:pink:white in a Mendelian ratio of 1:2:1. If two pairs of independently segregating alleles acted in the same way, a

12.22

A boy with Down's syndrome, or mongolism, caused by the action of a superfluous chromosome 21.* In the general population, about 14 live births per 10,000 have Down's syndrome, but the rate is much higher when the mother is over 35 years old. Apparently chromosome 21 suffers increasing tendency toward nondisjunction with advancing age. Many potential sufferers die before birth, so the actual number of trisomics of this kind is probably much higher than the live birth figures indicate. People with Down's syndrome are mentally retarded and have lower than average resistance to the diseases everyone is likely to have.

range of intensities would appear from deep red through shades of pink to white, again in a Mendelian fashion, with ratios of $\frac{1}{16}$ deepest red, $\frac{4}{16}$ definitely red, $\frac{6}{16}$ light red, $\frac{4}{16}$ pale pink, and $\frac{1}{16}$ white

---

* The reference here to chromosome 21 is an example of a conscious continuation, for practical reasons, of a known error. When after long search the cause of Down's syndrome was identified as a chromosomal aberration, it was thought that the chromosome involved was No. 21. Later research established that No. 22 was in fact the aberrant chromosome, but the descriptive identification of the condition, "trisomy 21," has become so entrenched in medical writing on the subject that there has been general agreement to continue to use it.

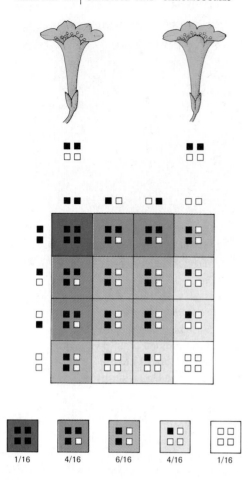

12.23

Inheritance of intergraded flower color as a result of interaction of two pairs of genes. The Punnett square shows the results of crossing two flowers of moderately intense pigmentation, each carrying two of a possible four pigment-producing genes. In the offspring, one out of 16 has a chance of receiving all four pigment-producing genes, and one out of 16 has an equal chance of receiving none, with intergrades having other chances, as shown in the squares.

(Figure 12.23). The appearance of only one in 16 with deepest red would be an indication of the presence of two pairs of alleles. If only one out of 64 had the most intense color and one of 64 was white, it would be correct to conclude that three pairs of alleles were involved. With more than about five pairs of alleles at work, analysis becomes impossible. Not only are the differences between colored flowers too slight to categorize, but there are also obvious effects of such environmental factors as light and temperature, both of which can affect flower color. Probably most quantitative inherited features are the result of multiple genes.

Skin color in humans is an excellent example of the expression of multiple genes. Just as people are generally neither giants nor dwarfs, so is their skin color neither pure black nor pure white. The degree of lightness or darkness depends on the thickness of the skin and the amount of melanin present. Melanin is a pigment in the outer layers of the skin that helps prevent ultraviolet light from reaching the more sensitive skin layers underneath. The more melanin, the darker the skin, and the amounts of melanin vary greatly. The skin of a person who has been tanned by the sun has become thicker, and the amount of melanin has increased.

Current thought is that there are probably four pairs of genes involved in the expression of skin color, with the resulting possibility of nine different skin shades. Figure 12.24 shows how medium-skinned parents may produce children with lighter or darker skin than their own.

GENETICS PROBLEMS

1. In fruit flies, males have XY sex chromosomes, females have XX, and white eye color is sex-linked. What are the expected sexes and eye colors from the following crosses?

   a) red-eyed (homozygous) female × white-eyed male

   b) red-eyed (heterozygous) female × white-eyed male

   c) white-eyed female × red-eyed male

2. In pigeons, as in birds generally, males have a matching pair of sex chromosomes, and females have an unmatched pair (the WW-WZ type). A sex-linked gene in pigeons affects the color of the head feathers. Gray-headed females, mated with cream-headed males, give equal numbers of cream-headed males, gray-headed males, and gray-headed females.

   a) Which allele is dominant?

   b) Offer an explanation for the lack of cream-headed females.

3. From the pedigree shown opposite, answer the following questions. (Circles indicate females; squares, males. Open figures indicate normal phenotypes; black, colorblind individuals.)

12.24

Quantitative inheritance of skin color in humans. If pigmentation is determined by four pairs of melanin-producing genes, then a couple with medium coloration might each have four melanin-producing and four neutral genes. Their children might have anything from none to eight melanin-producing genes, or nine possibilities, as shown in the diagram. It is thus possible for moderately pigmented parents to produce children that are very dark or very light or any of seven intermediate shades.

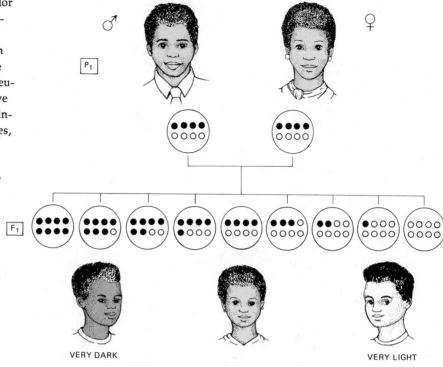

VERY DARK   VERY LIGHT

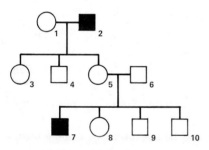

a) What is the probable genotype of 1?

b) Is the answer to (a) 100% certain?

c) What is the genotype of 5? of 9?

d) How certain are the answers to (c)?

e) If 8 marries a normal man, what are the chances that she will have a colorblind son? If she marries a colorblind man?

4. Gene A is linked to B with a crossover value of 6.6%; B is linked with C with a value of 9.8%; and A is linked to C with a value of 3.4%. Arrange the genes in order on a chromosome. Explain why 3.4 plus 6.6 does not equal 9.8.

5. In tomatoes, tall vine (T) is dominant over dwarf (t), and spherical fruit (S) is dominant over pear shape (s). A tall, spherical tomato was crossed with a dwarf, pear-shaped one. The results were:

Tall, Sphere	142
Tall, pear	34
dwarf, Sphere	36
dwarf, pear	138

a) What should the ratios have been if the two genes were on different chromosomes?

b) What is the map distance between the two loci?

c) Why is this answer probably a little lower than the correct map distance?

d) What experiments may be performed to improve the accuracy of the figure?

6. The Y chromosome in a human male carries a few known genes, such as the one for ichthyosis hystrix gravior (the "porcupine men" of eighteenth-century England). If a porcupine man has children, what can you expect with regard to genotypes and phenotypes and the distribution of sexes?

(a)

(b)

(c)

12.25

Gene action modified by the influence of the environment. Himalayan rabbits and Siamese cats have darker hair on the ears, noses, tails, and feet than on the warmer parts of their bodies (a). The fact that the hair on the lighter parts of the body is produced from cells genetically capable of making dark hair can be shown by removing the hair from a light part and keeping that part cool during regrowth of the hair (b). If that is done, the hair that grows in the cooled part is dark, like that on the ears (c). Similar effects occur in plants. A Macintosh apple is greener on the north side and redder on the south side, which received more sunlight.

7. If an organism is heterozygous for genes at four loci, all on different chromosomes, what is the possible number of kinds of gametes, at least with respect to those loci?

8. In male fruit flies, crossing over does not occur. Genes at loci A and B are linked, with a distance of 18 map units between them. What kinds of gametes can be produced by a male fly, with genotype AaBb?

9. In a disputed paternity case, a woman with blood type B had a child with blood type O, and she claimed that it had been fathered by a man with blood type A. What can be proved from these facts?

10. What are the possible parental genotypes of a person with blood type AB?

ANSWERS TO GENETICS PROBLEMS

1. a) All red-eyed, half male and half female.

   b) One-quarter red-eyed female, one-quarter white-eyed female, one-quarter red-eyed male, one-quarter white-eyed male.

   c) Half female, all red-eyed, and half male, all white-eyed.

2. a) Cream. b) "Cream" must be lethal. It can be tolerated only if counteracted by a "gray gene" on a homologous chromosome. Female pigeons have the WZ chromosome pair, and there is no chance for such a counteraction.

3. a) Number 1 may seem to be homozygous normal because she had three normal children by a colorblind man, but since she had a colorblind grandson from a normal son-in-law, she must be a carrier. b) Yes.

   c) Number 5 is a carrier, and number 9 is a normal male.

   d) Both genotypes are certain.

   e) Number 8 has a 50% chance of being a carrier, since her mother is definitely a carrier and her father is normal. She also has a 50% chance of passing the colorblind gene to a child, and a further 50% chance that a child will be male. Thus the chance of producing a colorblind son is $\frac{1}{2} \times \frac{1}{2} \times \frac{1}{2} = \frac{1}{8}$. The chance is the same whether her husband is normal or colorblind, since he will pass only a Y chromosome to any son.

4. $C \leftarrow (3.4) \rightarrow A \leftarrow (6.6) \rightarrow B$. Some double crossovers between B and C did not get counted.

5. The parent genotypes were TtSs and ttss.

   a) About equal numbers of each phenotype.

   b) 20 map units.

   c) Double crossovers not detected. This gives a misleadingly low percentage crossover.

   d) Find loci between the loci in question, and use the sum of the smaller, intermediate distances.

6. All sons will be phenotypic porcupine men. Daughters will not be affected either genotypically or phenotypically, because they receive no Y chromosomes.

7. One locus could give two kinds of gametes, or 2^1, and two loci give four kinds, 2^2. Therefore four loci could give 2^4, or 16 kinds.

8. If A and B are on one homolog and a and b on the other, then the only kinds of gametes are AB and ab. If A and b are on one homolog and a and B on the other, the only kinds of gametes are Ab and aB.

9. Since the man could have been genotypically AO and the woman BO, he *could* have been the father, but this is no proof that he actually was or was not the father.

10. AA × BB; AA × AB; AA × BO; AO × BB; AO × AB; AO × BO; AB × BB; AB × BO; AB × AB.

SUMMARY

1. An explanation of both Mendelism and Darwinian variations was proposed by Hugo de Vries during the period 1901–1903. De Vries called inherited changes *mutations*.

2. The so-called mutations of de Vries and other geneticists provided some useful information on patterns of inheritance and interactions among genes, but the chemical explanation for mutations had to wait for the demonstration of DNA as the genetic material.

3. Although the causes of natural mutations are not known, biologists can induce them artificially by a number of methods. Mutations can be induced by X-rays and other radiation that has sufficient energy to cause chemical changes of molecules in cells. Mutations can be induced by increasing the temperature and by introducing chemical mutagens, such as atmospheric pollutants and pesticides.

4. Most mutations are not only recessive but detrimental. However, occasional mutations have been advantageous; in fact, favorable mutations are the basis for the genetic variability that makes evolution work.

5. Only those mutations that occur in gonad cells in animals have any chance of being passed on to later generations.

6. In the early 1900s researchers discovered that in fruit flies chromosomes are responsible for the determination of sex.

7. *Nondisjunction* in fruit flies occurs when a pair of homologous chromosomes fails to separate during meiosis.

8. Numerous crosses with a variety of nondisjunctions demonstrated that in flies the sex of an individual is determined by the ratio of X chromosomes to the number of autosomes. This is *not* true of human sex determination.

9. When genes tend to be inherited together, they are said to be *linked*. Genes that are linked compose a *linkage group*, and the specific genetic traits carried by those genes are inherited as a group, not separately.

10. Chromosomes can break and exchange parts during meiosis, making genetic *crossovers*. These crossovers are expressed as recombinants. With information about nondisjunction, gene linkage, and crossing over, geneticists concluded that genes located on chromosomes are the carriers of hereditary traits.

11. Geneticists are able to construct *chromosome maps*, which locate genes along a chromosome, based on the percentage of crossovers.

12. Physical rearrangements in chromosomes are *deletions, duplications, inversions*, and *translocations*.

13. When there are more than two alleles at a specific locus in a population, the alleles are called *multiple alleles*. A well-known example of multiple alleles is the ABO blood-grouping system in humans.

14. A visible feature that is specifically associated with the sex of its possessor is a *sex-linked trait*. The best-known examples of sex-linked traits in humans are hemophilia and colorblindness.

15. Some effects of nondisjunction in human chromosomes are *Down's syndrome* and *cat's cry syndrome*. Examples of aberrations in human sex chromosomes are *Turner's syndrome* and *Klinefelter's syndrome*.

16. Some characteristics, such as body size or skin color, can be influenced by more than one pair of genes. When several pairs of genes act on a single characteristic, they are called *multiple genes*.

RECOMMENDED READING

Beadle, George W., "The Genes of Men and Molds." *Scientific American*, September 1948. (Offprint 1)

Clarke, C. A., "The Prevention of 'Rhesus' Babies." *Scientific American*, November 1968. (Offprint 1126)

Crick, F. H. C., "The Structure of the Hereditary Material." *Scientific American*, October 1954. (Offprint 5)

Grobstein, Clifford, "The Recombinant-DNA Debate." *Scientific American*, July 1977. (Offprint 1362)

Lerner, I. Michael, and William J. Libby, *Heredity, Evolution and Society*, Second Edition. San Francisco: W. H. Freeman, 1976.

McKusick, Victor A., "The Mapping of Human Chromosomes." *Scientific American*, April 1971. (Offprint 1220)

Mittwoch, Ursula, "Sex Differences in Chromosomes." *Scientific American*, July 1963. (Offprint 161)

Muller, H. J., "Radiation and Human Mutation." *Scientific American*, November 1955. (Offprint 29)

Ruddle, Frank H., and Raju S. Kucherlapati, "Hybrid Cells and Human Genes." *Scientific American*, July 1974. (Offprint 1300)

Srb, Adrian M., Ray D. Owen, and Robert S. Edgar (eds.), *Facets of Genetics* (Readings from *Scientific American*). San Francisco: W. H. Freeman, 1970.

Suzuki, D. T., and A. J. F. Griffiths, *An Introduction to Genetic Analysis*. San Francisco: W. H. Freeman, 1976.

Todd, Neil B., "Cats and Commerce." *Scientific American*, November 1977. (Offprint 1370)

Winchester, A. M., *Human Genetics*, Second Edition. Columbus, Ohio: Charles E. Merrill, 1975. (Paperback)

13

Reproduction in Animals

ANIMALS REPRODUCE BY SEXUAL AND ASEXUAL means. In either case, hereditary traits are passed on from generation to generation through the genetic code of DNA. There are enormous advantages to sexual reproduction, but most of all, it is preferable because it allows for variation and diversity through a *recombination* of parental genes. Animals that reproduce asexually can produce only offspring identical to themselves, and only through infrequent mutations can variations develop that might help the species adapt to a changing environment.

ASEXUAL REPRODUCTION IS USUALLY LIMITED TO THE SIMPLER ANIMALS

Asexual reproduction in higher multicellular animals is rare; it is limited almost entirely to the simpler types. One-celled animals, like one-celled plants, frequently undergo fission, although most of them use some sexual methods as well. Corals bud off small replicas of themselves, and many parasitic worms simply break into pieces, each of which is capable of developing into a complete worm.

Some female insects, such as aphids and bees, can lay viable eggs without fertilization. Although such reproduction is literally sexless, it is plainly derivative of a sexual method and usually serves only as a partial substitute for a sexual method.

VARIATIONS IN ANIMAL REPRODUCTIVE PROCESSES

The essential feature of sexuality is the union of unlike chromosome sets. In the unicellular (or as they are sometimes called, the noncellular) animals, the protozoa, sexual mating involves the fusion of two nuclei, but there is no increase in the number of individuals. Thus there is sex without actual reproduction, but *genetic recombinations* do take place, and that is the *essence of sexuality*. Table 13.1 outlines the variations of animal reproduction.

Many animals have fascinating methods of mating, and some of the more interesting reproductive

TABLE 13.1

Variations of Animal Reproduction

METHOD	TYPICAL ANIMALS USING METHOD	CHARACTERISTICS OF REPRODUCTION	NATURE OF OFFSPRING
Asexual			
Fission	Protozoans, some metazoans.	Body of parent divides into two equal parts. May be transverse or longitudinal.	Exact replicas of parent.
Budding	Hydra: external budding. Freshwater sponges, bryozoans: internal budding.	Unequal division. Bud develops organs like parent's and detaches itself. Buds are usually external; if internal they are gemmules.	The new individual arises as an outgrowth (bud) from the parent.
Fragmentation	Platyhelminthes, Nemertinea, Echinodermata.	Organism breaks into two or more parts.	Each new part capable of becoming a complete animal.
Sporulation	Protozoa.	Multiple fission. Many cells are formed, enclosed together in a cystlike structure.	Replicas of parent.
Sexual Modification			
Parthenogenesis	Rotifers, plant lice, some ants, bees, and crustaceans.	Modification of sexual reproduction. Unfertilized egg develops into complete individual. Usually occurs for several generations, followed by a biparental generation in which the egg is fertilized.	Fertilized bee eggs become females (queens or workers); unfertilized bee eggs become males (drones). Other examples various and complex.
Sexual			
Conjugation	Protozoans.	Two individuals fuse together temporarily, exchange micronuclear material. Union may occur between like cells (isogametes) or different cells (anisogametes). May be difficult to distinguish sex.	Contain nuclear material from both parents. Morphologically but not genetically like parents.
Biparental reproduction	Most vertebrates, invertebrates.	Familiar method involving separate and distinct male and female individuals. Each individual produces either sperm or ova, rarely both. Separate sexes (dioecious).	Similar to parents but not identical.
Hermaphroditism	Flatworms, hydra, tapeworms, some molluscs	Individuals have both male and female organs (monoecious condition). Self-fertilization usually prevented by producing gametes at different times.	Similar to parents but not identical. Contain equal amounts of DNA from each gamete.

Adapted from Cleveland P. Hickman, Sr., Cleveland P. Hickman, Jr., and Frances M. Hickman, *Integrated Principles of Zoology*, Fifth Edition (St. Louis: The C. V. Mosby Company, 1974) pp. 721, 722. Used with permission.

13.1

The starworm, *Bonellia viridis*. When a starworm hatches, it may become either male or female, depending on its luck. If while still undetermined it encounters a female, it becomes a male, (encircled in the picture) and it remains tiny, spending its life in the larger female. If while still undetermined it fails to meet a female, it becomes a female. The difference in size between the two sexes is as remarkable as the difference in habits. Two-thirds natural size.

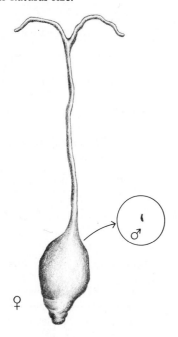

variations are presented here. One of the eight arms of a male octopus is actually a *hectocotylus*, a chamber for storing sperm. At the time of mating, the male inserts this special arm into the breathing cavity of the female, where eggs are already ripening. The arm breaks off inside the female (another arm will grow in time for the next mating season), and when the eggs are ripe, the arm ruptures and releases the sperm to meet the eggs. Female bristle worms eat the sexual parts of the males in order to fertilize their eggs. The sperm are liberated from their sacs when they reach the female's stomach. From there the sperm can travel to fertilize the eggs. The starworm, Bonellia, has an even more unusual reproductive pattern (Figure 13.1). The dwarf males spend their entire lives as parasites within the female's body, where they continue to fertilize eggs as the eggs become ripe. Honeybees have a female-oriented society, as we shall see in Chapter 29. After the queen bee mates with many male bees, their male reproductive organs break off, remaining inside the queen to prevent the loss of sperm. The male honeybees, having fulfilled their prime function, proceed to bleed to death.

Many aquatic animals simply shed gametes into the water they live in, allowing sperm to find eggs as best they can. Such a dependence on fertilization of free-floating gametes may seem to be a risky way to start new generations, but several factors increase the chances of fertilization and survival: (1) The eggs contain enough stored food to give the young animals a start in life until they can feed themselves. (2) The chances for successful fertilization are improved by the synchronized release of sperm and eggs, as well as by the huge numbers of gametes released. (3) Specialized behavior patterns tend to bring sperm close to eggs. (4) Some method of timing ensures that eggs and sperm have a good chance to meet; phases of the moon that affect the flow of tides are well correlated with spawning times, and some animals correlate their spawning times with the length of days, thus

13.2

Like many marine organisms, sponges shed sperm loose in water, depending to a considerable extent on chance that the sperm will come into contact with and fertilize an appropriate egg. In the sponge shown here, a cloud of sperm is being sent out from the mouth of the sponge like smoke from a chimney.

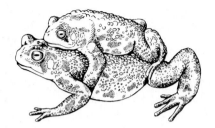

13.3
Mating between toads. Close association of the male and the female is involved, but fertilization is external. As the female lays eggs, the male deposits sperm over them.

liberating their gametes, each in its proper season (Figure 13.2). Thus aquatic invertebrates and fishes do breed successfully. The factors that contribute to successful fertilization and survival are the products of long evolutionary trends that ensure the survival of the species.

Almost all land-dwelling animals have adopted some form of mating that allows fertilization of eggs inside the body of the female. Even such diverse groups as insects, birds, and snails practice internal fertilization. In the lower vertebrates, some (sharks) may use internal fertilization, but most fishes and amphibia (frogs, toads, salamanders) have specialized behavior that guarantees that sperm will be deposited close to the eggs. In frogs, for example, the male clasps the female at egg-laying time and ejects sperm over the egg masses as the female liberates them (Figure 13.3). Fertilization and embryo formation occur in water, usually with no parental attention, although in some species of toads the female carries the eggs in pouches in the skin of her back (Figure 13.4). The wet environment there allows the embryos to grow until they can live independently. Some fishes have ritualized behavior that increases the chance for fertilization (Chapter 28).

Sex cells can survive and perform their task of fertilization only in a wet medium. This requirement provided no insurmountable problems for aquatic animals, but as animals migrated to the land, they needed to develop special reproductive structures that would somehow retain a wet environment for fertilization, even though mating itself took place on the dry land. In humans and most other land-dwelling vertebrates, the penis is inserted into the vagina during copulation to accomplish a wet, internal fertilization. Fertilization is *always* internal in mammals. Although the human species is not typical of mammals, it is understandably of special interest to students, so human reproductive structures and functions will be emphasized in this book.

13.4
An unusual instance of parental care for young animals. This Surinam toad, *Pipa pipa*, from northern South America, provides miniature pools in pockets on her back, where eggs, placed there by the male, hatch and go through the early stages of development.

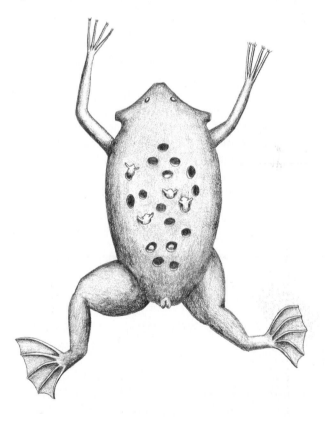

HUMAN SEX DETERMINATION

The male sex cell, the sperm, contains the haploid number of 23 chromosomes. The female sex cell, the ovum or egg, also contains 23 chromosomes. When a sperm fertilizes an egg, the male and female chromosomes enter into a single nucleus, and the diploid number of 46 chromosomes results. A new individual, with its full complement of hereditary material, begins its development. But what determines whether this new organism will be a boy or a girl? The answer is: the nature of the chromosomes. In both males and females, 22 pairs of chromosomes always match. The 22 dependably matching pairs (44 chromosomes) are called *autosomes*. The remaining two chromosomes are the *sex chromosomes*. The pair of sex chromosomes look alike in females, and they are designated *XX*. Sex chromosomes do not look alike in males, and they are designated *XY*.

After meiosis in males, half of the sperm contain 22 autosomes and an X chromosome, and the other half contain 22 autosomes and a Y chromosome. If an egg (always containing an X chromosome) is fertilized by the X-chromosome-bearing sperm, an XX zygote results and will develop into a female; an egg fertilized by a Y-bearing sperm produces an XY zygote and will develop into a male. So the father actually determines the sex of the child, since only his sperm cells contain the variable, the Y chromosome.

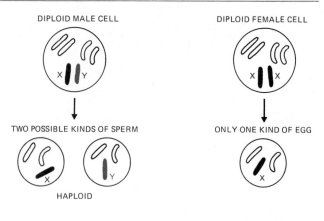

Sex is determined in many animals (including humans) and some plants by a special pair of chromosomes. In mammals and many insects, if the pair is an even match, XX, the animal is female; if the pair does not match, XY, the animal is a male. In such animals, males can produce two kinds of sex-chromosome-bearing sperm, either X or Y. Females can produce only one kind of egg, X. The sex of a new individual is determined at the time of fertilization of the egg by the sperm. An X-carrying sperm determines that the zygote will develop into a female; a Y-carrying sperm determines that the zygote will develop into a male.

HUMAN SEXUAL ANATOMY AND FUNCTION: MALE

The reproductive function of the male is the production of sperm cells and their delivery to the female (Figure 13.5). The organs that produce the sperm cells, *testes*, also produce the primary male hormone, *testosterone*, which aids sperm production. These duties, simple in comparison with the duties of the female, require only the *gonads* (testes), *accessory glands* to furnish a carrying fluid for the gametes, *ducts* for their transmission, and the *penis*.

The Testes

The paired *testes* are formed inside the body cavity, but they descend shortly before birth into an external sac between the thighs, the *scrotum*. The lower temperature outside the body is necessary for active sperm production; the sperm remain infertile if they are retained inside. For the same reason a fever is capable of killing hundreds of thousands of sperm cells. In warm temperatures the skin of the scrotum hangs loosely, and the testes are held in a low position. In cold temperatures the muscles under the skin of the scrotum contract, and the testes are pulled closer to the body. In this way, the temperature of the testes remains somewhat constant. Sweat glands also help to cool the testes.

The testes contain about 15 to 20 coiled *seminiferous tubules*, which produce the *sperm cells*, or *spermatozoa*, at a daily rate of a billion or so in young men. The actual combined length of the seminiferous tubules in both testes is about 225 meters (750 ft) (Figure 13.6).

Hormonal Regulation in the Male

Among the seminiferous tubules in the testes are small masses of *interstitial cells*, which secrete the male sex hormones, especially *testosterone*. Testosterone affects the production of spermatozoa, the development of the sex organs, and the appearance of such secondary male sex characteristics as a deep voice, facial and body hair, skeletal proportions, and distribution of body fat.

If there is a deficiency of testosterone, all the accessory male reproductive organs will decrease in size and activity. In most cases, both penile erection and volume of ejaculation will diminish markedly also. It is not yet certain how (or if) testosterone affects male behavior in general, but it is recognized that deficiencies of testosterone will produce an overall decrease in male sex drive and behavior. It is also known that secondary male sex characteristics will be affected by testosterone deficiencies. A castrated male will gradually acquire such female characteristics as fatty deposits in the breasts and hips, lack of facial hair, and smooth skin texture.

Besides testosterone, males also secrete the pituitary (hypophysis) sex hormones *FSH* (follicle-stim-

13.5

A median longitudinal section through a male human pelvis.

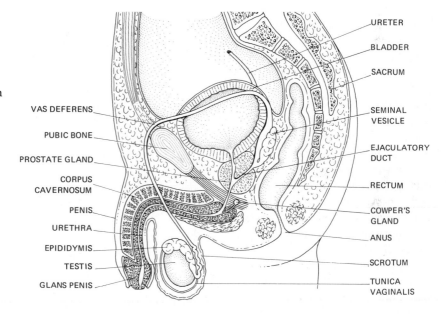

13.6

Structure of a mammalian testis, dissected to show the seminiferous tubules, within which sperm are produced. The sperm cells pass along the *seminiferous tubules* and are stored in the coiled tubules of the *epididymis*, from which they are passed to the outside.

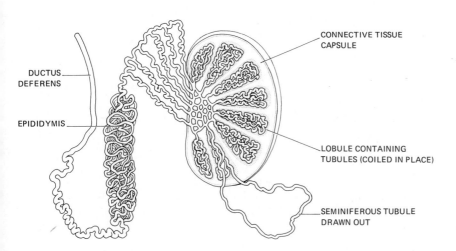

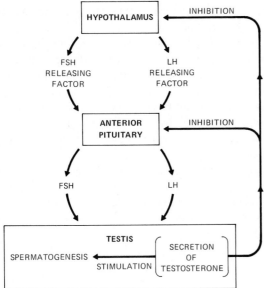

13.7

Scheme of sperm-production control by hormones. Releasing factors sent from the hypothalamus stimulate the anterior pituitary to release hormones that stimulate the secretion of testosterone. Testosterone stimulates sperm production but also acts as an inhibitor to both hypothalamus and pituitary in a negative feedback system.

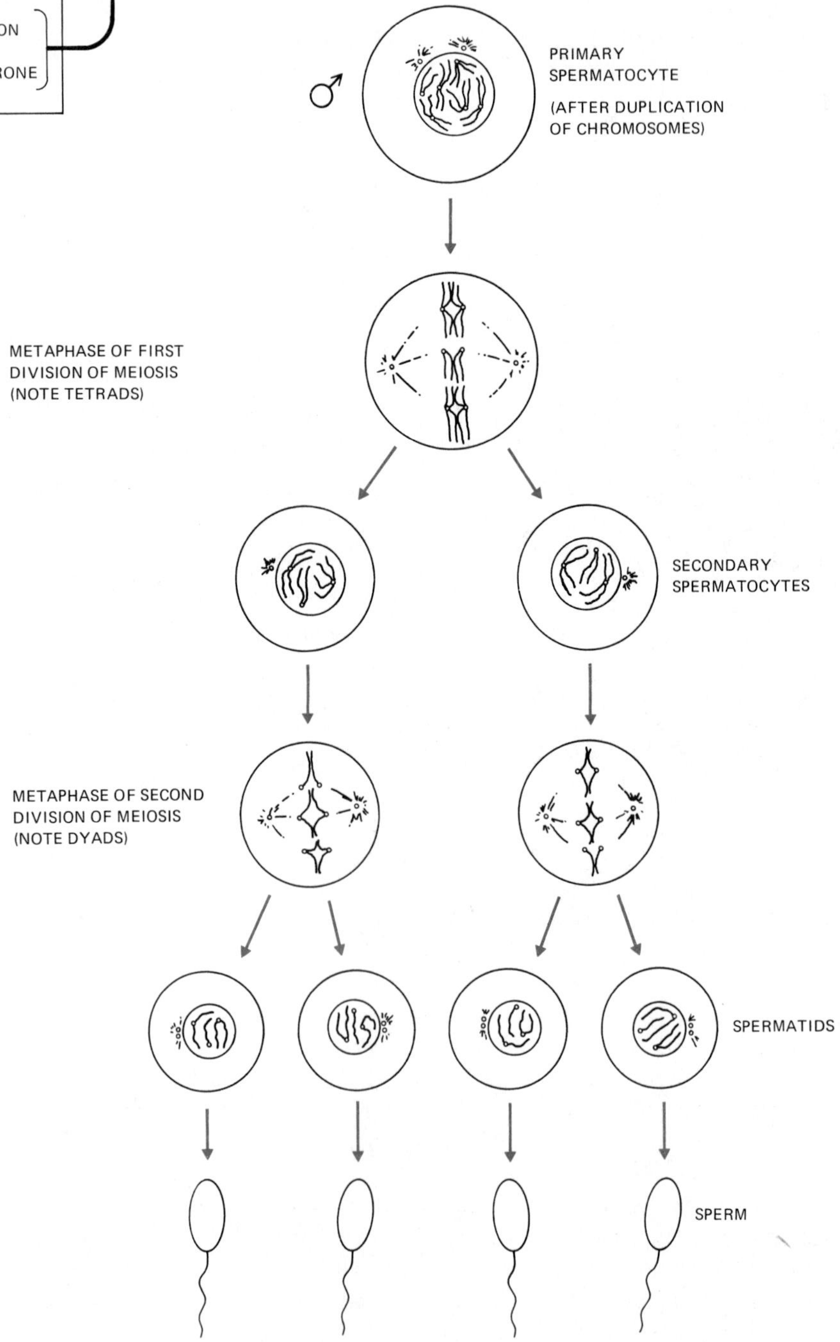

13.8

The cellular activity that leads to sperm production. A cell that is about to have its chromosome number reduced is a *primary spermatocyte*. It has a diploid chromosome number, and each chromosome itself is double. At the end of the first meiotic division, the cells resulting are *secondary spermatocytes*. After the second meiotic division, the cells are *spermatids*. Spermatids must undergo a reorganization in order to become sperm; this includes formation of a penetrating point, or *acrosome*, and a tail, or *flagellum*.

PRIMARY
SPERMATOCYTE
(AFTER DUPLICATION
OF CHROMOSOMES)

METAPHASE OF FIRST
DIVISION OF MEIOSIS
(NOTE TETRADS)

SECONDARY
SPERMATOCYTES

METAPHASE OF SECOND
DIVISION OF MEIOSIS
(NOTE DYADS)

SPERMATIDS

SPERM

ulating hormone) and *LH* (luteinizing hormone) also known as *ICSH* (interstitial cell-stimulating hormone). Both FSH and ICSH are chiefly responsible for the stimulation of spermatogenesis and testosterone secretion in the testes (Figure 13.7). As we shall see in Table 13.3, FSH and ICSH are also major female sex hormones.

Puberty and "Menopause"

A young male usually matures sexually when he is about 14 years old, although sexual maturation may take place a couple of years earlier or later. This period of sexual maturation is called *puberty* (from the Latin for "adult"). Apparently puberty is initiated by unknown factors that cause the hypothalamic portion of the brain to stimulate the pituitary gland to secrete LH and FSH. These hormones, in turn, stimulate secretions of testosterone from the testes and steroid secretions from the adrenal glands.

Although males usually experience a gradual decrease of testosterone secretion after they reach 40 or 50, there are no such drastic changes as occur with women at that age. (Those changes will be described later.) Indeed, any psychological problems during this period are probably caused not by a lack of testosterone but by a self-imposed fear of impotency and old age. Even with a decrease in testosterone, however, normal males may retain their sexual potency well into their eighties.

Spermatogenesis

Sperm cells are formed in the testes after a complex, precisely controlled series of events occurs in the seminiferous tubules. After repeated mitotic divisions, which increase the number of cells (the *spermatogonia*), meiosis results in quartets of haploid cells, the *spermatids* (Figure 13.8). Changes in shape that each spermatid undergoes include the development of an *acrosome* (Greek, "sharp body") at the front tip, which contains several enzymes that aid the sperm in penetrating an egg; a compact *nucleus* containing the chromosomes (and therefore all the genetic material, DNA); a *middle piece* containing mitochondria, which supply the energy for movement; and an undulating *tail*, which drives the swimming cell forward (Figure 13.9).

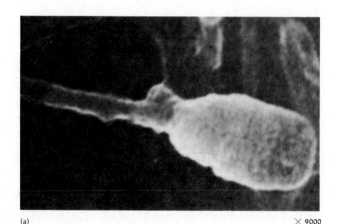

(a)   × 9000

13.9

(a) A scanning electron micrograph of a sperm cell as it would appear swimming. The compact, pointed head and the flexuous tail show to advantage. (b) Schematic drawing of a sperm cell showing its internal structure. The sperm head is mostly nucleus. The acrosome, which sits on top of the nucleus, aids the sperm in penetrating the surface of the egg. The middle piece, mainly made of a coil of mitochondria, furnishes energy for the sperm's movement. The actual movement is accomplished by the beating of the tail. × 5000.

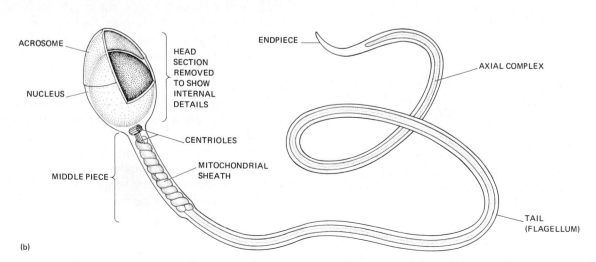

(b)

× 3000

13.10

A sperm about to penetrate the surface of an egg. Such dramatic photographs as this, with their three-dimensional effect, have been possible only in the last decade as a result of the development of the scanning electron microscope.

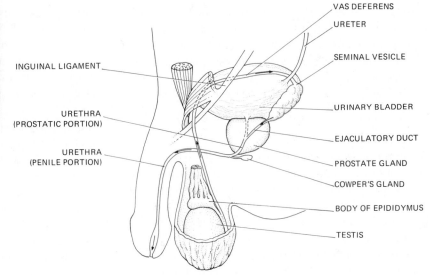

13.11

Dissection of the male human sexual organs. Compare this figure, in which only the testes, penis, and accessory parts are emphasized, with Figure 13.5, in which the general anatomy of the whole pelvic region is shown. The arrows trace the looped path of sperm from the testes through the urethra to the outside.

The production of sperm is periodic in most mammals, coinciding with seasonal breeding activity. But humans do not have a specific breeding season, and spermatogenesis proceeds on a continuous basis. Sperm cells at varying degrees of maturation are always present in the seminiferous tubules.

A sperm cell is one of the smallest cells in the body (the head of a sperm is about .005 mm long), and it is basically a nucleus with a tail.* Yet it requires almost three months for completion. If a small thimble were filled with sperm cells, it would contain more sperm than there are people in the world. (The population of the world is almost five billion.)

Accessory Ducts and Glands

Several structures are associated with the testes (Figure 13.11). Applied directly to each testis is an *epi-*

didymis (Greek, "upon the twins"; the "twins" are of course the testes), a coiled tube that would be about six meters (20 ft) long if it were stretched out. The haploid sperm cells are stored in the epididymis. All the seminiferous tubules of the testes lead into the epididymis, which then fuses into the major seminal duct, the *vas deferens*. The vas deferens receives secretions from the *seminal vesicles* and the *prostate gland* as it passes alongside them to the urethra. The *urethra* is the final tube, which leads through the penis and transports sperm outside the male body. During ejaculation, fluids from the seminal vesicles and prostate gland are secreted into the vas deferens, and these secretions plus the sperm cells make up the *semen* (Latin, "to sow"). With the onset of sexual excitement, the *bulbourethral* (Cowper's) glands secrete alkaline fluids into the urethra to neutralize the acidity of any remaining urine. These fluids also act as a lubricant.

Sperm cells make up only about 5–10 percent of human semen. The remainder is from the diluting medium of the accessory glands, which provides nourishing fructose for the sperm, a suitable alkaline

---

*   Placed end to end, about 600 sperm would equal 2.54 cm (1 in.), and if placed side by side, with heads touching, 6000 sperm would equal 2.54 cm. It would take about 1000 sperm cells to cover the period at the end of this sentence.

TABLE 13.2

Accessory Male Sex Glands

GLAND	EFFECT OF SECRETION	CONTENT OF SECRETION
Seminal vesicles	Provides nourishment for sperm; helps neutralize vaginal acidity.	Water, fructose, vitamin C; alkaline (about 60% of total semen).
Prostate gland	Makes sperm motile; helps neutralize vaginal acidity.	Water, cholesterol, buffering salts, phospholipids, prostaglandins; alkaline (about 30% of total semen).
Bulbourethral (Cowper's) glands	Neutralizes any urine in the urethra prior to ejaculation.	Alkaline mucus (about 5% of total semen).*

* The sperm constitute about 5 percent of the semen; the rest of the semen, as shown here, comes from the accessory sex glands.

medium to help neutralize vaginal acidity, and buffering salts and phospholipids that make the sperm motile (Table 13.2).

The Penis

The function of the penis is twofold: It carries urine through the urethra to the outside, and it transports semen through the urethra during ejaculation. In addition to the urethra, the penis contains three cylindrical strands of *erectile tissue*, mainly the two spongy *corpora cavernosa*. During erotic stimulation, which may be either tactile or mental, the spaces of the corpora cavernosa become engorged with blood under high pressure. As part of the overall response to stimuli from the nervous system, the arteries leading to the penis are dilated (enlarged), and the veins leading away are constricted. This dual action prevents the blood from escaping, and the penis becomes enlarged and firm in an <u>erection</u> (Figure 13.12). The erect penis can be inserted into the vagina during sexual intercourse. The tip of the penis is the *glans*, a sensitive area containing many nerve endings and therefore an important source of sexual arousal. The word "glans" is from the Latin for "acorn," which is the basic shape of the glans.

SOME CLINICAL CONSIDERATIONS

Ordinarily, there are 300–500 million sperm cells released during a normal ejaculation. If fewer than about 150 million normal spermatozoa are released per ejaculation, the semen is likely to be ineffective. A man who produces an inadequate number of sperm cells is said to be <u>sterile</u>, but sterility may also be

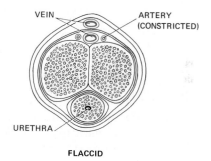

FLACCID

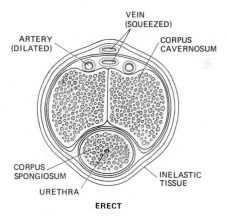

ERECT

13.12

Cross sections of a human penis, showing the difference between the flaccid and erect conditions. Erection is produced when blood under pressure enters the penis via the artery. As the *corpus cavernosum* enlarges, it squeezes shut the vein that normally carries blood back to the body. The resulting buildup of pressure creates a stiffening of the organ.

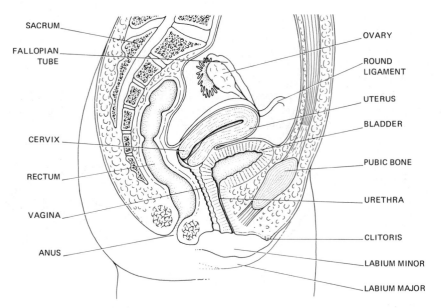

SACRUM
FALLOPIAN TUBE
CERVIX
RECTUM
VAGINA
ANUS

OVARY
ROUND LIGAMENT
UTERUS
BLADDER
PUBIC BONE
URETHRA
CLITORIS
LABIUM MINOR
LABIUM MAJOR

13.13
A median longitudinal section through a human female pelvis. This figure is the female counterpart of Figure 13.5, which shows a male pelvis. The uterus and vagina, which are central and single, are shown in section, but the ovary and oviduct (Fallopian tube), both of which are bilateral and double, are shown with only the left-hand one in place.

caused by other abnormalities. The removal of the testes, *castration*, used to be practiced on young boys to guarantee that they would remain sopranos to take female roles in musical productions or to make certain that harem guards were harmless. A castrated male will not produce testosterone, and so after a while there will be little or no male sex drive; however, erection of the penis may still be possible after castration. Castration is still used to produce tender roosters (*capons*) or steer steaks.

Sterility may also be caused by venereal diseases, which may interfere with sperm production in many ways, or by such diseases as mumps, which destroy the epithelium of the testes.

Impotence, which is simply the inability to produce or maintain an erection of the penis, is more likely to be a psychological problem than an anatomical one.

Circumcision (Latin, "to cut around") is the removal, for religious or health reasons, of the *prepuce*, or *foreskin*, a fold of skin over the end of the penis. Today the principal reason most circumcisions are performed is that the operation practically ensures freedom from cancer of the penis.

HUMAN SEXUAL ANATOMY AND FUNCTION: FEMALE

Sexuality in females is more complex than it is in males. Not only do the females have to produce eggs, but after fertilization they must nourish, carry, and protect the developing embryo. They must also nourish it for a time after it is born. Besides all of this, females have the added complication of a monthly rhythmicity imposed on the entire system.

The Ovaries and Oviducts

The egg-producing organs are the paired *ovaries*, elongated bodies about 5 centimeters (2 in.) long at maturity. Like testes, ovaries have a dual purpose. They release eggs (ova) on a fairly regular monthly schedule, and they produce female sex hormones. At birth a human female has about 400,000 potential eggs, but most of them never mature. In fact, no more than 400 of these potential eggs will mature. Beginning at puberty, the eggs mature one at a time (rarely more) in keeping with the monthly reproductive cycle, and they continue maturing during the reproductive years. Most women cease to be fertile during their forties, although some women in their late fifties may still bear children.

The ovaries contain *follicles*, the actual centers of egg production. Each follicle contains a potential ovum, and follicles are always present in several stages of development (Figure 13.14).

Each ovary is partially covered by a funnel-shaped *infundibulum*, into which the egg passes when it is released from the ovary. Apparently there is no particular pattern that selects one ovary over the other each month. The egg is effectively swept across a tiny gap into the infundibulum by feathery cilia (Figure 13.15), and few ova are ever lost in the abdominal cavity. From the infundibulum, an egg can

13.14

Photomicrograph of a section through a monkey ovary, showing various stages in the development of eggs. The *primordial follicle* is one that might have developed into a mature egg-containing follicle. It is necessary to say "might have" because only a few of the potential follicles actually do ripen. A *growing follicle* is just what it says. The *maturing follicle,* which is practically full size, shows the enormous difference in size between it and the primordial follicles. After the egg has been discharged from the follicle, the follicle is absorbed (unless pregnancy has resulted), as the *degenerating follicle,* smaller and irregular in shape, shows.

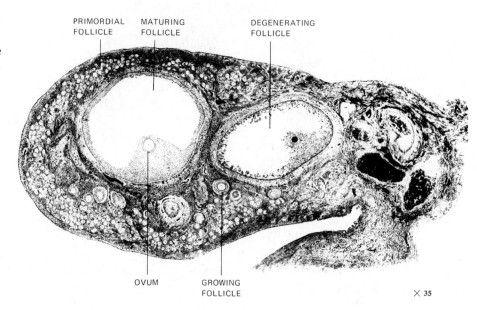

PRIMORDIAL FOLLICLE   MATURING FOLLICLE   DEGENERATING FOLLICLE

OVUM   GROWING FOLLICLE

× 35

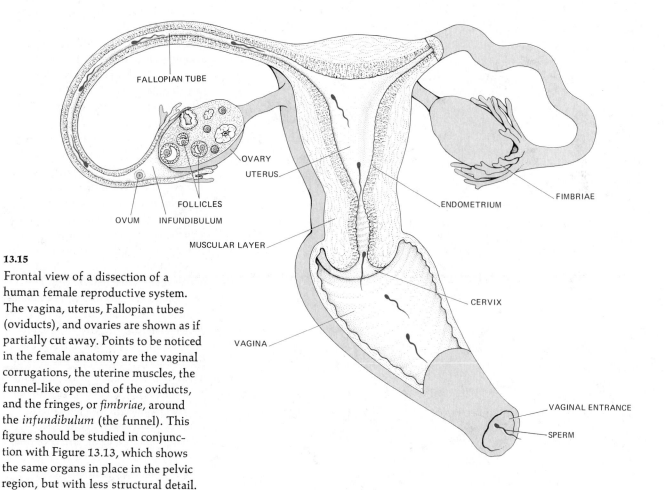

FALLOPIAN TUBE

OVARY
UTERUS

FOLLICLES

OVUM   INFUNDIBULUM

MUSCULAR LAYER

ENDOMETRIUM

FIMBRIAE

CERVIX

VAGINA

VAGINAL ENTRANCE

SPERM

13.15

Frontal view of a dissection of a human female reproductive system. The vagina, uterus, Fallopian tubes (oviducts), and ovaries are shown as if partially cut away. Points to be noticed in the female anatomy are the vaginal corrugations, the uterine muscles, the funnel-like open end of the oviducts, and the fringes, or *fimbriae,* around the *infundibulum* (the funnel). This figure should be studied in conjunction with Figure 13.13, which shows the same organs in place in the pelvic region, but with less structural detail. These figures are the female counterparts of Figures 13.5 and 13.11, which show the male reproductive system.

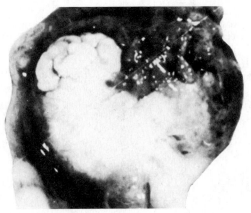

× 0.75

13.16
A human ovary, with its partial covering of *fimbriae* at the end of the oviduct. When an egg bursts from its follicle in the ovary, it drifts into the open end of the oviduct because fluid currents are created by the ciliary action of the oviduct lining. Eggs are passive and are free in the body cavity until they are caught in the current of liquid at the mouth of the oviduct.

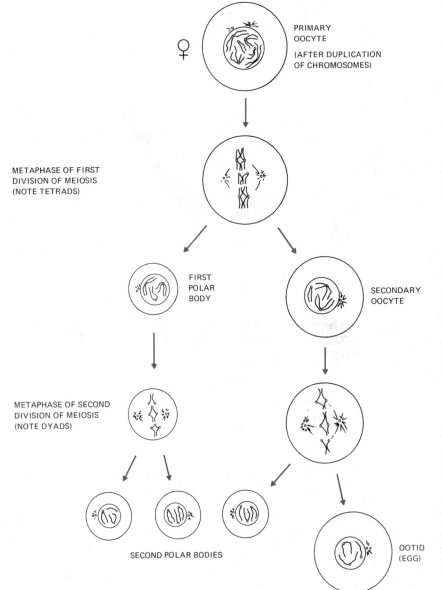

♀

PRIMARY OOCYTE
(AFTER DUPLICATION OF CHROMOSOMES)

METAPHASE OF FIRST DIVISION OF MEIOSIS (NOTE TETRADS)

FIRST POLAR BODY

SECONDARY OOCYTE

METAPHASE OF SECOND DIVISION OF MEIOSIS (NOTE DYADS)

SECOND POLAR BODIES

OOTID (EGG)

13.17
The cellular events in the maturation of an egg. The diploid *primary oocyte* (pronounced OH-oh-site) has two complete chromosome sets, and in every pair of chromosomes there are two *chromatids*; that is, each chromosome is double. As meiosis proceeds, a *secondary oocyte* is produced after the first meiotic division, but the two cells resulting from that division are not equal. One is the functional cell (the secondary oocyte), and the other is a *polar body*, destined for destruction. After the second meiotic division, reduction of the chromosome number is complete, and the result is a haploid *ootid* (pronounced OH-oh-tid), which will develop into a mature egg. The first polar body may or may not divide, so there may be only two final polar bodies instead of the three shown in the diagram. It does not matter, however, because the polar bodies are not functional. The bulk of the cytoplasm from the primary oocyte goes into the one functional egg. Compare this series of events with that of sperm formation (Figure 13.8), and note that for every primary spermatocyte, four viable sperm are produced, but for every primary oocyte, there is only one egg.

13.18

External view of the human female genitalia.

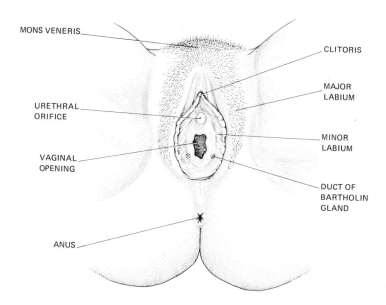

MONS VENERIS

CLITORIS

MAJOR LABIUM

URETHRAL ORIFICE

MINOR LABIUM

VAGINAL OPENING

DUCT OF BARTHOLIN GLAND

ANUS

pass down the *Fallopian tube,* or oviduct, which leads to the uterus, or womb.

Oogenesis

The maturation of eggs, *oogenesis,* differs from the maturation of sperm (Figure 13.8). Not only are the eggs matured one at a time under the rhythmic direction of nervous and hormonal control, but the cellular procedure is unlike that of sperm production. Where one primary spermatocyte yields four spermatozoa after meiosis, a primary oocyte yields only one egg (Figure 13.17). After the first meiotic nuclear division, which produces daughter cells with 23 chromosomes, the egg-to-be divides into two cells of unequal size. One is perhaps a thousand times as large as the other, and it contains almost all of the food-rich cytoplasm. The small one, a *polar body,* may divide again or disintegrate, but it never functions in reproduction. The large cell goes through a second meiotic division, and another tiny polar body is "pinched off." It, too, is destined to die. The remaining cell becomes the egg.

This phenomenon, in which meiosis produces four viable male (or malelike) gametes but only one viable female gamete, is common in many organisms. Most animals proceed thus, and so do such distantly related organisms as flowering plants and algae.

The Uterus, Vagina, and External Genitalia

The Fallopian tubes terminate in the *uterus,* a pear-shaped organ located directly behind the urinary bladder. Not only is the uterus pear-shaped, but it is pear-*sized* as well. However, it increases three to four times in size during the nine months of pregnancy.

Every month, in response to the secretion of the hormone estrogen, the lining of the uterus (*endometrium*) is built up in preparation for the possible implantation of a fertilized egg (the phenomenon of *pregnancy*). Secretions of another hormone, progesterone, aid in the development of the endometrium into an active gland rich in nutrients and ready to receive a fertilized egg. If pregnancy does not occur, the endometrium is sloughed off, and another monthly, or *menstrual,* cycle begins.

If fertilization and implantation do occur, the uterus houses, nourishes, and protects the developing fetus within its muscular walls. As the pregnancy continues, estrogen secretions develop the smooth muscle in the uterine walls, in preparation for the expulsive action of childbirth.

The uterus leads downward to the *vagina,* a muscle-lined tube about 8–10 centimeters (3–4 in.) long that receives sperm from the penis during sexual intercourse and allows the fetus to pass down from the uterus during childbirth. An inner mucus membrane secretes acids that help prevent infection (but also create an environment hostile to sperm). A fold of skin called the *hymen* partially blocks the vaginal entrance. The hymen is usually ruptured during the female's first sexual intercourse, but it may be broken earlier through other physical activities. The vaginal canal ends at the *cervix,* or neck of the uterus. The cervix is a prominent source of infection or cancer that should be checked regularly by a gynecologist.

The external genital organs are the labia major, the labia minor, vestibular glands, and the clitoris (Figure 13.18). These are collectively called the *vulva.*

13.19
Human mammae, or female breasts, shown front and side, illustrated as diagrammatic sections. Functional and supporting tissues are indicated. Mammary glands are variations of epidermal tissue.

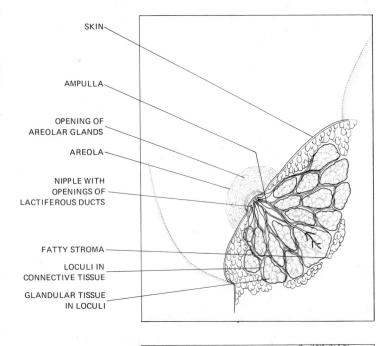

SKIN
AMPULLA
OPENING OF AREOLAR GLANDS
AREOLA
NIPPLE WITH OPENINGS OF LACTIFEROUS DUCTS
FATTY STROMA
LOCULI IN CONNECTIVE TISSUE
GLANDULAR TISSUE IN LOCULI

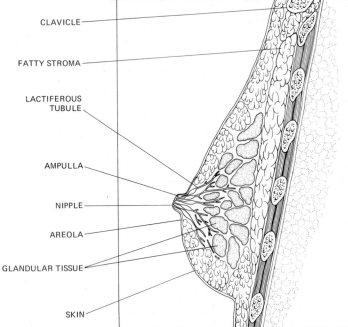

CLAVICLE
FATTY STROMA
LACTIFEROUS TUBULE
AMPULLA
NIPPLE
AREOLA
GLANDULAR TISSUE
SKIN

13.20
Scheme of follicle and ovum (egg) development, as influenced by the follicle-stimulating and luteinizing hormones FSH and LH). The presence of LH (from the pituitary) in the ovary enhances follicle and egg development. It also increases ovarian output of estrogen. Estrogen stimulates follicle and egg development and at the same time inhibits the output of both hypothalamus and pituitary. Thus gamete formation in females somewhat parallels that in males (see Figure 13.7).

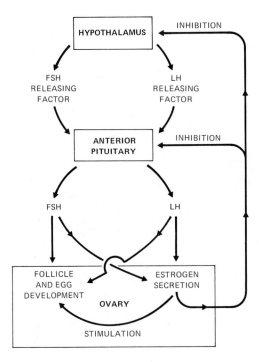

HYPOTHALAMUS
INHIBITION
FSH RELEASING FACTOR
LH RELEASING FACTOR
ANTERIOR PITUITARY
INHIBITION
FSH
LH
FOLLICLE AND EGG DEVELOPMENT
ESTROGEN SECRETION
OVARY
STIMULATION

The *labia* are folds of skin and mucus membrane, fatter in the major than in the minor, that surround and protect the vaginal and urethral openings. The major lubrication during sexual arousal comes from a "sweating" action of the vaginal walls and labia minor. The *clitoris*, just anterior to the vaginal opening, is structurally comparable to the penis; like the penis, it is composed almost exclusively of erectile tissue. During sexual excitement the clitoris becomes engorged with blood. However, although it is one of the major sources of sexual arousal for women, the clitoris plays no direct role in reproduction.

The Mammary Glands

Breasts are present in an undeveloped form in children and men, but during puberty in the adolescent female the breasts begin their development as mammary, or milk-producing, organs (Figure 13.19). During pregnancy, secretions of estrogen and progesterone cause the milk-secreting glands within the breasts to develop; the actual milk-producing hormone is *prolactin*.

The breasts contain an extensive drainage system comprising many lymph nodes. These nodes may be susceptible to breast cancer (the most common form of cancer in females), and frequent self-examination of the breasts should be a routine practice.

Hormonal Regulation in the Nonpregnant Female

Males are continuously fertile from puberty to old age, and throughout that period their sex hormones are secreted at a steady rate. The female, however, is fertile only during a few days of her monthly uterine cycle, and the complicated pattern of hormone secretion is intricately related to the cyclical release of an egg cell from the ovary.

The maturation of an egg in an ovary is timed by a cyclical production of hormones in the hypothalamic region of the brain. The timing is determined in an unknown manner in every individual even before birth. The action is indirect, in that the *gonadotropin-releasing hormones* from the hypothalamus act on the

TABLE 13.3
Major Human Reproductive Hormones

HORMONE	FUNCTION OF HORMONE	SOURCE OF HORMONE
Female		
FSH (Follicle-stimulating hormone)	Causes immature egg and follicle to develop; increases estrogen; stimulates new gamete formation and development of uterine wall after menstruation.	Pituitary gland (controlled by hypothalamus)
LH (Luteinizing hormone)	Stimulates further development of follicle and egg; stimulates ovulation; increases progesterone; aids development of corpus luteum.	Pituitary gland (controlled by hypothalamus)
LTH (Luteotropic hormone)	Causes the corpus luteum to secrete progesterone and estrogen.	Pituitary gland (controlled by hypothalamus)
Progesterone	Inhibits release of LH; stimulates thickening of uterine wall.	Corpus luteum (controlled by LH)
Estrogen	Stimulates thickening of uterine wall; stimulates development of female characteristics; decreases FSH; increases LH.	Follicle, corpus luteum (controlled by FSH)
CG (Chorionic gonadotropin)	Prevents corpus luteum from disintegrating; stimulates steroid secretion from corpus luteum.	Embryonic membranes, placenta
Prolactin	Allows mammary glands to secrete milk after childbirth.	Pituitary gland (controlled by hypothalamus)
Oxytocin	Stimulates uterine contractions during labor.	Pituitary gland (controlled by hypothalamus)
Male		
Testosterone	Increases sperm production; stimulates development of male characteristics; inhibits LH secretion.	Testes (controlled by LH)
LH (ICSH) (Interstitial cell-stimulating hormone)	Stimulates secretion of testosterone.	Testes (controlled by the pituitary)
FSH	Increases testosterone production; aids sperm maturation.	Testes (controlled by the pituitary)

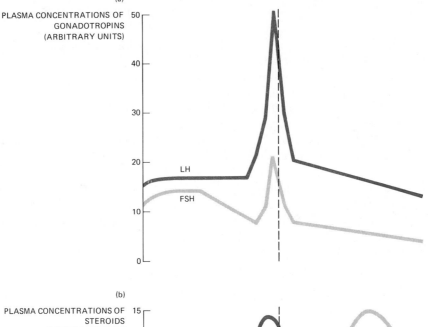

(a)

PLASMA CONCENTRATIONS OF
GONADOTROPINS
(ARBITRARY UNITS)

LH

FSH

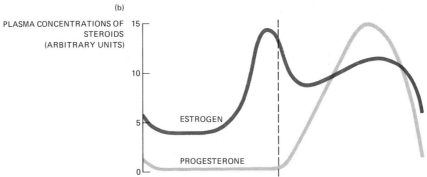

(b)

PLASMA CONCENTRATIONS OF
STEROIDS
(ARBITRARY UNITS)

ESTROGEN

PROGESTERONE

13.21

Correlation of hormonal and physical events in the human menstrual cycle. (a) Concentrations of LH and FSH in the blood. (b) Concentrations of the steroid hormones estrogen and progesterone in the blood. (c) Maturation of the follicle and egg, release of the egg, and degeneration of the follicle (no pregnancy resulting after ovulation). (d) Diagram of uterine activity through two menstrual periods and one intermenstrual period.

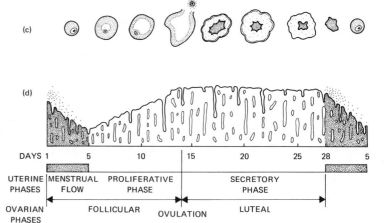

(c)

(d)

DAYS 1   5   10   15   20   25   28   5

UTERINE PHASES: MENSTRUAL FLOW | PROLIFERATIVE PHASE | SECRETORY PHASE

OVARIAN PHASES: FOLLICULAR | OVULATION | LUTEAL

pituitary gland. The pituitary gland then releases three additional hormones, which act in concert with the others to bring about the egg's maturation and release from the ovary, a process that is known as _ovulation_.

A simple way to understand the hormonal alterations of a normal monthly cycle may be to follow the development of the ovum. The approximate sequence (the _exact_ sequence cannot be precisely outlined) is as follows:

1. The follicle-stimulating hormone, _FSH_, promotes the development of the ovum and one of the immature follicles, the egg-producing center in the ovary.

2. An ovarian hormone, _estrogen_, is produced by the follicle and causes a buildup and enrichment of

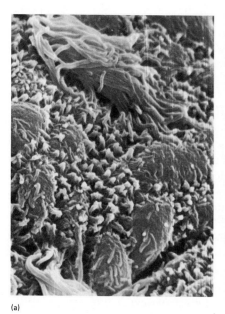

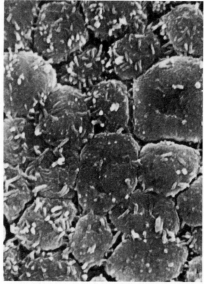

(a)

(b)

13.22
Scanning electron micrographs of the lining of human Fallopian tubes during different phases of the menstrual cycle. During the proliferation phase (a), some of the cells have cilia and some are secreting droplets. The secreting cells are covered with microvilli. In the secretory phase (b), microvilli are still present on the surfaces of the secreting cells, but the general surface of the lining has a more lumpy appearance than during the proliferation phase, and the cilia appear shorter and less numerous. The named phases refer to the uterine condition at the time the photographs were taken (see Figure 13.21).

the endometrium, as well as the inhibition of further production of FSH.

3. The elevated estrogen level triggers the pituitary secretion of <u>LH</u>, the luteinizing hormone, which causes the release of the ovum and also brings about the conversion of the collapsed follicle into an endocrine organ, the *corpus luteum.*

4. The corpus luteum secretes another hormone, *progesterone,* which completes the development of the endometrium and maintains this uterine lining for 10–14 days.

5. If the ovum is not fertilized and implanted in the endometrium, there is a disintegration of the corpus luteum, and progesterone production ceases (Figure 13.21).

6. Without progesterone, the endometrium breaks down, and *menstruation* occurs. (The Latin word for "month" is *mensis.*) During menstruation, the dead cells and blood released by hemorrhage of the uterine arteries are drained from the body through the vagina.

7. As the progesterone level decreases, the level of LH drops off also. This causes the pituitary to renew active secretion of FSH, which stimulates the development of another ovum. The monthly cycle begins again (Figure 13.21).

After ovulation, estrogen and progesterone act in the bloodstream to inhibit the release of LH and FSH by the pituitary. This is a feedback control that keeps more than one follicle at a time from maturing. There will be no more maturation of a follicle and no more ovulation until the next turn of the cycle *unless the egg is fertilized and pregnancy ensues.* Then the story changes and requires a separate treatment.

Hormonal Regulation in the Pregnant Female
Pregnancy sets a new series of events in motion. The ovaries themselves are affected because as the embryo develops, its covering membranes and the placenta release a hormone, *chorionic gonadotropin* (CG), that keeps the corpus luteum from disintegrating. Remember that if there is no pregnancy, the corpus luteum does disintegrate. If there *is* pregnancy, the implantation of the embryo triggers a feedback that causes the corpus luteum to remain during most of the period of pregnancy. This scheme ensures that the corpus luteum will continue to produce progesterone. If the amount of progesterone does not remain especially large, the uterine wall will break down, and the embryo will become separated from the uterus. This is one cause of *miscarriage.*

As the embryo develops, other hormones are secreted. *Prolactin* and *oxytocin* induce the mammary glands in the breasts to secrete milk after childbirth, and oxytocin also stimulates the uterine contractions that expel the baby from the uterus during childbirth.

Lest all this sound more certain than it really is, we should point out that the entire mammalian reproductive method is subject to great variability, not only from species to species, but even within the human species. For example, most girls living together in a sorority house will usually menstruate within the same two-week segment of the month. How can we explain such a synchronization of menstrual periods? (Apparently they are influenced by changing body odors.) Or to use an example from what we in our arrogance call the "lower animals," why does a female rabbit ovulate only after mating instead of on a given schedule? The hormones that are the direct regulators of sexual functions are plainly under the control of factors that may be described as psychological. This is a short way of saying we are far from knowing all that happens.

Puberty and Menopause

The onset of puberty in young girls usually occurs between the ages of 10 and 14 years. During puberty, the accessory organs, uterus, vagina, and breasts grow and mature under the influence of pituitary gonadotropins and estrogen from the ovaries. Hair begins to grow in the pubic region and the armpits, the pelvis widens and accumulates fatty deposits, and an active sex drive commences. Estrogens also stimulate a thickening of the skin and an increase in its water content, which tends to counteract somewhat the excessive secretion of adolescent oils and fats that may lead to acne. Finally, estrogens bring about an increase of calcium deposits at the ends of the long bones and cause the bone cells to stop dividing; the combination of these events ordinarily signals the end of skeletal growth.

The first menstrual period, the *menarche* (Greek, "beginning the monthly"), occurs during the latter stages of puberty, although actual ovulation may not occur until a year or so later. The final menstrual period, the *menopause* ("ceasing the monthly") usually occurs at sometime in the woman's fourth decade, signaling the end of reproductive ability. Menopause does not happen suddenly, and menstrual periods and ovulations will most likely begin to be irregular a few years before the final menstrual period.

After menopause, the breasts, uterus, and genitals begin to atrophy, but sex drive does not necessarily decrease, and in fact it may increase. Mental and physical pressures may be evident during menopause. Hot flashes of the skin, which are typical, are due to estrogen deficiency that somehow causes dilation of blood vessels in the skin. Some of the problems of menopause may be relieved by small doses of estrogen.

Prior to menopause, diseases such as hardening of the arteries occur infrequently in women, probably because estrogen reduces cholesterol in the blood. But as estrogen secretions continue to diminish after menopause, the incidence of cardiovascular disease becomes almost equal in men and women.

HUMAN MATING

Many animals have elaborate courtship rituals (Chapter 28), but few have anything like the variety of sexual practices that humans have. Men are sexually potent from the onset of puberty until late in age, and they have nothing comparable to the rutting season of, say, deer or the nesting season of birds. Women are affected by the cycle of ovulation and menstruation, but they are fertile throughout the year, and they remain so from about the first menstruation until the last.

In humans, the act of sexual intercourse, *coitus* (Latin, "to come together"), is usually preceded by a period of preliminary sexual excitement, in which physiological and psychological changes occur in both sexes as a preparation for actual *copulation*, or coupling. Until fairly recently, it was thought that female sexual response was different from and probably less intense than the sexual response of a male. Fortunately, an objective approach to the study of sexual response during recent years has shown that previous ideas about sexual response were not only narrow-minded but incorrect as well. We know now that males and females can experience relatively similar feelings during sexual intercourse.

Male and Female Sexual Response

The initial foreplay of sexual intercourse may be referred to as the *excitement phase*. (Of course, coitus cannot really be divided neatly into discrete stages, but we will describe it as if it were, hoping that increased clarity makes up for the clinical approach to an emotional activity.) During the excitement phase the penis becomes erect, fluids from the bulbourethral glands begin to be secreted, and the clear liquid oozes slowly from the penis. This viscous alkaline fluid neutralizes the acidity of any urine remaining in the urethra and acts as a lubricant later when the penis is inserted into the vagina. In the female, the clitoris and the nipples become erect, and the vagina secretes lubricating mucuslike fluids. Heart rate, breathing rate, and muscular tension increase in both sexes. A greater sensitivity to touch is apparent, and there is reduced sensitivity to pain and outside stimuli.

The excitement phase may last for as long as the couple want it to, or as long as they can refrain from the actual insertion of the penis into the vagina, which initiates the *plateau phase*. During this phase the penis is thrust rhythmically within the vagina. The feeling of sexual pleasure is increased for both partners, and both the glans of the penis and the clitoris are particularly stimulated. (The clitoris is massaged by the surrounding labia, not by the penis itself.) By now the semen has already been transported into the vas deferens by contractions of the genital ducts; the scrotum has also contracted and lifted the testes close to the body. Vaginal secretions continue, and pelvic thrusting finally reaches a point of involuntary movement as sexual pleasure continues to heighten. The next stage, *orgasm*, occurs.

Orgasm, the climax of sexual excitement, is accompanied in males by *ejaculation* of the semen, which is expelled through the penis via the urethra by massive contractions of the vas deferens, the seminal vesicles, and the other accessory glands, all of which contribute their secretions to the semen. Each normal ejaculation produces 3–5 milliliters (0.10–0.18 fl oz) of semen. The seminal fluid enhances the sperm's motility, and they begin active swimming at speeds up to 5 centimeters (2 in.) an hour, wiggling their tails erratically at 14–16 hertz (cycles per second). Some sperm may reach the uterus a few minutes after ejaculation and may reach the Fallopian tubes as soon as 30 minutes after ejaculation. (Sperm cells seem remarkably adept at progressing through the uterus into the Fallopian tubes, in spite of opposing currents generated by cilia.) Female orgasm does not necessarily accompany every copulation, and in humans, at least, female orgasm is not necessary for fertilization. In the female, any sexual climax is accompanied by the usual psychological effect of orgasm, with high breathing and heart rate and flushing of the skin. Although multiple orgasms may occur, there is nothing in the female climax that represents an ejaculation.

Following orgasm there is a latent or *resolution phase*. In men, a climax is usually followed by a period in which another climax is not possible for some time, but there is no such refractory period in women. After orgasm, both partners undergo a rapid return to normal pulse, breathing, and circulation, and there is a subjective feeling of lassitude and relaxation.

CONCEPTION

Once every 28 days or so, a single human egg oozes through a particularly fragile portion of the ovary wall and is carried by ciliary action through the Fal-

13.23   × 100

Scanning electron micrograph of the inner lining of a Fallopian tube. The irregular sphere is an egg.

lopian tube toward the uterus (Figure 13.23). The egg, moving much more slowly than the sperm, takes four to seven days to travel the 10 centimeters (4 in.) from the ovary to the uterus. During the first third of the journey, where fertilization is most advantageous, the egg slows its pace, as if awaiting the sperm. Fertilization must occur no more than 24 hours after mating, since the egg will remain viable for only that period of time. Sperm cells may remain viable in the female tract for about 72 hours. (If semen is stored in a laboratory at very cold temperatures the sperm will remain viable for many months.)

During mating, millions of flagellating sperm cells enter the female's vaginal canal.* If mating takes place at about the same time as ovulation, millions of spermatozoa enter the vagina, and some may travel toward the opposite-moving ovum, but only one sperm cell will eventually enter and fertilize the egg. If only one sperm may enter, why are millions dis-

---

* Allowing for enormous differences in size, a mature sperm swims about 50 percent faster than an Olympic swimmer. Put another way, a sperm can swim about 600 times its own body length in eight minutes, but the male Olympic swimmer travels about 400 times his body length in eight minutes. (And the Olympic swimmer does not have all the obstacles and unfriendly current to negotiate.)

13.24

Scanning electron micrograph of the meeting of sperm and egg. A sea urchin egg, the large sphere, is practically surrounded by hundreds of sperm. The outer covering of the egg is not its cell membrane but a special coating, the *vitelline membrane*, on which only certain areas are receptive to sperm. Within moments after the entry of a sperm, the membrane thickens. It is then called a *fertilization membrane*, which prevents other sperm from penetrating the egg, ensuring that each egg is fertilized by only one sperm.

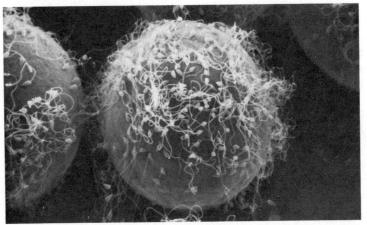

× 150

13.25

Scanning electron micrographs of the penetration of a sperm into an egg. The acrosome of the sperm touches the microvilli on the egg surface (a). The microvilli in the region of contact rise up to meet the sperm (b). The vitelline membrane rises and thickens, becoming the fertilization membrane, through which it is impossible to see. The fertilization membrane has been peeled away, disclosing the sperm, almost buried in the egg's cell membrane after three minutes (c). The sperm rapidly sinks into the egg cytoplasm, and a minute after the sperm body gets into the cell membrane, only the tail remains outside (d).

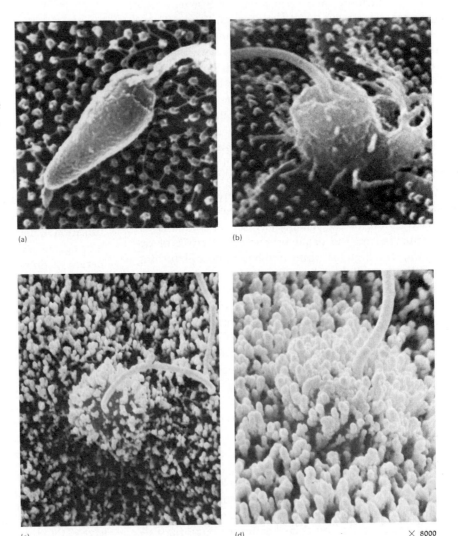

(a)

(b)

(c)

(d)

× 8000

13.26

Fertilization as seen inside an egg. Scanning electron micrographs, novel and impressive as they are, do not reveal the important part of the fertilization process: the union of male and female nuclei. In this series of pictures, photomicrographs of sectioned, stained eggs show the events. The specimen is Ascaris, a parasitic intestinal worm of swine and horses, which has long been a favorite subject for studies on fertilization. (a) The sperm has just penetrated the egg membrane. (b) The sperm loses some of its compactness as it approaches the egg nucleus in the center of the egg cytoplasm. (c) The heavy fertilization membrane surrounds the egg. The egg nucleus, not having undergone meiosis until the sperm is present, is undergoing its first reduction division. (d) The egg nucleus undergoes its second reduction division. (e) The egg nucleus has completed meiosis, and the two disintegrating polar bodies are outside the egg cytoplasm, visible as dark spots on the right-hand side of the figure. The egg and sperm nuclei are touching, but which is male and which is female is impossible to tell. (f) The egg and sperm nuclei have fused, and the zygote nucleus, the first diploid nucleus of the new generation, is in its first metaphase, on its way to producing an embryo.

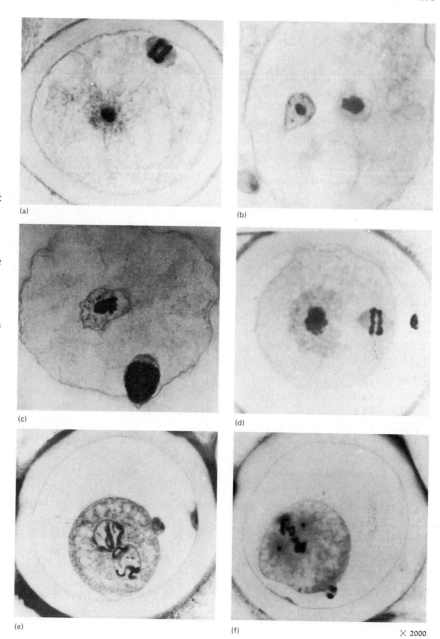

(a)

(b)

(c)

(d)

(e)

(f)

× 2000

patched? Most likely, more than enough sperm are discharged to ensure that at least one will successfully complete its mission. Not only must the sperm travel from the vagina all the way to the Fallopian tubes (a trip of 15–20 centimeters that may take as much as a few hours); they must also resist the spermicidal acidity of the vagina and overcome opposing fluid currents in the uterus and oviducts. So it is not surprising that the mortality rate of the sperm is enormously high. It is also postulated that the quantity of sperm discharged is large so that several

thousand will still reach the egg and act in unison to provide sufficient amounts of hyaluronidase. This enzyme chemically dissolves just enough of the outer wall of the ovum to allow a single sperm to enter. If more than one sperm somehow enters the ovum, the development of the zygote is almost always abnormal.

Some aspects of fertilization are illustrated in Figures 13.24 through 13.26. Fertilization usually takes place in the Fallopian tube. (If a fertilized egg should become implanted in the Fallopian tube, for

TEST-TUBE BABIES

On November 10, 1977, in Bristol, England, an egg was surgically removed at the time of ovulation from one of Mrs. Lesley Brown's ovaries and fertilized in a laboratory with sperm from her husband, Gilbert Brown. Two and a half days later, the cultured embryo was implanted in Mrs. Brown's uterus through her cervix. From that point on, Mrs. Brown had a normal pregnancy. On July 25, 1978, Mrs. Brown gave birth to an apparently normal 5-pound 12-ounce baby girl. Little Louise Brown's birth made the front page of practically every major newspaper in the world. Louise Brown was famous, but most biologists thought it was only a matter of time before technical problems would be overcome, making such procedures almost commonplace.

The unusual method of conception—not related in any way to cloning, which produces an individual genetically identical to its "parent"—was necessary because of a blockage in Mrs. Brown's Fallopian tubes that made it impossible for her eggs to be fertilized. Gynecologist Patrick C. Steptoe and physiologist Robert G. Edwards supervised the procedure. In several previous attempts over the past 12 years, they had been successful in inducing pregnancies, but the earlier pregnancies had ended in miscarriages.

Rabbits, mice, and cattle "test-tube babies" have been produced successfully for several years in the United States, but no "test-tube" conceptions and births of monkeys or apes

have been accomplished. The birth of Louise Brown represents a bold achievement without the intermediate successful experimentation with our closest evolutionary relatives. We now think toward the next development—growing an embryo, then a fetus, and finally a fully developed baby independent of a uterus.

Less publicized, but equally impressive technically, was the production of a mouse with six parents. Clement L. Markert and Robert M. Petters of Yale University cultured together a few cells from each of three embryos with genes for different coat colors (black, yellow, and white) and then implanted the aggregate embryo in a female. The result, shown above, had the three colors of hair that would be expected in a mouse with that genetic background.

---

instance, instead of properly in the uterus, an *ectopic pregnancy* ensues, and surgery is usually performed to remove the implanted embryo.) About an hour after fertilization, the haploid (23 chromosomes) nuclei of the egg and sperm fuse to form a single diploid cell with 46 chromosomes, the *zygote*. (Not surprisingly, it is in this context that the egg and sperm are called *gametes*, derived from the Greek word for "to marry.") Immediately after fertilization, a *fertilization membrane*, which sperm cannot penetrate, forms in an unknown manner around the zygote. In this way, only one sperm enters the ovum. Now embryonic development begins (Chapter 14).

CONTRACEPTION

All contraceptive methods have one aim: prevention of pregnancy. This can be achieved by preventing the production of eggs or sperm, by keeping eggs and sperm from meeting, or by preventing the implantation of an embryo in the uterus. All these methods are possible, but some are difficult to achieve, some are fallible, some are dangerous to the female, and some cause psychological or moral revulsion. Table 13.4 describes the many methods of contraception now available, and Table 13.5 lists several methods under investigation.

TABLE 13.4

Methods of Contraception (In decreasing order of effectiveness)

METHOD	MODE OF ACTION	EFFECTIVENESS IF USED CORRECTLY	ACTION NEEDED AT TIME OF INTER-COURSE	REQUIRES RESUPPLY OF MATERIALS USED	REQUIRES INSTRUC-TION IN USE	REQUIRES SERVICES OF A PHYSICIAN	SUITABLE FOR MENSTRUALLY IRREGULAR WOMEN	POSSIBLE SIDE EFFECTS OR INCONVENIENCES
Vasectomy*	Prevents release of sperm	Very high	None	No	No	Yes, operation	Yes	Usually irreversible
Tubal ligation†	Prevents egg cell from entering uterus	Very high	None	No	No	Yes, operation	Yes	Usually irreversible
Oral pill, 21-day adminis-tration	Prevents follicle maturation and ovulation	Very high	None	Yes	Yes, timing	Yes, prescrip-tion	Yes	Early: some water retention, breast tenderness, nausea. Late: possible blood clots, hypertension
Intra-uterine device (coil, loop)	Prevents implan-tation	High	None	No	No	Yes, to insert	Yes	Some women do not retain device. Possible menstrual discomfort, infection
Diaphragm with jelly	Prevents sperm from entering uterus. Jelly kills sperm.	High	Must be inserted before intercourse	Yes	Yes, must be inserted correctly each time	Yes, for sizing and instruction on use	Yes	None, but may cause over-lubrication. Cannot be fitted to all women
Condom (worn by male)	Prevents sperm from entering vagina	High	Yes, must be put on prior to intercourse	Yes	Not usually	No	Yes	Some reduction of sensation in male, interrupts foreplay
Tempera-ture rhythm	Determines ovulation time by noting body temperature at ovulation	Medium	None	No	Yes, must learn to interpret chart correctly	No, but physician should advise	Yes, if skilled in reading graph	Requires abstinence during part of cycle
Calendar rhythm	Abstinence during parts of cycle	Medium to Low	None	No	Yes, must know when to abstain	No, but physician should advise	No	Requires abstinence during part of cycle
Vaginal foams	Kill sperm	Medium to Low	Yes, requires application before intercourse	Yes	No	No	Yes	None usually, may irritate
Withdrawal	Remove penis from vagina before ejaculation	Low	Yes, withdrawal	No	No	No	Yes	Frustration in some
Douche	Washes out sperm	Lowest	Yes, im-mediately after	No	No	No	Yes	None, but must be done immediately after intercourse

* Vasectomy prevents the release of sperm but does not alter the production of male hormones.

† Tubal ligation prevents the passage of the mature egg cell to the uterus but does not alter the production of female hormones.

Adapted from James E. Crouch and J. Robert McClintic, *Human Anatomy and Physiology*, Second Edition (New York: John Wiley & Sons, Inc., 1976), p. 721. Used with permission.

TABLE 13.5
Other Methods of Contraception under Investigation

METHOD	MODE OF ACTION	EFFECTIVENESS IF USED CORRECTLY	ACTION NEEDED AT TIME OF INTER-COURSE	REQUIRES RESUPPLY OF MATERIALS USED	REQUIRES INSTRUC-TION IN USE	REQUIRES SERVICES OF A PHYSICIAN	SUITABLE FOR MENSTRUALLY IRREGULAR WOMEN	POSSIBLE SIDE EFFECTS OR INCONVENIENCES
"Minipill"; very low content of pro-gesterone	Inhibits follicle develop-ment	High	None	Yes	Yes	Yes, pre-scription and regular checkups	Yes	Irregular cycles and bleeding
"Morning after pill"	Arrests pregnancy probably by preventing implantation. Fifty times normal dose of estrogen.	By currently available data, high	None	Yes	No	Yes, pre-scription	Yes	Breast swelling, nausea, water retention. Possible cause of vaginal cancer in fe-male offspring
Vaginal ring: inserted in vagina, contains proges-teroid	"Leaks" progester-oid into blood-stream through vagina at constant rate. Thereby inhib-its follicle maturation	Studies are "promising"	None	Yes, perhaps at yearly intervals	Yes	Yes, to insert	Yes	Spotting, some discomfort
Once-a-month pill	Injected in oil base into muscle. Slow passage of birth control drug into cir-culation inhib-its follicle maturation	Said to be 100 percent	None	Yes, on a monthly basis	No	Yes, pre-scription and regular checkups	Yes	Similar to oral pill
Depo-Provera (DMPA) 3-month injection	Injected. Inhibits follicle development	By currently available data, high	None	Yes, on a 3-month basis	Yes	Yes, injection and regular checkups	Yes	Similar to oral pill

From James E. Crouch and J. Robert McClintic, *Human Anatomy and Physiology*, Second Edition (New York: John Wiley & Sons, Inc., 1976) p. 722. Used with permission.

VENEREAL DISEASES

Some parasites are transferred from host to host mainly by sexual contacts. Consequently, they are known as *venereal* parasites (from Venus, Greek goddess of love). Contrary to popular myths, venereal diseases (VD) are rarely contracted by casual, dry contact with persons or objects. The two most common venereal diseases in humans are gonorrhea and syphilis, both caused by bacteria.

Gonorrhea is primarily an infection of the reproductive and urinary tracts, but any moist part of the body, especially the eyes, may suffer. The causative organism, *Neisseria gonorrheae*, can be recognized by microscopic examination. If left untreated, gonorrhea becomes difficult to cure, causing painful inflammation of any mucus membrane, but it can be cured by antibiotics, especially if it is caught early. Before routine treatment of eyes in the newborn was established, thousands of babies were blinded by gonorrheal infection at the time of passage through the vaginal canal. Gonorrhea is commonly called *clap*.

Syphilis is a more dangerous disease, caused by a motile, corkscrew-shaped bacterium, *Treponema pallidum*. Its early symptom is a sore, the hard chancre, at the place where infection occurred. After perhaps a few months, various problems develop, including fever and general body pain. The symptoms frequently disappear, leaving the victim with the false impression that the disease is gone. But later, circulatory or nervous tissue may degenerate, so that paralysis, insanity, and death follow. Some individuals develop a tolerance for the parasites and can therefore carry and transmit the disease, even though they do not seem to have it. One of the most pitiful aspects of venereal infection is the transmission of the parasites from a mother to a baby. The syphilis bacterium is able to cross the placenta during pregnancy, whereas the gonorrhea bacterium seems unable to do so. Thus the developing child can contract syphilis early in its development and exhibit some of the terrible manifestations of the disease at birth. If the baby contracts syphilis during the actual birth process, it is not likely to exhibit any symptoms at the time of birth. However, a baby infected with syphilis will grow poorly, be mentally retarded, and die early.

Like gonorrhea, syphilis can be cured by antibiotics if treatment is started in time, and it seems likely that a syphilis vaccine for humans will be developed in the near future. Rabbits injected with dead syphilis bacteria have shown a partial resistance to subsequent injections of live syphilis bacteria, indicating that protective antibodies had been formed. But many more tests with animals such as chimpanzees must be concluded successfully before a vaccine can be used with humans. In the meantime, it is well to remember that the occurrence of venereal disease has reached epidemic proportions in this country, especially among teenagers and young adults, and false feelings of security about quick antibiotic cures should not be encouraged. It is true that venereal diseases can be cured with greater ease than ever before, especially if they are reported and treated early, but antibiotic-resistant strains of the gonorrhea bacterium are known. The most effective treatment is still an intelligent prevention.

SUMMARY

1. A number of methods can be used to regulate sexuality, but the end result is similar in all instances, with one animal producing small, motile gametes (spermatozoa), and the other producing larger, nonmotile food-filled gametes (eggs, or ova).

2. In protozoa there may be sex without actual reproduction, but genetic recombinations do take place, and that is the essence of sexuality. Almost all land-dwelling animals have adopted some form of mating that allows fertilization of eggs inside the body of the female. Most fishes and amphibia have specialized behavior that guarantees that sperm will be deposited close to the eggs.

3. The reproductive function of male animals is the production and delivery of sperm cells to the female.

These activities require the *testes, accessory glands* and *ducts,* and the *penis.*

4. Among the *seminiferous tubules* in the testes are small masses of *interstitial cells,* which secrete the male sex hormones, especially *testosterone.*

5. The formation of sperm cells in the testes (*spermatogenesis*) takes place after a complex, precisely controlled series of events occurs in the seminiferous tubules. After repeated mitotic divisions, which increase the number of cells (the *spermatogonia*), meiosis results in quartets of haploid cells, the *spermatids.*

6. Haploid sperm cells are formed in the testes and stored in the *epididymis,* which fuses into the *vas deferens.* The vas deferens passes alongside the *seminal*

vesicles and the *prostate gland* (and receives secretions from them) to the urethra. During ejaculation, fluids from the seminal vesicles and prostate gland are secreted into the vas deferens, and these secretions plus the sperm cells make up the *semen.*

7. The *penis* contains three cylindrical strands of erectile tissue, mainly the *corpora cavernosa,* which facilitate an erection during sexual excitement.

8. A man who produces an inadequate number of sperm cells for reproduction is *sterile.* Sterility may also be caused by castration, venereal diseases, and other diseases, such as mumps.

9. Sexuality in female mammals is more complex than it is in males. Females produce eggs, carry and protect the embryo, and care for the newborn offspring. Nervous, nutritional, and hormonal controls in females are more varied and intricate than they are in males.

10. The egg-producing organs in humans are the *ovaries,* which release eggs (*ova*) on a fairly regular monthly schedule. They also produce the female sex hormones *estrogen* and *progesterone.* The ovaries contain many regions called *follicles,* which are the centers of egg production.

11. In the maturation of eggs (*oogenesis*) a primary *oocyte* yields only one egg. After the first meiotic nuclear division, the egg-to-be divides into two cells of unequal size; the smaller one is a *polar body,* which never functions in reproduction. After the second meiotic division, another polar body is pinched off and dies. The remaining cell becomes the egg.

12. During pregnancy, the *uterus* houses, nourishes, and protects the developing fetus within its muscular walls. The *vagina* receives sperm during sexual intercourse and allows the fetus to pass down from the uterus during childbirth. The external female genital organs are collectively called the *vulva.*

13. The maturation of an egg in an ovary is timed by a cyclical production of *gonadotropin-releasing hormones* in the *hypothalamus.* These hormones act on the *pituitary gland* to release *FSH, LH,* and *LTH,* which stimulate *ovulation.* The *corpus luteum* produces *progesterone,* and the ovarian *follicles* secrete *estrogen.*

14. *Menstruation* occurs if the egg is not fertilized and implanted. The uterine wall breaks down, and the dead blood cells and tissues of the uterus are discharged through the vagina.

15. If the egg is fertilized, another hormone, *chorionic gonadotropin* (CG), prevents the corpus luteum from disintegrating. The secretion of progesterone continues, and other hormones aid pregnancy, childbirth, and milk production.

16. Few animals have the variety of sexual practices that humans have. Men are sexually potent from the onset of puberty until late in life, and women are likewise fertile throughout the year and remain so from the first menstruation until the last. In humans, the act of *coitus* is usually preceded by a period of preliminary sexual excitement, in which physiological and psychological changes occur in both sexes.

17. *Fertilization* usually takes place in the Fallopian tube, where the nuclei of the haploid sex cells fuse to form a single diploid cell, the *zygote.* Immediately after fertilization, a *fertilization membrane* forms around the zygote, preventing further penetration of sperm cells.

18. All *contraceptive* methods have one aim: prevention of pregnancy. This can be achieved by preventing the production of eggs or sperm, by keeping eggs and sperm from meeting, or by preventing the implantation of an embryo in the uterus.

19. *Venereal* parasites may be transferred from host to host through sexual contacts. The common venereal diseases in humans are *gonorrhea* and *syphilis,* both caused by bacteria.

RECOMMENDED READING

Bermant, Gordon, and Julian M. Davidson, *Biological Bases of Sexual Behavior.* New York: Harper & Row, 1974. (Paperback)

Epel, David, "The Program of Fertilization." *Scientific American,* November 1977. (Offprint 1372)

Harrison, Richard J., *Reproduction and Man.* New York: W. W. Norton, 1971. (Paperback)

Masters, W. H., and V. E. Johnson, *Human Sexual Response.* Boston: Little, Brown, 1966.

McCary, James Leslie, *Human Sexuality,* Second Edition. New York: D. Van Nostrand, 1973. (Paperback)

Nilsson, Lennart, Mirjam Furuhjelm, Axel Ingelman-Sundberg, and Claes Wirsén, *A Child Is Born,* Revised Edition. New York: Delacorte Press/ Seymour Lawrence, 1977.

Pengelley, Eric T., *Sex and Human Life,* Second Edition. Reading, Mass.: Addison Wesley, 1978. (Paperback)

Vander, Arthur J., James H. Sherman, and Dorothy S. Luciano, *Human Physiology: The Mechanisms of Body Function,* Second Edition. New York: McGraw-Hill, 1975. (Chapter 14)

14

Human Development

THE ADULT HUMAN BODY CONSISTS OF ABOUT 50 trillion (50 million million) cells, and each type of cell is a specialist with a specific job to do. Cells that are programmed to be heart cells, for instance, will never be anything else. In fact, cells are so specialized that the liver cells of a human and a horse are more alike than are human liver cells and other types of human cells. Cells are functional. They *do* something special, and if they are healthy and normal they will never do anything else. Organs do something specific, too. For instance, *all* vertebrate livers do basically the same thing, and therefore the liver cells of fishes, amphibia, reptiles, birds, and mammals are noticeably similar in structure. Structure and function are closely related.

Such specialization sounds simple enough, but no one has yet determined how it occurs (see Chapter 8). We do know that all the many specialized cells of an adult body are derived from a single fertilized egg cell. As the fertilized ovum divides again and again, the resultant cells begin to differ more and more from their original forms. They also become increasingly different from one another. What is known about DNA makes it likely that the information coded in each cell originates in DNA. One further possibility is that certain cells may secrete substances akin to enzymes or hormones that somehow provoke other cells to grow and differentiate. Even environmental

14.1
Seventeenth-century microscopists imagined that within sperm cells they could see little men, and drew pictures of sperm cells, like the one shown here, with human bodies inside. They thought that babies are simply the expansion of the *preformed* human beings. This idea of "preformation," was opposed to the presently accepted idea of "epigenesis," by which embryos are developed through growth and differentiation.

conditions may be important stimuli. But still, how does a cell "know" when and how to differentiate? And if mitosis creates daughter cells genetically identical to the parent cell, how is the pattern of similarity altered to produce specialized, different cells? Ideas about development have been altered through the centuries, and current theories have been helped immeasurably by our recent expansion of information about DNA.

About 2000 years ago, Aristotle proposed his hypothesis for the development of new organisms. He thought that menstrual fluid from the female provided the fundamental material for embryonic devel-

opment, that the male semen furnished the "dynamis," the life-giving energy, and that specialized body parts emerged during embryonic development. This concept of *epigenesis* was also advocated by some seventeenth- and eighteenth-century scientists as an alternative to another controversial theory, *preformation*. According to the theory of preformation, a fertilized egg contained a fully developed miniature animal, which merely needed the nourishment of a female uterus for it to grow into a normal-sized baby at birth (Figure 14.1). "Ovists" believed the preformed body originated in the egg, and "spermatists" thought the miniature body was in the sperm.

Even before Aristotle, it was understood that the act of sexual intercourse is a prerequisite for a mammal's conception and birth. But until the late nineteenth century it was not realized that both parents contribute genetic material (chromosomes) to the zygote through egg and sperm. With this knowledge it was relatively easy to understand that the embryonic development of the new organism was programmed by the genetic input of the parents.

A REVIEW OF MITOSIS AND MEIOSIS

Before we begin a discussion of embryonic development, it may be helpful to review the basic principles of cell division, originally presented in Chapter 8. In Figure 14.2 we can observe *mitosis*, the reproduction of nuclei and the equal distribution of DNA to each, and *meiosis*, the pair of nuclear and cellular divisions that reduce the chromosome number from diploid to haploid.

PRINCIPLES OF EARLY EMBRYONIC DEVELOPMENT

When a sperm meets an egg, it digests its way into the egg cytoplasm and sheds its tail. Meiosis in the egg, which had stopped at metaphase of the second division, is resumed after entry of the sperm. The second polar body is pinched off, leaving the maternal nucleus with a haploid set of chromosomes. The egg and sperm nuclei then fuse to form the diploid *zygote* nucleus. This nucleus contains all the genetic material, DNA, that subsequent mitotic divisions will distribute equally to all cells of the embryo. After one sperm has gained entrance to the egg, changes in the egg's surface take place to prevent additional sperm from entering.

After a sperm penetrates an egg, the egg quickly surrounds itself with an enveloping coat, the *fertilization membrane*, and a series of cell divisions called *cleavage* begins (Figure 14.3). Cleavage divisions, unlike later growth divisions, result in smaller and more numerous cells but not in increased size of the early embryo. Human cleavage is relatively slow. The first cleavage occurs about 24 hours after fertilization, and subsequent cleavages take place every 10 to 12 hours. In comparison, frog eggs may undergo cleavage every hour.

Early Cleavage in Vertebrates

Because early stages in human development are difficult to obtain and almost impossible to experiment on, most descriptive and experimental information comes from other animals, such as frogs, whose eggs are much larger than human eggs. The exact procedure of early embryonic development is somewhat different in animals such as reptiles and birds, with large, yolk-filled eggs, and in mammals, whose eggs have little yolk, but the essential principles hold for all vertebrates. Regardless of what kind of egg is

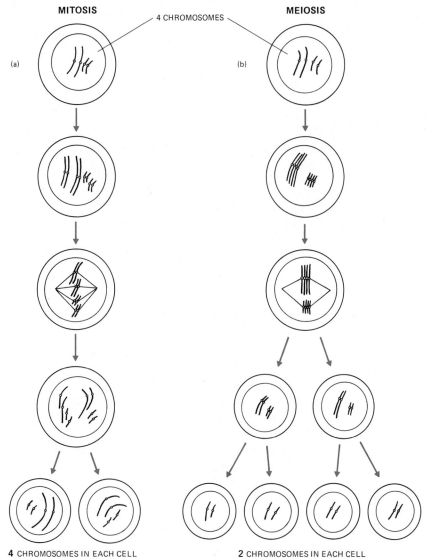

4 CHROMOSOMES IN EACH CELL   **2** CHROMOSOMES IN EACH CELL

14.2

Comparison of mitosis and meiosis. (a) In a mitotic division, each chromosome is duplicated once. When a diploid cell (shown here with four chromosomes) undergoes division, two cells result, each with four chromosomes identical to those in the original cell. (b) In meiosis, each chromosome not only duplicates but enters into a close pairing relation with its homolog, with the result that there is a four-stranded *bivalent*. There are four chromatids in a specific arrangement. The first meiotic division produces two cells, each with two chromatids. The second meiotic division results in four cells. The nucleus of each of the four cells has one chromatid (now a chromosome) from the old bivalent, and the total chromosome count is half what it was in the original diploid cells. A pair of reduction divisions produces four haploid cells from one diploid cell.

14.3

(a) The first divisions (*cleavage*) in an idealized embryo. The single-celled zygote still has two polar bodies attached to it. Successive divisions result in two, four, eight, sixteen cells, and so on, each new cell smaller than the zygote cell. These divisions continue until a *blastula*, a hollow ball of cells, is formed. The whole blastula, consisting of many small cells, is scarcely larger than the one original zygote cell. (b) Photographs of cleavage in a frog embryo. Compare the reality of these pictures with the diagrammatic quality of the drawings in (a).

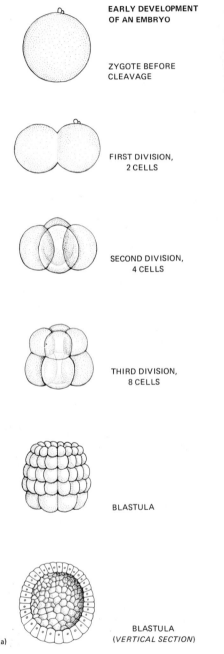

EARLY DEVELOPMENT
OF AN EMBRYO

ZYGOTE BEFORE
CLEAVAGE

FIRST DIVISION,
2 CELLS

SECOND DIVISION,
4 CELLS

THIRD DIVISION,
8 CELLS

BLASTULA

BLASTULA
(*VERTICAL SECTION*)

(a)

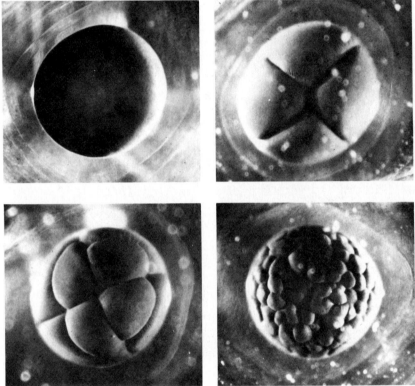

(b)

$\times$ 20

Courtesy Carolina Biological Supply Company

undergoing cleavage, the result is the formation of a number of cells that will start the embryo toward increasing developmental specialization.

Gastrulation

Cell divisions continue, but since the whole embryo remains the same size at this early stage, the cells are becoming smaller as cleavage progresses. The cells begin to move about on the surface of the young embryo, entering a reorganizing phase called *gastrulation*, during which the rearrangement of cells and growth of new ones result in an embryo with a developing body. The essential feature is the formation of layers of cells, with each layer capable of developing into special tissues (Figure 14.4).

14.4

The process of *gastrulation* in the lancelet, Amphioxus, a primitive chordate. Once a blastula has developed, one side of the growing embryo changes in such a way as to make the wall bend inward (invaginate). The result is a double-walled cup, a *gastrula* (Greek, "a little stomach"). The hole where invagination occurs is the *blastopore*. In Amphioxus, as in all deuterostome animals, the blastopore is at the posterior end of the embryo. The blastopore will become the anus of the mature animal. Later the mouth will break through at the opposite (anterior) end.

14.5

Origin of the three *germ layers* in a sea urchin. The *ectoderm* is the outermost layer. The *endoderm* is started when gastrulation starts. By the time gastrulation is complete, the endoderm is well established as the inner layer of the gastrula. At certain places within the embryo, cells begin to grow in the space between ectoderm and endoderm. The resulting tissue is *mesoderm*, which in most animals makes up the bulk of the mature body.

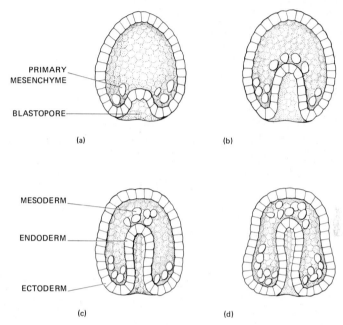

(a)   (b)

(c)   (d)

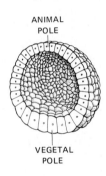

ANIMAL POLE

VEGETAL POLE

(a)

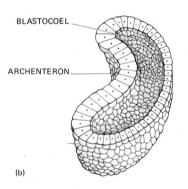

BLASTOCOEL

ARCHENTERON

(b)

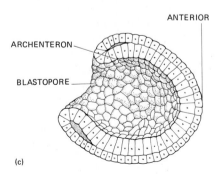

ANTERIOR

ARCHENTERON

BLASTOPORE

(c)

Once gastrulation has occurred, the whole embryo has an outer coating, the *ectoderm* (*ecto*, outside; *derm*, skin), and an inner lining, the *endoderm* (*endo*, inside) (Figure 14.5). The ectoderm cells will give rise to the outer layers of skin, the hair, the fingernails, the enamel of teeth, and the brain with its associated nervous tissue. The endoderm will form the lining of the alimentary tract, digestive glands, and lungs. Between the ectoderm and endoderm, new cells proliferate, filling the part between the two original layers and eventually building the bulk of the embryo. This is the *mesoderm* (*meso*, middle), the progenitor of connective tissue in lower skin layers, bone, muscle, blood, and the covering tissues of the internal organs.

THE DEVELOPMENT OF
THE SPINAL CORD AND THE EYE

As an embryo develops, it undergoes differentiation of its parts. In studying differentiation, biologists have asked what, where, when, how, and why. The questions of how and why are just beginning to be answered and the answers are still not satisfactory. Such questions are the stuff of developmental biology, one of the most active of present-day scientific fields. Once the ectoderm, endoderm, and mesoderm are formed, other organs and tissues of the body can begin to differentiate. The nervous system, for example, develops very early and actually plays a part in regulating the development of other parts of an animal body. The development of the vertebrate brain, spinal cord, and eye are presented here.

This series of pictures of developing frog embryos shows the origin and differentiation of the spinal cord and some other major body structures. Even though frog eggs, and consequently frog embryos, have features that are conspicuously different from their human counterparts, the general way that development proceeds is similar in both species.

In (a-1) gastrulation has occurred. A section through the midregion shows a lump at the top: the neural plate, destined to become part of the central nervous system. In (a-2) the ridge, which is on the dorsal side of the embryo, has risen more sharply. The section shows the neural plate thickened into two folds so heaped up that a narrow furrow has formed between them. Below these neural folds there is a rounded body (seen in cross section), the notochord, which is flanked on each side by developing masses of mesoderm. These masses will become muscle. In (a-3) the neural folds have grown together along the upper edges to form a hollow tube that will be the spinal cord. In (a-4) the dorsal ridge is more pronounced than it will ever be again. The body cavity, the coelom, shows as two slits in the mesoderm, right and left of the notochord. At its anterior end, the spinal cord will enlarge, retaining its essential hollowness, to become the brain. Nerves proliferate out from the brain and spinal cord, growing and branching until they form the nervous system. In this way, one entire set of internal structures has an ectodermal origin.

The series of events leading to the formation of a vertebrate eye is instructive because it illustrates an embryological principle of general applicability. That is the principle of *induction*, in which one tissue is caused, or *induced*, to change its developmental pattern as a result of the influence of a different tissue. The physical events are relatively simple. At the anterior end of the embryo, swellings of the enlarging brain grow out right and left toward the outer covering of ectoderm. They are the *optic vesicles*, which are the forerunners of eyeballs (b-1). As an optic vesicle approaches the outer covering, the ectoderm, it causes the ectoderm nearest the

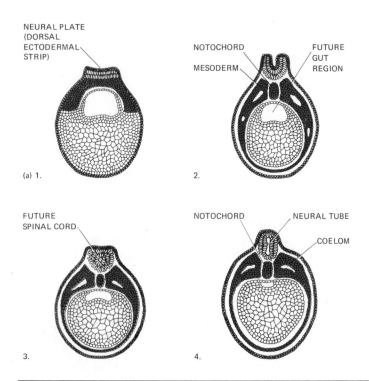

NEURAL PLATE
(DORSAL
ECTODERMAL
STRIP)

(a) 1.

NOTOCHORD

MESODERM

FUTURE
GUT
REGION

2.

FUTURE
SPINAL CORD

3.

NOTOCHORD

NEURAL TUBE

COELOM

4.

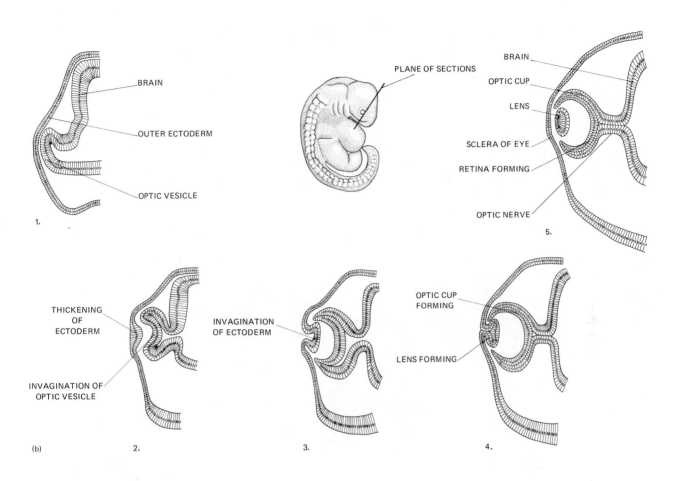

1. — BRAIN — OUTER ECTODERM — OPTIC VESICLE

PLANE OF SECTIONS

5. BRAIN — OPTIC CUP — LENS — SCLERA OF EYE — RETINA FORMING — OPTIC NERVE

(b) 2. THICKENING OF ECTODERM — INVAGINATION OF OPTIC VESICLE

3. INVAGINATION OF ECTODERM

4. OPTIC CUP FORMING — LENS FORMING

vesicle to thicken (b-2), then dimple inward toward the vesicle while the vesicle is folding in on itself (b-3). The dimple becomes a deep invagination, which sinks beneath the surface of the ectoderm and pinches off like a submerged bubble (b-4). This bubble, now surrounded by the collapsed vesicle, becomes the eye lens, and the vesicle becomes the outer wall of the eyeball, while the old ectoderm, healed over, becomes the outer layer, the cornea or sclerotic layer, of the eye (b-5).

In the induction of eye lens formation, it is obvious that at least two factors are at work. One can be generally called a genetic factor, since a vertebrate eye could not develop in, say, a spider. The proper genes must be present. The other factor is one of time and place, since the

eye lens develops only when the optic vesicle is close. Both these factors are contained within the embryo itself, but development can also be influenced by forces from outside the embryo. During the 1950s, many European women took a tranquilizer, thalidomide, during the early stages of pregnancy. The drug crossed the placenta into the embryonic blood and interfered with normal growth of the developing limbs. Many babies were born with deformed arms or legs, or even without limbs. Another dramatic cause of birth defects is the virus of German measles, or rubella, which can penetrate the unborn baby's defenses during the first three months of pregnancy, and last even past birth. Rubella virus can damage hearts, eyes, ears, and the blood.

14.6

Human triplets. Multiple births are always an exception in human beings. Twins are born once every 86 births, triplets once in 7400 births, and quadruplets once in 635,000 births. In most cases, one or more of the children born as quintuplets or sextuplets do not survive beyond infancy. The chances of quintuplets are astronomically low—about one in 55 million. Women who are older than 35 or who have had children previously may also produce multiple offspring at an abnormally high rate. This may be caused by irregular ovulation patterns in older women and the resultant release of more than one egg at a time. Apparently the tendency to release more than one mature egg at a time is an inherited trait. However, the cell separation that produces identical twins is not due to a hereditary factor, and its cause is not known.

The set of triplets shown here consists of two identical twin boys and their fraternal "twin" sister, who bears no more than a family resemblance to the identical-looking boys. In this case, more than one egg was fertilized, creating fraternal male and female twin embryos. Then one of the fertilized eggs (the male) split apart into two identical embryos, forming the identical boys in the picture. Triplets may also be formed in other ways. One fertilized egg could divide more than once, forming four identical embryos. If one of them died at the embryo stage, only three identical triplets would be born. Or three separate eggs could be fertilized. Numerous other combinations are possible.

14.7

Identical quadruplets, the brood of a mother armadillo (Dasypus). Since all four baby armadillos seem to be identical, even to minute details, they are thought to originate from a single egg. They thus are members of a small, well-defined *clone*. × 0.33.

DEVELOPMENT OF HUMAN EMBRYOS

As soon as a single sperm cell penetrates an egg cell, the egg undergoes its final meiotic division, and the polar body containing very little cytoplasm is formed. The haploid nuclei of the sperm and the egg unite in the act of fertilization, and the egg, now a diploid zygote, continues its journey toward the uterus. The process of ovulation has already caused the corpus luteum to initiate the monthly development of the uterine walls, and the fertilized egg will be accepted by the uterus, as an unfertilized egg would not be.

Implantation in the Uterus

The fertilized egg travels through the Fallopian tube until it reaches the uterus four or five days later. While it has been moving toward the uterus, the fer-

tilized egg has been dividing. After two or three more days of floating freely in the uterus, where it is nourished by uterine fluids, the embryo is ready to be received by the enriched uterine walls. By now the embryo is a fluid-filled hollow sphere of about 100 cells plus a covering mass of cells, the *trophoblast*, which will develop into a system of membranes outside the embryo. These membranes will transport nutrients to the embryo and will remove wastes from it. The whole mass, including trophoblast and the contained embryo, is a *blastocyst*.

The uterine wall and the blastocyst have continued to develop in the few days since fertilization occurred. When the uterus and the blastocyst have both reached compatible levels of development, the blastocyst sinks into the nutritious inner layer of the uterine wall, the *endometrium*. This process, known

14.8

Twins may be *identical* or *fraternal*. (a) *Fraternal twins* are formed when more than one egg is released from the ovary or ovaries and each egg is fertilized. Each fraternal human twin has its own placenta, umbilical cord, chorionic sac, and amniotic sac. Fraternal twins are the commonest of multiple births; 70 percent of all twins are fraternal, and 30 percent are identical. Aside from being the same age, fraternal twins resemble one another no more than any other siblings. (b) *Identical twins*, developed from the one fertilized egg, are genetically identical. Because of their common inheritance, identical twins are always the same sex. Identical twins are created when the embryo from a single egg breaks in two at a very early stage of development.

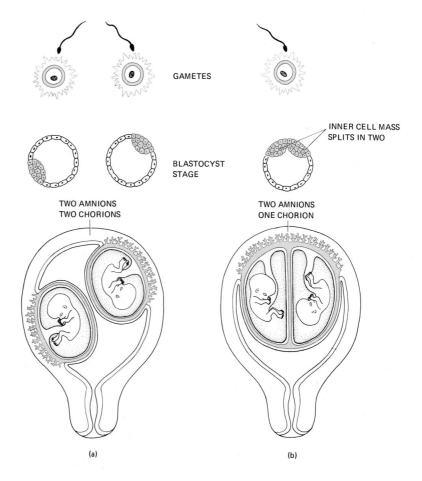

14.9

Implantation of an embryo in a uterine wall. The series of drawings shows the events in the production of an egg, elimination of polar bodies, fertilization, cleavage, the development of a blastocyst, and finally the sinking of the young embryo into the thick wall of the uterus. By the time implantation has occurred, the embryo is a multicellular body but still practically microscopic. The process of embedding seems to be a mutual effort on the part of both embryo and uterus. Once implantation has been accomplished, the uterine wall grows over the embryo and starts contributing the outer portion of the *placenta*; the embryo itself contributes the inner part.

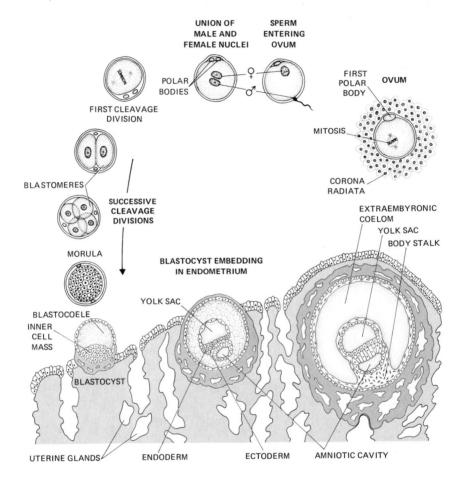

as *implantation* (Figure 14.9), is the point at which pregnancy begins.

The Fetal Membranes and the Placenta

As the trophoblast grows, it branches and extends into the tissues of the uterus, which increase along with it. The two kinds of tissue, the embryonic and maternal, grow until there is sufficient surface contact to ensure adequate passage of nourishment and oxygen from the mother, as well as removal of metabolic waste, including carbon dioxide, from the embryo. The embryo is soon completely covered by the envelope of *extraembryonic membranes*, one of which is the *amnion* filled with fluid (the *amniotic fluid*). The embryo floats in the amniotic fluid inside the amnion (Figure 14.10). At one side, the developed trophoblast tissue becomes the absorbing part of the embryonic mass, the *chorion*. At the site of implantation, the chorion joins intimately and intricately with uterine tissue to develop into the *placenta* (Greek, "*flat cake*"). A full-term human placenta is about 2.5 centimeters (1 in.) thick, and 22 centimeters (9 in.) across, and it weighs about 0.45 kilogram (1 lb).

The extensive mass of blood vessels and connective tissue that composes the chorion makes an inner lining of the placenta. These blood vessels are formed from the embryo and are connected to the embryo by way of the *umbilical cord*. The cord contains two arteries, which carry carbon dioxide and nitrogen

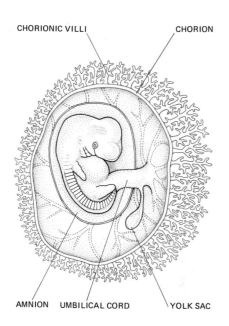

CHORIONIC VILLI   CHORION

AMNION   UMBILICAL CORD   YOLK SAC

wastes from the embryo to the placenta, and a vein, which carries oxygen and nutrients from the placenta to the embryo. A gelatinous cushion surrounds the vessels of the umbilical cord. This resilient pad, together with the pressure of blood and other liquids gushing through the cord, prevents the cord from twisting shut when the fetus becomes active enough to turn around in the womb.

There is normally no direct connection between the embryonic and the maternal tissue, at least no actual blood flow and no nerve connection (Figure 14.11). Sugars, water, oxygen, and hormones can cross the placental barrier, as can such poisons as lead, insecticides, and drugs. A newborn baby can show drug withdrawal symptoms if its mother used heroin during pregnancy. Nevertheless, the growing embryo is well insulated from most of the possibly harmful influences that the mother is subjected to.

14.10

Two important membranes form around the developing embryo at the time of implantation. The outermost membrane is the *chorion*, which makes up most of the placenta. The chorionic membrane contains many finger-like projections (*villi*), which secrete enzymes that help the developing embryo to adhere firmly to the uterine wall, and which also allow the exchange of nutrients, gases, and metabolic wastes. Because of the intricately folded system of chorionic villi, the total surface of the chorion is about 50 times the surface area of the skin of the newborn baby. The second membrane is the *amnion*, a fluid-filled sac that enables the developing embryo to float suspended in a relatively injury-free environment during pregnancy.

14.11

Circulation in the placenta. A section of uterine wall is shown at the place where the umbilical cord is connected. Two umbilical arteries in the cord carry blood, poor in oxygen and rich in carbon dioxide, from the fetus to the placenta. The connection between the fetal part of the placenta and the maternal part is close, with thousands of villi on the fetal part embedded in the maternal part, increasing the contact surface enormously. The fetal capillaries come close to the maternal capillaries and exchange their load of carbon dioxide for oxygen. In addition to exchange of respiratory gases, the placenta makes possible the elimination of metabolic wastes from the fetus and allows the entry of foods, vitamins, and salts.

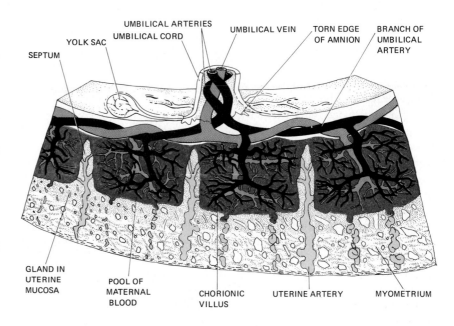

UMBILICAL ARTERIES
UMBILICAL CORD   UMBILICAL VEIN   TORN EDGE   BRANCH OF
OF AMNION   UMBILICAL
ARTERY
YOLK SAC
SEPTUM

GLAND IN
UTERINE
MUCOSA
POOL OF   CHORIONIC   UTERINE ARTERY   MYOMETRIUM
MATERNAL   VILLUS
BLOOD

14.12

Embryonic human sex organs appear the same for males and females until the embryo is about seven weeks old. After 10 weeks, the female genitals (top row) develop a distinct bud at the top that will become the clitoris. Swellings in the middle area will develop into the labia, and the vertical slit that is evident from the earliest stages will remain separated to become the opening for the urethra and vagina. The genitalia of the male fetus (bottom row) do not differ greatly from the female organs at 10 weeks. By the twelfth week it is evident that the bud is developing into the penis, and the swellings fuse together over the vertical slit to become the scrotum. The testicles will descend into the scrotum when the fetus is about seven months old. After 34 weeks, both the male and female sex organs look very much as they will at birth.

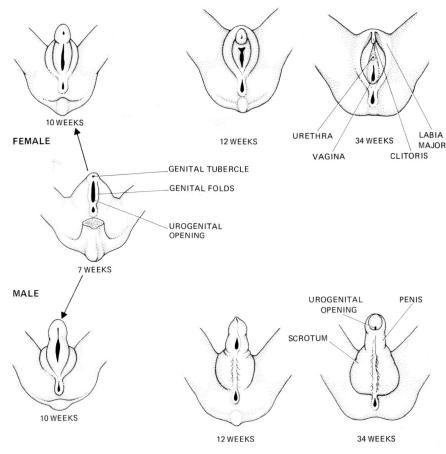

The placenta, which is the embryo's only contact with the outside world, is later shed as the *afterbirth.*

It has long been thought that oxygen passed through the placenta from the maternal blood to the fetus by *passive diffusion* (the random movement of highly concentrated gas molecules to areas of low concentration). There is new evidence, however, that molecules of oxygen may actually be shuttled across the placenta at an unusually fast rate by means of an unidentified carrier molecule. It also seems that certain chemicals, such as anesthetic gases, insecticides, tranquilizers, and the carbon monoxide in cigarette smoke, may interfere with the efficiency of the carrier. Fetuses that are deprived of this facilitated oxygen transport may develop into underweight children with a susceptibility to developmental defects. Severe oxygen deprivation may also cause miscarriages. It has been known for some time that female anesthetists and women who smoke heavily during pregnancy have more miscarriages and underweight children than do other pregnant women.

LATE EMBRYONIC DEVELOPMENT AND BIRTH

The human ovum is one of the largest cells in the body, just big enough to be seen with the unaided eye. After a week's growth, it reaches a diameter of about half a millimeter, but it is a formless group of cells, a blastocyst.* At the end of two weeks, it has begun tissue differentiation, but it is still mostly made

---

* The hereditary program is in effect from the moment of conception, and only the most drastic environmental conditions can alter it. For instance, although the mother should carefully restrict her weight gain during pregnancy, the baby will not be *permanently* larger if its mother has disregarded the obstetrician's advice and overeaten anyway. On the other hand, a baby will not necessarily be smaller just because the mother has not eaten enough during her pregnancy. Indeed, the fetus may be deprived of important nutritional elements, but if the mother does not supply enough food, the fetus will begin to consume its mother's body to obtain its basic nourishment.

up of extraembryonic cells essential for its nourishment. By three weeks, a 2.5 millimeter (0.1-in.) embryo has begun organ formation, including the primitive gut and nervous system. An incomplete but beating heart is present in a month, and the embryo reaches a diameter of about 5 millimeters. During the second month, organ building continues, with discernible features appearing, including limbs, facial contours, and sex organs (Figure 14.12). Once this stage is reached, the individual is no longer referred to as an embryo; it is called a _fetus_.

After three months in the uterus, a fetus is recognizably human. Although only about 50 to 60 millimeters (2–2.4 in.) long, it contains all the organ systems characteristic of the adult. The last six months of

pregnancy are devoted to increase in size and maturation of the organs developed during the first three months. By the time the fetus is 100 millimeters (4 in.) long, it can move and be felt by the mother, but it is thin, wrinkled, hairy, and watery. As it ages, the fetus loses most of its hair, its bones begin to harden, it picks up fat, and it becomes mature enough to be born. It is said to have come _to term_.

The Process of Childbirth

At term, about 280 days after the last menstrual cycle before conception, the muscles of the uterus begin to contract rhythmically. The exact cause of these initial contractions is still not known, but several reasonable possibilities exist. We know that estrogen

14.13

The birth of a human baby. (a) The internal events, as shown in a series of models. (Photographs of live births show only external events.) When the time has come for a baby to be born, strong, rhythmic contractions of abdominal and uterine muscles force the baby out through the birth canal. The amniotic sac breaks, releasing the amniotic fluid in which the baby has floated for most of its life, and the muscles of the cervix of the uterus and of the vagina relax. Most babies are born head first. If the head (the baby's largest part) does not come first, the baby can sometimes be turned around before delivery. However, it may be delivered breech first, although that makes for a difficult delivery. (b) How the birth of a baby looks to the woman bearing it.

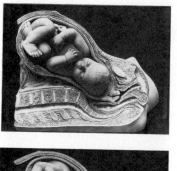

(a)

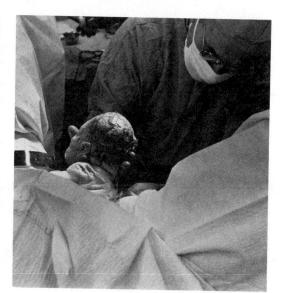

(b)

AMNIOCENTESIS: "VISITING" THE BABY BEFORE BIRTH

Amniocentesis is the technique of obtaining cells from an unborn infant. The procedure is to locate the fetus by bouncing high-frequency sounds off it and recording the echoes. This is a safe method, unlike the use of X-rays, which are potentially dangerous to embryos. Then a hypodermic needle is inserted into the amnion, usually directly through the abdominal wall. The embryo is surrounded by the amnion, and the amnion is filled with fluid in which a number of loose cells float. Some of the cells are from the embryo itself and some from the amnion, but they are all derived from the original zygote and are therefore genetically alike. Some 20 milliliters (0.6 fl oz) of the amniotic fluid are drawn into a syringe and may be grown as tissue cultures. They may be studied directly, but cultured cells can be made to yield more infor-

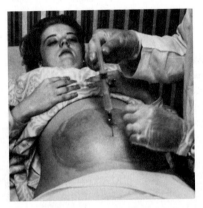

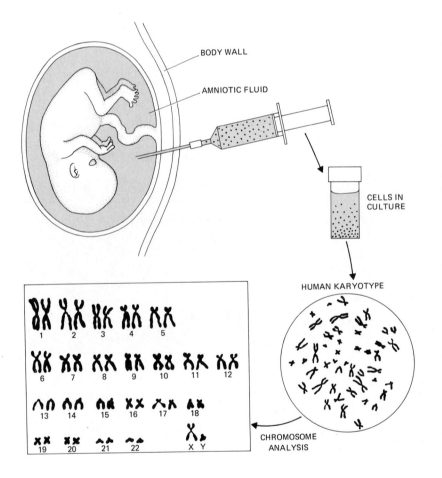

BODY WALL

AMNIOTIC FLUID

CELLS IN CULTURE

HUMAN KARYOTYPE

CHROMOSOME ANALYSIS

1 2 3 4 5
6 7 8 9 10 11 12
13 14 15 16 17 18
19 20 21 22 X Y

stimulates uterine contractions, and that progesterone inhibits them. Apparently, during the final days of pregnancy, the secretion of progesterone usually decreases. In addition, there is enough estrogen in the maternal blood by now to overcome the inhibitory effect of the progesterone, and contractions begin. (But the decrease in the level of progesterone does not occur in all women, and therefore the hormonal interaction of increasing estrogen and decreasing progesterone may not be the essential factor.)

Another possibility is that the growing fetus eventually reaches a size that distends the uterus in

mation. However, nearly a month is needed to establish a satisfactory culture, so the delay may be undesirable.

By amniocentesis the sex of the unborn can be determined, but despite possible curiosity of the parents about this, it has no medical importance. What is important is to find whether or not there are chromosome abnormalities, which might indicate possible deformities. Since the chromosome situation can be read with fair accuracy, it is possible to learn, for example, whether some duplication has occurred. If there is an extra chromosome of one of the middle-sized chromosome groups, the prediction can be made that the embryo, if allowed to come to term, will not survive infancy. The parents may decide that the best action is to terminate the pregnancy by having an abortion. On the other hand, if there is a history of abnormality in the family and consequent doubt concerning the unborn child, amniocentesis may show that the chromosomes are normal and that the child should be healthy.

Since the earlier an abortion is performed, the easier it is on the mother, amniocentesis is best done as soon as any question arises. The procedure of amniocentesis is reasonably safe for both embryo and mother, and it is advisable when there is any likelihood of serious fetal abnormality. This would be true if a mother is known to carry a chromosome aberration, or even if she is normal and over 40. Women over 40 have a disproportionately high number of babies with chromosome defects, probably because the eggs ovulated at that age are 20 or 25 years older than the eggs released in the time just after puberty, and the older eggs are more likely to be affected by multiple mutations than were the younger eggs. It is judicious for older expectant mothers to have the cells of their unborn babies checked.

such a way that uterine contractions are initiated. Some biologists speculate that the pressure of the child's head against the cervix stimulates the secretion of oxytocin by the pituitary, thus increasing uterine contractions. However, animals have been shown to begin typical uterine contractions on the expected date of birth even though the fetus had been removed weeks before. It seems that an actual full-term fetus is not necessary to start the contractions of childbirth. Also, since there is considerable variation in size of full-term fetuses, especially in relation to the sizes of their maternal uteruses, the size of the fetus appears to be too much of a variable to be a feasible cause of contractions.) On the other hand, it *is* recognized that human fetuses are born long before they are able to care for themselves—in a sense, they are always born prematurely. But if gestation continued for more than 9 or 10 months, the baby's relatively large brain could not pass through the vaginal canal.

Other suggestions for the initiation of childbirth are available. The female hormone *oxytocin* is known to be a powerful stimulant for uterine contractions— perhaps extra secretions of oxytocin initiate "labor pains." *Prostaglandins* are also effective stimulators of the smooth muscles of the uterus—secretions of prostaglandins increase during labor, and these chemicals may be the essential cause of precisely timed uterine contractions. It has also been speculated that uterine contractions are initiated (in sheep, at least) by secretions from the fetal hypothalamus portion of the brain, pituitary, and adrenal glands.

We can speculate that the uterine contractions that expel the baby during childbirth are caused by hormonal activity. But we cannot even speculate how these hormones are triggered with such impressive consistency within so many mammalian species.

At any rate, the initial contractions stimulate the liberation of oxytocin, which further stimulates even more powerful uterine contractions. Waves of muscular contractions spread down the walls of the uterus, forcing the fetus toward the cervix. By now, the cervix is dilated close to its maximum diameter of about 10 centimeters (4 in.). Early contractions occur about every half hour and last about one minute. Later the contractions become increasingly stronger, finally occurring every minute or so just prior to childbirth.

The amniotic sac may burst at any time during labor, or it may have to be ruptured by the attending physician. Either way, the ruptured sac releases the amniotic fluid. This phenomenon is known as "losing the waters," and if it occurs very early, it may signal the onset of labor. Another possible indication that labor has begun is the release of the cervical plug of mucus from the vagina. Ordinarily, either the bursting of the amniotic sac or the release of the cervical plug will occur early enough to provide ample time to prepare for childbirth, but great variations exist among pregnant women.

During pregnancy, the muscle cells of the uterus grow to as much as 40 times their former size, and

TABLE 14.1

Gestation Periods and Litter Sizes in Mammals

ANIMAL	GESTATION PERIOD IN DAYS (MEAN)	LITTER SIZE (MEAN)	(RANGE)
Opossum*	13	8	4–13
Mouse	20	4	1–9
Rat	22	8	6–12
Rabbit	31	8	1–13
Kangaroo*	33	1	1
Squirrel	42	4	1–8
Dog	60	6	1–12
Cat	63	4	2–6
Tiger	100	3	2–4
Pig	114	8	2–14
Lion	115	3	1–6
Sheep	148	2	1–3
Goat	151	2	1–3
Rhesus monkey	164	1	1–2
Baboon	187	1	1–2
Deer	203	2	1–3
Chimpanzee	227	1	1–2
Dolphin	255	1	1
Gorilla	260	1	1–2
Human	280	1	1–5
Cow	280	1	1–2
Horse	340	1	1–2
Humpback whale	360	1	1
Zebra	365	1	1
Camel	390	1	1
Sperm whale	480	1	1
Elephant	625	1	1

* Marsupial animals are born at a very immature stage. Immediately after birth each offspring makes its way to the mother's pouch, where it finds and clings to a nourishing nipple for about three months. During this nursing period the newborn's development continues to a more mature stage.

the number of muscle cells increases also, transforming the uterus into an enormously powerful muscle. The sturdy walls of the uterus harbor and nourish the fetus during pregnancy, and finally, through the muscular contractions during labor, the uterus forcefully expels the fetus outward through the vaginal canal. (The fetus does not assist its own birth passage at all.) Normally babies are born head first, with the soft bones allowing some deformation as the head is squeezed through the birth canal.

After it is born, the baby is still attached to the placenta by the umbilical cord. Immediately after the baby is expelled from the uterus, the attending physician clamps off the umbilical cord and cuts it. (In other mammals the mother severs this cord with her teeth.) Now the newborn is on its own, to a certain degree.

A few minutes after the baby is born, further uterine contractions expel the placenta, now called the *afterbirth*. These final uterine contractions also help to close off the blood vessels that were ruptured when the placenta was torn away from the uterine wall. Then the uterus returns to its original size.

The three stages of labor may be summarized as follows:

1. The first stage begins with the initial contraction of the uterus and ends when the cervix is dilated to its maximum.

2. The second stage is the delivery of the child.

3. The third stage is the delivery of the placenta (afterbirth).

The Breathing Process of the Newborn

Naturally, a fetus does not breathe air during its intrauterine life; it gets its oxygen from the placenta and delivers its carbon dioxide there. At birth, the baby must adapt quickly to life in air, and in order to do that, it must make several rapid physiological and anatomical adjustments. It must also shift from a fetal-type circulation, which depends on the placenta, to an essentially adult circulation, which depends on the lungs for gas exchange. Before birth, the fetal heart has an opening (the "oval window") between the atria that allows blood from the venous and arterial systems to mix. Also, the *ductus arteriosus* carries blood from the pulmonary artery to the aorta (Chapter 24), thus bypassing the lungs (Figure 14.14). Both the atrial opening and the pulmonary bypass close at birth, allowing the blood to pass through the pulmonary artery to the newly functioning lungs. If either the "oval window" or ductus arteriosis does not close off, the oxygen content of the blood will be low, and the baby's skin will appear slightly blue. A newborn child with this condition is therefore called a "blue baby." Usually a defective ductus arteriosus must be corrected surgically, but nothing can be done about that immediately after birth. The start of actual breathing, however, can be helped. Fluid can be cleaned from the nose and mouth of the *neonate* (newborn), the baby can be held upside down to help drain the breathing passage, and if necessary, it can be given a smack on its bottom to shock it into crying, its first gasp of air. More commonly these days, breathing is stimulated by a small whiff of carbon dioxide.

Lactation

Lactation includes both milk secretion and milk removal from the breasts. Throughout pregnancy the hormone *prolactin* stimulates milk production in the mammary glands, but it is not until after the baby is born that milk is actually secreted from the mother's breasts. It appears that placental secretions of estrogen and progesterone inhibit the release of milk from the breasts, but after the placenta has left the uterus there remains no deterrent to milk production, and nursing can begin.

Although several other hormones, including oxytocin, play a part in milk production and secretion, the most important agent is prolactin. It is not known what causes the secretion of prolactin during pregnancy, but it is clear that the continued production of prolactin during breast feeding is stimulated by the infant's sucking action on the nipple; breasts that are not emptied soon cease to produce milk. It also appears that nursing of the child aids in the return of the uterus to normal size and shape by the continued production of oxytocin, which is also responsible for the release of milk from the breast. Because of this stimulation by the suckling infant, lactation may be prolonged for two years or more if nursing continues. Usually, however, after about a year the nursing baby needs additional nutrients, such as iron, which the mother's milk cannot provide.

On the average, nursing mothers produce about one liter of milk each day. Multiple births result in larger secretions, usually about one liter for each baby engaged in suckling.

Although menstruation may not take place while the mother is nursing her baby, ovulation may occur anyway. In spite of tales to the contrary, a woman may indeed become pregnant during the lactation period. Nursing is certainly not an effective method of birth control.

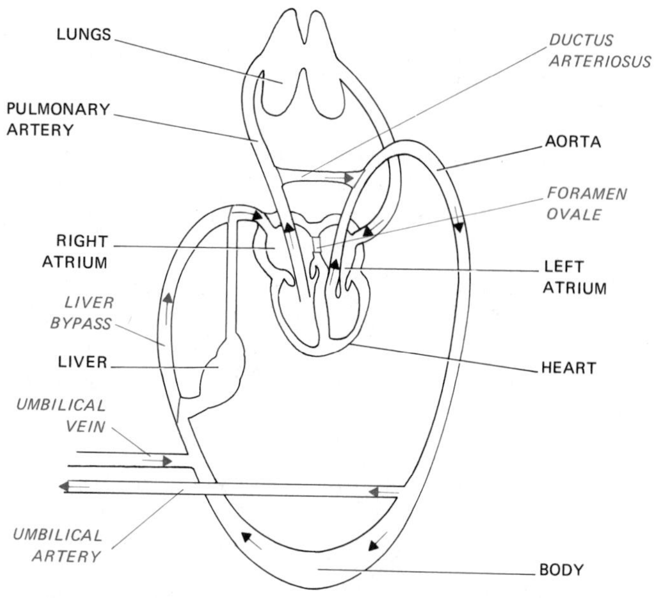

14.14
Human fetal circulation. The lungs of an unborn baby, being airless, cannot function in gas exchange. The respiratory exchange takes place in the placenta instead. Since the lungs of the fetus are useless until it is born, they receive only enough blood to allow them to grow, and most of the fetal circulation is detoured. One detour is the *ductus arteriosus*, a connecting vessel between the pulmonary artery and the aorta. A second is the *foramen ovale*, an opening between the right and left atria, which allows some of the blood from the body to recirculate directly. There is also a vessel that temporarily bypasses the liver. In a fetus, the liver is not completely functional, and the fetus depends on the maternal liver. The colored arrows and colored labels in the figure show which parts of the circulatory system are strictly fetal; the others will continue to function after birth as they did before. After birth, the fetal detours are closed: The umbilical vessels are cut, and the other bypasses are blocked by growth processes.

SUMMARY

1. Each type of cell is a specialist with a specific job to do. But it is not yet totally clear how such specialization occurs. Aristotle proposed his theory of *epigenesis* about 2000 years ago, and the theory of *preformation* was suggested in the seventeenth century. Neither theory is completely correct, although both contain some measure of accuracy. Not until the late nineteenth century was it realized that both parents contribute genetic material to the zygote through egg and sperm.

2. *Mitosis* is the reproduction of nuclei and the equal distribution of DNA to each. *Meiosis* is the pair of nuclear and cellular divisions that reduce the chromosome number from diploid to haploid.

3. When a sperm meets an egg, it digests its way into the egg cytoplasm and sheds its tail, and its nucleus fuses with the egg nucleus. The egg quickly surrounds itself with the *fertilization membrane*. The resultant diploid cell is the *zygote*.

4. The series of cell divisions initiated after fertilization is *cleavage*. The directional forces set up at the beginning of cleavage are essential to the proper completion of the organism, and they will be maintained throughout its entire life.

5. Once a hollow ball of microscopically small cells, a *blastula*, is formed, a new growth phase ensues. The cells begin to move about on the surface of the blastula, next entering a reorganizing phase of *gastrulation*. The shifting of cells includes an invagination (bending in) at one place on the sphere of cells. The region where invagination occurs is the *blastopore*.

6. Once gastrulation has occurred, the embryo has an outer coating, an *ectoderm,* and an inner lining, an *endoderm*. The *mesoderm* fills the part between the two original layers of ectoderm and endoderm, and it eventually builds the bulk of the embryo.

7. The human embryo begins to develop in the oviduct, but it passes down toward the uterus, cleaving as it goes, until after five or six days, it is ready for implantation in the uterine wall. By then the zygote is a ball of some 100 or more cells, which will become the embryo proper, plus a growing mass of cells on one side, the *trophoblast*, which will develop into a system of membranes outside the embryo. The whole mass, including trophoblast and embryo, is a *blastocyst*.

8. The *placenta* consists of the absorbing part of the embryonic mass (the chorion) and uterine tissue; it is the embryo's only contact with the outside world. The *umbilical cord* carries wastes from the embryo to the placenta, as well as oxygen and nutrients from the placenta to the embryo.

9. As an embryo develops, it undergoes *differentiation* of its parts. We know that cellular specialization and growth can result in functional organs, even though the exact regulating forces are not understood.

10. An early embryo, having a future hind end and a future head end determined, has a layer of ectoderm on the outside. Where the backbone will later grow, a strip of the ectoderm thickens to make the *primitive streak,* underneath which the *notochord* develops. The spinal cord, the brain, and the entire nervous system begin with the ectodermal primitive streak and notochord.

11. The series of events which leads to the formation of a vertebrate eye illustrates the principle of *induction,* in which one tissue is caused or *induced* to change its developmental pattern as a result of the influence of a different tissue.

12. After two months of development, the human embryo is called a *fetus*. After three months in the uterus, a fetus is recognizably human. The last six months of pregnancy are devoted to increase in size and maturation of the organs developed earlier. When the fetus is mature enough to be born, it has come *to term*.

13. At term, about 280 days after the last menstrual cycle, the muscles of the uterus begin to contract rhythmically, and the "labor" preceding actual childbirth begins. The exact cause of these initial uterine contractions is not known.

14. *Lactation* includes both milk secretion and milk removal from the breasts. Although several other hormones play a part in milk production and secretion, the most important hormone is *prolactin*.

RECOMMENDED READING

Balinsky, B. I., *Introduction to Embryology*, Fourth Edition. Philadelphia: W. B. Saunders, 1975.

Friedmann, Theodore, "Prenatal Diagnosis of Genetic Disease." *Scientific American*, November 1971. (Offprint 1234)

Gordon, Richard, and Antone G. Jacobson, "The Shaping of Tissues in Embryos." *Scientific American*, June 1978.

Nilsson, Lennart, Mirjam Furuhjelm, Axel Ingleman-Sundberg, and Claes Wirsén, *A Child is Born*, Revised Edition. New York: Delacorte Press/Seymour Lawrence, 1977.

Rugh, Roberts, and Landrum B. Shettles, *From Conception to Birth: The Drama of Life's Beginnings*. New York: Harper & Row, 1971. (Also available in paperback)

Saunders, John W., Jr., *Patterns and Principles of Animal Development*. New York: Macmillan, 1970.

Tanner, James M., and Gordon Rattray Taylor, *Growth,* Revised Edition. New York: Time-Life Books, 1973. (Life Science Library)

Wessells, Norman K., and William J. Rutter, "Phases in Cell Differentiation." *Scientific American*, March 1969. (Offprint 1136)

HOW A HUMAN EMBRYO DEVELOPS

*The First Week**
Initial cell divisions take place. The embryo in the oviduct is in a four-cell stage; by the time it reaches the uterus four or five days later, it has grown to about 32 cells. Blastocyst and trophoblast form, and implantation in the uterine wall takes place.

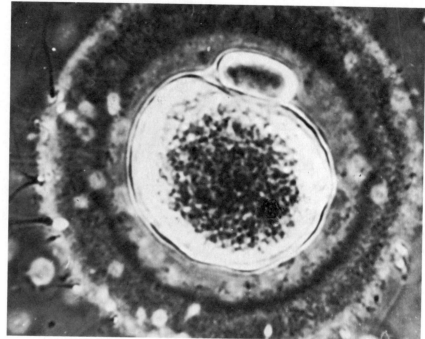

Sperm and egg fuse in fertilization.

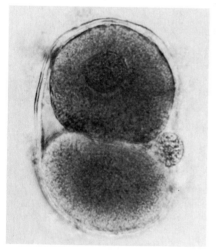

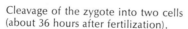

Cleavage of the zygote into two cells (about 36 hours after fertilization).

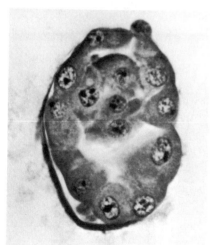

Cleavage and morula formation (about the third day after fertilization).

The blastocyst (about four days after fertilization).

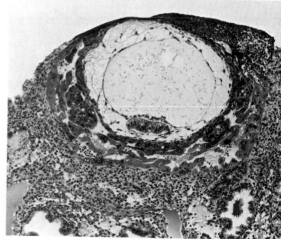

Implantation of the blastocyst (about 7–10 days after fertilization).

*Adapted from James E. Crouch and J. Robert McClintic, *Human Anatomy and Physiology,* Second Edition (New York: John Wiley, 1976), pp. 82–85. (Originally from Arey.) Used with permission.

Second Week
The extra-embryonic membranes, the chorion and the amnion, begin to develop; blood vessels grow within developing chorionic villi. The placenta is formed from villi and uterine tissues. The connection between placenta and embryo is the umbilical cord. Formed at this time are the three embryonic tissues from which all other mature tissues derive. (about 1.5 mm; all measurements from crown to rump)

Third Week
Body, head, and tail are distinguishable. The neural tube begins to form and will eventually develop into the brain, the spinal cord, and the rest of the nervous system. Primitive blood cells and vessels are present. (about 2.5 mm)

Fourth Week
The U-shaped heart pumps blood through a simple system of vessels. Placenta is functioning, and definitive villi appear. Limb buds are indicated. Umbilical cord begins to develop. Body is tubular and C-shaped. Intestine is a simple tube. Liver, gall bladder, pancreas, and lungs begin to form. Eyes, nose, and brain begin to form. (about 5.0 mm)

A three-layered embryo after about three weeks.

EMBRYO

UTERINE WALL

Actual size

THE DEVELOPMENT OF THE HUMAN HAND

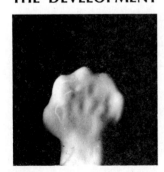

At about five weeks the forearm is shorter than the hand, and finger ridges begin to form.

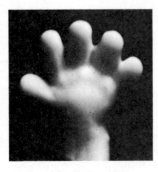

At six weeks the hand has begun to develop clearly delineated fingers.

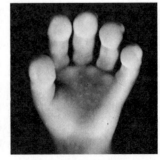

At 11 weeks fingertip buds are developed, and the muscles are active.

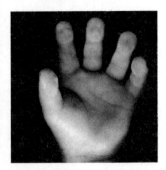

At five months the hand is practically complete, and thumb sucking may start.

Fifth Week
Primitive nostrils are present. Tail is prominent. Heart, liver, and primitive kidney begin to form. Jaws are outlined. Intestine elongates into a loop. Genital ridge bulges. Primitive blood vessels extend into head and limbs. Spleen is indicated. Spinal nerves are formed, and cranial nerves are developing. Stomach begins to form. Pre-muscle masses appear in head, trunk, limbs. Epidermis is gaining a second layer. Cerebral hemispheres are bulging. It is at this stage that drugs taken by the mother may affect the embryo; also such diseases as measles may be transmitted from the mother to the embryo. (about 8.0 mm)

Actual size

The embryo at about five weeks.

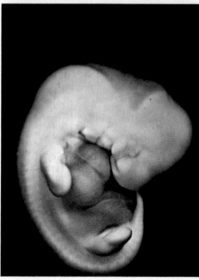

Actual size

The embryo at the end of five weeks.

Sixth Week

Upper jaw components are prominent but separate; lower jaw halves are fused. Head becomes dominant in size. External ear appears. Arms and legs are recognizable. Simple nerve reflexes are established. Heart and lungs acquire definitive shapes. Eye continues to develop and indicates color. (about 12.0 mm; ½ in.)

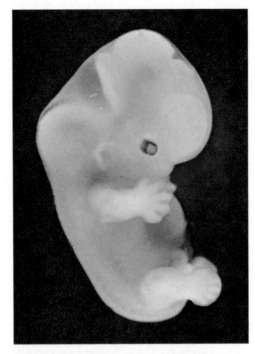

Actual size

The embryo at about six weeks.

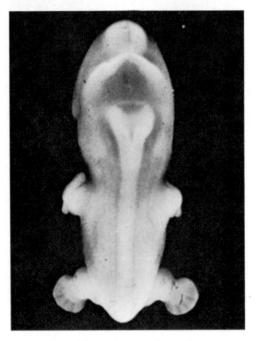

A rear view of the embryo at about six weeks.

Seventh Week
Face and neck begin to form. Fingers
and toes are differentiated. Back
straightens. Tail is regressing. Three
segments of upper limb are evident.
Jaws are formed and begin to ossify.
Stomach is attaining final shape and
position. Muscles are differentiating
rapidly throughout the body and assuming final shapes and relations.
Brain is becoming large. Eyelids are
forming. (about 17.0 mm; ¾ in.)

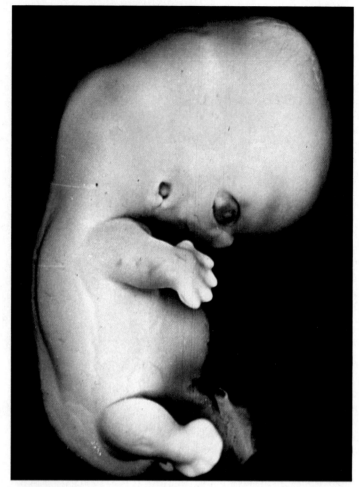

The seventh week.

Actual size

Eighth to Ninth Week

The embryo is now called a *fetus*. Nose is flat; eyes are far apart. Fingers and toes are well formed. Growth of intestine makes body evenly round. Head is elevating. Tongue muscles are well differentiated; earliest taste buds are indicated. Ear canals are distinguishable. Small intestine is coiling within umbilical cord; intestinal villi are developing. Liver is relatively large. Testes or ovaries are distinguishable as such. Main blood vessels assume final plan. First indications of skeletal bone formation. Muscles of trunk, limbs, and head are well represented, and the fetus is capable of some movement. All five major subdivisions of brain are formed, but brain lacks convolutions that are characteristic of later stages. External, middle, and internal ear are assuming final form. (about 23.0 mm; 1 in.)

Tenth to Eleventh Week

Head is erect as in later life. Limbs are nicely modeled. Fingernail and toenail folds are indicated. Lips are separate from jaws. Intestines withdraw from umbilical cord and assume characteristic positions. Anal canal is formed. Nasal passages are partitioned. Kidney is able to secrete. Bladder expands as sac. Vaginal sacs are forming. Rudimentary sex ducts form. Early lymph glands appear. Enucleated red cells predominate in blood. Earliest hair follicles begin developing on face. Spinal cord attains definitive internal structure. Eyelids are fused. Fetal pulse is detectable. Tear glands are budding. (about 40.0 mm; 1½ in.)

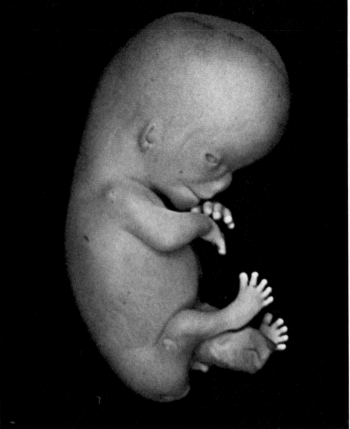

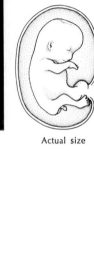

The fetus at about eight weeks.

Actual size

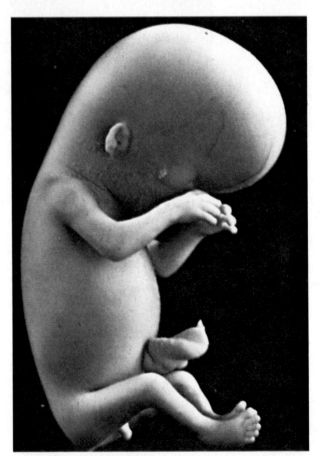

The fetus at about nine weeks.

Twelfth to Fifteenth Week
Head is still dominant. Nose gains bridge. External genitalia attain distinctive features. Tooth buds and bones form. Cheeks are represented. Nasal glands form. Lungs acquire definitive shape. Prostate and seminal vesicles appear. Blood formation begins in bone marrow. Some bones are well outlined. Epidermis is three-layered. Brain attains its general structural features. Characteristic organization of eye is attained. Mother feels uterus enlarging. All vital organs are formed. (about 56.0 mm; 2¼ in.)

Sixteenth to Nineteenth Week
Face looks "human." Hair appears on head. Muscles become spontaneously active. Body is outgrowing head. Hard and soft palates are differentiating. Gastric and intestinal glands begin to develop. Kidneys attain typical shape and plan. Testes are in position for later descent into scrotum. Uterus and vagina are recognizable as such. Blood formation is active in spleen. Most bones are distinctly indicated throughout body. Stretching movements, not felt by mother. Epidermis begins adding other layers to form skin. Body hair starts developing. Sweat glands appear. Eyes, ears, and nose approach appearance more characteristic of adult human form. General sense organs begin to differentiate. (about 112.0 mm; 4½ in.)

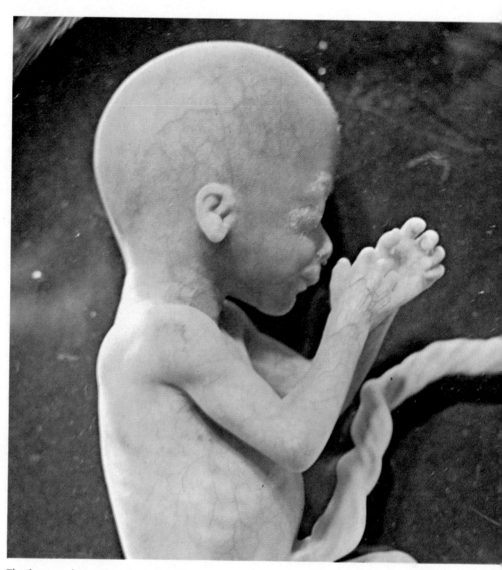

The fetus at about seven lunar months.

Fifth Lunar Month*
Body is covered by downy hair called lanugo. Nose bones begin to harden. Vaginal passageway begins to develop. Until birth, blood formation continues to increase in bone marrow. Fetal heart sounds are audible with stethoscope. Heart beats at twice adult rate. Lungs are formed but do not function. Gripping reflex of the hand begins to develop. Kicking movements and hiccuping may be felt by mother. (about 160.0 mm; 6½ in.)

*Because women's reproductive cycles are closer to the lunar month of 28 days than to the calendar month, fetal development is conventionally described in terms of lunar rather than calendar months.

Sixth Lunar Month
Body is lean but better proportioned. Colon becomes recognizable. Nostrils reopen. Hairs emerge. Cerebral cortex of the brain is layered, as is more typical of adult. Internal organs occupy normal positions. Thumb sucking may begin.

Seventh Lunar Month
Fetus is lean, wrinkled, and red. Eyelids reopen; eyebrows and eyelashes form. Nervous system is developed sufficiently so that fetus practices controlled breathing and swallowing movements. Uterine glands appear. Scrotum is solid until sacs and testes descend, from seventh to ninth month. Lanugo hair is prominent.

Brain enlarges, cerebral fissures and convolutions appearing rapidly. Retinal layers of the eye are completed, and light can be perceived. If delivered at this stage, the baby has at least a 10 percent chance of survival.

Eighth Lunar Month
Testes settle into scrotum. Fat is collecting, wrinkles are smoothing, and body is rounding from eighth to tenth month. Taste sense is present. Weight increase slows down. If delivered at this stage, the baby has about a 70 percent chance of survival.

Ninth Lunar Month
Nails reach fingertips. Fetus still cannot hear.

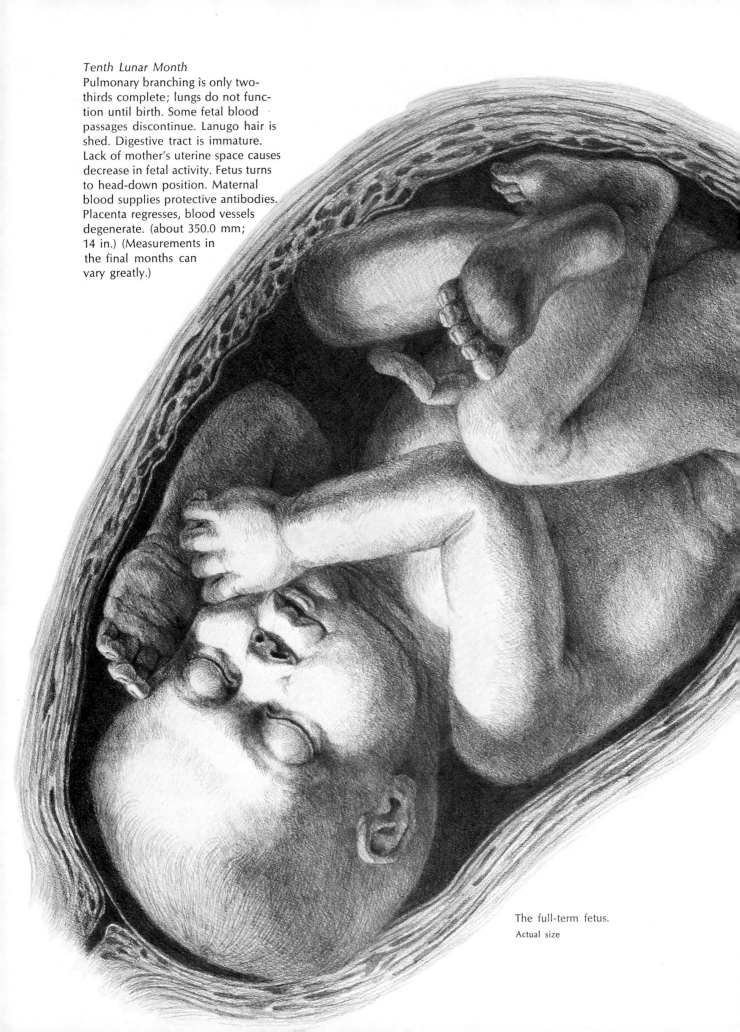

Tenth Lunar Month
Pulmonary branching is only two-
thirds complete; lungs do not function until birth. Some fetal blood
passages discontinue. Lanugo hair is
shed. Digestive tract is immature.
Lack of mother's uterine space causes
decrease in fetal activity. Fetus turns
to head-down position. Maternal
blood supplies protective antibodies.
Placenta regresses, blood vessels
degenerate. (about 350.0 mm;
14 in.) (Measurements in
the final months can
vary greatly.)

The full-term fetus.

Actual size

THE LIFE OF
PLANTS

PART FIVE

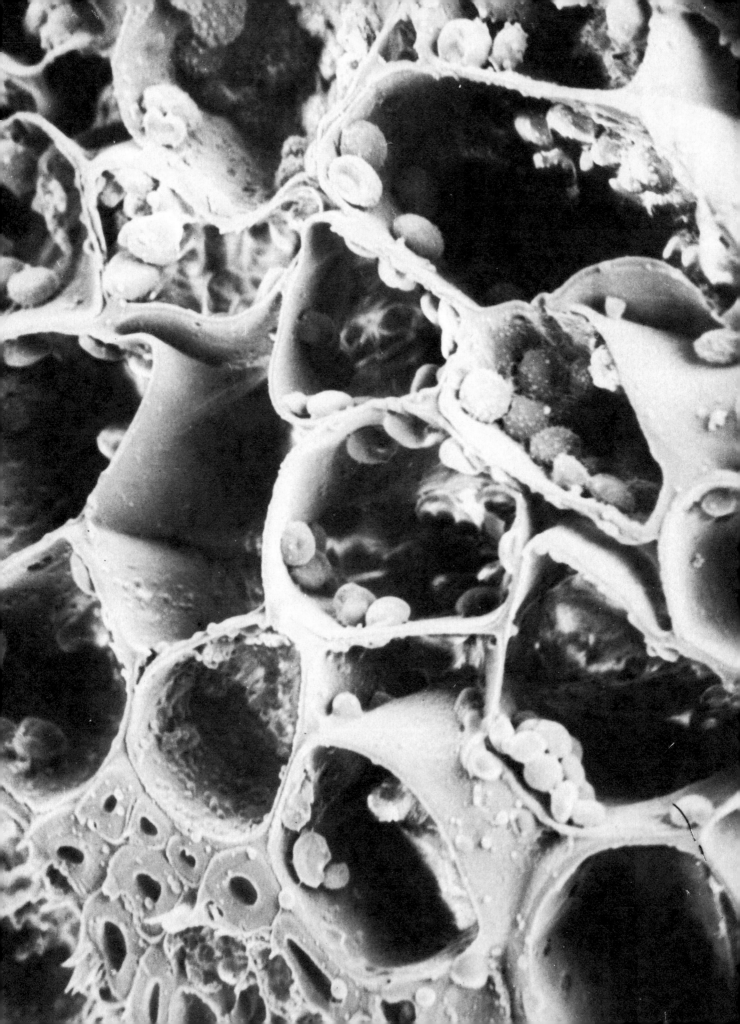

15

Reproduction in Plants

Scanning electron micrograph of
a cut across a vascular bundle of a
pitcher plant, Sarracenia. The
larger central circles are water-
carrying xylem cells, and the
smaller ones are food-carrying
phloem.

THE CONTINUITY OF LIFE ON EARTH DEPENDS ON
the ability of organisms to perpetuate their own
kind with enough accuracy to maintain their identity,
while providing enough variability and flexibility to
allow for changing conditions. During the course of
organic history, organisms have developed such a
range of reproductive methods that no general scheme
can be made to include them all. But in every method,
the end result of reproduction is the production of
new individuals, all more or less like their parents.

Patterns of reproduction have been evolving ever
since the first prokaryotes were formed several billion
years ago. At first, there was presumably no sex, no
recombination of genes through mating; there was
a mere pinching in two of a single cell. Then came
simple cellular fusions, such as those still used in
some unicellular algae and protozoa. Later, specialization led to distinction between male and female
gametes. In time, certain parts of an organism were
set apart as being peculiarly adapted to gamete production, and these parts evolved ever greater specialization until the highly differentiated sex organs of
multicellular plants and animals were formed. Along
with them there evolved the behavior patterns of animals, including the sexually related social activities
of human societies and the elaborate pollinating
mechanisms of the flowering plants. In this chapter,
we will consider plants and such plantlike organisms
as the fungi and lower algae. The plants currently

inhabiting the earth can be arranged in a series in which the structurally and functionally simple ones are placed at the beginning and the complex ones are at the end. Those in between are arranged in as reasonable an order as our present information allows.

The most commonly accepted idea is that algae are the simplest and most ancient plants. Then, omitting some minor groups of plants, come the mosses, the ferns, the conifers, and finally the flowering plants. The last are the most complex and the most recent. Keep in mind that when complexity is mentioned, it is the entire life history of the plant that is being considered, not any particular part. Indeed, as we describe the several types of plants, it will become apparent that some parts of a plant may be more complex than those of a supposedly more primitive type, but that other parts may actually be simpler. This is notable in a comparison of the relatively more primitive mosses and the more highly developed ferns. The sexual generation of a fern is structurally simpler, smaller, and shorter-lived than the sexual generation of a moss, but the simplification is more than offset by the fern's enormously greater complexity in the asexual generation. Similarly, the sexual generation of a flower is still simpler than it is in ferns, but the asexual generation is so complex that the plant as a whole is clearly the most advanced of the series.

The examples presented in this chapter will illustrate a progression from the simplest asexual types, through a primitive sexuality in some algae, to the specialized reproductive methods of the higher plants. The evolution of the diploid sporophyte is especially instructive. In those ancient plants that have undergone limited evolutionary change, the haploid phase is more prominent than the diploid. However, in plants of more recent origin, the haploid has decreased in complexity until in flowering plants it occupies only a small portion of the flower.

ASEXUAL REPRODUCTION IS USED BY SOME SPECIES OF PLANTS

If reproduction is sexual, there must be a union of two sets of chromosomes. In asexual reproduction, no such union is required, because new individuals are made from a single parent. Most multicellular animals are strictly sexual, but plants commonly use both sexual and asexual methods, varying from species to species, and occasionally varying within a single species. Some seed plants seem to require a sexual phase, but some lower organisms (for example, blue-green algae) appear to be completely asexual.

Reproduction without sex, that is, without union of two sets of chromosomes, is simpler than sexual reproduction. Asexual reproduction suffers a disadvantage in comparison with sexual reproduction in that it results in a slower distribution of variation through a population. Asexual reproduction does, however, have several advantages, and many organisms, especially microscopic ones, commonly use it. Asexual reproduction bypasses the need for an organism to find a mate, an advantage of considerable importance for survival when the population is sparse or when new territory is being invaded. It can also be exceedingly fast, as one can see when a bacterial population dramatically increases in an infected host, or an unrefrigerated piece of meat spoils under the action of microbial growth. Furthermore, except for an occasional mutation, all the offspring from an asexually reproducing organism are genetically identical. Having a whole population of individuals all genetically alike can be useful, especially in a restricted environment, such as in one animal body or a piece of decaying matter. The asexual method is so efficient that it is the common method in many plants.

Fission

The simplest example of asexual reproduction is *fission*, the division of a single cell and the separation of the two resultant cells. Many small multicellular plants break readily into smaller pieces, and the pieces

15.1

Scanning electron micrograph of the asexual spores (*conidia*) of black bread mold, Rhizopus.

× 3000

15.2

Asexual (vegetative) reproduction in a flowering plant. A lateral stem, or runner, branches out from a parent strawberry plant, and when it touches earth at its far end, it sprouts roots and makes an entire new plant.

15.3

Multiplication of mushrooms. From a central starting point, an ever-expanding underground mass of mushroom tissue sends up fruiting bodies around the periphery, making a "fairy ring."

grow independently. The massive "blooms" of green algae in ponds are the result of such fragmentation.

Spores

Most plants produce special cells that are reproductive but involve no nuclear fusion, although they may follow either meiotic or mitotic divisions (Chapter 8). These cells are usually called _spores_, but the term is a rather vague and general one with little technical precision. Enormously variable in form and manner of distribution, spores are common reproductive cells in algae and molds, which are capable of giving them off in astronomical numbers (Figure 15.1).

The air, the earth's waters, and most exposed surfaces carry a load of microscopic spores. Spores can survive in harsh environments for long periods by remaining dormant while conditions such as temperature and moisture remain at their extremes. When moisture and temperature are suitable, spores can germinate. Because some bacteria, such as the ones that cause botulism and tetanus, can form spores, they are able to endure conditions such as high temperatures that would kill non-spore-formers.

Vegetative Reproduction

Higher plants may reproduce asexually, or _vegetatively_, through processes in which a new complete plant is produced from the roots, stems, or leaves of the parent plant. This may occur naturally or through human interference by cutting, grafting, or layering. In higher plants, prostrate stems or even masses of tissue spread beneath or on the surface of the soil, branching at the advancing ends and decaying at the back, with many separate individuals resulting. Strawberry plants, for instance, may produce _runners_, or _stolons_, which grow along the ground away from the parent plant. Runners form roots and shoots that develop into new plants (Figure 15.2). Ferns and mushrooms may also spread, forming conspicuous circles of plants, which may be several meters in diameter (Figure 15.3).

15.4

Asexual reproduction by means of a cutting, or "slip." A stem of a mint plant, removed from its parent plant, has sprouted roots from the lower end of the stem. When planted, the rooted cutting will grow as an independent individual.

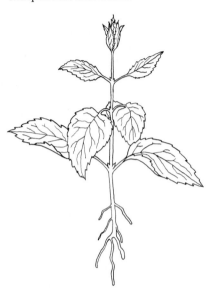

15.5

Asexual reproduction by grafting. A branch of a lilac bush has been connected to a stem of privet hedge. Care is taken to have the growing tissue of the *scion* (the lilac) connected to the growing tissue of the *stock* (the privet), and the joint is covered with wax and wrapped (arrow) to hold the graft in place and prevent drying. Grafts survive only if stock and scion are closely related. Lilac and privet, both in the olive family, are similar enough to ensure a successful graft.

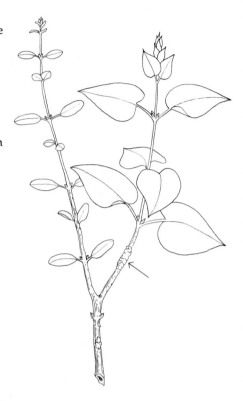

Humans sometimes desire genetically identical copies of plants in species that have special combinations of characteristics: apples, for example, or roses. These plants do not "come true" from seeds, because seeds involve sexual fusions and are therefore genetically different from the parent plants. To guarantee biological uniformity, people use _clones_, or groups of individuals all having the same genetic background, obtained without an intervening sexual fusion. Obtaining clones is easy with plants, because new plants can be grown from slips or cuttings (Figure 15.4). A piece of stem may be cut from a desirable variety and grafted onto a stem of any closely related seedling (Figure 15.5). With this procedure, one can have an orchard of uniform peaches or cherries or mangoes, or a garden of identical lilac bushes, or even several kinds of apples on a single tree.

The principle of cloning has been known for centuries, and practical gardeners have known how to produce multiple copies of plants almost as long as they have been cultivating them. Even cloned animals have been known for years, especially in armadillos, which routinely give birth to identical quadruplets. The recently developed technique for cloning frogs has sent a shiver of apprehension through some people, who see no threat in making a million identical copies of an organism with leaves and stems but foresee terrible results in doing something similar with an organism possessing eyes and legs.

THE ORIGIN OF COMPLEX SEXUALITY

The essential feature of any sexual reproduction is the union of two nuclei, usually of different genetic makeup, resulting in the formation of a single new nucleus.

In the simplest and presumably most ancient organisms, mating involves the complete fusion of two unicellular individuals (Figure 15.6). In some bacteria a strand of DNA may migrate from one bacterium across a protoplasmic bridge to a recipient (Figure 15.7). The result is a new set of gene combinations, even though reproduction in the sense of making additional individuals has not occurred. Biologists regard such conjugations as sex acts, whether or not actual reproduction results.

By observing a number of *simple* organisms, biologists have found a series of sexual types that may illustrate the way in which *higher* organisms came to have special sexual procedures. The simplest sexual process is the joining of *any* two free-living cells, with cytoplasmic union followed by nuclear fusion. A slight advance over such a primitive procedure occurred when special mating types evolved—a foreshadowing of differentiation into maleness and fe-

maleness. This evolution of mating types happens in some species in which certain individuals, not visibly different, can mate *only* with compatible individuals. Populations of these two types of individuals are called *mating strains*. A species in which no separate mating strains exist is called *homothallic*; a species in which special mating strains do exist is *heterothallic*. Even in heterothallic species there is no discernible "maleness" or "femaleness."

The next step in sexual complexity came with a visible difference between the sex cells, a slight difference in size or pigmentation. Still another advance

15.6

Sexual fusion of entire organisms. (a) A unicellular green alga, Chlamydomonas. (b) Two compatible cells meet. Each is haploid (N). (c) A single cell, a *zygote*, results from the fusion of the two separate cells. The zygote is diploid (2N). After a period of time, meiosis occurs, and four haploid cells are formed. × 1000.

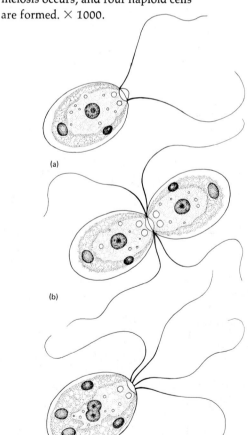

(a)

(b)

(c)

15.7

Mating in bacteria. Instead of fusing two cells into one, bacteria form a protoplasmic bridge from one cell to another, and bacterial nuclear material is sent through the connection, called a *pilus*. Since the mating time of a bacterium can be about an hour and a half, and the life span of a vegetative bacterium, in good growing conditions, is only about half an hour, mating lasts three times as long as the ordinary life of the individual.

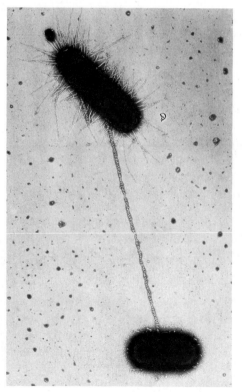

× 10,000

SEEDS AND CIVILIZATION

Early humans wandered about the earth in small groups, killing such animals as they could overpower, finding fruits in season, digging up starchy rhizomes, and even eating leaves when they could find nothing better. Because large societies are impossible to feed when everything must be hunted down or collected by chance, the tribal size was limited, probably to a few dozen at most. The very smallness of hunting tribes made protection difficult. Wild animals and marauding human enemies made life risky. The bulk of everyone's time was spent in searching for something to eat, and even so, humans must have been hungry more often than not. When societies were tiny and the bulk of a human's life was taken up in self-protection and scrounging for a bit of food, there was neither time nor power nor technique for building large permanent settlements.

Change came when it was discovered that fruits and seeds could be stored and then planted to produce food crops that could be harvested. Discoveries came slowly and independently in scattered places. In Jordan and Israel flint sickles about 10,000 years old indicate that grains, probably wheat and barley, were then being harvested. Knowledge of farming seems to have spread westward to Europe about 5000 B.C., with wheat and barley as the main crops. Meanwhile, skill in cultivation moved eastward to China, where there was wheat, but where another grass, rice, became the staple grain. From eastern Asia, people came through what is now Alaska and migrated to South America. The hunters in the cold territories of the north remained hunters, but in the gentler climates where rich valleys offered the opportunities, humans found wild grasses, which they learned to grow, to keep, to select, and to improve. By about 4000 B.C., the Mexicans were cultivating varieties of maize so

specialized that their wild counterparts are still unknown.

A person whose full effort is taken up with day-to-day survival has little time or energy for nonessentials. An occasional genius may paint animals on rock walls or carve figures in stone or bone, but cities and armies are possible only when there is leisure. A half-starved savage, prowling through the wilderness, must work 12 months a year merely to keep alive, but a Peruvian Indian with a maize patch can grow enough grain in three months of labor to feed himself and his family in relative comfort. Thus, when humans made seeds their servants, they gained time for leisure, and with that leisure they began to make buildings that were more than simply shelters. These were built with elegance and ornamented with carving and color, and they served for governmental or religious functions or for personal ostentation. No essential changes in the pattern have occurred since.

The great civilizations of the world have depended on the ability of seeds to store and concentrate a durable source of human food. The arts, the sciences, literature, architecture, politics, philosophy, even wars—all became possible because of grains. Figuratively speaking, the temples of Canton were built on rice, the cathedrals of Paris on wheat, and the pyramids of Chichen-Itza and the fortresses of Macchu Picchu on maize. The greatest achievements of humans are founded on grass.

Dates of the Cultivation of Some Important Domestic Plants

Maize (corn)	Years before Present
South central Mexico	7000
Peruvian highlands	6300–4800
Southwestern United States	4000

came with a change in motility. One sex cell, the larger one, lost its ability to move, while the smaller remained active. Since in higher organisms the larger nonmotile cell is the female gamete, and the smaller motile one is the male, the terms "male" and "female" are sometimes used even in simple microscopic species.

All the sexual methods described above can be observed in one genus of one-celled green algae, Chlamydomonas (Figure 15.8). But additional innovations have resulted in still more complex organisms. One was the development of multicellular bodies in which only a special portion of the organism is devoted to gamete production. Such a development

Wheat	
Southwestern Iran	9500–8750
Syria	9000
Eastern Palestine	9000
Jordan	9000
Eastern Turkey	9000
Northern Iraq	8750
Crete	8100
Thessaly (Greece)	8000–7000
Bulgaria	7000
Flax	
Syria	10,050–9500
Northern Iraq	7800–7600
Southwestern Iran	7500–7000
Lentils	
Syria	9000
Eastern Palestine	9000
Eastern Turkey	9000
Northern Iraq	8750
Western Macedonia	8200
Thessaly (Greece)	8000–7000
Southwestern Iran	7500–7000
Bulgaria	7000
Peas	
Eastern Palestine	9000
Eastern Turkey	9000
Northern Iraq	8750
Thessaly (Greece)	8000–7000
Barley	
Syria	9000
Eastern Palestine	9000
Eastern Turkey	9000
Southwestern Iran	8750–8000
Crete	8100
Thessaly (Greece)	8000–7000
Northern Iraq	7800–7600

Table reprinted by permission from Charles B. Heiser, Jr., "Origins of Agriculture," *1975 Yearbook of Science and the Future* (Chicago: Encyclopedia Britannica, 1974), p. 61. Used with permission.

15.8

Three gradations of sexuality in one genus, Chlamydomonas. (a) Two cells, structurally alike, meet and mate. This is the simplest kind of sexuality imaginable. (b) In a different species, one partner is larger than the other, making a sexual differentiation. The smaller is usually considered the male. (c) In still a third species, one partner not only is larger than the other but has lost its motility, although the smaller one is still able to swim. Having a larger nonmotile cell (usually called an egg) mate with a smaller motile cell (usually called a sperm) is typical of the sexual methods of most plants and animals.

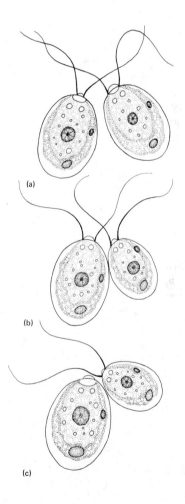

must have occurred early in organic history, as shown by the fact that some relatively primitive species have it, and it is now used by most living things.

The most recent complication was the provision for parental care of the new generation. In plants, either the egg is provided with enough reserve food to ensure a reasonable chance for survival of the new plant, or the new plant is kept in the parent plant during its early life. Protection of the young plants (or animals) has been the common procedure for most species, yet the more primitive sexual methods still persist in some plants alive today. By comparing sexual processes, we can see how complicated procedures may have developed, one step at a time, from homo-

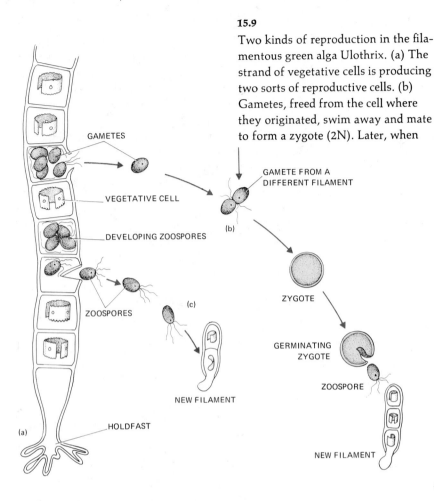

15.9
Two kinds of reproduction in the filamentous green alga Ulothrix. (a) The strand of vegetative cells is producing two sorts of reproductive cells. (b) Gametes, freed from the cell where they originated, swim away and mate to form a zygote (2N). Later, when meiosis occurs and the zygote germinates, swimming zoospores (N) will emerge from it, and they can grow into new Ulothrix filaments. (c) Ulothrix can also produce asexual swimming spores (zoospores), each of which can grow into a new filament. × 500.

GAMETES

VEGETATIVE CELL

DEVELOPING ZOOSPORES

ZOOSPORES

(c)

NEW FILAMENT

HOLDFAST

(a)

(b)

GAMETE FROM A DIFFERENT FILAMENT

ZYGOTE

GERMINATING ZYGOTE

ZOOSPORE

NEW FILAMENT

15.10
Reproduction by both asexual and sexual methods in the green alga Oedogonium. Asexual zoospores (N) break out of the parent filament, swim away, and later settle down to produce new individuals. When Oedogonium reproduces sexually, one cell grows into an enlarged egg while others are producing sperm cells. The sperm escape, swim to a hole in the cell containing the egg, and fertilize the egg. Later, the fertilized eggs undergo meiosis and germinate to produce new haploid individuals. × 400.

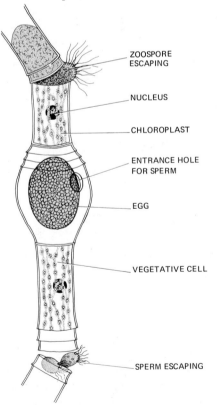

ZOOSPORE ESCAPING

NUCLEUS

CHLOROPLAST

ENTRANCE HOLE FOR SPERM

EGG

VEGETATIVE CELL

SPERM ESCAPING

thallism in the one-celled Chlamydomonas to seed production in a highly specialized flower.

SEXUAL REPRODUCTION IN LOWER PLANTS

In the heterogeneous assemblage of plants popularly known as algae, reproductive methods are as various as the plants themselves. Indeed, algae are so diverse that the current opinion among biologists is that they are not a single evolutionary group, that they instead represent several independent lines of development (Chapter 9). Giving a comprehensive treatment of the reproductive methods of even the main groups of algae is not possible in a small space, because these methods vary from the simplest homothallic Chlamydomonas to the complex life cycles of such sea weeds as the red algae.

Representative examples of algal reproduction illustrate some common procedures in microscopic, freshwater types. Ulothrix (Figure 15.9), Oedogonium (Figure 15.10), and Spirogyra (Figure 15.11) are common freshwater genera, each with its characteristic reproductive method.

Fungi

Fungi constitute another varied group of organisms. They are indeed so diverse and so different from other organisms that they are excluded from the plant kingdom by many biologists (Chapter 9). However, they have been traditionally considered as plants, and for convenience they will be treated along with plants in this book.

Reproduction in fungi is as varied as it is in algae. In fact, some fungi, such as those causing rust dis-

eases in plants, exhibit the most complex reproductive patterns to be found anywhere in nature.

In primitive water fungi, eggs are formed in single cells, which receive male nuclei delivered through protoplasmic threads. In spite of their structural simplicity, such water molds as the Achlyas have an elaborate system of hormones (Chapter 17) that regulate the growth and activity of the sexual apparatus (Figure 15.12). Other molds, such as the black bread mold, Rhizopus, not only have asexual spores but

15.11

In the green alga Spirogyra, mating procedure is by conjugation. (a) Two filaments lie side by side. (b) Tubes are sent out from each cell to a neighboring cell. (c) All the cellular material passes from one cell into the mating partner. (d) The cells fuse into a single *zygospore*. (e) The zygospores go into a period of dormancy, later to undergo meiosis and give rise to a new filament. × 500.

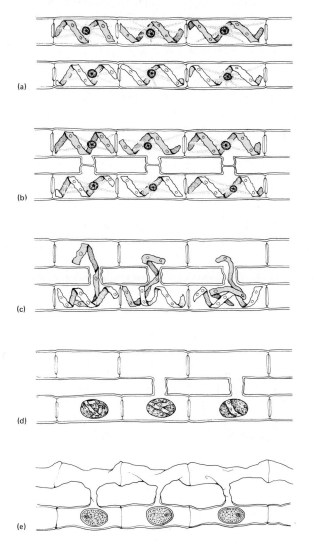

15.12

Sexual reproduction in the water mold Achlya. (a) Fungus threads of compatible mating types (left and right), approach one another. (b) A hormone, antheridiol, from one individual (left), stimulates the other to sprout *antheridial hyphae* (right). (c) The antheridial hyphae secrete a second hormone, which stimulates the production of *oogonial hyphae* (left), which attract the antheridial hyphae. (d) The antheridial hyphae further stimulate the maturation of eggs in the oogonial hyphae, whereupon nuclei from the antheridial hyphae pass into the oogonial hyphae, and fertilization of the eggs occurs. × 25.

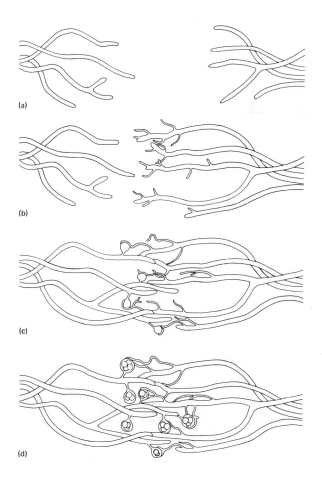

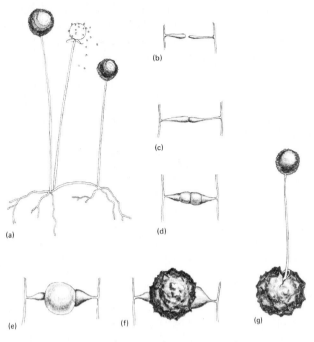

(a) (b) (c) (d) (e) (f) (g)

15.13

Reproduction in the black bread mold, Rhizopus. (a) The fungus threads send up hyphae, *sporangiophores*, bearing spore cases, *sporangia,* full of spores (see Figure 15.1). These asexual spores are the usual means of reproduction, unless compatible mating types meet. If they do, mating hyphae grow toward one another (b), meet (c), and form a bridging pair of cells (d), which fuse to make a single zygospore (e) containing hundreds of compatible nuclei. The nuclei mate, two by two, and the zygospore develops a thick covering (f). After a period of dormancy, the zygospore undergoes meiosis and then germinates, to send out a sporangiophore (g) with spores that are haploid (N). × 50.

practice a kind of conjugation, in which adjacent cells meet, fuse their cytoplasm and nuclei, and grow a heavy protective wall around the resultant zygospore (Figure 15.13). Later, meiosis occurs and the zygospore germinates to produce a new fungus.

Mushrooms, commonly the most familiar fungi, produce sexual spores. The structural and functional features of mushroom spore production are shown in Figures 15.14 and 15.15. As might be expected, many variants are known, but this procedure is a common one, and it is used by hundreds of species of mushrooms.

Mosses and Ferns

Sexual reproduction in mosses and ferns illustrates the general direction of plant sexuality, which is distinctly different from the sexual methods of animals. In practically all animals, a diploid organism possesses specialized cells that undergo meiosis, and the resulting haploid cells become eggs or sperm cells. Then the union of an egg and a sperm makes a diploid zygote. Plants rarely proceed so. Instead, plants have a whole haploid phase during which gametes are produced without meiotic divisions (Figure 15.16). That phase is the *gametophyte generation*, or the gamete-producing plant (*gamete*, Greek, "marry"; *phyte*, Greek, "plant"). The union of gametes produces a diploid zygote, which may grow mitotically into a diploid plant.

At maturity some of the diploid plant's cells undergo meiosis, and the resultant haploid cells are spores. This phase is the *sporophyte generation*. Since the haploid gametophyte, or sexual phase, is followed by the diploid sporophyte asexual phase and then by the gametophyte again, the whole phenomenon is called "*alternation of generations.*"

The moss plants usually seen on earth or rocks are sexual plants. Flask-shaped <u>archegonia</u> produce eggs on this moss, and egg-shaped <u>antheridia</u> produce sperm, all by mitotic nuclear divisions, that is,

15.14

Reproduction in a mushroom. The layers of tissue under the cap of the mushroom are the gills (shown enlarged in the inset), which bear the spore-producing cells.

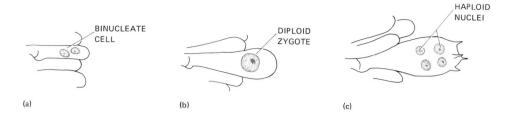

(a)   BINUCLEATE CELL   (b) DIPLOID ZYGOTE   (c) HAPLOID NUCLEI

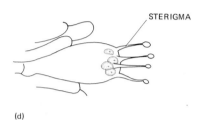

(d) STERIGMA

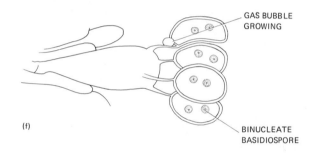

(e) UNINUCLEATE BASIDIOSPORE — MATURE BASIDIUM

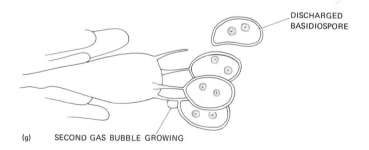

(f) GAS BUBBLE GROWING — BINUCLEATE BASIDIOSPORE

DISCHARGED BASIDIOSPORE

(g) SECOND GAS BUBBLE GROWING

15.15

Spore production in a mushroom. (a) The cells of a mushroom have two nuclei each. One binucleate cell is shown. The nuclei fuse. This is the act of mating. (b) The result of the fusion is a diploid zygote nucleus (2N). (c) Meiotic or reduction divisions follow, resulting in four haploid nuclei. Four slender processes grow from the end of the cell, which can now be called a *basidium*. (d) Spores begin to swell at the tips of the sterigmata, and a nucleus moves into each one. (e) The basidiospores grow to full size. (f) When complete, the spores are ready to be discharged. A bubble (or droplet?) oozes out at the base of each spore. (g) When the bubble reaches a critical size, it pops the spore free, catapulting it into space. × 2000.

15.16

Generalized scheme of *alternation of generations*. The diploid sporophyte generation begins with fertilization and the formation of a zygote, and it ends with meiosis at the time of spore production. The haploid gametophyte generation begins when meiosis results in spore production, and it lasts until egg and sperm meet in fertilization.

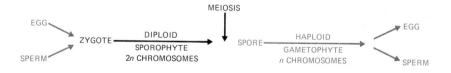

EGG / SPERM → ZYGOTE → DIPLOID SPOROPHYTE 2n CHROMOSOMES → MEIOSIS → SPORE → HAPLOID GAMETOPHYTE n CHROMOSOMES → EGG / SPERM

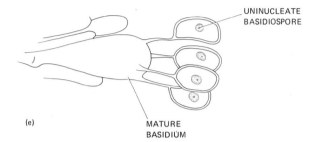

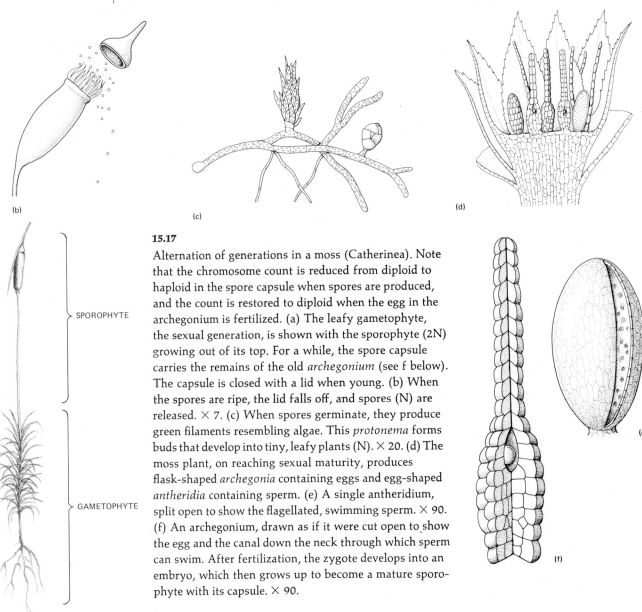

(b)

(c)

(d)

SPOROPHYTE

GAMETOPHYTE

(a)

15.17

Alternation of generations in a moss (Catherinea). Note that the chromosome count is reduced from diploid to haploid in the spore capsule when spores are produced, and the count is restored to diploid when the egg in the archegonium is fertilized. (a) The leafy gametophyte, the sexual generation, is shown with the sporophyte (2N) growing out of its top. For a while, the spore capsule carries the remains of the old *archegonium* (see f below). The capsule is closed with a lid when young. (b) When the spores are ripe, the lid falls off, and spores (N) are released. × 7. (c) When spores germinate, they produce green filaments resembling algae. This *protonema* forms buds that develop into tiny, leafy plants (N). × 20. (d) The moss plant, on reaching sexual maturity, produces flask-shaped *archegonia* containing eggs and egg-shaped *antheridia* containing sperm. (e) A single antheridium, split open to show the flagellated, swimming sperm. × 90. (f) An archegonium, drawn as if it were cut open to show the egg and the canal down the neck through which sperm can swim. After fertilization, the zygote develops into an embryo, which then grows up to become a mature sporophyte with its capsule. × 90.

(e)

(f)

without meiosis (Figure 15.17). Since the sperm are motile, they can swim from the antheridia to the archegonia, where fertilization occurs. The diploid zygote grows first into a few-celled embryo inside the archegonium. This new sporophyte bursts the archegonial walls to become a long, thin body with its lower end embedded in the old gametophyte and with a swollen bulbous capsule at the upper end. In the capsule, some cells divide meiotically, yielding quartets of haploid cells that become spores. The haploid spores are released from the capsule, and if they fall into a proper environment, they germinate into new haploid gametophyte moss plants.

Mosses are the simplest living plants in which the gametes are produced in special multicellular organs,

the antheridia and archegonia. Algae and fungi, in contrast, produce their gametes inside single cells. The mosses are also notable in that they nourish embryonic sporophyte plants in the archegonia. That arrangement or some other that is similarly protective is typical of all plants from mosses on up through the evolutionary scale of plants.

In mosses the gametophyte is the larger and more actively photosynthetic generation, but in ferns the arrangement is reversed. A haploid fern spore germinates to develop into a flat, heart-shaped plant about five millimeters across, with rootlike hairs growing into the soil, archegonia with eggs, and antheridia with sperm (Figure 15.18). The haploid motile sperm can swim from its antheridium to an

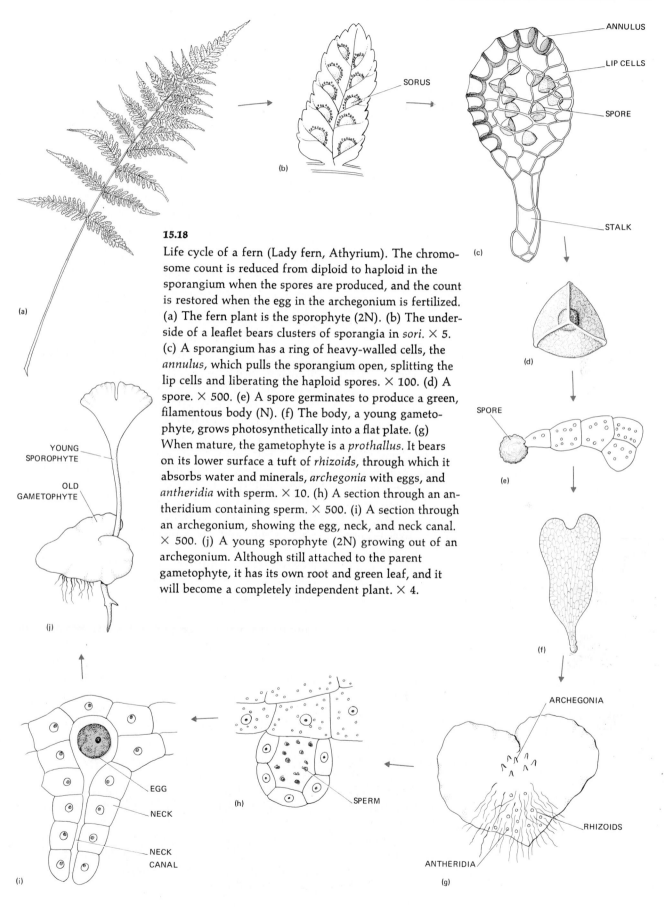

15.18

Life cycle of a fern (Lady fern, Athyrium). The chromosome count is reduced from diploid to haploid in the sporangium when the spores are produced, and the count is restored when the egg in the archegonium is fertilized. (a) The fern plant is the sporophyte (2N). (b) The underside of a leaflet bears clusters of sporangia in *sori.* × 5. (c) A sporangium has a ring of heavy-walled cells, the *annulus,* which pulls the sporangium open, splitting the lip cells and liberating the haploid spores. × 100. (d) A spore. × 500. (e) A spore germinates to produce a green, filamentous body (N). (f) The body, a young gametophyte, grows photosynthetically into a flat plate. (g) When mature, the gametophyte is a *prothallus.* It bears on its lower surface a tuft of *rhizoids,* through which it absorbs water and minerals, *archegonia* with eggs, and *antheridia* with sperm. × 10. (h) A section through an antheridium containing sperm. × 500. (i) A section through an archegonium, showing the egg, neck, and neck canal. × 500. (j) A young sporophyte (2N) growing out of an archegonium. Although still attached to the parent gametophyte, it has its own root and green leaf, and it will become a completely independent plant. × 4.

(a)   (b)

15.19

Comparison of haploid and diploid mitoses. In the haploid mitosis (a), characteristic of the gametophyte generation of a plant, there is only one representative of each chromosome type; in this species there are six types. In the diploid mitosis (b), there are two representatives of each chromosome type, as there would be in the sporophyte generation of the plant. By counting chromosomes, one can determine where the meiotic divisions occur in the life of a species.

archegonium and there fertilize a haploid egg. The resultant diploid zygote first grows into an embryo, then sends a root into the soil, and later sends an upward-thrusting leaf into the air. This is the young stage of the familiar fern plant, and it grows to become a mature fern plant. At maturity, fern leaves produce spore capsules, known as *sporangia* (singular, sporangium), in which meiosis occurs, followed by the rounding up of quartets of haploid spores. After release from a sporangium, a spore can germinate to yield a new fern gametophyte.

SPORE PRODUCTION

By the time plants had developed the habit of reproducing by means of seeds, they had increased the complexity and relative size of their diploid asexual sporophyte stage. They had also reduced the complexity and relative size of their haploid gametophyte sexual stage so much that biologists would probably not know what these stages mean in seed-bearing plants if it were not for the more readily understood models provided by the lower plants. Microscope observations of cells during all the phases of a plant's life reveal the time and place when the chromosome number is reduced by half, thus establishing when

meiosis takes place. Further investigation shows which structures are spores and where and when sexual union occurs. By comparing the chromosome counts and the manner of spore and gamete production in seed plants with those phenomena in ferns and mosses, one can learn how cones, flowers, fruits, and seeds function (Figure 15.19).

Heterospory

Ferns and fernlike plants make spores that, having undergone meiosis, grow into haploid gametophyte plants, usually capable of yielding both eggs and sperm. At many times and places, however, plants have evolved a method of producing two kinds of spores: those that germinate to form female gametophytes only or male gametophytes only. Since spores that are destined to produce female gametophytes usually are relatively large, they are called *megaspores* (Greek, "large seeds"). The smaller ones, which will develop into male gametophytes, are *microspores* (Greek, "small seeds"). The phenomenon of making two kinds of spores is *heterospory* (Greek, "different seeds"), in contrast to *homospory* (Greek, "same seeds"), in which only one kind of spore is produced. Heterospory is typical of seed-bearing plants.

In seed plants, spores are formed in a sporangium following meiosis, but they are not immediately released from the parent plant to begin an independent life as they would be in mosses and ferns. Instead, the spores begin developing while they are still in the sporangium, and the kind of development depends on the kind of spore. Megaspores are produced in megasporangia, which in seed plants are commonly called *ovules*. Megaspores develop into female gametophytes, which are haploid, egg-producing plants. They are typically nongreen, and they live parasitically on the parent sporophyte. Microspores are produced in microsporangia, which are called *anthers* in flowers. Microspores develop into male gametophytes. They, too, are parasitic, but they remain small, undergo only a few mitotic nuclear divisions, and when shed from the microsporangium are called *pollen grains*. When pollen grains are transported to the vicinity of an egg nucleus, they germinate, forming a long pollen tube that carries sperm nuclei directly to the egg nuclei.

The familiar seed plants of the world are *diploid sporophytes*. Since they produce haploid megaspores and microspores, they are essentially asexual. The sexual generation is the microscopic mega- or microgametophyte. When the gametophytes have produced eggs and sperm, and the resulting zygotes have begun

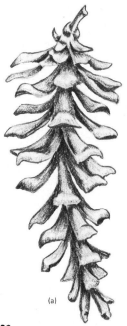

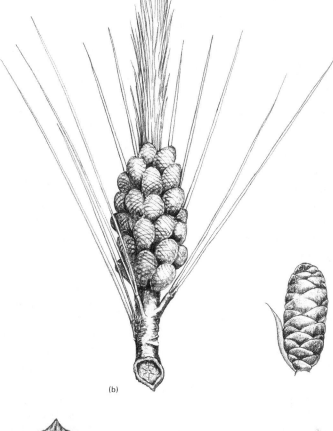

15.20

(a) An ovulate cone of white pine, *Pinus strobus.* × 0.8. (b) A cluster of staminate cones of pine, as seen in early spring, with old needles below and young needles at the end of the stem, natural size. One cone is shown enlarged, × 5.

(a)

(b)

15.21

The lower surface of a single scale from an ovulate cone, with two seeds loosely attached to it. The roundish lumps are the ripened ovules containing embryos, and the flat wing attached to the seed helps in wind dispersal. × 2.

to be embryos, new sporophyte generations have begun. These, with their surrounding tissues, are *seeds*. The embryo sporophyte in a seed can be thought of as the grandchild of the plant that bears it because the immediate parents of the embryo were the gametophytes.

Conifers

Coniferous plants are seed plants, but they have no flowers, and they are less advanced in an evolutionary sense than flowering plants. They are woody, and most bear needle-shaped evergreen leaves.

Pine trees are representative conifers, showing one common method of reproduction. They produce two kinds of cones: ovulate ones, which are the conspicuous familiar pine cones, and the small staminate cones, which appear for a few weeks in spring (Figure 15.20).

The ovulate or female cone, produced at the apex of a branch, is a cluster of scales arranged spirally around a central axis. Each scale, probably a specialized branch, bears on its upper surface a pair of *ovules* (Figure 15.21). An ovule is functionally a megasporangium covered by an integument, or seed coat. In the tissue of the ovule, one cell is destined to undergo meiosis. This is the *megaspore mother cell*, and it is recognizably larger than the surrounding cells. As is typical of spore formation, this megaspore mother cell undergoes meiosis, producing four haploid megaspores, three of which abort. The surviving

15.22

A long section through a young ovulate cone. × 12. The insert shows a portion of a single cone scale. × 50. Under each cone scale is a sterile bract. On the upper surface of the cone scale there is a pair of ovules, one of which is shown sectioned. The ovule is covered with *integuments*, part of the cone, which will form the seed coats. The female gametophyte (haploid), having grown from a megaspore, has two archegonia, each with an egg. An opening in the integuments, the *micropyle*, allows a pollen grain to come in close to the eggs. The pollen grain germinates, sending a pollen tube into the ovule, where haploid sperm nuclei are deposited in the archegonium and the haploid eggs are fertilized. The diploid embryo then develops in the ovule.

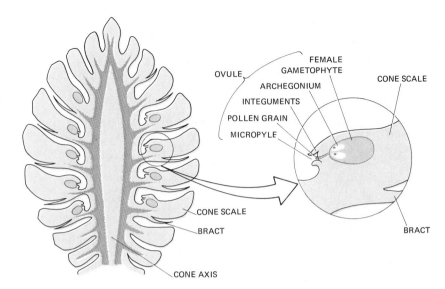

15.23

A long section through a staminate cone. The mature pollen sacs are shown full of mature pollen grains. × 25.

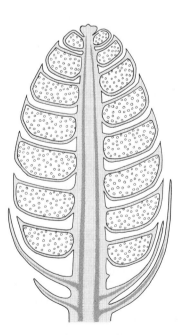

one follows the pattern of any megaspore: It grows by mitotic divisions into a haploid female gametophyte. When mature, the female gametophyte is a tiny mass of tissue with several archegonia (Figure 15.22). The archegonia are located at the end of the gametophyte nearest the *micropyle*, an opening through the integument. They are not such definitely constructed organs as those of mosses; rather, they consist only of a vesicle with a few neck cells and a conspicuous egg nucleus. The whole pine archegonium is recognizable as such only by comparison with that of lower plants.

In early spring, while the ovulate cones are being formed, staminate cones are growing on other branches. A staminate cone, less than a centimeter long, has a series of thin scales in a spiral around its central axis (Figure 15.23). Each scale bears on its lower side a pair of microsporangia that produce hundreds of diploid microspore mother cells, each of which divides meiotically into four haploid microspores. As microspores normally do, they grow mitotically into male gametophytes, but even when complete, they remain microscopic.

A microspore of a pine produces by mitosis two nonfunctional cells called prothallial cells, which are presumably remnants of a once-prominent vegetative gametophyte. The microsporangia break open and shed the *pollen grains*, which are incomplete male gametophytes, containing only the two vestigial prothallial cells and two functional nuclei. These are a tube nucleus, and a second nucleus, which will later divide to form two sperm nuclei (Figure 15.24).

Pollen drifts in air to the young ovulate cones, which open their scales and temporarily allow pollen to come close to the micropyles (Figure 15.25). Pollen lies quiet in the ovulate cone for a year until the eggs in the archegonia mature. In the second spring in the life of the ovulate cone, the pollen grains send out a germ tube containing sperm nuclei, one of which enters an archegonium to fertilize an egg nucleus. From the resulting zygote there grows an embryo with an axis and several embryonic leaves, the *cotyledons*.

The embryo is contained in and nourished by the female gametophyte.

The complete pine seed consists of the embryo embedded in the old female gametophyte, which is in turn surrounded by the integument, or seed coat (Figure 15.26). When the seed is mature at the end of the second summer, the cone scales spread, releasing the seed along with a piece of the scale, which peels off, giving the seed a wing that helps it drift in moving air.

15.24

A mature pine pollen grain, with conspicuous extensions or "wings." × 325.

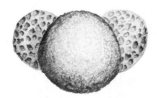

15.25

A germinated pine pollen grain. This is the complete male gametophyte (haploid). It has a pair of crushed *prothallial cells,* remindful of a time when the gametophyte was a more complex generation than it is now. It also has the functional parts: the *stalk cell,* which will disintegrate; the *body cell,* which will form two sperm nuclei; the *tube nucleus;* and the *pollen tube,* which will penetrate into the ovule. × 250.

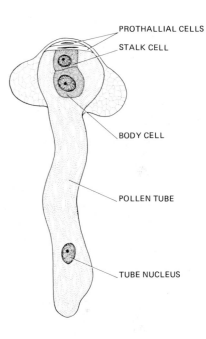

PROTHALLIAL CELLS

STALK CELL

BODY CELL

POLLEN TUBE

TUBE NUCLEUS

15.26

Pine seed and seedlings. (a) A long section through a pine seed. The embryo is nourished by the *endosperm,* which is the old haploid female gametophyte. The diploid embryo is a small sporophyte, with all its main parts established: an axis with a growing point at each end and a ring of seed leaves, or *cotyledons.* × 15. (b) When the seed germinates, the radicle

of the embryo stretches, and new cells at the tip contribute to the growing root. As the *hypocotyl* elongates, it drags the seed above ground. × 4. (c) When the top of the seedling reaches light, the seed coats and old gametophyte fall off, the hypocotyl straightens, and the cotyledons expand. The little pine tree is established and ready to grow on its own. × 4.

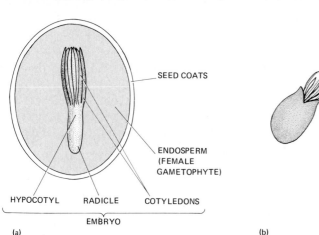

SEED COATS

ENDOSPERM (FEMALE GAMETOPHYTE)

HYPOCOTYL   RADICLE   COTYLEDONS

EMBRYO

(a)

(b)

(c)

Flowering Plants

Flowering plants are the most recent and most complex plants on earth. Since the first flowers on earth apparently were never fossilized, the transition forms that led to modern flowers are not known. One possibility is that flowers evolved in rather dry uplands, where conditions were not favorable for preservation, with the result that the links are indeed missing. Once having developed, flowering plants came to dominate the land, and they are now the most conspicuous plants on the surface of the earth.

FLOWERS: STRUCTURE AND VARIATION

A complete flower has all floral parts present: the usually green *sepals* under the usually colored *petals*, plus the reproductive parts (Figure 15.27). A ring of sepals makes a *calyx*, and a ring of petals makes a *corolla*. In most flowers there is a ring of *stamens* inside the ring of petals, and each stamen has a pollen-bearing *anther* at the end of a supporting filament. In the central part of the flower, a *pistil* containing *ovules* carries a columnar *style*, at the top of which is a *stigma*, which receives pollen.

"Textbook flowers," as described above, seldom occur, although poppies and wild roses approach the theoretical type. Most real flowers vary in one or more details, and every imaginable possibility can be found somewhere in nature (Figure 15.28). Sepals come in many shapes and sizes, and they may be green and leaflike, brightly colored and petal-like, or lacking. Petals may be large or small, colored or green, or missing entirely; they may be all alike, as in buttercups, or different within the same flower, as in orchids. Petals may be attached below the ovary or above it, and they may be individually produced, as in apples, or all fused into a single tube, as in honeysuckle. When petals and sepals are arranged with radial symmetry, as in a wheel, the flower is said to be regular (lilies), but if the symmetry is bilateral, the flower is irregular (mints).

Reproductive parts of flowers, too, may be lacking. An *imperfect flower* lacks one of the reproductive organs; that is, such a flower is either *staminate* or *pistillate*. Many trees (hollies and willows) have imperfect flowers. (If petals or sepals are lacking the flower is simply called *incomplete.*) Stamens and pistils may be few or many, arranged in a spiral or a whorl, separate or grown together in all possible combinations. Ovaries may contain one or few or many seeds, and the tissues to which the seeds are

15.27

A section through a representative flower. The stem end bearing the flower is the *receptacle*. The sepals, usually green, collectively make the *calyx*. The petals, frequently brightly colored, collectively make the *corolla*. The *pistil*, containing the *ovules*, has a more or less elongated *style*, topped by a sticky *stigma* on which pollen lands and germinates.

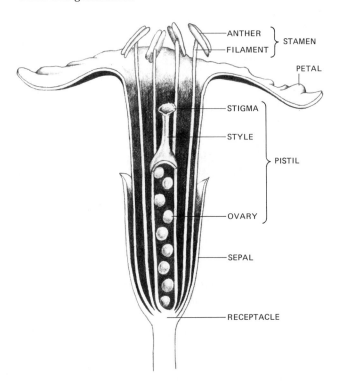

attached may be around the periphery of the ovary or at the top or bottom or along a central shaft.

In size, flowers range from the almost microscopic blooms of the floating Wolffias to the great, heavy magnolias (see Figure 12.1). They may be solitary, as in tulips, or collected into *inflorescences* that contain a few (cherries, for example) to hundreds or even thousands of individual flowers in a single head (sunflowers).

Spore Production in Flowers

A flower is not primarily a sexual structure, nor are pollen grains comparable to sperm cells. The main

15.28

Representatives of diverse families of flowering plants. (a) A jack-in-the-pulpit (Arum Family). (b) Oats (Grass Family). (c) Apple flowers (Rose Family). (d) Dog-tooth violet (Lily Family).

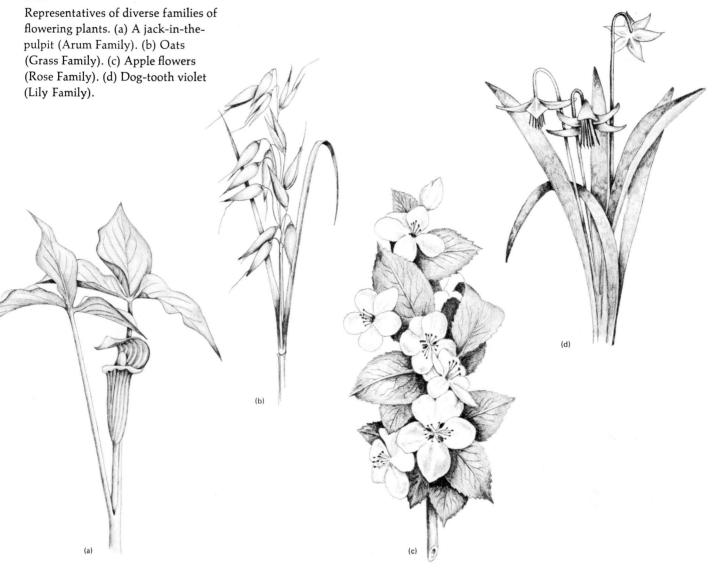

(a)

(b)

(c)

(d)

plant, the part that is conspicuous and does the main metabolic work, is the diploid asexual sporophyte, and the primary function of the flower is to produce spores. The mechanism of spore production is basically similar in all flowers, although enormous variation occurs in the external appearance of different families of flowering plants.

Like conifers, flowers make two kinds of spores: *microspores*, which can develop into male gametophytes, and *megaspores*, which can develop into female gametophytes. Megaspores are produced in ovules. Ovaries of flowers are not truly comparable to the ovaries of animals, because they do not produce

eggs following meiosis. The ovule in an ovary of a flower is a sporangium, in which meiosis results in spores, specifically four megaspores. One of these haploid megaspores grows into a female gametophyte inside the ovule, but even at maturity it is microscopic and hidden from external view. Since the ovules are enclosed within the ovaries, the entire group of flowering plants is called the *angiosperms*, which means "seeds in a vessel."

In most flowers, the megaspores divide mitotically, and this division is followed by two more mitoses, so that the mature female gametophyte finally comes to have eight nuclei. It is then called an *em-*

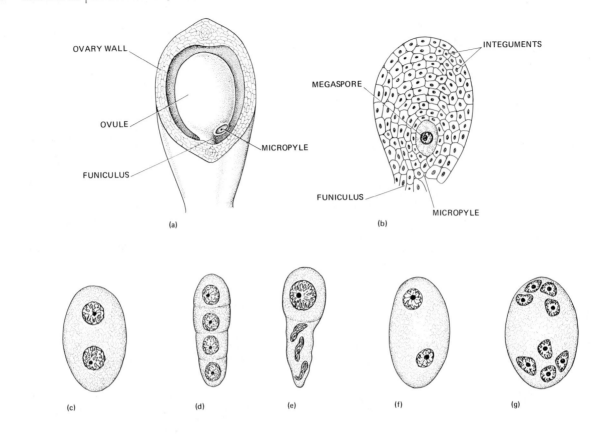

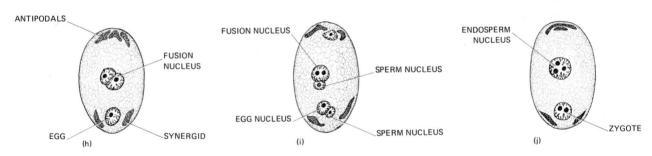

15.29

Development of the embryo sac in the ovule of a flower. (a) A sectional view into a cut ovary, showing one ovule. (b) A section through an ovule, attached to its placenta by the *funiculus.* The *micropyle*, through which the pollen tube will pass later, and the *integuments,* which will form the seed coats, are visible but (as in real material) not conspicuous. The largest cell in the ovule is the diploid *megaspore mother cell.* The megaspore mother cell undergoes the first of the meiotic divisions, resulting in a binucleate cell (c). The second meiotic division results in a row of four haploid megaspores (d). One megaspore survives; the other three die (e). The surviving megaspore nucleus divides mitotically to form a binucleate cell (f). Two more mitotic divisions produce an eight-nucleate *embryo sac* (g). Two nuclei, one from each end of the embryo sac, fuse to make a *fusion nucleus.* One nucleus near the micropyle becomes the egg. The other nuclei (three at one end of the embryo sac and two near the egg) disintegrate. The embryo sac is ready for fertilization (h). The pollen grain has deposited two sperm nuclei in the embryo sac (i). One is fusing with the egg nucleus and the other with the (now diploid) fusion nucleus. The act of *double fertilization* is a special feature of flowering plant reproduction. In (j) a haploid sperm and a haploid egg fuse to make a diploid *zygote,* which is ready to develop into an embryo. The fusion of a haploid sperm with a diploid fusion nucleus makes a *triploid endosperm nucleus*, which will produce the triploid endosperm tissue, a nourishing mass surrounding the developing embryo.

bryo sac (Figure 15.29). Three of the eight nuclei are of special importance: one that will be the egg nucleus, or female gamete, and two that fuse to make a new kind of nucleus, the diploid *fusion nucleus,* which is to function during the development of the seed. At this stage the embryo sac is complete and ready for fertilization.

While the ovules are producing megaspores and eventually embryo sacs, the anthers are in the process of forming pollen. An anther when young usually has four elongated lobes, each filled with *sporogenous tissue* composed of diploid cells. Each diploid cell is a *microspore mother cell.* When these cells undergo meiosis, each microspore mother cell yields four haploid microspores. The microspores, which round up into almost spherical free-floating cells, represent the first stage of a short-lived gametophyte. The microspore nuclei divide mitotically to form a binucleate cell that is soon covered by a firm double coating. This is the pollen grain, and at this stage, many plants split open their anthers to release the pollen. Pollen grains are variously shaped and covered with such distinctive markings that they can be identified, frequently with specific accuracy (Figure 15.30). Once the anthers are open, the pollen is ready to be carried to the stigma of the flower.

Pollination

Pollination is not fertilization; it is only the movement of pollen from anther to stigma. Actual fertilization occurs later. Pollination, however, must precede fertilization, and it is a requirement for seed setting. Flowers have evolved many methods of ensuring pollination, some of them complicated and involving structural and behavioral modifications of the pollinating agent. For example, flowers pollinated by hummingbirds produce sweet nectar in glands deep in the throat of the flower, and hummingbirds have the hovering flight and the long, thin beaks needed for obtaining the nectar (Figure 15.31). Each species needs the other.

(a)   × 4000

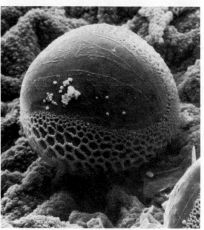

(b)   × 1300

15.30

Scanning electron micrographs of pollen grains. (a) Pollen of Cosmos, a member of the Composite Family, which usually has rough, spiny pollen. (b) Pollen of a lily, belonging to the monocotyledons, which usually have rather smooth, rounded grains.

15.31

Flowers pollinated by hummingbirds usually have nectar deep in long, narrow floral tubes where most animals cannot reach it. The long, slender beaks of the hummingbirds are well suited to obtaining the nectar, and as the birds feed, they also pollinate the flowers.

15.32

Pollination of the sage, Salvia. (a) A bee, entering the flower for the nectar at the base of the tube, runs into an anther blocking the tube entrance. The swiveled anther can seesaw like a rocker, so when the bee's head strikes the lower end, the upper end is pushed down against her back, dusting it with pollen, as shown in (b). Note that the stigma is at first high and closed. Later the style elongates and the stigma spreads, as in (c). When a bee, with her head and back loaded with pollen, enters an older flower, she brushes her back against the opened stigma, rubbing pollen onto it. Not only is pollination accomplished, but self-pollination is avoided, and cross-pollination is practically ensured.

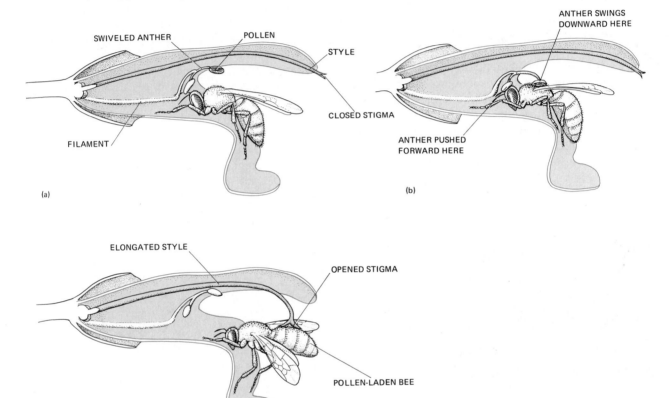

Most flowers, perhaps the earliest insect-pollinated ones, are pollinated by beetles, but other insects are important. Each species is likely to be adapted to the visits of some particular insect, whether ants, bees, wasps, butterflies, or moths.

Several instances are known in which the evolution of both flower and pollinator has resulted in carefully controlled mutual benefits. In wild sage, Salvia, hinged anthers with movable levers dust the hairy backs of feeding bees, and the downward pointing stigmas catch pollen as bees enter the blossoms (Figure 15.32). Yucca flowers provide a hatching and feeding ground in their ovaries for yucca moths, in compensation for which the egg-laying female moth collects pollen and applies it to the inverted cuplike stigma (Figure 15.33). Both the insect and the plant benefit from the association, and indeed neither could survive without the other. Some orchids (Ophrys) have flowers that resemble female bees both in appearance and odor. Male bees, in seeming to attempt mating with a female bee, will bring about the pollination of the orchid plant (Figure 15.34).

It cannot be inferred that either the plant or the pollinator has evolved such elaborate symbioses by an effort of will or intelligent design. There is no evidence that any biological change has ever been so

15.33

Pollination in the rock lily, Yucca. The flower has its stigmatic surface on the underside of a concave cup, which could hardly receive pollen by wind or gravity. The pronuba moth has special palps for the handling of pollen. She drills several holes in the flower ovary and lays eggs among the ovules. Then she gathers balls of pollen and crams them up into the concave stigmatic surface. This action ensures fertilization of the yucca eggs and provides ripening ovules for moth larvae to eat. After the larvae are grown, they eat their way out of the ovary, let themselves down to earth by a spun thread, and pass their pupal stages below ground. The following summer, the adult moths emerge in time for the flowering of the yucca, and the story is repeated. Everything must be integrated: the structural adaptations of moth and flower, the behavior of the moth, and the timing of both organisms. This is coevolution and mutual dependence.

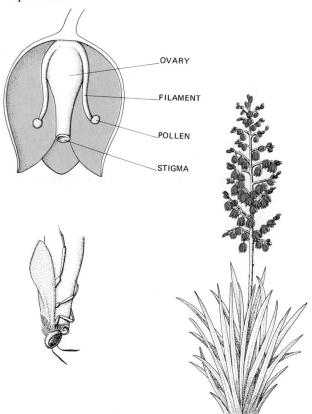

15.34

The fly orchid, Ophrys, has flowers that look and smell so like a female bee that male bees try to mate with them. This *pseudocopulation* effects pollination. × 3.

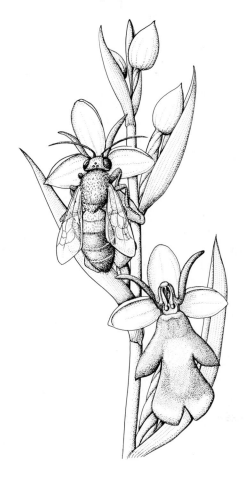

made, regardless of our human habit of ascribing purposeful action to plants and animals. The explanation most satisfactory to contemporary biologists is that adaptations that happen to function are selected in the competition for survival, even when the adaptations lead to relationships so complex that they seem humanly intelligent.

One aspect of pollination is its side effect on humans: It causes hay fever in susceptible victims. Wind-pollinated plants yield great quantities of light, air-borne pollen, which can be inhaled and can cause human allergies. Ragweeds, grasses, conifers, and catkin-bearing trees are the most troublesome (Fig-

FRUITS WITHOUT SEEDS

Bananas do not have seeds; at least those commonly sold in markets do not. Neither do pineapples or some varieties of grapes, oranges, and grapefruit. Where do such fruits come from? And how is it possible to produce them in commercial quantities?

The answer to the first question is: They come from seeded parents, which underwent some mutation that prevented ripening of the ovules but allowed the development of ovary walls and accessory parts. Most fruits abort when seeds are not formed, but some, known as *parthenocarpic* (Greek, "virgin fruit"), grow almost normally even without seeds. Most parthenocarpic fruits sold commercially were found by chance, appreciated by their discoverers, and saved from extinction by human interference.

Once obtained, a seedless fruit may be propagated in several ways. Seedless grapes and citrus fruits, such as seedless oranges, are propagated by grafting seedless branches on ordinary rootstocks. After a few plants begin growing and branching, practically unlimited numbers of grafted seedless plants can be produced, so that extensive vineyards or orange groves are possible.

Bananas and pineapples are propagated by cutting away and transplanting side shoots.

Even seedless fruits have ovules, but they do not mature. Ovules can be seen as black flecks in bananas or as pinpoint-sized white granules in pineapples. In some varieties of both fruits, real seeds do mature. Banana seeds are shiny, brown, and flattened, like small beans. Pineapple seeds, when ripened, look like apple seeds. Seed-bearing individuals may be used in plant-breeding stations to yield new varieties, adding to the stock of variability and offering a chance to combine mutations in novel, possibly desirable, fruits. In this way, many seedless varieties can be produced.

15.35
Wind-pollinated flowers. (a) The *catkins* of the aspen (Populus) hang free in the air and liberate masses of pollen in early spring. (b) Ragweed (Ambrosia) grows in waste places, producing great quantities of pollen from the small, green flowers. Ragweed is one of the worst hay fever plants in the northern temperate zone. In general, plants that are wind-pollinated have voluminous, light, waxy pollen. × 0.5.

(a)

(b)

ure 15.35). In general, such plants have small, inconspicuous flowers. Brightly colored flowers like goldenrods, even though they are popularly accused of causing hay fever, are insect-pollinated and do not make enough pollen to affect people.

Seed and Fruit Formation

When a pollen grain arrives on a receptive stigma, the secretion of the stigma stimulates it to send a slender germ tube out through one of the pores in the

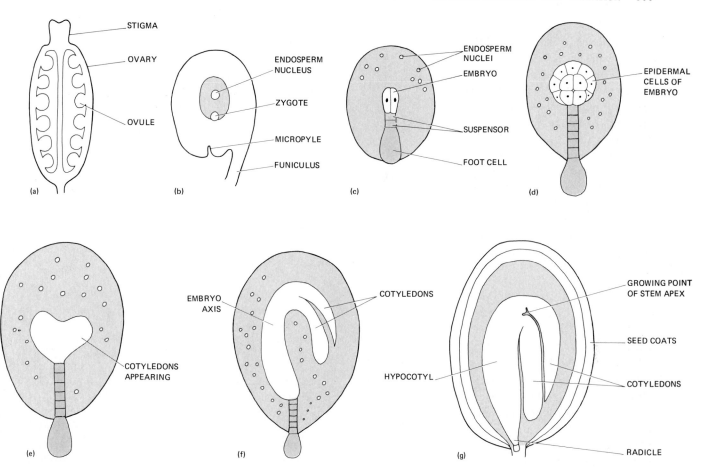

15.36

Development of a seed. (a) A long section of an ovary with young ovules. (b) An ovule, sectioned longitudinally. Fertilization has just taken place. There is a diploid zygote and a triploid endosperm nucleus. (c) The endosperm nuclei are dividing and making a loose mass of tissue around the growing embryo. The embryo has one enlarged cell at the base (the *foot cell*), a short row of cells (the *suspensor*), and the embryo proper, which at this stage has only two cells. (d) Further cell divisions in the embryo have produced an outer layer of cells that will form the epidermis of the future plant. (e) The endosperm continues to grow into a diffuse covering around the embryo, which has grown into a heart-shaped mass, indicating the future cotyledons. (f) Although there is still some endosperm, most of the space in the ovule is taken up by the growing embryo, which has an axis and two cotyledons that have been bent around. (g) The mature seed has seed coats, derived from the integuments of the ovule. The endosperm has been absorbed, and the embryo practically fills the entire seed. The axis of the embryo consists of a *hypocotyl* with a root growing point (the *radicle*) at the lower end and a large pair of cotyledons filling the ovule. An apical growing point between the cotyledons will make the stem apex of the future plant.

grain coat. This is the *pollen tube,* which grows down through the style of the flower, guided by unknown forces, until it comes to an ovule. There it finds its way into the ovule and penetrates the embryo sac. By this time, pollen tubes generally have three nuclei: a tube nucleus, which has no further function (if indeed it ever had one), and two sperm nuclei. In most flowers, the nucleus that results from the fusion of the fusion nucleus and the second sperm nucleus is a triploid nucleus, that is, with three sets of chromosomes. The union of one sperm nucleus with the fusion nucleus and the other with the egg constitutes the phenomenon of *double fertilization,* which is characteristic of flowering plants but of no other plant group.

After double fertilization has occurred, the zygote nucleus and the endosperm nucleus divide repeatedly. From the zygote comes the embryo, an immature plant of the new sporophyte generation (Figure 15.36). An embryo in a seed, even when the

15.37

Sections through seeds of two of the world's most important food plants. (a) Corn, or maize (Zea). The most conspicuous part is the starchy endosperm. Its outermost cells are heavily impregnated with protein. The embryo, too, contains protein, as well as quantities of fat, the source of corn oil. (b) Grain of wheat, generally similar to that of corn, but with a relatively larger embryo. The endosperm is the source of wheat flour. Note that a grain is not simply a seed, but a complete fruit covered by the ovary wall.

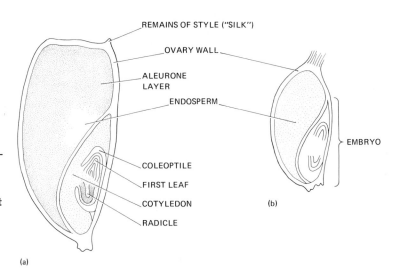

REMAINS OF STYLE ("SILK")
OVARY WALL
ALEURONE LAYER
ENDOSPERM
EMBRYO
COLEOPTILE
FIRST LEAF
COTYLEDON
RADICLE

(a)

(b)

seed is ripe, is an incompletely formed plant, but it usually has a rudimentary stem with an apical growing point at the "upper" end and a root growing point at the other, along with one or two special structures called seed leaves, or *cotyledons*. Meanwhile, the endosperm nucleus produces a formless mass of triploid tissue, the *endosperm*. The endosperm functions as a nourishing tissue for the growing embryo, sometimes becoming larger than the embryo itself, as in coconuts or the cereal grains. However, the endosperm may be used up by the time the seed is mature, so that the finished seed contains an embryo but no endosperm. Seeds of peanuts and beans are examples of seeds without endosperm.

Since the bulk of the world's food supply consists of flour or meal from wheat, rice, corn, millet, barley, and oats, and since the bulk of each of these grains is made of endosperm tissue, it is evident that humans live largely on triploid tissue (Figure 15.37). In all these grains, the embryo is a relatively small object, popularly known as the "germ," and it is usually removed from the grain during the process of making flour or when the grain is being finished for commercial sale.

As the endosperm and embryo grow, so do the integuments of the ovule and the walls of the ovary. At maturity the ovule has become a seed, the integuments have become seed coats, and the ovary wall has become the outer covering of the fruit. A fruit is in fact a ripened ovary, although it may include some nonovarian tissue in some plants (Figure 15.38). A

pea pod is simply an ovary with seeds in it, but apples and strawberries are quite different. The ovary of an apple flower is only the core, and the edible part of the apple consists essentially of stem. In a strawberry, the ovaries are what are usually called "seeds," and the juicy part of the berry is, once more, a swollen, sweet, juicy stem.

Remember that the embryo in a seed is not the direct offspring of the plant that bears it. It is rather the offspring of a pair of gametophytes contained in the parent plant, or more accurately, the grandparent plant. A seed contains tissues of three generations: (1) The seed coats belong to the original plant. (2) The pollen tube and embryo sac belong to the next, or gametophyte, generation although by the time a seed is mature, those two parts are usually destroyed. (3) The embryo belongs to the new sporophyte generation (Figure 15.39).

A seed is essentially an embryo in a protective coat, waiting in suspended animation until proper warmth, moisture, and sometimes sufficient aging make germination possible. Seed production is an effective way for plants to wait out cold or dry periods safely. The higher animals have no such dormancy period; when a mammalian embryo starts life, it must go ahead and be born and grow up. The only alternative is death.

Flowering plant seeds are of two possible kinds: those with one embryonic leaf (*monocotyledons*, such as grasses, lilies, and palms) and those with two embryonic leaves (*dicotyledons*, such as beans, walnuts,

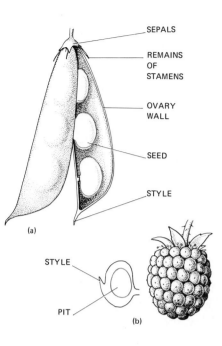

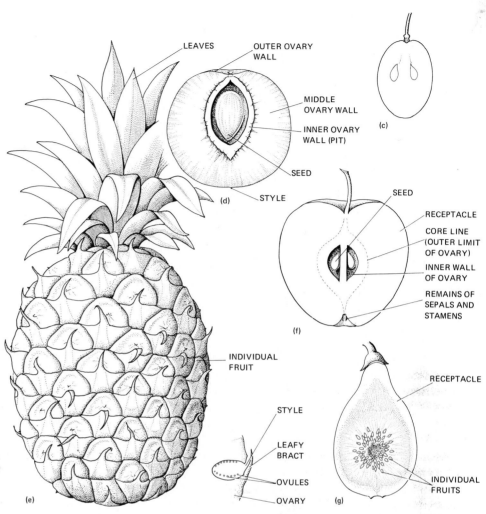

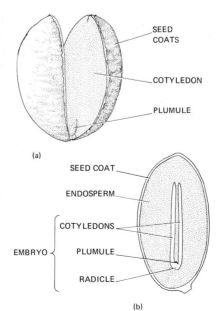

15.38

Representative fruits. (a) A *simple* fruit, developed from a single *carpel*. Lima bean (Phaseolus). (b) *Aggregate* fruit of blackberry (Rubus), with a number of *drupelets*, each developed from a separate ovary. A single drupelet is shown in section with its pit inside. (c) *Berry* of grape (Vitis). The fleshy part is ovary wall, and the seed coats are hard. (d) *Drupe* of peach (Prunus). The skin is from ovarian epidermis, the edible part is the middle ovary wall, and the hard pit is the inner ovary wall. The seed, covered with brown seed coats, is inside the pit. Almonds, nectarines, cherries, plums, and prunes are similar. (e) *Multiple fruit* of pineapple (Ananas). This is a mass of fruits, each derived from a separate flower. A single flower is shown in section, with tiny aborted ovules inside. (f) An apple (Pyrus) is mostly receptacle (stem) tissue. The ovary makes the core of the apple, and the papery inner layer of the ovary covers the seeds. (g) A fig (Ficus) is a special case of a multiple fruit, with the so-called seeds, structurally individual fruits, each derived from a separate flower.

15.39

Two types of seeds. (a) Seed of peanut (Arachis), with brown, papery seed coat, a small embryo axis with a growing point (*plumule*), and large cotyledons, which practically fill the seed coat. The endosperm, which was temporarily present when the ovule was developing, is completely absorbed by the time the seed is mature. (b) A castor oil "bean" (Ricinus), with hard seed coats, abundant endosperm, and thin, leafy cotyledons, shown edge on. The bulk of the stored food is in the endosperm.

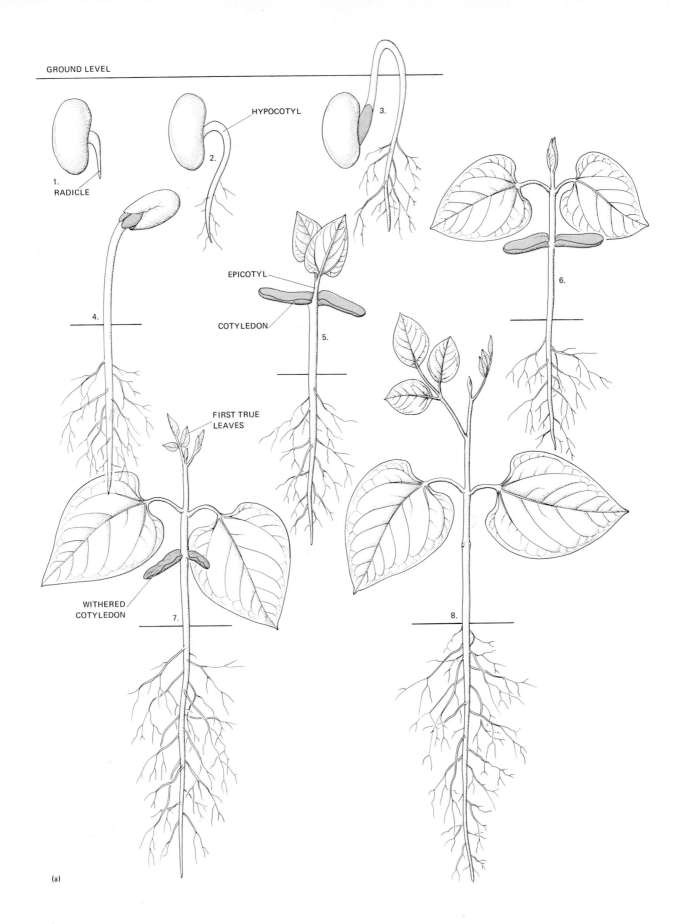

GROUND LEVEL

1.
RADICLE

HYPOCOTYL

2.

3.

4.

EPICOTYL

COTYLEDON

5.

6.

FIRST TRUE
LEAVES

WITHERED
COTYLEDON

7.

8.

(a)

oaks) (Figure 15.40). When a seed sprouts, it first grows a root downward (Chapter 17) and then a shoot upward. Once the shoot reaches the light, it develops chlorophyll and begins manufacturing its own food, but until that time it depends for its energy and material supply on nutrients stored either in the embryo itself or in the endosperm.

The ability of plants to pack a rich supply of carbohydrates, fats, proteins, vitamins, and minerals into a small volume of a seed has been important to animals as well as to the seeds themselves. Animals find seeds such a nourishing, concentrated food supply that in many instances animal survival depends on an adequate seed crop. It is probably no coincidence that the evolution of the warm-blooded animals came only after the evolution of seeds, since the availability of a constant high-energy source is essential to furnishing the warmth that such animals require.

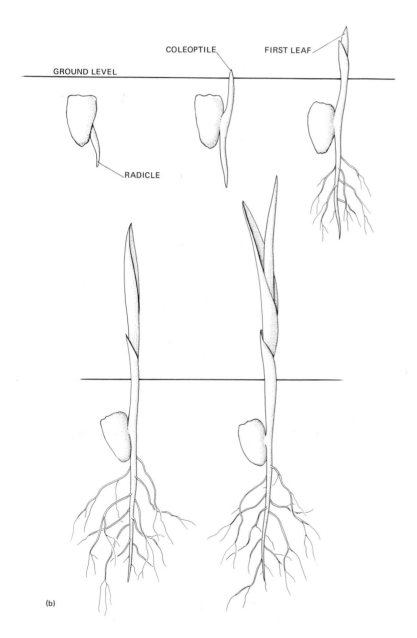

COLEOPTILE   FIRST LEAF

GROUND LEVEL

RADICLE

(b)

15.40

Germination of seeds. (a) Like all seeds, beans start germination by putting forth a root by extension of the *hypocotyl* and growth of its tip: the *radicle*. As the hypocotyl stretches, it arches up, shoves its way through the soil, and pulls the *cotyledons* along. Once into the light aboveground, the arch straightens, and the cotyledons expand. The growing point puts forth a pair of *juvenile leaves*, unlike any that will develop later, and then produces the first typical bean leaves. Meanwhile secondary roots branch from the first root, or primary root, and the plant is established. (b) Corn grains also start germination by sending out a root. The first part of the upward growing shoot to emerge is a slender, thimble-shaped *coleoptile* which covers the first leaf. The *cotyledon* remains inside the grain, acting as an absorbing organ that transports food from the bulky endosperm to the growing seedling. After the first leaf breaks through the coleoptile and uncurls, other leaves follow, and the seedling becomes independently photosynthetic. The root system consists not only of the primary and secondary roots but of *adventitious* roots, which sprout from the base of the stem and will eventually become the main root system of the mature plant.

SUMMARY

1. Most multicellular animals use strictly sexual reproduction, but plants commonly use both sexual and asexual methods, varying from species to species and occasionally within a single species.

2. Asexual reproduction results in a slower distribution of variation through a population than does sexual reproduction, but it has several advantages. It bypasses the need for an organism to find a mate, it can be exceedingly fast, and all the offspring are genetically identical, a situation that often is desirable.

3. The simplest example of asexual reproduction is *fission*, the division of a single cell and the separation of the two resultant cells.

4. Most plants produce special reproductive cells called *spores*, which do not involve nuclear fusion.

5. Higher plants may reproduce asexually, or *vegetatively*, through processes by which a new complete plant is produced from the roots, stems, or leaves of the parent plant, sometimes with human interference in cutting, grafting, or layering.

6. To guarantee biological uniformity, people use *clones*, or groups of individuals all having the same genetic background, obtained without an intervening sexual fusion.

7. The simplest sexual process is the joining of *any* two free-living cells, with cytoplasmic union followed by nuclear fusion. A slight advance over such a primitive procedure happens in some species in which certain individuals can mate *only* with compatible individuals. Populations of these two types of individuals are called *mating strains*. A species in which no separate mating strains exist is called *homothallic*; a species in which special mating strains do exist is *heterothallic*. The next step in sexual complexity came with a visible difference between the sex cells, a slight difference in size or pigmentation. Still another advance came with a change in motility. The larger sexual cell lost its ability to move while the smaller remained active. Another innovation was the development of multicellular bodies in which only a special portion of the organism is devoted to gamete production. The most recent complication was the provision for parental care of the new generation.

8. In the heterogeneous assemblage of plants popularly known as *algae*, reproductive methods are as various as the plants themselves.

9. Reproduction in *fungi* is as varied as it is in algae, and some fungi exhibit the most complex reproductive patterns to be found anywhere in nature.

10. Plants have a whole haploid phase during which gametes are produced without meiotic divisions; that phase is the *gametophyte generation*. The union of gametes produces a diploid zygote, which may grow mitotically into a diploid plant. At maturity, some of the diploid plant's cells undergo meiosis, and the resultant haploid cells are spores; this is the *sporophyte generation*. The whole phenomenon is called "*alternation of generations*."

11. *Mosses* are the simplest living plants in which the gametes are produced in special, multicellular organs, the *antheridia* and *archegonia*. Algae and fungi, in contrast, produce their gametes inside single cells. The mosses are also notable in that they nourish embryonic sporophyte plants in the archegonia.

12. In mosses the gametophyte is the larger and more actively photosynthetic generation, but in *ferns* the arrangement is reversed.

13. Spores that are destined to produce female gametophytes are usually relatively large, and they are therefore called *megaspores*. The smaller spores, which will develop into male gametophytes, are *microspores*. The phenomenon of making two kinds of spores, called *heterospory*, is typical of seed-bearing plants. Megaspores are produced in megasporangia, which in seed plants are called *ovules*. Microspores are produced in microsporangia, which are called *anthers*. When the gametophytes have produced eggs and sperm, and the resulting zygotes have begun to be embryos, new sporophyte generations have begun.

14. *Coniferous plants* are seed plants, but they have no flowers, and they are less advanced in an evolutionary sense than flowering plants. They are woody, and most bear needle-shaped evergreen leaves.

15. *Flowering plants* are the most recent and most complex plants on earth. Flowering plants are now the most conspicuous plants on the surface of the earth.

16. An ideal complete flower has all floral parts present: the usually green *sepals* under the usually colored *petals*, plus the reproductive parts. A ring of sepals makes a *calyx*, and a ring of petals makes a *corolla*. In most flowers there is a ring of *stamens* inside the ring of petals, and each stamen has a pollen-bearing *anther* at the end of a supporting *filament*. In the central part of the flower, a *pistil* containing *ovules* carries a columnar *style*, at the top of which is a *stigma*, which receives pollen.

17. A flower is not primarily a sexual structure, nor are pollen grains comparable to sperm cells. The main plant, the part that is conspicuous and does the main metabolic work, is the diploid asexual sporophyte, and the primary function of the flower is to produce spores. Like conifers, flowers make two kinds of

spores, microspores and megaspores. The entire group of flowering plants is called the *angiosperms*.

18. *Pollination* is not fertilization; it is only the movement of pollen from anther to stigma. Actual fertilization occurs later. Pollination, however, must precede fertilization, and it is a requirement for seed setting.

19. A seed contains tissues of three generations: The *seed coats* belong to the original plant, the *pollen tube* and *embryo sac* belong to the next, or gametophyte, generation, and the *embryo* belongs to the new sporophyte generation. A seed is essentially an embryo in a protective coat, waiting in suspended animation until proper warmth, moisture, and sometimes sufficient aging make germination possible.

20. Flowering plant seeds are of two possible kinds: those with one embryonic leaf, *monocotyledons,* and those with two embryonic leaves, *dicotyledons.*

RECOMMENDED READING

Bold, H. C., *The Plant Kingdom,* Fourth Edition. Englewood Cliffs, N.J.: Prentice-Hall, 1977.

Ditmer, Howard J., *Modern Plant Biology.* New York: Van Nostrand-Reinhold, 1971.

Echlin, Patrick, "Pollen." *Scientific American,* April 1968. (Offprint 1105)

Galston, A. W., *The Life of the Green Plant,* Third Edition. Englewood Cliffs, N.J.: Prentice-Hall, 1968.

Greulach, Victor, *Plant Structure and Function.* New York: Macmillan, 1973.

Grant, Verne, 'The Fertilization of Flowers." *Scientific American,* June 1951. (Offprint 12)

Mayer, A. M., and A. Poljakoff-Mayber, *The Germination of Seeds.* New York: Pergamon Press, 1963.

Raven, Peter, Ray F. Evert, and Helena Curtis, *Biology of Plants,* Second Edition. New York: Worth, 1976.

Ray, Peter Martin, *The Living Plant,* Second Edition. New York: Holt, Rinehart and Winston, 1972.

Sigurbjornsson, Bjorn, "Induced Mutations in Plants." *Scientific American,* January 1971. (Offprint 1210)

Torrey, John G., *Development in Flowering Plants.* New York: Macmillan, 1967.

van Overbeek, Johannes, "The Control of Plant Growth." *Scientific American,* July 1968. (Offprint 1111)

Weier, T. Elliott, C. Ralph Stocking, and Michael G. Barbour, *Botany: An Introduction to Plant Biology,* Sixth Edition. New York: John Wiley, 1974.

Wents, Frits W., *The Plants,* Revised Edition. New York: Time-Life Books, 1971. (Life Nature Library)

Wilson, Carl L., Walter E. Loomis, and Taylor A. Steeves, *Botany,* Sixth Edition. New York: Holt, Rinehart and Winston, 1971.

16

Development of Higher Plants

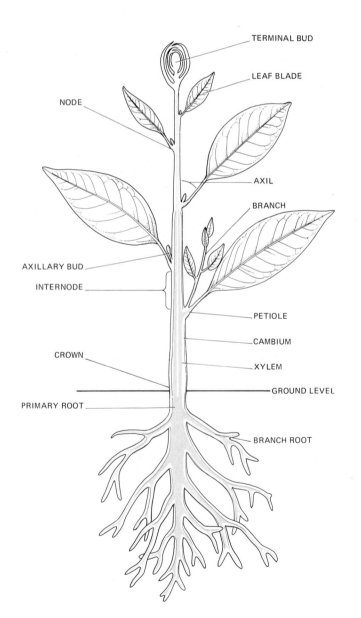

TERMINAL BUD

LEAF BLADE

NODE

AXIL

BRANCH

AXILLARY BUD

INTERNODE

PETIOLE

CAMBIUM

CROWN

XYLEM

GROUND LEVEL

PRIMARY ROOT

BRANCH ROOT

WHEN PEOPLE SAY "PLANTS," THEY ALMOST IN-variably mean vascular plants, those with woody conducting tissues, including the ferns and fernlike plants, the cone-bearing trees, and the woody and herbaceous flowers. In the present chapter, we are concerned with the general body plan of vascular plants, with the developmental processes by which

16.1
Schematic representation of a dicotyledonous plant. The underground root and the aboveground shoot meet at the *crown*. The terminal or *apical bud* is mitotically active, adding new cells at the top and thus adding to the height of the plant. In the terminal bud, *leaf primordia* and *axillary bud primordia* are initiated, later to become leaves with blades on *petioles* and *axillary buds* in the axils of the leaves. The place on a stem where a leaf is borne is a *node,* and the length of stem between nodes is an *internode.* A branch originates when an axillary bud begins independent growth. The central woody cylinder is composed of xylem surrounded by a sleeve of cambium, the tissue that makes the stem increase in girth. Stem branches originate only in axils of leaves, but root branches may occur anywhere along a root.

16.2

Schematic representation of a monocotyledonous plant (grass). The *apical growing point* adds new cells at the tip, which elongate and increase the height of the plant. Leaf primordia and axillary bud primordia are formed just below the growing point. Leaves have their bases practically encircling the stem before sending the blade away from the stem. The sheathing leaf bases add strength to the stem, which has no cambium to add girth. *Axillary buds* can begin growing as independent apices to become *axillary branches*. There is no central vascular cylinder as there is in dicotyledonous plants, but numerous separate vascular bundles extend from root to apex, with branches to the veins of the leaves. The fibrous root system develops from the lower end of the shoot.

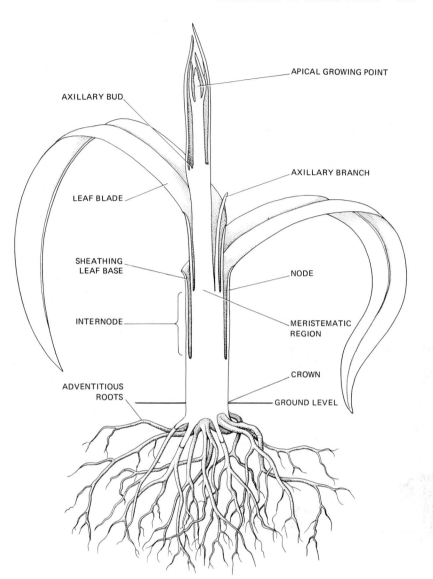

APICAL GROWING POINT

AXILLARY BUD

AXILLARY BRANCH

LEAF BLADE

SHEATHING LEAF BASE

NODE

INTERNODE

MERISTEMATIC REGION

CROWN

ADVENTITIOUS ROOTS

GROUND LEVEL

mature form is achieved, and with some of the variations that have evolved from a presumably common ancestor.

Superficially a head of cabbage is enormously different from a tropical vine stretching almost into invisibility to the top of a mahogany tree, and the fine turf of a putting green seems to have little in common with a giant bamboo, but the way these dissimilar plants grow and their structural similarities, once discovered, are clearly fundamental. There is one basic plan for all vascular plants. It consists of a *stem*, with a terminal growing point, bearing leaves along its length, and a *root*, also with a terminal growing point. Both stem and root contain internal tissues with conducting ability, the downward-flowing *phloem* and a more central upward-flowing stiffening

tissue, the *xylem*. Two major variations on that theme have proved successful. One is that of the nonflowering vascular plants and the dicotyledons, exemplified by most common herbs and woody trees, and the other is that of the monocotyledons, exemplified by grasses, lilies, and palms. (For their evolutionary position, see Chapter 9.) In spite of numerous and sometimes bizarre variations in some plant families, these two main types are so fundamental to the general scheme of higher plant structure that if they are understood, then the structure of almost any plant met in ordinary experience will be recognizable. Figures 16.1 and 16.2 offer a schematic comparison of the monocotyledonous and dicotyledonous plant bodies. Botanists usually shorten these terms to _monocot_ and _dicot_.

GERMINATION IS THE RECOMMENCEMENT OF THE LIFE OF A SEED

A mature, viable seed, ready to germinate, provides a reasonable start for a treatment of vascular plant development and anatomy. In order to start growing, a seed must be physiologically prepared. This may simply mean that there is a temperature above freezing, an adequate oxygen supply (because a germinating seed has a high respiratory rate), and enough water. Some seeds have additional special requirements, such as light or a ripening time to allow enzymes to act, or decay of a hard seed coat, or the presence of specific external stimulating compounds. But *water*, *warmth*, and *oxygen* are universally needed, and indeed the best way to preserve living, dormant seeds is to keep them dry, cold, and airless.

When the embryonic plant in a seed begins to grow, having obtained enough water, the first activity is in the *radicle*, the lower tip of the embryonic axis. Cells expand, and mitotic divisions begin at the extreme tip, causing the first or *primary root* to push through the seed coat. In a dicot seed such as a radish, the two embryonic leaves, the cotyledons, begin to expand. With the primary root growing down and that portion of the shoot below the cotyledons (the *hypocotyl*) expanding, the cotyledons are pushed up until they are above the ground surface. They then open out, turn green, and expose the apex of the stem. As the primary root grows, it sends out branches, *secondary roots*, which firmly anchor the seedling. At the same time the stem apex is mitotically active, growing upward and beginning to produce the first true leaves. The length of stem above the attachment of the cotyledons, the *epicotyl*, elongates further, and the seedling is on its way to becoming a self-sustaining plant.

A radish is an *epigeous* type (Greek, "above the earth") because it raises its cotyledons above the soil surface and uses them for photosynthesis. Many dicots are *hypogeous* (Greek, "under the earth"); that is, they keep their cotyledons below the soil by not stretching the hypocotyl. Instead, only the epicotyl elongates, and the first visible leaves are true leaves. Meanwhile, the cotyledons serve as a temporary food supply, which keeps the embryo fed until part of it can reach light and start photosynthesizing.

Monocot seeds, such as grasses, begin germination by sending out a primary root, but the single cotyledon acts as an absorbing organ that presses into the seed's main food supply in the endosperm (Chapter 15). The apical growing point is covered by a *coleoptile*, inside which the first true leaf, like a dagger in a sheath, grows upward. The leaf pushes through the coleoptile, splitting it and leaving it as a temporary basal collar. With a primary root established and a true leaf extending into the light, a monocot seedling is established.

Needless to say, different species of plants, both monocots and dicots, have variations on these methods of germination. One of the improbable methods is that of the red mangrove (*Rhizophora mangle*) of tropical beaches. In that species, the seed germinates while it is still on the parent tree, sending down a radicle several centimeters long (Figure 16.3). When

16.3

Vivipary in the red mangrove (*Rhizophora mangle*). The seed germinates while still attached to the parent plant, and the growing root hangs down. The entire fruit falls from the branch. If the seed lands in the soft mud below the tree, it plants itself and is on its way to becoming a mangrove tree.

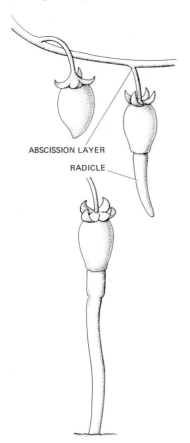

ABSCISSION LAYER

RADICLE

the seedling eventually falls, it sometimes sticks upright in the mud and thus is effectively planted. It then needs only to send out secondary roots and expand its leaves. Plants as well as animals can be *viviparous;* viviparous organisms have embryos that develop within the mother's body and derive most of their nourishment from the mother's tissues.

ROOTS ARE THE ANCHORING AND ABSORBING ORGANS OF A PLANT

When an expanding radicle breaks out from a germinating seed, part of the increase in size is simply the swelling of cells, but such increase is limited, and soon mitosis commences at the growing root tip. As new cells are formed, some remain as a permanent part of the root, but others are pushed forward (that is, usually downward) beyond the actual root tip, and they form a loose covering over the tip. The thimble-shaped cover is the *root cap,* whose wet, slimy cells make it possible for the elongating root to ease its way among the soil particles. As roots keep growing, root caps are continually worn away and replaced. At the very apex of the root tip is a *quiescent zone,* in which mitosis is minimal. Experiments with radioactive tracers show that most mitotic divisions take place just back of the quiescent zone.

Once a cell has divided into two cells, the new cells increase in size, not only taking in water but synthesizing new proteins and other protoplasmic compounds. Simple increase in size is accompanied by little or no apparent increase in complexity, but once full cell size is reached, the cells begin to *differentiate.* Some round up to become storage cells, some elongate and thicken to become conducting cells, and some send out extensions to become *root hairs.* The regions where these events occur can be seen in thin sections of roots: (1) mitoses in the region of cell division, (2) an obvious increase in cell size in the region of elongation, and (3) increase in diversity of shape and wall thickness in the region of differentiation. Long sections of root tips are frequently used in studies of differentiation because they show plainly the events through which all cells pass from first formation to maturity (Figure 16.4).

Some of the epidermal cells back of the region of elongation are capable of pushing out extensions as long tubular growths, the root hairs. These structures, which grow thickly enough to give a young root a fuzzy appearance, increase the absorbing surface of the root by about 20-fold. They squeeze their

16.4
Diagrammatic view of a median long section through a root tip. Just behind the root cap, mitoses increase the number of cells, which elongate as they age, pushing the root tip forward through the soil and driving the root cap ahead of the tip. Within the space of a few millimeters, cells can be found in all stages of development, from the perpetually embryonic mitotic region, through the region of cell enlargement, and into the region of final differentiation into mature cells of xylem, phloem, cortex, and epidermis.

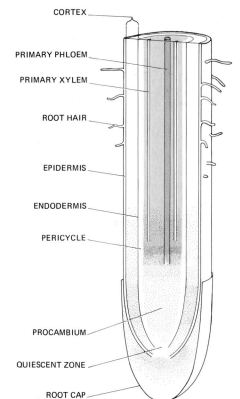

way among the soil particles, coming into such close contact that they can absorb almost all the water. Root hairs generally last only a few days, dying as new ones are grown near the growing point of the ever-elongating root (Figure 16.5).

By the time a root reaches functional maturity, it has a number of differentiated tissues, each with its own work to do. The outermost layer of epidermis not only bears root hairs but does some absorbing of water and mineral salts. Inward from the epidermis a usually voluminous cylinder of cortex is made up of rather large, simple cells, which pass water from the epidermis to the conducting tissues and also store reserve food, usually starch. The central conducting cylinder is sheathed in a sleeve of specialized cells, the *endodermis* (Greek, "inside skin"). The endodermis is lined with still another layer of structurally simple cells of *pericycle*. The bulk of the conducting tissue is <u>*xylem*</u>, whose elongated, heavy-walled cells are open-ended. Running lengthwise in the fluted

16.5

The development of root hairs. When cells in a root tip have reached their maximum size, some of the epidermal cells send out extensions into the soil. These extensions, the root hairs, may reach a length that is hundreds of times their diameter, thus increasing the surface of the root in contact with the soil. Root hairs are usually short-lived, usually lasting only a few days, but they are the main absorbers of water and minerals. × 1000.

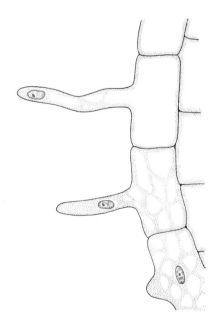

grooves of the xylem cylinder are strips of <u>*phloem*</u> whose elongated, thin-walled cells are mainly transporters of dissolved food. In most roots, the cells between the xylem and phloem, the *cambium* (Latin, "change"), which retains its *meristematic* (mitotic) ability, is capable of increasing the girth of the root by producing new xylem cells inward and new phloem cells outward. The central portion of dicot roots is usually filled with xylem, but in monocots there is frequently a core of thin-walled, relatively undifferentiated pith (Figure 16.6).

Root branches may originate anywhere along a root except at the growing tip. Some pericycle cells may become mitotically active, and by divisions they begin to form a new root apex, which swells and pushes outward through the endodermis, then through the cortex, finally bursting through the epidermis. Meanwhile, the cells just back of this growing tip are becoming like those behind the growing point of any root, with a region of mitosis, one of cell elongation, and finally one of differentiation. After a branch root has grown out from the parent root, xylem and phloem are differentiated in the new root, and these form a functional connection with the old xylem and phloem. Thus there is a continuous connection of the vascular system throughout the entire root system (Figure 16.7).

In woody plants, as roots grow older, the cambial cells come to make a complete cylinder around the xylem, and they keep adding more xylem year after year until an old root is mostly xylem, with only a thin covering of phloem and its associated tissues. Since the original epidermis can cover only a small root, it splits when a root swells in growth and is soon lost, but it is replaced by new, corky cells, which make an effective covering over the phloem. Old roots support large plants and conduct water, minerals, and food, but they are not absorptive.

Roots are more extensive than most people think. Because they are out of sight and appear small when plants are pulled out of the soil, root systems are not generally appreciated, but they are in fact more extensive than the aboveground plant parts. Tree roots generally extend well beyond the tips of branches. One of the best-known inventories of root growth is that of the botanist H. J. Dittmer, who measured all the roots of a single rye plant and reported a total of more than 600 kilometers (380 mi) of roots, with a surface area of 230 square meters (2500 ft²), not counting the extra area provided by root hairs. With the leaves measuring only 4.7 square meters (51 ft²), the roots had more than 50 times as much surface as the leaves. Although grasses (rye is a grass) may

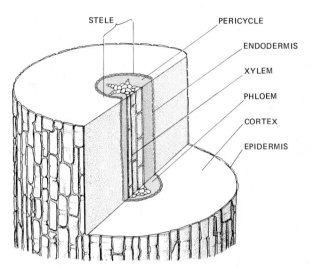

16.6

Diagrammatic cross section of a young dicot root. All the tissues here are primary, having come from the apical growing point. The solid central core of xylem in this example is like a fluted column, appearing somewhat star-shaped in cross section. In the "arms" of the xylem are phloem strands, separated from the xylem by a layer of cambium that will eventually contribute secondary xylem and phloem. A sleeve (a ring in cross section) of undifferentiated pericycle surrounds the entire vascular cylinder, or stele. Outside that is a second sleeve of specialized endodermis. A voluminous cortex surrounds the stele, usually making up the bulk of the young root. The outermost layer is the epidermis, some of whose cells grow into root hairs.

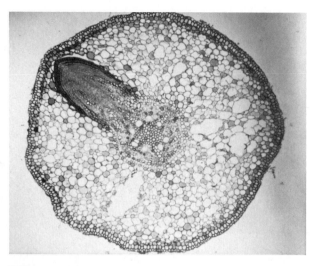

16.7

The origin of branch roots. In this cross section of an old root, a branch can be seen pushing its way from the central stele out through the cortex and bursting through the epidermis. The first cells to begin dividing when a branch starts are in the pericycle of the old root. They become meristematic and start a growing point, which becomes the apex of the branch root. Branches originate at points where xylem strands are close to the pericycle, but there do not seem to be any predestined places where such renewed mitotic activity commences. In stems, in contrast, branches originate superficially from branch buds in the axils of leaves, and their distribution is predetermined by the position of the leaves.

have relatively more root than woody trees do, the comparison is probably fair for most plants.

The original primary root grows out from the radicle of the seed, but in mature plants, roots may grow from any part. A root produced from an organ other than the radicle or another root is an *adventitious root*. Most of the roots of grasses, with maize (corn) as a clear example, are adventitious, coming from the base of the stem. Stems readily produce new roots, as anyone knows who has made cuttings or "slips" of garden plants. In such plants as African violets and begonias, even leaves are capable of generating roots.

Since they are in an underground environment, which is generally more stable than the aerial environment, roots are less variable than either stems or leaves. Botanists say that roots are "conservative."

Nevertheless, considerable variations can be found. Grass roots, for example, are numerous, thin, and fibrous, but sweet potato roots may become swollen with reserve starch. Aerial roots of climbing vines grow out from stems and attach themselves to trees or buildings. A few orchids have green roots that carry on the photosynthetic work of the plant. One of the most striking root systems is that of the "strangling fig" of the American tropics, of which many species exist. The fig seed germinates in the branches of a host tree and sends its roots down along the trunk until they reach the ground. The roots can eventually grow together, surrounding the entire host trunk and finally killing it. The banyan tree, another species of tropical fig, sends roots down from its branches until they reach and enter the soil. Then they grow thicker until they make stout props that hold up the branches

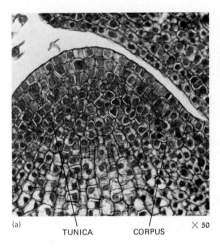

(a) TUNICA   CORPUS   × 50

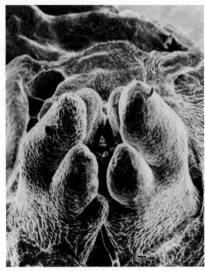

(b)   × 160

16.8

Apical growing points of stems. (a) In this median long section through the apex of a Chrysanthemum stem, the outer layer of tunica (which will give rise to later epidermal cells) covers the surface. Under the tunica, the corpus consists of a mass of small cells with thin walls, relatively large nuclei, and little cytoplasm. The corpus will give rise to all the other primary tissues of the leaves and stem. (b) Scanning electron micrograph of the shoot apex of a buckeye (*Aesculus californica*). The apex is just visible in the center of the picture. The points sticking up above the apex are leaf primordia.

and allow further spread. In this way, a single banyan tree can come to cover acres of ground, supported by a forest of roots.

STEMS SERVE MAINLY FOR SUPPORT AND CONDUCTION

The epicotyl of a germinating seed, with a meristematic apex at its tip, gives rise to the shoot. At first, a small increase in length comes from simple cell enlargement, but afterward, stem growth in length is dependent on mitotic activity in the apex. A *stem apex* is a soft, dome-shaped mass of tissue that may be a fraction of a millimeter thick in a small plant and up to several centimeters thick in some palm trees. It is covered with a layer of meristematic cells that divide in such a way as to keep the cover complete as growth continues. A stem apex does that by orienting its mitotic spindles parallel to the surface, so that the new cell walls are perpendicular to the surface. Such divisions are typical of the divisions in that outer apical layer, known as the *tunica*. Under the tunica is a mass of cells that divide in such a way as to make the stem increase in length. No one knows how the orientation of mitoses is regulated, but the result is a continuing growth of the apex, with the *corpus*, covered by its tunica, moving ever upward (Figure 16.8).

Just below the apical growing point, usually only a fraction of a millimeter, bumps grow out at precisely regulated places. These are the incipient leaves,

known as *leaf primordia*. Inside the young stem, events are similar to those back of a root growing point. Once produced, new cells first become enlarged and then begin to differentiate into conducting xylem and phloem. The first indications of differentiation are increases in the protein content of the cells and slight thickening of the cell walls, which can be seen in tissue sections in strands of cells extending from near the apex downward. As leaf primordia grow, they extend in length and become separated from one another by the elongation of the stem. The place on a stem that bears a leaf is a *node*, and the intervals between nodes are *internodes*. Nodes elongate little, but internodes elongate greatly, especially in dim light, causing a separation of the nodes and consequently a separation of leaves along the stem. As leaves mature, strands of xylem in them differentiate, and the path of differentiation is such that a meeting of the conducting tissues of leaf and stem results (Figure 16.9).

Soon after a stem apex produces a leaf primordium, it begins to produce a second smaller mass of cells just above the leaf primordium. The new growth is a *lateral bud primordium*. When the leaf is mature, the axillary bud will have developed in the axil of the leaf. The axillary bud is a copy of the terminal apex, and given the proper stimulation, it can begin to grow just as the terminal apex did; that is, it can expand, grow out, and become a side branch of the old shoot. Stem branches in general originate in this way. The manner of stem branching is different from that of root branching. Stem branches are produced specifically at nodes and come from superficial cells that originate in the stem apex. Root branches, in contrast, may appear anywhere along a root and are the result of meristematic cells originating deep inside the root in the pericycle. Roots do not have nodes and internodes, and this is one of the surest ways of distinguishing between roots and stems, which sometimes may closely resemble one another.

The patterns of maturation in monocot and dicot stems are somewhat different (Figure 16.10). We will follow first the dicot pattern. As the stem cells mature and the vascular tissues differentiate, a number of xylem strands form either a continuous cylinder inside the stem or a ring of separate strands. Then, outward from the xylem cylinder (or strands), phloem cells begin to differentiate. Anywhere from a few millimeters to several centimeters below the apex, depending on the species and environmental conditions, a fully functional stem develops. At such a place, before a cambium has begun to make new growth, all the cells are initially formed in the apex. Tissues composed of such cells are _primary tissues_, in contrast to secondary tissues, which will be derived

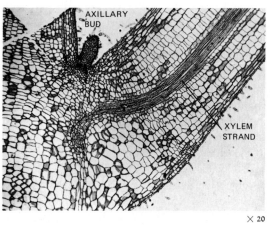

× 20

16.9

A median petiole section through a node of a Coleus stem. The xylem strand in the petiole of the leaf, stretching toward the upper right side of the picture, connects to the xylem of the stem. The bump in the angle formed by the stem and the petiole is a young axillary bud.

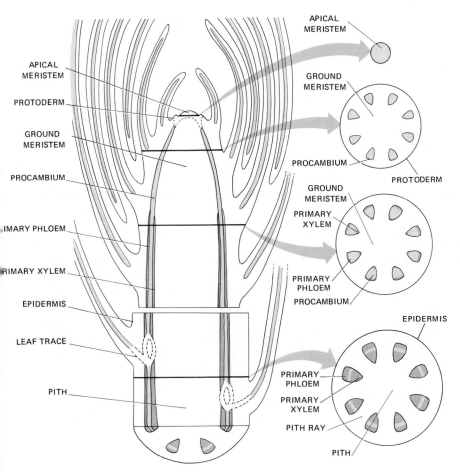

16.10

Diagrams of cross and long sections through a stem apex showing maturation of the tissues. The very tip of the stem, which is completely embryonic, is composed of undifferentiated cells. However, as new cells enlarge, they push the apex higher and higher, leaving behind the cells that differentiate to form the _vascular tissues_ and the _ground tissues: cortex_ and _pith_. Later, when the _vascular cambium_ has become the main growing tissue, the bundles will enlarge until they make a complete cylinder of wood.

16.11

A cross section through a woody stem: American sycamore (*Platanus occidentalis*). The central pith is surrounded by a cylinder of xylem, through which rays radiate out to the encircling bark. The innermost layer of cells of the bark, just outside the xylem, is the cambium. The pale splotches in the bark are masses of phloem, outside which is the cortex. In a young stem such as this (one year old), the outer epidermis is still complete. × 25.

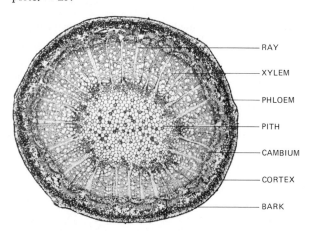

RAY
XYLEM
PHLOEM
PITH
CAMBIUM
CORTEX
BARK

later from cambium. In monocot stems, the early xylem strands are not developed in a ring or cylinder around a central pith. They are numerous and scattered throughout the "background" tissue.

As the apex grows upward and leaves the older part of the stem farther and farther behind, the older part of the stem begins secondary growth. The cells between a xylem strand and a phloem strand become meristematic, dividing so as to produce more xylem inward and new phloem outward. Thus the xylem is increased and the stem grows in girth. As this growth continues, it pushes the phloem outward like a man with a fat belly stretching his belt, and the old outermost phloem is first crushed and then split off. By the time a stem is a few years old, it will have a remnant of the old primary cells down the center (the pith) and a small amount of primary xylem, surrounded by a thick cylinder of secondary xylem, which is in turn surrounded by a thin cylinder of cambium. Outside the cambium is the bark, containing phloem and a replacement for the long lost epidermis. This replacement is the *cork*, produced by a second meristematic cylinder, the *cork cambium* (Figure 16.11).

16.12

Enlarged portions of a cross section through an old woody stem: basswood (Tilia). A corky layer at the left is produced by a cork cambium, which keeps making more cork as the stem grows. Inward from the cork is a layer of undifferentiated cortex. The vascular tissues, phloem and xylem, are complex tissues, each consisting of several kinds of cells. In the phloem are phloem tubes, rays, and fibers. The cambium between phloem and xylem, which is the main growing tissue, produces phloem toward the outside and xylem toward the inside. In the xylem are water-conducting cells, fibers, rays, and parenchyma. (Parenchyma is a term applied to any essentially undifferentiated plant tissue.)

CORK   CORTEX   PHLOEM FIBERS   CAMBIUM   SPRING WOOD   SUMMER WOOD   RAY   VESSEL

ANNUAL RING

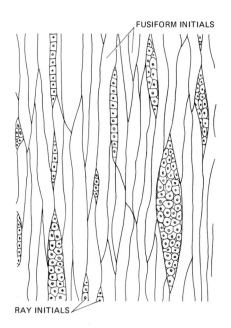

FUSIFORM INITIALS

RAY INITIALS

16.13

A tangential view of cambium of black walnut, *Juglans nigra*, as it would appear if a piece of bark was stripped down to the wood. The two kinds of cells are (1) *fusiform initials*, which divide lengthwise to yield long xylem cells toward the inner side or phloem cells toward the outside; and (2) *ray initials*, the short, rounder cells in groups, which divide to produce rays. × 150.

16.14

Tree-ring dating. A slender cylinder of wood, called a core, is cut from a tree (or log or sawed piece of lumber) with an increment borer. The spacing between rings is carefully measured. A sequence of spacings is established and can be compared with that of other samples. In any particular locality, all the trees will have thin rings in poor growing years and thicker rings in better years. A sample from a living tree provides a starting date. The older part of that sample will overlap an older sample of unknown date. More and more overlapping samples have made it possible to date wood hundreds of years old with great accuracy.

The cambium, only one or two cell layers thick, has two kinds of cells: long slender ones, which are to yield long slender xylem or phloem cells, and clusters of smaller and rounder cells, which are to yield rays. These two cell types are known as *fusiform initials* and *ray initials*, respectively (Figure 16.13). The cambium is especially active at the beginning of each growing season in the spring of the year, making a new ring of large xylem cells around last year's old xylem. As the season progresses, the cambium produces smaller and smaller xylem cells until by late summer, when it ceases growth, it is making relatively small ones. Each year, then, a new cylinder of xylem is laid down around last year's cylinder, and when a tree is cut across, these *annual rings* can be easily seen. Since normally only one ring is laid down each year, the number of rings equals the number of years the tree has lived. As the xylem increases in girth, the cambium expands, leaving behind it in its progress not only the annual rings but also the strips of ray tissue, which appear in cross cuts of wood as radial lines running from the central pith toward the bark (Figure 16.14).

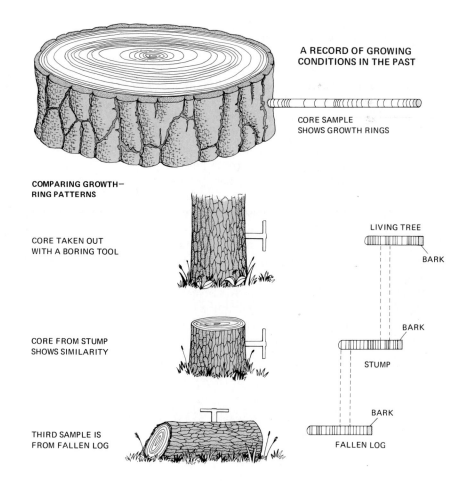

A RECORD OF GROWING CONDITIONS IN THE PAST

CORE SAMPLE SHOWS GROWTH RINGS

COMPARING GROWTH-RING PATTERNS

CORE TAKEN OUT WITH A BORING TOOL

CORE FROM STUMP SHOWS SIMILARITY

THIRD SAMPLE IS FROM FALLEN LOG

LIVING TREE

BARK

BARK

STUMP

BARK

FALLEN LOG

HOW BANANAS GROW

A banana plant looks like a tree. It raises its huge flat leaves five or six meters into the air and has a smooth, waxy trunk as thick as a man's leg. But internally it is nothing like the ordinary trees of the temperate zones. There is no bark and no central woody column. It is soft and tender, certainly softer and more tender than most of the weeds that invade temperate waste places.

During most of its life, a banana plant has practically no stem. Such stem as it does possess is little more than a lump of tissue at its base, with a growing point projecting upward a few centimeters and producing leaf primordia and eventually leaves. As each leaf grows, it stretches up through the central part of the so-called trunk, forming an ever-thicker leaf base, which is like a tube. As more and more inner leaves are formed, the older outer ones become larger and tougher until there is an elongated, stalklike structure made up of concentric rings, similar to the rings of leaf bases in an onion bulb. At its upper extremity, a leaf puts out a winglike expansion of its midrib, making the main part of the photosynthetic leaf. What looks like a leaf and, indeed, functions as a leaf is structurally an expanded petiole, with the morphologically "true" leaf, comparable to a blade of grass, represented by a tiny thread of tissue at the tip of the expanse. That little tip, twisted like a dried pig's tail and only a couple of centimeters long, is scarcely noticeable at the end of a two-meter leaf. Most people look-ing at a banana plant never see the true leaves. They see only the stout leaf bases and the ex-panded wings along the main midvein.

A cut across a banana stalk shows the rings of leaf bases, rings within rings, as spongy as a roll of corrugated cardboard (b). A longitu-dinal cut through the center of a stalk reveals the real stem, like an elongated heart of a cab-bage, with leaf bases of various ages attached at the sides (c). It is apparent that structurally

(a)

(b)

the stalk is not comparable to the stalk of such a familiar tree as a maple.

When a banana plant is ready to flower, the apical growing point of the stem begins to grow, forcing its way up through the center of the rings of leaf bases until it reaches the top. Then it bends over, first forming clusters of pistillate flowers and then forming clusters of staminate ones. The pistillate flowers, which have ovaries but no stamens, are destined to become banana fruits. Each flower cluster becomes a "hand" of bananas, and the mass of clusters becomes the entire bunch of bananas. Banana fruits point upward. If the apex is allowed to grow, it will make a long string of staminate flower clusters (with stamens but no ovaries), each covered by a waxy purple bract. The part of the stem that bears staminate flowers is inedible and never finds its way to markets.

A banana stalk bears fruit only once. When harvest time comes, a stroke of a machete brings the whole plant down, the staminate part of the stem is cut off, and the fruit completes its ripening after it is removed from the rest of the plant. Then side buds grow out from the base of the cut stump, and these little plants, the suckers, can be dug out and transplanted (c). In about 18 months they are ready to bear a new crop.

(c)

Annual rings vary in thickness with weather; a dry year gives a thin ring and a wet year a thicker one. By making careful measurements of successive rings in trees of known age and overlapping a number of samples, we have been able to draw a number of conclusions about the past from the science of *dendrochronology* (Greek, "tree timing"). Tree-ring dating has made possible the establishment of accurate dates for many historical edifices, especially the buildings of the Indians of the American southwest. Also from tree-ring measurements, climates can be described as far back as 8000 years. One unexpected development of dendrochronology is the demonstration that dating by the use of carbon-14 is not accurate without a correction. Carbon-14 dating, based on the assumption that the amount of carbon-14 has remained constant through the years, does not agree with tree-ring dating, but tree-ring dating is the more convincing method, and therefore the carbon-14 method has to be adjusted.

Wood has so many uses and is such common material that an understanding of its various characteristics should be a part of everyone's fund of information. The differences between different kinds of wood are determined by the species of trees, the climates in which they grew, the part of the trunk they were taken from, the direction of the saw cut, and individual peculiarities. The hardness of wood is mainly determined by the species. *Softwoods* are officially derived from conifers, whether the wood is actually soft or not, and *hardwood* comes from flowering trees. Although linden wood from a flowering tree is softer than yellow pine wood from a conifer, linden is sold as a hardwood and yellow pine as a softwood, but in general the distinction is well founded. There is a difference, too, in woods of different ages, because some trees deposit pigments in the older parts of the trunk more than in the younger parts. The central cores (the *heartwood*) of walnut, cherry, and red cedar (*Juniperus virginiana*), for example, are deeply colored, but the outer rims (the *sapwood*) are pale.

The *grain* of wood is determined by the way the cambial cells lay down xylem and by the final differentiation of the xylem cells. Cambium may produce uniform masses of cells, all neatly lined up parallel to each other and with little difference in cell size or wall thickness, and the result, as in white pine, is a homogeneous, smooth, almost characterless wood. Or the cambium may be twisted and give rise to xylem cells contorted in various ways, and these may mature into tissues with many sizes of cells with different wall thicknesses, as happens in curly or tiger-

16.15

The difference between spring wood and summer wood in white pine, *Pinus strobus*. In this cross section, the spring wood with its larger cells is at the left. As summer approaches, the cells do not grow as large, and when fall comes, growth stops. In the following spring (extreme right), growth starts up again, and large, spring wood cells are once more produced. The alternation of cell size makes rings of softer (large cells) and harder (smaller cells) tissue.

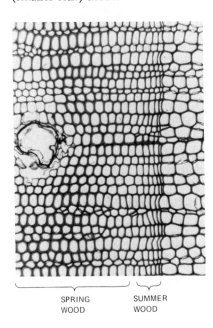

SPRING WOOD   SUMMER WOOD

16.16

Cross sections of hardwoods. (a) In oak (Quercus), the vessels produced in spring are conspicuously larger than those produced later in the season. In an annual ring in such wood, there is a ring of the large vessels in-side a ring of smaller ones. Such wood is *ring porous.* (b) In linden (Tilia), the vessels are about the same size throughout any annual ring. Such wood is *diffuse porous.*

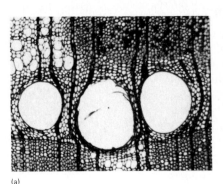

(a)

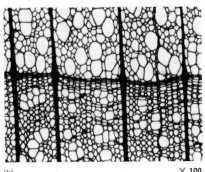

(b)   × 100

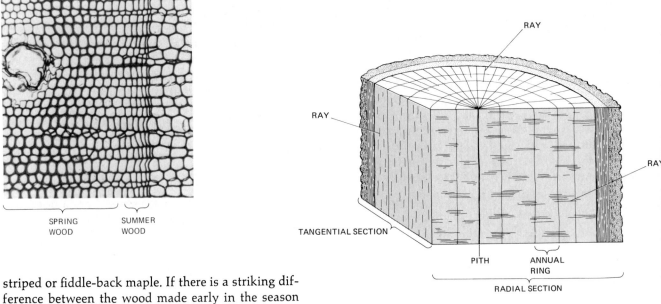

RAY

RAY

RAY

TANGENTIAL SECTION

PITH   ANNUAL RING

RADIAL SECTION

16.17

The effect of the direction of cut on the appearance of wood grain. In a cross section, the annual rings appear actually as rings, and the rays as narrow lines crossing the annual rings. In a radial section, which passes directly through the center of the stem, the annual rings appear as long streaks, and the rays as cross-hatching across the annual rings. In a tangential section, which passes anywhere except through the center, the annual rings appear as long streaks, and rays as short lines.

striped or fiddle-back maple. If there is a striking difference between the wood made early in the season (*spring wood*) and that made later (*summer wood*), the annual rings will be conspicuous, as they are in yellow pine (Figure 16.15). Some woods produce a few large water vessels in the spring but few or none later on, as in oak. When oak wood is seen in cross section, the spring wood shows a ring of holes, and it is called *ring porous*. In contrast, yellow poplar wood has vessels formed throughout the year and is *diffuse porous*. Since softwoods have no water vessels, only tracheids, they are nonporous. Rays, too, affect the appearance of wood. If they are vertically shallow, they will not be visible to the naked eye, as in birch, in which rays are only a fraction of a millimeter high. If they are extensive, as in oak, cherry, and maple, they may be conspicuous (Figure 16.16).

The way in which wood is cut makes a great difference in the appearance of the cut surface. If it is cut across, the annual rings appear as circles or parts of circles, with rays running radially. If a cut is directly through the pith, that is, along a radius of the cylinder of the tree trunk (a *radial section*), the annual rings appear as more or less parallel lines, provided that the tree was reasonably straight. The rays then appear as shining strips of various widths, depending on the species, running across the lines of the annual rings. If the cut is through a log anywhere else, as it usually is (a *tangential section*), the annual rings will show up as wavy, irregular streaks, and the rays will usually show poorly or not at all (Figure 16.17).

With wood becoming ever more scarce and expensive, fine woods are commonly cut in thin sheets and glued over a base of less desirable wood to make *veneer*. Veneers have an added advantage in being dimensionally more stable than solid blocks, because the grain of the veneer is frequently glued across the grain of the base, so that both pieces hold each other in place when changes in humidity might otherwise cause swelling, shrinking, or warping. *Plywood* is made by gluing sandwiches of three to seven thin wooden sheets together with the grains alternately crossing each other. Since plywood can be made with the inner layers of flawed wood, it can be made cheaply. Besides, if made with weatherproof glue, it is not affected by moisture, and it is unsplittable.

Woods are prized for many special qualities: red cedar for its aromatic resins, mahogany for its color, lignum vitae for its hardness, hickory for its resilient flexibility and for the flavor of its smoke, rosewood for its color patterns, white pine for its homogeneity,

16.18

A tree dissected. The corky outer bark is peeled out, as are the phloem (the inner bark), the thin cambium layer, and a mass of the living sapwood. The sapwood, which is living wood, is wet and relatively softer than the dead central heartwood.

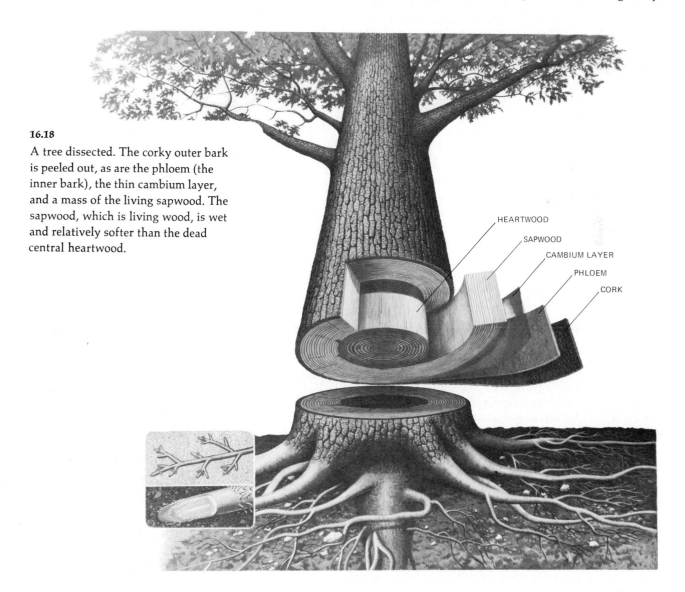

HEARTWOOD

SAPWOOD

CAMBIUM LAYER

PHLOEM

CORK

16.19

Variation in the form of stems. (a) An onion stem is a short, broad lump at the base of the onion bulb, with an apex, no cambium, and roots growing from the bottom. The bulk of the onion bulb is made up of succulent leaf bases. (b) A cabbage head has a longer but still stubby stem. (c) A tree with one main stem and minor side branches. (d) A tree with no one main stem but many almost equal side branches.

(a)

(b)

(c)

(d)

spruce for its ability to resonate in musical instruments, pernambuco for its springiness in violin bows, teak for its durability, beech for its resistance to water damage, white oak for its retention of liquid in wine casks, redwood for its resistance to decay, dogwood for its resistance to wear. The list could go on. In recent times, the cost of fine wood has driven wood technologists to making partially synthetic substitutes, molding boards from wood chips bonded in plastic, or printing photographs of wood onto plastic sheets, as in wall panel boards and table tops.

The general structure of stems is subject to as much variability as is their tissue structure (Figure 16.19). The differences depend on the relative amounts of growth of apices and cambium. If apical growth is limited and leaf growth is voluminous, the result can be a compact mass, like a head of cabbage. Or the stem may be almost completely suppressed, with fleshy leaf bases surrounding the apex. If the whole complex is under the soil, with only leaf tips emerging, the result is a bulb, such as an onion. If there is one main apex, the plant grows in a pattern like a conventional Christmas tree, but if many apices are about equal, a spreading twiggy pattern, like that in a maple tree, is formed. A squat, heavy tree results when cambial growth is vigorous and apical growth restricted, but when cambial growth is minimal and apical growth unlimited, the plant grows as a slender vine that must be supported. Irises have horizontal stems that lie on or just under the soil surface. Potato plants have conventional stems above ground, but some of the stems below ground become starchladen and swell to make the edible potato *tubers*. In hawthorns, some woody branches remain short but hard and sharp, but the prickles on rose stems are only epidermal outgrowths not connected by any vascular elements. Thorns cannot be broken off without tearing wood, but prickles snap off cleanly and easily. Just as some things that do not look like stems (potato tubers) really are, so some that look like stems are not. The apparent stem of a banana plant is a

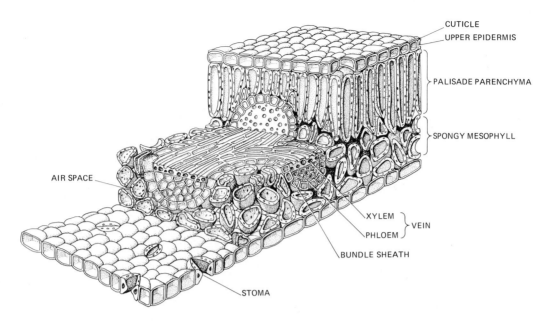

16.20

A dissected leaf. A mature leaf has a layer of *epidermal cells* over its surface. On the outside these cells have a waterproof coating of waxy *cuticle.* *Stomata* are usually more common in the lower than in the upper epidermis. The stomatal pore leads into an air space inside the leaf. The *spongy*

mesophyll (Greek, "middle leaf"), located above the lower epidermis, consists of chloroplast-bearing cells that are partially pulled away from one another as the growing leaf expands, with the result that they are loosely connected and have a labyrinth of air channels throughout the tissue.

Above the spongy mesophyll, the cells of the *palisade parenchyma*, also chloroplast-bearing, are neatly compacted like the nap of velvet. A system of *veins* extends through leaves. Veins are coated with cells, a *bundle sheath*, making a covering like the insulation on an electrical wire. × 75.

mass of leaf bases, and the morphological stem, except when the plant is flowering, is a short nub of tissue near the base, deeply hidden in the leaf bases.

LEAVES ARE THE MAIN PHOTOSYNTHETIC ORGANS OF PLANTS

The leaves of vascular plants are thought to have evolved from the growing together of flattened branches. In their individual development now, however, they are known to arise as outgrowths of tissue just below the growing point of the stem apex. As a young leaf increases in size, it begins to flatten out laterally and to undergo differentiation of its internal tissues. While still in the terminal bud, it is folded in a neat, compact package, but when it opens, its cells swell and some of them continue for a while to divide until the leaf achieves its mature form.

The tissues of leaves vary from species to species. The amount of cuticle ranges from none in aquatic leaves to a heavy covering in such succulent leaves as those of Sedums. Stomata occur on both sides of some

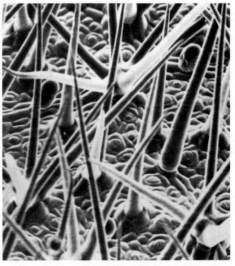

× 100

16.21

A scanning electron micrograph of a leaf surface. Spiky extensions of epidermal cells give the leaf a bristly appearance.

16.22

A portion of a cross section through a corn leaf (*Zea mays*), a C-4 plant. The chloroplasts in the ordinary mesophyll cells do not build up starch, but they do have grana. Carbon dioxide is fixed in the mesophyll. In the bundle sheath cells, those surrounding vascular bundles, the chloroplasts *do* form starch grains (indicating carbohydrate synthesis) but have atypical grana. × 350.

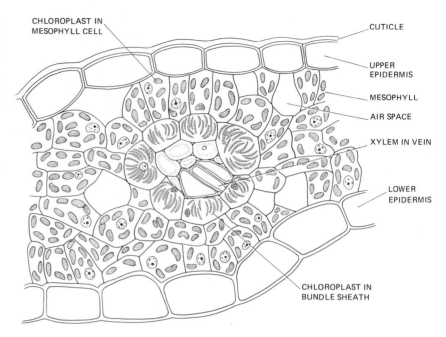

CHLOROPLAST IN MESOPHYLL CELL

CUTICLE

UPPER EPIDERMIS

MESOPHYLL

AIR SPACE

XYLEM IN VEIN

LOWER EPIDERMIS

CHLOROPLAST IN BUNDLE SHEATH

16.23

Environmental effects on leaf growth. Long sections of pine needles (*Pinus contorta*) grown in dry and moist habitats. (a) In a dry place, the needles are thinner and the cells that make up the tissues are more compact. (b) In the moist-grown needles, there is much more air space between the chlorophyll-bearing cells of the mesophyll.

(a)   × 40

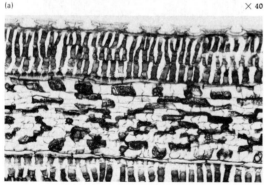

(b)

leaves, or only on the upper epidermis, as in floating water lily leaves. The epidermis may be one thin layer, or it may have several thick layers of cells, as in *Ficus elastica*, the popular rubber plant of public foyers. The palisade layers may be one, two, or three layers deep, or, as in many ferns, not present at all. The mesophyll may be so thin that the leaves wilt on any warm, dry day, as squash leaves do, or it may be so thick, juicy, and filled with mucilaginous materials that the leaves resist drying for weeks, as in the succulent "air plants" (Kalanchoe) (Figure 16.23).

The arrangement of leaves on stems and their gross structure are determined by genetic forces that are incompletely understood. In the apical bud of a mint plant, leaf primordia grow opposite each other in pairs, and when the leaves mature, they occur two at a node, one on each side of the stem. A few plants, such as bedstraw (Galium), produce a circlet of leaves at a node, but most plants bear only one. The solitary leaves, however, are so placed that one can follow an imaginary spiral twining up a stem and passing through the base of each leaf stalk, or *petiole*. The geometry of leaf arrangement, called *phyllotaxy* (Figure 16.24), is so precise and so predictable in most species that it has excited the curiosity of botanists and mathematicians for many years. It is possible to express phyllotactic arrangements with formulas that can be applied as well to other natural structures or

16.24

A spiral arrangement of leaves on a stem. In this example, every leaf is directly above the second one below. The arrangement, which is dependable for a species, is determined by the way the apical growing point starts its leaf primordia.

16.25

Stems that look like leaves. In the butcher's broom (Ruscus), the broad, flattened branches are green and have such a leafy aspect that one could easily mistake them for leaves. But real leaves have buds in their axils, and these "leaves" have no buds. In-stead, *they* occur in the axils of tiny, colorless scales, which in this instance are the morphological but functionless true leaves. Another indication that these leaf-like branches are branches is the fact that they bear flowers in the middle of the surface. Natural size.

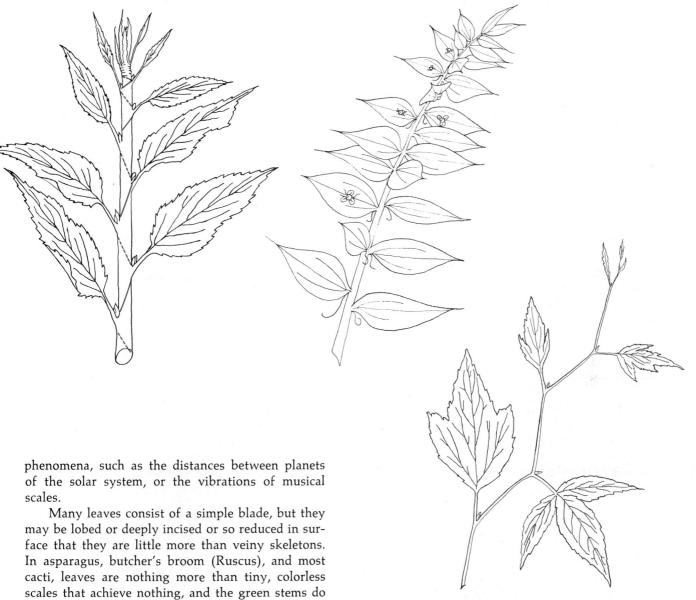

phenomena, such as the distances between planets of the solar system, or the vibrations of musical scales.

Many leaves consist of a simple blade, but they may be lobed or deeply incised or so reduced in surface that they are little more than veiny skeletons. In asparagus, butcher's broom (Ruscus), and most cacti, leaves are nothing more than tiny, colorless scales that achieve nothing, and the green stems do all the photosynthetic work (Figure 16.25). Mulberry and sassafras leaves are not all alike even on one plant (Figure 16.26). When a leaf blade is all in one piece, it is said to be *simple,* but when it is subdivided into *leaflets,* it is *compound.* The difference becomes apparent when one finds the position of the axillary bud, which is produced at the base of the petiole of

16.26

The inconstancy of leaf shape in ivy (Cissus). Leaf shape varies from merely lobed to three-parted. A number of plants, notably sassafras and mulberries, have such a variety of leaves on any one plant. × 0.5.

16.27

Simple and compound leaves. (a) A simple leaf of privet (Ligustrum), in which the leaf blade is not divided. × 1. (b) A compound leaf of mountain ash (Sorbus), in which the leaf blade is divided into two rows of leaflets, with the midrib of the leaf naked between the leaflets. × 0.5. (c) The doubly compound leaf of the Kentucky coffee tree (Gymnocladus), in which each leaflet is in its turn subdivided into still smaller leaflets. × 0.25. In all these examples, the whole leaf grows from a stem at a node, and it has an axillary bud just above the point of attachment of the leaf. There is no bud in the axil of a leaflet. The vascular structure in a leaf stalk, or petiole, is quite different from that of a stem.

a *leaf* but not at the base of a *leaflet* (Figure 16.27). Leaflets themselves may be compounded, making a *doubly compound* leaf, and many ferns are triply compounded. One genus, Davallia, which is five times compounded, has leaves like a fine filigree (Figure 16.28). Leaf form helps in determining species of plants, but it is of little use when applied to larger taxa, such as genera and families, because leaves are the least stable in form of all plant parts.

In all but a few rare species, leaves are temporary organs. At the end of the growing season in herbaceous plants and deciduous trees, leaves are shed because of a weakening of the cells across the base of the petiole. The timing of leaf drop, or abscission, is regulated by the amount of the growth-control compounds, *auxin* and *abscisic acid*. In most instances, abscission is facilitated by the alteration of a few

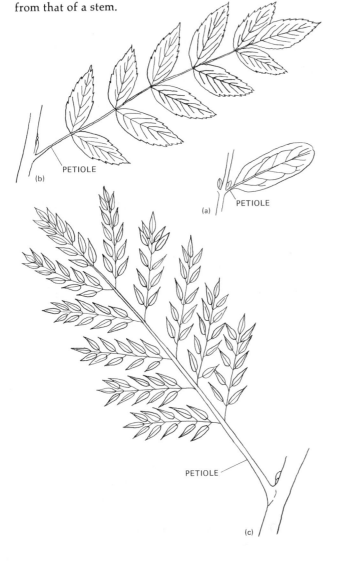

(b) PETIOLE

(a) PETIOLE

PETIOLE

(c)

16.28

A leaf of the rabbit's foot fern, *Davallia canariensis*. The blade is divided, the divisions are subdivided, and those subdivisions are still again subdivided until the compounding reaches the fifth order. Davallia seems to have reached the ultimate in compounding. Such a lacy structure may be useful to a plant in allowing light to penetrate through the upper leaves and shine down to the lower ones. × 0.25.

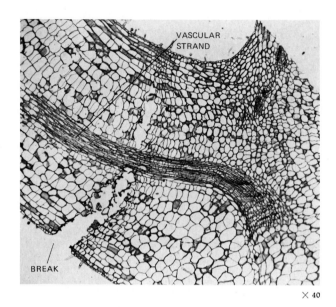

VASCULAR
STRAND

BREAK

× 40

16.29

How a leaf is cut off from a branch: the *abscission layer.* At the end of a growing season, or when a leaf grows old or is damaged, the cells across the base of the petiole are weakened so that a slight movement of the leaf will break it off. This section cut through a node of a coleus stem just before leaf fall shows the crack appearing across the petiole. The vascular strand, the dark streak in the photograph, is the last tissue to break. Abscission layers are generally formed in temperate climates at the approach of cool weather, or in tropical climates at the beginning of a dry season.

cells that make an *abscission layer* across the petiole (Figure 16.29). The cells are *suberized,* that is, impregnated with a corky, waterproofing, waxy material, which seals the scar where the leaf was. Evergreen leaves remain in place at least through a winter, but many are cast off at the end of the second growing season. In some unusual species, such as the bristlecone pine of the California mountains, individual leaves have been found to remain in place as long as 30 years.

SUMMARY

1. Water, warmth, and oxygen are universally needed for *seed germination.*

2. When the embryonic plant in a seed begins to grow, the first activity is in the *radicle.* Then mitotic divisions cause the *primary root* to push through the seed coat. As the primary root grows, it sends out *secondary roots.* The *epicotyl* elongates further, and the seedling is on its way to becoming a self-sustaining plant.

3. All seeds begin germination by sending out a *primary root,* but in monocot seeds, a single cotyledon acts as an absorbing organ that presses into the seed's main food supply in the endosperm. The first true *leaf* pushes through the *coleoptile,* splitting it. With a primary root established and a true leaf extending into the light, a monocot seedling is established. Variations exist in both dicot and monocot seed germination.

4. When an expanding radicle breaks out from a germinating seed, mitosis soon commences at the growing root tip. A *root cap* makes it possible for the elongating root to ease its way among the soil particles. Once elongation has ceased, the cells *differentiate,* with epidermal cells sending out extensions to form *root hairs.*

5. By the time a root reaches functional maturity, it has a number of differentiated tissues: *epidermis, cortex, endodermis, pericycle, xylem, phloem, cambium,* and in some species *pith.*

6. After a branch root has grown out from the pericycle of the parent root, xylem and phloem are differentiated in the new root, and they form a functional connection with the old xylem and phloem. There is a continuous connection of the vascular system throughout the entire root system.

7. A root produced from an organ other than the radicle or another root is an *adventitious root.*

8. Since they are in an underground environment, roots are less variable than either stems or leaves. Nevertheless, considerable variations can be found.

9. The epicotyl of a germinating seed gives rise to the shoot. Practically all stem growth in length in dicots is due to mitotic activity in the *stem apex.* No one knows how the orientation of mitosis is regulated, but the result is a continuing growth of the apex, with

the *corpus*, covered by its *tunica*, moving ever upward.

10. Incipient leaves are *leaf primordia*. Soon after a stem apex produces a leaf primordium, it begins to produce a second smaller mass of cells, a *lateral bud primordium*, just above the leaf primordium.

11. Both phloem and xylem are complex tissues, each consisting of several kinds of cells. In the phloem are *sieve cells*, *companion* cells, and *phloem fibers*. In the xylem are water-conducting cells, either open-ended *tracheae* or closed *tracheids*, *xylem fibers*, *rays*, and *xylem parenchyma*.

12. The *cambium* has two kinds of cells, *fusiform initials* and *ray initials*.

13. Each year a new cylinder of xylem is laid down around last year's cylinder, producing *annual rings*. Tree-ring dating (*dendrochronology*) has made possible the establishment of accurate dates for historical edifices and the description of past climates.

14. *Softwood* comes from conifers and *hardwood* from angiosperms. The central core of wood, the *heartwood*, may be more dense and pigmented than the outer *sapwood*. *Spring wood* has larger cells and is softer than *summer wood*. *Rays* are strips of cells streaking across the grain of wood in a radial direction toward the *bark*, which contains all tissues outside the cambium.

15. The *leaves* of vascular plants are thought to have evolved from the growing together of flattened branches. In their individual development now, however, they are known to arise as outgrowths of tissue just below the growing point of the stem apex.

16. A completed leaf has a layer of *epidermal cells* over its surface. These cells are coated on the outside with waxy *cuticle*. *Stomata* are usually more common in the lower than in the upper epidermis; the stomatal pore leads into an air space inside the leaf. The internal tissues of leaves are the *spongy* and *palisade parenchyma* (or *mesophyll*) and the *veins*, consisting of *xylem* and *phloem* and sometimes *cambium*, covered by a *bundle sheath*.

17. In all but a few rare species, leaves are temporary organs. Leaf drop at the end of the growing season, *abscission*, is regulated by the amount of the growth-control compounds, *auxin* and *abscisic acid*.

RECOMMENDED READING

Adams, P., J. J. W. Baker, and G. E. Allen, *The Study of Botany*. Reading, Mass.: Addison-Wesley, 1970.

Albersheim, Peter, "The Walls of Growing Plant Cells." *Scientific American*, April 1975. (Offprint 1320)

Arnett, R. H., and G. F. Bazinet, Jr., *Plant Biology*, Fourth Edition. St. Louis: C. V. Mosby, 1977.

Bell, P. R., and D. E. Coombe, *Strasburger's Textbook of Botany*, Thirtieth Edition. New York: Longman, 1976.

Biddulph, Susann, and Orlin Biddulph, "The Circulatory System of Plants." *Scientific American*, February 1959. (Offprint 53)

Bierhorst, D. W., *Morphology of Vascular Plants*. New York: Macmillan, 1971.

Cutter, Elizabeth G., *Plant Anatomy*. Reading, Mass.: Addison-Wesley, 1978.

Epstein, Emanuel, "Roots." *Scientific American*, May 1973. (Offprint 1271)

Esan, Katherine, *Anatomy of Seed Plants*. New York: John Wiley, 1960.

Foster, A. S., and E. M. Gifford, Jr., *Comparative Morphology of Plants*. San Francisco: W. H. Freeman, 1974.

McMahon, Thomas A., "The Mechanical Design of Trees." *Scientific American*, July 1975.

Raven, Peter H., Ray F. Evert, and Helena Curtis, *Biology of Plants*, Second Edition. New York: Worth, 1976.

Wents, Frits W., *The Plants*, Revised Edition. Time-Life Books, 1971. (Life Nature Library)

17

Nutrition, Growth, and Regulation in Plants

PRACTICAL BIOLOGISTS—FOOD HUNTERS AND farmers—have known for centuries that warmth, water, and soil are necessary for plant growth, but it was only with the advent of experimental procedures, beginning in the nineteenth century, that we began to understand in some detail how plants use the environment and how they regulate their own internal affairs. Energy capture by photosynthesis (Chapter 4) and energy release by respiration (Chapter 5) have already been described. The input of materials other than carbon, hydrogen, and oxygen will be treated in the present chapter, along with those biologically synthesized materials with which plants determine their own rate, direction, and form of growth.

MINERAL NUTRITION OF PLANTS

Soil is the source of all the mineral nutrients taken in by plants and consequently (with some negligible exceptions) by animals. Soil is therefore of prime importance in the continuation of life. Far from being just an aggregation of rock particles, soil is such a changing and variable medium that pedologists, the students of soil, frequently think of it as something almost living. Besides the rock particles that make up the bulk of most soils, there are water, gases, organic matter, and living organisms.

Soils

The chemical composition of the parent rock from which any soil was derived naturally determines much of the nature of the soil (Figure 17.1). Even though all rock particles are soluble to some extent in water, some are more resistant than others. Silicates are less soluble than carbonates or phosphates. Some soils are rich in magnesium, others in iron, others in calcium. Occasionally some uncommon element, such as selenium, is present in concentrated amounts, and it can make a difference to plants (and the animals that eat the plants, too, and in that instance the incorporated element is poisonous). The size of soil particles is as important as their composition. Large particles fail to hold much water, but extremely small particles can become hard and compact. The most satisfactory soils for plant growth are of mixed size, such as occurs in a *sandy loam*. Particles ranging in size from 2 to 0.02 millimeters (mm) in diameter are classified as *sand*. *Silt* ranges from 0.02 down to 0.002 mm, and *clay* particles are smaller than 0.002 mm. Mixtures of these different-sized particles, named according to the proportions of each, are so variable that no simple general classification of the world's soils has been devised, and pedologists now try to computerize soil types in an effort to make some sense out of the multiplicity of kinds. Soils are not just loose mixes of particles, like the free-flowing sand of a dry sandy beach. Rather, they are made up of loose aggregates of particles, called *crumbs*, held in place by organic material. The crumb structure of a soil affects many of its qualities as a medium for plant roots, especially its water-holding power, its softness, and the amount of air it can contain.

The acid-base reaction of soil is important for soil fertility. If a soil solution is too acid, the crumb-producing ability of calcium is damaged, organisms in the soil are inhibited, and general productivity is impaired. If the soil solution is too alkaline, iron becomes insoluble, and organisms cannot live without iron. The "acid-loving" plants, such as the heaths (blueberries, azaleas, rhododendrons, heather), are not so much lovers of acid as lovers of iron, which they seem to need more than do most plants, and which is made available by acidity. On the pH scale (Chapter 1), with neutrality at 7, soils are generally most productive at pH 5 to 6, or slightly on the acid side. A few crops, such as legumes (clover, peas, alfalfa), do best in slightly alkaline soils.

Water may simply trickle through soil after rain without remaining long enough to be of much use to plants. Or it may be present, stuck to soil particles as a thin layer. Such water is too tightly bound to soil

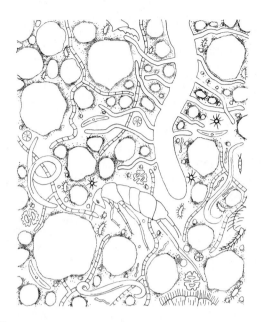

17.1

A worm's-eye view of a pinch of soil. In soil are roots with root hairs, diatoms, various green and blue-green algae, protozoa, round worms, earthworms (too large to show in such an enlarged view), bacteria, unicellular and filamentous fungi, water, dissolved minerals, air rich in carbon dioxide, such insects as springtails, and rock particles of all sizes.

to be available to roots. In clay soils, with their high proportion of extremely small particles, there may be as much as 20 percent water, but it is so firmly attached that it is useless to plants. *Capillary water*, however, is water that is present in liquid form but held firmly enough so that it does not drain away to the *water table* below. Capillary water is the important source of water to growing plants, and it must be renewed periodically, or roots must grow to it. In life, roots do indeed grow into the moist places. As a root tip advances, followed by its region of root hairs (Chapter 16), it removes the water from the soil particles it can reach, making a dry cylinder of soil around the root. This does not imply that roots actively seek water, for they do not, but it does mean that much water uptake is the result of active growth. If one recalls that Dittmer's rye plant produced nearly 600 kilometers of roots (Chapter 16) in one season, it is easy to understand that daily growth of a root system can be considerable.

Soil must be provided with abundant air spaces, or most plants cannot grow in it. If air spaces are eliminated by trampling, as on paths, or by heavy machinery, if they are filled with water by flooding, or if gas exchange with the free atmosphere is prevented by piling on of soil (as happens in construction projects), plants usually die. Corn in a flooded field or a tree whose roots are covered by a bulldozer in a building lot is doomed because the roots, which are alive and require an adequate oxygen supply, will be killed. Even under the best of conditions, air pockets in soil contain higher concentrations of carbon dioxide than are in the atmosphere because of the respiratory action of roots and soil organisms, with no compensating photosynthesis. (It is, after all, dark down there.)

Organic material makes up varying amounts of soils. In the black soils of swampy regions, the soil may be almost entirely organic and will burn if ignited, but drifting sand dunes may be practically devoid of organic material. Good growing soils have a mixture of inorganic matter, which provides lightness and crumb texture, and organic material, which holds water and helps maintain crumbs. The organic material comes from dead and partially decomposed plant and animal parts and from the excreta of animals, worked on by soil animals, bacteria, and fungi.

The familiar dark color of a good organic soil is due to the presence of *humus*, a hygroscopic, decay-resistant mix of protein and lignin derived largely from woody plant parts.

Because its inhabitants are mostly too small to see, soil organisms are rarely noticed but actually the soil is alive with bacteria, protozoa, fungi, various worms, insects, plant roots, and even large burrowing animals. A handful of a rich agricultural soil can contain as many bacteria as there are people living on the earth. As a result of the biological activity in the upper levels of the soil (living things are not common below the top meter or so), the soil is being constantly stirred up, acidified by carbon dioxide production, and altered by the living, dying, eating, and excreting of its inhabitants.

Macronutrients

Some elements are needed in rather large quantities by plants and animals. Excluding oxygen, carbon, and hydrogen, which are obtained from air and water, all the required elements must be absorbed by plants from soil. Those needed in greater amounts are calcium, nitrogen, phosphorus, potassium, sulfur, magnesium, and iron. These are the *macronutrients* (Table 17.1). To get into the living world, they must be absorbed into plant roots, usually via root hairs,

TABLE 17.1

Mineral Elements Necessary for Growth of Plants, Listed in the Order of Quantities Needed

MINERAL ELEMENTS	LOCATION, FUNCTION
Macronutrients	
Nitrogen	Part of almost every organic molecule except neutral fats and hydrocarbons.
Potassium	Required by many plant activities but with poorly understood function.
Sulfur	Part of amino acids (cystine, cysteine) and some enzymes.
Calcium	Required for firm intercellular middle lamellas; presence affects permeability of cell membranes; used in enzymes.
Magnesium	Part of chlorophyll molecules; cofactor in enzyme actions.
Phosphorus	Part of DNA, RNAs, membrane lipids, ATP and related compounds, coenzymes.
Iron	Part of molecule in enzymes, essential in cytochromes.
Micronutrients (trace elements)	
Boron	Probably concerned with sugar transport.
Chlorine	Unknown.
Sodium	Unknown.
Manganese	Coenzyme action in photosynthesis, respiration, and nitrogen activities.
Zinc	Amino acid synthesis, coenzymatic activity.
Copper	Part of respiratory enzymes.
Molybdenum	Part of nitrate-nitrite reduction in bacteria.

17.2
Determination of the need for mineral elements. Tobacco seedlings were grown in a solution containing all the elements needed (on the right) and other solutions lacking some elements. The differences in growth are apparent.

and then incorporated into plant protoplasm, where they remain until the plant dies and its components are returned to the soil by the bacteria and fungi of decay or until the plant is eaten by some other organism. Eventually, the consuming organism itself is eaten or dies. In either instance, the elements originally incorporated into some plant must find their way back to the soil to be used over and over. At any one moment, the chances are good that any human individual possesses some of the iron that once was part of the hemoglobin of the world's greatest geniuses or heroes or scoundrels.

Every element is recycled, even those that are not metabolically useful. If an element is in the soil, it will dissolve to some extent, and a plant will take it in. Traces of such useless elements as aluminum, lead, and gold can be found in plants, and their concentration is an indication of the concentration of the elements in the soil. Gold analysis of plants, for example, can lead to an activity called geobotanical prospecting, by which people hope to be able to locate precious metals. Biologists, however are mainly interested in biologically useful elements and have traced the recycling of them all through many possible pathways. Carbon and sulfur cycles are familiar, but here we will present only one especially instructive cycle, that of nitrogen (Figure 17.3).

Micronutrients
Besides the macronutrients, there are the *micronutrients*, sometimes called *trace elements*. They are required in such small concentrations that they were slow to be recognized, because efforts to demonstrate their necessity were hindered by the difficulty of ob-taining pure enough chemicals. Even now, except in the most precise cultural experiments, it is not necessary to add trace elements to ordinary culture media because the usual shelf reagents contain, as contaminants, enough zinc and boron and cobalt to satisfy at least the tougher laboratory organisms.

TRANSLOCATION IN PLANTS

Plants do not have a circulatory system in the sense that animals do, with fluid continually going and returning to a starting place. Instead, water and dissolved salts or ions usually enter plants through roots. The water mostly passes through the plant and is evaporated from the leaves, and the nonvolatile materials are left behind in the plant. The minerals are incorporated into protoplasm or into foods, and some of the water is used in photosynthesis. The resulting foods are either stored in leaves or transported to other storage organs or to active regions where they are used metabolically. There is, then, upward and downward movement of materials, but the transporting system, especially as far as water is concerned, is an open system, and it is better compared to a pipeline than to a circulating system.

Water enters roots mainly through root hairs, the driving force being primarily the diffusion pressure of the water molecules. Living roots contain many dissolved materials: sugars and other organic compounds; and salts, which are kept inside the cells by the activity of plasma membranes. The water in the soil is more nearly pure water than the water inside the root cells, and there is a difference in water con-

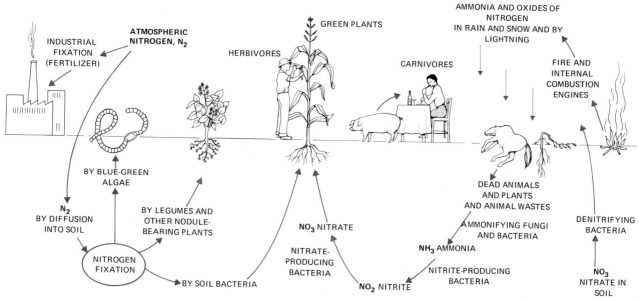

17.3

The flow of nitrogen in earth and organisms. The inorganic aspects of nitrogen flow include the commercial fixation of atmospheric nitrogen into fertilizers and the washing of nitrogen compounds (ammonia and nitric and nitrous oxides) from the air by rain and snow. Such compounds are released by fires and, increasingly, by gasoline motors. Although vascular plants cannot make protein from the nitrogen of the air, they can use nitrogen that has been oxidized to nitrate by such microorganisms as the root-nodule bacteria that live mostly in legumes, by a number of blue-green algae, and by some free-living soil bacteria. Once plants have made proteins, or at least amino acids, animals can eat the plants and get protein from them. The nitrogen in dead plants and animals and in their wastes is generally as useless to plants as atmospheric nitrogen is, but such organically bound nitrogen can be released by the action of microorganisms, especially the ammonia-producing bacteria and fungi of decay. Ammonia, in turn, can be oxidized to nitrate via nitrite (NO_2) in soil. Except in special instances (such as fossilized animals still containing some of their original amino acids), nitrogen is on the move from atmosphere to plant to animal to soil in an endless string of possibilities. Although nitrogen is the most voluminous component of the atmosphere, nitrogen-rich foods are the scarcest and most expensive of human nutrients.

centration between the soil water and cell sap. Since the cell membranes are freely permeable to water but not to the solutes in the cell sap, they allow water to diffuse in. Once in a root, water can diffuse from cell to cell in the cortex of the root, always following a "downhill gradient," that is, moving continuously from the place where there is more water (less solute) to where there is less water (more solute). Once across the cortex, the water can enter the water vessels of the root xylem. It is then free to pass up the roots and to the leaves by way of the stems, carrying some dissolved minerals with it.

Rise of Water

Water pours through stems with such speed and volume and to such heights as cannot be explained by the usual mechanical methods of water rise. Neither capillarity nor suction, aided by atmospheric pressure, could raise water to the top of even a 25-meter (80-ft) tree or vine, much less one of the tall eucalyptus or Douglasfir trees. The most widely accepted explanation for the flow of water up trees is the tension-cohesion theory of Dixon and Joly, which holds that evaporation from the leaves pulls the water up from the roots, taking advantage of the tensile

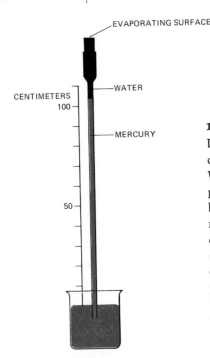

EVAPORATING SURFACE

WATER

CENTIMETERS
100

MERCURY

50

17.4
Demonstration of the pulling power of evaporation on a water column. Water cannot be pushed up by air pressure (we usually say "sucked up by a vacuum") higher than 76 centimeters (30 in.). A continuous water column, however, has enough tensile strength to allow it to pull a mercury column higher than air pressure can push it. With an apparatus like that shown here, a mercury column can be pulled up over 100 centimeters (40 in.). A plug of some porous material, such as plaster or wood, is made thoroughly wet, and the glass tube is filled with water and dipped in a mercury reservoir. As evaporation proceeds, the mercury column is pulled up until it passes the limit imposed by the weight of air on earth (atmospheric pressure). This experiment establishes the principle of "transpiration pull," by means of which evaporation from leaves can pull water from the soil to the tops of tall trees.

strength* of water in a closed tube (Figure 17.4). The theory is supported by many observations and experiments on living plants, as well as by mechanical models that can raise water higher than atmospheric pressure can account for. So long as leaves are actually transpiring, there is a flow of water from soil to roots to stems to leaves, pulled along by a water deficit in the leaves. Other suggestions, such as the active pumping of water by xylem along the way, have not stood up to experimental checking.

The transpiration pull that exists when a tree is in full leaf could not exist in a deciduous tree in early spring before leaves have appeared. At such a time, a pushing force must be acting, and the most likely source of such a push is in the roots. The tender, threadlike roots of tomato plants have been shown to be capable of pumping water from the tip toward the shoot and doing so at a high rate. One difficulty is that such pumping is demonstrable only in isolated roots in culture, not in intact plants. So far, no one has devised a convincing demonstration of effective root pressure in a normal, functioning plant. Yet water does rise in leafless trees, and even the limited demonstration of root pressure is indicative of a push-

ing force that could help water move upward during times when transpiration pull is lacking.

Phloem Transport

The movement of materials in the xylem can be followed with a fair degree of certainty. Both the usual transport of water and mineral salts upward during times of active transpiration and the spring transport of dissolved sugars, especially notable in sugar maple trees, can be readily verified. The movement of substances in the phloem cannot be so easily followed. Phloem tissues are thin, and they are protected by outer cells that are frequently thick and corky. The phloem cells themselves are so delicate that any experimental treatment is likely to interfere so seriously with their functioning that little has been learned from direct interference with them. Radioactive tracers that leave the living cells intact have been used to show that phloem can carry water and dissolved foods both up and down stems, but the forces that determine the direction of flow and provide the driving power remain unknown. Unlike water-conducting vessels in xylem, conducting tubes in the phloem are alive, although the main conducting cells, the *sieve tubes*, are strangely lacking in nuclei. The fact that the sieve tubes are perforated at ends and sides but provided with active cell content has prompted the suggestion that sieve cells somehow act as living

---

*   The tensile strength of a material is the amount of stretching force it can stand without breaking.

pumps. The smaller, closed companion cells (Chapter 16), which accompany sieve tubes, may have some effect on phloem translocation, but their actual use is not known.

One ingenious approach to the study of phloem transport is to use aphids (Figure 17.5). These tiny insects can drive their suckers through the epidermis of a stem and into a living phloem cell, then suck out the juice as if each had a microscopic hypodermic syringe. The puncture does not seem to harm the plant cell. The aphid beak, called the stylets, can be neatly severed with a razor blade, and phloem exudate will continue to flow from the cut stump until droplets can be collected for analysis. A series of such analyses shows that dissolved sugars, amino acids, and various other substances can be transported by phloem either from leaves or from roots, but the mechanisms of transport remain obscure.

GROWTH REGULATION IN PLANTS

Plants have no recognizable glands like the glands of animals; nor do they have a circulatory system. They do, however, have the ability to synthesize compounds that are transported from one part of the organism to another and that can result in specific physiological reactions. Such compounds in animals are called _hormones,_ and when regulatory compounds were found in plants, they were also called hormones. The term hormone is still used for some plant substances, but many botanists prefer "_growth-regulating substance,_" of which a number of natural and synthetic examples are known.

Tropisms and Auxin

The early work on plant growth-regulating substances was done in an effort to understand tropisms. _Tropisms_ are growth movements whose directions are determined by the direction from which the stimulus comes. They are _positive_ when the growth is toward the stimulus and _negative_ when growth is away from the stimulus. A plant that is illuminated from one side and grows toward the light is showing _positive phototropism_. If a shoot grows away from the gravitational pull of the earth, it is showing _negative geotropism_. Other tropisms are _thigmotropism_ (response to touch) and _chemotropism_ (response to chemicals).

The original observations on phototropism were made by Charles Darwin, who investigated plant movement, earthworm habits, barnacles, and animal behavior, to name his lesser contributions (Figure

17.5

Using aphids to collect samples of plant sap. An experimenter trying to obtain sap from a living tree would have to use such a gross needle that the phloem would be damaged, thus invalidating the experiment. But the slender sucker of an aphid can enter a phloem cell apparently without harming it. Then the aphid exudes the fluid from its back end, where it can be collected for analysis. In the upper picture, a section through the bark of a basswood twig shows an aphid's stylets in the phloem cells. The lower picture shows an aphid sucking on a twig. A droplet of exudate, called honeydew, is hanging from the aphid's abdomen. Thus the content of phloem fluid can be determined.

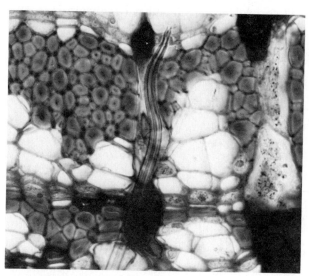

× 200

× 35

SOMATIC EMBRYOS

In the early 1960s, scientists learned how to grow plant embryos from mature, differentiated cells without meiosis, gamete formation, or fertilization. The trick is to use the techniques of plant cell culture. Mature cells are cleanly removed from a plant organ, such as a root or leaf, and placed in a sterile solution in which a particular blend of plant hormones induces cell division. Soon a mass of rapidly dividing, undifferentiated cells floats in the liquid medium. When some of the cells are moved to another medium with a different hormone mixture, they begin to grow in an organized fashion and eventually become embryos. They look remarkably similar to zygotic embryos, but since they grow from somatic (nonreproductive) tissues, they are called *somatic embryos*.

The growth of somatic embryos is an impressive feat, showing that it is possible to bypass the usual sexual cycle and still generate functional embryos. It also demonstrates that mature cells retain a full complement of genetic information; that is, they are *totipotent*. The mature cells have everything necessary to form an entire new plant.

Somatic embryos can be grown quickly and in large numbers. Over 100,000 carrot embryos can be grown in a single Petri dish in about four weeks. The embryos are wonderful subjects for studying the processes of embryonic growth and differentiation, because, unlike ordinary *zygotic embryos*, they are not buried in maternal tisue.

In the photograph, a number of somatic embryos of the caraway plant (*Carum carvi*) are floating in a medium containing sugar,

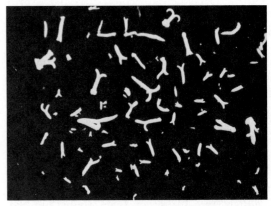

×1

mineral salts, vitamins, and equal concentrations of the plant growth regulators zeatin, gibberellic acid, and abscisic acid. The Y-shaped body has a normal, single hypocotyl and two bent cotyledons, and it looks like any embryo from a typical seed.

Because there is no genetic segregation or recombination, each somatic embryo, as well as the plant that grows from it, has exactly the same genetic makeup as the plant that originally provided the cells. Growing multiple somatic embryos therefore provides a quick and efficient method of asexual reproduction, or *cloning*. Such observations have spawned much speculation on the possibility of cloning animals, even humans. Although no human cloning has been accomplished, the technique offers promise in the rapid multiplication of special one-of-a-kind plants, such as food crops or ornamental flowers, and it proves that cloning can be a practical reality.

17.6). He used grass seedlings to show that the light-sensitive part of the seedling is the tip of the coleoptile (Chapter 16), but that the active part is *below* the coleoptile. This finding indicated that something in the tip was affected by light, but that "something," whatever it was, moved down lower to cause a curvature at some distance from the receiving source. Later experiments showed that if the tip was removed and the seedling was illuminated from one side, there was no response, but if the tip was removed and replaced *off center*, the seedling would bend even when uniformly illuminated all around or not illuminated at all. Thus was born the idea of a plant hormone.

The actual substance responsible for the observed curvatures was found after techniques for its separation were developed. First it was discovered that coleoptile tips could be cut off and placed on gelatin sheets, and small portions of gelatin, stuck to the side of a growing coleoptile, could cause bending (Figure 17.7). The active substance was extracted, concen-

17.6

Darwin's demonstration of the light receptor site in a seedling. (a) A grass seedling grown in the dark or in uniform illumination sends its coleoptile straight up. (b) If illuminated from one side, the coleoptile bends toward the light source. (c) If the tip is cut off, the coleoptile will not bend, even when unilaterally illuminated. (d) If a cap is put over the tip, the coleoptile is insensitive to unilateral illumination. (e) If the shaft of the coleoptile is kept dark by a sleeve, the same part of the coleoptile that bent in (b) will bend. Therefore the tip is the sensitive part, but the shaft of the coleoptile is the responding part.

17.7

Hormonal control of tropisms. (a) Auxin and coleoptile bending. Coleoptile tips are cut and placed on small squares of agar to allow diffusion of material from the tips into the agar. Auxin is the effective component. If an auxin-containing agar block is placed on top of a decapitated coleoptile, the coleoptile will grow at a rate equal to that of an untreated seedling, but if a coleoptile is decapitated and no auxin is provided, growth stops. For still another proof, when an auxin-treated agar block is placed off center atop a decapitated coleoptile, the auxin diffuses down one side of the coleoptile, causing it to grow faster on that side and making the shaft bend. (b) One seedling was placed upright in the center, and other seedlings, one on each side, were fastened horizontally. They were photographed repeatedly on the same film for two hours. The images show the upward bending of the horizontally placed seedlings. Since auxin is pulled downward by gravity, it accumulates on the lower side of the horizontal seedlings and causes faster growth there. The result is the upward bending of the shaft.

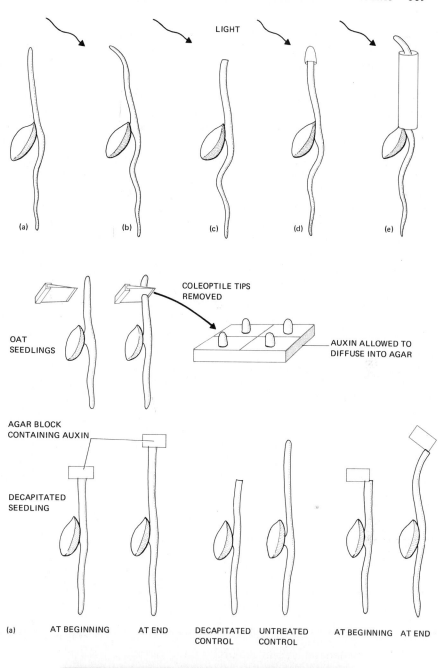

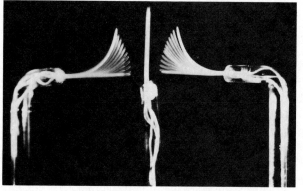

× 0.67

trated, and identified as indoleacetic acid, or IAA, now commonly called _auxin_. It can be shown experimentally that auxin is synthesized in the coleoptile tip of grass seedlings and that it moves away from its source under several influences: (1) It moves geographically downward under the influence of gravity; (2) it moves away from a one-sided light source; and (3) it moves morphologically downward, that is, "downward" with respect to the plant axis. Just as animals have a head end and a tail end, so plants have an essentially upward-growing shoot and a downward-growing root, and that _polarity_ seems to be irreversible. It can also be shown that in shoots, higher concentrations of auxin can cause cellular enlargement.

With this much information, we can begin to understand how a plant turns toward light: (1) Auxin is present in a coleoptile tip, where it is synthesized. (2) It moves downward. (3) A light source at one side drives the auxin toward the dim side. (4) The auxin, which is more concentrated on the dim side, causes the cells on that side to elongate. (5) The cells on the illuminated side do not elongate as much. (6) The shoot bends toward the light. The plant is not "trying to get to the light"; it simply cannot do anything else. A similar series of events "explains" upward growth away from gravitational pull. (The word "explains" is in quotation marks because this is only a partial explanation, and no full explanation is available.) (1) Auxin moves away from the tip, but this time it moves _morphologically downward_, regardless of the position of the seedling; that is, it moves toward the base of the plant. (2) If the seedling happens to be lying on its side, the auxin will move not only morphologically downward but also toward the actual downward side of the shoot, and it will accumulate along the lower side. (3) Cells on the lower side will elongate faster than those on the upper side. (4) The shoot will bend upward.

These explanations seem acceptable until we look at the action of roots. When a primary root emerges, it shows _positive geotropism_ and turns down. If auxin causes cells to elongate, a root on its side should grow upward, as a shoot does. The answer is that auxin causes cell enlargement only in the proper concentrations, and above a certain concentration, it can act as an inhibitor. Roots are much more sensitive to auxin than shoots are, and a concentration that can cause elongation in a shoot can cause inhibition in a root. Besides, roots have root caps, which seem to be receptors. Root caps in some way receive the gravitational stimulus and cause hormonal regulation of directional growth, involving auxin and other regulators, such as abscisic acid, to be discussed later. With

this much understood, the ability of a whole seedling to right itself becomes clear. Neither in nature nor in agriculture are seeds planted so that their radicles point down and their epicotyls up, yet when they germinate they end up properly. The auxin, moving from the shoot apex, determines that the shoot will grow upward. Auxin plus other regulators make the root grow downward until the seedling is established.

The amount of auxin used in one geotropic response is too small to measure directly. Only a millionth of a gram can be extracted from about 10,000 coleoptile tips. With concentrations as low as they are in living plants, no present-day chemical analytical methods are sensitive enough to measure the actual amounts. Under such circumstances, biologists use a technique that is effective in many instances when vanishingly small amounts of an active substance have to be determined: _bioassay_. In a bioassay, some organism of known ability is tested. Bioassays are routinely used in the measurement of quantities of vitamins, hormones, antibiotics, and of course, plant-growth substances. To perform a bioassay on auxin, one starts with a known amount of auxin. Then by successive dilutions, one reaches suitable concentrations, so that the auxin can be incorporated in a tiny block of agar, which is stuck to the side of a decapitated oat seedling grown in the dark. The amount of resultant curvature is measured and related to the amount of auxin present, and a standard curve is obtained. When an unknown quantity of auxin has to be measured, the curvature of a new set of oat seedlings is observed, and the unknown becomes the known. (Although the principle of all bioassays is the same, such determinations for vitamins, for example, are usually made on microorganisms.)

Once IAA was found to be _the_ auxin, several chemically related compounds were found to be capable of producing similar effects. Naphthalene acetic acid, indole propionic acid, and indole butyric acid, all synthetically produced, can act as auxins. Another related compound, 2,4-dichlorophenoxyacetic acid (2,4-D), is such a potent growth stimulator that, when applied in concentrated form, it can cause a plant to grow itself to death. Consequently, it is useful as a cheap weed killer. In spite of this information, however, we still do not know what there is about the nature of some molecules that gives them an auxin-like quality or how, at a molecular level, auxin affects cells.

Nontropistic Auxin Action

Auxin is involved in more than tropisms. It enters into the regulation of plant life from seed to fruit. It stimulates root initiation. Application of auxin to the

A SPECTRUM OF FLOWER COLOR

Chlorophyll is so predominantly the pigment of plants that the plant world is essentially a green world. Plants do, however, synthesize enough other pigments to produce practically any of the thousands of colors recognizable by human perception. Some flowers reflect all the light that falls on them, resulting in white. The whiteness of a flower is like the whiteness of snow. Light is reflected from countless microscopic particles—cells in a flower or crystals in snow—with air spaces in between. Flowers have no white pigment. If a flower absorbed *all* the sunlight, it would appear black. Few plant parts are black; those that approach blackness have such a concentration of green or red pigments that they appear black. Some fungi and many animals produce the dark pigment melanin, which can be quite black when concentrated. And tree bark, especially when wet, can be almost totally absorptive of light.

As a painter can make an endless variety of colors from three primary colors, so plants can make flowers of all colors from a few pigments: the reds, blues, and violets mainly from anthocyanins and related compounds, and the yellows and oranges from carotenes and carotenelike compounds. These, with the browns of tannins, the greens of the chlorophylls, and white light reflected from air spaces, yield colors of the entire spectrum. The colors of anthocyanins, shown in the flowers of Fuchsia, Meconopsis (a blue poppy), gentians, and Iris, can be changed by differences in concentration, in molecular structure, and in acidity. They are water-soluble and they accumulate in cell sap. The orange in the Gazania flowers and the yellow of Wormia (a tropical shrub) are due to carotenes, which are fat-soluble pigments in plastids of the cytoplasm.

Few plant colors are spectrally pure; virtually all are mixtures, making an array of hues, including pinks, magentas, mauves, purples, scarlets, crimsons, and blues from the intense blue of larkspurs to the sky blue of forget-me-nots.

GENTIANS

MECONOPSIS

CAMPANULA

HICKORY

WALLFLOWER

FUCHSIA

WORMIA

HICKORY

Leaves are more spectacular in fall than in spring because they are larger and easier to see, but many leaves are brilliantly colored in spring before chlorophyll becomes so concentrated that it turns them green. At the end of a growing season, chlorophyll may disintegrate, leaving the yellow and orange carotenes exposed in leaves. In addition, cool weather and bright sun contribute to the synthesis of the same anthocyanins that give flowers color. The golden yellow of hickory leaves in fall is due to carotenes, the flaming oranges of maples to a combination of carotenes and anthocyanins, and the bronzes of oaks to anthocyanins and tannins. All the variations in pigments and concentrations of pigments make possible the multicolored tapestry of an autumn forest.

OAK

MAPLE

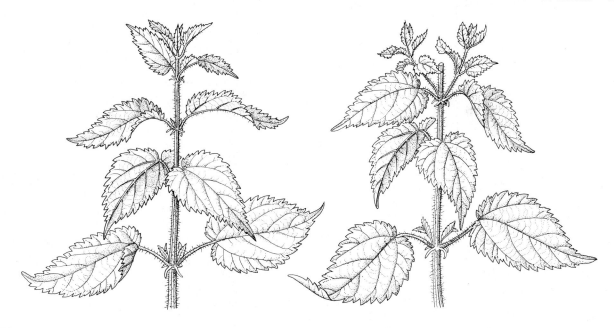

17.8

Apical dominance and the effect of removing the apex. In the intact plant, the main growing point is growing steadily, and the lateral buds in the axils of the leaves are held in check by the auxin draining down from the apex. When the apex is cut off, the auxin source is removed and the lateral buds begin to grow. In their turn, they now exert an inhibiting influence.

outside of a stem can cause it to put forth a veritable beard of roots, and synthetic auxins are commercially used to make roots grow on such woody cuttings as holly. Auxin can inhibit as well, as is shown when the auxin produced in an apical bud moves down a stem and keeps nearby axillary buds from breaking into growth. This phenomenon is called *apical dominance* (Figure 17.8). If a strong apical bud is cut off, one of the side buds will soon send out a branch and become the main apical bud, *inhibiting* in its turn the buds below it. All gardeners know that nipping off terminal buds will make a bush bushier.

Leaf abscission is prevented by auxin. If most of a leaf is cut away, an abscission layer will form and the petiole will fall off, but if auxin is artificially applied to the cut surface, the petiole will remain. The same kind of action occurs on fruit stalks, and it frequently causes fruit to drop off before ripening. Auxin sprays on apple orchards can make the trees hold the fruit instead of allowing the massive loss known as "June drop," when many little apples are likely to fall prematurely. Auxin influences the differentiation of cambium cells or cells in tissue cultures into xylem cells.

A last action of auxin is the stimulation of fruit development. When a pollen grain fertilizes an egg nucleus in an ovule, it brings enough auxin with it to start growth of the ovary wall, making the fruit grow. Most fruits will not grow if the ovules are not provided with pollen-borne auxin, although bananas and pineapples are exceptions. As we will see, however, auxin frequently acts together with other growth-regulating substances, and it is not the only one or even the most potent one.

Seed Germination and Phytochrome

As investigations on tropisms led to the discovery of auxin and its various activities, so investigations on seed germination and flower formation led to *phytochrome* (Greek, "plant color") and its activities. Phytochrome is a light-sensitive pigment that was first postulated and later isolated as a result of study of two seemingly unrelated phenomena. One is the ability of seeds to remain dormant for years and to germinate only when brought to the soil surface by plowing or washing. The other is the timing of flowering as determined by the length of day.

Not all seeds are light-sensitive. However, those of lettuce of the Grand Rapids variety do not germinate, even when warm and moist, until they have had a period of illumination. Because white light is a mixture of wavelengths, and because light-triggered re-

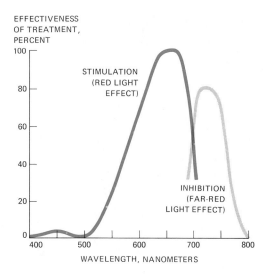

17.9

The action spectra of stimulation (colored line) and inhibition (gray line) of germination in the Grand Rapids variety of lettuce seed. The curves are made by plotting the effectiveness of the treatment against the wavelength. Light from a bright white source is passed through a prism or across a diffraction grating, which separates the light into a spectrum. Samples of seeds are placed so that each sample of, say, 100 seeds receives light of a specific wavelength. The number of germinating seeds is counted. If active germination is sought, the number of germinated seeds is plotted for each wavelength. The resulting curve is called an action spectrum. As the colored line shows, when seeds were irradiated with light at 500 nm, there was no germination, but at about 650 nm, all the seeds germinated. That gives the action spectrum for stimulation of germination.

If inhibition is being considered, the fewer the seeds germinating, the more effective the inhibition. The gray line shows a decline in germination at longer wavelengths (700–750 nm). At about 720–750 nm few seeds germinated, and the radiation inhibition was highly effective. If the shapes of the curves, especially the peaks, are compared with the absorption spectrum curves for phytochrome in Figure 17.12, the similarities are apparent. Such similarity between absorption and action spectra supports the supposition that phytochrome is the light-sensitive pigment involved in light stimulation of germination.

actions are usually the result of differential light absorption by a pigment, it was necessary to obtain an *action spectrum* of lettuce germination by exposing seeds to a series of lights of different wavelengths Figure 17.9). The first discovery was that the critical wavelength was in the red portion of the spectrum, at about 660 nanometers (a nanometer is one-billionth of a meter), a specific wavelength that stimulated germination. A second discovery was that the effect of red light could be counteracted by somewhat longer wavelengths in the so-called far red, at 730 nm. If lettuce seeds were irradiated with alternating wavelengths of 660 and 730 nm, the response was determined by the last dose received. Thus it seemed that there was an absorbing pigment that was able to undergo alteration, but the pigment was left in a physiologically active condition only after being irradiated with light of 660 nm.

Photoperiodism

While some investigators were working on seed germination and light, others were concerning themselves with the flowering habits of certain plants that had specific day-length preferences (Figure 17.10). The habit of some organisms to respond to certain lengths of time of illumination is called *photoperiodism*. The earliest subject of extensive research was a strain of tobacco known as Maryland Mammoth, which flowers only when kept in greenhouses over the winter, when the days are short. When subjected to long days, Maryland Mammoth remains vegetative. It remains vegetative even during short days if the nights are interrupted, even for a few minutes with a dim light. This kind of information, like that on seeds, suggested the activity of a pigment, and the action spectrum of flowering was obtained. If the midnight interruption was with light at 660 nm, flowering was inhibited in Maryland Mammoth even during short days, but if the interruption was with light at 730 nm, the plants flowered. Further, if the two wavelengths were alternated, flowering or lack of it was determined by the last dose received (Figure 17.11).

The fact that seed germination and flowering had so many features in common led to a search for the sensitive pigment. It appeared likely that the same pigment was acting in both phenomena, since it was reversible at the same wavelengths in both. The pigment, however, is too dilute in living plants to be visible, or even to be detected with a spectrophotometer. To complicate matters, the pigment was expected to be blue or green because it absorbs in the red portion of the spectrum; and of course chloro-

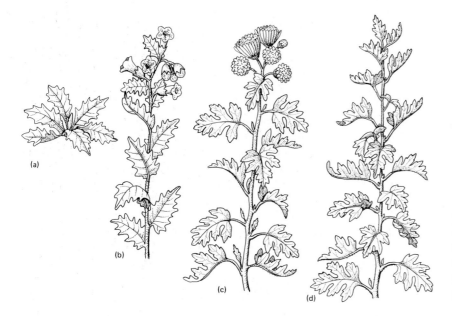

17.10

The effects of day length on flowering. Black henbane (*Hyosyamus niger*) is a long-day plant, which flowers only when days are longer than a critical period. (a) A plant of black henbane kept artificially on a regime of short days remains vegetative. (b) The same species flowers when provided with long days and short nights.

Chrysanthemum varieties, however, are frequently short-day (or long-night) plants. (c) A chrysanthemum plant kept on a short-day (long-night) regime flowers. (d) A similar plant kept on a long-day schedule remains vegetative.

17.11

The red-far-red reversal effect in the plant on which much of the original photoperiodism work was done: Maryland Mammoth tobacco. The bars indicate a 24-hour day, with the white portion representing the relative amount of light and the colored portion representing darkness. When grown in long days and short nights, the plant remains vegetative. When grown in short days and long nights, the plant flowers. However, if the long night is interrupted by a flash, only a few minutes long, of red light (or white light containing red), the plant behaves as if it had had only a short night and stays vegetative. But the effect of the red light can be obliterated by a flash of far-red light. If that is provided, the plant reverts to its short-day behavior and flowers.

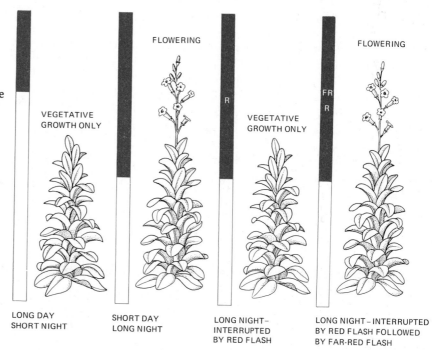

VEGETATIVE GROWTH ONLY

FLOWERING

VEGETATIVE GROWTH ONLY

FLOWERING

LONG DAY SHORT NIGHT

SHORT DAY LONG NIGHT

LONG NIGHT – INTERRUPTED BY RED FLASH

LONG NIGHT – INTERRUPTED BY RED FLASH FOLLOWED BY FAR-RED FLASH

phyll, being similar, would act as a cover for the proposed pigment, which was named phytochrome before it was actually isolated. By growing thousands of corn seedlings in the dark to prevent chlorophyll formation, biologists at the United States Department of Agriculture in Maryland isolated a pigment that gave an absorption spectrum practically duplicating the action spectra obtained from lettuce seeds and tobacco flowers. It absorbed maximally at 660 nm, but when it was so irradiated, it underwent a change and absorbed maximally at 730 nm. Then, however, if irradiated by light at 730 nm, it shifted back to 660 nm again. The pigment is somewhat similar chemically to already known algal pigments, but it also has

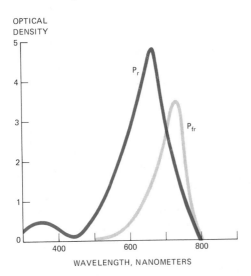

OPTICAL
DENSITY

WAVELENGTH, NANOMETERS

17.12
The absorption spectra of P_r (P_{660}) (colored line) and
P_{fr} (P_{730}) (gray line). The points on the curves are obtained from readings on a spectrophotometer, an instrument that can be used to measure the amount of light of
any given wavelength passing through a sample. The
amount of light that is absorbed at each of a series of
wavelengths is plotted with the amount of absorption, or
optical density, along the ordinate and the wavelength
along the abscissa. The resulting absorption spectrum is
characteristic of the sample, each pigment having its own
"profile." Absorption spectra are used not only to aid in
the identification of unknown samples but to find which
pigments are involved in physiological processes. Note
that P_r has its maximum absorption at 660 nm, and P_{fr}
has its maximum at 730. Compare Figure 17.9.

a protein component. As was predicted, it is bluish
green, about the color of a Go light in a traffic signal
(Figure 17.12). When phytochrome is in the form
that absorbs mainly at 660 nm, it is called P_{660} or
P_{red} or P_r; when it is absorbing at 730, it is P_{730} or
$P_{far-red}$ or P_{fr}.

Because the pigment is reversible, as are the phenomena of germination and flowering, and because
the same wavelengths cause both the photochemical
effect in the pigment and the biological effects in the
living plants, the pigment is presumed to be the
photoreceptor and to be responsible for the observed
effects. This is a general procedural method for biologists: If an absorption spectrum matches an action
spectrum, the absorbing pigment is assumed, though
not proved, to be the agent triggering the physiological event.

An additional feature of phytochrome is its ability to revert spontaneously in the dark from the P_{fr}
form to the P_r form. This feature seemed to offer an
explanation for the flowering effect in a short-day
plant. Perhaps during a long night, enough conversion occurred from P_{fr} to P_r to allow the initiation
of flower buds, but if the night was interrupted by
a flash of white light containing light at 660 nm, the
P_r would absorb it and be sent back to the P_{fr} state.
Unfortunately, this supposition is too simple, and at
present the molecular method of function of phytochrome is still unknown, especially since there are
not only *short-day* (or more properly long-night)
plants but *long-day* ones, as well as indifferent ones
that may flower under any day-night regime. The
latter two types would require a different explanation. Besides, there is a whole range of flowering
types, as shown in Table 17.2. No unified theory
explains these variations satisfactorily.

"Florigen"

Phytochrome itself is not thought to initiate the setting of flower buds directly. Instead, it causes some
as yet undiscovered metabolic change, probably by
altering the DNA activity of meristematic cells,
with resultant changes in enzyme (protein) content.
Grafting experiments indicate that some diffusible
substance is directly responsible, with the phytochrome mainly in leaves acting as the receptor (Figure 17.13). If a short-day plant is exposed to short
days, so that it is stimulated to flower, and is grafted
through a hole in an opaque partition to another
plant that is kept on a nonflowering schedule, the
plant on the nonflowering schedule will flower. This
result is taken to mean that something like a flower-

inducing hormone migrates from the short-day side to the long-day side and makes the long-day side bloom, regardless of the light duration. Repeated efforts to isolate this postulated hormone have failed, but it already has a name: _florigen_ (Latin, "flower maker"). Recently, other complex grafts between short-day and long-day plants have produced evidence of still another hormone, this one an antiflorigen that inhibits flowering. The switch from a leaf-producing to a flower-producing bud is a complex change, but progress is constantly being made toward understanding it.

Other Phytochrome Activities

With the discovery of photoperiodism and its connection with phytochrome came a whole series of discoveries of other biological phenomena that are associated with day-length and with the red-far-red reversal. These include bud dormancy, plastid orientation, root development, leaf abscission, xylem formation, differentiation of stomata, anthocyanin synthesis, and a couple of dozen more plant activities, not to mention such photoperiodic responses in animals as egg-laying behavior in mites, snails, trout, and pheasants, migration directions in birds, and mating times in horses and sheep.

The phytochrome effects in postgermination action of seedlings are an especially instructive set of

TABLE 17.2

Categories of Flowering Plants, According to Photoperiod Requirement

TYPE	REQUIREMENT, IF ANY	EXAMPLES
Long-day	Flowers only during long days (short nights).	Sugar beets Winter barley
Long-day	Long days increase flowering but are not necessary.	Lettuce Bluegrass
Short-day	Short days (long nights) are necessary	Pigweed Tobacco
Short-day	Short days increase flowering but are not necessary.	Cosmos Cotton
Day-neutral	No particular photoperiod required.	Cucumber Corn
Long-short-day	Flowers when short days follow long days.	Bryophyllum Night-blooming jasmine
Short-long-day	Flowers when long days follow short days.	Winter rye Candytuft

17.13

Grafting and flowering. The cocklebur, Xanthium, is a short-day plant. (a) If plants are grown in light-tight boxes and provided with short days, they flower; if given long days, they do not. (b) If a plant on a short-day regime is grafted through a hole in its box to another plant that is kept on a long-day regime, the second plant will flower even though the light it is receiving would normally prevent flowering. This indicates that some diffusible substance is passing through the point of the graft and causing the second plant to overcome the inhibition that its own light regime should dictate.

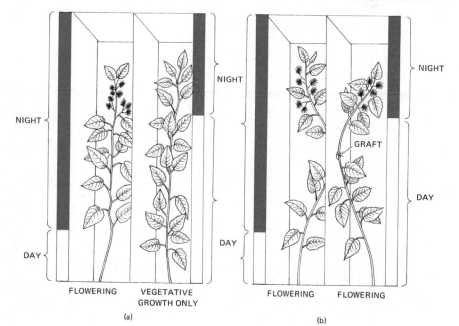

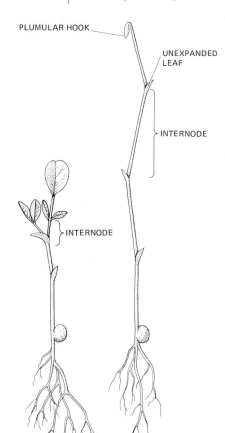

PLUMULAR HOOK

UNEXPANDED LEAF

INTERNODE

INTERNODE

17.14

The red-far-red reaction in a seedling. When illuminated with red light (or white light containing red), the seedling grows normally as shown at the left. The internodes are restricted in length, the leaves expand, and the plumular hook at the apex straightens. When illuminated with far-red light only, the seedling has much longer internodes, the leaves remain unexpanded, and the plumular hook stays bent over. Because the seedling reacts to the same wavelengths of light as it would if flowering (and certain other activities) were involved, it seems likely that the same compound—phytochrome—is responsible for the behavior in those related activities.

phenomena (Figure 17.14). When a dicot seed germinates, it sends its negatively geotropic epicotyl upward, but it keeps its first true leaves, all folded and unexpanded, bent down in such a shape as to "elbow" its way through the soil. As long as the seedling is pushing its way through soil, it is of course in the dark, and it maintains the habit described until it reaches light. Then, under the specific influence of light at 660 nm, it straightens out its plumular hook, expands its first leaves, and stops the internodal elongation. No buried seedling could get up to that

all-essential photosynthetic light without its auxin-controlled geotropism and its phytochrome-controlled structural development. The whole system functions dependably; if it did not, a seedling would starve before it reached light, and that would be the end of the line for that particular genetic combination.

Other Growth-Regulating Substances

Continuing investigation has revealed the presence of several additional growth-regulating substances. As was true of auxin and phytochrome, the discoveries came as a result of research on something other than hormones. Studies on leaf fall, on tissue cultures, on plant diseases, on dormancy of buds, and on ripening of fruits have shown that all these activities, plus a number of additional unexpected ones, are influenced by an array of compounds that can act either alone or in concert to regulate the general form and behavior of plants.

Before World War II, Japanese botanists, unknown to Western scientists, were working on a rice disease caused by a fungus, *Gibberella fujikuroi*. The main symptom was spindly growth, which gave the name "foolish seedling disease" to the trouble. When a substance named *gibberellin*, obtained from cultures of the fungus, was applied to healthy rice plants, it produced the typical symptoms, but those results were not thought to have general significance until similar compounds were isolated from normal, healthy plants. About 40 gibberellins and gibberellin-like compounds are now known and can be isolated from all parts of many green plants. Among many developmental and physiological actions that gibberellins can affect, two are especially well substantiated: growth in height and starch digestion in seeds. Some plants are genetic dwarfs (Mendel's peas, for example), with a short, bushy growth habit. Upon treatment with gibberellins, they can be made to grow tall and thin like their nondwarf counterparts. Other plants, like headed cabbage, make compact rosettes of leaves until day-length or temperature changes make them shoot upward, or "bolt," as horticulturalists say (Figure 17.15). Instead of forming a typical cabbage head, a cabbage plant dosed with gibberellic acid grows upward like the weedy wild ancestor from which the heading variety was derived. Further, when a seed begins to germinate, the stored starch in the endosperm (in barley, for example) is digested to soluble sugar. The rate of hydrolysis is determined by the gibberellic acid concentration. Gibberellins can also substitute for low temperature requirements in breaking dormancy of buds or in promoting flowering. In their ability to cause increase in cell size, gibberellins are like auxins, and the two hormones probably work together in living plants to make their final effects.

In the 1950s, F. Skoog and C. O. Miller, culturing plant tissues and trying a variety of compounds to enhance growth, found a new class of growth-regulating substances, which have been called *cytokinins* (Greek, "cell movers") because they stimulate cell divisions. Auxin can make cells enlarge, but it has no direct ability to promote mitosis. The most striking characteristic of the cytokinins is just that. It was not surprising that the cytokinins are chemically similar to a component of DNA: the nitrogen base adenine. Several natural cytokinins have been found, including zeatin and kinetin, besides some 40 adeninelike compounds. These not only can affect mitotic rates but are important in the normal development of embryos, and they have the surprising ability to prolong the durability of chlorophyll in isolated bits of leaf. Like the gibberellins, the cytokinins work in collaboration with auxin in maintaining normal plant growth.

Ethylene, C_2H_4, is the simplest and the longest known of growth regulators, although it was not at first recognized as such. It was thought of as a toxic component of industrial gases, and it was found to be a natural product of plant metabolism only years later. One of its most readily observable features is its ability to hasten the ripening of fruits. In fact, people have known for centuries that hard pear fruits would soften if they were individually wrapped in paper. Each fruit, so enclosed, kept its self-generated ethylene in the package and so brought about the desired ripening. Now ethylene is used commercially to encourage artificial softening of hard fruits. On the other hand, if softening is not wanted in stored fruits, the ethylene can be washed away by a flow of inert gas, keeping fruit firm longer. Ethylene hastens leaf abscission, inhibits internode elongation, causes increase in stem thickness, and interferes with geotropism. These actions are antagonistic to those of auxin, and it is likely that ethylene and auxin work together, as other growth-regulating substances do, to make plants conform to workable patterns. The procedure resembles the use of motor power coupled with brakes to keep a motor vehicle under control. With one force working to speed up an activity and another force working to slow it down, the activity can be kept at a functionally successful rate.

The most recent plant growth regulator to be discovered is *abscisic acid*, or ABA (Figure 17.16). It was discovered independently by several investigators working on totally different problems. Some were

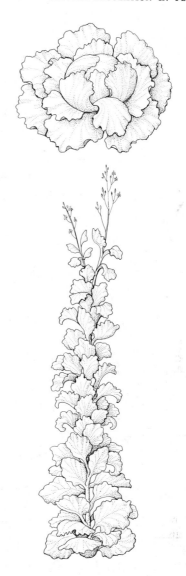

17.15

The effect of gibberellin on the growth habit of lettuce. The untreated plant at the top remains short and vegetative. The plant on the bottom has been treated with Gibberellin A_3, and it has responded by growing tall and going into its flowering phase.

17.16

Structural formula for a molecule of abscisic acid.

WITCHWEED

A North Carolina farmer plants corn. It comes up and grows, but along with it is a small, red-flowered weed, and the corn crop that year is so poor that the farmer decides not to plant corn again next year but to try peanuts instead. During the next season, a peanut crop is weed-free, so the encouraged farmer in the third year goes back to corn. There is the troublesome weed again. It attaches its roots to the roots of corn and saps the strength of the corn, seriously lowering the productivity of the crop. Its seeds lie in the soil ungerminated until the proper host begins to grow. Then the weed seeds sprout, and the seedlings send their parasitic roots to meet the host roots. After that, the weeds thrive at the expense of the host.

The parasite is one of the witchweeds of the Genus Striga, from a Greek word for "a bird that screams in the night." Our species is *Striga asiatica*, but the other species attack other grasses, such as sorghum, and they are equally destructive and specific.

The question of how witchweed seeds are able to remain dormant until corn grows nearby prompted one listener to this story to ask, "Do they have little eyes?" The answer is that a substance required for germination is diffused into the soil by corn roots but not by roots of other crops. When that substance reaches a witchweed seed, then and only then does it begin growth. By testing witchweed seeds in culture dishes and adding various compounds isolated from corn roots, experimenters identified the critical substance as kinetin (6-furfuryl-amino purine):

Witchweed seeds can be made to germinate by adding kinetin to the water in which the seeds are soaked or by growing corn roots near them, but most crops apparently do not exude kinetin into the soil, and witchweed seeds can lie dormant for years until a suitable host grows nearby.

The original problem of how to control witchweeds is still unsolved, although we know at least how the seeds regulate their own germination time. As so frequently happens, however, in seeking one answer, experimenters found a new question. Using radioactive carbon and phosphorus tracers, biologists found that water and minerals pass more readily from corn to witchweeds than in the other direction, as though a one-way valve were operating, but manufactured food does not pass readily in either direction. How does the parasite determine the direction of flow? The answer, when found, will be a straightforward physical or chemical one, *mysterious* when unknown, but no more *mystical* than the once secret reason for dormancy.

The practical aspect of parasitism of witchweeds is still a problem. How can you get rid of the parasite and leave the host undamaged?

asking why some dormant buds cannot be induced to begin growth during winter. Others were concerned with leaf fall, especially for the practical reason that commercial harvesting of cotton bolls is easier when the leaves of the cotton plant have dropped. The regulating compound involved in these and other plant activities proved to be a single substance. It is responsible for the dormancy of such woody plant buds as those of maples. It is well known that oak or maple branches brought into a warm place in midwinter will not open. Abscisic acid holds them tightly shut. Abscisic acid also acts as an antagonist to auxin in such matters as cell elongation. Here, to continue the comparison with a machine, auxin is the motor power and abscisic acid is the brake. Another feature of abscisic acid, its ability to cause stomata of leaves to close, suggests that it plays a part in controlling water loss in times of stress. Although the exact uses of all the plant growth regulators have not been clarified, enough has been learned to indicate that they act together in maintaining the integrity of entire plants.

The actual or supposed uses of the various plant hormones or growth regulators have been briefly

(a)

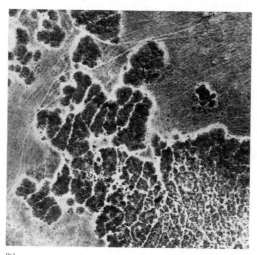

(b)

17.17
Spacing in plants. The purple sage (*Salvia leucophylla*) of the southwestern United States exudes volatile oils that are toxic to other plants. When a sage bush is established and gives off its plant-repellant compounds, other plants cannot grow close to it. (a) A sage bush (left) is surrounded by an area of bare sand in which plant growth is inhibited. (b) An aerial view of patches of sage, each clearly surrounded by white margins devoid of plants.

indicated but not their distribution within a plant. Auxin occurs throughout a plant, but it is most concentrated in the apex, young leaves, and root tips. Gibberellins are like auxins in their distribution. Cytokinins are restricted to actively growing regions where mitosis is occurring. Ethylene is most actively produced in maturing fruits, and abscisic acid in mature leaves. In our state of incomplete knowledge of these compounds, we can only say that there appears to be a balance, achieved through centuries of evolution, that keeps plants working effectively through the seasons and through the vicissitudes of cold and drought and disease.

Growth Retardants

Besides the growth regulators, there are a number of known compounds, both natural and synthetic, that inhibit vegetative growth of plants. One of these, *maleic hydrazide*, is used commercially to keep tobacco plants within desired limits, and it has been tried as a control for lawn and roadside grasses to reduce the need for mowing. Nothing in the molecular structure of maleic hydrazide or the other inhibitors gives any clue as to the reasons for their activity; we know only that they work.

Some plants regulate not only their own lives but those of others. The ability of black walnut trees to hold down the growth of plants near them has been in the folklore for centuries. In fact, a compound called *juglone* that is present in walnut bark and hulls is toxic to other plants. In regions where the living for plants is precarious, as in the arid American Southwest, competition for water is intense, and some plants have evolved means of killing off competitors by exuding inhibitors and thereby saving the little precious moisture for themselves. One plant especially effective at protecting its own territory in this way is a shrub, Adenostoma, whose exudates include a number of phenolic compounds (related to carbolic acid). Another is a desert sage, *Salvia leucophylla*, which exudes camphor and other turpentinelike products that are poisonous to seedlings. The regular spacing of these plants, obvious even to casual observers, is the result of such exudates and helps guarantee living space for a necessarily limited number of individuals (Figure 17.17). The phenomenon of keeping competitors at bay in this manner is known as *allelopathy* (Greek, "making others suffer"). For examples of territorial protection in animals, see Chapter 29, and for competition among plants, see Chapter 31.

SUMMARY

1. *Soil* is the source of all the mineral nutrients taken in by plants and consequently by most animals. Soil is therefore of prime importance in the continuation of life. Besides the rock particles that make up the bulk of most soils, there are water, gases, organic matter, and living organisms.

2. The acid-base reaction of soil is important for soil fertility. If a soil solution is too acid, the crumb-producing ability of calcium is damaged, soil organisms are inhibited, and general productivity is impaired. If the soil solution is too alkaline, iron becomes insoluble, and organisms cannot live without iron.

3. *Capillary water* is present in the soil in liquid form, but it is held firmly enough so that it does not drain away to the *water table* below. Capillary water is the water available to plants.

4. Soil must be provided with abundant air spaces, or most plants cannot grow in it.

5. Good growing soils have a mixture of inorganic matter, which provides lightness and crumb texture, and organic material, which holds water and helps maintain crumbs. *Humus* is a hygroscopic, decay-resistant mix of protein and lignin derived largely from woody plant parts.

6. The soil is alive with bacteria, protozoa, fungi, various worms, insects, plant roots, and burrowing animals that constantly stir up the soil, acidifying it by carbon dioxide production and altering it by living, dying, eating, and excreting.

7. Some elements are needed in large quantities by plants and animals—these are the *macronutrients*. Excluding oxygen, carbon, and hydrogen, which are obtained from air and water, all the required elements must be absorbed by plants from soil. Besides the macronutrients, there are the *micronutrients*, sometimes called *trace elements,* which are required in very small amounts.

8. Plants do not have a circulatory system in the sense that animals do, but there is upward and downward movement of materials in an open transporting system. Water and dissolved salts or ions enter plants usually through roots. The water mostly passes through the plant and is evaporated from the leaves, and the nonvolatile materials are left behind in the plant.

9. Water enters roots mainly through root hairs, the driving force being primarily the diffusion pressure of the water molecules.

10. The widely accepted tension-cohesion theory of Dixon and Joly states that water is able to flow up trees because the evaporation from the leaves pulls the water up from the roots, taking advantage of the tensile strength of water in a closed tube. The mechanism of transport in the phloem, however, remains obscure.

11. Plants have the ability to synthesize compounds called *growth-regulating substances,* which stimulate specific physiological actions.

12. *Tropisms* are growth movements whose directions are determined by the direction from which the stimulus comes. The actual substance responsible for tropisms in plants is *auxin*. In grass seedlings auxin is synthesized in the coleoptile tip, and it moves away from its source under several influences. Besides causing tropisms, auxin enters into the regulation of plant life from seed to fruit, inhibiting as well as stimulating.

13. *Phytochrome* is a light-sensitive pigment involved in seed germination and flowering. The action of phytochrome is influenced by its ability to change from a red-light–absorbing form (P_r) to a far-red-light–absorbing form (P_{fr}).

14. *Photoperiodism* is the response of organisms to specific lengths of time of illumination.

15. The name "florigen" has been given to a hormone not yet isolated, which seems to be responsible for inducing flower buds.

16. Other growth-regulating substances besides auxin and phytochrome have been discovered. *Gibberellins* are hormones that affect growth in height and starch digestion in seeds; *cytokinins* stimulate cell divisions; *ethylene* hastens the ripening of fruits and is generally antagonistic to auxin; and *abscisic acid* regulates leaf fall and other plant activities.

17. Besides the growth regulators, several known natural and synthetic compounds inhibit vegetative growth of plants.

RECOMMENDED READING

Frieden, Earl, "The Chemical Elements of Life." *Scientific American*, July 1972.

Hendricks, Sterling B., "Light." *Scientific American*, September 1968.

Janick, Jules, Carl H. Noller, and Charles L. Rhykerd, "The Cycles of Plant and Animal Nutrition." *Scientific American*, September 1976.

Levitt, J., *Introduction to Plant Physiology*, Second Edition. St. Louis: C. V. Mosby, 1974.

Naylor, Aubrey W., "The Control of Flowering." *Scientific American*, May 1952. (Offprint 113)

Noggle, G. R., and G. J. Fritz, *Introductory Plant Physiology*. Englewood Cliffs, N.J.: Prentice-Hall, 1976.

Strobel, Gary A., "A Mechanism of Disease Resistance in Plants." *Scientific American*, January 1975. (Offprint 1313)

van Overbeek, Johannes, "The Control of Plant Growth." *Scientific American*, July 1968. (Offprint 1111)

Weier, T. E., C. R. Stocking, and M. G. Barbour, *Botany*, Sixth Edition. New York: John Wiley, 1979.

Wilson, Carl L., Walter E. Loomis, and Taylor A. Steeves, *Botany*, Sixth Edition. New York: Holt, Rinehart and Winston, 1979.

THE
PHYSIOLOGY
OF
ANIMALS

PART SIX

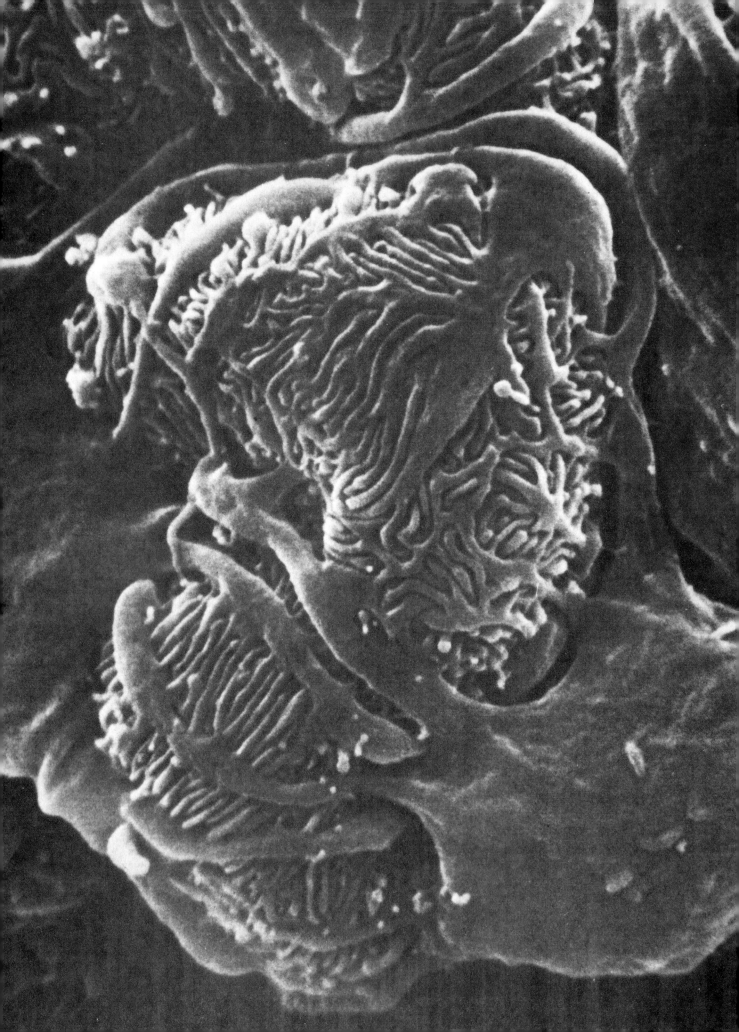

18

Animal Hormones and the Endocrine System

GLANDS ARE TISSUES OR ORGANS THAT SECRETE chemical substances. They are usually classified according to their system of secretion. Ductless glands, or _endocrine_ glands, secrete their substances directly into the bloodstream, and _exocrine_ glands, such as salivary and sweat glands, secrete into ducts that lead to specific sites. The endocrine system is one of two great regulating systems in the body; the other is the nervous system. The endocrine system also regulates such long-term processes as growth and reproductive ability, whereas the nervous system controls short, quick responses to stimuli. In fact, both systems are so interrelated that they might very well be considered a single regulatory agency. For example, the hypothalamic portion of the brain controls the secretions of many endocrine glands, and the secretions of the endocrine glands, called _hormones_ (Greek, "to arouse"), have a direct effect on many aspects of nervous control. Hormones regulate body metabolism, growth and development, stress responses, and many other critical functions. In general, hormones are indispensable for the maintenance of homeostasis.

Hormones are chemical compounds, usually either steroids (a type of lipid) or proteins. They are synthesized from raw materials in the cells and secreted into the bloodstream. When liberated into the blood, hormones are carried throughout the entire body, but they are effective only in those "target areas" where specific cells are programmed to re-

Scanning electron micrograph of mammalian kidney tissue inside the filtering capsule, showing the blood vessels from which wastes are taken, covered with fine filaments of the capsule cells.

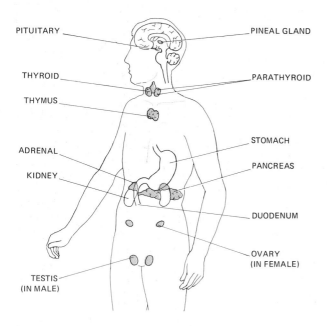

18.1

The endocrine glands of the human body.

spond to certain hormones. Hormones are effective in extremely small amounts. Only a few molecules of a given hormone may produce a dramatic response in a target cell.

Hormones generally work on a negative *feedback system* (Figure 18.2). For example, the metabolic rate of an animal body must be kept steady, and a hormonally regulated feedback system ensures steadiness. If the metabolic rate drops, a pituitary hormone, the *thyroid-stimulating hormone*, causes the thyroid gland to secrete another hormone, *thyroxine.* Thyroxine causes an increase in metabolic rate. On the other hand, if the metabolic rate rises, production of thyroid-stimulating hormone by the pituitary is slowed down or stopped, the thyroid secretes less thyroxine, and the metabolic rate is lowered. In normal individuals, the thyroid-pituitary feedback system maintains a steady state of thyroxine secretion and consequently a constant metabolic rate. Throughout this chapter we will see again and again how the endocrine glands harmonize not only with each other but with the nervous system as well.

THE VERTEBRATE PITUITARY GLAND: THE "MASTER GLAND"?

The *pituitary* gland, sometimes called the *hypophysis* (Greek, "undergrowth," because the pituitary is located directly under the brain), is a two-lobed gland about the size of a bean. The pituitary is connected by a stalk of neurons and blood vessels to the hypothalamus, facilitating an integrated coordination

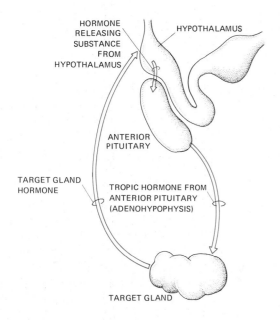

18.2

Diagram of how the thyroid gland controls metabolism and is itself controlled. The hypothalamus produces a hormone-releasing substance that causes the anterior pituitary to release a hormone into the blood. When the hormone reaches the thyroid gland, that gland releases thyroxin, which brings about an increase in metabolic rate. High metabolic rates cause a decrease in the production of the hormone-releasing substance by the hypothalamus. This self-balancing system keeps the metabolic rate of the body at a constantly fluctuating but safe level.

INVERTEBRATE HORMONES

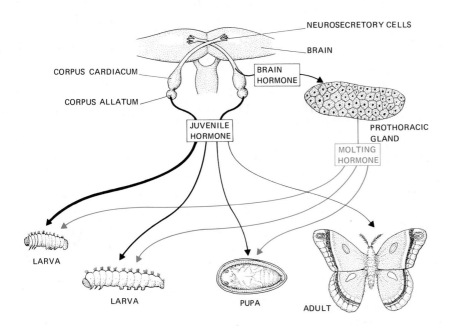

Hormone control of insect metamorphosis.

In the previous chapter we saw how the hormones of a plant regulate its growth. The growth and development of invertebrates, insects for example, are also regulated by hormones. Other activities, such as molting, coloration, and reproductive ability, are also controlled by hormonal action. Most of the hormones produced by invertebrates originate in cells of the nervous system. Although hormones have recently been identified in such invertebrates as crustaceans and mollusks, more explicit information is available on insect hormones, especially *ecdysone* and *juvenile hormone*.

In insects, a set of organs produce hormones that activate the genetic program of precisely timed and regulated developmental stages. Many insects, such as butterflies, ants, bees, and wasps, undergo a developmental transformation from egg to larva to pupa to adult. (The word "larva" comes from the Latin word for ghost, and "pupa" derives from the Latin word for doll.) This progression of developmental stages is called *complete metamorphosis*. Each stage requires the formation of a new exoskeleton as the previous one is shed, or *molted*. Molting and metamorphosis are both controlled by hormones.

Ecdysone, the molting hormone, is a steroid produced by a prothoracic gland of the larva. Ecdysone, under the control of a hormone from the brain, apparently acts directly on chromosomes to stimulate molting. Ecdysone is balanced by the second insect hormone, *juvenile hormone*, which is produced by two small organs near the brain. Both ecdysone and juvenile hormone control insect growth and development. Ecdysone favors the formation of the pupa, and juvenile hormone prolongs the larval stages through several molts. As a result of decreased production of juvenile hormone throughout the larval and pupal stages, the adult insect finally emerges.

Promising experiments with juvenile hormone have indicated that it may be effective as an insecticide spray because of its ability to prolong the larval stage. If juvenile hormone could be used in this way to inhibit insect development, it would be a welcome alternative to such environmentally harmful chemicals as DDT.

TABLE 18.1

Major Human Endocrine Glands and Their Functions

GLAND	HORMONE	FUNCTION OF HORMONE	MEANS OF CONTROL	TARGET AREA
Anterior Pituitary	Growth hormone (somatotropin, STH)	Stimulates growth, increases protein synthesis, affects metabolism	Hypothalamic growth-hormone-releasing factor	Bone and muscle
	Thyroid-stimulating hormone (TSH)	Stimulates production and secretion of thyroxine by the thyroid gland	Hypothalamic TSH-releasing factor and thyroxine	Thyroid gland
	Prolactin (lactogenic hormone, LTH)	Stimulates lactation	Hypothalamic prolactin-inhibiting factor	Mammary glands
	Adrenocortico-tropic hormone (ACTH)	Stimulates adrenal cortex secretion	Hypothalamic ACTH-releasing factor and cortisol	Adrenal cortex, skin, liver, mammary glands
	Luteinizing hormone (LH)	Stimulates development of corpus luteum, release of mature ovum, and production of progesterone and estrogen in female, secretion of testosterone in male	Hypothalamic LH-releasing factor and testosterone in male, progesterone in female	Ovaries, testes
	Follicle-stimulating hormone (FSH)	Stimulates ovulation in female, increases sperm production in male	Hypothalamic FSH-releasing factor, estrogen	Ovaries, testes
	Melanocyte-stimulating hormone (MSH)	Uncertain in humans	Unknown	Melanocytes
Posterior pituitary (Oxytocin and vaso-pressin are actually synthesized in the hypothalamus and released from the posterior pituitary.)	Oxytocin	Stimulates uterine contraction, milk ejection	Action potentials in hypothalamic secretory neurons	Uterus, mammary glands
	Vasopressin (anti-diuretic hormone, ADH)	Inhibits urine formation by controlling water reabsorption from kidneys, contracts arterioles and causes a rise in blood pressure	Action potentials in hypothalamic secretory neurons	Kidneys, smooth muscle
Thyroid	Thyroxine	Controls rate of metabolism and growth	Thyroid-stimulating hormone (TSH) of pituitary	Most cell types
	Thyrocalcitonin	Lowers calcium level in blood	Blood calcium concentration	Bone, kidneys, other cells

GLAND	HORMONE	FUNCTION OF HORMONE	MEANS OF CONTROL	TARGET AREA
Parathyroid	Parathormone (PTH)	Increases calcium level and decreases phosphate level in blood	Blood calcium concentration	Bone, intestine, kidneys, other cells
Adrenal cortex	Cortisol, cortisone, and related hormones	Regulates storage of glycogen by liver and conversion of protein into glucose; anti-inflammatory	ACTH	Many cell types
	Aldosterone	Controls retention of sodium and loss of potassium in urine	Angiotensin and blood potassium concentration	Kidneys
Adrenal medulla	Adrenalin (epinephrine)	Increases pulse and blood pressure	Sympathetic nervous system	Heart and smooth muscle
		Controls breakdown of glycogen, increases blood sugar		Liver and muscle
	Norepinephrine	Constricts blood vessels, increases metabolic rate	Sympathetic nervous system	Artery walls
Pancreas (Islets of Langerhans)	Insulin	Lowers blood sugar, increases liver and muscle glycogen	Blood glucose concentration	Muscle, liver, and adipose tissue
	Glucagon	Decreases liver glycogen, increases blood sugar	Blood glucose concentration	Liver and adipose tissue
Ovary (follicle)	Estrogens	Development of sex organs and female characteristics	FSH	Reproductive tract and other parts of the body
Ovary (corpus luteum)	Progesterone, estrogens	Influences menstrual cycle, stimulates growth of uterine wall, maintains pregnancy	LH	Uterus
Placenta	Estrogen, progesterone, chorionic gonadotropin (CG)	Maintains pregnancy	Unknown	Ovary, mammary glands, uterus
Testes	Testosterone (androgens)	Development of sex organs and male characteristics	LH	Reproductive tract and other parts of the body
Digestive tract	Secretin	Stimulates release of pancreatic juice to neutralize stomach acid	Acid in small intestine	Cells of pancreas
	Gastrin	Produces digestive enzymes in stomach	Food entering stomach	Stomach lining
	Cholecystokinin (CCK)	Stimulates release of pancreatic enzymes and gall bladder contraction	Food in small intestine	Pancreas, gall bladder

NEW IDEAS ON
BIOLOGICAL CONTROL

Hormone research in the last half of the twentieth century is enormously productive. From the discovery of a single hormone, secretin, the work on chemical controls of biological action has grown faster and faster, until dozens, even hundreds of hormones and hormonelike substances have been found in all kinds of organisms, and new ones are being described with increasing frequency. Unlike investigations on vitamins, which were productive for a while (no new vitamin has been found for about 25 years), hormone studies are continuing to yield useful results. In fact, vitamin D is now thought to act much like a hormone in the regulation of calcium metabolism.

One indication of the vitality of any scientific area is the awarding of Nobel Prizes. The 1977 Nobel Prize in Physiology or Medicine was awarded to Rosalyn S. Yalow of the Veterans Administration Hospital in New York, Andrew V. Shally of the Veterans Administration Hospital in New Orleans, and Roger C. L. Guillemin of the Salk Institute in La Jolla, California. All three were cited for their "formidable development" of research in the field of protein hormones.

Yalow, who worked with Solomon Berson until he died in 1972, received half of the $145,000 award for her development of radioimmunoassays (RIAs). The RIA technique measures hormone levels in the blood with such sensitivity that physicians can now diagnose hormonal conditions that would have escaped them previously. Use of the RIA technique has demonstrated, for example, that although diabetic children show insulin deficiencies in their bodies, adult diabetics do not. However, adult diabetics seem incapable of utilizing their insulin to control glucose levels in the blood.

The RIA technique operates by setting up a competition between a body hormone and its antibody. By introducing a small amount of a radioactive form of the hormone and measuring how much of it survives the action of the antibody, the researcher can determine how much of the natural hormone was present.

Guillemin and Shally shared the remaining half of the prize for their discovery that the hypothalamus region of the brain secretes protein hormones that control the secretions of the anterior pituitary gland. Guillemin and Shally were aided by Yalow's RIA technique, which considerably expedited their research. In this work as well, another scientist, the English anatomist Geoffrey Wingfield Harris, who died in 1971, deserves credit for the original research indicating that the hypothalamus actually controls pituitary function via connecting neurons and blood vessels. Unfortunately for Berson and Harris, Nobel Prizes are not awarded posthumously.

Several hormones of the hypothalamus have been identified so far, all of them important in controlling human body functions, and all potentially useful in medical treatment of human malfunction. Among them are the following four. (1) CRF (corticotropin-releasing factor) stimulates release of adrenocorticotropin (ACTH) from the anterior pituitary, which in turn controls the function of the adrenal cortex. (2) TRH, consisting of equal amounts of the amino acids histamine, proline, and some form of glutamic acid, may be a neurotransmitter that can influence behavior, and it may be useful in treating such diseases as schizophrenia and Parkinson's disease. (3) LHRH acts on the anterior pituitary to release both LH (luteinizing hormone) and FSH (follicle-stimulating hormone). Since LH and FSH subsequently control the functions of both the ovaries and the testes, LHRH may be useful in establishing new approaches to birth control. (4) GIF (growth-inhibiting factor), also called somatostatin, indirectly inhibits secretions of TRH from the hypothalamus, glucagon and insulin from the pancreas, and gastrin and hydrochloric acid from the stomach. GIF may prove helpful in treating diabetes, ulcers, and growth disorders.

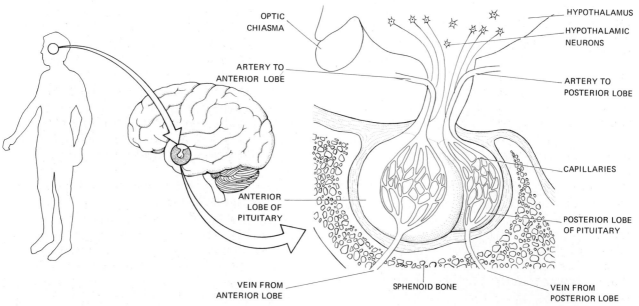

OPTIC
CHIASMA

ARTERY TO
ANTERIOR LOBE

ANTERIOR
LOBE OF
PITUITARY

VEIN FROM
ANTERIOR LOBE

SPHENOID BONE

HYPOTHALAMUS

HYPOTHALAMIC
NEURONS

ARTERY TO
POSTERIOR LOBE

CAPILLARIES

POSTERIOR LOBE
OF PITUITARY

VEIN FROM
POSTERIOR LOBE

18.3

Structure and control of the pituitary gland. This small but all-important gland has two parts, the anterior and posterior, and is located just below the hypothalamus in the brain. Secretions from the hypothalamus are carried to the two portions of the pituitary by a pair of small arteries. Hormones from the pituitary may act directly (influencing growth rates) or indirectly by causing other glands to secrete *their* hormones.

between the nervous and hormonal systems (Figure 18.3).

The pituitary is considered the most important endocrine gland (even though its daily secretions are less than one-millionth of a gram) and it is sometimes referred to as the "master gland" because of its control over most of the other endocrine glands. (In truth, the hypothalamus might justifiably be called the "master gland," since hypothalamic hormones control the secretions of the anterior pituitary.) This master control system is derived from the anterior lobe of the pituitary, which in turn is controlled by the hypothalamus. Some of the regulatory hormones secreted by the anterior lobe are the thyroid-stimulating hormone (TSH), which stimulates the thyroid to secrete *thyroxine;* the adrenocorticotropic hormone (ACTH), which means "the hormone that stimulates the cortex (outer part) of the adrenal glands"; and the gonadotropic hormones, follicle-stimulating hormone (FSH), and luteinizing hormone (LH), which control the secretions of the ovaries and testes.* The anterior pituitary also secretes a growth hormone, *somatotropin* (STH), which regulates the growth of the skeleton; and *prolactin,* a hormone that stimulates milk production in the mammary glands.

The posterior lobe of the pituitary secretes *oxytocin,* which stimulates uterine contractions during childbirth and milk secretion after childbirth; and *vasopressin* (ADH), which aids water reabsorption in the kidneys and stimulates the smooth muscles of the arteries. Emotional strain can cause increased secretions of vasopressin, which in turn produces a rise in blood pressure and an inhibition of urine formation. Drugs such as nicotine and ether stimulate the secretion of vasopressin, but alcohol has the opposite effect. For example, if you consume large quantities of beer, you will stimulate your urine production, not only because you increase the liquid content of your body, but also because the alcohol in the beer inhibits vasopressin secretion, which then stimulates urine production. Besides, another component of beer, lupulin from hops, is a *diuretic,* or urine stimulator.

---

*   We can understand how one hormone is able to stimulate the same kind of action in two very different glands (ovaries and testes) if we remember that both the male and the female sex organs have the same embryonic origin.

18.4

Growth control by pituitary hormone. A normal-sized individual is compared with a "giant," whose pituitary was overactive, and a dwarf, whose pituitary was underactive.

Some Disorders of the Pituitary

One of the most familiar results of a disorder of the pituitary gland is *giantism*, caused by oversecretion of growth hormone during the period of skeletal development (Figure 18.4). In this case, the body grows beyond the normal size range, and it may reach more than 2½ meters (8 ft) and 180 kilograms (400 lb). Death usually occurs before the individual is 30 years old. *Acromegaly* is a form of giantism that usually occurs in adults after skeletal development is complete. If too much growth hormone is secreted after maturity, the skeleton does not lengthen any further, but some cartilage and bone will thicken. Such thickening makes the face, hands, and feet widen considerably (Figure 18.5). Acromegaly may be treated with some success by using radiation treatments.

Insufficient secretion of growth hormone produces *pituitary dwarfs*, or *midgets*, persons who have normal intelligence and body proportions, but whose overall height does not progress beyond that of a normal six-year-old child. Premature senility is common, however, and pituitary dwarfs usually die before they reach 50. Although sexual development is usually minimal and reproductive ability is probably impaired, some pituitary dwarfs are capable of pro-

18.5

The effects of overproduction of pituitary growth hormone after maturity. If the pituitary recommences secretion of growth hormone after full growth is reached, only the bony extremities are affected, so that hands, feet, and jaws increase. The onset of the afflic-tion, called acromegaly (Greek, "large peaks"), is shown in this series of photographs taken of an individual who was normal until she reached adulthood but then developed the acromegalic symptoms apparent in the last picture.

18.6
Position and anatomy of the thyroid
and parathyroid glands.

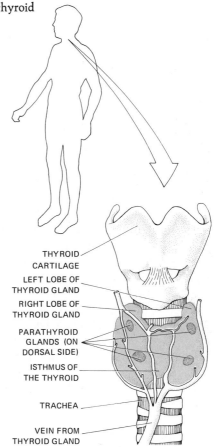

THYROID CARTILAGE
LEFT LOBE OF THYROID GLAND
RIGHT LOBE OF THYROID GLAND
PARATHYROID GLANDS (ON DORSAL SIDE)
ISTHMUS OF THE THYROID
TRACHEA
VEIN FROM THYROID GLAND

ducing normal children. As we shall see in the next
section, some dwarfism is caused by deficiencies of
thyroid hormone rather than by deficiencies of pituitary growth hormone.

THE THYROID GLAND REGULATES
METABOLIC FUNCTIONS

The *thyroid* gland has two lobes, one on each side of
the junction between the larynx (voice box) and trachea (windpipe). The two lobes of the thyroid are
connected by a bridge of thyroid tissue called an
isthmus (Figure 18.6). The thyroid hormone is *thyroxine*, an amino acid containing iodine. Thyroxine
controls *metabolism*, the rate at which food and oxygen are used to generate heat and energy throughout
the body. The synthesis of thyroxine and its release
are controlled by the thyroid-stimulating hormone
(TSH) of the pituitary gland.

Some Disorders of the Thyroid
The thyroid is concerned mainly with such metabolic
processes as growth and cellular respiration, and any
deviation from the normal rates of secretion will affect many body functions. This is especially true because thyroxine is not directed at a particular target
area but is effective throughout the body.

Overactivity of the thyroid gland results in *hyperthyroidism* (Grave's disease), a condition caused
by an oversecretion of TSH by the pituitary or by a
thyroid tumor. Symptoms include nervousness, irritability, increased heart rate and blood pressure,
weakness, weight loss, elevated use of oxygen at
rest, and bulging eyes (exophthalmus), caused in
part by an increase in the fluid behind the eyes (Figure 18.7). Drug therapy that inhibits thyroxine production and administrations of radioactive iodine
have successfully replaced surgery as treatments for
hyperthyroidism.

Underactivity of the thyroid is *hypothyroidism*,
usually associated with *goiter* (Latin, "throat"), an
enlarged thyroid. Such a swelling in the neck is
caused when insufficient iodine in the diet forces the

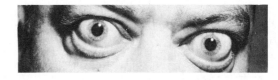

18.7
External effect of thyroxine overproduction. One of the results of heightened metabolism and increased blood
pressure is a bulging of the eyes,
which gives the sufferer a startled
expression.

thyroid to expand in an "effort" to produce more thyroxine (Figure 18.8). Most cases of goiter used to be found in areas away from the ocean, where iodine content in the soil and water supply is low (Figure 18.9). With the addition of minute amounts of iodine to ordinary table salt and drinking water in recent years, goiter has practically disappeared as a common ailment.*

Symptoms of hypothyroidism include decreased heart rate, blood pressure, and body temperature; lowered basal metabolism rate; and underactivity of the nervous system.

Underactivity of the thyroid during infancy causes *cretinism*, which results in mental retardation and irregular development of bones and muscles. The skin is dry, eyelids are puffy, hair is brittle, and the shoulders sag. Because of similar physical characteristics, cretinism is often confused with Down's syndrome (mongolism), a genetic disease (Chapter 12). Ordinarily, cretinism cannot be cured, but early diagnosis and treatment with thyroxine may arrest the disease before the nervous system is damaged. If the thyroid becomes underactive during adulthood, *myxedema* results, producing swollen facial features, dry

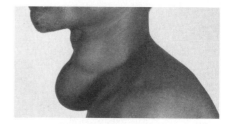

18.8

The effect of thyroxine underproduction: a *goiter*. If the thyroid gland does not secrete sufficient thyroxine, usually because of iodine deficiency in the diet, the gland becomes enlarged, although not so much in most cases as that shown here. The condition of *hypothyroidism* causes a whole series of difficulties, including mental retardation. It was formerly thought that people suffering from hypothyroidism were in some way supernatural. Hence they were called *cretins*, a variant pronunciation of the French word *chrétien*, or Christian.

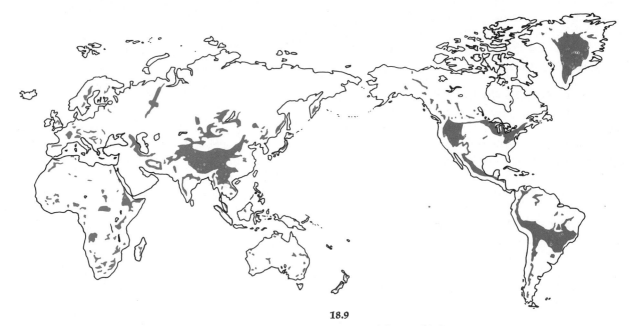

18.9

A map of the world showing the regions where goiter is (or was) most common. It is notable that none of the colored areas touches the sea. Because seawater contains iodine, people who eat food from the sea (fishes or seaweeds) obtain enough iodine and rarely suffer goiter.

---

* Almost 4000 years ago, the Chinese treated goiters (they did not call them that) by adding the ashes of burned seaweed and sponge to the diet of the afflicted individual. Although the cause of the disease was not known, and iodine in the seaweed and sponge was not consciously being prescribed, the "cure" was usually at least partially successful.

18.10

One of the causes of dwarfing is underactivity of the thyroid gland. This
circus dwarf is an example of that
phenomenon.

18.11

Position and anatomy of the adrenal
glands, the source of a number of
hormones.

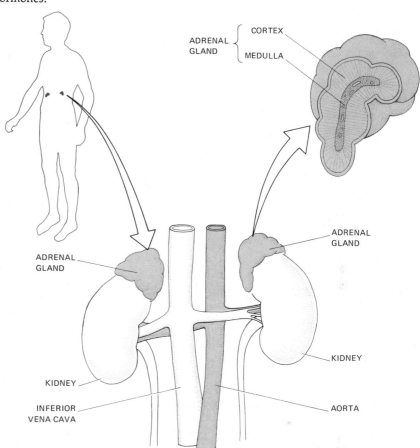

skin, low basal metabolism rate, tiredness, possible
mental retardation, and intolerance to cold in spite of
increased body weight. Like cretinism, myxedema
may be corrected if it is treated early.

The familiar "circus dwarf" is the irreversible
result of an underactive thyroid gland (Figure 18.10).
Such dwarfs are usually intelligent and active, and
they have stubby arms and legs, a relatively large
head, and flattened facial features. It has been suggested by some scientists that Attila the Hun (406?–
453) was such a dwarf, but no firm evidence exists for
this captivating claim.

THE PARATHYROID GLANDS

The four *parathyroids* are tiny glands embedded in
the thyroid gland. The parathyroid hormone, *parathormone* (PTH), regulates the use of several chemicals, especially calcium. Improper balance of calcium
in the blood may cause faulty transmission of nerve

impulses, destruction of bone tissue, and hampered
bone growth, and it may produce a paralysis of the
muscles (tetany). Small as the parathyroid glands are,
their removal invariably causes death. Before their
existence and location became known in 1850, the
parathyroids were frequently excised during goiter
surgery, and the patients died "mysteriously."

THE ADRENALS

The two *adrenal* (Latin, "upon kidneys") glands rest
like berets on top of the kidneys. Each adrenal is composed of an inner portion, the *medulla* (Latin, "marrow," meaning inside), and an outer portion, the *cortex* (Latin, "bark," referring to the outer portion of
something, just as the bark of a tree is the outer covering) (Figure 18.11).

The adrenal cortex is divided into three specific
zones, each providing a different type of steroid hormone. More than 50 adrenal cortical hormones have

JFK'S "SUNTAN" AND NAPOLEON'S NAPS

It is unlikely that the president of the United States, especially such a vigorous man as John Fitzgerald Kennedy, would want to admit to having an incurable disease. But during the campaign of 1960 it was "leaked" that Kennedy had Addison's disease, caused by an underactivity of the adrenal cortex. Ironically, the symptoms of Addison's disease include tiredness and an inability to deal with stress—characteristics not usually associated with the public JFK.

Basically, Addison's disease results from the failure of the adrenal cortex to produce its hormones. Although the full extent of Kennedy's case was never disclosed, it was known that he received oral doses of cortisol frequently. Without cortisol, Kennedy probably could not have synthesized enough blood glucose between meals, and his metabolism might have been sluggish because of an inability to mobilize

Napoleon at 34.

John Fitzgerald Kennedy.

Napoleon at about 50.

been identified, but the most widely known are the mineralocorticoid group (including aldosterone), which affect sodium metabolism, glucocorticoids (such as cortisol), which affect metabolism of foods, especially carbohydrates, and the cortical sex hormones (mostly male androgens). Cortisone and cortisol work as anti-inflammatory drugs in such diseases as arthritis, bursitis, and allergies by temporarily shifting the balance from the inflammatory mineralocorticoid hormones toward the anti-inflammatory glucocorticoids.

The adrenal medulla chiefly secretes *epinephrine*, commonly known as *adrenalin*. Adrenalin is stimulated by the nervous system to produce a condition

proteins and fats from the tissues. Even with special doses of cortisol, it is likely that President Kennedy's muscles still would have been weak.

But Kennedy looked healthy. His constant suntan, for instance, made some envious of his life. He did sail and play touch football, but his "suntan" may have been partially due to Addison's disease, since one of its characteristics is a bronzing of the skin, apparently caused when excess ACTH produces a darkening of the skin pigment.

An even more dramatic example of a public figure with a hormonal defect is Napoleon, who supposedly suffered from a disorder of the hypothalamus that affected his pituitary, adrenals, thyroid, and sex organs. (And as if that wasn't enough, he contracted syphilis in later life.) As a young man, he was thin, virile, and so energetic that he rarely needed sleep. Napoleon was only 52 when he died, and by the time he was 46, his mind and body had deteriorated so much that, according to a report of the time, he actually fell asleep during the Battle of Waterloo, which he lost.

It is easy to speculate that Napoleon's indecision and fits of hysteria caused his late military failures, but there is disagreement over the report that his autopsy provided definite proof of his brain disease. One of the forms of a hypothalamic hormonal imbalance produces feminine characteristics in a male. When Napoleon died in 1821, he had acquired many of the secondary sex characteristics of a woman, even to the widening of hips and narrowing of shoulders. He was reputed to have been proud of his full breasts.

sometimes called the "fight or flight" effect, which permits the body to react quickly and strongly to emergencies. The secretion of adrenalin causes extreme effects that last a very short time; in fact, enzymes in the liver inactivate adrenalin in about three minutes. After adrenalin has been secreted, some blood vessels constrict and others dilate, redistribut-

ing blood flow to such organs as the brain and muscles, where it is most needed. Blood is moved away from the skin, reducing the danger from a possible surface wound, blood pressure rises, and the time needed for blood clotting is reduced. Circulation increases, and enzymes are activated in the liver to release sugar from glycogen for increased energy. Digestion, an energy-consuming function, is halted during the emergency by the diversion of blood from the stomach and intestines to the muscles, where extra energy may be needed. In each case, the secretion of adrenalin favors the body's survival in times of stress.

Some Disorders of the Adrenal Glands

Apparently, the adrenal medulla is never underactive, and though overactivity is very dangerous, it is quite rare. In contrast, disorders of the adrenal cortex do occur occasionally. Underactivity of the cortex produces *Addison's disease*, whose symptoms include anemia (deficiency of red blood cells), weakness and fatigue, increased blood potassium, and decreased blood sodium. Skin color becomes bronzed because excess ACTH, produced by the pituitary in an effort to restore a normal cortical hormone level, induces abnormal deposition of skin pigment. (See the essay at left.) Until recently, Addison's disease was usually fatal, but now it can be controlled with regular doses of cortisol and aldosterone.

Two diseases caused by overactivity of the adrenal cortex are *Cushing's disease* and *adrenogenital syndrome*. Cushing's disease is usually caused by a cortical tumor that overproduces glucocorticoids. Symptoms include fattening of the face, chest, and abdomen (the limbs remain normal), accompanied by abdominal striations and a tendency toward diabetes caused by increased blood sugar. Protein is lost, and the muscles become weak. Adrenogenital syndrome is also caused by an overactive adrenal tumor, which stimulates excessive production of the cortical male sex hormones. These androgens cause male characteristics to appear in a female, and they will accelerate sexual development in a male. Hormonal disturbances during fetal development of a female child may cause a distortion of the genitals, so that the clitoris and labia become enlarged and resemble a penis and scrotum. In a mature woman, an extreme case of adrenogenital syndrome may produce a bearded lady, so frequently found in a circus sideshow, although such pronounced male characteristics in a female may be caused by other hormonal malfunctions besides defects in the adrenal cortex.

18.12

(a) Location and anatomy of the pancreas, containing the islets of Langerhans. (b) Photomicrograph of a section of pancreatic tissue, showing an islet (the central mass) containing the darker stained alpha cells and lighter beta cells.

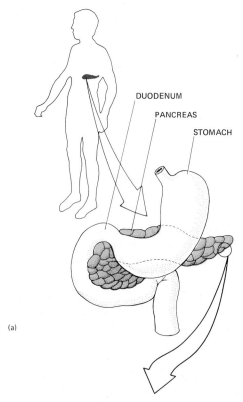

(a)

DUODENUM

PANCREAS

STOMACH

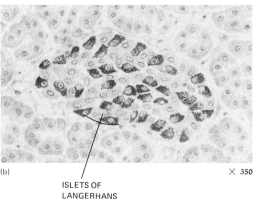

(b)   × 350

ISLETS OF
LANGERHANS

THE PANCREAS

The *pancreas* is tucked directly under the stomach (Figure 18.12). About 99 percent of its weight is involved in producing digestive enzymes, which are transported through a duct into the small intestine. The true endocrine portion of the pancreas is found in its remaining one percent. This portion of the pancreas secretes two hormones from a cluster of cells called the *islets of Langerhans*. The islets contain two special groups of cells, designated as alpha and beta. Alpha cells secrete *glucagon*, and beta cells secrete *insulin*. Insulin is secreted when the level of blood sugar rises, such as right after a meal. The most important effect of insulin is to facilitate glucose transport across cellular membranes. Insulin also enhances the conversion of glucose to glycogen, which then can be stored in the liver as a source of sugar.*

The liver acts as a regulator of blood sugar by taking up excess glucose when insulin and blood sugar are high, and by releasing sugar back into the blood when insulin and blood sugar are low (Figure 18.13). Glucagon has functions opposite to those of insulin. Glucagon's most important job is to increase blood glucose concentration by stimulating the liver to convert glycogen to glucose.

Some Disorders of the Pancreas

About four percent of the United States population will develop *diabetes mellitus* at some time in their lives. Diabetes, an inherited disease (recessive), is caused when beta cells do not produce enough insulin. When this happens, glucose accumulates in the blood and is deposited in the kidneys, but cannot enter the cells. Because the cells are unable to use the accumulated glucose (the most readily available energy source in the body), the body actually begins to starve. Appetite may increase, but eventually the body consumes its own tissues anyway, literally eating itself up.

Because the removal of glucose from the kidneys requires large amounts of water, the diabetic person produces excessive sugary urine. In response to the increased urine production, with the possibility

---

* Insulin was first extracted as a workable antidiabetic compound in 1922 by two Canadian scientists, Frederick G. Banting and C. H. Best, who worked at the Toronto laboratory of John J. R. MacLeod. Banting and MacLeod received the 1923 Nobel Prize. Insulin was the first protein for which the complete amino acid sequence was discovered. That feat was accomplished in 1954 by Frederick Sanger at Cambridge University. Sanger won the Nobel Prize in 1958.

18.13

General scheme of blood sugar regulation. The liver is the center for storage of carbohydrate, as well as the center for the change of soluble glucose to stored glycogen, and vice versa. Whether glucose is changed to glycogen or glycogen is changed to glucose is determined by the relative concentrations of the hormones insulin and glucagon in the blood. Insulin makes the liver convert glucose to glycogen, and glucagon makes it convert glycogen to glucose. These hormones are regulated by adrenal hormones, which in turn are regulated by the adrenocorticotropic hormones (ACTH), which in their turn are regulated by the brain. The system, generally automatic, can be altered by conscious states.

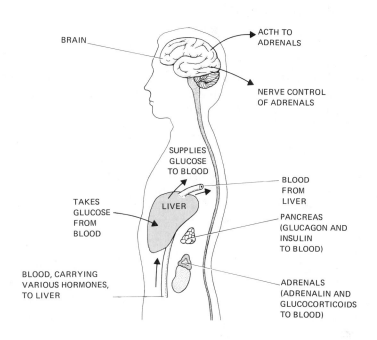

of serious body dehydration, diabetics become extremely thirsty, drink huge amounts of liquids, and may excrete as much as 20 liters of urine per day.*

Diabetes is incurable in any form, but mild diabetes can usually be controlled by strict dietary regulation. More serious cases may require treatment with regular injections of insulin. If the disease is untreated, almost every part of the diabetic's body will be affected, and gangrene, hardening of the arteries, other circulatory problems, and further complications may occur.

Medicinal insulin is generally obtained from animals, but experimentally produced insulin can be extracted from bacteria. Bacteria-produced insulin is not yet available commercially, but rat genes for insulin production have been implanted in the bacterium *E. coli*, and at least a precursor of insulin has been harvested. Genetic engineers hope that by the

---

* The word "diabetes" comes from the Greek word for "siphon," or "to pass through," referring to the seemingly instant elimination of liquids. The word was actually used by the Greeks as early as the first century. In the seventeenth century, the sweetness of diabetic urine was discovered, and the name of the disease was lengthened to diabetes mellitus. "Mellitus" comes from the Greek word *meli*, which means honey. In England, the disease had already been called "the pissing evil."

time you are reading this text, techniques for bulk production of therapeutic insulin may be available.

Low blood sugar, *hypoglycemia*, may be caused by excessive secretions of insulin. Sugar is not released from the liver, and the brain is deprived of its necessary glucose. Hypoglycemia can be controlled by regulating the diet, especially the carbohydrate intake. On a short-term basis, the sugar in a glass of orange juice may restore the normal glucose balance in the blood, but a long-term treatment would usually consist of a reduction of carbohydrates in the diet. Carbohydrates tend to stimulate large secretions of insulin, thereby removing sugar from the bloodstream too quickly.

THE GONADS

In Chapter 13 we saw how the gonads secrete hormones that control reproductive functions. Besides the ovaries and testes, other sex organs, such as the placenta and corpus luteum, also contribute to the complex feedback mechanism of hormonal control, especially in the female. We refer the reader to Chapter 13 for a review of the intricate interrelationships of hormones that regulate reproductive mechanisms in humans.

OTHER SOURCES OF HORMONES

Besides the major endocrine organs already described, other glands and organs carry on hormonal activity. We will consider briefly the pineal gland, the thymus gland, the digestive system, and the prostaglandins. The last are recently discovered hormones that are found in many parts of animal bodies.

The Pineal Gland

The pineal gland is a pea-sized body located in the midbrain. Several chemicals have been isolated from the pineal, especially melatonin. Melatonin affects skin pigmentation and seems to react to changes of light through the eye. Lizards and some amphibia appear to be able to camouflage themselves through this interrelationship of light and skin color. Although the function of the pineal gland in humans is uncertain, it may act as a "biological clock" that affects secretions of sex hormones.

The Thymus Gland

The thymus gland is located under the breastbone in the center of the chest. It is large and active only during childhood, and it reaches its maximum effectiveness during early adolescence. After that time, the thymus atrophies because of the action of sex hormones. It finally ceases activity altogether after 50 years or so, and it may therefore play an important role in the process of aging. The main function of the thymus in humans seems to be the formation of a permanent immunity system of antibodies. There is speculation that the thymus may be a true endocrine gland that secretes a hormone (thymosin) into the bloodstream to affect other regions of antibody production, such as the lymph nodes and the spleen.

The Digestive System

The best known of the several digestive hormones are the polypeptides gastrin and secretin. *Gastrin* stimulates the production of hydrochloric acid when food enters the stomach; thus gastrin is produced by the same organ, the stomach, that is its target organ. *Secretin*, the first hormone to be discovered (by British scientists Bayliss and Starling in 1903) and the first substance to actually be called a "hormone," stimulates a pancreatic secretion that neutralizes stomach acid that passes to the small intestine.

Prostaglandins

Prostaglandins are fatty-acid hormones that can be found in tissues of the intestines, liver, kidney, pancreas, heart, lung, thymus, brain, and both male and female reproductive organs. Human semen contains the highest concentration of prostaglandins in the body; however, in spite of their name, prostaglandins are mostly secreted from the seminal vesicles, not the prostate gland, as was once thought.

Some of the many effects of prostaglandins are the contraction of uterine smooth muscle, inhibition of progesterone secretion by the corpus luteum, lowering of blood pressure, enhancement of fertilization, reduction of gastric secretion, and reduction of infections through stimulation of bacterium-destroying blood cells. In addition, these hormones may be useful in relieving asthma, arthritis, ulcers, and hypertension.

Not for the first (or the last) time in this book, we must say, "The exact mechanism is not known." In fact, little is known about how prostaglandins work. These "super hormones" may work inside cells to turn chemical reactions on and off, or they may do their job by changing the composition of cellular membranes.

WHAT MECHANISM CONTROLS THE ENDOCRINE SYSTEM?

In the late 1950s, Earl W. Sutherland, Jr., and his colleagues at Case Western Reserve University isolated a substance called *cyclic AMP*. Because Sutherland found a rise in cyclic AMP following increased hormone levels, he proposed that cyclic AMP played a major role in regulating hormonal activity, and in 1971 he received the Nobel Prize in recognition of his *second-messenger concept*.

According to Sutherland's hypothesis, a hormone acts as a "first messenger," carrying its chemical message from the cells of the endocrine gland to the entire body. At the cells of the target organ, the first messenger binds to receptor sites on the cell membranes and reacts with an enzyme (adenyl cyclase), which then acts on ATP inside the cell. ATP is converted by the enzyme into cyclic AMP, which becomes the "second messenger." The second messenger then diffuses throughout the cell and finally causes the cell to respond with its specific endocrine function (Figure 18.14).

The second-messenger concept has been shown to apply to some hormones (notably the protein hormones) but not all. Another possible mechanism for the regulation of hormonal activity concerns the stimulation of protein synthesis. Several hormones, such as the thyroid hormones (and ecdysone, the molting hormone of insects and other invertebrates), seem to

18.14
Hormonal action at the cellular level. A hormone is carried by the blood all over the body (1), but it is effective only in certain places, known as "target organs." Adrenalin, for example, can cause the pancreas to respond, but not retinas or spleens. Once arrived at a target organ that recognizes the hormone (2), that organ's cells convert some of its ATP to cyclic adenosine monophosphate, cyclic AMP (3–5). That compound is called the "second messenger," as if the hormones were working as a relay team. The cyclic AMP then induces intracellular changes (6–7), such as the release of glucagon, with the resultant release of glucose from glycogen in the liver (8). When you are frightened, the recognition of danger is in your brain. That starts a chain of events: hypothalamus to pituitary to adrenals to pancreas to liver to blood to muscles. If you are scared enough, your body is quickly put into a condition of readiness, either to run with maximum speed or to dive into battle.

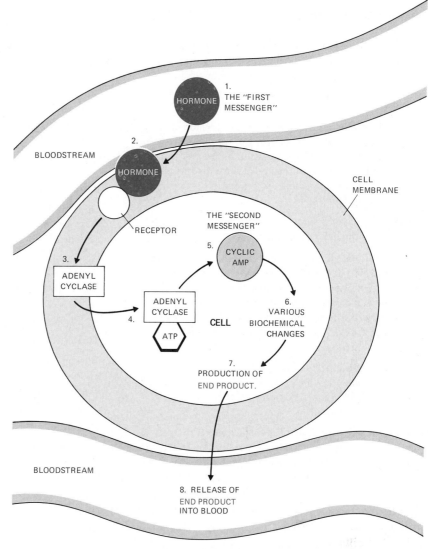

act directly on specific genes by stimulating the synthesis of proteins (enzymes) and causing the transcription of certain kinds of messenger RNA. It is possible that the hormone inhibits the repressor substance, so that the DNA of structural genes is turned on, allowing mRNA to start building enzymes. (See the discussion of the Jacob-Monod scheme in Chapter 8.) The enzymes or their end products are then available for a specific activity, such as the release of glucose into the blood.

For the time being, these two mechanisms are considered the only credible ones. Much remains to be learned about how hormones work.

SUMMARY

1. Ductless glands, or *endocrine* glands, secrete *hormones* directly into the bloodstream, whereas exocrine glands secrete into ducts that lead to specific sites. The nervous system and endocrine system are interrelated. Hormones generally work on a *feedback system*, and they are crucial for the maintenance of homeostasis.

2. The *pituitary* gland is connected by a stalk of neurons and blood vessels to the hypothalamus, facilitating an integrated coordination of nervous and hormonal activities.

3. The pituitary is considered an especially important endocrine gland because it controls most of the other

glands. Some hormones of the anterior lobe are TSH, ACTH, FSH, LH, STH (the growth hormone), and prolactin. The posterior lobe secretes oxytocin and vasopressin. Some disorders of the pituitary are *giantism, dwarfism,* and *acromegaly.*

4. The *thyroid* gland secretes *thyroxine,* which controls the rate at which food and oxygen are converted into heat and energy. Thyroxine contains the greatest concentration of iodine in the body. Some disorders of the thyroid are *hyperthyroidism, hypothyroidism* (goiter), and *cretinism.*

5. The *parathyroids* secrete *parathormone,* which mainly regulates the balance of calcium in the blood.

6. Each *adrenal* gland is composed of the *medulla* and the *cortex.* The medulla chiefly secretes *adrenalin,* and the cortex secretes many hormones, especially *aldosterone* and *cortisol,* which affect metabolism. Some disorders of the adrenal cortex include *Addison's disease, Cushing's disease,* and *adrenogenital syndrome.*

7. The *pancreas* (islets of Langerhans) produces *glucagon* and *insulin,* which respectively increase and decrease blood sugar. A common disorder of the pancreas is *diabetes.*

8. The *gonads,* especially the female gonads and secondary sex organs, secrete steroid hormones, which are involved in complex and integrated feedback systems.

9. The *pineal* gland secretes *melatonin,* which affects skin color in amphibia and may affect sex hormones in humans.

10. The *thymus* gland atrophies after adolescence. It probably produces a permanent immunity system.

11. The *digestive system* produces several hormones, including *gastrin,* which stimulates the production of hydrochloric acid when food enters the stomach, and *secretin,* which influences the neutralization of acid in the small intestine.

12. *Prostaglandins* are distributed throughout the body. The effects of prostaglidins are many, including contraction of uterine muscle, inhibition of progesterone secretion by the corpus luteum, and lowering of blood pressure. Prostaglandins may prove useful in relieving asthma, arthritis, ulcers, and hypertension.

13. Two mechanisms have been proposed for the control of the endocrine system. According to the *second-messenger concept,* a hormone causes the conversion of ATP to cyclic AMP inside a target cell. This causes the cell to respond in its specific way. Another possibility is that some hormones may inhibit repressor substance and thereby allow the buildup of enzymes (proteins) that cause the target cells to begin their activity.

RECOMMENDED READING

The following books are helpful references for the entire physiology section, Chapters 18–24.

Anthony, Catherine P., and Norma J. Kolthoff, *Textbook of Anatomy and Physiology,* Ninth Edition. St. Louis: C. V. Mosby, 1975.

Crouch, James E., and J. Robert McClintic, *Human Anatomy and Physiology,* Second Edition. New York: John Wiley, 1976.

Guyton, Arthur C., *Basic Human Physiology: Normal Function and Mechanisms of Disease,* Second Edition. Philadelphia: W. B. Saunders, 1977.

Hickman, Cleveland P., Sr., Cleveland P. Hickman, Jr., and Frances M. Hickman, *Integrated Principles of Zoology,* Fifth Edition. St. Louis: C. V. Mosby, 1974.

Langley, L. L., Ira R. Telford, and John B. Christensen, *Dynamic Anatomy & Physiology,* Fourth Edition. New York: McGraw-Hill, 1974.

Nilsson, Lennart, and Jan Lindberg, *Behold Man: A Photographic Journey of Discovery Inside the Body.* Boston: Little, Brown, 1974. (Available in paperback)

Schmidt-Nielsen, Knut, *Animal Physiology,* Third Edition. Englewood Cliffs, N.J.: Prentice-Hall, 1970. (Paperback)

Schottelius, Byron A., and Dorothy D. Schottelius, *Textbook of Physiology,* Eighteenth Edition. St. Louis: C. V. Mosby, 1978.

Vander, Arthur J., James H. Sherman, and Dorothy S. Luciano, *Human Physiology: The Mechanisms of Body Function,* Second Edition. New York: McGraw-Hill, 1975.

The following works are recommended specifically for Chapter 18.

Guillemin, Roger, and Roger Burgus, "The Hormones of the Hypothalamus." *Scientific American,* November 1972. (Offprint 1260)

LeBaron, Ruthann, *Hormones: A Delicate Balance.* New York: Bobbs-Merrill, 1972. (Pegasus paperback)

McEwen, Bruce S., "Interactions between Hormones and Nerve Tissue." *Scientific American,* July 1976. (Offprint 1341)

Nathanson, James A., and Paul Greengard, "Second Messengers in the Brain." *Scientific American,* August 1977. (Offprint 1368)

O'Malley, Bert W., and William T. Schrader "The Receptors of Steroid Hormones." *Scientific American,* February 1976. (Offprint 1334)

Pastan, Ira, "Cyclic AMP." *Scientific American,* August 1972. (Offprint 1256)

Pike J. E., "Prostaglandins." *Scientific American,* November 1971. (Offprint 1235)

19

The Action of Nerve Cells

ONCE ANIMALS EVOLVED BEYOND THE SIMPLE one-celled stage, it became necessary to coordinate the actions of all parts of a single animal. Two methods developed, one chemical and one electrical, or nervous. As we saw in the last chapter, the chemical secretions of the endocrine glands go directly into the bloodstream and to all parts of the body. In contrast, the nervous system works by means of electrochemical impulses conducted by specialized nerve cells from one specific part of the body to another.

In an organism that must find food and shelter and avoid danger, an internal information system, coupled with an awareness of the external world, must be dependable enough to ensure survival. And such an integrated system must work as fast as the food supply that might escape or the predator that might kill. At the same time, the effects of this information system must be short-lived to allow for rapid adjustment to change. A chemical control system does not meet these requirements, but an electrical one does. (The endocrine system and the nervous system have been compared to a large city's postal system and telephone system, respectively.) But is either system really in charge? Or do nerves work closely with hormones to keep the body functioning properly? Most likely it is the latter.

NEURONS ARE THE FUNCTIONAL UNITS OF THE NERVOUS SYSTEM

The fundamental unit of the nervous system is the nerve cell, or _neuron_. The human nervous system is basically an organized network of about 15 billion neurons, and branches of one neuron may touch hundreds of other neurons. Actually, most of the cells in the nervous system are not neurons at all, but _glia_, nonconducting cells that protect and nourish the neurons.* Glial cells may also function to store memory traces, and they may overexcite neurons, a condition that could bring about epileptic seizures.

The two basic properties of neurons are _excitability_, the capacity to respond to stimuli, and _conductivity_, the ability to conduct a current along a nerve cell. These properties combine chemical and electrical characteristics. We know much about the electrical characteristics of nerve impulses, but substantial information about the chemical action of nerve cells is lacking. However, we do know that when our neurons are operating, they secrete certain chemicals that affect glands, muscles, or even other neurons. These chemicals (which will be described later) are basically the same in all vertebrates.

Parts of the Neuron

Like other cells, a neuron contains a nucleus and cytoplasm, but its cytoplasm extends far beyond the main body of the cell (Figure 19.1). A neuron is made up of three identifiable parts, each associated with a specific function:

1.   The _cell body_ (soma) contains a nucleus and several organelles responsible for maintaining the metabolism, growth, and repair of the neuron. Some of the organelles in the neural cytoplasm are nissl bodies, which contain RNA and apparently synthesize protein necessary for such physiological purposes as memory tracing. Although a neuron may grow and be repaired to some degree, it is not capable of mitosis after very early childhood, and therefore it cannot reproduce after that time. Once a mature nerve cell of the brain or spinal cord dies, it cannot be replaced, and after the age of 30, human beings lose neurons in the brain at the rate of about 30 cells a minute.

---

*   In fact, glial cells protect the neurons from foreign substances so efficiently that antibiotics usually distribute poorly in the brain tissue. Similarly, white blood cells that enter the brain to fight a brain infection may merely group together in ineffective clumps.

19.1

Diagram of a generalized nerve cell. The cell body, with its nucleus and cytoplasm, has many fine, branching extensions that carry impulses toward the cell body (dendrites) or away from the cell body (axons). In this drawing, the axon is longer than the dendrites, but that is not always so. An insulating sheath of fatty myelin surrounds some of these cellular extensions. Constrictions in the myelin sheath, the nodes of Ranvier, occur at intervals. A nerve cell and its branches may be only a few micrometers from one end to the other, or the branches may be meters long, as in a giraffe's leg.

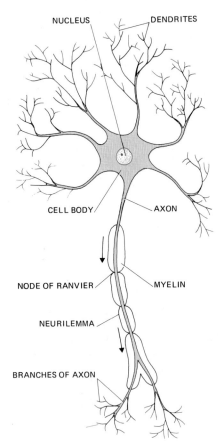

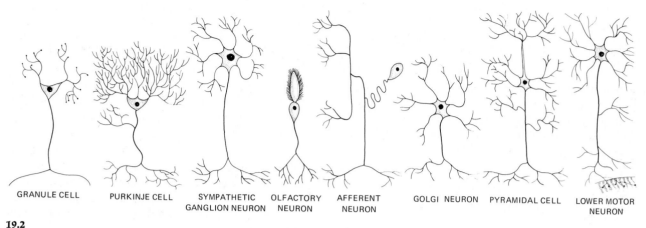

GRANULE CELL   PURKINJE CELL   SYMPATHETIC GANGLION NEURON   OLFACTORY NEURON   AFFERENT NEURON   GOLGI NEURON   PYRAMIDAL CELL   LOWER MOTOR NEURON

19.2

Examples of some of the variations in structure of nerve cells. In spite of their different appearance, they all have the same parts (cell body, axons, dendrites) and serve one basic function: the transmission of impulses.

2. The _dendrites_ (Greek, "tree"), are numerous, short, threadlike branches (processes) extending out from the cell body. Dendrites carry impulses *toward* the cell body. A cell may have as many as 200 dendrites.

3. Each nerve cell has a single _axon_, a long process extending as much as one or two meters from the cell body. (The diameter of the axon determines the velocity of the impulse conduction; thick axons conduct faster than thin ones.) An axon usually carries impulses *away* from the cell body, and it usually ends in a spray of numerous axon terminals. An axon may be enclosed by a fatty insulation sheath called _myelin_, which aids in the conduction of electrical nerve impulses. Another covering, the _Schwann sheath_, may also be present outside the brain and spinal cord (the central nervous system); it enhances the regeneration of peripheral nerve tissue, which does not include the brain and spinal cord. Multiple sclerosis is the degeneration of myelin sheaths in the brain and spinal cord; the cause is unknown, and there is no cure.

Types of Neurons

There are three types of neurons, classified according to their functions:

1 _Afferent_ neurons carry impulses *toward* the central nervous system. Some afferent neurons are also known as *sensory* neurons because they carry sensory information to the brain and spinal cord from distant parts of the body where actual stimuli (such as a stubbed toe) are received. There is no sensation until the message has been relayed to the central nervous system. Dendrites of afferent neurons receive impulses from *receptor cells*, which pick up stimuli (environmental changes).

2. _Efferent_, or *motor* neurons, carry impulses *away* from the central nervous system to the glands or muscles where the physical activity actually takes place. Axons of motor neurons terminate in *effectors*, which bring about the relevant action in the effector organ. Your bare foot jerks away from a sharp carpet tack after effectors, in the form of leg muscles, receive an impulse from efferent neurons.

Poliomyelitis (infantile paralysis), which is a viral infection, sometimes causes the destruction of the neurons leading to skeletal muscle. The result is paralysis of muscles, sometimes including those needed for breathing.

3. _Interneurons_ lie entirely within the brain and spinal cord and constitute well over 90 percent of all nerve cells. Interneurons carry impulses from sensory neurons to motor neurons, and they are involved in the processing of input information and in such complex activities as learning, emotions, and language. The number of interneurons between afferent and efferent neurons depends on the complexity of the activity. Such intricate actions as learning and memory may require thousands of interneurons. In other less complicated actions (such as a simple reflex), sensory and motor neurons may be directly connected, with few or no interneurons involved.

Neurons classified according to their structure and the number of their processes may be seen in Figure 19.2.

A *nerve* is simply a bundle of fibers (the Latin root word for nerve means "sinew") enclosed in connective tissue like many telephone wires in a cable (Figure 19.3). Nerves composed of sensory fibers that detect environmental changes are *sensory nerves*, and those made up of only motor fibers that stimulate muscles and glands are *motor nerves*. Nerves that contain both sensory and motor fibers, as most body nerves do, are called *mixed* nerves.

HOW IS A NERVE IMPULSE TRANSMITTED?

A *nerve impulse* is not the same as a flow of electricity. Instead, it is an electrochemical impulse, and it travels at a much slower rate than electricity. The actual rate is about 100 meters (330 ft) a second, as compared with about 300,000 *kilo*meters a second (the speed of light) for an electrical current. Another difference between a nerve impulse and an electrical current is that the fiber impulse does not diminish in power as it moves, as an electrical current would.

A minimum stimulus (*threshold*) is required to instigate an impulse, but an increase of intensity of the stimulus does not increase the strength of the impulse. The phenomenon is like firing a gun. If insufficient force is applied to the trigger, nothing happens. But once the sufficient minimum force is applied, pulling the trigger harder will not make the gun fire harder. The phenomenon of having a nerve cell fire at full power or not at all is an example of the *all-or-none law*.

To a person who can perceive the difference between a light touch and a strong one, or between a soft sound and a loud one, it may seem wrong to say that the all-or-none law applies. Actually, such differences can be perceived when the *frequency* of the impulse on the tissue is changed. Fibers may conduct impulses at a rate of a few a second or as many as 100 a second. The more frequent the impulses, the higher the level of excitation. Also, the *number of neurons* involved makes a difference in the way the intensity of the stimulus is perceived. For example, consider the difference you feel between getting a strong push and getting a light touch. Even though the same hand may be in contact with the same part of your body, more of your neurons are affected by the strong push. But in an experimental situation, one can show that the all-or-none law does hold.

19.3

A nerve in cross section. (a) A photomicrograph, showing bundles of fibers. Of the thousands of fibers in a nerve, some may be carrying impulses in one direction while others are carrying impulses the other way. (b) A simplified diagram of a nerve, drawn to show the individual fibers more clearly than a photograph at such a low magnification can show them.

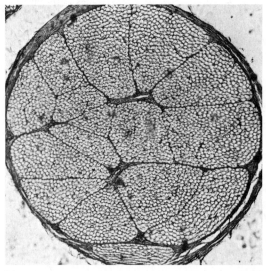

(a)   × 200

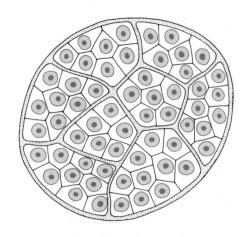

(b)

19.4

Diagram of a simple reflex arc. In the scheme as shown here, only three neurons are involved: a receptor, a connector, and an effector. An actual instance as simple as this is rare in such a complex animal as a human being, but it does illustrates the principle on which a reflex works.

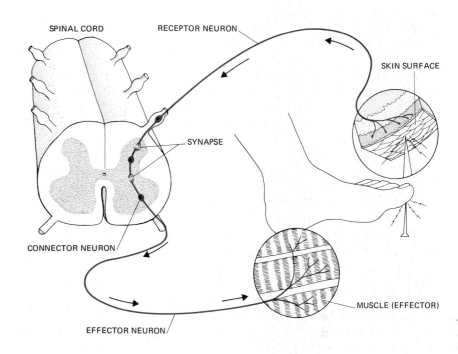

The Reflex Arc

If you hurt a finger, you pull back your hand before making a conscious decision to move. Or if a doctor taps the tendon just below your knee cap, your leg jerks up without your conscious control. Why? What has happened is a fairly simple reflex activity, conducted by two or three neurons connected in a *reflex arc* (Figure 19.4). A simple reflex arc makes possible an automatic reaction to a stimulus. It bypasses the brain, processing the incoming information in the spinal cord. In this way, the muscles can react faster than if they had to wait for instructions from the brain. Such reflex actions explain how a chicken whose head has been cut off can still run around awhile before it falls down dead.

A simple reflex arc usually involves (1) the reception of a stimulus, (2) transmission of the nerve impulse to a sensory nerve cell in a cluster of cells (a *ganglion*) in a spinal nerve, (3) transmission of the impulse to a connector neuron in the gray matter of the spinal cord, (4) transmission of the impulse to a motor nerve cell in the spinal cord, (5) transmission of the impulse out an efferent fiber to an effector, usually a muscle, and (6) contraction of the muscle. Even in most of the simplest organisms there occurs specialization of cells that results in *sensory cells* that receive outside stimuli, *nerve cells* that conduct impulses, and *effector cells* that contract and move the organism. Such a simple system is all that is necessary to make up the mechanism of a reflex arc, but such simple reflexes are rare in reality, and a reflex arc in humans is certainly much more complex. In higher animals and humans the more usual complicated reflexes are conducted by a chain of reflexes. Not only is there the possibility of several muscles being activated almost simultaneously, but the person experiencing the reflex action is aware of it and may even shift his balance—an enormously complicated action. One great difference between the complex and flexible nervous systems of human beings and the relatively rigid nervous systems and behavior patterns of simpler animals is the complexity of the human reflex arc.

The Mechanism of Nerve Action

Although little is known of how stimuli such as sound and touch can be changed into a nerve impulse, some information is available concerning the way impulses are transmitted along nerve fibers. It appears that a nerve impulse is a *wave of electrochemical change* in the distribution of electrically charged ions on both

THE GIANT AXONS OF SQUIDS

The diversity of living types does more than provide biologists with entertainment and a source of speculative philosophy. Knowing that organisms have an enormous variety of structures and ways of doing things, experimenters are always looking for special creatures that may be particularly useful as research tools. For the work Morgan wanted to do, he could scarcely have invented a better animal than fruit flies. For biochemical genetics, Beadle and Tatum found an almost ideal organism in the fungus Neurospora. An object of choice in neurological research in 1936 by J. Z. Young was a squid.

A startled squid can shoot itself backward through the water by a sudden contraction of the muscular cover over its body, the mantle. The spurt of water coming out of its siphon tubes provides jet propulsion, accompanied by a cloud of black fluid. (It has been facetiously suggested that biologists, like squids, progress by going backward as fast as possible and covering their route with ink.) Squids need and have a rapid system for sending impulses to the mantle muscles, and that system includes a so-called giant axon. Most nerve fibers are visible to the naked eye, but they are too slender for the electrodes and micropipettes that were available to neurobiologists of the 1940s. However, the big squid axon is several centimeters long, even in a small specimen, and it is up to a millimeter thick, easily thick enough to receive a thin electrode. The axon, which is formed from the fusion of many axons, is physically a tubular sheath filled with axoplasm. With measuring electrodes placed both inside and outside the axon, one can determine changes in potential as impulses pass along the axon. The giant axon can be emptied and then refilled with different solutions. When Young did so, he showed that an experimental axon can continue to transmit impulses so long as the inside solution contains potassium and the outside is bathed in a sodium solution. Further, the speed of transmission is affected by the size of the axon. An ordinary squid axon, one of the small ones, carries impulses at only about 4.3 meters per second, but the giant axons are five to ten times faster. Since the days of the classic experiments of A. L. Hodgkin and H. F. Huxley at Plymouth, England, finer electrodes have been developed and researchers can now work on ordinary nerve fibers, but the fundamental demonstrations of nerve action were made with squid axons. Such axons are important to squids because a rapid transmission system is useful in escaping danger, and to us because we can take advantage of their specialized structure to learn biological principles.

sides of the membrane of the nerve fiber (Figure 19.5). A nerve impulse is conducted as follows:

1. In a resting cell, potassium ions are concentrated inside the cell, and sodium ions are concentrated outside the cell membrane. Although potassium and sodium both have positive electrical charges, the differing amounts of ions set up a *relatively* negative charge inside the cell.* An external stimulus changes the permeability of the membrane, and a *permeability change* passes along the nerve fiber like a moving wave.

2. During the short period that the permeability change lasts, about 1/1000 of a second, sodium ions rush into the cell at the point of stimulus.

3. During that brief time, the inside and the outside electrical charges of the membrane become reversed at the point of impulse. The outside becomes negative with respect to the inside, and the inside becomes relatively positive.

4. The point of impulse on the membrane stimulates an adjacent point, and the electrical charges there become reversed as they did at the original point. This passage of the impulse is the *action potential*.

---

* Electric eels maintain noticeably high voltages (600–700 volts) because the cell membranes of their muscles are stacked one on top of the other, and all the grouped membranes work together to discharge electricity.

5. The impulse ends when some potassium ions rush to the outside of the cell.

6. The return to the original resting state is brought about by the pumping of sodium outward. Sodium ions are pumped out of the cell, and potassium ions are pumped in. As a result of the pumping action and the shifting of places by sodium and potassium, positive ions accumulate on the outer membrane, and the original condition, called the *resting potential*, is restored. The outside membrane returns to a positive charge, and the inside of the membrane becomes relatively negative once again. The membrane is said to be *repolarized*.

The sodium-potassium "pump," powered by ATP, is self-renewed at every point along the fiber; for this reason the nerve impulse does not decrease in strength with distance, as an electric current in a typi-cal wire would. After each firing, the nerve cell must rest for about 1/500 of a second. During this *refractory period* the fiber membrane is repolarized. Afterwards the fiber is "recharged" and is ready to carry out another impulse. Human beings are limited to about 300 impulses per second.

How Fast Does a Nerve Impulse Travel?

The speed of passage of an impulse along a fiber is related to the thickness of the axon. The larger the diameter of the axon, the faster is its rate of conduction. Also, thickly myelinated fibers will conduct more rapidly than fibers with only a thin sheath. The speed at which the human nerve impulse travels is relatively slow—about 100 meters a second (300 km/hr), or three times faster than a car on a highway, but slower than a jet plane. If it were any faster, it would not be synchronized with the functional ability of the

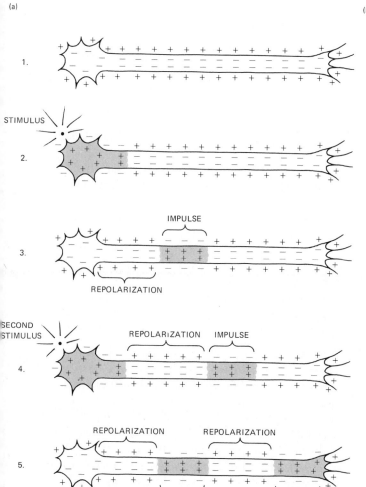

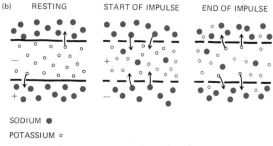

SODIUM ●
POTASSIUM ○

19.5

Diagrammatic representation of the passage of a nerve impulse. (a) An impulse as an electrical phenomenon. In (1), which shows a resting cell, the outside of the fiber membrane is positively charged with reference to the inside of the membrane. In (2) the nerve cell has received a stimulus, and the relative charges are reversed. Each point along the fiber that has reversed its charge causes the next point along the line to do the same. The movement of the charge reversal along the fiber is the impulse. (3) After the impulse has passed, the original polarized condition is reestablished. Repolarization must occur before a second impulse can be transmitted. In (4) a second impulse is received and can be transmitted because the fiber has had time for repolarization to be established. In (5) the first impulse has reached the end of the fiber, and the second one is on its way. (b) The passage of an impulse along a fiber, represented as an exchange of ions instead of a change of electrical charge. Both methods of representation are correct; they show the same phenomenon with different symbols.

19.6

Transmission of an impulse across a synapse. (a) A low-magnification view of a nerve cell, touched by an end bulb of an axon ending (encircled on the drawing). (b) An enlarged view of the end bulb, containing tiny droplets called synaptic vesicles. (c) At still greater enlargement, only a portion of one side of the end bulb is shown. The sequence of events, from left to right: a synaptic vesicle approaches the membrane, makes contact with and breaks through the membrane, and empties its contents into the cytoplasm of the nerve cell. The material in the vesicle carries the message across the synaptic gap.

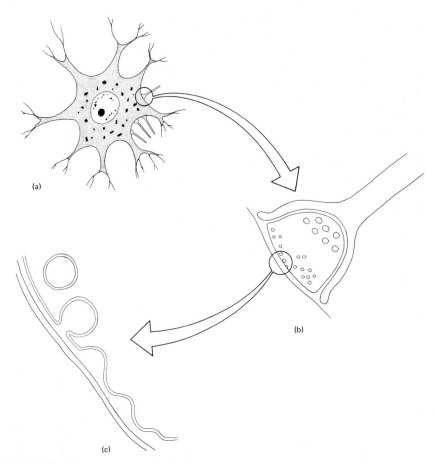

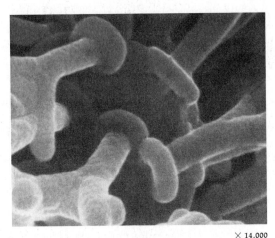

× 14,000

19.7

A scanning electron micrograph of a nerve cell (filling the center of the picture) with a number of end bulbs, or synaptic knobs, touching it. The three-dimensional effect of scanning pictures contrasts with the flat view provided by thin sections photographed in transmission electron micrographs. This material is from a mollusc, Aplysia.

bones and muscles, and the body would tangle hopelessly.

Fishes usually conduct impulses about half as fast as mammals, and snakes and earthworms may be one-third as fast as mammals, but no other animals are even close to the mammalian rate of motor conduction in nerves. Of course, *reaction* time to a stimulus is something else again. Some other animals—cockroaches, for example (see the essay on p. 512)—react faster than mammals because some of their axons are thicker than ordinary ones, and the impulse has to travel a shorter distance in a small animal.

The Synapse

There is always a narrow gap (about 500 times thinner than a strand of your hair) between one neuron and the next. How do nerve impulses travel from one neuron to another? The junction between the axon terminal of one neuron and the dendrite or cell body of the next neuron is a *synapse* (Greek, "union") (Figure 19.6). If an impulse is to reach a functional site, such as the brain or a muscle, it must pass across the chemical bridge of the synapse. The current belief is that the tips of axons release a chemical substance that crosses the synaptic gap into the dendrite or body of the next cell (Figure 19.7). One such substance is *acetylcholine* (about 10 other such "transmitters" are known).

Acetylcholine, when passed to the adjacent neuron, causes changes in the cell membrane that start the same kind of depolarization that the previous cell experienced, and a wave of excitation passes along the second nerve fiber. The impulse is on its way again after a temporary delay at the synapse. The acetylcholine, after crossing the synaptic junction, is quickly destroyed by an enzyme, *cholinesterase*. Some chemical substances, such as some insecticides and "nerve gases," inhibit the chemical action of cholinesterase. Acetylcholine is not destroyed, but no new nerve impulses can be transmitted. Muscle cells remain in a state of depolarization, and the breathing muscles are paralyzed. Death by asphyxiation results.

The synapse acts as a one-way valve, with acetylcholine accumulating in the axon ends only, not in the dendrites. Impulses can pass from axon ends across the synapse, over to dendrites of other neurons, not

the other way. Because of this built-in directional control by synapses, impulses do not travel in all directions at once. If they did, an organism with a complex nervous system would be thrown into uncontrollable spasms by every strong impulse.

Transmission across a synapse, like transmission along a nerve fiber, requires the input of energy from ATP. Details of how such transferring chemicals work are incomplete, and even the nature of some of the compounds is uncertain, especially in the central nervous system.

It is possible for one neuron to have as many as 200,000 synapses. Such a great flexibility of possible connections and pathways makes possible the brain's unique complexity and unpredictability. Of course, the pathways to be followed for any given impulse are not chosen at random. They have been carefully programmed to avoid crossed circuits and garbled messages. In the next chapter we will see how nerve impulses are actually transmitted by the nervous system to cause muscular and glandular activity.

SUMMARY

1. The fundamental unit of the nervous system is the nerve cell, or *neuron*. The two basic properties of neurons are *excitability* and *conductivity*.

2. A neuron is made up of three parts, each associated with a specific function: the *cell body*, the *dendrites*, and the *axon*.

3. There are three types of neurons, classified according to their functions: (1) *afferent* neurons carry impulses *toward* the central nervous system; (2) *efferent* neurons carry impulses *away* from the central nervous system to effector organs; (3) *interneurons* carry impulses from sensory neurons to motor neurons and they are involved in the processing of input information and in such complex activities as learning, emotions, and language.

4. A *nerve impulse* is an electrochemical impulse, not a flow of electricity. A nerve cell is either conducting an impulse at full power or not conducting at all, a phenomenon known as the *all-or-none law*.

5. A *reflex arc* is a group of neurons regulating a fairly simple activity. It includes a receptor, at least one afferent neuron, usually an interneuron, at least one efferent neuron, and an effector. Reflex arcs permit body movements without conscious participation of the brain.

6. A nerve impulse is a wave of electrical change that passes along the membrane of a nerve fiber, with inward movement of sodium ions and outward movement of potassium ions. The nerve fiber potential is restored to its resting state by the action of a "sodium pump," which moves sodium ions out of the axon.

7. A neuron conducts impulses at speeds proportional to the diameter of the axon. Human nerve impulses are conducted at speeds of about 100 meters per second.

8. A *synapse* is the junction between the axon terminal of one neuron and the dendrite or cell body of the next neuron. "Transmitter" chemicals such as acetylcholine cross the synaptic gap and excite the next nerve fiber.

RECOMMENDED READING

Helpful references for the entire physiology section, Chapters 18–24, are listed at the end of Chapter 18.

Baker, Peter F., "The Nerve Axon." *Scientific American*, March 1966. (Offprint 1038)

Lester, Henry A., "The Response to Acetylcholine." *Scientific American*, February 1977. (Offprint 1352)

Patterson, Paul H., David D. Potter, and Edwin J. Furshpan, "The Chemical Differentiation of Nerve Cells." *Scientific American*, July 1978.

Shepherd, Gordon M., "Microcircuits in the Nervous System." *Scientific American*, February 1978.

20

The Nervous System

THERE IS PROBABLY NOTHING IN THE WORLD more complex and more beautifully designed than the human nervous system. Comparisons with computers, space machines, and communications systems fail (even though we will succumb to such comparisons in this chapter). The human nervous system, and especially the human brain, is just too far ahead of the competition.

The nervous system is a single communications network, but for the convenience of explaining and learning, it is usually divided into two parts. The *central nervous system* is the brain and spinal cord, and the *peripheral nervous system* consists of those nerve cells and fibers that lie outside the brain and spinal cord. It includes the cranial and spinal nerves, and it is concerned with the transmission of sensory and motor information to all parts of the body. The special *sensory system* is involved with vision, hearing, and the other senses.

Although the nervous system of humans differs from that of other animals in the development of the brain, it is basically similar to that of the other vertebrates (Figure 20.1), and the main subject of this chapter will be the human nervous system.

THE CENTRAL NERVOUS SYSTEM

The central nervous system, consisting of the spinal cord and the brain, is protected within the backbones (vertebrae) and skull. It may be likened to a central control system, a home base for a far-reaching communications network.

The Spinal Cord

We have already seen how the spinal cord operates in reflex actions. It is also the connecting link between the brain and most of the body. The spinal cord is about 45 centimeters (18 in.) long in adults and about as thick as your index finger, and it extends through the protective vertebrae of the spinal column from the base of the brain to the second lumbar vertebra (about waist level). In an infant, the spinal cord is about the same length as the spinal column, but the column eventually outgrows the cord.

The space in the spinal column not occupied by the cord is filled by connective tissue and the meninges, three protective membranes enveloping the

20.1

Nervous systems of four animals. A coelenterate animal, Hydra, is simple, compared with a mammal, but it does have tissue differentiation, including a network of specialized cells that function as a primitive nervous system. A flatworm, Dugesia, has a definite head end, containing ganglia, and a pair of readily visible nerves, one running down each side of the body, with branches to all parts. An insect, such as a honeybee, Apis, has a complex, well-developed nervous system, with ganglia and a central nerve cord running the length of the animal on the ventral side, with branches to all sensory and moving parts. Mammals, especially the human species, have the most highly specialized nervous systems of all animals.

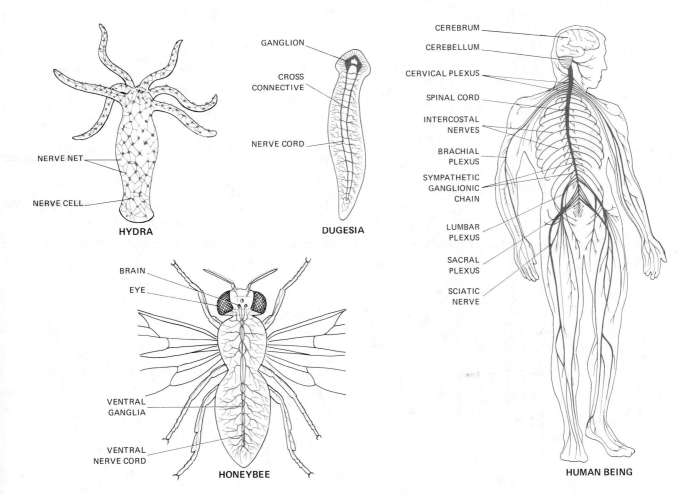

HYDRA · NERVE NET · NERVE CELL

DUGESIA · GANGLION · CROSS CONNECTIVE · NERVE CORD

HONEYBEE · BRAIN · EYE · VENTRAL GANGLIA · VENTRAL NERVE CORD

HUMAN BEING · CEREBRUM · CEREBELLUM · CERVICAL PLEXUS · SPINAL CORD · INTERCOSTAL NERVES · BRACHIAL PLEXUS · SYMPATHETIC GANGLIONIC CHAIN · LUMBAR PLEXUS · SACRAL PLEXUS · SCIATIC NERVE

20.2

Gross anatomy of the human spinal cord and the bones that protect it. Each vertebra, precisely fitted to the ones on each side of it, has a passageway for the spinal cord and openings between the spinal processes through which pass the pairs of *spinal nerves*. As the figure shows, spinal nerves do not leave the spinal cord as single, solid outgrowths; instead, they have two parts, a *ventral* and a *dorsal root*. A tiny canal, too small to see in a picture of this size, passes the length of the cord. The central region of the cord contains the *gray matter*, which in cross section is shaped somewhat like the letter H. The *white matter* surrounds the gray matter. The entire cord is covered by the *meninges*, known to many people because infection of the meninges is spinal meningitis. Note that the bulk of the vertebral column is on the ventral side of the cord, and that the bony covering of the cord is thinner on the side toward the skin of the back than it is on the side toward the belly. The bumps that can be felt in the midline of a human back are *spinal processes* sticking out from the vertebrae.

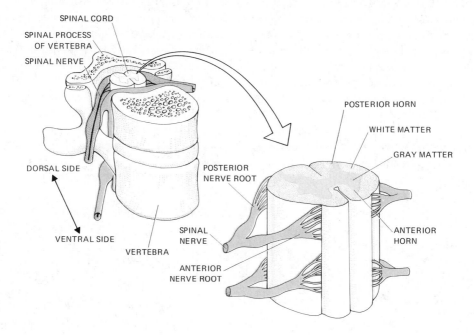

SPINAL CORD

SPINAL PROCESS OF VERTEBRA

SPINAL NERVE

DORSAL SIDE

VENTRAL SIDE

VERTEBRA

POSTERIOR NERVE ROOT

SPINAL NERVE

ANTERIOR NERVE ROOT

POSTERIOR HORN

WHITE MATTER

GRAY MATTER

ANTERIOR HORN

cord. The meninges, from the outside in, are the dura mater, arachnoid layer, and pia mater.

When cut across, the spinal cord shows a tiny central canal, a dark portion of H-shaped "gray matter," surrounded by "white matter." The gray matter consists mostly of cell bodies and synapses, and the white matter consists mainly of white myelinated nerve fibers, which extend up and down the cord (Figure 20.2). Inside the white matter are several pathways, or tracts. The descending tracts carry motor impulses to skeletal muscle fibers, and the ascending tracts carry sensory impulses to the brain. If the spinal cord is severed, there will be no motor or sensory function below the cut.

There are 31 pairs of spinal nerves branching to each side of the body, from the neck all the way down the column. Spinal nerves are mixed, containing both sensory and motor fibers that, with the cranial nerves, form the peripheral nervous system.

The Brain

The human adult brain weighs about 1.4 kilograms (3 lb),* contains more water (about 85 percent) than any other organ in the body, and has the consistency of scrambled eggs. Yet it is the most complex and organized instrument ever evolved.

The brain consumes about one-quarter of the oxygen being used in the entire body, and it is so sensitive to shortages of either oxygen or glucose that lack of these substances will cause extremely rapid damage to brain tissue. Proper and continuous nourishment is so important to the brain that it has built-in regulating devices that make *impossible* a

---

*   The human brain is almost completely grown by the time we are seven years old, but soon after birth, nerve cells begin to die. And they continue to die throughout our lives, never again equaling in number what we had at birth.

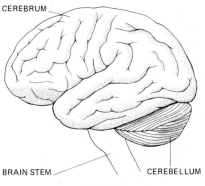

CEREBRUM

BRAIN STEM   CEREBELLUM

SIDE VIEW

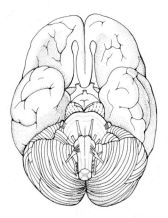

FROM BELOW

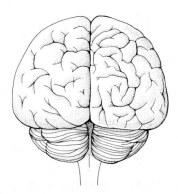

BACK VIEW

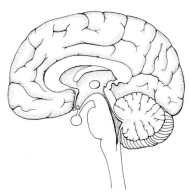

LONG (SAGITTAL) SECTION

20.3
Side, bottom, back, and sectional
views of the human brain.

constriction of blood vessels that would reduce the incoming blood supply, even though it is relatively easy to dilate (enlarge) these same vessels through drug use.

The brain is covered by the same three meninges that protect the spinal cord. Between the innermost layer and the middle membrane is a space filled with cerebrospinal fluid. The main function of the fluid and meninges is to cushion and protect the brain. The cerebrospinal fluid supplies a special "floating" environment, so the very soft brain is relieved of gravitational pull and is cushioned against hard blows on the skull and sudden movements that might force the brain against the skull. Such damage to the brain may cause concussion, which means "shake violently" in its Latin form.*

The three main parts of the brain are the brainstem, cerebellum, and cerebrum (Figure 20.3). The brainstem is the stalk of the brain, relaying messages between the spinal cord and the brain. It is further divided into the medulla, pons, and midbrain. Several groups of neurons in the *medulla* control such vital functions as heart rate, breathing rate, and blood pressure. The medulla also regulates vomiting, sneezing, coughing, and swallowing. The *pons* forms a bulge at the forward portion of the brainstem. It consists mainly of fibers from the medulla to the cerebrum, and of connections to the cerebellum. The pons controls certain respiratory functions. The *midbrain,* at the anterior end of the brain stem, is directly connected to the pons. Reflex centers are located toward the rear of the midbrain. Some relatively minor visual, auditory, and postural reflexes are monitored in the midbrain.

---

* Woodpeckers seem to be prime candidates for brain damage or concussion. They may bang their beaks against trees 40–45 times in less than three seconds, and they usually drum several hundred times a day, more during the mating season. Concussion and other injury are avoided, however, because woodpeckers have a narrow air space between the brain and its outer membrane; this air space probably reduces the fluid transmission of shock waves. In addition, a woodpecker's brain is cushioned with spongy bone and thick muscles that also act as shock absorbers.

20.4

A generalized brain, compared with brains of selected vertebrate animals. The five main sections of a brain vary with the degree of specialization an animal has evolved. In a fish, the olfactory lobes are larger than the cerebrum, but in an amphibian, that relationship is reversed. In a human the olfactory bulb is reduced to an insignificant level whereas the cerebrum has grown to make the bulk of the brain. In the diagrams in the right-hand column, the cerebrum is made conspicuous in all examples because it is used as a measure of the behavioral complexity of the animal. From fish through reptile to human, the relative size of the entire brain has increased, but the part of the brain taken up by the cerebrum increased proportionately even more. Compare the size of the cerebrum and cerebellum in the human brain with the same parts in a bird.

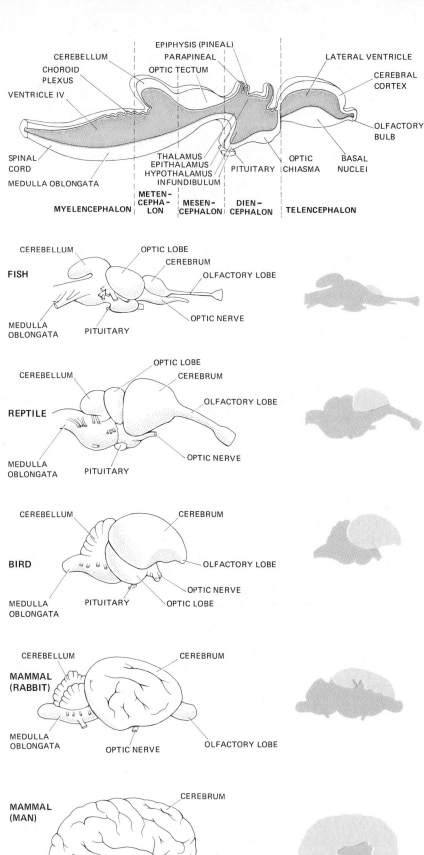

EPIPHYSIS (PINEAL)
PARAPINEAL
CEREBELLUM
OPTIC TECTUM
CHOROID PLEXUS
VENTRICLE IV
LATERAL VENTRICLE
CEREBRAL CORTEX
OLFACTORY BULB
SPINAL CORD
MEDULLA OBLONGATA
THALAMUS
EPITHALAMUS
HYPOTHALAMUS
INFUNDIBULUM
PITUITARY
OPTIC CHIASMA
BASAL NUCLEI

MYELENCEPHALON | METENCEPHALON | MESENCEPHALON | DIENCEPHALON | TELENCEPHALON

CEREBELLUM
OPTIC LOBE
CEREBRUM
OLFACTORY LOBE
FISH
MEDULLA OBLONGATA
PITUITARY
OPTIC NERVE

CEREBELLUM
OPTIC LOBE
CEREBRUM
OLFACTORY LOBE
REPTILE
MEDULLA OBLONGATA
PITUITARY
OPTIC NERVE

CEREBELLUM
CEREBRUM
OLFACTORY LOBE
BIRD
OPTIC NERVE
MEDULLA OBLONGATA
PITUITARY
OPTIC LOBE

CEREBELLUM
CEREBRUM
MAMMAL (RABBIT)
MEDULLA OBLONGATA
OPTIC NERVE
OLFACTORY LOBE

CEREBRUM
MAMMAL (MAN)
CEREBELLUM
MEDULLA

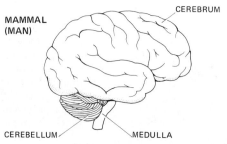

20.5

Areas of the human brain in which specific functions are localized. The information on which such a map is based has come largely from accidents in which loss of a particular sense or function can be correlated with damage to a known part of the brain. In experiments with animals other than humans, selected places in the brain can be electrically stimulated, and the resulting activity or lack of it can be observed.

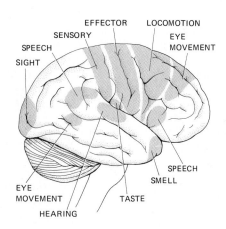

The *cerebellum* is the second largest portion of the brain (the cerebrum is the largest). It is the coordinating center for muscular movement. The cerebellum transforms cerebral impulses into actual, coordinated muscular movement. It does not initiate muscular movement, but motor impulses from the cerebrum would be meaningless without the cerebellum to carry them out. Balance, precision, timing, and body positions are concerns of the cerebellum in humans. In birds, which are generally more agile than mammals, the cerebellum is relatively more important and seems to be one of the major centers of learning.

The *cerebrum* is the largest portion of the human brain. It is divided by a deep fissure into two lateral halves, or hemispheres, and each hemisphere is further separated into four lobes, the frontal, parietal, temporal, and occipital. Each lobe contains special functional areas, including speech, vision, movement, learning, and memory (Figure 20.5). A tough bridge of nerve fibers, the *corpus callosum*, connects the hemispheres and relays nerve impulses between them. (See the essay on p. 417.) In humans, the cerebrum is essential for survival, although exactly which parts or functions are critically necessary is not known. Popularly, the cerebrum is considered the region where thinking is done, but no localization of consciousness or intellectual learning has been established. The cerebrum has been mapped for specific regions that perceive sensory stimuli from specific parts of the body, such as fingers, lips, trunk, and genitalia. It has also been mapped for the localization of regions that control specific parts of the body.

If the cerebrum is damaged, there will be loss of function, depending on the specific area of the injury. For instance, damage to the speech area in the temporal lobe may produce *aphasia* (inability to understand written or spoken words).

The outer portion of the cerebrum, the *cortex*, is a thin, convoluted covering containing about 12 billion neurons.* The folded convolutions increase the surface area about three times. The gray color of cortex nerve cells gives the outer brain its common designation as "gray matter." The inner, medullary portion of the cerebrum shows the whiteness of myelinated efferent, afferent, or connective fibers. Also buried within the cerebrum are the cell bodies of cell clusters called basal ganglia, which help to coordinate muscle movements.

The *thalamus*, below the cerebrum, has connections to all the activities of the central nervous system. It acts as a monitor to all incoming and outgoing impulses and modifies some of them. The thalamus contains part of a net of nerve cells, the *reticular formation*, which filters all incoming stimuli, discarding unimportant ones and sending others to the proper decoding centers in the brain. Damage to the thalamus causes Parkinson's disease, which is characterized by involuntary tremors.

The *hypothalamus*, below the thalamus, controls many physiological and endocrine activities, and it contains centers that regulate basic behavioral patterns, such as those involved in feeding, fighting, reproduction, or escape. It is in the hypothalamus that experimentation has revealed "pleasure centers," "hunger centers," and "fighting centers."

---

*   These 12 billion brain neurons are constantly producing an electric current that uses about as much power as a flashlight.

TABLE 20.1

Major Differences Between the Somatic and
Autonomic Nervous Systems

Somatic

1. Has axons that go from the central nervous system to
   the effector muscle without synapse.
2. Activates skeletal muscles only.
3. Always excites the effector organ.

Autonomic

1. Has fibers that synapse in ganglia outside the central
   nervous system.
2. Affects smooth muscle, cardiac muscle, or glands.
3. May inhibit *or* excite effector organ.

THE PERIPHERAL NERVOUS SYSTEM

The peripheral nervous system enables the brain and
spinal cord to communicate with the entire body. It
may be further divided into the *somatic* nervous system and the *autonomic* nervous system. The somatic
system involves "voluntary" movements of the skeletal muscles, and the autonomic system controls such
"involuntary" activities as breathing and glandular
secretion.

The Somatic Nervous System

The somatic nervous system is composed of nerve fibers that lead from the brain through the spinal cord
to the skeletal muscles. These fibers pass from the
brain directly to the cells of the skeletal muscles without any synapses. The action of such motor neurons
is always to contract muscles.

20.6

The autonomic nervous system and
its relation to the brain and spinal
cord. Although it is less conspicuous
than the large central parts of the nervous system, the autonomic nervous
system, with two pairs of nerve cords
paralleling the central cord on each
side, is an essential regulator of the
internal organs. Acting as controls
against one another, the sympathetic
and the parasympathetic systems
maintain effective control over such
activities as heart rate, secretion,
peristaltic movement of intestines,
and blood flow through arteries.

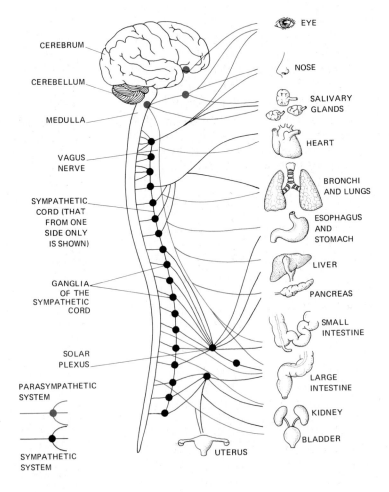

THE SPLIT BRAIN IN HUMANS

Recently the left and right hemispheres of the cerebrum have come to be thought of as two separate brains. Such an idea was proposed intuitively in the past, but now there seems to be enough experimental data to substantiate a claim for the split-brain theory in humans. Apparently the left hemisphere (the "left brain") dominates such activities as speech, writing, and logical reasoning. The right brain seems to dominate intuition, spatial perception, and creative and athletic functions. In normal instances the right brain controls the left side of the body, and the left brain controls the right side. Connecting the two brains is the *corpus callosum*, a bridge of 200 million nerve fibers carrying nerve impulses from one hemisphere to the other.

Experiments with the corpus callosum solidified the case for the split brain. In the early 1950s Ronald E. Myers and Roger W. Sperry found that when the corpus callosum of a cat was severed, both hemispheres functioned independently, almost as if they were two complete brains. Subsequently, the corpus callosum operation was performed on several patients suffering from epilepsy, and these individuals were usually cured as a result of the separation of their cerebral hemispheres.

The speech center is located in the left hemisphere. When split-brain patients were asked to view objects across their total left-right field of vision, they could point to the objects they saw in their total field, but could *orally describe* only the objects in the left side of the field. Obviously, the patients could "see" with both sides of their brain, but if they had to talk about what they saw, their left brains were "blind." Remember, the speech center is in the left brain, which relates to the *right* side of body.

Can the right brain learn to "speak" when the speech center in the left brain is damaged? Apparently it can, and apparently other functions can be transferred ("relearned," if you will) from one hemisphere to the other, especially in young children before the activity becomes totally specialized. But many adults have demonstrated the ability for fully developed language skills in their right hemispheres.

Individuals with split brains do not suddenly become less intelligent or less mentally alert, and there is no reason to believe that right-handed people are more intelligent than left-handers, or vice versa. Western society has downgraded left-handedness and consequently the right brain for many centuries. ("Sinister" and "gauche" are both derived from words meaning "left," whereas words like "adroit" and "dextrous" are related to "right.") Now we must finally give the right brain its due. We still do not know why the left brain is usually dominant, but we do know that the right brain can be just as "adroit" as the left brain when it has to be.

The Autonomic Nervous System

The main differences between the somatic nervous system and the autonomic nervous system are shown in Table 20.1.

The autonomic nervous system helps the somatic system to regulate normal body functions and to control the use of body resources (Figure 20.6). It is composed of two parts, the *sympathetic* and the *parasympathetic systems*, each complementing the other. The sympathetic system, including nerves from the thoracic (chest) and lumbar (midback) region of the spinal cord, can cause relaxation of intestinal wall muscles and an increase in sweating, heart rate, and blood flow to voluntary muscles. The parasympathetic system, containing nerves from both the brain and sacral (low) end of the spinal cord, acts in opposition to the sympathetic, causing contraction of intestinal wall muscle and decreasing sweating, heart rate, and blood flow to voluntary muscles.

Both systems include paired ganglia, to the right and left of the spinal cord. The ganglia of the parasympathetic system are in the effector organs, somewhat removed from the spinal cord; the ganglia of the sympathetic system are in two rows near the spinal cord. All the ganglia are connected to the spinal cord

BIOFEEDBACK: THE BRAIN CONTROLS THE BODY

The so-called involuntary activities of the body, such as breathing and regulation of heart rate, are in fact not completely involuntary. The method of control of such "automatic" body functions is popularly called *biofeedback*, and it can be learned. One important factor in biofeedback training is knowing what is happening in one's cerebral and visceral functions. For instance, some migraine-headache sufferers, aware that their headaches are associated with certain dilated blood vessels, have learned how to prevent the vessels from expanding when a headache seemed to be approaching.

Recently, with a growing popular interest in Yoga and other Eastern mysticism, biofeedback training of one kind or another has developed rapidly. Soothing alpha waves are called up at will, heartbeats are slowed or steadied, blood is diverted into the hands away from a potentially aching head, and skin temperature is changed.

Of course, some chicanery accompanies most fads, especially profitable ones. Prescribed body positions, incantations, special costumes, smells, sounds, and decorative environment are all unnecessary, though some may aid a faltering beginner by strengthening his resolve. The one factor that all the religious and other methods have in common is an insistence on unswerving thought, whether it is called prayer, meditation, concentration, or scheming. Directed thought is more difficult to achieve than most people expect; it is rare for anyone not disciplined in concentration to attend strictly to a single idea for more than a few seconds. Legitimate efforts with biofeedback training may provide some long-sought relief for such diseases as epilepsy and other brain-related problems. Indeed, perhaps illness need not be in the brain to be treated with biofeedback techniques.

---

by preganglionic fibers, and to effectors by postganglionic fibers.

The postganglionic fibers secrete *norepinephrine* at their endings to transmit impulses from the fiber to the effector organ. Both the sympathetic and parasympathetic systems release acetylcholine as the chemical transmitter for ganglionic synapses.

The sympathetic system is the only nerve supply to the adrenal medulla, and it is capable of stimulating the flow of adrenalin from the adrenal gland in response to a typical stress situation, commonly called a "fight-or-flight" condition.

Although the autonomic system regulates many important body functions almost independently of the brain and spinal cord, it is basically incorrect to refer to it as an "involuntary" system. It is becoming increasingly obvious to experimenters that animals (including humans) can learn to control such activities as heart rate and breathing rate if there is a motivation to do so. (See the essay above.)

THE SENSES: PERCEPTION OF STIMULI

All knowledge and awareness depends on the reception and decoding of stimuli from the outside world. Such a process of assimilating afferent information is called *perception*. Practically all organisms, except perhaps bacteria, have some perceptive mechanisms, and there is some evidence that bacteria do also.

Organisms must maintain homeostasis, and one of the ways they do so is through a sensory system that permits the body to adjust to changing environmental conditions (including internal, visceral conditions). Structures that are capable of perceiving such changing stimuli are _receptors_. One method of classifying receptors is according to the kinds of energy they respond to, as follows: chemoreceptors (taste, smell), mechanoreceptors (touch, pain, hearing), photoreceptors (vision), thermoreceptors (heat, cold), and proprioceptors (balance and other information about one's own body). In terms of the physiology of

WHAT ANIMALS SEE

Light and color are such common features in the world of human experience that they are almost taken for granted. But even so, we should consider two basic principles that make vision possible: absorption and reflection of light. Although sunlight has a white appearance, it actually consists of many different colors. A glass prism will separate sunlight into reds, oranges, yellows, greens, blues, and violets, the full spectrum of colors visible to the human eye. Sunlight also contains other "colors" not visible to the human eye, infrared and ultraviolet. When sunlight falls on an object, the color components of the light may be reflected or absorbed. If all the light is absorbed, the object will appear as black to the color-sensitive eye. "Color" is seen when light is reflected from an object. Green plants are perceived as green because their green pigment, chlorophyll, reflects the green portion of the spectrum to our eyes; the green plant absorbs the other colors in the light. To "see" a color, the light receptor of the organism must be sensitive to the particular colors being reflected from the object to the eye. On the following pages we will see how several different animals perceive color.

Predatory birds, such as hawks, falcons, and eagles, are superb hunters because they fly faster than any other animal can. But to complement their flying skill, birds need exceptional vision so they can see prey at a distance. In fact, birds not only fly faster than any other animal; they also see more sharply. The resolution of some birds' eyes is about eight times that of human eyes because birds have about 1.5 million cone cells per fovea compared with our 200,000. Because of the abundance of cone cells, birds can resolve a distant object well enough to identify it clearly. A human would see the same object as a blur. The pictures here demonstrate how a ground squirrel at a distance would look to a hawk and to a human.

It should not be surprising that birds can also see close objects clearer than humans can. Birds are able to change their focus instantly, and they can see close objects as well as they can see distant ones. Visual acuity is so important to birds that many have eyes that are larger than their brains.

A distant ground squirrel as seen by a hawk.

The same ground squirrel as seen by a human.

If animals could talk to us and tell us what they see, we would know more about their color vision than we do now. But we still could not know everything, because there is no guarantee that one animal could successfully communicate its view of the world to another animal. Even a human being does not know exactly what another person sees. The senses are so subjective that we can only describe them approximately to other people. It is fairly easy to know if someone is seeing *no* color, or if a person mistakes one end of the spectrum for the other, but we cannot say that we know absolutely what another person or another animal sees. The photographs on these pages come as close as possible to showing the world as seen by a human, a dog, a horse, and an insect. Likewise, the color spectra on the following page are at best approximations of the world of color as seen by several animals.

Adapted from an idea in *Life Before Man*, a Time-Life Book.

A human being sees the world in full color, from violet
to red, but certain colors, such as ultraviolet, are beyond
the human's normal range of visible light. The field of
vision is about 180 degrees, and the overall picture is exceptionally clear.

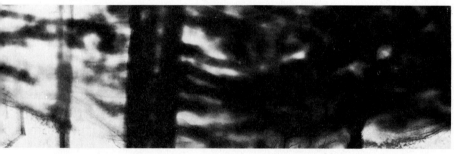

A dog does not see color; it sees various shades of gray. The field of vision is wider
than a human's, but objects in the background tend to be out of focus because animals such as dogs and cats are nearsighted. Note that the foreground is sharper than
in the picture above of the human's world.

Because a horse's eyes are located on the sides of its head, it can see everything in a wide field of almost 360 degrees without
turning its head. Very close objects would be blurred, but everything else appears in focus, even faraway objects and those on
the extreme edges of the picture. A horse can probably distinguish finer gradations of gray than a dog, but it still does not
perceive color.

A honeybee sees a field as wide as that of a horse, and it perceives it in some color. A bee can see ultraviolet light (a human
cannot) and fine divisions of violet, blue, and blue-green, but it perceives the green-yellow-orange range only indistinctly. Dark
orange and red are probably perceived as black. Because of its compound eyes with thousands of separate facets, the bee probably
perceives a blurred picture made up of individual pieces, like a mosaic (although it is possible that the bee sees an image
totally beyond our realm of perception and imagination). The honeybee's eyes do not move, and individual lenses cannot be
focused. For this reason, only a small portion of a bee's entire field of vision will be in focus at any one time as the lens points
directly at it.

HUMAN

Humans see a full range of color from violet to red but are unable to see ultraviolet and infrared light. As far as we know, humans have the most discriminating sense of color perception of all animals.

BIRD

A bird's eye is almost ten times as acute as a human eye and responsive to the same portion of the spectrum. Apparently, birds are partially blind to blue but very sensitive to red. Some birds of prey can detect infrared rays emanating at night from the bodies of mice and other prey. Some birds may not see violet. Many birds show no interest in green, and in fact they may not see green at all.

Color vision apparently exists in most insects, fishes, reptiles, and birds, but it is absent in most mammals. The primate mammals, including humans, can perceive color, and it is possible that among mammals *only* the primates are able to distinguish color. Other mammals, such as cats, dogs, horses, cows, lions, and elephants, are almost certainly color-blind. Color perception is highly developed in most fishes, reptiles, and birds, and in such insects as dragon-flies and bees. Amphibians probably lack a discriminating sense of color vision, although frogs seem to be able to distinguish between blue and green. If a trapped frog is given a choice, it will jump toward blue and avoid green, presumably heading for the open areas of sky and water rather than the more confining environment of grass.

The different visual abilities of animals and the specific visual perceptions of each species were not randomly evolved. Rather, they are biologically sensible and appear to be related to specific problems and conditions in the environment. Bees, for example, have developed color sensitivities in areas where such sensitivities are useful. Bees usually do not pollinate red flowers because they cannot distinguish red. In contrast, bees *do* pollinate blue flowers and some flowers that reflect ultraviolet light. In these wavelengths, bees have a highly developed color sense. The natural selection of the evolutionary process has been effectively at work.

FISH

It is not known with certainty what colors fishes perceive, but they do seem to react to the bright colors exhibited during mating. One group of biologists presents the spectrum above, with the full range of colors, including ultraviolet light. Another school believes that fishes are attracted to yellow and green, and there is some evidence that they do not see the blue and violet areas at all. This group of biologists also thinks that fishes react to orange and red light as if it were black, as shown below. In either spectrum, the colors are probably blurred by the fish's murky environment.

FISH

TURTLE

Turtles and most reptiles have a well-developed color vision. The most important colors seem to be orange, green, and violet. Yellow and yellow-green are apparently seen as orange. Red seems to appear as violet, thereby closing the color circle. It is not certain that snakes can distinguish color, but apparently lizards can.

HONEYBEE

Bees cannot make sharp distinctions between green, yellow, and orange, but they can discriminate between blue and blue-green. Bees cannot see dark orange and red (red probably appears as black), but they do see ultraviolet light, which humans do not see.

CAT

Cats, dogs, horses, and most other mammals (primates are the main exceptions) see only shades of gray instead of color.

the nervous system, a receptor may be thought of as the peripheral end of a sensory neuron. All stimulated receptors generate the same type of nerve impulse, and the different sensations are brought about when nerve fibers connect with specialized portions of the central nervous system. As with nerve impulses in general, the intensity of "sensory" stimuli may be altered by increasing or decreasing the frequency of the stimulus, or by stimulating more or fewer receptors at one time.

In summary, it appears that all receptors are capable of changing (*transducing*) stimuli into nerve impulses. Subsequently, localized regions of the brain translate unspecialized impulses into specific, conscious sensations. In this way, even mechanical pressure on an eye stimulates impulses in the optic nerves that will ultimately be interpreted as light. If you jar your eyes hard enough, even in total darkness, you will "see stars" because the brain recognizes impulses coming through optic nerves only as light.

The Eye and Vision

Visible light is a portion of the electromagnetic spectrum within the range of about 380–760 nanometers (a nanometer is one-billionth of a meter). The human eye is sensitive to wavelengths within this range, and so we all "see" objects that emit or are illuminated by waves in our receptive range. Other organisms (insects, for example) are sensitive to shorter wavelengths in the range of ultraviolet, X-rays, and gamma rays, and some organisms can "see" longer wavelengths, such as infrared waves (Figure 20.7). Of course, no eye actually *sees*; it merely receives light. It is the brain that "sees" by converting nerve impulses into information. If you were to shine a light directly onto the optic center of the brain, no light would be "seen," and in fact, no other response would be recorded either. To go one step further, if you could switch the nerves that carry light impulses to the optic centers of the brain with the ones that carry sound impulses to the auditory centers, it would

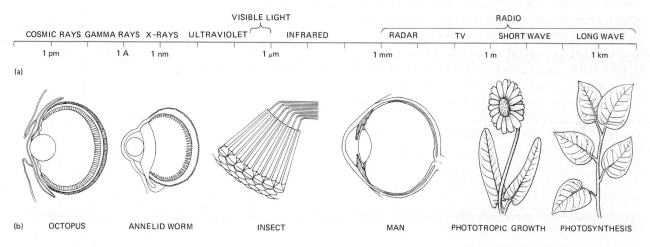

(a)

| COSMIC RAYS | GAMMA RAYS | X-RAYS | ULTRAVIOLET | VISIBLE LIGHT | INFRARED | RADAR | TV | RADIO SHORT WAVE | LONG WAVE |

1 pm   1 A   1 nm   1 μm   1 mm   1 m   1 km

(b) OCTOPUS   ANNELID WORM   INSECT   MAN   PHOTOTROPIC GROWTH   PHOTOSYNTHESIS

20.7
The electromagnetic spectrum and some biological radiation receptors. (a) A scale showing the wavelengths of electromagnetic radiation, from the shortest cosmic rays to long radio broadcast waves. That part of the spectrum commonly absorbed and used by living things is about midway between the extremes. It is called "visible light" because it is the part that affects human retinas. The limits vary from species to species, and in human eyes from individual to individual, but the range is generally bounded by the near ultraviolet at about 300 nm and the near infrared at about 700 nm. (b) Some representative light-absorbing parts: eyes of animals and leaves (generally) of plants. Regardless of the differences in form, all depend on having a light-gathering structure and a selective light-absorbing pigment. The octopus eye and the human eye, eons apart in evolutionary history, are notably similar, but the insect eye is built on quite a different plan. Plants do not usually have any means of concentrating light, although some mosses and flowering plants have lenslike thickenings that do effectively achieve such concentration.

20.8

Comparison of a human eye with a camera. There are several structural similarities. Light from an illuminated object passes through a hole in the front of the apparatus (pupil of an eye, aperture of a camera) and it is focused by a lens on a sensitive field (retina of an eye, film in a camera). In both the image is reversed, but one can turn a photographic film over to look at it, and in much the same way a brain can interpret the image to understand which way is up. In an eye, the actual image is both upside down and reversed. One important difference between the camera and the human eye is that the eye focuses on nearer or more distant objects by changing the shape of the lens, but the camera (like some fish eyes) focuses by moving the lens toward or away from the film.

(a)   × 240

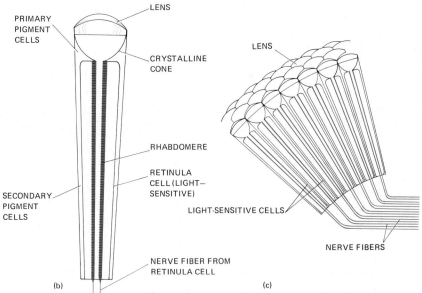

(b)   (c)

20.9

The compound eye. (a) A scanning electron micrograph of part of a fly's eye, showing the hundreds of simple eyes, *ommatidia*, that constitute the whole eye. (b) A diagram of a long section of a single ommatidium. Light is collected by a front lens and passed down a long tunnel to the receptor cell at the bottom. The field of view of an ommatidium is narrow. (c) A diagram of a compound eye, cut away to show the arrangement of the ommatidia. An insect or other animal with such compound eyes probably does not form a picturelike image; rather, it must have some sort of mosaic perception. Although some insects obviously have effective vision, we cannot imagine how an insect brain interprets the stimuli it receives. No one knows how the world looks to a dragonfly, even though it is possible to take a photograph through the myriad lenses of such an animal and look at the picture.

20.10

Diagram of a section through a human eye. The outer coat, the tough *sclera,* is transparent across the front, where it is called the *cornea.* Back of the cornea is a chamber filled with clear liquid, the *aqueous humor.* The *iris* is a muscular disk, in the center of which is a hole, the *pupil.* The *lens* not only helps form a visual image but can change to accommodate the eyes to near or distant objects. The lens is connected to radiating fibers from the *ciliary body,* the active focusing organ. The inner layer of the eye is the dark-pigmented *choroid coat,* which cuts down on reflection and glare by absorbing stray light. The bulk of the eye is filled with a semisolid *vitreous humor,* which sometimes has particles floating in it. If you look at a neutral, featureless background, such as a uniform gray sky, you can see them drifting about. They are called *musci volantes* (Latin, "flying flies"). The sensitive layer of the eye, the *retina,* has a slight depression, the *fovea,* where the retinal elements are so distributed as to give maximum sharpness of vision. An eye can be moved in its socket by a set of six muscles.

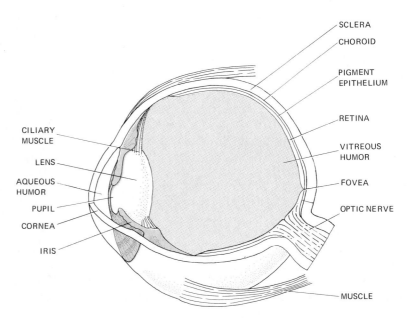

SCLERA
CHOROID
PIGMENT EPITHELIUM
RETINA
VITREOUS HUMOR
FOVEA
OPTIC NERVE
MUSCLE
CILIARY MUSCLE
LENS
AQUEOUS HUMOR
PUPIL
CORNEA
IRIS

actually be possible for the brain to "hear" light and "see" sound. The brain is impartial. It will convert *all* sufficiently strong action potentials into nerve impulses. The brain doesn't "know" the differences between light, sound, touch, and the other senses. If the sight center of the brain is stimulated, the brain "sees," even if the external stimulus is actually the loud backfire of a truck.

Practically all animals respond to light in some way, and light receptors vary from the image-forming, color-sensitive camera-type eyes of vertebrates (Figure 20.8) to the simple photoreceptors of unicells, which can distinguish only intensity. The nocturnal earthworm has no eyes, but its body is covered with light-receptor cells that cause it to move away from bright light. In contrast, predator birds like eagles have developed extremely efficient eyes that outweigh their brains. And a frog, which has perfectly good eyesight, will nevertheless respond only to moving objects. A frog surrounded by insects will starve to death if the insects do not move. (The subject of perception in animals is discussed in detail in Chapter 28.)

Insects and other arthropods have compound eyes, with dozens or even thousands of separate, tubular *ommatidia* (Figure 20.9). Each of these units has its own lens and light-sensitive receptor cells, and each ommatidium views a single section of the total picture. The compound eye is the most usual animal eye. It probably is better suited to detecting motion than to perceiving a clear image of a visual field. The greater the number of ommatidia, the more distinct is the image. The eyes of bees contain about 15,000 ommatidia, and some predatory insects have twice that number. (A dragonfly has about 30,000 separate ommatidia.)

Within a human eyeball, about 2.5 centimeters (1 in.) in diameter, are over 100 million neurons that convert light waves into electrochemical impulses that are decoded by the brain (Figure 20.10). As we have seen, the eyes of vertebrates (including humans) work like a camera, the image being brought to a

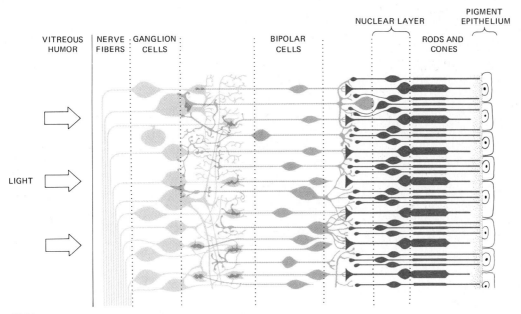

VITREOUS HUMOR | NERVE FIBERS | GANGLION CELLS | BIPOLAR CELLS | NUCLEAR LAYER | RODS AND CONES | PIGMENT EPITHELIUM

LIGHT

20.11

Diagram of a section through a verte-brate retina. Light, coming from the left through the vitreous humor, passes through three layers of cells before reaching the light-sensitive *rods* and *cones*. Beyond the rods and cones there is a pigmented epithelial layer which prevents back scattering. When light energy stimulates a rod or cone, that energy is transduced into the electrical energy of a nerve impulse. The impulse is sent from the receptor cells through an intermediate set of *bipolar cells* and finally to gan-glion cells before passing on to the optic nerve. The function of the eye is to receive light stimuli, change the stimuli into nerve impulses, and send those impulses to the brain; interpret-ing the impulses is the work of the brain.

20.12

Scanning electron micrographs of (a) rods and (b) cones from amphibian retinas. The difference in shape is ob-vious; the difference in function is that the rods are more effective in dim light but not sharp image formers, and the cones are used in color vision and in accurate focusing on an image.

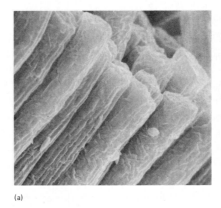

(a)

(b)   × 5000

focus on the sensitive *retina* (Figure 20.11). The retina is composed of layers of slender cells, the under-neath layer consisting of photoreceptors, the *rods* and *cones*, and a covering layer of sensory neurons (Figure 20.12). The sensory neurons send axons to the brain via the optic nerve, a bundle of axons.

There are about 125 million rods and 7 million cones in each human eye. The rod cells, which are not color-sensitive, function mainly in dim light and in peripheral vision. The cone cells are concentrated in the center of the retina directly behind the lens, es-pecially in a small area called the *fovea*, where image formation is sharpest and color vision is most acute. The fovea contains only cones, and animals with the most cones in the fovea have the sharpest vision. (The human fovea has about 200,000 cones but some birds have up to eight times that number.) Humans can usually see most sharply by looking directly at an object so that the image falls on the fovea, but they can see better in dim light by *not* looking directly at an object, so that the image falls on the periphery of the retina, where the highly sensitive rod cells are (Figure 20.13).

The biochemistry and biophysics of vision are incompletely known. We know that each rod contains a light-sensitive pigment called rhodopsin, which consists of a colorless protein, opsin, and a colored carotenoid molecule, retinal (or retinene), a derivative of vitamin A. One photon—that is, one unit of light energy—is enough to affect one molecule of rhodopsin, and in humans visual stimulation can be achieved by as few as 5 to 10 molecules of affected rhodopsin. What is called the *primary light effect* occurs when light is absorbed by rhodopsin, which then separates into opsin and retinal. This reaction somehow causes the receptor cell to fire, and the impulse is relayed to the brain. Afterward, the retinal is enzymatically recombined with opsin to generate rhodopsin again. For decades before these reactions were discovered, a deficiency of vitamin A was known to

20.13

Finding your blind spot. Hold the page directly in front of your right eye, keeping your left eye closed. While staring steadily at the target, move the page closer. When the page is about 20 to 30 centimeters (8 to 12 in.) from your eye, the X should disappear, because the image of the X will at that point fall on the nonsensitive part of the retina, where the optic nerve and blood vessels connect to the eyeball. The presence of the blind spot is not annoying, because two eyes are usually working in cooperation, and any object projected on the blind spot of one eye is projected on a sensitive spot of the other, so the object is constantly visible.

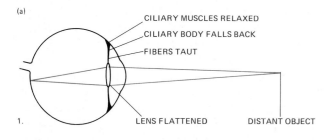

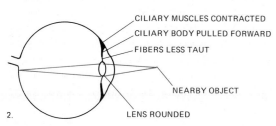

20.14

Making a sharp image on the retina. (a) How the eye *accommodates*, or focuses on near and distant objects. When the ciliary muscles relax, they fall back, tightening the fibers that suspend the lens. The tightened fibers pull the lens into a flatter shape, making distant objects come into focus on the retina. As shown in (1), a relaxed eye automatically focuses on distant things. Effort is needed, even if it is thoughtless effort, to focus on near objects. The ciliary muscles contract, pulling the whole ciliary body forward and reducing the size of the opening. This lessens the tension on the lens suspension fibers and allows the lens to assume a rounder shape, making near objects come into focus on the retina. As shown in (2), with the scale greatly exaggerated, nearby objects can be seen in sharp focus when the ciliary muscles are under tension. That is why looking at a distant scene, even across a room, is restful to the eyes of a person doing long-term close work. (b) Correcting vision in eyes that are myopic (nearsighted), or hyperopic or presbyopic (farsighted). In a myopic eye the focal point is in front of the retina, but placing a reducing lens in front of the eye will make the focal point coincide with the retina, as shown in (1) and (2). If the focal point is behind the retina, placing an enlarging lens in front of the eye makes the correction, as shown in (3) and (4).

cause the reduced sensitivity to light known as night blindness.

Night vision is almost totally rod vision, since the color-sensitive cones require 50 to 100 times more stimulation than rods do. The human eye can adjust to night or day vision, but nocturnal animals, such as bats and owls, have only rods, and daytime animals, such as some birds, have only cones, and these specialized animals lack the visual adaptability of humans (Figure 20.15).

The mechanism of color vision is not well understood. According to one theory, colors are perceived when red-sensitive, green-sensitive, or violet-sensitive cells (cones) in the retina are stimulated. The final color is determined by the combining of the different levels of excitation of each cell type (each cone). With the exception of primates and a few other species, most mammals are thought to be color-blind. (Can an animal with cones be color-blind, as many seem to be, or do we simply not know how to test animals accurately for color perception?)

The Touch Receptors

By "touch" we refer to different physical stimuli (such as heat, pressure, and pain) and different sense receptors (Figure 20.16). For example, there are microscopic receptors for direct touch in the skin, called *Meissner's corpuscles.* Touch stimuli are also received by hairs, which affect the skin when they are bent. (A bug crawling along a hairy arm will be felt even if its feet never touch the skin.) Firm pressure affects other large receptors called *Pacinian corpuscles.* Warmth is perceived by the *Ruffini endings,* and cold by the *end bulbs of Krause.* Although the nerve impulse is identical for each type of receptor, the brain interprets the sensation according to affected nerve ending in the specific receptor. If a sufficiently intense stimulus is applied, more than one receptor may be

stimulated, and the brain may be confused. When that happens, the person may be stimulated by intense cold but interpret it wrongly as heat. There are more than half a million touch receptors on the surface of the human body, and most of them are on the lips, tongue, and fingertips.

That indefinable phenomenon known as pain is initiated by the action of free nerve endings that are present in most parts of the body, although intestines and brain tissue have no pain receptors. We do not know whether pain receptors respond directly to injured tissue or indirectly to some substance released by damaged cells. The perception of pain is to some extent a matter of attention. Who hasn't discovered an already dry cut without having any knowledge of when the wounding happened? There are about three million pain receptors distributed over the surface of the body. Such a large number is probably required to offer maximum warning and protection against harmful injury.

Taste and Smell

Taste and smell, which are both chemically activated senses, are similar in their action. In fact, much of what we normally call taste, or *gustation,* is actually a function of our sense of smell, or *olfaction.* In order for a substance to be smelled, it must first be dissolved in the surface liquid on the membrane of the olfactory area. Thus taste and smell complement one another. A person whose nasal passages are blocked by a cold cannot "taste" his food. But in spite of some similarities, taste and smell are separate sensations, and they will be so treated here.

The specialized receptor cells for taste in humans are located in the *taste buds* on the top and underside of the tongue, and to a lesser degree on the larynx and pharynx (Figure 20.17). Taste buds are stimulated by four basic taste sensations—sweet, sour,

20.15
Extremes in the development of mammalian eyes. A mole rat lives mostly underground in the dark and has tiny, almost useless eyes. A flying squirrel is active mostly at night and must make the most of whatever light is available. Its eyes are relatively large, making their light-gathering power greater than that of human eyes. No animal can see in the dark, but some can see better in dim light than others can.

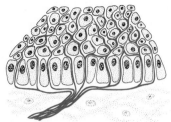

FREE NERVE ENDING (PAIN)

CORPUSCLE OF RUFFINI (HEAT)

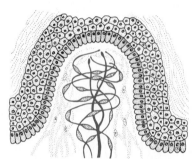

MEISSNER'S CORPUSCLE (TOUCH)

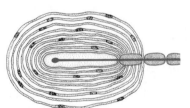

PACINIAN CORPUSCLE (DEEP PRESSURE)

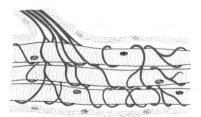

MUSCLE SPINDLE (PROPRIOCEPTION)

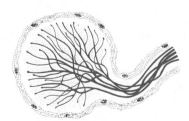

KRAUSE'S END-BULB (COLD)

20.16

Receptors in skin and muscle. Specialized nerve endings in the skin are generally but not uniformly distributed over the body. Meissner's corpuscles, which are sensitive to light touch, are numerous in fingertips, but because fingertips are used extensively, they may have enough toughened dead skin on the surface to blunt the sensitivity of the corpuscles of Ruffini, which are sensitive to heat. If a stimulus is intense enough, it may affect receptors that would not normally be sensitive to a stimulus of that type.

For example, one may get a sensation of cold by putting a hand in very hot water. Deep pressure is less likely than a creasing of the skin to stimulate pain-sensitive nerve endings; a tiny pinch hurts more than a big one. Nerve endings in muscles and joints give information on the position of various body parts. Without looking, a pianist knows how far to reach to find an exact note, and a tennis player watches the moving ball, not his own arms.

20.17

Taste buds on a tongue. (a) Scanning electron micrograph of a taste bud. (b) Diagrammatic section through a taste bud, showing the arrangement of the nerve endings. × 30.

(a)

× 20

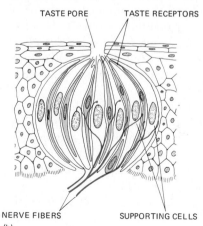

TASTE PORE   TASTE RECEPTORS

NERVE FIBERS   SUPPORTING CELLS

(b)

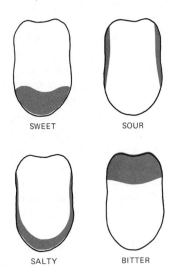

20.18

Distribution of various kinds of taste buds on a tongue. The number of "flavors" that can be detected strictly by taste is limited, but these, in combination with thousands of perceptible odors, give the impression of many tastes. It is common knowledge that when one has a cold in the head, food seems tasteless. The olfactory receptors are blocked, and the sufferer is left with only taste buds working.

salty, and bitter. Although the areas of response to these tastes are located on specific parts of the tongue (Figure 20.18), there is no conclusive evidence that there are different types of taste buds for each taste sensation. In fact, the tastes we perceive are probably the result of a combination of different intensities of more than one basic sensation. Taste buds have a life of only about 10 days, and they are replaced with decreasing frequency as we get older, usually hampering our sense of taste somewhat.

Odors are more difficult to classify than tastes, but human sensitivity to odors is greater than it is to taste. There are millions of receptor cells in the nasal passages. Even so, human olfactory sensitivity is less than that of most animals and tremendously less than that of insects. (See Chapter 28 for a more comprehensive discussion of chemoreceptors, especially smell, in animals.)

Human smell receptors are associated with the mucus-secreting membrane lining the upper nasal cavity. Smell receptors in humans are slender cells with thin filaments attached. The basal ends of these filaments are connected to the cranial nerve that leads to the olfactory lobes in the brain.

Of all the senses, the least understood is smell. One theory suggests that many different odors can be

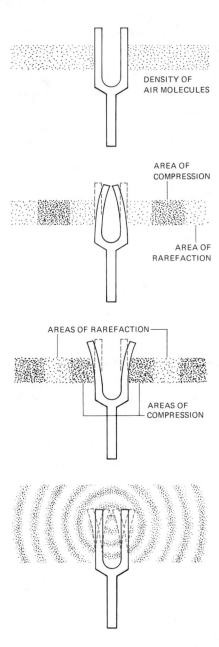

20.19

Sound generation by a tuning fork. When the prongs of a fork vibrate, they rock back and forth, sending into the surrounding air alternating compressions and rarefactions that spread out from the source like ripples in a pool. When a compression reaches an animal's eardrum, it pushes against the eardrum. When a rarefaction follows, it allows the eardrum to spring back. A tuning fork gives out a regular succession of waves producing a "pure tone" of a set frequency (vibrations per second), but most tones are mixtures of frequencies, whether they are generated by a vibrating string, a drum head, explosions in a gasoline motor, animal vocal cords, resonant air columns in organ pipes, or the splashing of a stream. Loudness, or intensity, is a matter of the energy put out by the sound generator.

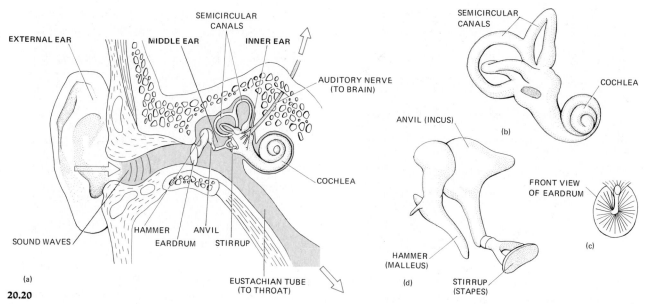

20.20

(a) Anatomy of the human external, middle, and inner ear. Sounds transmitted through air enter the auditory canal and set the eardrum to vibrating. The vibration is then transmitted to the auditory bones, the hammer, the anvil, and the stirrup. The bones, acting as levers, effectively amplify the vibrations, so that the fluid in the cochlea becomes agitated, causing selected hairs to be moved. If the hairs are bent, nerves are stimulated, and the message is sent to the brain for interpretation. Not all sound is heard through air. One's own voice is heard largely by bone conduction, and that is why most people are astonished when they first hear themselves speaking from a recording. (b) Enlarged view of the semicircular canals, the organ of balance, and the cochlea, which contains the nerve endings sensitive to sound. (c) The eardrum as seen from the inside of the ear. (d) The auditory bones, which transmit vibrations from the eardrum to the cochlea.

detected because of the many possible interactions between the molecules of the odor substance and the several kinds of receptor cells. One way to explain such odor variations: If odor molecules have specific chemical shapes that can fit into receptor sites in many combinations, the better the fit, the greater the firing rate of the afferent nerve fiber will be. Odor discrimination then results from the simultaneous but varying stimulation of receptor cells.

Hearing and Balance

Two such seemingly unrelated stimuli as sound and positional change are considered together because they are received in the same organ: the inner ear. Physically, sound is the alternating compression and rarefaction of the medium (usually air) through which the sound is passing. In air that is disturbed by sound, waves of compression, in which air molecules are pushed together, are followed by waves of rarefaction, in which the air molecules are farther apart (Figure 20.19). Without a carrying medium such as air, there can be no sound. The waves in the air affect the receptor cells in the ear. The physical changes in the receptor cells are somehow changed to electrical impulses in the nervous system. The impulses are transmitted to the brain, where they are interpreted as sound. Psychologically, sound is the interpretation of nerve impulses from the auditory neurons.

Human ears are generally responsive to frequencies from about 50 cycles per second to about 20,000 cps, but some people can hear from 16 cps to 30,000 cps, especially during their early years. Hearing ability declines steadily from early childhood, because tissues in the inner ear lose their elasticity. In addition, vibrating hair cells begin to degenerate, and harmful calcium deposits may form.

Very few naturally vibrating bodies produce simple vibrations or a "pure" tone, such as that produced by a vibrating tuning fork. Natural objects produce instead combinations of frequencies. When the combinations are subjectively interpreted by our nervous system, they give various sounds their "quality," or timbre.

There are four basic parts of the auditory receptor (Figure 20.20). (1) A thin membrane, the tympanum, or eardrum, vibrates in response to air vibrations in

(a)

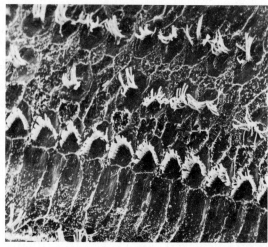
(b)

$\times$ 500

20.21

Injury to the inner ear by intense noise. (a) A scanning electron micrograph of a normal guinea pig organ of Corti, showing a regular pattern of hair cells, and (b) an injured organ of Corti after exposure to 24 hours of loud rock music (about 120 decibels). The injured ear shows most of the hair cells irreversibly shriveled and destroyed. Once hair cells in the organ of Corti are destroyed, they can no longer transduce sound, and continued exposure to dangerously loud noises will inevitably produce a per-

manent hearing loss. Current estimates are that noise is twice as bad now as it was in the 1960s, and it is expected to continue to double every 10 years. Studies have shown that factory workers have twice as much hearing loss as white-collar office workers. Approximately 10 million Americans use hearing aids, and many of these individuals may have suffered hearing impairment through prolonged exposure to sounds that may not immediately be thought of as excessively loud.

the same way that a drum vibrates in response to the beat of a drumstick. (2) Three tiny, paper-thin bones, the hammer, the anvil, and the stirrup, transmit vibrations and amplify them 22 times by lever action from the eardrum to (3) the oval window, a membrane-covered opening in the bone. (4) The inner ear, or labyrinth, contains a coiled tube called the cochlea (Latin, "snail shell"). These parts of the ear work together in the following way.

1. External sound causes the eardrum to vibrate.
2. The eardrum moves the hammer.
3. The hammer moves the anvil.
4. The anvil moves the stirrup.
5. The stirrup produces vibration in the oval window.
6. Vibrations in the oval window cause waves of compression in the inner ear.

In the cochlea is the *organ of Corti*, which consists of membranes and some 30,000 sensory cells.

Vibrations in the liquid in the labyrinth displace the receptor cells, and the displacement, in some unknown manner, causes the cells to discharge nerve impulses. Sounds of different frequencies stimulate different cells, and the brain interprets the stimuli in terms of perceived pitch. Differences in loudness are perceived as a result of differences in the frequency of impulses along a neuron, with more frequent impulses (as many as 1000 per second) indicating louder sound.

It is not totally accurate to refer to the ear as the organ of balance, since the sense of balance is only partly seated there. Our sense of equilibrium is aided substantially by other senses, especially vision and muscle receptors, which constantly help us to locate our positions. (Try standing on your toes with your eyes closed. Without your eyes to guide your body, you will invariably begin to fall forward. Now stand on your toes with your eyes open and note the difference.) Nevertheless, the inner ear does contain specific parts that help the body to cope with changes in position and acceleration (Figure 20.23). The main receptors for equilibrium are the utricle, the saccule, and three fluid-filled semicircular canals. The two basal reservoirs of the canals, the *utricle* and the *saccule,* are filled with a liquid called endolymph. They also contain hair cells that are bent by the weight of attached crystals as the position of the head changes. The bending of the hair cells stimulates the vestibular nerve, which ultimately relays the message to the cerebellum, which in turn directs the necessary muscular adjustment. Receptors in the utricles and saccules regulate equilibrium in responses to changes in the *position* of the head (posture), whereas responses to *movements* of the head are initiated in the ampullae, swellings that join the ear canals with the utricle.

The ampullae are also filled with endolymph and contain cilia. The plane of each of the _three semicircular ear canals_ in each ear is at right angles to the other two, so that at least one canal will be affected by every head movement. When the head

moves, the fluid in the canals lags because of inertia. The relative movement of the endolymph stimulates the cilia in the ampullae, and proper nerve impulses are initiated to compensate for the head and eye movement. You may become dizzy or motion-sick when your head and eyes are rotated in one direction and the lagging endolymph rotates in the same direction but slower, and there is no opportunity for self-regulating countermovements. Motion sickness does not occur in deaf people or infants, neither of whom have functional receptor cells in the balance organs.

CONSCIOUSNESS AND BEHAVIOR

No other animal is so much a creation of its brain as is the human animal. In fact, one of the greatest differences between humans and other animals is the exquisite complexity of the human brain—no other

20.22

Decibel scale and common noise sources.

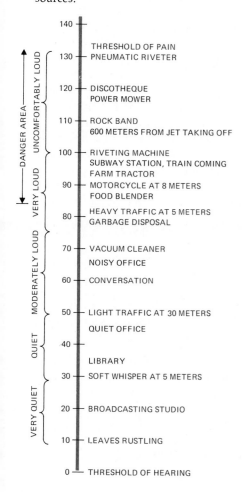

20.23

(a) The organ of equilibrium: the *semicircular canals* of the inner ear. They properly should be called the "circular canals" because they allow the internal fluid to move around completely. One is vertical and oriented front-to-back, one is vertical and oriented side-to-side, and one is horizontal. Any movement of the head can cause relative movement of the canal fluid over hairs in the bulbous swellings (*ampullae*). When the hairs are bent by the flow of the fluid over them, nerve impulses are set up and transmitted to the brain. (b) The passage of fluid over the hairs in an ampulla.

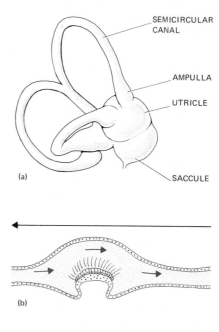

animal can think, learn, or communicate as effectively as we can. Of course, other animals *can* learn and remember, sleep and dream, and be mentally affected by drugs and other stimuli, but in the following sections we will emphasize these functions as they relate to the *human* brain, remembering that when we speak of the nervous system, and particularly the brain, we are stressing differences of *degree*. The human brain is simply the most complex one there is.

Learning and Memory

Little is known about the processes of learning and memory. Specific areas of the brain have been located for different types of learning, however, and it now seems certain, contrary to what was once thought, that no single area of the brain controls learning. In fact, long-term studies of children with learning disabilities have shown that alternate areas of the brain can be trained to "relearn" what had once been learned by a portion of the brain now damaged. Apparently the brain can make such learning adjustments as long as there is a memory trace of what had been learned in the first place.

One theory of learning suggests that when one nerve circuit is used over and over again, synaptic connections are improved, and "preferred pathways" are created. These pathways establish a learning experience. (See Chapter 28 for a discussion of learning patterns in animals.)

Synapses may also be related to memory, but no definite information is available. It *has* been shown that the hippocampal area of the brain is crucial for recent memory storage, and if the hippocampus is destroyed, there is no retention of recent events. A more serious effect is produced by the removal of the medial portion of the temporal lobes of each cortex. As the British neurologist P. W. Nathan has said, "When certain parts of the temporal lobes are cut out, the record of a life is cut out with them." But Nathan goes on with a questionable analogy. "It is as if one's memory is in a filing cabinet and someone has taken [the files] away." More likely, the temporal lobes represent a directory that permits the retrieval of information stored in the cabinets. It is not known in which part of the brain memory is stored.

Another provocative theory suggests that memory is the result of specific RNAs, which are in turn responsible for specific proteins. The speculation is sometimes contradictory but interesting. It has been shown that memory traces can be erased by chemicals that are known inhibitors of protein synthesis. It may be logical, then, to suppose that RNA is a coding element for memory. Indeed, RNA increases in propor-

tion to brain activity and decreases when brain activity slows down. The force of this argument is lessened by the fact that upsurges of RNA also occur during most normal activity of neurons, and organs other than the brain (such as the kidneys, thyroid, and pancreas) also show increases in RNA content whenever they are stimulated by their appropriate hormones.

Sleep and Dreams

The brain is always active, and in fact, the human brain is more active in some areas when the body is asleep than when it is awake, and the brain requires a greater blood supply during sleep. Whether a person is asleep or awake, brain activity may be recorded on an instrument called an electroencephalograph. Both the machine and the record it produces, the electroencephalogram, are referred to by the initials *EEG*. The EEG shows the changing levels of consciousness in the brain (Figure 20.24), and it is frequently used to detect brain damage by locating areas of altered wave patterns. The four distinct wave patterns in the brain are known as alpha, beta, delta, and theta waves. Alpha waves, evident in relaxed adults whose eyes are closed, usually indicate a state of well-being. Beta waves are typical of an alert, stimulated brain. Delta waves are seen during deep sleep, in damaged brains, and in infants. Theta waves are usually found in children or in adults under stress.

It is generally agreed that sleep is one of the more important activities of human beings (lack of sleep will cause death faster than lack of food), but the exact benefits of sleep are not known, and the causes of sleep are equally unknown.* But the EEG has made possible the detection of at least four separate stages of sleep, which are described below.

1. During *Stage 1 Sleep* the rhythm of alpha waves slows down, and the individual experiences a floating sensation. This stage is usually not classified as true sleep, and individuals awakened from Stage 1 are quick to agree, insisting that they were merely "resting their eyes."

2. *Stage 2 Sleep* is indicated by the appearance on the EEG of short bursts of waves known as "sleep spindles." Sleep is not deep.

3. Delta waves appear during *Stage 3 Sleep*. Intermediate sleep is characterized by steady breathing,

---

* It is particularly ironic that so little is known about sleep, because if you are 20 years old, you have already spent about eight years of your life asleep. By the time you are 60, you will have slept about 20 years.

slow pulse rate (about 60), and decline of temperature and blood pressure.

4.   *Stage 4 Sleep* is the deepest stage, also known as oblivious sleep. It usually begins about an hour after falling asleep. Although the EEG indicates that the brain acknowledges outside noises and other external stimuli, the sleeper is not awakened by such disturbances.

Sleep proceeds in cycles 80–120 minutes long. An important condition known as <u>REM</u> (rapid eye movement) <u>sleep</u> takes place during the re-entry into a new Stage 1 phase. It is during the REM period that dreams occur, and the brain is then almost completely separated from the outside environment. Breathing and heart rate may increase or decrease, testosterone secretion increases and penile erections occur, and twitching body movements and perspiration are common. (Naturally, such physiological changes during REM sleep do not affect males only. Preliminary studies show that the clitoris also becomes erect during the REM period, and estrogen secretion probably increases as well.)

The REM period is thought to be essential for a general relaxation of normal buildups of stress and tension, and it is probably necessary for brain maturation.* Babies, whose central nervous system is not fully developed, spend more than half of their sleeping hours in the REM state, but adults spend only about 15 percent of their total sleep in the REM condition. Growth hormone is secreted by children during deep sleep, so apparently sleep is essential for the normal growth of young children. It has been found that mentally retarded patients have less REM sleep than normal people, and in fact, the more severe the retardation, the shorter is the period of REM sleep.

Sleep is a highly individual process. No two people seem to require the same amounts of sleep (gen-

---

*   Dreams may serve to release tension or fear that builds up during the day. This theory is given credence by studies that show that seven out of eight adult dreams are somewhat unpleasant, and about 40 percent of the dreams of children are actually nightmares. Another theory suggests that dreams are an attempt by the brain to bring coherence to the accelerated firings of neurons in the brainstem during REM sleep. According to this theory, dreams occur when the arbitrary firings of the neurons are translated into images as close to our actual experiences as possible. The brain is attempting to take random impulses and direct them into themes with a "plot." The more bizarre dreams may be caused when the brain is unable to connect the neuron activity with any logical experiences or memories.

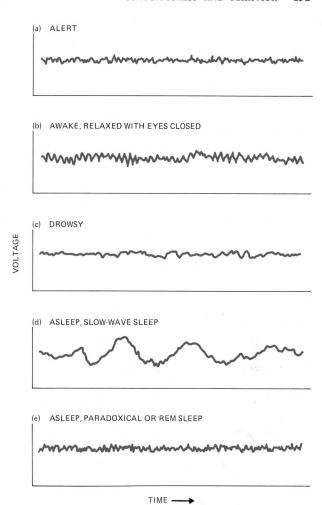

(a) ALERT

(b) AWAKE, RELAXED WITH EYES CLOSED

(c) DROWSY

VOLTAGE

(d) ASLEEP, SLOW-WAVE SLEEP

(e) ASLEEP, PARADOXICAL OR REM SLEEP

TIME ⟶

20.24
Electroencephalograms. Electrodes are connected to the outside of the skull, and variations in the electrical activity of the brain are fed into an amplifier with a recording chart. The chart moves at a steady rate, and as a pen moves in accordance with the electrical changes, it draws a line showing the brain activity. An experienced EEG reader can tell by the traced patterns whether the activities are normal.

TABLE 20.2

The Effects of Drugs on the Nervous System

DRUG	EFFECTS	MECHANISM OF ACTION
Stimulants		
Caffeine (coffee, tea, cola drinks)	Mild reaction: acceleration of heart rate, dilation of pupils, increase in blood sugar.	Stimulates sympathetic nervous system, facilitates synaptic transmission.
Nicotine (tobacco)	Medium reaction: acceleration of heart rate, dilation of pupils, increase in blood sugar.	Stimulates sympathetic nervous system, facilitates synaptic transmission.
Amphetamines (Dexedrine, Methylampheta-mine, or "speed," Benzedrine)	Powerful reaction: acceleration of heart rate, dilation of pupils, increase in blood sugar, reduced sense of fatigue. Depressant effect on appetite.	Stimulate sympathetic nervous system.
Cocaine	Temporary sense of well-being and alertness.	Stimulates central nervous system, inhibits uptake of norepinephrine.
Depressants		
Ethyl alcohol	Small amounts have a stimulant, then a sedative effect as the lower brain centers are depressed. Loss of dexterity, insensitivity to touch. Distorted vision, interference with hearing, difficulty in speaking. Loss of coordination and balance. Unconsciousness, coma.	Reduction of neuron function in the brain. Motor and sensory regions of cortex inhibited. Depression of visual, auditory, speech centers of cortex. Cerebellum inhibited. Depression of reticular formation.
Barbiturates (Seconal, Nembutal, Amytal)	Promote sleep; high doses induce respiratory failure. Barbiturates and alcohol combine to produce extreme depression and may cause suicide.	Depression of reticular formation. Depression of medulla oblongata in large doses.
Tranquilizers (meprobamate: Miltown, Equanil; chlorpromazine: Thorazine; chloriazepoxide: Librium; diazepam: Valium)	Reduce anxiety and tensions.	Block receptors of epinephrine and acetylcholine, depress reticular-formation activity.
Opiates (opium, morphine heroin, codeine, methadone)	Reduce pain, induce muscle relaxation, lethargy. Highly addictive.	Depress thalamus.
Anesthetics (ether, chloroform, benzene, toluene, carbon tetrachloride)	Induce unconsciousness. Volatile hydrocarbons produce effects similar to alcohol intoxication.	Depress central nervous system. Block transmission of electrical impulse that triggers contraction of ventricles.
Hallucinogens		
Mescaline, psilocybin, LSD, dimethoxy-methyl-amphetamine	Distort visual and auditory perceptions. Enhancement of emotional responses. Hallucinations.	Mimic molecular structure of serotonin, a transmitter substance in parts of the brain. Duplicate effects of nervous system activity.
Marijuana, hashish	Produce a sense of well-being.	Unknown.

Adapted from John W. Kimball, *Biology*, Fourth Edition (Reading, Mass.: Addison-Wesley, 1978), pp. 498–502.

20.25

The effect of a drug (caffeine) on the web-building skill of a spider, *Araneus diadematus*. (a) An untreated spider can build a typical orb-shaped web with regular spokes and neat connecting lines. (b) A spider that has been fed sugar water containing caffeine tries to build a web but fails to keep to the standard plan and makes a mess.

(a)

(b)

erally, we need less as we get older), and no two cycles proceed according to the same pattern. But even though some people do not remember their dreams, everyone *does* dream every time a full cycle progresses into the REM stage. We almost always awake directly from REM sleep, and almost all sleepers awakened during their REM period will recall having been in a dream state.

Most people will experience extreme psychological discomfort after a few days of sleep (and dream) deprivation, although occasional rare individuals can live for years with only two or three hours of sleep per day. Without sufficient sleep, the ATP reserve in the body declines precariously, adrenal stress hormones are secreted into the blood steadily, and mental and physical levels of performance falter markedly. Large amounts of a chemical similar to LSD appear in the blood, and this may be the cause of hallucinations (such as those that happen to sleepy drivers) and psychotic behavior. Sleep is the only relief.

Drugs and the Nervous System

Throughout this chapter and the previous one we have seen examples of how important chemical reactions are for the proper functioning of the nervous system. Chemicals may also *interfere* with the normal workings of nerve impulses. In the last chapter we saw how a synapse might be "short-circuited" by the action of chemical substances. Neuromuscular functions may be altered at the synaptic level by different chemicals, several of which are described here.

Curare Curare, a potent drug still used in parts of South America to poison arrowheads, functions by binding to the nerve receptor site normally occupied by acetylcholine. But unlike acetylcholine, curare

does not allow a change in the permeability of the cell membrane, and it is not destroyed by cholinesterase. Consequently, curare blocks the normal action of acetylcholine (which is still released) at the junction of nerve and muscle. The muscles, including the breathing muscles, are unable to contract, and death from asphyxiation follows. (Poisonous mushrooms have a similar effect on neuromuscular transmissions.) A minute dose of curare is sometimes used as a muscle relaxant in controlled surgical situations.

Pesticides and "nerve gases" Some organic phosphate pesticides (such as malathion) and "nerve gases" also inhibit the action of cholinesterase. When that enzyme is inhibited, acetylcholine is not destroyed, and muscle cells remain in a state of depolarization. As with curare, no new nerve impulses can be transmitted properly, and the muscles are paralyzed. Asphyxiation results.

Botulinus toxin Botulinus toxin, produced by the bacterium *Clostridium botulinum*, is responsible for a type of food poisoning known as botulism. Botulinus toxin blocks the release of acetylcholine from presynaptic junctions. Naturally, if acetylcholine is not present, the muscle cells cannot be stimulated by nerve impulses, and the typical result is death by asphyxiation. Botulism may be on the rise with an increase in home canning in this country. Unfortunately, botulinus toxin is an extremely powerful poison, and it has been said that about one-half pound of it would be enough to kill the human population of the entire world.

Many other alterations of the nervous system mechanism may be caused by drugs, and Table 20.2 outlines some of them.

SUMMARY

1. The *central nervous system* consists of the spinal cord and the brain.

2. The three main parts of the brain are the brainstem, cerebellum, and cerebrum. The *brainstem*, the stalk of the brain, relays messages between the spinal cord and the brain. The *cerebellum* is the coordinating center for muscular movement. The *cerebrum* is popularly considered the region where thinking is done, and its control functions are numerous.

3. The *peripheral nervous system* enables the brain and spinal cord to communicate with the entire body. It is further divided into the *somatic* nervous system and the *autonomic* nervous system. The somatic system involves "voluntary" movements of the skeletal muscles, and the autonomic system controls such "involuntary" activities as breathing and glandular secretion.

4. The autonomic nervous system is composed of the *parasympathetic* system and the *sympathetic* system. The first is mainly inhibitory, and the second is mainly stimulating.

5. Structures that are capable of receiving stimuli are *receptors*. All receptors are capable of changing energy into nerve impulses, a process called *transduction*.

6. Practically all animals respond to light in some way, and light receptors vary from the image-forming, color-sensitive *camera-type eyes* of vertebrates to the simple photoreceptors of unicells. Insects and other arthropods have *compound eyes*.

7. The retina is composed of layers of slender cells, one layer consisting of photoreceptors (*rods* and *cones*) and a covering layer of sensory neurons. The sensory neurons eventually send axons to the brain via the optic nerve.

8. Direct touch receptors are *Meissner's corpuscles*; firm pressure affects receptors called *Pacinian corpuscles*; warmth is perceived by the *Ruffini endings*, and cold by the *end bulbs of Krause*.

9. Much of what we call taste, or *gustation*, is actually a function of our sense of smell, or *olfaction*. The specialized receptor cells for taste in humans are located in the *taste buds* on the tongue. The tastes we perceive are probably the result of a combination of different intensities of more than one basic sensation.

10. Of all the senses, the least understood is smell. One theory suggests that many different odors can be detected because of the many possible interactions between the molecules of the odor substances and the several kinds of receptor cells.

11. Both hearing and balance have receptors in the inner ear.

12. One theory of learning proposes that when one nerve circuit is used over and over again, synaptic connections are improved, and these "preferred pathways" establish a learning experience. Synapses may also be related to memory, but no definite information is available. It is possible that RNA and protein synthesis are connected to memory.

13. Sleep can be divided into four distinct stages and a transition stage, *REM sleep*, when dreams occur. Neither the exact benefits nor the causes of sleep are known.

RECOMMENDED READING

Helpful references for the entire physiology section, Chapters 18–24, are listed at the end of Chapter 18.

Bailey, Ronald H., *The Role of the Brain*. New York: Time-Life Books, 1975. (Human Behavior)

Blakemore, Colin, *Mechanics of the Mind*. New York: Cambridge University Press, 1977.

Gazzaniga, Michael S., "The Split Brain in Man." *Scientific American*, August 1967. (Offprint 508)

Geschwind, Norman, "Language and the Brain." *Scientific American*, April 1972. (Offprint 1246)

Haber, Ralph N., "How We Remember What We See." *Scientific American*, May 1970. (Offprint 528)

Heimer, Lennart, "Pathways in the Brain." *Scientific American*, July 1971. (Offprint 1227)

Horridge, G. Adrian, "The Compound Eye of Insects." *Scientific American*, July 1977. (Offprint 1364)

Kimura, Doreen, "The Asymmetry of the Human Brain." *Scientific American*. March 1973. (Offprint 554)

Kupchella, Charles E., *Sights and Sounds: The Very Special Senses*. Indianapolis: Bobbs-Merrill, 1976. (Paperback)

Lassen, Niels A., David H. Ingvar, and Erik Skinhøj, "Brain Function and Blood Flow." *Scientific American*, October 1978.

Mueller, Conrad G., and Mae Rudolph, *Light and Vision*, Revised Edition. New York: Time-Life Books, 1970. (Life Science Library)

Pappenheimer, John R., "The Sleep Factor." *Scientific American*, August 1976. (Offprint 571)

Routtenberg, Aryeh, "The Reward System of the Brain." *Scientific American*, November 1978.

Stevens, S. S., and Fred Warshafsky, *Sound and Hearing*, Revised Edition. New York: Time-Life Books, 1972. (Life Science Library)

HOW ANIMALS MOVE

Animals move. This may seem too obvious to mention, but it acquires significance when we remember that plants do *not* move about. All animals use the same basic principle to move. Whether they creep, run, swim, jump, or fly, they move forward by pushing backward against the ground, water, or air, or they move backward by pushing forward. In so doing, the animal must change its shape—the moving animal must have moving parts.

HOW BIRDS FLY

Animal movement has always fascinated humans, but the flight of birds has had a special attraction. Some magnificent attempts to fly have been made by humans, but none has equalled the splendor of a bird in flight—or its economy. The 747 and the Concorde can fly, and so can bats and many insects, but birds are unique because they have feathers.

Birds are built to fly. Their thin, hollow bones and powerful breast muscles provide light weight and high power, two essential requirements for any flying machine. The refinements are remarkable. Teeth and heavy jawbones have been replaced by the gizzard, which does the work of teeth grinding food. Significantly, the gizzard is located near the bird's center of gravity, where it places little demand on the flying apparatus. The special structure of a bird extends to the inner organs and the entire mechanism that regulates homeostasis. A bird's heart is large and strong to cope with the strain of flying, and the blood has a high concentration of energy-producing sugar. The respiratory system of a bird is even more specialized, and in proportion to body weight it is about four times more extensive than a human respiratory system. Besides lungs, birds have air sacs that extend throughout the body. In addition to improving the breathing system, the air sacs form an effective cooling mechanism. But most of all, it is the distinctive feathers of a bird that allow it to soar, glide, race, and hover.

A bird flies by applying power on the downstroke of its wings. As the wings pull down, the edge of each wing feather curls upward to produce a propellerlike action that lifts the bird up and pushes it forward. During the powerful downstroke the feathers overlap and present the maximum thrusting surface against the air. The downstroke ends on a level with the bird's bill. As the upstroke begins, the primary feathers at the tip of the wing twist and separate to permit air to pass through. This action makes it easier for the bird to lift its wing. Some forward propulsion and lift continue to occur as the wings move up and back against the air, but most of the driving force is produced as the downstroke begins at the start of a new cycle.

A hummingbird hovers with the aid of a special shoulder joint that permits the wing to swivel at the shoulder (many insects have such swivel joints). The forward stroke looks much like the conventional downstroke of other birds, except that because the hummingbird's body is held almost vertically, the wing stroke is parallel to the ground instead of downward. This provides lift but no forward propulsion. As the wing moves into the backstroke, it swivels almost 180 degrees at the shoulder. The edge of the wing is turned backward in a "mirror image" of the forward stroke, so that once again the hummingbird achieves lift without forward propulsion. By using its swivel action and a wingbeat rate of 50–70 beats per second, the hummingbird is able to hover like a helicopter while it sips nectar from a flower.

Adapted from an idea in *The Birds,* a Time-Life Book.

FEATHERS AND FRAMES:
FORM FOLLOWS FUNCTION

Not only are feathers beautifully designed for flying; they also provide low weight and insulation against the weather. In fact, their properties of insulation are so effective that birds can usually live in climates that are too cold for other animals. The gross structure of large contour feathers provides the bird with its overall sleekness. These feathers make streamlined flight possible. But the most impressive engineering is shown in the finer parts of a flight feather. The shaft is strong but lightweight, with a flexible tip so the bird can twist the feather at the end of a downstroke. Extending diagonally from the shaft are the parallel barbs of the feather, each made up of many smaller barbules that overlap and are hooked together by even smaller barbs. The result is an interwoven pattern that produces the strongest of all wings (including airplane wings), while still retaining its lightness. But as strong as feathers are, they are not totally resistant to wear. Most birds have oil glands that help offset the abuse that feathers take. Birds smear the oil on feathers with their beaks—one form of preening. Other birds preen their feathers with a fine powder derived from the feathers themselves. Feathers are replaced about once a year, usually in late summer, after the nesting season. Some birds also acquire new brightly colored feathers *before* the nesting season, in time for their ritual courtship displays. Typically, most birds molt a little at a time so that their flying is not curtailed, but many water birds shed all their feathers at once, and they cannot fly until the new feathers grow in.

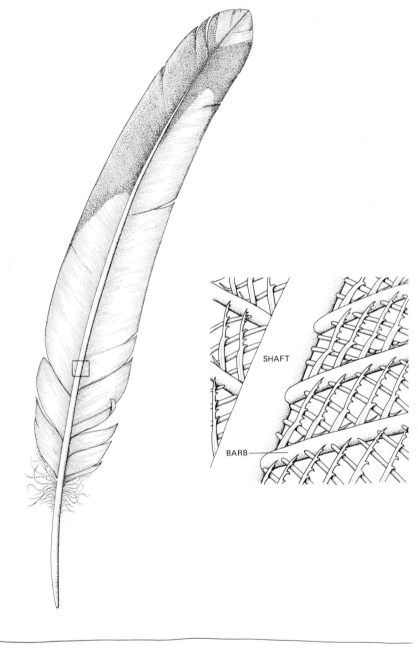

SHAFT

BARB

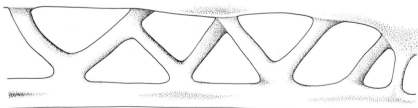

A long section of a bird's wing bone.

Besides being strong and light, some of the hollow bones of birds reveal an internal trusslike structure that makes them even stronger. The pattern of these struts has been duplicated in airplane wings and steel beams used in the construction of buildings. As strong as such bones are, the skeletons of some birds actually weigh less than their feathers. X 10.

HOW SNAKES SLITHER

Snakes have evolved from animals that lost their limbs when they began to burrow. In the course of adaptive evolution, the entire body of a snake became an organ of locomotion. Instead of legs, snakes use the backbone and muscles to move. They have 100–400 vertebrae; most mammals have about 16. Snakes cannot move forward on totally smooth surfaces because there is nothing for them to push against. Also, in order to move forward, snakes must have room to change the curvature of their bodies, allowing their muscles to shorten on one side of the body while complementary muscles lengthen on the opposite side. For this reason, most snakes can move easily through a wavy glass tube, but they cannot move through a circular tube or a narrow straight one. There are exceptions and variations, as shown in the drawings on the next page, which illustrate the four basic kinds of snake movement. Most snakes are not limited to just one movement. In fact, they can usually use more than one kind of movement at the same time.

The most familiar method of snake movement is *lateral undulation*. The snake moves by pressing its curved body against stones, roots, or any other upraised surfaces along the ground. It pushes off from these contact points, each portion of its body following the portion ahead of it to trace a single path. A minimum of two contact sites on alternate sides of the body must be maintained simultaneously in order for a snake to move forward. The efficiency of lateral undulation decreases as the number of contact points increases, because the snake expends more energy for the forward motion it produces. The lateral undulation of a snake is basically the same movement used by a swimming fish.

Large snakes, such as pythons and boa constrictors, break the rule of not being able to travel in a straight line. These loose-skinned snakes have a powerful muscle that runs the length of their bodies. The muscle is attached at the sides to scutes, wide abdominal scales that look somewhat like tractor treads. As the series of scutes are held to the ground, the snake moves the muscle between them. By constantly removing the forward scutes from the ground and pressing down with the rear ones, the snake is able to propel itself forward. This *rectilinear* movement allows the snake to advance toward its prey in a straight line or to move across flat surfaces. Such movement is slow, and when the snake wants to move quickly it uses lateral undulation.

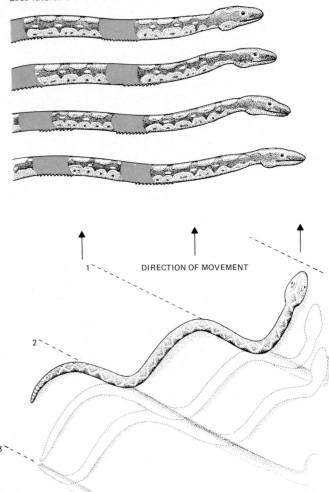

DIRECTION OF MOVEMENT

If a snake is enclosed within a narrow straight space, it cannot use lateral undulation and resorts to a *concertina movement* instead. With this method the midbody is folded into several S loops, which are pressed against the sides of the channel. As the snake moves its head forward, it also pushes against the channel to move the upper half of its body forward. Then it forms additional loops behind its head and pulls up its tail while pressing the forward loops against the channel walls.

When a snake needs maximum speed it uses the *sidewinding* technique. A sidewinder like a desert viper literally moves sideways. Only three points of the snake's body touch the ground, at the neck, midbody, and tail. The head points in the direction of travel, and the rest of the body is inclined at about 60 degrees to the direction of travel. With the body in this wide-open position, the snake finds more contact points than it would if traveling in a straighter line, and progress is swift. The snake moves by forming two loops with its body and lifting its head off the ground. The neck was in one track (1), a point at the center of the body was in a second track (2), and the tail was in the third track (3). If the head moves toward a new track, the body will move into track 1 and the tail will shift into track 2. Between the tracks, the looped body forms bridges that do not touch the ground. These arches undulate back and forth as the snake progresses, but they remain off the ground, leaving only three straight tracks, with a little hook at the end where the head once rested.

Sometimes snakes do more than slither. Some poisonous snakes can rise up and strike forward, and the African desert viper *Bitis caudalis* can even jump to avoid hot ground surfaces. But in the sequence shown here, this tree snake seems to have outdone itself.

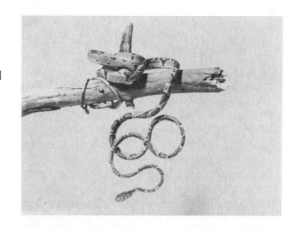

Preparing to move to a higher branch, the snake loosens its coils enough to extend its body, but it retains a firm grip with its tail.

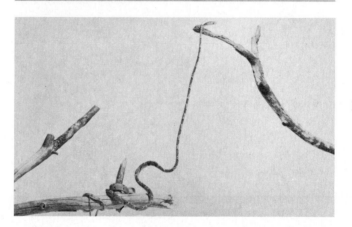

Tree-dwelling snakes have the ability to stiffen their entire bodies, as this one is doing as it stretches toward a higher branch.

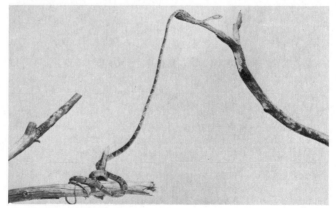

Using its coils as an anchor, the snake is able to raise its head and reach the new branch.

Moving forward, the snake begins to pull up its body as it uncoils from the branch below.

HOW ANIMALS RUN

The fastest land animals have the greatest combination of length of stride and rate of stride. But the real clue to speed is the ability to move as many joints as possible in the same direction at the same time. The cheetah, the acknowledged champion sprinter, has long legs in proportion to its body, a supple spine that speeds up the flexing motion of the body to go along with the legs, long and flexible feet, and a very small collarbone that is situated well forward on the body to allow a free

swivel movement of the shoulder. Hooved animals, such as antelopes and gazelles, have no collarbone at all, and thus their shoulder flexibility is increased even further. Their speed is increased by their single-digit feet, which correspond to standing on tiptoe. Animals such as cats and dogs walk on their "fingers." Slower-moving animals, such as bears and human beings, keep the soles of their feet in contact with the ground when they walk. The tiptoe posture of hooved animals helps

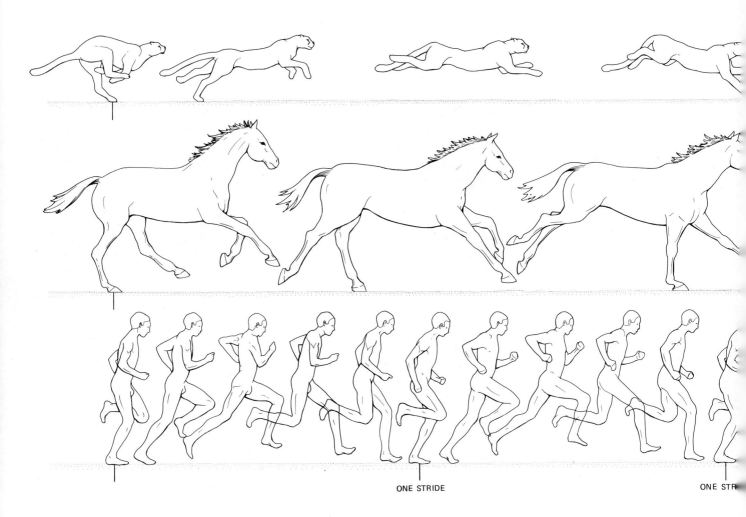

ONE STRIDE   ONE STR

The strides of a cheetah, a horse, and a man are shown in these sequential drawings. The cheetah and the horse are in full gallop, their most powerful stride, and the man is running at top sprinting speed. Both the horse and the cheetah have strides of about 7 meters (23 ft), but the cheetah completes 3.5 strides per second, whereas the horse completes only 2.5. This ratio coincides exactly with the top speeds of the animals: The cheetah, a sprinter, can run short distances at about 112 km/hr (70 mph) and the horse may be able to

sprint as fast as 80 km/hr (50 mph) for a short period. A horse can maintain a steady pace of about 24 km/hr (15 mph) for distances even greater than 50 kilometers (30 mi). An Olympic sprinter can run as fast as 36 km/hr (22 mph) for the 100-meter dash. Both the cheetah and the horse depend on their powerful shoulders and legs for their speed, but the cheetah is also aided by a supple spine that extends the length of its stride and complements the thrusting movement of its legs. The human spine, like the horse's, is too rigid to increase the length of the leg significantly, and elastic ligaments in the feet of horses and other hooved animals give an extra springiness and lift. All these factors contribute greatly to an animal's speed. Equally important, especially over long distances, is the ability of the heart and lungs to keep the muscles well supplied with oxygen. Without energy-producing fuel, even the most streamlined system of levers and pendulums (bones and muscles) would be useless.

ONE STRIDE

ONE STRIDE

ONE STRIDE

ONE STRIDE

aid substantially in running, and both the horse and the human sprinter rely mostly on the backward push of their legs on the ground to produce forward momentum. Humans are also assisted by the balancing action of the swinging arms. All three runners take all feet off the ground at some time during the sprint, and the cheetah is airborne almost half the time. The horse and the man leave the ground completely for about one quarter of their stride, the horse's unsupported leap being about four times the length of its body.

The European common hare (*Lepus europaeus*) uses its outstanding speed to run, hop, and zigzag away from predators. An adult hare is larger than a rabbit and can accelerate to 72 km/hr (45 mph) by using its extra-long hind legs. But some streamlined hunting birds, such as the falcon, are the fastest moving animals of all. They can fly at about 160 km/hr (100 mph) and can dive accurately at speeds up to 290 km/hr (180 mph).

When the Australian crested dragon wants to move faster, it switches from four legs to two. Consequently, it is called the "bicycle lizard." The long tail remains on the ground for balance.

21
The Skeletal and Muscular Systems

IN GENERAL, AS ANIMALS BECAME MORE COM-plex they also became bigger. And the larger they became, the more important it was for them to develop a reliable means of supporting their bodies, protecting their internal organs, and aiding movement. The skeleton provided support and protection, and together with the muscles, it permitted the body to move. Most animals have some firm supporting structure. Even protozoa may secrete shell-like tests (Latin, "pot") or cover themselves with grains of sand. Sponges make stiff internal fibers of geometrical elegance. Invertebrates commonly have external coverings, such as the shells of molluscs and the tough outer skeletons (*exoskeletons*) of crustaceans and insects. Bony, jointed internal skeletons (*endoskeletons*) were not produced until fishes appeared on earth, but from then on, all the vertebrates have had bones, except the sharks and their like, which have skeletons made of relatively elastic cartilage.

Animals with exoskeletons, like the crayfish, cannot grow any larger than their outer covering. They must shed their old skeletons and grow larger ones during the periods of molting. In contrast, animals with endoskeletons have fewer growth restrictions, as demonstrated by elephants, whales, and dinosaurs.

BONES DO MORE THAN SUPPORT THE BODY

The vertebrate skeleton is contained within the body and is composed of bone and cartilage (connective tissue) surrounded by soft tissue. The main function of the vertebrate skeleton is to provide support and protection for the internal organs, but bones are also the storehouse and main supply of reserve calcium and phosphorous, and they aid movement by providing a point of attachment for muscles. In the higher vertebrates (including humans), the bone marrow manufactures red blood cells and some white cells. Bone is not dry, brittle, or dead. It is a living, changing productive tissue.

The Structure of Bone

Bone contains only about 20 percent water, as compared with other tissues of the body, which may have as much as 90 percent water. The solid portion of bone is composed mainly of inorganic minerals, especially calcium phosphate. The rest is organic material made up of bone cells (osteocytes) and bone collagen, a fibrous protein. The collagen and minerals are held together by a cementlike substance in a way that resembles the meshing of reinforced concrete. The strength of bone has been compared to cast iron, even though bone is much lighter and more flexible.

Besides having strength and flexibility, bone is alive and dynamic. The balance of calcium and other minerals in the bones, blood, and muscles depends largely on the ability of the minerals in the bones to move back and forth between the bones and other parts of the body. For example, most of the body's calcium is found in the bones, but calcium is vital for such functions as muscle contraction, the beating of the heart, and blood clotting. When the calcium supply in the body drops below the required level, calcium from the bones is deposited in the blood for transport to the site of the shortage. When the body is deprived of food for long periods, the blood withdraws large quantities of minerals from the bones. Expectant mothers use some of their reserve supply of bone minerals to provide the fetus with the minerals it needs to build its skeleton.

The skeleton of the human embryo is almost entirely made up of cartilage for about two months, at which time it begins to harden. Cells of the flexible cartilage are replaced by bone cells, which remove calcium phosphate and other minerals from the blood and deposit them to form the bones. This process is ossification. We finally reach our adult height when all the bones have replaced the cartilage, and the bones no longer lengthen. This process occurs earlier in women than in men, accounting for the relative tallness of men. Some cartilage remains in the adult body, in the external ear and the tip of the nose, for example.

Usually our bones begin to deteriorate by the time we are 40. Until then our bodies have continued to build bone, even after the full skeletal height is reached, but once the deterioration starts, we lose bone cells. As vertebral disks soften and weaken and the back begins to arch, the characteristic "humpback" of old age may result. Women are affected

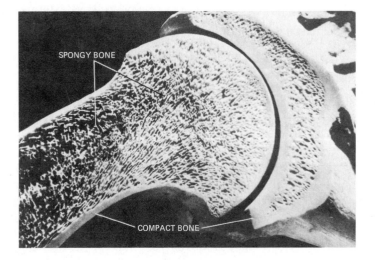

SPONGY BONE

COMPACT BONE

21.1

A section through a human hip joint. The shell-like, solid outer part of the bones contrasts with the porous inner part. Large bones commonly have such separation into *compact* and *spongy* regions. The streaked appearance of the spongy bone is due to the strands of hard material that are built in the direction of greatest pressure, achieving lightness without loss of strength.

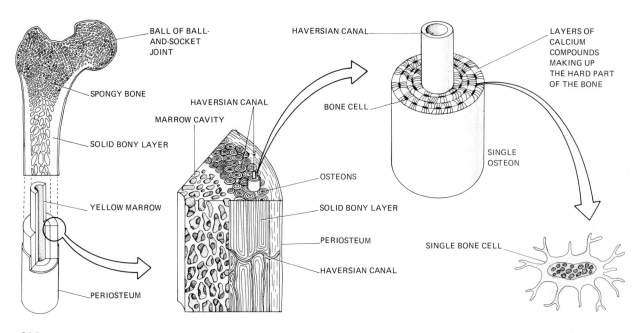

21.2

Gross and microscopic anatomy of a bone. An enlarged view of a portion of the compact bone shows the covering layer, the *periosteum*, and a number of cylindrical subunits, the *osteons*, penetrated by channels. Blood vessels pass through *Haversian canals*. Living cells are dispersed through the hard part of the bone (calcium carbonate and calcium phosphate).

more than men, but sooner or later most of us are destined to shrink and creak.

The Classification of Bone

Bones may be classified according to their external shape as long, short, flat, or irregular, or according to their internal composition, as follows:

1. *Spongy (cancellous) bone* consists of an open, interlaced pattern of bony tissue designed to support maximum stresses and shifts in weight distribution (Figure 21.1). Spongy bone is found inside many bones, including ribs and the ends of long bones. The thigh bone (femur) is the longest, thickest, and strongest bone in the body; its latticed interior is so adaptable that it can shift its stress lines along the ridges if body stresses change. Because bones grow strongest where the stress is greatest, broken bones heal faster when some limited stress is placed on them during the healing process.

2. *Compact bone* is very dense, and contains concentric rings of calcified bone known as osteons (Figure 21.2). Within the osteons are cavities, in the center of which are *Haversian canals*, elongated tubes that carry blood vessels throughout the bone. The shafts of long bones are composed of compact bone, as in the midregion of the femur.

THE VERTEBRATE SKELETON IS ESSENTIAL FOR SUPPORT, MOVEMENT, AND PROTECTION

The growth of a verebrate skeleton depends on the proper balance of minerals, hormones, and vitamins. As we saw in Chapter 18, too much or too little growth hormone from the pituitary gland can produce giants or midgets, and dietary deficiencies can hamper proper skeletal development or even damage mature bones. Some bones normally become fused together as children grow, so that of the original 270

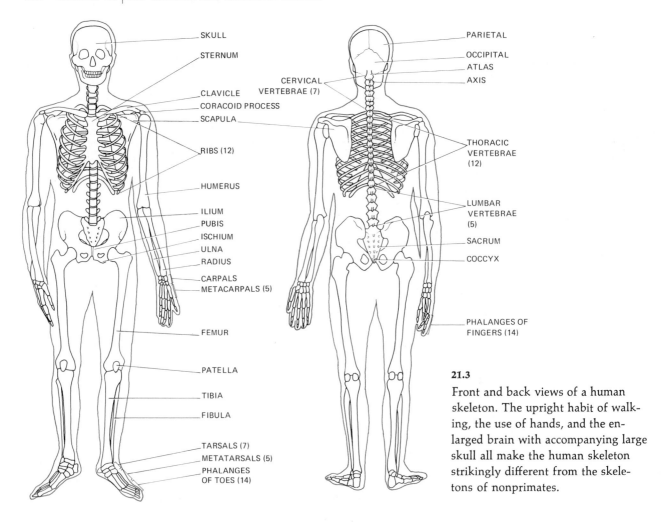

SKULL
STERNUM
CERVICAL VERTEBRAE (7)
CLAVICLE
CORACOID PROCESS
SCAPULA
RIBS (12)
HUMERUS
ILIUM
PUBIS
ISCHIUM
ULNA
RADIUS
CARPALS
METACARPALS (5)
FEMUR
PATELLA
TIBIA
FIBULA
TARSALS (7)
METATARSALS (5)
PHALANGES OF TOES (14)

PARIETAL
OCCIPITAL
ATLAS
AXIS
THORACIC VERTEBRAE (12)
LUMBAR VERTEBRAE (5)
SACRUM
COCCYX
PHALANGES OF FINGERS (14)

21.3

Front and back views of a human skeleton. The upright habit of walking, the use of hands, and the enlarged brain with accompanying large skull all make the human skeleton strikingly different from the skeletons of nonprimates.

or so bones in the infant human body, only about 206 are present in the adult human skeleton (Figure 21.3). Of our 206 bones, 106 are found in the hands, wrists, feet, and ankles, where an intricate combination of bones allows the most varied movement.

The Organization of the Vertebrate Skeleton

The vertebrate skeleton, of which ours is not a typical example, consists of a central spinal column, which is a flexible stack of a varying number of bones called vertebrae; a skull, which is basically designed to protect the brain; two limb girdles to support the forelimbs and hindlimbs; and the limbs themselves. The skull, vertebrae, and rib cage make up the *axial skeleton,* and the forelimbs, hindlimbs, shoulders, and pelvis form the *appendicular skeleton.* In terms of evolution, the axial skeleton was the earlier one.

In humans, the skull is composed of 28 bones, and only the jaw is freely movable. The head is about

12 percent of our total body weight in infancy, but only about 2 percent when we are fully grown. The spinal column contains seven cervical (neck) vertebrae, which permit a broad range of lateral and vertical head movement; 12 thoracic vertebrae, which are attached to the ribs; five lumbar vertebrae in the small of the back, which support most of the body's weight; the sacrum, whose five small fused bones connect the pelvic girdle to the backbone; and a tail bone, or coccyx, of four small fused bones. (If a backache occurs, it will most likely happen in the lumbar or sacral areas.)

The rib cage has 12 pairs of flexible ribs, 10 of which are united by elastic cartilage in front to the breastbone, or sternum, to allow for expansion during breathing. Besides supporting the trunk, the backbone protects the spinal cord, which passes lengthwise through holes in the vertebrae. There are also openings between vertebrae through which pass the spinal nerves. Pads of elastic cartilage called intervertebral disks, located between the vertebrae, not only absorb shocks but allow the vertebrae to move smoothly against each other. The disks are filled with a jellylike substance that may leak out of a ruptured disk and cause great pain by irritating nearby nerves, producing muscle spasms that serve to protect the spinal cord by restricting movement. Severe physical injuries may actually crush the disk altogether, and major surgery may be required to repair the damage.

Backaches probably cause as much trouble for the average person as any other type of chronic physical problem. The most important function of the backbone is to protect the spinal cord, and it performs that function admirably. "Backache" trouble doesn't originate with the cord, however. It starts with any of the 33 vertebrae and their supporting structures. Problems with the kidneys, prostate gland, or liver can cause back pain. Even emotional problems

SUNLIGHT AND SKELETON BUILDING

When children or any young vertebrates (except fishes) are first developing and hardening their bones, they need a supply of the hormone *calciferol*. Without it they become literally rickety. They suffer pain and lassitude, their long bones are weak and ill formed, their heads are too large and their rib cages misshaped. The malady is acute in northern latitudes, where winter sun is weak, and it is especially bad in smoggy cities, where children are kept indoors away from sunlight. Calciferol was for many years called Vitamin D and is still so labeled on the milk sold in the United States and most of Europe, but it is in fact a steroid that influences the absorption of calcium from the gut and its deposition in bones. Calciferol is therefore more properly called a hormone than a vitamin. Land-dwelling animals can synthesize the precursor of calciferol as a form of cholesterol, but they lack the enzyme necessary for the small but essential molecular change from the precursor to the active molecule of calciferol. That change, however, can occur if the cholesterol is irradiated with ultraviolet light, such as is present in sunlight. Children and baby animals exposed to enough sun do not develop rickets. The disease is unknown in regions where light is plentiful and growing infants are normally kept outdoors, nor does rickets develop even in such northern dwellers as Eskimos, who obtain their calciferol from the fish they eat. (Since fish are underwater animals, they receive no ultraviolet light, but they have the ability to make their own calciferol.)

Proper bone development can be guaranteed (other factors being adequate) if an infant is supplied with calciferol in its diet or is kept in enough sunlight to allow its cholesterol in the skin to be changed to calciferol, which can be absorbed. Although some calciferol is essential to the health of vertebrates, too much (like too much of any hormone) can be detrimental, and artificial intake of large amounts is dangerous.

Even now, a common sight in the high Andean villages of Ecuador is a tiny Indian mother carrying a large child in a sack on her back. The round, black eyes peering cheerfully out of the sack are frequently those of a two-year-old who in other parts of the world would be running about on its own legs. If you ask about the practice, you will learn that if the children are put down to walk before their bones are sufficiently developed, their legs will go crooked. The Indian diet is severely lacking in meat, and although Ecuador is on the earth's equator (hence the name of the country) and ultraviolet light is plentiful, the high altitude makes the air very cold. As a consequence, people cover themselves with clothes, and presumably they know from experience that rickets is a threat.

During the eighteenth and nineteenth centuries, when northern Europe was living through the Industrial Revolution, coal smoke so beclouded the skies of cities during the long dark days of winter that many of the factory workers, crowded by cold and poverty into slums, were permanently disabled simply because they did not know enough to get outside. Here is one of the saddest effects of human ignorance of basic biological information.

may manifest themselves as backaches by producing muscle tension that leads to dull back pain. Overweight people may have backaches when their abdominal muscles weaken so much that the back muscles must pull in the opposite direction to hold the body erect. This is the same problem that pregnant women have when they suffer increasingly severe backaches as the pregnancy progresses.

Joints

To be effective, bones must not only be stiff enough to support muscles and internal organs; they must permit movement. This they do by means of the *joints*, or articulations. A joint is the place where two separate bones meet, or where a cartilage joins a bone.

The structure of a joint depends on its function. Some joints are immovable, as in skulls; some, like the hinged joints of knees, bend essentially in one plane; some allow movement in several planes, such as the gliding joints of wrists; and some are free in all planes, such as the ball-and-socket of the shoulder (Figure 21.4). Joints are held together by tough *ligaments* (there are about 1000 ligaments in the human body), and covered by special membranes, the synovial membranes. The moving parts of joints, where contact between two bones is made, are bathed in a viscous synovial ("egg white") fluid, which, along with the texture of the bone covering, makes a joint as nearly frictionless as any moving parts known to humans.

21.4

Joints. (a) Diagram of a long section through a joint. The two bones that move against one another are held together by tough, slightly elastic ligaments, which cover the joint and enclose the fluid in the synovial cavity. Where the bones approach each other, they are covered by marvelously slick articular cartilage, which

offers minimal friction between the moving parts. (b) An elbow is an example of a hinge joint. Movement is restricted to one plane. (c) In the hip, a ball-and-socket joint, the thigh bone (femur) can work against the pelvic girdle freely up and down, sideways, and with a rotating motion.

The shoulder joint is similar. (d) Wrists (and ankles) have gliding joints. Gliding joints do not have as great an angular range as hinge or ball-and-socket joints, but they do have freedom of direction, and they allow a variety of diverse bendings and turnings.

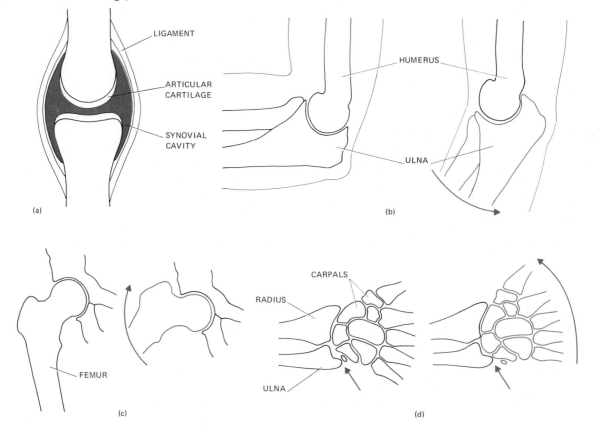

THE INTEGRATION OF BONES AND MUSCLES

Joints make a skeleton potentially movable, but the bones cannot move by themselves. But muscle cells, because they have the ability to contract, specialize in movement.* Bones (and exoskeletons, too) move as levers, with one part rotating about or being hinged to another. A muscle, capable of contraction, is attached to one bone at one end and across a joint to the next one at its other end. When the muscle contracts, the two bones are pulled toward one another, as when an elbow bends, or around on one another, as when a neck twists.

During contraction, a muscle grows shorter by about 15 percent of its resting length, as well as thicker. A contracted muscle is brought back to its original condition by the contraction of its opposing muscle, and it can be stretched to 120 percent of its resting length. A muscle does no work while being stretched. Because muscles work by contracting, not by expanding, they must work in pairs, one altering the relative position of two bones in one way, and its opposite muscle pulling them back.

The Lever Action of Bones and Muscles

Usually, a muscle is attached to a bone by a *tendon* (Greek, "to stretch"), connective tissue composed of collagen. When a muscle contracts, one of the bones attached to it remains stationary while the other bone moves along with the contraction. The junction point where the muscle is attached to the stationary bone is the point of *origin*, and the point of attachment of the muscle to the bone that moves is the *insertion*.

A contracting muscle pulls the attached bone toward the muscle in a movement called *flexion*. Another complementary movement, *extension*, moves the bone away from the contracting muscle. Flexion bends the joint, and extension straightens it. The biceps muscle contracts to pull your forearm and hand toward your shoulder, and the *antagonistic* muscle, the triceps, contracts to straighten your arm (Figure 21.5). Most muscles or groups of muscles have an antagonistic muscle or group to permit movement in opposite directions, or to allow a limb to be rotated. Basically, bones operate as a system of levers.

Although opposing muscle pairs usually work antagonistically, they are not necessarily equal. The biceps muscle of the upper arm is used to bend an elbow more than the opposing triceps is used to extend

---

*   The human body contains about 600 muscles, which weigh about 18 kg (40 lb) in a woman and 30 kg (65 lb) in a man.

21.5

A pair of muscles working in extension and flexion. The forearm is extended when the triceps muscle contracts, and the biceps relaxes and is passively pulled longer. The forearm is flexed when the contraction and relaxation are reversed. Muscles do not extend themselves. They can only contract and then be pulled back to the elongated condition by the contraction of some antagonistic muscle.

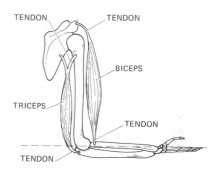

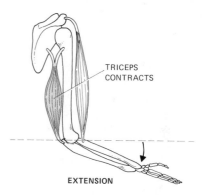

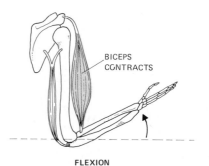

21.6

Position of muscles and tendons. (a) The anatomy of a real hand at a keyboard, with the tendons exposed. Except for some thumb and little finger muscles, hands are almost devoid of muscle. Finger movement is mainly achieved by contraction of muscles in the forearm. (b) How a hand would have to be built if the finger-activating muscles were attached directly to the bones they move. Such a hand would be not only clumsy but probably weaker because it could scarcely have muscles as massive as the ones now present in the forearm.

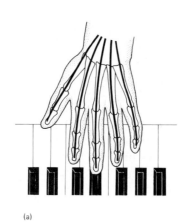

(a)

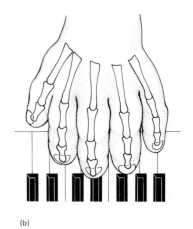

(b)

it. The boy who wants to show off his muscle exhibits his biceps, never his triceps. Jaw muscles are even more strikingly unbalanced. The closing muscles of a jaw are short, heavy, and powerful, but the opening muscles are slight and weak. Anyone can hold an alligator's mouth shut, but trying to hold one open is not wise.

The working muscle is not always close to the moving bone. The muscle used for standing tiptoe, which works by pulling up the heel bone, is not in the ankle but behind the tibia, and it is called the calf muscle, or gastrocnemius. Hands are almost devoid of muscles except for thumb muscles and extensors of the little finger. Piano players do not get tired hands, no matter how strenuously they play (Figure 21.6). Their forearms, however, can become fatigued because that is where the finger-working muscles are located.

SKELETAL MUSCLE

Skeletal muscle (also called striated or voluntary muscle) is considered a "voluntary" type of muscle because it is under the control of the somatic nervous system, and it contracts when we consciously want it to. It is attached to and causes the movement of the skeleton. It is also responsible for such voluntary actions as the elimination of feces and urine. The fibers of skeletal muscle, composed of multinucleate cells, extend along the full length of the muscle.

The Cellular Structure of Skeletal Muscle

The organ called a _muscle_ (from the Latin, "little mouse," because the movement of muscles under the skin resembles a running mouse) is actually an aggregate of muscle fibers (Figure 21.7a, b), just as a nerve is a bundle of nerve fibers. The basic cellular unit of muscles is therefore called the *muscle fiber* (Figure 21.7c). The adult human body probably has about six billion fibers in its 600 or so muscles. Ordinarily, one muscle is larger than another because the larger muscle contains more bundles of fibers.

Within the muscle fiber are smaller units called _myofibrils_. The long, thin myofibrils are made up of a fairly large protein, *myosin*, and a smaller protein, *actin*. The arrangement of these large and small proteins produces the alternating light and dark bands characteristic of skeletal ("striated") muscle (Figure 21.7c). The light bands, called I bands, contain actin; the darker bands, called A bands, contain myosin. Cutting across each I band, like a dime in a stack of pennies, is a disk called the Z line. Within the A band there is a paler H band that also contains a disk, the M line. A section of one Z line to the next, that is, one whole A band and two halves of the two adjoining I bands, constitutes a _sarcomere_, the functional unit of muscle contraction (Figure 21.7d).

The Molecular Mechanism of Muscle Contraction

The thin actin filaments and the thick myosin filaments are connected by molecular bridges, usually called cross bridges (Figure 21.8). During contraction the cross bridges slide back and forth across the filaments in a ratchetlike movement. The _sliding-filament theory_ of muscle contraction, proposed in 1960 by H. E. Huxley, states that muscles are shortened when the actin and myosin filaments slide past each other without changing their lengths. The I bands shorten during contraction, but the A bands do not

21.7

The anatomy of muscle, from the gross structure of an arm to the molecules of muscle protein. A whole muscle is made of bundles, which in turn are made of muscle fibers. Each fiber can be frayed into fibrils, which show the banding that makes striated muscle striated. The striations are an expression of the position of different compounds distributed in the fibril. The two main muscle proteins are *actin* and *myosin,* and their overlapping or failure to overlap gives the muscle its banded appearance. The thicker protein filaments are myosin and the thinner ones are actin.

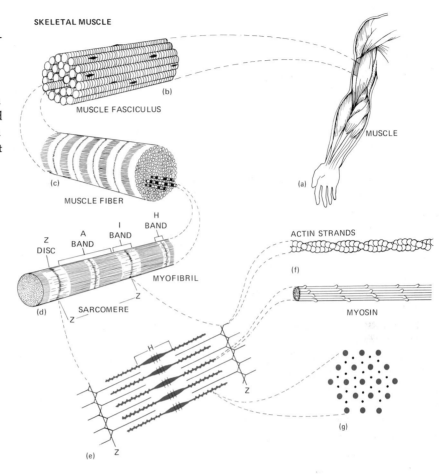

21.8

Electron micrograph of a thin cross section of muscle fibrils, showing the ends of the protein filaments. The larger spots are myosin and the smaller ones are actin. The diagram shows the spatial arrangement of the actin and myosin filaments, with cross bridges.

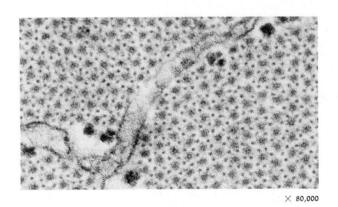

× 80,000

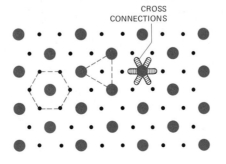

21.9

How the muscle protein filaments slip past each other as a muscle moves. The thin actin filaments and the thicker myosin filaments glide in and out. In (a), which shows a stretched muscle, the two Z lines marking the two ends of a single sarcomere are far apart, and the actin and myosin filaments overlap only for a short portion of their tips. (b) A muscle fibril in its resting state shows the Z lines closer together, the actin and myosin filaments deeply overlapped, and no H zone. (The H zone is visible in the stretched condition as a place where no actin is present.) The thin actin filaments do not overlap. (c) When a muscle contracts, the Z lines come even closer together, the actin and myosin filaments overlap still more, and the actin filaments even overlap themselves. The ratchetlike cross bridges that reach from myosin to actin are not shown. The whole action is known as the sliding-filament model of muscle contraction.

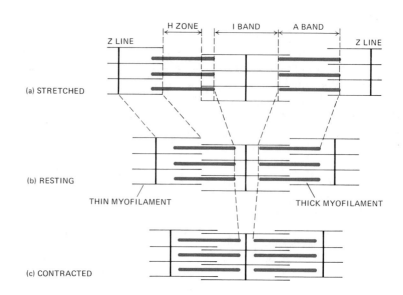

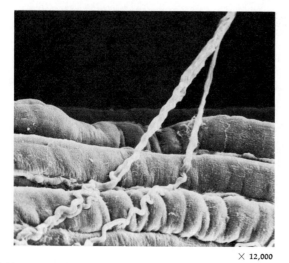

× 12,000

21.10

Scanning electron micrograph of nerve endings on muscle fibrils. The job of a muscle is to contract, but contraction cannot be useful unless it is controlled, and the control comes by way of the nerves.

(Figure 21.9). Striated muscle contractions occur so rapidly that each cross bridge may attach and release the active sites in the filaments as many as 100 times per second.

The action potential of muscle cells is similar to the action of nerve cells, described in Chapter 19. The chemical process of muscle contraction, known as *excitation-contraction coupling*, is initiated by an electrical stimulus in the membrane of the muscle cell (Figure 21.10). Like the activity of nerve cells, muscular activity is mediated by the action of ions. Two regulator proteins, troponin and tropomyosin, act as natural inhibitors of muscle contraction by preventing actin from combining with myosin in a resting muscle, but when calcium ions are introduced into the muscle fibers, the muscle contracts immediately. This happens because the inhibitory effects of troponin and tropomyosin are offset by calcium ions, which bind to troponin and permit a restoration of activity in the cross bridge. The source of these calcium ions is the *sarcoplasmic reticulum*, a specialized form of the endoplasmic reticulum found in other cells (Figure 21.11).

Shortly after death, rigor mortis ("death stiffness") sets in and causes the corpse to become stiff. This phenomenon is caused by the loss of ATP in the dead muscle cells. Without ATP, actin-myosin cross bridges continue to be formed, but the resultant bonds cannot be broken. The dead muscles, therefore, are held in a locked, rigid position. Rigor mortis disappears as the dead body decomposes.

The Nervous Control of Muscle Contraction

As we saw in the previous chapter, afferent nerves carry messages from sensory receptors in the muscles to the brain and spinal cord, and efferent nerves from the brain and spinal cord provide motor impulses that lead to muscle contraction. Skeletal muscle fibers are packed together into bundles averaging about 150 fibers, and each bundle is controlled by a single motor nerve fiber. Such a group of skeletal muscle fibers activated by the same nerve fiber is called a *motor unit*. The motor unit makes contact with the appropriate nerve fiber through a synapse, a chemical bridge described in detail in Chapter 19. The synapse fulfills the possibilities for electrical stimulation of both the nerve fibers and the muscle fibers, and it makes body movement possible.

The nervous system transmits impulses to skeletal muscles continually, especially to the small motor units, where a constant state of weak muscle contraction is maintained. This slight tension, called muscle tone, serves to keep the muscles ready to react to the full stimulus of a contraction. Muscle tone also helps to keep the muscles from getting flabby through disuse. As discussed in Chapter 18, in a stressful situation secretions of adrenalin from the adrenal glands will prepare the muscle for extremely quick action. So-called nervous or jumpy people are usually in a state of such muscular readiness that they may actually twitch.

Energy for Contraction

Energy release is the most obvious feature of muscle contraction. Aside from the mechanical work done, there is considerable heat generated, which indicates still further energy release. If energy is being released during muscle contraction, it must come from some biological source. As cellular energy usually does, it comes from ATP, which yields some of its energy as it is broken down to ADP and phosphate. The immediate reaction, substantiated by firm experimentation, involves an actin-myosin-ATP association, with release of phosphate and energy, which is used in the contractile process. Some of the ATP comes from the usual aerobic oxidation of foods, principally sugars and fats, via the Krebs cycle. Some comes from a special ATP-generating process involving a high-energy compound, *creatine phosphate*, which can form ATP from ADP by having its own phosphate transferred.

During times of heavy muscular activity, much ATP comes from anaerobic respiration, or *glycolysis*, which means "the loosening of sugar" (Chapter 5). When anaerobic oxidation is the source of ATP, the incompletely oxidized result is lactic acid. Since lactic acid is toxic, it cannot accumulate indefinitely in muscles. It must be removed by the circulating blood and then carried to the liver. Once lactic acid reaches the liver, it is converted back to glycogen. During strenuous exercise, lactic acid is generated faster than oxy-

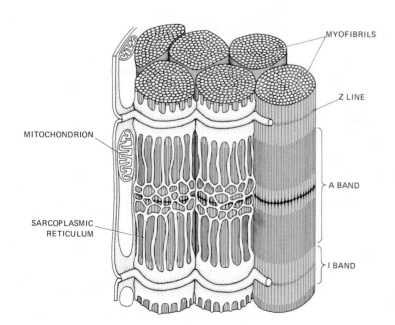

MYOFIBRILS

Z LINE

MITOCHONDRION

SARCOPLASMIC RETICULUM

A BAND

I BAND

21.11

A three-dimensional drawing of a bundle of muscle fibrils, reconstructed from electron micrographs of cross and long sections of muscle. The fibrils run vertically. The right-hand myofibril is represented as having had its sarcoplasmic reticulum peeled off to show the bands and the Z line, a disk separating adjacent sarcomeres. Vertebrate muscle from a frog. × 17,000.

SPRAINS, STRAINS, DISLOCATIONS, AND FRACTURES

Practically all of us have had a sprained ankle or a sprained wrist. It hurts for a while but usually returns to normal after a few days, even without special treatment. But what is a sprain, and what is a strain? A *sprain* is an injury to the ligaments around a joint, usually resulting in some tearing and making movement of the joint painful. In a slight sprain, the ligaments heal within a few days, but severe sprains (which are usually referred to as "torn ligaments") may take weeks or even months to heal. In some cases the ligament must be repaired surgically. A *strain* is an injury to a muscle or its tendon, and it may range from a simple "charley horse" to a violent wrenching that requires an operation to repair the torn tissue.

A *dislocation* is such a severe strain of the muscles, tendons, and ligaments that a bone is pulled out of its socket and fails to return to its proper position. The disconnected bones must be repositioned, usually by slipping the dislocated bone back into place by hand, a process called "reducing." If a dislocation can be reduced by simple manipulation, the after-effects may be slight, but if surgery is required, the damage can take months to heal. Hip and shoulder joints are commonly dislocated, but any movable joint, such as knuckles, knees, and elbows, may be "thrown out of joint."

A *fracture* is a broken bone. Broken bones that are reset soon after an injury have an ex-

cellent chance of healing perfectly because the living tissue and adequate blood supply at the fracture stimulate a natural repositioning. It follows that the supple, healthy bones of children mend faster and better than the more brittle bones of older people. In either, the fracture is almost always encased in a plaster cast until the bone sets permanently. In elderly people, bones contain relatively more calcified and less organic material, and consequently old bones break more easily. A fall that a child hardly notices can be serious in an older person. "To fall and break a hip," a common disaster among the elderly, could frequently be better stated, "to break a hip and fall," because the fragile old bones may crack merely under the strain of walking, making the legs give way. The hip is broken before the body hits the ground.

There are many different types of fractures. The simplest fracture is the *greenstick*, suffered mainly by children whose elastic bones may bend as well as break. Accompanied only by slight, longitudinal cracks, greenstick fractures are difficult to detect even in X-ray pictures, but they heal readily. Technically, a *simple fracture* is one in which the skin is not broken. In a *compound fracture*, broken ends of bone push through the skin, because the crushed or shattered parts may be sharp. The most serious fractures require not only surgery but physical therapy afterward.

gen can be brought to the respiring tissues. The result is that muscles "feel tired" until adequate oxygen can be restored. Because of this temporary shortage of oxygen, heavy breathing continues even after exercise while the *oxygen debt* is being made up; that is, the excess lactic acid is completely removed, and the ATP and creatine phosphate store is replenished.

SMOOTH MUSCLE

Smooth muscle is found mostly in the internal organs of the body (but not the heart), and for that reason it is also known as visceral muscle. Smooth muscle cells are long, tapered cells with one nucleus in the fiber of each cell (Figure 21.12). Smooth muscles contain no striations as in skeletal muscle, and they have only a few myofibrils; smooth muscle does contain actin and myosin filaments, however, and therefore the basic elements of muscle contraction may be accomplished.

One of the dominant features of smooth muscle is its slow and rhythmic action, which makes it suitable for the contractile control of the walls of the intestinal tract, the uterus, blood vessels, respiratory passages, and urinary and genital ducts. Smooth muscle not only helps to move the contents of tubes such as the intestines but also regulates the diameter of

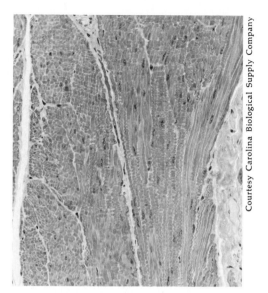

Courtesy Carolina Biological Supply Company

× 150

21.12
Photomicrograph of a section of
smooth muscle from a snake. Smooth
muscle is so called because it lacks
the banding of striated muscle. It is
slow and steady in action and as such
is well suited to the organs in which
it occurs. Intestines, for example, have
smooth muscles in the walls, and they
are effective in forcing food along.

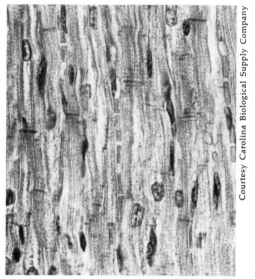

Courtesy Carolina Biological Supply Company

× 700

21.13
Cardiac muscle. Besides striated and
smooth muscle, animals have a third
type, which is especially suited to the
repeated contractions a beating heart
must provide. Cardiac muscle does
have striations, much like those of
ordinary striated muscle, but the arrangement of the sarcoplasmic reticulum is special. Like smooth muscle
but unlike the usual striated muscle,
cardiac muscle is not under direct
voluntary control.

blood vessels, for example, by slow and steady contractions that do not upset the normal circulation of
the blood or the systematic beating of the heart.

CARDIAC MUSCLE

Cardiac muscle is found only in the heart. Meant to
resist prolonged wear, it is more useful for endurance
than for speed. Contractions of cardiac muscle last
longer than those of smooth muscle. This type of action is necessary to give the heart enough time to
pump blood from its chambers. Because of its strenuous and continuous pumping activity, cardiac muscle
must have a constant supply of oxygen, and oxygen
deprivation for more than 30 seconds or so may cause
the cells to stop contracting, with resultant heart
failure.

Cardiac muscle is striated like striated muscle and
contains the same type of myofibrils and actin and
myosin filaments (Figure 21.13). Until recently it was
thought that cardiac muscle was composed of one
mass of branching fibers called a *syncytium*, but observations with the electron microscope now suggest
an organization of closely packed, separate cells, each
with its own nucleus. This principle can be illustrated
when a frog's heart is surgically removed and then
cut into little pieces; each piece will go on beating for
as long as half an hour.

SUMMARY

1. Invertebrates commonly have external coverings, such as the tough *exoskeletons* of crustaceans and insects. Bony, internal skeletons (*endoskeletons*) were not produced until fishes appeared on earth, but after that all vertebrates except sharks and their like have had bones.

2. The main function of the vertebrate skeleton is to support and protect the body, but bones are also the main supply of reserve calcium and phosphorous, and the marrow produces blood cells.

3. Bones may be classified according to their internal composition as either *spongy* or *compact*.

4. The human skeleton is made up of an *axial* skeleton (skull, vertebrae, ribs) and an *appendicular* skeleton (arms, legs, and the structures that attach them to the axial skeleton).

5. The point where two separate bones meet or where a cartilage joins a bone is a *joint*, or *articulation*. The structure of a joint depends on its function. Joints are held together by *ligaments* and covered by *synovial membranes*.

6. A muscle is usually attached to a bone by a *tendon*. The junction point where the muscle is attached to the stationary bone is the point of *origin*, and the point of attachment of the muscle to the bone that moves is the *insertion*. Most muscles or groups of muscles have an *antagonistic* muscle or group to permit movement in opposite directions, or to allow a limb to be rotated. Basically, bones operate as a system of *levers*.

7. Muscle fibers are made up of *myofibrils*, which in turn are made up of the proteins *myosin* and *actin*. Within the myofibrils is a *sarcomere*, the functional unit of muscle contraction.

8. The *sliding-filament theory* of muscle contraction states that muscles are shortened when the thin actin and thick myosin filaments slide past each other without changing their lengths. The chemical process of muscle contraction, known as *excitation-contraction coupling*, is initiated by an electrical stimulus in the membrane of the muscle cell.

9. A group of skeletal muscle fibers activated by the same nerve fiber is called a *motor unit*.

10. The biological source of energy release during muscle contraction is ATP. The immediate reaction involves an actin-myosin-ATP association, with release of phosphate and energy.

11. The three types of muscles are *skeletal* (also called striated or voluntary muscle), *smooth* (found mostly in the internal organs of the body), and *cardiac* (found only in the heart).

RECOMMENDED READING

Helpful references for the entire physiology section, Chapters 18–24, are listed at the end of Chapter 18.

Cohen, Carolyn, "The Protein Switch of Muscle Contraction." *Scientific American,* November 1975. (Offprint 1329)

Gans, Carl, "How Snakes Move." *Scientific American,* June 1970. (Offprint 1180)

Gray, James, *How Animals Move.* Baltimore: Penguin Books, 1964. (Paperback)

Gray, James, "How Fishes Swim." *Scientific American,* August 1957. (Offprint 1113)

Hildebrand, Milton, "How Animals Run." *Scientific American,* May 1960.

Hoyle, Graham, "How is Muscle Turned On and Off?" *Scientific American,* April 1970. (Offprint 1175)

Huxley, H. E., "The Mechanism of Muscular Contraction." *Scientific American,* December 1965. (Offprint 1026)

Lester, Henry A., "The Response to Acetylcholine." *Scientific American,* February 1977. (Offprint 1352)

Loomis, W. F., "Rickets." *Scientific American,* December 1970. (Offprint 1207)

Merton, P. A., "How We Control the Contraction of Our Muscles." *Scientific American,* May 1972. (Offprint 1249)

22

Nutrition and
the Digestive System

THINK OF THE DIGESTIVE SYSTEM AS A TUBE within a tube. The inside tube is the digestive tract, including the stomach and intestines, and the outer tube is the body itself. Food that is in the stomach is still not inside the body proper, just as a raisin lying in the middle of the hole in a doughnut is not *inside* the doughnut. Food enters the digestive tract and undergoes *digestion*, the process of breaking down large food molecules into small food molecules. But the small molecules are useless unless they can get into the individual cells of the body. This is accomplished when the small food molecules pass through the cell membranes of the intestines into the bloodstream. This process is called *absorption*.

The twofold process of digestion/absorption has the same results in all animals: (1) it reduces large molecules of food to small molecules that can pass through cell membranes; (2) the small molecules can be used as a source of energy; and (3) the small molecules, such as amino acids, can be used by cells to rebuild, repair, and reproduce themselves. Of course, a constant source of energy is vital for all organisms, just as it is vital for an organism to be able to rebuild, repair, and especially reproduce its cells. Digestion and absorption provide the usable fuel that helps keep an animal alive. In this and the following chapters we will see how the body obtains and assimilates nutrients, how waste products are removed, and how oxygen and carbon dioxide are transported into and out of the overall body system.

WHAT ARE THE ESSENTIAL NUTRIENTS?

During the course of evolutionary adaptation to their environments, humans have lost the ability to synthesize many of their cellular constituents. To remain alive, therefore, humans must include certain essential components in their diets. These essential compounds are carbohydrates, lipids (fats), and proteins, as well as smaller molecules, such as vitamins that are essential to the proper functioning of enzyme systems.

Carbohydrates

Glucose is the body's main energy source. It provides about two-thirds of the body's daily calorie requirement, and it is needed for cellular respiration. Glucose is produced from other carbohydrates, such as sugars and starches. The conversion of starches and such complex sugars as sucrose (table sugar) takes place in the mouth and small intestine. Glucose may be used quickly for the synthesis of ATP, or it may be converted into glycogen and stored in the liver, where it is readily available when cells require energy.

Other sugars that can be easily used by the body are fructose (found in most fruits), maltose (malt), and lactose (milk).

Fats

Fats are an important part of every animal cell. They are essential structural components of the plasma membrane, the endoplasmic reticulum, mitochondria, and myelin sheaths around nerves. Animal cells can usually synthesize their lipid constituents, but there are some limitations. Rats, for example, are unable to synthesize certain fatty acids. Young rats on a fat-free diet do not develop normally. They stop growing and do not mature sexually.

When fats are digested, they are broken down into their component parts, one of which is fatty acids. Fatty acids are long, thin molecules primarily consisting of hydrogen and carbon. If all the carbons have the maximum number of hydrogen atoms attached to them, they are called _saturated_ fatty acids. If some of the carbons contain fewer hydrogen atoms, they are spoken of as "unsaturated" fatty acids. The human body can synthesize certain fatty acids, but others, labeled _essential_ fatty acids, must be obtained from food. Most of the essential fatty acids are of the

unsaturated type. Fatty acids, as well as the various kinds of fats they make up, are important to the body in a number of ways. They may serve as energy sources, but they may also play very important roles in the structure of cell membranes.

Cholesterol is a complex fat with an entirely different molecular structure from fatty acids. It is an important constituent of all healthy living cells and an important component of any proper diet. However, too much cholesterol in the diet may contribute to circulatory diseases, such as the obstruction of arteries by fatty deposits (atherosclerosis). Animal fats and butter contain large amounts of this complex fat. It is believed that cholesterol and other fatty deposits on the walls of the arteries reduce the space available for blood flow. In its extreme forms this blockage can cause very serious circulatory defects, heart disease, and death. Unfortunately, the precise relationship of diet cholesterol to heart disease is not completely clear. Factors other than cholesterol intake may play an important part.

Proteins

The word "protein" comes from the Greek word that means "primary" or "holding first place." It is an apt name for such an important nutrient, since proteins, specifically _amino acids_, are required by all animals. Without proteins our bodies would waste away and die. No new cells would form, and no oxygen could be carried through the blood by the protein _hemoglobin_. The human body contains 20 different amino acids, which can be joined in practically unlimited combinations in protein molecules. The structure and amino acid composition of proteins vary from plant to animal and from animal to animal, even from human to human. For example, differences in protein complexes make it unlikely that one person can accept a skin graft from another. Because of structural differences, proteins must be broken down by the digestive system and rebuilt to meet the specifications of each organism. When you eat beef, you are not absorbing _beef_ proteins. Rather, you digest the beef, hydrolyzing the proteins to yield amino acids (the structural units of proteins), which are then absorbed. When your body synthesizes new protein, the amino acids are rearranged according to the genetic message of human DNA to make _human_ proteins—more specifically, _your_ proteins. In a similar way, a builder can construct many different types of houses with the same building materials.

Proteins are digested in the stomach and small intestine. The digested protein will eventually become skin, hair, fingernails, enzymes, hormones, and in general the tissues of the body. Those amino acids

HOW ANIMALS EAT

Death comes swiftly in the natural world. To eat means to kill, and some animals may actually resort to cannibalism. Lions have been known to eat their cubs rather than begin a hunt, and in the frenzy of a group kill, hyenas may wound and then eat members of their own pack. Animals that hunt in packs usually spend much of their time trying to identify a weak or isolated member of the group they are stalking. Once the selection is made, the predators will chase the chosen victim until it is caught and killed, unless it manages to get away. If the prey has been chosen well, its handicap will be too much for it to overcome, and the hunters will eventually have their feast.

Cheetahs are so fast that few animals have a chance to escape unless the cheetahs are seen early, or unless inexperienced cubs such as these mishandle the chase. These young cheetahs stay close to their mother as they pursue an antelope and trap it as the mother watches. The kill is finally made by the mother, and only then do the apprentice hunters begin their meal.

Hyenas are not scavengers, but ferocious hunters that use the combined skills of the pack to subdue large animals like this African wildebeest. Once the prey has been wounded enough for them to bring it down, the hyenas begin to eat it alive. The hyena uses its powerful jaws to crush the bones and hide of the wildebeest. Only the skull and horns will be left. Lions are not nearly so thorough, and they are apt to abandon an animal before it is completely eaten. The scene on the next two pages usually follows as the vultures move in.

There is probably no more efficient hunter than a swooping killer bird. The types of prey and the devices for killing are almost as diverse as the birds themselves, but one factor is constant: When a killer bird strikes, it strikes with overpowering speed.

There are about 130 species of owls. About one-third of them are night hunters, aided by an acute sense of vision and a radarlike system of hearing. This barn owl probably did not need more than the light of the stars to make its catch, but it could have done just as well in total darkness if the mouse had made even the slightest sound.

Predatory birds like falcons, hawks, and eagles are some of nature's best-designed hunters. The peregrine falcon is streamlined for pursuit, diving at 290 kilometers per hour (180 mph), the fastest speed attained by any animal. This male osprey feasts on the head of a sea trout he has just caught in the shallows of Florida Bay. After the osprey has devoured the fish's head, he will carry the remainder of the carcass back to the nest in his powerful talons. His mate will tear the fish into little pieces for herself and the young ospreys.

Anhingas are also known in the southern United States as snake birds or darters. The anhinga spears a fish with its pointed beak and then deftly tosses the fish into the air and catches it in the mouth. The fish is always caught head first so the scales and fins will not get stuck in the anhinga's throat.

LIFE Nature Library/Animal Behavior
Published by Time-Life Books Inc.

An eastern garter snake captures a toad and slowly maneuvers it into its mouth. The toad will be settled in the snake's gut, where powerful digestive juices will dissolve the limbs of the toad after three or four days. Most snakes have flexible skins and elastic jaws that allow them to swallow large objects. A snake's jaw can be opened relatively wider than a human jaw because the hinge where a snake's jawbone joins the skull is free to drop down, making the mouth enormous. A man with a snake's jaw and throat could swallow a football.

TONGUES AND . . .

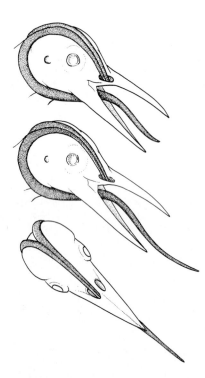

Woodpeckers may have perfected their hunting and eating skills as far back as 50 million years ago. In addition to a jackhammer beak and a specialized tail and claws that hold it securely in a vertical position against the tree trunk, the woodpecker has a flexible and retractable tongue that dips deep into bored holes to dislodge grubs and other food. The barbed tongue continues under the woodpecker's jaw, up around the back of the head, and is attached to the right nostril, leaving the left nostril clear for breathing. Because the tongue is held by elastic tissue, it can be retracted or protruded at will.

An anteater's tongue is usually about 60 centimeters (2 ft) long. In some species, like the giant anteater shown above, the muscles of the tongue are anchored on the rib cage, providing the strength and elasticity the tongue needs to snap at prey and pull back into its mouth at flashing speed. In the scaly anteater, the long muscles attach even farther back, on the pelvis. The anteater breaks into termite mounds and ant hills with its claws and then uses its sticky tongue to reach into crevices to catch the insects.

Frogs, toads, and many other reptiles catch insects by flicking out a sticky tongue faster than the insect can act. Large insects that cannot be thrown neatly down a frog's gullet into its stomach are lined up lengthwise by the frog's forelegs and swallowed. A frog's pointed tongue is coated with viscous saliva. The tongue is usually held flat against the floor of the mouth, but it is unrolled and thrust outward when the frog sees a moving insect nearby. If the insect remains motionless, the frog will not strike.

Like frogs, chameleons also possess a long, sticky tongue that catches prey. But the chameleon's tongue is longer than a frog's, and the thrusting mechanism is more complicated. When not in use, the chameleon's tongue is folded up, accordion-style, at the back of the mouth. When an insect comes within range, the chameleon contracts one set of muscles that unfurl the tongue slightly, and then relaxes another set of muscles that run the full length of the tongue. This second action shoots the tongue forward at least the length of the chameleon's body, and the insect is trapped by the sticky bulb on the tip of the tongue.

The tongue of a honeybee is hollow, like a drinking straw. The bee sips nectar from a flower through the tube in its tongue and stores the liquid in its stomach. (Actually, enzymes begin converting the nectar to honey even before the bee returns to the hive.) Once back in the hive, the bee deposits the nectar in a special reservoir, where it is transformed into honey.

This African okapi is able to wrap its tongue around a leafy branch. It eats the leaves by pulling its tongue along the branch and stripping the leaves into its mouth.

. . . TEETH

A tiger's sharp teeth are typical of those of a carnivore that tears at its food. Long, pointed incisors at the front and strong canines are used for seizing and tearing, and the molars and premolars have been modified to shear like scissors.

Herbivores characteristically have flat molars and premolars that are suitable for grinding plants and for thorough chewing.

Rodents have long, curved front incisors that are used for gnawing; they have no canines at all. Because of their constant use, the incisors continue to grow, compensating for wear. The teeth remain sharp because only the front surfaces are coated with enamel, allowing the back side to wear down more quickly and creating a beveled edge.

The teeth of sharks are notorious. Not only are they long and razor-sharp, but they are lined up in several rows like soldiers, the young ones waiting to replace the old ones when they wear down or fall out. The new teeth are always larger than the ones they replace, keeping pace with the shark's overall growth. In ten years a tiger shark may use up about 24,000 teeth.

that the human body cannot synthesize and that must therefore be included in the diet are called the *essential*, or *required*, amino acids. Those that can be synthesized are equally necessary for survival, but they are not included in the technical term "essential amino acids." Humans require eight different essential amino acids in their diet (threonine, methionine, valine, leucine, isoleucine, lysine, phenylalanine, tryptophan). Children need two others (arginine, histidine) during their developing years. There are large amounts of protein in meat, eggs, fish, seeds, and dairy products.

Vitamins

Vitamins play such an important role in supporting the work of a number of enzymes, especially those involved in cellular respiration, that vitamin deficiencies cause many different diseases (Table 22.1). Because the body cannot synthesize them from other food sources, vitamins must be included in the diet constantly. To be effective, vitamins must not be altered during the digestive process.

Vitamins are usually classified as either water-soluble or fat-soluble. They are ordinarily required in minute amounts, but their importance is enormous.

Minerals

At least 14 inorganic substances are essential for the maintenance of a healthy body, and their benefits range from the development of good bone structure to the activation of enzymes. Table 22.2 outlines the essential inorganic substances (*minerals*) and relevant facts about them.

THE DIGESTIVE PROCESS REDUCES LARGE INSOLUBLE MOLECULES TO SMALL SOLUBLE ONES

In its simplest terms, digestion is the process of breaking down large, complex, insoluble molecules of food into small, soluble molecules that can be absorbed and used by the cells.

The digestive process, assisted by enzymes, takes place in the digestive tract, or *alimentary canal*. In humans, this muscular tube starts at the mouth and ends at the anus, and it is usually about nine meters (30 ft) long. The major compartments of the alimentary canal are the buccal cavity (commonly but not quite correctly called the mouth), the esophagus (a slender tube connecting the throat with the stomach), the stomach, and the intestines. The glands associated with the digestive tube contribute mucus for lubrication, enzymes for chemical breakdown, agents for

emulsification of fats, and ions for the control of acid-base relationships in the local environment.

ADAPTATIONS IN DIGESTIVE SYSTEMS

The process of reducing food particles in size and complexity is fundamentally the same in all organisms, but of course some differences occur. Some animals, such as earthworms and birds, who eat almost continuously, have a separate storage segment, the *crop*, from which food is released at controlled intervals (Figure 22.1). Another adaptation, the *gizzard*, is a muscular sac containing sand or small stones, which the toothless bird or worm has swallowed to help crush and grind food that would be broken down by teeth in other animal groups. Without such stones in the gizzard an animal such as a chicken starves

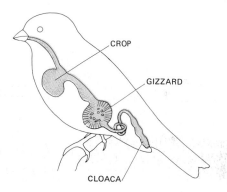

22.1
The digestive system of a bird. Food, swallowed whole, can be stockpiled in a swelling at the lower end of the esophagus, the *crop*. From the crop, it is passed into the *gizzard*, morphologically the stomach but with heavy muscular walls that rub the food with the gravel that the bird takes in with the food. The abrasive action in the gizzard takes the place of chewing in toothed animals. Once ground up in the gizzard, food is passed to the small intestine, where digestion and absorption occur. Gravel and undigested wastes are passed out through the *cloaca* (Latin, "sewer"), an enlargement at the terminus of the alimentary canal through which move feces, waste from the kidneys, and eggs or sperm.

TABLE 22.1

Vitamins

DESIGNATION LETTER AND NAME	MAJOR SOURCES	FUNCTION	DAILY REQUIREMENT	EFFECTS OF DEFICIENCY
A (Carotene)	Egg yolk, green or yellow vegetables, fruits, liver, butter	Formation of visual pigments, maintenance of normal epithelial structure	5000 I.U.*	Night blindness, skin lesions
B-Complex Vitamins (B₁ through Biotin)				
B_1 (Thiamine)	Brain, liver, kidney, heart, whole grains, yeast	Formation of enzyme involved in Krebs cycle	1.5 mg	Stoppage of carbohydrate metabolism at pyruvate; beriberi, neuritis, heart failure, mental disturbance
B_2 (Riboflavin)	Milk, eggs, liver, whole cereals, spinach	Cytochromes in oxidative phosphorylation (hydrogen transport)	1.5–2.0 mg	Photophobia, fissuring of skin
B_6 (Pyridoxine)	Whole grains, liver, milk	Coenzyme for amino acid metabolism and fatty acid metabolism	1–2 mg	Dermatitis, nervous disorders
B_{12} (Cyanocobalamin)	Liver, kidney, brain; bacterial synthesis in gut	Nucleoprotein synthesis (RNA); prevents pernicious anemia	2–5 mg	Pernicious anemia, malformed red blood cells
Folic acid	Meats, vegetables, eggs	Nucleoprotein synthesis; formation of red blood cells	0.5 mg or less	Failure of red blood cells to mature, anemia
Niacin	Whole grains, liver, chicken, yeast	Coenzyme in hydrogen transport (NAD, NADP)	17–20 mg	Pellagra, skin lesions, digestive disturbances, dementia
Pantothenic acid	Yeast, liver, eggs	Forms part of coenzyme A (CoA)	8.5–10 mg	Neuromotor disorders, cardiovascular disorders, gastrointestinal distress
Biotin	Egg white; synthesis by intestinal bacteria	Concerned with protein synthesis, CO_2 fixation, and nitrogen metabolism	150–300 mg	Scaly dermatitis, muscle pains, weakness
C (Ascorbic acid)	Citrus fruits, tomatoes, potatoes, butter	Vital to collagen synthesis and intracellular substance	75 mg	Scurvy, failure to form connective tissue fibers
D_3 (Calciferol)	Fish oils, liver, milk, sunlight	Increases calcium absorption from gut, important in bone and tooth formation	400 I.U.	Rickets (defective bone formation)
E (Tocopherol)	Green leafy vegetables, wheat germ oil	Maintains resistance of red blood cells to hemolysis (bursting of red blood cells)	Not known	Increased fragility of of red blood cells
K (Naphthoquinone)	Liver, leafy vegetables; synthesis by intestinal bacteria	Aids in prothrombin synthesis by liver	Not known	Failure of blood to coagulate
Inositol	Fruits, nuts, vegetables	Aids in metabolism, prevents fatty liver	Not known	Fatty liver

Adapted from James E. Crouch and J. Robert McClintic, *Human Anatomy and Physiology*, Second Edition (New York: John Wiley, 1976), pp. 656–657. Used with permission.

* International units

TABLE 22.2
Major Minerals

MINERAL	MAJOR SOURCES	FUNCTION	DAILY REQUIREMENT	EFFECTS OF EXCESS	EFFECTS OF DEFICIENCY
Calcium	Dairy products, eggs, fish, soybeans	Bone structure, blood clotting, muscle contraction, excitability, synapses	About 1 g	None	Tetany of muscles, loss of bone minerals
Chlorine	All foods, table salt	Acid-base balance, osmotic equilibria	2–3 g	Edema	Alkalosis, muscle cramps
Cobalt	Meats	Necessary for hemoglobin formation	Not known	None	Pernicious anemia
Copper	Liver, meats	Necessary for hemoglobin formation	2 mg in adults	Toxic	Anemia (insufficient hemoglobin in red cells)
Fluorine	Fluoridated water, dentifrices, milk	Hardens bones and teeth, suppresses bacterial action in mouth	0.7 part/ million in water is optimum	Mottling of teeth	Tendency to dental caries
Iodine	Iodized table salt, fish	Synthesis of thyroid hormone	0.1–0.2 mg in adults	None	Goiter, cretinism
Iron	Liver, eggs, red meat, beans, nuts, raisins	Oxygen and electron transport	16 mg in adults	May be toxic	Anemia
Magnesium	Green vegetables, milk, meat	Bone structure, co-factor with enzymes, regulation of nerve and muscle action	About 13 mg	None	Tetany
Manganese	Bananas, bran, beans, leafy vegetables, whole grain	Formation of hemoglobin, activation of enzymes	Not known	Muscular weakness, nervous system disturbance	Subnormal tissue respiration
Phosphorus	Dairy products, meat, beans, grains	Bone structure, intermediary metabolism, buffers, membranes, phosphate bonds essential for energy (ATP), nucleic acids	About 1.5 g	None	Unknown; related to rickets; loss of bone mineral
Potassium	All foods, especially meats, vegetables, milk	Buffering, muscle and nerve action	1–2 g	Heart block	Changes in ECG, alteration in muscle contraction
Sodium	Most foods, table salt	Ionic equilibrium, osmotic gradients, excitability in all cells	About 6 g for adults	Edema, hypertension	Dehydration, muscle cramps, kidney shutdown
Sulfur	All protein-containing foods	Structural, as amino acids are made into proteins	Not known	Unknown	Unknown
Zinc	Meat, eggs, legumes, milk, green vegetables	Part of some enzymes	Not known	Unknown	Unknown

Adapted from James E. Crouch and J. Robert McClintic, *Human Anatomy and Physiology,* Second Edition (New York: John Wiley, 1976), p. 658. Used with permission.

to death, unable to begin the digestive process. Conscientious bird feeders supply sand as well as grain for wild birds during snowstorms.

Organisms like tapeworms, parasites that live in vertebrate intestines, do not have digestive systems (Figure 22.2). Tapeworms simply absorb the predigested food of their hosts and therefore do not need any digestive mechanisms of their own. The cow and some other grass-eating mammals have a large organ at the base of the esophagus called the *rumen*. The rumen contains protozoa and bacteria that break down cellulose, one of the consitutents of cell walls, and synthesize the necessary vitamins. Food is also stored and fermented in the rumen, and it can be regurgitated and rechewed several times before it passes on to the stomach.

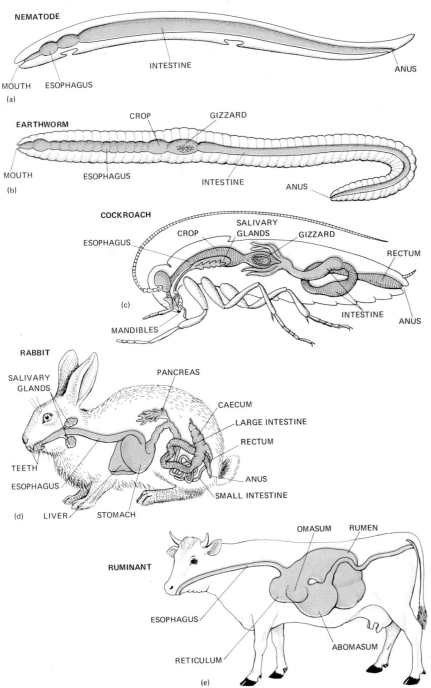

22.2

Comparison of the digestive tracts of several invertebrate animals and a mammal. Some of the simpler animals, such as jellyfish, do not have any special structures that make a definite digestive tract. The main features of a digestive tract include a mouth for ingestion, in most animals a storage pouch, glands for the production of digestive enzymes, a more or less convoluted tube from which nutrients can be absorbed, and an anus through which wastes can be expelled. There is usually some means of reducing the size of ingested particles: a filter, a rasp, chewing teeth, or an internal grinder.

Some animals (the ruminants, such as cattle) have a double system, in which food is swallowed with minimal chewing and passed into the first two enlargements of the alimentary tract: the rumen and the reticulum. After it has been partially broken down by bacterial action, it is regurgitated and chewed further. The food is swallowed again and sent to two more containers, the omasum and the abomasum, where digestive enzymes act on it before it is passed to the small intestine for absorption of the nutrients. Most of the digestive action in a ruminant is accomplished by bacteria.

22.3

The human digestive tract. The drawing shows some parts, notably the small intestine, out of realistic proportion for the sake of clarity.

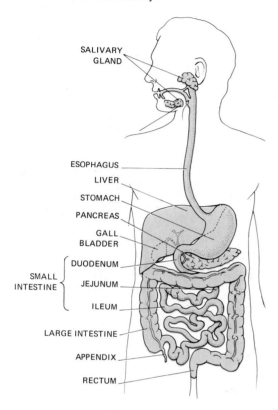

SALIVARY GLAND

ESOPHAGUS

LIVER

STOMACH

PANCREAS

GALL BLADDER

DUODENUM

SMALL INTESTINE

JEJUNUM

ILEUM

LARGE INTESTINE

APPENDIX

RECTUM

THE HUMAN DIGESTIVE SYSTEM IS COMPARTMENTALIZED

Humans are _omnivores_; that is, they can and do eat practically any type of food. Because of this varied intake of foods, the overall human digestive system is less specialized than that of some other animals, whose systems are designed for only meat (carnivores) or only vegetation (herbivores), for instance (Figure 22.3). However, much specialization occurs within the different parts of the human digestive system. Probably the most impressive aspect of the system is its sequential coordination, which involves a network of controls, regulated by hormones and the central nervous system. These controls produce the required activities at the appropriate time and under the proper conditions. Most of the functions of the digestive system are autoregulatory; that is, they take place without our conscious control.

The human digestive system is compartmentalized, and each part is adapted to a specific function (Figure 22.4). The esophagus (Greek, "I shall carry"

22.4

Compartmentalized digestion in the human body. Each compartment of the digestive system has a specific function in the overall process of digestion. Digestion begins in the mouth and ends in the small intestine, and carbohydrates, fats, and proteins are broken down at specific locations throughout the system. In the _mouth_, some starches and sugars are broken down into simple sugars by the enzymatic action of ptyalin in saliva, but most food molecules are still too large to be absorbed into the bloodstream. In the _stomach_, the digestion of carbohydrates continues, and proteins are partially digested. Fats remain relatively unaffected. Some small molecules and inorganic salts may be absorbed from the stomach. In the _duodenum_, fats are emulsified by bile from the gall bladder, and secretions supplied by the pancreas break down carbohydrates, fats, and proteins. In the rest of the _small intestine_, the breakdown of carbohydrates, fats, and proteins continues, and most of the absorption to the rest of the body takes place.

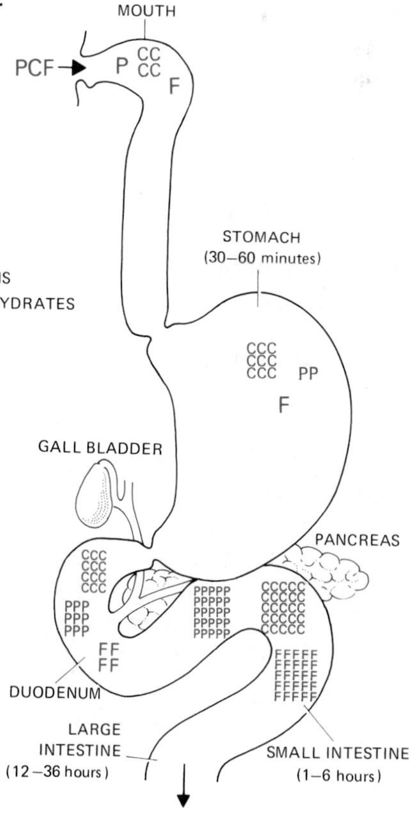

MOUTH

PCF→ P CC CC F

STOMACH (30–60 minutes)

CCC CCC CCC PP F

P PROTEINS
C CARBOHYDRATES
F FATS

GALL BLADDER

PANCREAS

CCC CCC CCC CCC PPP PPP PPP FF FF

PPPP PPPP PPPP PPPP

CCCCC CCCCC CCCCC CCCCC

FFFFF FFFFF FFFFF FFFFF

DUODENUM

LARGE INTESTINE (12–36 hours)

SMALL INTESTINE (1–6 hours)

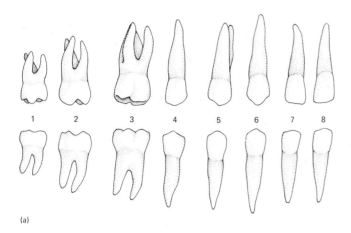

(a)

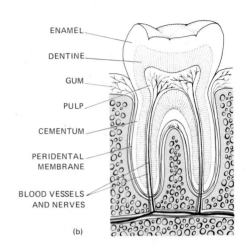

ENAMEL

DENTINE

GUM

PULP

CEMENTUM

PERIDENTAL
MEMBRANE

BLOOD VESSELS
AND NERVES

(b)

22.5

Human teeth. (a) Half a set of teeth: upper and lower from the right side. The short, prong-rooted molars (1, 2, 3) and premolars (4 and 5) are for grinding; the pointed canines (6) are tearing teeth; the chisel-edged incisors (7 and 8) are for nibbling and gnawing. After a few years' use, a set of teeth becomes so individual and distinctive, with its cracks, losses, shifts of position, and dental fillings, that it can be used for identification after other special peculiarities, such as fingerprints, are gone. Teeth are the hardest parts of the body. When hominid fossils are found, the teeth are usually virtually intact. (b) Diagram of a long section of a tooth. The hard, brittle outer covering is the enamel. Inward from the bony *dentine* is a soft *pulp*, containing nerves and blood vessels. The vulnerable parts of a tooth are the enamel and the peridental membrane, which covers the tooth below the gum line.

and "to eat") passes the food from the mouth to the stomach; the stomach stores food and breaks it down with acid and some enzymes; the small intestine is the site of most of the chemical digestion of large molecules into smaller ones, and the place where most of the nutrients pass into the bloodstream; the large intestine carries undigested food, removes water from it, and releases solid waste products through the anus.

If we follow a mouthful of food through the digestive tract, we can observe the digestive process in greater detail.

Digestion in the Oral Cavity

In the mouth, where it all begins, the food is masticated (chewed) by the ripping, crushing, and grinding action of the teeth (Figure 22.5). As much pressure as 500 kilograms per square centimeter can be exerted by the molars, considerably greater than the pressure under the tires of a large Cadillac. At the same time, the food is moistened by saliva, which intensifies its taste and eases its passage down the delicate tissue of the esophagus. Saliva, secreted by the salivary glands, contains an enzyme, amylase, that converts large molecules of starch into small molecules of malt sugar. *This conversion from large molecules to small ones will continue throughout the digestive process.* The specific amylase in saliva, also known as ptyalin, is the first of the many digestive enzymes at work in the digestive tract.

Now the moistened, softened parcel of food, known as a bolus, is ready to be swallowed. The first stage of swallowing is voluntary, and it is the final voluntary digestive movement until the waste products are expelled during defecation. The tongue serves as a digestive organ at this time by helping to push the bolus into the pharynx, or throat. The esophagus, a muscular tube about 25 centimeters (10 in.) long connects the pharynx to the stomach. When swallowing begins, a valve called the epiglottis presses down

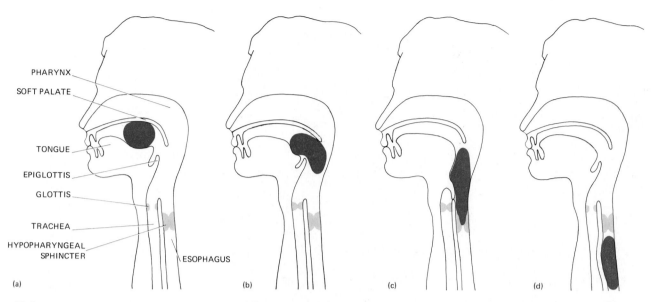

PHARYNX

SOFT PALATE

TONGUE

EPIGLOTTIS

GLOTTIS

TRACHEA

HYPOPHARYNGEAL
SPHINCTER

ESOPHAGUS

(a)   (b)   (c)   (d)

22.6

Action of the epiglottis in swallowing. Air breathed in through the nose and pulled down through the trachea must pass across the passageway through which food goes from mouth to stomach, and yet choking (erroneous entry of food into the trachea) is relatively rare. As food starts to go by the opening to the trachea (the glottis) (b), a flap of tissue, the *epiglottis,* folds down over the opening (c), allowing the food to slip past and into the esophagus (d). The act of swallowing is involuntarily followed by an exhalation to clear the glottis; to inhale directly after a swallow requires conscious determination and feels unnatural.

and prevents the food from entering the trachea (windpipe) or the larynx (voice box). If a tiny bit of food slips into the wrong channel, you will cough and choke until the passage is cleared, or you will die of suffocation.

The complete first stage of swallowing consists of four simultaneous movements (Figure 22.6): (1) The soft palate is raised to prevent food from entering the nasal cavity. (2) The muscles at the rear of the mouth form a thin slit, which keeps large objects from passing into the pharynx. (3) The vocal cords tighten, and the epiglottis swings back over the opening of the larynx. (4) The main muscle of the pharynx contracts and initiates a peristaltic wave, which pushes the food through the esophagus and into the stomach in 4–10 seconds (Figure 22.7).

Digestion in the Stomach

The stomach has been alerted that food is on the way, and secretions of hydrochloric acid and the stomach

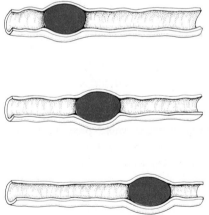

22.7

Peristalsis. Once past the pharynx, a mass of food (a *bolus*) is beyond voluntary recall. From there on through the alimentary tract, it is propelled by the reflex action of waves of muscular contractions and relaxations. (Astronauts in a gravity-zero situation have no trouble swallowing.)

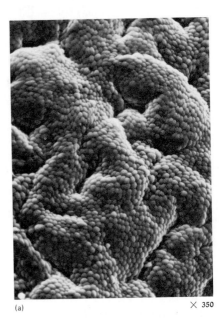

(a)   × 350

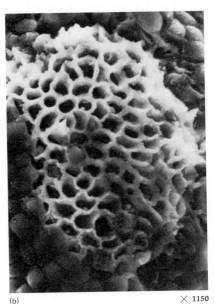

(b)   × 1150

22.8
Normal and damaged stomach linings, as seen in scanning electron micrographs. (a) A normal stomach lining, raised into ridges and furrows and pebbled with convex cell surfaces, has a continuous mucus coating that protects the living tissue from the digestive action of the stomach contents. (b) If the protection is destroyed, some of the cells can be killed, and the result is an open sore or ulcer. Although the stomach wall itself has no direct pain perceptors like those in the outer skin, an ulcer can rub against the opposite stomach wall and cause an ill-defined but definite discomfort, which may seem to be coming from somewhere other than the real site of trouble.

enzyme pepsin have already begun in some of the 35 million glands lining the stomach (Figure 22.8). Pepsin and hydrochloric acid, the main gastric juices, convert protein to smaller molecules (amino acids, proteoses, and peptones) but do not harm the stomach lining, which is coated with a thick layer of protective alkaline mucus. It is thought that stomach ulcers are formed when the heavy coating of mucus is absent. (Actually, ulcers are 10 times more common in the initial portion of the small intestine–the duodenum—than in the stomach.) Also, the food in the stomach helps absorb and dilute some acid. The acidic environment in the stomach is necessary to make pepsin effective. Hydrochloric acid acts on an otherwise inactive protein called pepsinogen to produce active pepsin. The acid condition is changed to an alkaline (basic) one in the small intestine, but if some stomach acid is inadvertently pushed upward, the result is a feeling known as "heartburn."

The secretion of gastric juices in the stomach occurs in three phases:

1. The *cephalic* ("head") phase. When food is seen, smelled, or tasted, the stomach is stimulated by vagus nerves from the brain, and a small amount of juice is secreted.

2. The *gastric* phase. Gastric juice is secreted when incoming food stretches the stomach. Further secretions occur when the hormone gastrin enters the stomach through the bloodstream and causes the production of large amounts of gastric juice.

3. The *intestinal* phase. Additional small amounts of gastric juices are secreted when the partially digested food enters the opening section of the small intestine.

Besides producing the gastric juices, the stomach stores large quantities of food, uses a churning action to mix food into a soupy slurry called chyme, and slowly empties the partially digested food into the small intestine. In fact, the most important function of the stomach is to regulate the flow of chyme into the small intestine. The stomach is well suited for storage because its muscles have little tone, allowing for enlargement up to four liters as necessary.

Food enters the stomach by way of the cardiac sphincter (so named because it is located near the heart), an opening regulated by a ringlike muscle. When the food is ready, it leaves the stomach through the pylorus, (Greek, "gatekeeper").

Digestion in the Small Intestine

After spending approximately two to four hours in the stomach (liquids pass through almost immediately), the food moves into the small intestine, where further contractions continue to mix the soupy liquid. It takes anywhere from one to six hours for the food to move through the 6-meter-long (20 ft) small intestine, where the digestion of all three basic types of

HUNGER AND STARVATION

Hunger may be nothing more than the desire for food, simple "appetite." One's appetite may be stimulated by the sight or smell of food, or even merely reading about it. A person who feels hungry for dessert after an adequate meal clearly has no physiological need to eat. When the need for food exceeds intake, however, a different hunger, a real and physiological hunger, sets in. Hunger pangs, one sign of hunger, are caused by the rubbing of the walls of an empty stomach, and they can be temporarily relieved by a drink of water. When grim starvation occurs, however, not much water is needed to maintain water balance in a body, because little is lost in excretion. A starving man can subsist on a quarter of a liter of water a day, as compared with 20 liters a day for a well-fed diabetic.

If the lack of food continues, body reserves must be used because there are unceasing demands, not merely for energy but for energy that can be supplied only by specific compounds. To keep up body heat and to provide for movement, a number of kinds of molecules can be used, and they can be readily derived from carbohydrates, fats, or proteins, but a living brain demands a constant glucose supply except in really severe starvation, when some alternatives do become acceptable.

Continuing hunger results in shakiness of muscles and loss of mental alertness. A sometimes popular notion that fasting sharpens the mind is based in part on the observation that overeating causes lassitude, as indeed it does, and in part on a self-delusion similar to that which accompanies partial drunkenness. The half-starved or the half-drunk mind may feel acute because its most effective critical abilities are dulled, and it is really more stupid than normal, so stupid, indeed, that it cannot recognize its own stupidity. In long-term starvation, the sufferer first uses reserves of fat, but when those are exhausted, muscle proteins are drawn on. Meanwhile, the starving one thinks constantly of food, finding reminders of it where a nourished person would never see them. Waste papers are seen as having once wrapped food, and birds flying are perceived as edible meat.

If some but not enough food is available, muscle waste and body fluids accumulate in the abdomen, causing skeletal spindliness of limbs and a puffy belly. In absolute starvation, the abdominal swelling does not occur. It must be emphasized that both quality and quantity of food affect the body reactions during times of insufficiency. Simple lack of calories has its own problems of inadequate energy and brain glucose, but a complication occurs when calorie intake by way of carbohydrate is sufficient, but protein content is too low. Then comes a disease of protein deficiency, *kwashiorkor*, with abdominal edema, splotchy colored skin, loss of hair pigment, and general weakness.

The effects of insufficient food and of unbalanced food are of special concern in childhood. Children who have too few calories do not grow to full size, and even if they get abundant nourishment in later life, they never overcome the initial stunting. Brains, too, are adversely affected both by too few calories and by too little protein. Studies have shown that the number of brain cells in an undernourished child or young animal is smaller than normal. Surprisingly, however, a smaller quantity of brain cells does not necessarily mean lowered intelligence. A famous investigation of Dutch army inductees showed that young men who had been born in the midst of the World War II famine had overcome their prenatal and immediately postnatal malnutrition so completely that there was no observable difference between them and their better-fed siblings. The starving ones and their mothers had lived for about half a year while sharing fewer than 500 Calories a day, but having survived to manhood, the young men were normally bright.

Assuming a varied diet that includes carbohydrates, proteins, fats, vitamins, and minerals, an average adult can use 3000–4000 Calories a day. A hardworking lumberjack cutting timber in the freezing woods may need twice that, and an inactive person can survive, though miserably, on half that.

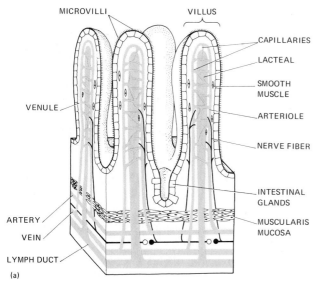

MICROVILLI VILLUS

CAPILLARIES

LACTEAL

SMOOTH MUSCLE

ARTERIOLE

NERVE FIBER

INTESTINAL GLANDS

MUSCULARIS MUCOSA

VENULE

ARTERY

VEIN

LYMPH DUCT

(a)

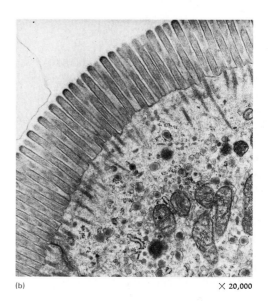

(b) × 20,000

22.9

Lining of the small intestine. (a) Schematic drawing of a section through a small portion of the inner lining of a small intestine. Four villi are shown projecting into the central cavity of the intestine. The villi, which increase the absorptive surface, are capable of slow, limited movement, and they are provided with abundant blood vessels. The surface epithelial cells, which are subject to constant abrasion, are being renewed by growth of new cells. Each villus is covered with smaller projections over the surface, the microvilli, which are just visible as a fuzzy edge to the villus in an ordinary microscope. × 250. (b) A thin section, shown in an electron micrograph, shows a small portion of an epithelial cell with the microvilli well defined.

food is completed. It is there also that practically all of the digested molecules of food are absorbed into the bloodstream. Alcohol may be absorbed through the stomach walls, but the rest of the food mixture moves into the intestine.

The absorbing surface of the small intestine is increased by millions of fingerlike protrusions called *villi* (Latin, "shaggy hairs"), which give the appearance of the tufts of a turkish towel or the velvety pile of a rug (Figure 22.9). The villi aid digestion further by adding their constant waving motion to the peristaltic movement already occurring throughout the small intestine. The thousands of cells making up each villus are the units through which absorption from the intestine takes place. The absorptive surface of each of these cells contains thousands of *microvilli*, which further increase the surface area for absorption. An entire small intestine exposes some 250 square meters (300 yd²) of surface area.

Enzymes for Digestion in the Small Intestine

Increased surface area and waves of motion are important to digestion in the small intestine, but even more important is the action of pancreatic enzymes and bile secreted into the small intestine by the pan-

creas and liver through a common duct. Bile salts break down fat droplets into particles small enough for the enzymes to attack. The physical breakdown of fats is called *emulsification*, the same process by which soap breaks down dirt. The liver secretes bile, which is collected, stored, and modified in the gall bladder.

These secretions of pancreatic juices and bile, together with intestinal enzymes, help to change the environment in the duodenum and entire small intestine from acidic to alkaline. The action of pepsin is stopped by the new alkaline environment that allows intestinal enzymes to function properly. Enzymes furnished by the pancreas and intestine break down carbohydrates, fats, and proteins. These enzymes are the *carbohydrases*, the *lipases*, and *proteases*. Note that enzymes almost always end with the "ase" suffix; the prefixes, such as "prote" in protease, usually indicate quite clearly what substance is being acted upon (Table 22.3).

Digested carbohydrates and proteins are eventually transported to the liver, and digested fats are carried into the lymphatic channels (Chapter 24). Ultimately, almost all the digested food molecules will be passed into the bloodstream, to be circulated throughout the body as energy-producing and tissue-building nourishment.

Action of the Large Intestine

Food passes from the small intestine into the large intestine in a watery condition. By now, digestion is complete, and the large intestine functions to remove water and salts from the liquid matter. Removal of water converts liquid wastes into *feces*, a semisolid mixture that is stored in the large intestine until ready to be eliminated through the anus during defecation. The final products of digestion remain in the large intestine from 12 to 36 hours.

Few bacteria are present in the stomach while food is being digested there, but many bacteria inhabit the large intestine. These bacteria are harmless as long as they remain in the large intestine, and in fact they are useful in synthesizing vitamins K and B_{12}. Besides containing bacteria, the feces are composed of water, undigested food, such as cellulose (from cabbage or celery, for example), and other waste products from the blood or intestinal wall. The brownish color of the feces is caused by the presence of degradation products of red blood cells in the form of bile pigments (bilirubin).

The elimination of most of the body waste products is conducted not by the digestive tract but by the kidneys and lungs. The next two chapters will examine these organs, as well as two other extremely important organs, the liver and the heart.

TABLE 22.3
The Major Digestive Enzymes

ENZYME	SOURCE	SUBSTRATE	FUNCTION
Carbohydrases			
Salivary amylase (ptyalin)	Salivary glands	Starches	Begins carbohydrate digestion; breaks down starch into maltose
Pancreatic amylase	Pancreas	Starches	Converts starch to disaccharides
Maltase	Intestinal glands	Maltose	Converts maltose to glucose
Lactase	Intestinal glands	Lactose	Converts lactose to glucose and galactose
Sucrase	Intestinal glands	Sucrose	Converts sucrose to glucose and fructose
Lipases			
Gastric lipase	Gastric glands	Fats	Converts fats to fatty acids and glycerol
Pancreatic lipase	Pancreas	Fats	Converts fats to fatty acids and glycerol
Proteases and Peptidases			
Pepsin	Stomach mucosa	Proteins, pepsinogen	Converts proteins to polypeptides
Trypsin	Pancreas	Proteins, chymotrypsinogen	Converts proteins to polypeptides
Chymotrypsin	Pancreas	Proteins	Converts proteins to polypeptides
Carboxypeptidase	Pancreas	Proteins	Releases terminal amino acid from end of protein
Enterokinase	Duodenal mucosa	Trypsinogen	Converts trypsinogen to trypsin
Peptidases	Intestinal glands	Peptides	Converts peptides to amino acids
Nucleases			
Ribonuclease	General	RNA	Splits nucleic acids into free mononucleotides
Deoxyribonuclease	General	DNA	Splits nucleic acids into free mononucleotides

TABLE 22.4
The Major Digestive Hormones

HORMONE	SOURCE	STIMULUS	FUNCTION
Gastrin	Stomach	Food entering stomach, vagus nerve, protein	Stimulates secretion of gastric juices
Enterogastrone (Gastric Inhibitory Peptide, GIP)	Small intestine	Fats, glucose, and acids in intestine	Inhibits secretion of gastric juices and stomach contractions; stimulates secretion of insulin from B cells of pancreas
Secretin	Small intestine	Acid and peptides in duodenum	Stimulates pancreatic juices containing bicarbonate; stimulates liver to produce bile
Pancreozymine (cholecystokinin)	Small intestine	Peptides and fat in duodenum	Stimulates gallbladder to release bile; stimulates pancreas to release enzymes
Enterocrinin	Small intestine	Chyme in intestine	Stimulates release of intestinal juices

SUMMARY

1. Food cannot be used until it has been reduced in size enough to enter the cells of the body, not just the digestive tract. Large, insoluble food molecules are broken down into small, soluble molecules through the process of *digestion*. The small molecules produced by digestion are taken into the bloodstream across the cell membranes of the intestines through the process of *absorption*.

2. Food provides energy and the material to build and repair cells. Humans must include in their diets certain essential components, such as *carbohydrates*, *fats*, and *proteins*.

3. *Vitamins* play an important role in supporting the work of enzymes, especially those involved in cellular respiration. At least 14 *minerals* are also essential to the human body.

4. The digestive process takes place in the digestive tract, known as the *alimentary canal*.

5. Humans are *omnivores*, and their digestive system is designed to digest and assimilate almost any type of food.

6. Digestion begins in the mouth, where enzymes in saliva convert large molecules of starch into small molecules of sugar. The act of swallowing is the final voluntary digestive act until defecation. *Peristalsis* helps to propel food along the digestive tract.

7. *Gastric juices* in the stomach act on protein to convert it into smaller molecules of amino acids. Besides mixing the food and producing gastric juices, the stomach stores chyme and regulates its flow into the small intestine.

8. Enzymes furnished by the pancreas and the intestine and *bile* secreted by the liver break down carbohydrates, fats, and proteins in the small intestine. Practically all the digested molecules of food are absorbed into the bloodstream from the small intestine.

9. By the time food reaches the large intestine, digestion is complete. The large intestine functions to remove water and salts from undigested food and to produce feces.

RECOMMENDED READING

Helpful references for the entire physiology section, Chapters 18–24, are listed at the end of Chapter 18.

Davenport, Horace W., "Why the Stomach Does Not Digest Itself." *Scientific American*, January 1972. (Offprint 1240)

Fernstrom, John D., and Richard J. Wurtman, "Nutrition and the Brain." *Scientific American*, February 1974. (Offprint 1291)

Lieber, Charles S., "The Metabolism of Alcohol." *Scientific American*, March 1976. (Offprint 1336)

23

The Excretory System and Homeostasis

ABOUT A HUNDRED YEARS AGO, THE GREAT French physiologist Claude Bernard said, "All the vital mechanisms, however varied they may be, have only one object, that of preserving constant the conditions of life in the internal environment." Bernard was speaking of the body's ability to maintain an inner stability (homeostasis), no matter what the external conditions are. Bernard compared our cells to flowers in a greenhouse. The conditions outside the greenhouse may be sweltering or frigid, and yet the environment of the flowers must remain constant. In particular, life on earth needs to survive in a variety of quantities and concentrations of water. At one extreme are the freshwater fishes, which must keep sufficiently concentrated body fluids in a highly dilute environment. At the other extreme are desert animals, which have practically no liquid water available and must obtain and conserve every molecule of water they can. Most terrestrial mammals are in between. They must conserve water, but at the same time they must remove excess salts and body wastes, especially nitrogenous ones. About two-thirds of the body weight of animals is water, and varying amounts of water are constantly entering and leaving the cells (through metabolic activity) and the body. The balance of water must be maintained, and only slight variations are tolerable.

Humans lose water by evaporation from moist lung surfaces and through the skin. Water, salts, and organic substances are given off in sweat, but the prime regulator of water balance and waste elimination is the kidney system.

SALT AND WATER BALANCE IN ANIMALS

Most animals living in water environments have specialized methods of maintaining salt and water balance. But because the conditions of the open sea do not change much, oceanic invertebrates have little need for complex internal regulators. The bodies of these animals have the same salt concentration as the sea water that bathes them, and their permeable body surfaces allow their interior fluids to conform to any changing concentrations of sea water (Figure 23.1). The lack of internal regulatory processes in such animals can be demonstrated by placing a specimen in fresh water, where salts are rapidly lost and water is

taken up. If oceanic invertebrates are forced to remain in fresh water, they will soon die.

Animals living in fresh water have a great need to prevent salt loss and to limit water uptake. In the freshwater environment, simple physical diffusion will favor the loss of salts from the animal's body, and osmosis will favor the uptake of water. Internal regulatory mechanisms counter these tendencies and permit animals to live in a low-salt environment of fresh water, where they maintain a salt concentration higher than that of the environment.

Freshwater invertebrates keep their salt concentrations higher than that in the water by pumping water molecules out, by absorbing salt and food, and by actively transporting sodium and chloride from the environment into the body. If it is not possible to limit the water uptake, the active excretion of water will maintain the balance necessary to sustain life.

Terrestrial animals have other problems. If they are to get rid of body wastes and excrete excess salts, they must lose water by evaporation or through uri-

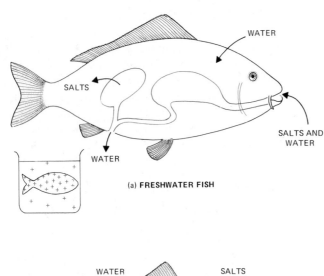

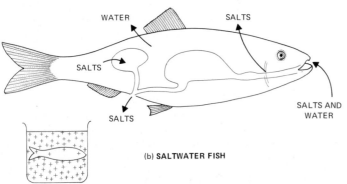

23.1

Water regulation in fishes. (a) A freshwater fish must maintain a salt concentration in its body fluids higher than that in the surrounding water. If it loses too much salt or takes in too much water, it dies. The water that does come in through the skin and gills is passed out in plentiful urine, and what little salt does come in is filtered out and retained. (b) A saltwater fish has a different requirement. It must maintain a salt concentration in its body fluids lower than that in seawater. It drinks a lot, urinates little, and excretes concentrated salt solution from glands in the gills. It also loses water through the skin.

nation. For the terrestrial animal water is always a scarce commodity. Terrestrial animals may acquire water in three ways: by eating food with water in it; by releasing water from food by oxidation; and by drinking water. Some terrestrial animals will use all three methods, but others will rely almost entirely on one form only.

Some animals, such as marine birds, turtles, and lizards, are unable to produce highly concentrated urine the way mammals can, and these animals excrete salt through special salt glands located near the eyes. As part of their overall adaptation to flight, birds have no bladder for the storage of urine. Their kidneys excrete a semisolid uric acid as part of the intestinal waste.

INVERTEBRATE EXCRETORY MECHANISMS

Excretory mechanisms vary greatly among invertebrates. Many protozoans and freshwater sponges have simple intracellular organelles called *contractile vacuoles*, which are essentially reservoirs of excess water. Because freshwater protozoa contain more salt than the water they live in, they absorb water through osmosis. This excess water is accumulated in the expanding contractile vacuole until it is finally emptied back into the outside water. Protozoan waste products are actually eliminated directly from cell membranes into the outside water, so the contractile vacuole is more accurately described as an organ of water balance, not excretion.

Many invertebrates have excretory organs called *nephridia*. Mechanical differences exist from one organism to another, but a nephridium is basically a tubular system that carries wastes to the outside. Flatworms have an arrangement of branched tubes called a *flame cell system*, which removes liquid wastes from intercellular spaces. The wastes are swept into excretory ducts by flagella (which reminded early observers of a flickering flame), and the liquids are finally excreted through excretory pores on the surface of the body. In this way, the flame cell system maintains water balance at the same time that it excretes wastes. Segmented worms have a system of so-called true nephridia, which adds a network of blood vessels and a collecting bladder to the system. The process of urine formation is basically the same in both types of nephridia.

The excretory systems of insects and spiders are called *Malpighian tubules*, an arrangement well suited for land animals in dry environments. As salts

TABLE 23.1

Water Content of Animal Bodies

ORGANISM		PERCENTAGE OF BODY COMPOSED OF WATER
Pea weevil		48%
Kangaroo rat		65%
Herring		67%
Chicken		74%
Frog		78%
Lobster		79%
Earthworm		80%
Jellyfish		95%
Human		65%
Bone	22%	
Brain	74.5%	
Muscle	75.5%	
Kidney	82.5%	
Blood	83%	

and wastes are secreted into the tubules, an osmotic gradient is created that allows water to flow into them. A urinelike fluid is then drawn into the intestine, where rectal glands reabsorb water and most of the potassium in the salts. The remaining waste, usually uric acid, is then excreted through the rectum. Such a system was probably the forerunner of the excretory systems of land-dwelling vertebrates.

THE VERTEBRATE KIDNEYS AND EXCRETORY SYSTEM

Human adults take in about 2500 milliliters (2½ qt) of water each day, most of it through foods and liquids they eat and drink.* Under normal conditions, the same amount of water is given off daily, about a third of it evaporated from the skin and lungs, and

---

* Only about 47 percent of a person's daily water intake comes from drinking. As much as 39 percent is supplied by so-called solid food (meats are 50–70 percent water, most vegetables contain more than 90 percent water, and even "dry" bread is about 35 percent water), and the body's own metabolism produces about 14 percent of the daily water requirement.

TABLE 23.2

Average Normal Routes of Water Gain and Loss in Adults

	MILLILITERS PER DAY
Intake:	
In liquids (drunk)	1200
In food (eaten)	1000
Metabolically produced	350
Total	2550
Output:	
Evaporation (skin and lungs)	900
Sweat	50
Feces	100
Urine	1500
Total	2550

From Arthur J. Vander, James H. Sherman, and Dorothy S. Luciano, *Human Physiology: The Mechanisms of Body Function*, Second Edition, p. 320. Copyright 1975 by McGraw-Hill Book Company, New York. Used with permission of McGraw-Hill Book Company.

more than half excreted in urine (Table 23.2). The remarkable organs that regulate this constant water balance are the *kidneys,* paired, bean-shaped organs, each one about the size of a large bar of soap. The kidneys lie behind the lining of the abdominal cavity, one kidney on each side of the spinal column, at about waist level (Figure 23.2). Each kidney contains more than one million *nephrons,* the functional units of excretion. By "the functional units of excretion" we mean that each nephron is an independent urine-making machine.

Each nephron contains a long, thin excretory tubule that originates as a cupped sac called *Bowman's capsule.* The excretory tubules are coiled and winding, and if straightened out, they would be about 80 kilometers (50 mi) long in an adult. Each of the million or so excretory tubules leads into a large collecting duct that eventually drains into the central cavity of the kidney. This cavity is connected to the ureter, a long tube that transports urine from the kidney to the urinary bladder, where the urine is held until it is excreted through the urethra to the outside.

23.2

The human urinary system. The kidneys are the main blood filters. Blood from the aorta passes through the kidneys, where potentially dangerous waste materials are removed, and returns to the main bloodstream by way of veins to the inferior vena cava. Water, salts, urea, pigments, and other wastes are constantly accumulated in the bladder, which is emptied periodically.

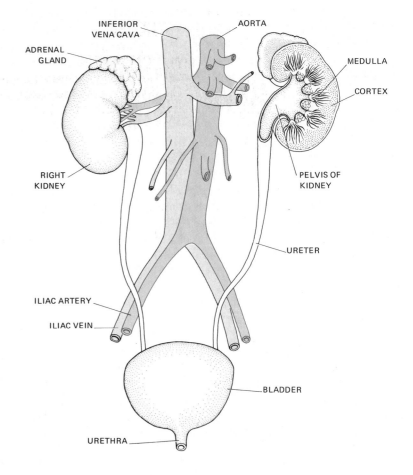

ARTIFICIAL KIDNEYS

If the kidneys fail, toxic wastes build up in the body until cells and organs begin to deteriorate and eventually die. Many tens of thousands of Americans suffer from kidney failure, but more than 16,000 of them are leading relatively normal lives because of the successful use of the artificial kidney, or *dialysis* (diffusion).

The principle of diffusion is basic to the working mechanism of the artificial kidney. Very simply, the artificial kidney is a mechanical device that pumps 5–6 liters of blood from the body through a membrane of cellophane. The blood is rinsed by a briny solution, and waste products, such as urea, uric acid, excess water, creatinine, sodium, and potassium, diffuse through the pores of the cellophane by osmotic pressure. The cleansed blood is then rerouted back into the body.

This procedure is as follows: One cellulose tube is permanently linked to an artery (usually in the arm), and another is attached to a nearby vein. The two tubes are connected by a rubber shunt when not in use. During dialysis the tubes are hooked up to the machine. Several shunt variations are also used, including a bovine graft (part of a cow vein) that links the artery and vein when dialysis is not taking place. Blood is pumped from the artery through an oxygenated salt solution similar in ionic concentration to body plasma. Since the concentration of wastes is higher than the normal concentration of the plasmalike fluid, the wastes will automatically diffuse through the permeable membrane of the tubes into the rinsing fluid. The membrane is porous to all blood substances except proteins and red blood cells. The wastes are eliminated from the body, and the purified blood is free to flow back into the

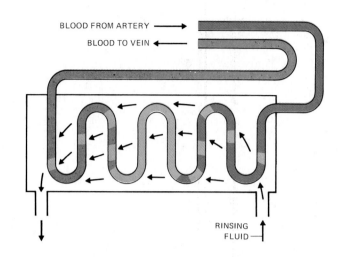

BLOOD FROM ARTERY →

BLOOD TO VEIN ←

RINSING FLUID —

body. Dialysis is sometimes used to add nutrients to the blood. For instance, large amounts of glucose may be added to the salt solution, so that the glucose may be diffused into the blood at the same time that wastes are being removed.

It takes about six hours and 20 passes through the bathing fluid to complete a full cycle of dialysis, and most patients receive treatment two or three times a week. Unfortunately, even this incredibly successful machine (it can remove wastes from the blood 30 times faster than a natural kidney) provides only partial relief for kidney-failure victims. All patients remain uremic (Greek, "urine in the blood") to some degree, and some victims are unable to receive enough cleaning of the blood to make the treatments worthwhile. In such cases kidney transplants are sometimes attempted, but so far the usual problems of tissue rejection have minimized the success of the transplant operations.

Kidneys are filters. Although the human kidneys account for less than one-half of one percent of the body weight, each day they filter about 1900 liters of blood. It is the kidneys, not the intestines as one might think, that are the body's main regulators of waste disposal. Not only do the kidneys regulate water balance; they also clean and filter the blood and control its chemical concentration.

How do the blood vessels and the nephrons work together to filter the blood and produce urine? Blood enters each kidney through a renal artery, which branches into smaller vessels called arterioles. Each

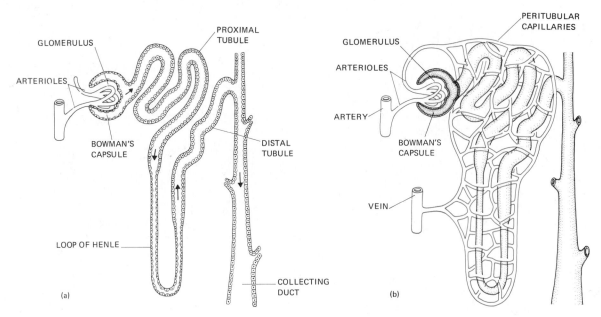

23.3

Microscopic anatomy of the urine collecting unit. (a) A *nephron* in outline, with blood supply removed. Fluid enters the central canal of the nephron in *Bowman's capsule,* passes through the proximal tubule, the loop of Henle, and the distal tubule to the collecting duct, by which time it has been changed into urine. (b) A nephron with its accompanying blood supply. An artery sends a twisted ball of capillaries into the cup of Bowman's capsule. From the ball, the *glomerulus,* blood fluids are forced under pressure into Bowman's capsule: water, sugars, salts, and various wastes. From Bowman's capsule there passes a vein that breaks up into capillaries surrounding the tubules, and these reunite after sending blood past the tubules. After having been filtered through the nephron, the purified blood leaves by way of a vein to rejoin the general circulation.

arteriole enters a Bowman's capsule and subdivides even further to form a tiny ball of capillaries, a clump of small blood vessels known as the *glomerulus* (Figure 23.3). The blood in the glomerulus is separated from the Bowman's capsule by a thin layer of tissue. Some of the fluid portion of the blood passes from the glomerulus into Bowman's capsule and along the excretory tubule. The blood vessels that leave the glomerulus are intimately associated with the excretory tubule. As we will see in the following sections, the blood returns to the body circulation through a renal vein after it has passed through the network of blood vessels leading from the glomerulus. The intertwining of these blood vessels with the excretory tubule facilitates the crucial exchange of materials to and from the blood.

THE MECHANISM OF URINE COLLECTION

When amino acids from proteins are oxidized, most of the final breakdown products are in the form of carbon dioxide (exhaled from the lungs) and water. But the nitrogen from amino acids is removed in the form of urea, which must be excreted in urine. If nitrogen wastes accumulate in the tissues, they cause poisoning, starvation, and eventual suffocation of the tissues. The maintenance of the body's salt content by extraction of excess salt from the blood is another crucial process accomplished by the excretion of urine.

Glomerular Filtration, Tubular Reabsorption, and Tubular Secretion

In the formation of urine a series of events occurs, and the end result is the elimination of wastes and the conservation of necessary water, salts, and other useful compounds. Three basic processes (Figure 23.4) take place during the formation of urine from the blood. (1) When blood circulates from the arteries into the glomerulus of the kidneys, it is under high pressure. This pressure forces some of the plasma fluid into Bowman's capsule, and the blood cells and large proteins are filtered out. This process is known as *glomerular filtration.* (2) As fluid moves through the nephron, useful substances, such as water, some salts, glucose, and amino acids, that were initially lost from the blood during filtration are returned to the blood by active transport. This process is called *tubular reabsorption*. (3) Certain substances are secreted *into* the fluid as it passes through the nephric tubules. In this way, such waste products as hydrogen and potassium ions, drugs, and foreign organic materials may be excreted. This process is *tubular secretion*.

Urine and Urination

While the kidney is forming urine, it is also regulating the concentration of substances in the extracellular fluid by removing excess wastes and conserving substances that are present in normal or subnormal quantities. About 190 liters of water filter through the glomeruli each day, but only about 1½ liters are excreted in urine. The difference, about 99 percent, is reabsorbed. Urine is composed mostly of chloride, sodium, potassium, urea, creatinine, and uric acid. The volume and concentration of urine are regulated by two hormones and an enzyme. The hormones are *ADH* (antidiuretic hormone), or vasopressin, and *aldosterone;* the enzyme is *renin* (not rennin). Renin is important for the regulation of sodium balance, and apparently it also plays a role in water balance by stimulating the thirst reflex in the brain. Thirst is also indicated by other signals, such as a dryness of the mouth. ADH controls the permeability of the walls of the collecting ducts. When dehydration occurs, more ADH is released and more water is withdrawn from the urine. The opposite effect is achieved during overhydration. Aldosterone, an adrenal cortical hormone, stimulates the reabsorption of sodium from the kidney and thereby reduces the amount of sodium in the urine.

Urine from the kidneys flows through the ureters to the *urinary bladder*, a thick, muscular sac with a capacity of about one liter. However, the bladder rarely fills to its full capacity, as we shall see. Every 10 or 20 seconds, peristaltic waves force small amounts of urine from the renal pelvis into the bladder. The bladder becomes distended as it fills with urine, and when an adult accumulates about 300 milliliters (9 fluid ounces), the sensory endings of the pelvic nerve in the bladder wall are stimulated, and there is an urge to urinate. In an adult, this urge can be counteracted by consciously controlling the external urethral sphincter, but an infant lacks the development of the higher brain centers and cannot control the sphincter. Urination in an infant is a spinal reflex action initiated by the distention of the bladder. In an adult, those same impulses generated by stretch receptors in the bladder are mediated by control centers in the cerebrum.

An adult normally eliminates about 1½ liters of urine a day, but several conditions may increase the output considerably. Fortunately, urine production decreases while the body is asleep, so that we are usually not disturbed during the night. Increased stress or fear may cause increased urine production. Sometimes the production of urine may not actually increase during stress or excitement, but the urge to urinate may occur anyway. Pregnant women know that as the fetus develops it puts more and more pressure on the woman's bladder, causing frequent urinations. Cold weather also increases the urge to urinate. This happens because the blood is moved away from the skin and into the internal organs in an effort to conserve heat. The extra blood entering the organs

23.4

Simplified scheme of urine production in the kidney, specifically in the nephron. (1) In *filtration*, fluid passes from the blood into Bowman's capsule. (2) In *tubular secretion*, materials pass from the blood of the capillaries into the nephric tubules. (3) In *reabsorption*, materials (especially water and sugar) pass from the tubules and the loop of Henle back into the blood of the capillaries. The remaining fluid in the collecting tubule is urine. Tubular secretion and reabsorption are carried on simultaneously.

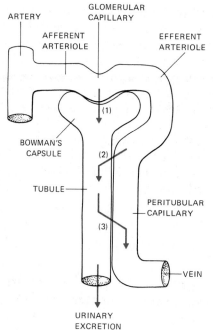

subsequently causes the kidneys to filter more blood and produce more urine. Certain foods, such as mustard, pepper, tea, coffee, and alcohol, increase urine production. In contrast, nicotine decreases the production of urine.

THE VERTEBRATE LIVER

The liver is the largest gland in the human body, weighing about 1.5 kilograms (3 lb) in an adult. An organ of vital importance, it has many functions besides the synthesis of bile salts described earlier. The liver may be considered an excretory organ because of some of its excretory functions, but in deference to this versatile organ, we list here all of its *major* functions (the liver actually performs more than 500 separate functions).

A. *Secretion and Excretion*

   1. Forms bile acids and excretes them as bile salts.

   2. Breaks down hemoglobin.

B. *Metabolism*

   1. Withdraws excess sugar from the blood and stores it as glycogen.

   2. Converts many end products of digestion (amino acids, lactic acid, glycerol) into glycogen.

   3. Reconverts stored glycogen into glucose and releases it into the blood.

   4. Makes galactose available for tissue utilization by converting it to glucose in a series of enzymatic reactions.

   5. Converts lactic acid from muscles into glycogen.

   6. Metabolizes protein.

   7. Synthesizes amino acids into structural, enzymatic, or plasma proteins.

   8. Metabolizes amino acids to provide energy through conversion to carbohydrate intermediates.

   9. Synthesizes the nonessential amino acids.

  10. Synthesizes creatine, purines, pyrimidines.

  11. Regulates amino acid content of the body.

  12. Plays an important role in lipid, cholesterol, and phospholipid metabolism.

  13. Synthesizes and degrades fatty acids.

DESERT RATS, CAMELS, AND "WARM-BLOODED" FISHES

Most desert animals need water as regularly as we do (their bodies contain about the same percentage of water as ours—65% of body weight), and they usually get it from the water in vegetation. Cactus plants, the most likely source of liquid refreshment, contain about 90 percent water. But desert rats, usually called kangaroo rats, eat dry seeds and dry plants that contain very little water. And yet the kangaroo rat has adapted successfully to desert life. Where does its water come from? Mostly from oxidation of food and the tiny amounts of moisture in the air, hardly enough to keep the little rodent supplied with enough water unless its water loss is kept low. The key to the balance between low water intake and low water loss is that the kangaroo rat is a nocturnal animal. It remains in its relatively cool and humid burrow all day and avoids the dangerous heat of the sun, which would cause water to evaporate from the body. Besides, the kangaroo rat's kidney is so efficient that it can excrete urine as concentrated salt water. Under experimental conditions kangaroo rats drank salt water and successfully excreted both the excess urea and the salt it contained. So far as is known, no other mammal can drink large quantities of salt water and survive. When most mammals (including humans) drink salt water, they actually increase their thirst, because the body becomes dehydrated in the process of getting rid of the excess salt.

  14. Produces all plasma proteins except immune globulins.

  15. Synthesizes fibrinogen and prothrombin for blood coagulation.

  16. Maintains proper circulating levels of proteins necessary for blood clotting.

C. *Storage*

   1. Stores vitamins A, riboflavin, pyridoxal, B_{12}, folic acid, and pantothenic acid.

Another famous desert animal is the camel. So legendary is the camel's ability to conserve water, in fact, that as many myths as facts about the camel exist. In the first place, the camel does *not* store water in its hump. Nor does it possess an extra stomach for that purpose. Instead, the camel can survive, in the somewhat milder winter months at least, on water-filled plants, and it rarely needs to drink water directly. Apparently the camel does not "store" water at all but merely consumes enough to retain a healthy water-to-body-weight ratio. A thirsty camel will lose weight, which is replaced almost instantly when it drinks water. Camels can drink almost 103 liters (27 gallons) of water in 10 minutes. Like the kangaroo rat, its urine contains huge quantities of urea and salt and very little water, and camels may even be able to drink water containing large amounts of salts.

Besides all these survival abilities, the camel's body temperature fluctuates with shifts of outside temperature, and in this way perspiration is reduced. The camel's temperature may go as high as 40°C (105°F) during the day and as low as 34°C (93°F) at night. Finally, this remarkably adaptable animal has a twofold insulation system. Its hair insulates against the heat, and its fatty hump concentrates the body fat in one place instead of distributing it all over the body. Thus, a lack of insulative material between the hair and skin permits heat to flow outward, and a cooling effect results.

One more example of unusual methods of temperature regulation concerns the "warm-blooded" or homeothermic fishes, such as the bluefin tuna. This time, the circulatory system, not the kidney, is the crucial mechanism. The *rete mirabile* ("miracle net") is a tissue containing a rich network of closely packed veins and arteries that not only permits the transport of oxygen through the blood but also creates a countercurrent heat exchange system that warms the tissues and increases the power of muscles. This is accomplished by a circulatory bypass that prevents body heat from escaping through the gills as usual, directing it instead back to the muscle.

As body temperature increases within an acceptable range, the muscle power also increases, because warmer muscles can contract more rapidly than cooler ones. The bluefish tuna and the mackerel shark, two fishes equipped with a rete mirabile system, are predators, and their increased muscle power enables them to swim faster, thereby becoming more effective predators. Interestingly, the bluefish tuna (and the other tunas with a rete mirabile) and the mackerel shark are not related. They have independently evolved similar means of adapting to an environment where controlled body temperature makes for successful living. This is a vivid example of parallel, or convergent, evolution, which we will examine in some detail in Chapter 26.

---

2. Stores iron in the form of ferritin.

3. Acts as a blood reservoir.

D. *Detoxification*

1. Attacks and detoxifies hydrocarbons.

2. Degrades many drugs.

3. Produces urea and other compounds by removing nitrogen (as ammonia) from amino acids, thus eliminating toxic compounds.

REGULATION OF BODY TEMPERATURE

Since the designations "cold-blooded" and "warm-blooded" are too general and misleading to be suitable for scientific use, we will use instead the terms *poikilothermic* (Greek, "varied heat") and *homeothermic* (Greek, "same heat"). Animals such as reptiles, fishes (there are exceptions; see the essay above), and amphibia are poikilothermic; that is, they are basically unable to stabilize their body tempera-

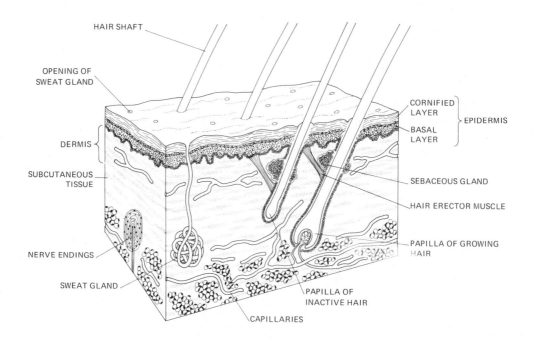

HAIR SHAFT

OPENING OF
SWEAT GLAND

DERMIS

SUBCUTANEOUS
TISSUE

NERVE ENDINGS

SWEAT GLAND

CORNIFIED
LAYER   } EPIDERMIS
BASAL
LAYER

SEBACEOUS GLAND

HAIR ERECTOR MUSCLE

PAPILLA OF GROWING
HAIR

PAPILLA OF
INACTIVE HAIR

CAPILLARIES

23.5

Diagram of a block of human skin, magnified about 20 times. Skin covers the entire human body. It meets and is continuous with the inner lining of the body at external openings: the mouth, nostrils, and ears, and the anal, urinary, and genital openings. A 70-kg (155-lb) man has about 11 kg (25 lb) of skin. Skin protects against injury and bacterial infection; it keeps in warmth during cold times (especially in furry animals) and dissipates heat during hot times by losing water through evaporation; it helps the kidneys regulate water balance; it protects against injurious radiation by absorbing ultraviolet light and uses that same ultraviolet for conversion of cholesterol, secreted by the skin, into essential vitamin D; it produces hair, fingernails, and toenails (or those variants called hooves) and horns in animals that have them; it produces sweat and such variants thereof as milk and ear wax; and it yields odors that are believed to be important, even in people, as sex attractants. It is well known that the ridges on fingertips can be used to identify individuals, but it is not well known that skin from any other part of a human body could be used just as well.

Skin is flexible; it can be folded and straightened indefinitely without cracking or wearing out. It is adaptable; it can be as thin as 0.07 mm over the inner side of an arm, or up to several millimeters on soles of feet. Even unborn babies have thickened skin on their soles and palms. Skin is sensitive; from its nerve endings come impulses that inform the owner of pressures, wrinkles, temperatures, textures, or injury. Skin is complex in its origin; the outer layer is derived from the embryonic ectoderm, the inner layers from mesoderm.

Structurally, skin is one of the more complex organs in the body. The outer layer, the *epidermis*, has layers of flattened, cornified (horny) cells, which are continually being worn away and replaced by new ones from the basal layer, where protein synthesis is especially active. The second layer, the *dermis*, contains an array of tissues: hair follicles that push out a continuous filament of hair shaft (up to several meters if not cut or worn off); sweat glands, simple, coiled ducts that give off water and salts to the outside; sebaceous glands, like microscopic bunches of grapes, secreting the oily substance that keeps skin soft, or fur and feathers waterproof; muscles that move hair shafts; nerves; blood vessels; and an insulating layer of fatty tissue.

Skin is reactive. It can darken in response to ultraviolet light by the creation of melanin from the amino acid tyrosine; or in response to hormonal changes, the melanin can be intensified in special areas, such as a woman's face darkened by the "mask of pregnancy." It can release sweat when the skin is warm or when fright stimulates a "cold sweat." Cold or excitement can make the hairs stand on end, or the muscles that erect the hair shafts can draw so tight as to pucker the surface, making "goose pimples."

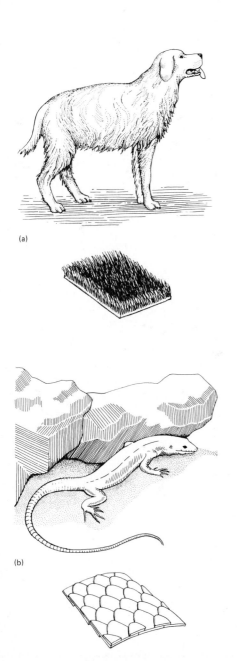

tures, which change along with the surrounding temperatures. Birds and mammals, however, are homeothermic; that is, they are able to maintain fairly constant body temperatures independent of external temperatures (Figure 23.6).

Birds and mammals vary their physical activities greatly, and yet their body temperatures usually stay within a range of 2°C. Mammals have body temperatures of 36°C to 38°C (37°C is normal for humans), and birds range between 40°C and 42°C. Homeothermic animals produce heat through metabolic activity, such as oxidation of food, cellular metabolism, and muscular contraction. Shivering is merely an increase of muscular activity designed to increase heat production through muscular contraction. Cooling is accomplished mostly by evaporation of water through the skin (perspiration) and lungs. (Desert animals have elaborate cooling mechanisms. See the essay on p. 470.)

Adaptations to Cold

Homeothermic animals must adapt to temperatures ranging from about −60°C to 60°C. Although most homeotherms have body temperatures close to 38°C, they still manage to adapt to these extreme tempera-

23.6

Temperature controls. (a) Being homeothermic ("warm-blooded"), mammals can remain active in places where other animals would be immobilized. A hairy skin is an important insulator and keeps the heat in. That is helpful in cold times and places, but it works as effectively in warm times and places, in which there must be some means of cooling off. Most mammals have cooling devices; sweat glands are one common type. Animals that do not sweat (dogs, for example) hang out their tongues and pant, letting evaporation help cool them. (b) Poikilothermic ("cold-blooded") animals have little or no internal temperature-regulating systems, and they generally lack elaborate insulating covers. Snake and lizard skins are scaly, not hairy, and dry, not moist. Such animals help themselves keep a somewhat even temperature by moving to a warmer place or to a cooler place, as the occasion demands.

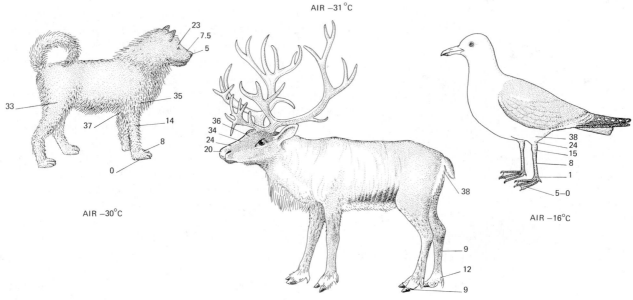

AIR −31°C

AIR −30°C

AIR −16°C

23.7
Even the best-insulated arctic animals have unprotected areas that must be left unencumbered by heavy fur or feathers if they are to function properly. Feet and noses, for instance, are usually naked to the cold. A *countercurrent heat exchange* between warm arterial blood to the foot and cold venous blood from the foot permits heat to pass directly from the artery to the vein. This exchange rechannels the heat back to the body, away from the extremities. As a result, the extremities remain colder than the body, usually close to the temperature of the outside environment. (This *rete mirabile* or "miracle net" system of heat retention is also discussed in the essay on p. 470). If the webbed feet of arctic gulls were not as cold as the water in which the gulls swim, the feet would lose more heat than the body could supply. Furthermore, if a gull with warm feet stood still on the ice, its feet would melt the ice. Before the gull could move again, the ice might refreeze and trap the bird. Having cold feet has its advantages.

ture ranges extending from the arctic zone to the tropic zone (Figure 23.7). For instance, an Eskimo's body temperature is about 37°C, and yet his external environment may be 50 to 100 degrees cooler. What are the main methods of adapting to the cold?

Homeothermic animals can cope with extreme cold by adapting to it, escaping from it, or enduring it. Besides increasing muscular activity and generating body heat through the metabolism of food, some animals retain body heat by using some form of insulation (fur or clothes, for example). An arctic fox with winter fur can live comfortably in external temperatures more than 85°C colder than its own body temperature. In contrast, tropical animals would have difficulty adjusting to temperatures 15 or 20 degrees lower than their body temperature. In such conditions, the tropical animals would have to increase their metabolic rate to produce heat. (A naked human has little natural insulation, and temperatures lower than 20°C are uncomfortably cold.)

Small animals, such as mice, have a relatively larger body area exposed to heat loss than larger animals do, and they cannot rely on insulative heat alone. Instead, such animals as arctic hares and mice live under the snow, where the temperature usually drops no lower than −5°C.

Many birds and other homeothermic animals counter the cold of winter by removing themselves from it altogether. These animals migrate to warmer climates, an activity discussed further in Chapter 28. Also discussed in that chapter is hibernation, a method used by some animals to endure the seasonally cold weather without actually leaving their home sites.

Adaptations to Heat

Desert animals have evolved ways to live successfully in their extremely hot and dry environments. Camels are the best-known examples, but many other animals are able to maintain homeostasis in hot climates. The main consideration for desert animals is to reduce water loss by evaporation. Many small desert animals are nocturnal; they avoid the sun by living in burrows during the day. Where drinking water is not avail-

(a)

(b)

able, animals derive enough water from solid foods. Desert animals are also able to conserve water by excreting concentrated urine and dry feces.

Animals that are too large to live in burrows cope in other ways. Light-colored fur reflects the sun's rays and keeps heat from penetrating the body. Hairless portions of the body, usually unexposed to the sun, contain capillary networks that permit heat loss, and animals with horns or large ears (such as gazelles and African elephants) use these parts of the body as cooling centers through the process of convection (Figure 23.8).

Many animals, including humans, are effectively cooled by the evaporation of perspiration, but dogs do not sweat at all. Instead, they lose heat by panting, an increased rate of breathing that facilitates water evaporation from the tongue and upper respiratory tract. The normal breathing rate for a dog is 15 to 30 breaths per minute. The accompanying excessive loss of carbon dioxide does not cause unconsciousness in a dog as it would in humans.

23.8

A contrast in cooling systems. (a) An elephant (*Loxodonta africana*) has ears so large that one of them can hide its whole head. The thin flaps, richly supplied with blood vessels, act as radiators that dissipate heat. The ears are one of an elephant's most useful cooling devices. (b) At the other extreme of ear size, the European pigmy shrew (*Sorex minutus*) has minute external ears. Such a small animal has so much surface in proportion to its bulk that it needs to conserve all the heat it can. In spite of its name, *S. minutus* is not the least of its kind; another pigmy shrew from the Mediterranean region, *Suncus etruscus*, is only about half as long and weighs about as much as a healthy peanut.

SUMMARY

1. Seawater animals lack internal regulatory processes that maintain a salt and water balance in their bodies, and they soon die in fresh water. In contrast, freshwater invertebrates can maintain the concentration of their blood and their body fluids even in diluted seawater.

2. Terrestrial animals must acquire large amounts of water, and they must also eliminate salts and other body wastes through the lungs and skin, and through excretion of urine and feces.

3. Excretory mechanisms vary greatly among invertebrates. Many protozoans and freshwater sponges have simple intracellular organelles called *contractile vacuoles*. Many invertebrates have excretory organs called *nephridia*, each of which is basically a tubular system that carries wastes to the outside. Flatworms

have a dual arrangement of branched tubes called a *flame cell system*, which maintains water balance at the same time that it excretes wastes. The excretory systems of insects and spiders are called *Malpighian tubules*, a system that was probably the forerunner of the excretory systems of land-dwelling vertebrates.

4. Humans lose water by evaporation from moist lung surfaces and from the skin. Water, salts, and organic substances are given off in sweat, but the prime regulator of water balance and waste excretion is the *kidney system*.

5. Urine formation essentially consists of forcing into nephric tubules all the blood except the proteins and blood cells (*glomerular filtration*), and then retrieving from the tube everything except some water, salts, urea, and other wastes (*tubular reabsorption*).

6. While the kidney is forming urine, it is also regulating the concentration of substances in the extracellular fluid by removing wastes and conserving substances that are present in normal and subnormal quantities. Urine is formed from the capillaries of the *glomerulus*.

7. The vertebrate *liver* has many important functions, including the storage of foods, secretion of bile salts, conversion of stored glycogen to glucose, breakdown of hemoglobin, metabolism of proteins, and the synthesis of blood plasma products associated with blood clotting.

8. *Poikilothermic* animals are basically unable to stabilize their body temperatures, which change along with the surrounding temperatures. *Homeothermic* animals are able to maintain fairly constant body temperatures independent of external temperatures.

RECOMMENDED READING

Helpful references for the entire physiology section, Chapters 18–24, are listed at the end of Chapter 18.

Carey, Francis G., "Fishes with Warm Bodies." *Scientific American*, February 1973. (Offprint 1266)

Heller, H. Craig, Larry I. Crawshaw, and Harold T. Hammel, "The Thermostat of Vertebrate Animals." *Scientific American*, August 1978.

Kappas, Attallah, and Alvito P. Alvares, "How the Liver Metabolizes Foreign Substances." *Scientific American*, June 1975. (Offprint 1322)

Leopold, A. Starker, *The Desert.* New York: Time-Life Books, 1969. (Life Nature Library)

Leopold, Luna B., and Kenneth S. Davis, *Water.* New York: Time-Life Books, 1966. (Life Science Library) (Chapter 5)

Montagna, William, "The Skin." *Scientific American.* February 1965. (Offprint 1003)

Schmidt-Nielsen, Knut, "The Physiology of the Camel." *Scientific American*, December 1959. (Offprint 1096)

Schmidt-Nielsen, Knut, and Bodil Schmidt-Nielsen, "The Desert Rat." *Scientific American*, July 1953. (Offprint 1050)

Smith, Homer W., "The Kidney." *Scientific American*, January 1953. (Offprint 37)

24

The Respiratory and Circulatory Systems

MOST ANIMALS CAN LIVE FOR DAYS WITHOUT food or water, but it takes only a few minutes without oxygen for the muscles, the heart, and the brain to fail. The intake of oxygen is called *respiration*, an unfortunate term because it has two related but different biological meanings: (1) Respiration at a cellular level means the oxidation of food with the consequent release of energy (Chapter 5). (2) Respiration at the gross anatomical level means *breathing* (air containing oxygen enters the lungs, and air containing waste gases is expelled from the lungs), and it is the breathing or respiratory system that will be discussed here. This chapter will also deal with the circulatory system, which is related to the respiratory system because in many animals oxygen is carried (circulated) by the blood to all parts of the body.

THE EVOLUTION OF GAS EXCHANGE IN ANIMALS

Although most animals do not breathe as humans do, they must have some sort of mechanism for obtaining oxygen and eliminating carbon dioxide. All animals exchange gases across a respiratory surface. In some animals this may be the entire body surface. Other animals accomplish gas exchange by using gills,

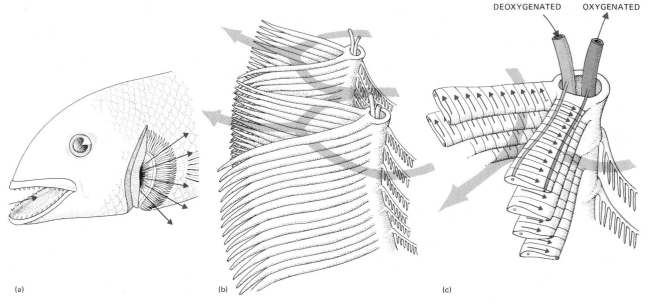

DEOXYGENATED   OXYGENATED

(a)

(b)

(c)

24.1

The mechanism by which a fish obtains oxygen from water. (a) The fish gulps water into its mouth and makes the oxygenated water pass over the gills and then out the gill slits in the sides of the pharynx. (b) The gills are finely divided, offering abundant surface for gas exchange, and they are provided with an array of blood vessels close to the surface. (c) The blood entering the gills from the fish's body is lower in oxygen and higher in carbon dioxide than the water flowing past. Since each gas tends to diffuse to a region of lower concentration, oxygen moves from the water into the blood, and carbon dioxide moves from the blood into the water. The exchange is helped by *countercurrent flow;* that is, the blood and water are moving in opposite directions. Such an arrangement results in an efficient system because at any place along the countercurrent exchanger the difference in concentration of a diffusing substance is maximized. If blood and water flowed along together in the same direction, even if there were concentration differences, the relative concentrations in blood and water would be less, and the rate of exchange would be lower.

lungs, or tracheae. These methods will be considered in the following sections.

Respiration by Surface Diffusion

Many small organisms, such as protozoa, earthworms, and small crustaceans, respire by the direct diffusion of gases through the surfaces of their bodies. Because of this direct diffusion, no other special respiratory apparatus is necessary. Animals smaller than one millimeter in diameter are most successful with surface diffusion. However, other larger aquatic animals, such as eels and frogs, supplement lung or gill respiration with gaseous diffusion through the skin, especially when they are totally immersed in water.

Gills

Used by water-dwelling animals, especially fishes, gills are the most efficient means of respiration in water. But efficient as gills are, fishes still expend about 20 percent of their total energy extracting oxygen from water. In contrast, mammals use only about two percent of their energy for breathing. There are only small amounts of dissolved oxygen contained in water. In order for a fish to supply sufficient oxygen to its blood, it must constantly take in water through the mouth. The water is forced over the gills, and it passes out of the body through the gill slits. Fish *gills* are soft, blood-filled filaments that allow the blood to flow in a direction opposite to the flow of water, thus establishing an efficient *countercurrent flow* (Figure 24.1), which is aided by a gill pump and the swimming motions of the fish. A gill may be defined as an enlarged, outward-turning respiratory surface, whereas a lung is a cavity formed by the turning in of the respiratory surface.

Lungs

Land-dwelling animals developed *lungs,* which are moist, saclike organs containing a fine network of

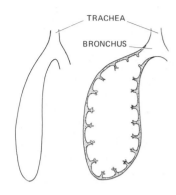

MUD PUPPY

ELONGATED

UNLOBULATED

HYDROSTATIC

FROG

DEEP ALVEOLI

ROUNDED

MORE VASCULAR

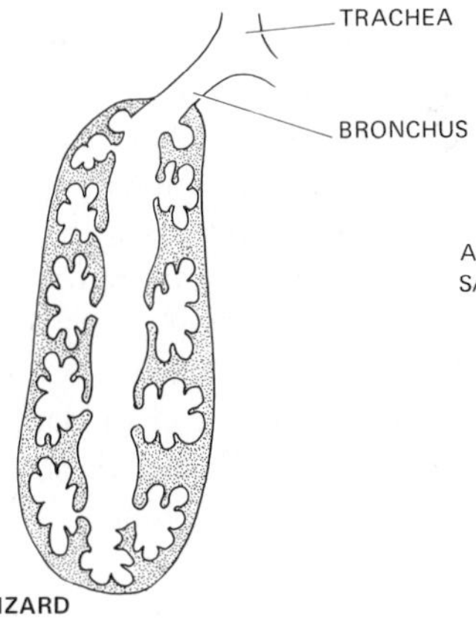

LIZARD

HIGHLY VASCULAR

CENTRALLY LOCATED
ALVEOLI

BEGINNING OF
BRONCHIOLES

SEPTA

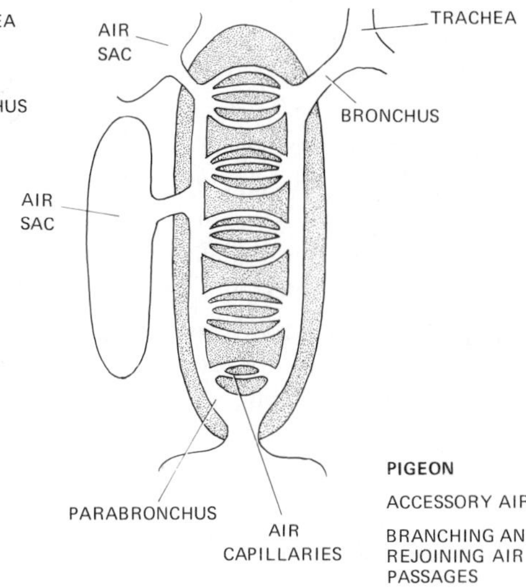

PARABRONCHUS

AIR
CAPILLARIES

PIGEON

ACCESSORY AIR SACS

BRANCHING AND
REJOINING AIR
PASSAGES

MANY SEPTA

NO BLIND ALVEOLI

EXCHANGE BY AIR
CAPILLARIES
ON WALL
PASSAGEWAYS

EMPTIED WITH
EACH BREATH

HIGHLY VASCULAR

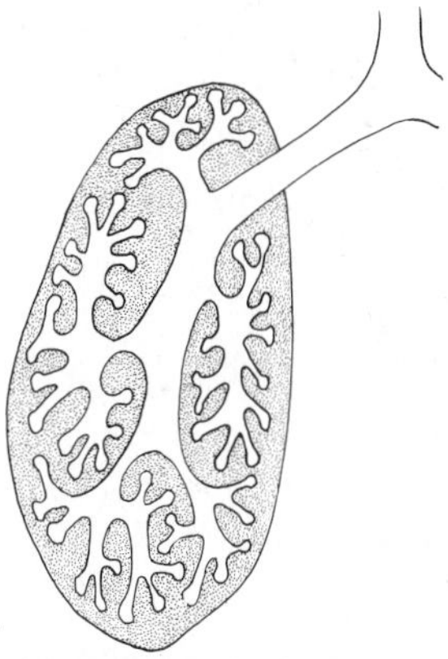

24.2

Stages in the evolution of lungs. In
the more primitive amphibians, such
as the mud puppy, the lung is only
a simple sac. In more advanced amphibia, such as frogs, the lung has
the beginnings of subdivisions, and
it is provided with more blood vessels. Reptiles, such as lizards, have
deeply lobed *alveoli* and abundant
blood vessels in their lungs, and there
is a hint of the development of
branched air tubes, bronchioles.

Birds have highly branching and rejoining air tubes throughout the
lungs, as well as an extensive system
of connected air sacs. Since mammals, like birds, are homeothermic,
they have high oxygen requirements.
The gas-exchange surface of mammalian lungs is increased maximally
by branched bronchioles leading to
clusters of elastic, thin-walled alveoli,
whose walls are well supplied with
capillaries.

RAT

COMPLEX LOBES

TERMINAL ALVEOLI (EXCHANGE)

MULTIPLE SUBDIVIDED BRONCHIOLES

NO CENTRAL AIR CAVITY

HIGHLY VASCULAR

MANY SEPTA

blood vessels. The first lungs probably appeared in
animals like lungfishes, and amphibians still retain
simple lungs. In the more highly developed organisms, the internal structure of the lung is more complex, having evolved from simple sacs to highly
folded and lobulated mammalian lungs, which contain millions of tiny air sacs called *alveoli*. Each alveolus is surrounded by a network of rich blood vessels
(Figure 24.2).

The lung adequately serves the respiratory needs
of terrestrial vertebrates, but it is an inefficient mechanism nevertheless, especially in comparison with
gills. Water flows directly and efficiently past a gill,
but air must enter *and* leave the lung through the
same opening. Without a free one-way flow of air,
problems of lung ventilation occur, and they are further complicated by the fact that the lungs are not
directly in contact with the outside air sources. Instead, a series of airways connects the lungs to the
outside, thus setting up an area of unventilated and

24.3

The respiratory system of a bird. (a) Diagram of the anatomy of a bird's trachea, lungs, and air sacs. The air sacs, which are directly connected with the lungs, branch throughout many parts of the bird's body, including the hollow bones, thus delivering oxygen to respiring tissues more effectively than would be possible with lungs only, as in mammals. The requirements of flight, including a need for lightness and a high rate of muscular activity, plus the demands of a relatively high temperature, make it necessary for a bird to have an efficient method of delivering oxygen to the tissues and removing carbon dioxide. (b) A scanning electron micrograph of the interior of a bird's lung. The network of air tubes ensures continuous passage of air close to the capillaries. In a bird's lungs, the air flows *through* the tubes.

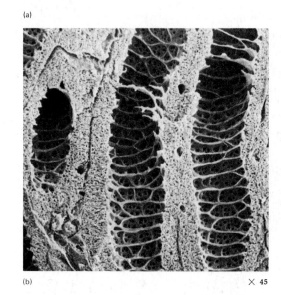

(a)

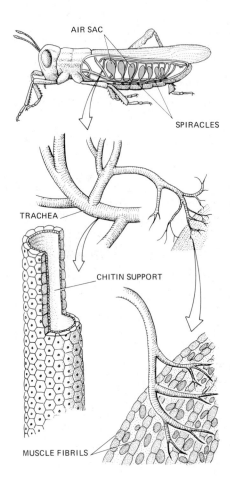

(b)   × 45

24.4

Breathing mechanism in an insect. Air goes in and out through small openings along the sides of an insect's abdomen. These *spiracles* lead into air tubes, the *tracheae*, which branch and rebranch until fine endings reach the tissues. Pulsations of the insect's body create alternating pressure and suction. Insects with aquatic larval stages have various gill-like structures that provide gas exchange, but an adult insect can be drowned only if its abdomen, not its head, is kept under water.

24.5

The human respiratory system. Inhaled and exhaled air passes though the pharynx (the swallowing apparatus), then the larynx (the Adam's apple, or voice box, containing vocal cords), the tubular trachea (the wind pipe), the two bronchi, which branch into ever smaller bronchioles and finally into the alveoli (the terminal air sacs) of the lungs. The thin walls of the alveoli are richly supplied with capillaries and are kept moist. Respiratory gases are exchanged through the walls of the alveoli and the capillaries.

uncirculated air that never reaches the lungs. Only about 70 percent of the human lung is filled with fresh air after each inhalation. Bird lungs are more efficient, as shown in Figure 24.3.

Tracheae

Insects and some other land-dwelling arthropods have a system of air tubes called *tracheae* (not to be confused with human tracheae, or windpipes), which branch out from surface openings called spiracles to all parts of the body (Figure 24.4). The spiracles are made operative by the movement of the body. The internal ends of the tracheae are air capillaries that carry oxygen directly to the body cells. Carbon dioxide is transported from the body cells through the tracheae to the outside.

RESPIRATION IN HUMANS

Homeothermic animals, with their need for large amounts of oxygen, evolved a system of air breathing. One of the advantages air breathers have over aquatic animals is that oxygen diffuses out of the air about 300,000 times faster than it diffuses out of water. In this section we will consider the human breathing mechanism and the way in which gases are exchanged in the lungs prior to being transported by the blood.

The Organization of the Respiratory System

The respiratory system in mammals is made up of the lungs, the several passageways leading from the outside to the lungs, and the muscular mechanism that moves air in and out of the lungs (Figure 24.5). Before the air we breathe reaches the lungs, it must enter the nose or mouth and pass through the trachea (windpipe). The trachea branches into tubes called bronchi, which enter the lungs and divide into smaller and smaller tubules, finally terminating in air sacs called *alveoli*, the sites of gas exchange between the lungs

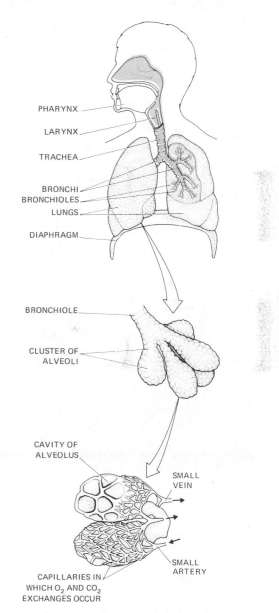

PHARYNX

LARYNX

TRACHEA

BRONCHI
BRONCHIOLES
LUNGS

DIAPHRAGM

BRONCHIOLE

CLUSTER OF
ALVEOLI

CAVITY OF
ALVEOLUS

SMALL
VEIN

SMALL
ARTERY

CAPILLARIES IN
WHICH O_2 AND CO_2
EXCHANGES OCCUR

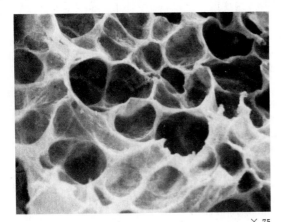

× 75

24.6
Scanning electron micrograph of the inside of a rat lung. The dark holes show alveoli and alveolar ducts, which lead into deeper alveoli. This picture gives an idea of the convolutions that add to the amount of surface available for gas exchange.

× 2500

24.7
Scanning electron micrograph of the interior of a hamster trachea. The upstanding objects are cilia, which wave continuously, carrying mucus and inhaled solid particles, such as dust, upward from the direction of the lungs. This cleaning operation keeps the breathing passages clear. Most of the lining of mammalian respiratory surfaces is provided with cilia, including the bronchi, tracheae, and nasal passages.

and the bloodstream (Figure 24.6). Each lung contains about 300 million alveoli, each surrounded by many blood capillaries. Gas exchange is facilitated by the thin membrane walls (only one cell thick) that separate the alveoli from the blood vessels and also by the enormous surface area of the alveoli, which amounts to about 70 square meters. Unlike many smaller animals, humans receive oxygen indirectly, through the blood.

Throughout the upper respiratory tract the air is filtered, warmed, and moistened. (Air breathed through the mouth receives fewer of these benefits than air taken in through the nose.) Epithelial glands in the passages secrete mucus (about a liter each day) that traps foreign particles that are swallowed and eventually eliminated in the feces. The mucus is kept moving, and a new film of mucus is secreted about every 20 minutes. An important role in this purifying mechanism is played by the cilia that sweep the mucus downward along the tract (Figure 24.7). Air pollutants, such as cigarette smoke, can paralyze these cilia and cause the mucus to clog, a reaction frequently resulting in "smoker's cough." Phagocytes, cells that engulf foreign substances, are also neutralized by air pollutants. In contrast, when cilia in the nose are partially paralyzed by cold air, there is an overproduction of mucus, which is not swept back into the throat as usual. When this happens, the mucus drips forward and causes a runny nose.

The lungs, which are composed mainly of the elastic tissue of the alveoli, contain very little muscle. They are covered by a layer of epithelium called pleura and held within the bony chest cavity. The expansion and contraction of the lungs is accomplished by the *diaphragm*, a muscular partition between the thoracic and abdominal cavities, and by abdominal muscles and muscles between ribs.

The Mechanics of Breathing

The intake of air, *inspiration*, occurs when air rushes into the lungs to equalize a reduction in air pressure in the thoracic cavity (Figure 24.8). The mechanism operates as follows:

1. Rib muscles contract and elevate the ribs.

2. The muscles of the diaphragm contract, lowering the diaphragm and increasing the volume of the chest cavity.

3. Abdominal muscles relax to compensate for the compression of abdominal organs.

4. The increased size of the chest cavity causes the pressure in this cavity to drop below the atmospheric pressure, and air rushes through the res-

24.8

The unconscious mechanisms of breathing. When the carbon dioxide level in blood reaches a certain concentration, the respiratory center in the medulla sends an impulse to the diaphragm, causing the diaphragm muscles to contract. This contraction, shown in (b), flattens the diaphragm, pushes out the abdominal muscles, and enlarges the chest cavity. The resulting reduced pressure in the lungs makes air, which is at higher pressure outside, rush into the lungs. Diaphragm contraction is aided by the contraction of the *intercostal muscles* between the ribs. When these contract, the ribs are raised in a hingelike action, further expanding the rib cage. When lungs are filled with air, stretch receptors in the lungs send impulses to the respiratory center, which sends back impulses causing the diaphragm to relax. Pressure of the abdominal wall against the internal organs makes the diaphragm rise into a dome shape as shown in (a), expelling air from the lungs. Forcible expulsion of air is accomplished by contraction of the abdominal muscles. Although breathing can be consciously controlled to some extent, no amount of determination can override the automatic drive to exhale and inhale.

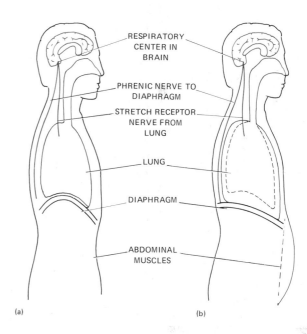

piratory passages into the lungs, equalizing the pressure.

Expiration, or the expulsion of air from the lungs, occurs in the following way:

1. Muscles of the ribs and diaphragm relax, allowing the thoracic cavity to return to its original, smaller size.

2. Abdominal muscles contract, pushing the abdominal organs against the diaphragm.

3. The elastic lungs contract as the air is expelled.

An ordinary breath will move about 500 milliliters (1 pint) of air in and out of the lungs, but a deep breath may increase the lung capacity to about 4000 milliliters (4 liters). Trained athletes may have a lung capacity of more than six liters.

The Control of Breathing Rate

The regulation of the rate of breathing is controlled by the amount of carbon dioxide in the blood. Even a slight increase of carbon dioxide stimulates the breathing rate until enough carbon dioxide has been released to restore the normal concentration of carbon dioxide in the blood. When the carbon dioxide level of the blood rises, the respiratory center in the brain sends out nerve impulses that increase the rate of operation of rib muscles and the diaphragm, resulting in faster breathing.

Thus a high carbon dioxide content of the blood causes an *increase* in breathing, and a low carbon dioxide content causes a *decrease*, a classic example of negative feedback control. Note that it is mainly the carbon dioxide concentration, not the oxygen concentration, that regulates the breathing rate. A person breathing in an atmosphere low in oxygen may not feel discomfort as long as the carbon dioxide proportion does not rise. However, although the main regulator of the breathing rate is the level of carbon dioxide in the blood, the breathing rate is also increased somewhat by a drop in the level of oxygen in the blood.

Gas Exchange in the Lungs

The blood coming to the lungs is low in oxygen. Through the process of diffusion, oxygen from the air in the alveoli will move into the blood, and carbon dioxide and water in the blood will move into the

24.9

Scheme of gas exchange in lungs and tissues. When blood passes through the lung capillaries and comes close to the alveoli of the lungs, the concentration of oxygen in the alveoli is high and that of carbon dioxide is low. The opposite is true of the concentrations in the blood. Since diffusion favors a movement from a region of high concentration to a region of low concentration, the two respiratory gases are exchanged. The blood passing through the lungs loses carbon dioxide and gains oxygen while the lungs lose oxygen and gain carbon dioxide. The oxygenated blood is distributed over the body, and when it reaches capillaries in tissues, the situation is reversed: The high concentration of oxygen in the blood favors diffusion into the respiring cells, where the oxygen is low. Carbon dioxide moves in the opposite direction. Thus living cells are constantly supplied with oxygen from the lungs, and the waste carbon dioxide is eliminated.

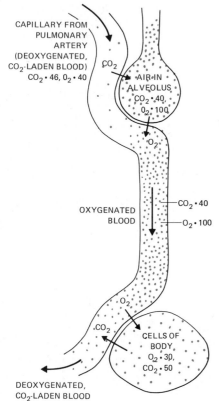

CAPILLARY FROM PULMONARY ARTERY (DEOXYGENATED, CO_2-LADEN BLOOD)
$CO_2 \cdot 46, O_2 \cdot 40$

CO_2

AIR IN ALVEOLUS $CO_2 \cdot 40$, $O_2 \cdot 100$

O_2

OXYGENATED BLOOD

$CO_2 \cdot 40$
$O_2 \cdot 100$

O_2

CO_2

CELLS OF BODY $O_2 \cdot 30$, $CO_2 \cdot 50$

DEOXYGENATED, CO_2-LADEN BLOOD

alveoli (Figure 24.9). Oxygen and carbon dioxide move in different directions, but both move from areas of relative abundance (high concentration) to areas of lower abundance (lower concentration). Both gases move freely through the walls of the capillaries and the alveoli.

THE GAS-CARRYING ABILITY OF THE BLOOD

On its own, the water of blood can carry only about one percent of an adult's oxygen requirement. But *hemoglobin*, the respiratory pigment in the red blood cells, has such an affinity for oxygen that it binds about 98 percent of the oxygen available in the lungs. In this way, hemoglobin acts as a carrier and makes possible the effective transport of oxygen throughout the body. Hemoglobin is a protein; each molecule contains five percent *heme*, an iron-containing compound, and 95 percent *globin*, a protein. Although the heme makes up only a small portion of this molecule, it is absolutely essential to the function of hemoglobin as an oxygen-carrying pigment. It is the heme that gives hemoglobin its red color.

The Transport of Oxygen in the Blood

Oxygen is abundant in the lung alveoli, and it combines readily with hemoglobin to form oxyhemoglobin. The oxyhemoglobin is then transported in the red blood cells to the tissues, where oxygen is less abundant. Because oxygen is constantly being used up by cellular oxidation, functioning tissues have less oxygen than the lungs. Because of the low concentration of oxygen in the tissues, the oxygen in oxyhemoglobin is free to diffuse into the cells of the tissues. Varying amounts of oxygen are released, depending on cellular needs.

Unfortunately, some toxic agents bind to hemoglobin even more readily than oxygen does. Air pollutants such as insecticides and sulfur dioxide bind to hemoglobin to prevent its effective carrying of oxygen. Such poisons cause a numbness or dizziness that is characteristic of a lack of oxygen. Probably the best known hemoglobin poison is carbon monoxide, found in automobile exhaust fumes. Carbon monoxide binds to hemoglobin about 210 times faster than oxygen and forms a stable compound. Carbon monoxide at concentrations of 0.1 or 0.2 percent in the air is dangerous, and increased amounts may induce death by blocking the uptake of oxygen by hemoglobin. When this happens, tissues die of asphyxiation.

One of the most serious lung problems is *emphysema*, a condition that develops when alveoli lose

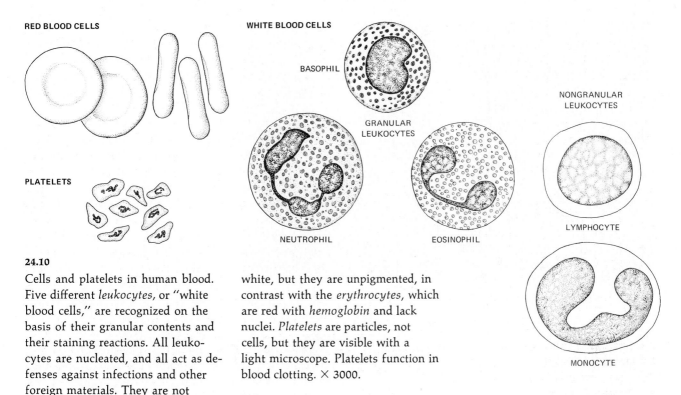

RED BLOOD CELLS

PLATELETS

WHITE BLOOD CELLS

BASOPHIL

GRANULAR LEUKOCYTES

NEUTROPHIL

EOSINOPHIL

NONGRANULAR LEUKOCYTES

LYMPHOCYTE

MONOCYTE

24.10
Cells and platelets in human blood. Five different *leukocytes*, or "white blood cells," are recognized on the basis of their granular contents and their staining reactions. All leukocytes are nucleated, and all act as defenses against infections and other foreign materials. They are not white, but they are unpigmented, in contrast with the *erythrocytes*, which are red with *hemoglobin* and lack nuclei. *Platelets* are particles, not cells, but they are visible with a light microscope. Platelets function in blood clotting. × 3000.

their elasticity, usually because of extreme air pollutants that clog breathing passages and destroy sensitive tissues. When this happens, the alveoli do not retain enough elasticity to expand and collapse rhythmically. Waste carbon dioxide cannot be expelled properly, and it is trapped within the alveoli. At the same time, oxygen is not distributed freely to the blood. Fortunately, a normal adult has about eight times as many alveoli as he needs for routine activities, and acute cases of emphysema are not so prevalent as they might be otherwise.

The Transport of Carbon Dioxide in the Blood
Carbon dioxide is more soluble in water than oxygen is, and it diffuses easily through capillary walls. Once in the blood, carbon dioxide is transported in three ways:

1.   About 67 percent of the carbon dioxide reacts with water to form carbonic acid. In the lungs, where carbon dioxide is less abundant than in the blood, this reaction will operate in the reverse direction, and carbon dioxide will be lost from the blood.

2.   About 25 percent of the carbon dioxide reacts with hemoglobin and is carried from the tissues to the lungs. When hemoglobin carrying carbon dioxide arrives in the lungs, the carbon dioxide is exchanged for oxygen.

3.   The remaining 8 percent of carbon dioxide is dissolved directly in the blood as molecular carbon dioxide (CO_2).

BLOOD SERVES MANY FUNCTIONS IN THE HUMAN BODY

Although animals have evolved various ways to transport food, gaseous waste products, and regulatory substances throughout their bodies, the most common method uses a circulating fluid. _Blood_ is the term used for circulating fluids in general, even though some blood, as in invertebrate animals, may be blue or yellow and perform functions somewhat different from those of blood in humans. Blood in the human body serves a number of functions. It carries dissolved foods, transports carbon dioxide and oxygen, distributes chemical regulators (hormones), acts as a defense against infection, helps maintain an even temperature, and removes waste products.

Composition of Blood
The liquid portion of mammalian blood is called _plasma_ (about 55%), and the solid portion is known as _formed elements_ (about 45%), mostly blood cells suspended in the plasma (Figure 24.10). The follow-

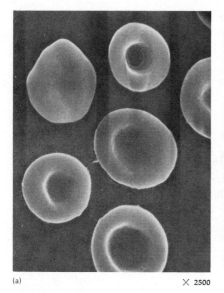

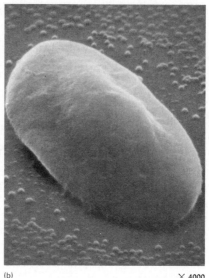

(a)   × 2500   (b)   × 4000

24.11
Scanning electron micrographs of erythrocytes. (a) Mammalian erythrocytes, showing the typical concavity. This does not show the absence of nuclei. A myth that camel blood has nucleated erythrocytes, once started, was repeated in textbooks for many years, long after it was refuted by repeated observation. In fact, camel erythrocytes are no more nucleated than those of any other mammal, but they are somewhat elliptical. (b) Frog erythrocytes, much larger than those of human blood, are nucleated (the nucleus is the bump in the middle of the cell) and elliptical.

ing tabulation summarizes the composition of mammalian blood.*

PLASMA

1. *Water*, 90 percent.
2. *Dissolved solids*, consisting of the plasma proteins (albumin, globulins, fibrinogen), glucose, amino acids, electrolytes, various enzymes, antibodies, hormones, metabolic wastes, and traces of many other organic and inorganic materials.
3. *Dissolved gases,* especially oxygen, carbon dioxide, and nitrogen.

FORMED ELEMENTS (solid components)

1. *Red blood cells* (erythrocytes) transport oxygen to body cells and remove carbon dioxide.
2. *White blood cells* (leukocytes) serve as scavengers and immunizing agents by destroying bacteria at infection sites.
3. *Platelets* (thrombocytes) help blood to coagulate.

Each cubic milliliter of human blood contains about five million *erythrocytes* (red blood cells), and there are about 25 trillion erythrocytes in the human body; 40 to 50 red blood cells in a row would reach across the period at the end of this sentence. (De-

creased production or increased destruction of red blood cells produces a condition known as *anemia.*) Erythrocytes make up about half the volume of human blood. The erythrocyte is a disk, slightly indented on both sides, providing a larger surface for gas diffusion than a flat disk or a sphere (Figure 24.11). An erythrocyte exists for about 120 days in the bloodstream before it finally fragments. It is then engulfed by scavenger cells called macrophages in the liver, bone marrow, or spleen. Every *second* about 7–10 million erythrocytes are destroyed, and the same number are created. The iron from hemoglobin is retained and used again. The rest of the heme is converted to bilirubin, a bile pigment, which is excreted in feces. If the liver is faulty, bilirubin may accumulate in abnormally high amounts and cause the skin to turn yellow—a condition known as *jaundice.* During its brief life span an erythrocyte may travel 1100 kilometers (700 mi) through the bloodstream (about 75,000 round trips from the heart to other parts of the body and back to the heart), bending to squeeze through the tiny capillaries.

The term *leukocyte* (or white blood cell) is used to cover a number of slightly different cell types that circulate in the blood. Leukocytes are slightly larger than erythrocytes. Unlike erythrocytes, leukocytes have nuclei, and they are capable of independent ameboid movement to pass through blood vessel walls into tissue spaces. In adults, there are about 700 erythrocytes for every leukocyte.

Leukocytes in adults may increase as a result of an infection from 8000 to 25,000 per cubic millimeter, about the same number that newborn children have.

---

* Adapted from Cleveland P. Hickman, Sr., Cleveland P. Hickman, Jr., and Frances M. Hickman, *Integrated Principles of Zoology*, Fifth Edition (St. Louis: C. V. Mosby, 1974), p. 607.

Some leukocytes are actively engaged in engulfing and digesting foreign material, including bacteria. Specialized leukocytes called lymphocytes play an important role in producing antibodies to establish an immune response to foreign substances.

   Leukemia is a disease characterized by uncontrolled leukocyte production. The white blood cells tend to use the oxygen and nutrients that normally go to other cells, causing the death of otherwise healthy cells. Because almost all the new white blood cells are immature and incapable of normal function, victims of leukemia have little resistance to infection. In most cases, the combination of starving cells and lack of an immune system is enough to cause death. The cause of leukemia is unknown, although a viral infection is considered likely.

   Platelets (thrombocytes) are minute disks that serve as starters in the process of blood clotting, to be described below.

Coagulation of Blood

The fluidity of blood is delicately balanced. If blood was not liquid, it could not function, and yet if it could not solidify, any small wound would allow the blood to escape unhindered, as it does in victims of "bleeder's disease," or hemophilia. When a tissue is damaged in a normal person and blood leaves a vessel, the sensitive platelets disintegrate and release thromboplastin (Figure 24.13). This enzyme activates the protein prothrombin in the presence of calcium ions, converting it into the active enzyme thrombin. Thrombin then catalyzes the conversion of fibrinogen into the insoluble, stringy protein fibrin. The fibrin threads entangle the blood cells and create a clot. This whole process may be summarized as follows:

1. prothrombin $\xrightarrow[\substack{+ \\ \text{calcium}}]{\text{thromboplastin}}$ thrombin

2. fibrinogen $\xrightarrow{\text{thrombin}}$ fibrin

   Why doesn't blood clot in the blood vessels? Because the enzyme thrombin is not present in the circulating blood. Besides thrombin and the other coagulation agents mentioned above, as many as 35 compounds may be required for blood coagulation, a necessary precaution to prevent clotting when no bleeding has occurred. Mosquitoes and other blood-sucking animals are able to counteract the clotting action of thrombin by having an anticoagulant present in their saliva. Without this mechanism the victim's blood would clot in the delicate mouth parts of

24.12
Scanning electron micrograph of human erythrocytes from a patient suffering from sickle-cell anemia. Instead of being normal biconcave disks, these cells are variously distorted. Some are stretched out so long and thin that they bear some resemblance to a sickle blade—hence the name of the disease.

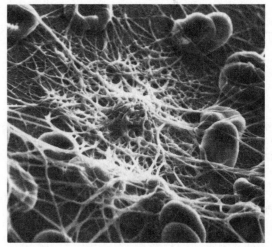

× 5180

24.13
A scanning electron micrograph of a portion of a blood clot. The tangled threads are *fibrin*, which binds the clot into an insoluble mass, and the circles are blood cells, enmeshed in the fibrin filaments.

the mosquito. On the other hand, the entry of too much anticoagulant into an animal's bloodstream can be fatal. A recent plague of blood-sucking bats (*Desmodus rotundus*) in Central and South America threatened to ruin the cattle industry until a special anticoagulant was injected into the rumens of the cattle. The blood of the specific vampire bat was unable to clot after a full meal of the blood of the treated cattle. As a result, the small capillary breaks that normally occur in a bat's wings during flight did not seal as usual. The bats died of internal bleeding or were weakened to the point where they were unable to hunt for food, and they starved to death. The anticoagulant works by binding the plasma protein and preventing the utilization of Vitamin K, which is needed for the synthesis of prothrombin. The anticoagulant apparently has no adverse effect either on the cattle or on the consumer who eventually eats the beef.

THE HEART IS A DOUBLE PUMP

In vertebrate animals, including humans, blood flows in a *closed system*, remaining essentially within the carrying vessels. Most animals, however, have *open systems*, in which blood oozes generally throughout the body in relatively large, loosely jointed sinuses, and it is kept moving partly by rather simple hearts and partly by body movement (Figure 24.14). Such circulation is typical of crustaceans and insects. In the mammalian circulatory system, which will be treated here, blood is pumped from a muscular heart out through elastic arteries, then through microscopically fine vessels, the capillaries, and finally through veins back again to the heart.

The Quantity of Blood Circulated

A heart begins beating in a human embryo within a few weeks, and it must continue through the life of

24.14

Comparison of circulatory systems. (a) The two-chambered fish heart is a single pump, forcing deoxygenated blood (darker gray) from body tissues out through the gills, from which oxygenated blood (darker brown) goes to the body tissues. (b) In a frog (amphibian), deoxygenated blood (darker gray) enters the three-chambered heart, where it is mixed (lighter gray) with blood from the lungs (darker brown). A mixture of oxygenated and deoxygenated blood (lighter gray) is sent out over the body. (c) A turtle (reptile) has two incompletely separated ventricles. Deoxygenated blood from the body (darker gray) is partially mixed with oxygenated blood from the lungs (darker brown) before being sent to the lungs (lighter gray). Blood sent from the left ventricle to the body is a mixture of deoxygenated and partially deoxygenated blood (lighter brown). (d) Birds and mammals separate deoxygenated blood (darker gray) from oxygenated blood (darker brown). The right side of the heart is only an accessory pump that sends blood to the lungs to be reoxygenated.

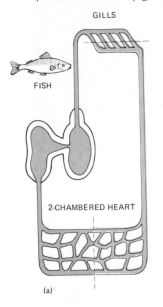
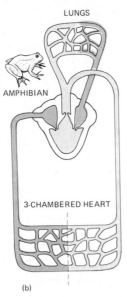
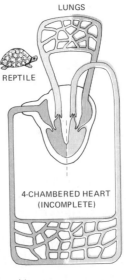
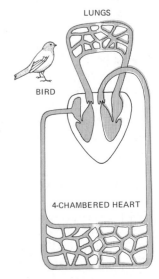

(a) GILLS — FISH — 2-CHAMBERED HEART

(b) LUNGS — AMPHIBIAN — 3-CHAMBERED HEART

(c) LUNGS — REPTILE — 4-CHAMBERED HEART (INCOMPLETE)

(d) LUNGS — BIRD — 4-CHAMBERED HEART

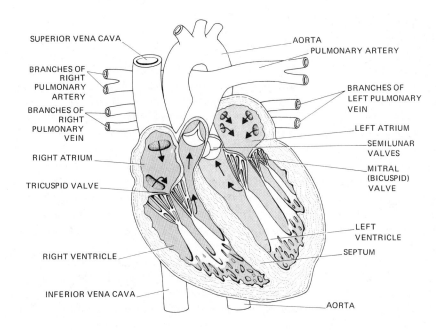

24.15

Dissection of a human heart, as seen from the front, with the ventral part of both atria and both ventricles removed.

the individual at a rate of about 70 times a minute, 100,000 times a day, or about 2½ billion times during a 70-year lifetime.* The heart rarely misses a beat, but if it skips four or five in a row, the body will probably fall down dead.

The human body contains about 5½ liters (12 pints) of blood, and the heart takes slightly more than a minute to pump a complete cycle of blood through the body. In times of strenuous exercise the heart can quintuple this output. And yet, the heart is a relatively small organ, no larger than your fist, and it weighs only about 300 grams (11 oz).

The Mechanics of Heart Action

The heart is actually a double pump, separated into two halves by a thick vertical wall called a septum. The heart is partitioned into two sections, the left somewhat larger than the right section, each acting as an individual pump. However, the two sections are synchronized in their pulsations. Each half of the heart consits of an *atrium* (formerly called an auricle), and a larger, stronger, thicker-walled *ventricle* (Figure 24.15). The atria serve as reservoirs, and both

---

* The specialized muscle tissue that performs this extraordinary feat is contained only in the wall of the heart. The only comparable muscular competence occurs in the uterus of a woman in preparation for childbirth and during actual labor and delivery, but the uterus does not have to perform at this level for 70 years or so. The heart does.

contract at the same time to force blood into the ventricles. Atrioventricular (A-V) valves permit blood to flow from the atria to the ventricles but prevent the blood from seeping back into the atria.

Blood enters the right atrium from the veins. At this point the blood is low in oxygen and high in carbon dioxide, since it has just returned from the organs and tissues of the body. The blood is squeezed from the right atrium into the relaxed right ventricle by means of a rather weak contraction of the atrial walls (this is the first pump). Backflow is prevented by the right A-V valve. Then the right ventricle contracts, forcing blood out into the pulmonary arteries, one to each lung. The blood cannot surge back into the ventricle because of another valve, the pulmonary semilunar (S-L) valve. The right side of the heart is thus strictly a pump that keeps blood flowing to the lungs.

After the blood has been sent past the oxygen-rich air sacs of the lungs, the blood returns to the heart via the left atrium. The left atrium contracts and forces blood into the relaxed left ventricle. Then the left ventricle contracts (this is the second pump) and forces the blood into the largest artery in the body, the aorta. The aorta branches into the arteries that carry oxygen-rich blood to all parts of the body. Although the left and right ventricles pump almost simultaneously, so that equal amounts of blood enter and leave the heart, the pumping pressure of the thick-walled left ventricle is greater, because it must

THE UNHEALTHY HEART

"Heart disease" may be defined as any disease that affects the heart, but it can also be a disease of the blood vessels. A more appropriate term is "cardiovascular diseases," which includes both heart and blood vessel disorders. About 29 million Americans have cardiovascular diseases, which are responsible for more deaths than all other causes of death combined. Although it may be decreasing, the American death rate from cardiovascular disease is still one of the highest in the world. Among the disorders of the cardiovascular system are atherosclerosis, coronary artery disease, stroke, and high blood pressure.

Atherosclerosis is characterized by deposits of fat, fibrin (a clotting material), cellular debris, and calcium on the inside of arteries. These built-up materials stick to the inner walls of the arteries, narrowing the space for passage of blood and reducing the elastic stretchiness. The condition is dangerous because it cuts down blood flow, and if the artery is closed off entirely, no blood can flow. If the closed artery is in the heart, a *heart attack* occurs. If the artery takes blood to the brain, a *stroke* occurs.

Although no one really knows the basic cause of atherosclerosis, several factors are known to increase its progress. Among them are cigarette smoking and the amount of animal fat and cholesterol in the diet. Other factors that may have an effect are hypertension, diabetes, age, stress, heredity, and the male sex hormones.

There are three ways in which atherosclerosis can cause a heart attack. (1) It can completely clog a coronary artery. (2) It can provide a rough surface where a blood clot (*thrombus*) can form and grow to the point where it closes off the artery and causes a heart attack called a *coronary thrombosis*. (3) It can partially block blood to the heart muscle and cause the heart to stop beating rhythmically.

The usual warning signs of a heart attack (they are not always the same) are (1) a heavy, squeezing pain in the center of the chest; (2) pain that may spread into the shoulder, arm, neck, or jaw; (3) sweating; and (4) nausea, vomiting, and shortness of breath. Sometimes these symptoms go away and come back later. Quick medical attention is imperative.

Angina pectoris (Latin, "chest pain") occurs when not enough blood gets to the heart muscle. Sometimes exercising or stress will cause angina. Stopping the stress may relieve the pain, drugs may be prescribed, or surgery may be necessary to replace the damaged artery. Angina does not always lead to heart attack. Sometimes collateral circulation develops, more blood reaches the heart muscle, and the pain decreases. Angina may even disappear altogether if the heart muscle is receiving enough blood.

A *stroke* occurs when something cuts off the brain's blood supply. There are several ways a stroke may happen. (1) Atherosclerosis in the arteries of the brain or neck may block the flow of blood. (2) A blood clot (thrombus) may form in the atherosclerotic vessel, closing off the artery and causing a *cerebral thrombosis*. (3) A traveling blood clot (*embolus*) can become wedged in a small artery of the brain or neck; this kind of stroke is an *embolism*. (4) A weak spot in a blood vessel may break; this is a *cerebral hemorrhage*. (When the weak spot bulges, it is called a cerebral *aneurysm*.) (5) In rare cases, a brain tumor may press on a blood vessel and shut off the blood supply.

Because the brain controls the body's movements, any part of the body can be affected by a stroke. The damage may be temporary or permanent. If a brain artery is blocked in the area that controls speaking, speech will be affected. Muscles or vision may be affected. Even memory can be affected. "Little strokes," which may be warning signs of an impending major stroke, should be treated by a physician.

High blood pressure (hypertension) is blood pressure that is too high all the time, instead

---

supply blood to all parts of the body, whereas the right ventricle supplies only the lungs.

A complete heart cycle, or beat, consists of a ventricular *contraction*, or *systole* (sis-toe-lee), and a ventricular *relaxation*, or *diastole* (die-ass-toe-lee), when the ventricle receives blood from the contracting atrium (Figure 24.16). The only rest the heart gets is about half a second during the brief period between

of going up or down in a normal way. Hypertension may be an inherited problem. Usually there are no external signs that blood pressure is high, and only regular medical examinations can detect the condition.

During hypertension, the artery walls are hard and thick, and the important stretchiness of arteries is reduced. Once the stretch is gone, the heart must work harder to pump enough blood, and if hypertension persists, the heart may become enlarged, a condition called *hypertensive heart disease*. High blood pressure can also cause a stroke if the extra force of the blood breaks an artery in the brain and a cerebral hemorrhage occurs.

More than one million Americans die annually from diseases related to hypertension. About 25 million Americans have high blood pressure and almost that many are borderline cases. Most forms of treatment involve the reduction of sodium intake (hence the need to stop using table salt), but no matter what drug therapy is prescribed, the purpose is to dilate the arteries so that blood pressure is lowered.

An *artificial pacemaker* is an electronic device that takes over the job of the heart's natural pacemaker. Before the heart muscle will contract, it must receive an electrical impulse, which normally comes from the heart's natural pacemaker, a center of specialized muscle tissue. As long as these impulses are sent regularly, the heart pumps at a steady pace. But if something interferes with the electrical impulses, the natural pacemaker cannot do its job. The heart then pumps too quickly or too slowly or irregularly. The result is that not enough blood—carrying oxygen and nourishment—gets to the body cells. The answer to this problem is an artificial pacemaker, which is placed in the body (in a rather simple operation) to make the heart beat normally. If checked regularly by a physician, an artificial pacemaker can do its job indefinitely.

24.16
One cycle of a heartbeat. (a) Blood from the body comes into the right atrium at the same time that blood comes into the left atrium from the lungs. (b) Both atria contract at the same time, forcing the valves between atria and ventricles open and emptying blood into the ventricles. (c) The ventricles, gorged with blood, hesitate for an instant. (d) On signal from the atrioventricular node (see Figure 24.17), the ventricles contract. The resulting pressure closes the valves between atria and ventricles and opens the valves leading out of the ventricles. The right ventricle forces deoxygenated blood (lighter brown) into the pulmonary artery toward the lungs, and the left ventricle forces oxygenated blood (darker brown) into the aorta toward the body. By this time, the atria are ready to be refilled for another heartbeat.

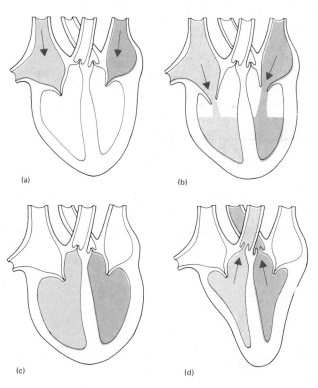

(a)   (b)

(c)   (d)

ventricular contractions; also, the heartbeat usually slows down during sleep because most of the capillaries are inactive then. The characteristic "lub, dup" heart sounds are caused not by the contractions of the heart, but by the sudden closing of the heart valves. When a valve leaks, the condition is known as a heart "murmur," and the backflow of blood may be heard as a whooshing sound.

24.17

How heartbeat is controlled. When the atria are filled with blood, an impulse is generated in the sinoatrial (S-A) node. As the impulse spreads throughout both atria, their muscles contract. The impulse from the S-A node also stimulates the atrioventricular (A-V) node. This second node, upon sending out its own impulse over the ventricle, causes contraction of the ventricular muscles. The two nodes serve as the "pacemaker," keeping the heart beating rhythmically. The numbers on the diagram give the time, in fractions of a second, required for the impulses from the S-A node to pass through the heart. The electrocardiogram shows the changes in electrical potential as the impulse progresses. The peak occurs about 0.16 seconds after the initiation of the impulse, when ventricles are starting their maximum contraction.

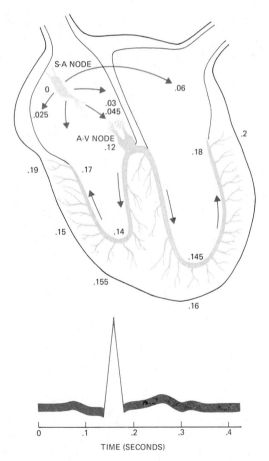

TIME (SECONDS)

The lower the metabolic rate of an animal, the fewer heartbeats it will produce. Usually, the smaller the animal, the higher will be its heart rate. An average human heart rate is about 72–84 per minute, for example, but other animals have the following average rates per minute: shrew (the smallest mammal), 800; robin, 600; mouse, 400; rabbit, 200; cat, 125; fish, 60; elephant, 30.*

The Control of Heart Muscle Activity

The origin of a heartbeat is a specialized section of muscle tissue, called the *sinoatrial* (S-A) *node*, near the entrance to the right atrium. The S-A node is known as the *pacemaker* of the heart, and its original contraction reaches across both atria to the *atrioventricular* (A-V) *node* and spreads along the walls of both ventricles to specialized fibers in the tips of the ventricles (Figure 24.17). Now the simultaneous contractions may begin at the tips of the ventricles and squeeze the blood out through the pulmonary artery and aorta.

The strength of heart contractions can be changed as body needs vary. More blood is pumped during exercise, for instance. Likewise, the *rate* and *rhythm* of cardiac muscle activity can be altered by several factors, including blood temperature, concentration of ions, and the chemical environment. But the main factor is nervous activity, and the control center is located in the medulla of the brain.

When the control center in the medulla stimulates the *cardiac* (sympathetic) nerves, norepinephrine is liberated at the nerve endings, and the heart rate increases. In contrast to these cardiac accelerator nerves, the *vagus* (parasympathetic) nerves are *inhibitory* nerves. When the vagus nerves are stimulated, acetylcholine is liberated at the nerve endings, and the heart rate decreases. This slowdown is caused by the hyperpolarization of the nerve membranes. The accelerator and inhibitory nerves are continually active, both acting in a feedback system. Both sets of counterbalancing nerves terminate in the S-A node, directing the activity of the pacemaker.

---

* Animals with slow pulse rates usually live longer than animals with rapid heartbeats. A rabbit, for instance, lives only about five or six years, and an elephant lives about 60 years. Human beings seem to be an exception. We have a faster pulse rate than elephants, and yet we generally live about 10 years longer. Most animals, no matter what size, average about one billion heartbeats per lifetime, but humans, the exception, average about 2½ billion beats per 70-year lifetime.

THE CIRCULATORY SYSTEM HELPS TO MAINTAIN HOMEOSTASIS

In order to perform its many functions properly, blood must circulate continuously throughout the body. The center of the circulatory system is the heart, but it is the circulatory vessels—veins, arteries, and capillaries—that help maintain cellular homeostasis by supplying cells with the substances they need for metabolism and regulation, and by carrying off their waste products for excretion in kidneys and lungs (Figure 24.19). A "second circulatory" system is the lymphatic system, which returns to the blood excess extracellular fluid, protein, and other substances that are lost from the blood (see p. 496).

Arteries, Capillaries, Veins

Blood moves through a system of three types of vessels: *Arteries* carry blood *away* from the heart; *capillaries* are small, thin-walled vessels that permit exchange of nutrients, oxygen, and carbon dioxide

24.18

William Harvey (1578-1657). When Harvey proved that the heart is a pump that recirculates blood through the body, he introduced an idea of a living body as a mechanical object, and he did so with such thorough, logical, and convincing demonstrations that there was no argument against it.

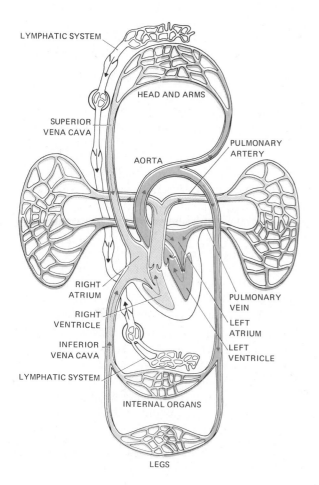

24.19

The human blood and lymphatic systems. In this schematic diagram, the heart is drawn disproportionately large in order to show its internal structure, and the major branches of arteries and veins are indicated merely as networks of capillaries. Blood vessels are in black and color, and the lymphatic system is in white. The blood moves in a closed system, with at least the formed elements —cells and platelets—remaining mostly in the blood vessels. The lymphatic system, in contrast, is not a closed system. The fluid from the blood oozes out from capillaries into the tissues and is worked into the lymphatic system. The direction of flow of lymph is not determined by a lymphatic heart; it is aided by one-way valves, the general movement of the body, and the directional flow of blood. (See also Figure 24.23.)

BLOOD PRESSURE

Arterial blood pressure depends on the volume of blood in the arteries and the elasticity of the arterial wall, as well as on the rate and force of heart contractions. If the arteries are supple, they can be stretched by large volumes of blood without an appreciable rise in blood pressure. Conversely, blood pressure will rise easily in arteries that have become hard, internally coated, or inelastic. When blood is ejected into the arteries by the ventricles during systole, an equal amount of blood is not simultaneously released out of the arteries. In fact, only about one-third of the blood leaves the arteries during systole, and the excess volume raises the arterial pressure. After systole, when the ventricular contraction is over, the arterial walls return to their unstretched condition as blood continues to leave the arteries. Pressure slowly decreases, but before all the blood has left the artery, the next ventricular contraction occurs, and the pressure begins to build up again. Because of this consistent rhythm, the arterial pressure never reaches zero, and there is always enough pressure left to keep the blood flowing.

Blood pressure levels are expressed by two numbers, both representing the height (in millimeters) of a column of mercury. The high number is the systolic pressure (when the heart contracts), and the low number represents the interval between heartbeats (the diastolic pressure). The lower figure is critical, because the higher the diastolic pressure is, the less rest a heart will get. An overworked heart simply does not last as long as a well-rested one. A normal adult blood pressure is 140/90 or *less*. Blood pressure is considered high, or hypertensive, in an adult when the systolic reading exceeds 160 and the diastolic reading is higher than 95. The *pulse pressure* is the difference between systolic and diastolic pressure. In the example above of "normal" blood pressure, 140-90, the pulse pressure equals 50. (This is not the same as the pulse *rate*, or the number of beats per minute, which usually ranges from 72–84 beats per minute in normal adult humans. The pulse rate can easily be felt at the inside of the wrist.)

The instrument used for measuring blood pressure is the easily recognized but not easily

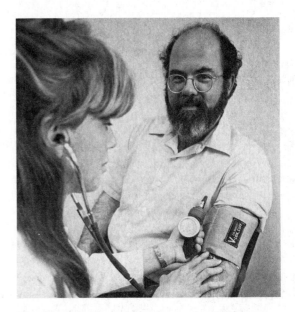

The determination of blood pressure. With a sphygmomanometer (Greek, "thin pulse measurer"), the maximal (systolic) and minimal (diastolic) pressures in arteries can be determined without injury.

pronounced *sphygmomanometer*. The sphygmomanometer consists of an inflatable cuff, a rubber bulb, and a column of mercury. The hollow cuff is wrapped around the upper arm and inflated to a pressure above the systolic pressure (the tightened cuff closes the arteries in the upper arm and prevents blood from flowing to the lower arm). A stethoscope is placed over the artery just below the cuff, and the pressure in the cuff is slowly released until the arterial pressure is greater than the pressure of the cuff. At that point, a recognizable sound can be heard through the stethoscope. This sound signals the high-velocity release of blood, and the figure on the mercury column at which it occurs represents the *systolic* blood pressure. When the cuff pressure is lowered further, a louder sound is heard, and then the sound gradually becomes softer. When the sound stops altogether, the *diastolic* blood pressure is noted on the column. The absence of sound indicates a free-flowing gush of blood through the open artery.

between the blood and tissues; _veins_ carry blood _toward_ the heart (Figure 24.20).

Of all the types of blood vessels, the arteries have the greatest pressure, and the arterial walls, which are appropriately strong and elastic, adjust to the great pressure of the contraction of the ventricle during systole. Blood travels through the aorta at a speed of about 40 cm/sec (0.9 mph). In people who suffer from hardened arteries (_arteriosclerosis_), the arteries are too rigid to expand properly, and the systolic blood pressure rises higher than the normal range. The arteries must also be able to squeeze down during diastole, the period of least arterial pressure. Arteries branch into smaller arterioles, whose walls contain smooth muscle. By contractions of this muscle, the arterioles control the varying flow of blood as organs require it.

Arterioles enter the body tissues and branch out further to form _capillaries_, the link between arteries and veins. Capillaries are the smallest and most numerous blood vessels. If all the capillaries in an adult human body were connected, they would stretch out about 96,000 kilometers (60,000 mi). Such an abundance of capillaries makes available an enormous surface area for the easy exchange of substances between the blood and nearby cells. Capillaries are so narrow (less than 0.01 mm in diameter) that red blood corpuscles must squeeze through in single file (Figure 24.21). Because the capillaries are so narrow, a high-fat diet may present circulatory problems.

24.20

Varieties of mammalian blood vessels. _Arteries_ are rather firm-walled but elastic, with an internal lining, a coating of smooth muscle, and a sheath of connective tissue. The walls expand slightly with each pulse of pressure from a heartbeat, or they can be either relaxed or constricted by nervous control. The _veins_ are generally softer and more flexible than arteries. They have the same kinds of tissues as the arteries, but they lack the stiffness, and they collapse if blood pressure is not maintained. The _capillaries_ are microscopically fine vessels, with walls mostly only one or a few cells thick. They allow passage of water and dissolved materials, and even some blood cells (especially leukocytes). The capillaries are distributed everywhere except in the dead outer layers of skin and in such special places as eye lenses.

24.21

A microscopic view of blood flowing through capillaries. The smallest capillaries have such a narrow passageway that the blood cells move through in single file, allowing time for the necessary gas and nutrient and waste exchanges to take place between blood and functioning cells. Blood working its way along in capillaries can be seen with only slight magnification in the web between a frog's toes.

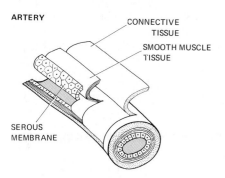

ARTERY
CONNECTIVE TISSUE
SMOOTH MUSCLE TISSUE
SEROUS MEMBRANE

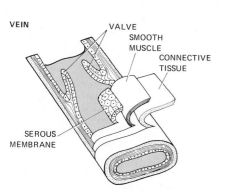

VEIN
VALVE
SMOOTH MUSCLE
CONNECTIVE TISSUE
SEROUS MEMBRANE

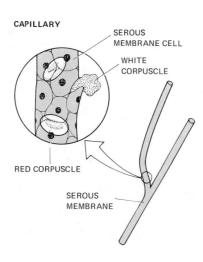

CAPILLARY
SEROUS MEMBRANE CELL
WHITE CORPUSCLE
RED CORPUSCLE
SEROUS MEMBRANE

× 500

After a high-fat meal, fatty globules are present in the blood. These globules may adhere to red blood cells, making it difficult for the heart to push the blood through the capillaries.

The thin wall of the capillary is *selectively permeable*; that is, it allows some dissolved substances to filter through and holds back others (Figure 24.22). Such substances as gases, waste products, salts, sugars, and amino acids pass freely through the capillary walls, but the large proteins present in the fluid part of the blood pass through only with difficulty. Likewise, the cells of the blood cannot pass through the walls of the capillaries.

As capillaries leave the individual cells of the body, they join together to form *veins*. The venous walls are thinner and less elastic than arterial walls, and their diameter is greater. Blood pressure in the veins is low, and the venous blood is helped along by surrounding muscles and prevented from flowing backward by one-way valves. These valves are especially important in the legs, where gravity cannot assist the return of blood to the heart. If a person stands still for a long time, the leg muscles are not able to push the venous blood up. In such cases, the venous blood returns more slowly to the heart and tends to accumulate in the veins. Reduced return of blood to the heart may result in poor blood flow to the rest of the body, including the brain. This could cause fainting. Another problem is that the veins in the feet and legs may be stretched somewhat as tissue water accumulates there. If the one-way valves in the veins in the legs weaken and become permanently dilated, *varicose veins* result.

The Lymphatic System

The *lymphatic system* is sometimes called the "second circulatory" system (Figure 24.23). However, it is quite different from the system that transports blood around the body. The lymphatic system is not a closed, circular system, nor does it have a pump. It is made up of a network of thin-walled vessels that carry a clear fluid called *lymph*. The system begins with very small vessels, lymph capillaries, in contact with the tissue. These vessels join together to make larger ducts, which pass through specialized structures called lymph nodes and continue on to drain their contents into a vein in the neck. The composition of lymph is similar to that of blood. Lymph contains water, plasma, some protein, electrolytes, and white blood cells (lymphocytes), but it lacks red blood cells and most of the blood proteins. Lymph is derived from the fluid portion of the blood that passes from the arterial ends of capillaries out into the spaces around cells. The fluid lymph surrounds and bathes the cells of the body.

Some of the fluid derived from the blood returns to the blood in the venous end of the capillaries. What remains may enter the lymphatic system and move slowly from remote regions of the body toward the lymph nodes in the groin, arms, armpits, and neck. The lymph nodes are specialized tissues that act as sieves, removing cellular debris, old cells, and foreign particles from the lymph fluid before the fluid rejoins the venous blood (Figure 24.24). When infection or injury occurs, lymph nodes may enlarge as they collect dead tissue cells or many lymphocytes. This produces the familiar "swollen glands" that result from some diseases and are evident frequently in the neck. Basically, the lymphatic system performs three major functions: (1) It returns to the blood excess fluid and proteins from the spaces around cells. (2) The lymph nodes filter and destroy bacteria before returning lymph fluid to the blood. (3) The lymphatic system plays a major role in the transport

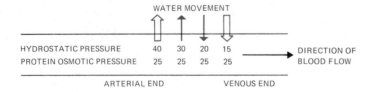

24.22

Movement of water out of and into capillaries. As blood enters a capillary, it is under pressure from the pumping of the heart, and the pressure is sufficient to force water and small molecules through the capillary walls into the surrounding tissues. As blood moves further along through the capillary, the pressure drops and no more water leaves. The proteins in the plasma, however, exert an osmotic influence on the fluid, and by the time the blood comes to the venous end of the capillary, there is enough difference between the osmotic value of the fluid in the capillary and that in the surrounding tissue to cause water to move back into the capillary.

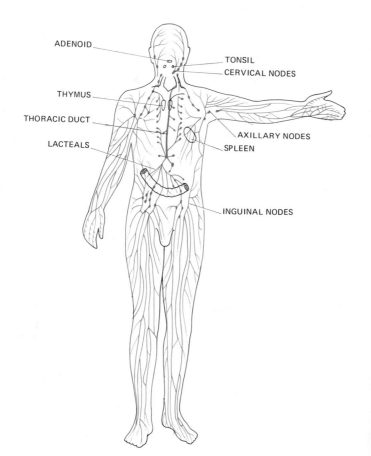

ADENOID

TONSIL
CERVICAL NODES

THYMUS

THORACIC DUCT

AXILLARY NODES

LACTEALS

SPLEEN

INGUINAL NODES

24.23
Distribution of the lymphatic system in a human body. Although lymphatic vessels are distributed throughout practically all living tissue, they are concentrated in some regions, especially the neck, upper chest, armpits, viscera, and lower abdomen. If an infection occurs, lymph *nodes* may become so enlarged that they can be felt as lumps in the neck or groin. (See also Figure 24.19.)

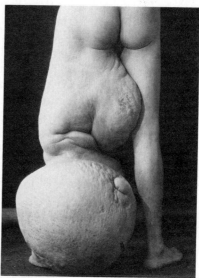

24.24
The effect of stoppage of lymph flow. A tropical roundworm, Filaria, can be injected into the human bloodstream by mosquitoes. The parasite burrows into tissues and obstructs the return of lymph to the lymphatic ducts. The result is a swelling of the affected part, an affliction known as *elephantiasis*. An unusually severe case is shown here. The damage is not locally painful, but continued interference with the movement of lymph can be fatal. No cure is known.

of fat materials from the tissue surrounding the intestine to the blood. This action is accomplished by a portion of the lymphatic system that leads from the small intestine to an opening in the vena cava.

IMMUNOLOGY

Several observations on the reactions of people and animals to disease led to the discovery of the *immune systems* and hence to the whole science of *immunology*. One observation was that even during the worst of epidemics—for instance, during the plague—not everyone caught the prevalent disease. Some individuals were able to mingle freely with the dying and walk away untouched. Some people are immune to certain diseases. Another observation is that one who has once had a disease, say mumps, is highly unlikely to suffer from the same disease again. One has, so to speak, earned one's immunity. Empirically, then,

IMMUNITY FROM NAZI SLAVERY

During World War II, when the Germans were killing or enslaving Poles and Jews as racially inferior people, an immunological trick was employed to save lives in a political way. E. Lazowski, now at the Illinois Children's Hospital in Chicago, and S. Matulewicz, now at the Université Nationale in Zaire, knew that an infection by a relatively harmless bacterium, *Proteus* OX-19, could cause a blood reaction resembling that of epidemic typhus, the dreaded louse-borne "disease of human misery." The Weil-Felix reaction had been known since an earlier epidemic in Poland in 1916, and it was routinely accepted as a diagnostic test for true typhus. The doctors reasoned that a blood test of a healthy person who had been injected with a dead suspension of *Proteus* OX-19 would indicate typhus.

For more than 25 years there had been no epidemic typhus in Germany, and consequently the Germans had lost much of their immunity to the disease. Knowing that, German doctors had no desire either to import the disease into their country or to go themselves into places where they could come into contact with it. Thus any Pole who showed a positive Weil-Felix reaction was presumed to be carrying typhus and was not wanted in Germany.

A Polish slave laborer, allowed a 14-day home leave, volunteered to be the first test case of an artificially induced immunological experiment. Although the trial carried risks, some known and some unknown, both to the Polish doctors and the worker, he preferred those risks to a return to slavery. He took the *Proteus* injection, officially disguised as "protein stimulation therapy." His blood sample, sent to the German laboratories, prompted a telegraphic reply: Weil-Felix positive. The local German officers, interpreting the telegram to mean typhus, excluded both the worker and his family from Germany in order to keep out typhus-carrying lice.

With the technique established, Lazowski and Matulewicz began injecting many Poles with *Proteus* OX-19, thus producing false typhus-positive blood tests, until a dozen villages were officially declared an epidemic area, relatively exempt from German interference. Suspecting that Polish doctors might send mislabeled blood samples, the German doctors tried to take samples only in their own laboratories, but of course their results were also positive because the reaction was a true biological one, even though it had been artificially induced. As Lazowski and Matulewicz reported in 1977, "Our private immunological war gave us a deep satisfaction because we knew that we had saved the lives of people who would otherwise have been killed."

---

medical doctors learned that in a practical way they can confer immunity artificially by administering a weakened form of a disease producer. Edward Jenner is frequently credited with originating the practice of giving a vaccination (Latin, "vaca," cow) with cowpox to keep a possible victim from getting deadly smallpox. In fact, by the seventeenth century various kinds of vaccinations had been in use in many places for years. Louis Pasteur later applied the same principle when he inoculated sheep against anthrax, and today such immunizing methods as treatment with the Salk vaccine against polio are still based on the same idea. It was not until there had been extensive study of blood chemistry and of the nature of cell membranes that the biological reason for immunity was partially understood. Even yet, with all the research being carried on, the bulk of the knowledge of immune reactions probably remains to be discovered.

The essential action in the development of an animal's immunity to any specific outside influence—and that influence is not limited to microorganisms—is the synthesis of a specific protein, known as an *antibody*. After the entrance of a foreign material, whether it is a bacterium, a virus, or an inhaled or swallowed substance, an animal's body may start the production of an antibody. The foreign material, known as an *antigen*, somehow stimulates the synthesis of a particular kind of molecule, the antibody, whose exact molecular shape is such as to match the molecular shape of the antigen. The antibody can tie up the antigen firmly enough to render it ineffective by coagulating it in a way roughly equivalent to what happens to the white of an egg when it is boiled. If an invading bacterium provides the antigen, and the antibody can be produced fast enough to keep ahead of the growth of the invader, the animal will destroy the bacterium. This would be a case of naturally developed immunity. In human medicine, a person who has been exposed to a possibly contagious infection may, as a precaution, receive an injection of a mild, perhaps weakened or heat-killed pathogen. The hope is that the body system will produce enough antibody of the correct type to avoid disease. Thus *active immunity* may be achieved. However, if a disease is found only after it has progressed to an uncomfortable or dangerous stage, the sufferer may be given a dose of actual antibody. Such an antibody is often prepared by injecting the causative bacteria into an animal such as a horse, and allowing the horse to manufacture the proper antibody. Serum containing the antibody is taken from the horse and administered to the human patient.

The same system that protects an animal from infection can be annoying or even deadly. Developing an immunity against typhoid fever is desirable, but inadvertently developing an immunity or sensitization to wheat flour makes eating difficult. A person with a sensitivity to what most people consider ordinary food is *allergic*, and allergic reactions are expressions of the same kind of antigen-antibody activity that helps an animal body defend itself against bacteria and viruses (see the essay on page 500).

Blood transfusions must be strictly monitored because of the antigen-antibody system. In one series of blood groups (many are known), there are four possible combinations of factors (blood types), based on the kinds of antigens in red blood cells and the kinds of antibodies in blood serum. The basic principle of blood transfusion is that a transfusion must not be made if the cells of the donor's blood will be *agglutinated* (clumped together) by the recipient's plasma.

Now that organ transplants are mechanically feasible, one of the critical details that must be attended to is the possibility of an immune reaction that may cause the recipient to reject the new organ from a different person with a different set of proteins. In a related way, some women chronically have miscarriages because they lack a factor that prevents the immunological rejection of the embryo in the mother's uterus. Because the embryo has proteins that resemble its father's as well as its mother's proteins, the mother's body may react to the embryo as though it were a "foreign" antigen unless the mother possesses the proper antibody. Obviously, most women have an immunological mechanism that makes chronic miscarriages unusual.

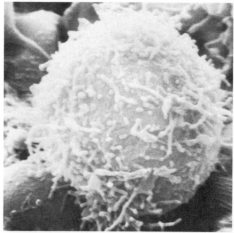

(a)   × 8000

(b)   × 3500

24.25
Scanning electron micrographs of lymphocytes. (a) Unlike red blood cells, lymphocytes have nuclei and are capable of undergoing mitosis, increasing enormously in number when an infection occurs. (b) The photograph of this flowerlike cluster has in the center a mouse spleen lymphocyte containing proteins (immunoglobulins) specific for sheep red blood cells. When such a cell comes into contact with sheep red cells (the surrounding "petals" in this photograph), the two kinds of cells are firmly bound. Once a lymphocyte has developed such a specific ability, it can transmit that ability to the cells resulting from its divisions. All the lymphocytes descended from such a cell retain its immune reaction; they are said to have an immunological memory.

ALLERGY: WHAT IS IT?

Most people have some kind of allergy, but few understand what an allergy is. We blow our noses, wipe our eyes, and take "anti-allergy" pills, but even if we deduce that an "antihistamine" is a drug that opposes a "histamine," who knows what a histamine is? To start with, hay fever is not a fever and is not caused by hay. And antihistamines do not attack the allergy; they fight against the work your own body is trying to do to defend itself against the foreign substance that starts the allergic reaction.

What we call hay fever is generally a reaction to pollen in the air, a seasonal phenomenon that is usually caused by tree pollen in the spring, grass pollen in summer, and weed pollen in the late summer and fall. Ragweed does the most damage because it produces huge amounts of pollen. A single ragweed plant produces more than 10 billion pollen grains. (Such a prodigious output is necessary to offset the hazards encountered by plants that depend on the wind to carry pollen to the female flower. The output can be compared to the large amounts of sperm released at each ejaculation.)

An _allergy_ is an abnormal body reaction, and the agent that causes the unusual reaction

(coughing, sneezing, headache, hives, etc.) is an _allergen_ or _antigen_. The most common allergen is dust, and other well-known allergens are pollen, food, and chemicals. When you are allergic to something (chocolate, for instance), your body reacts to it as though it were a foreign substance. Your body is triggering a typical immune reaction, the same sort of procedure that produces beneficial immunities against infections or that causes troublesome rejections of organ transplants.

Once the allergen is recognized as a foreign substance, your body produces specific chemicals that stimulate the formation of an _antibody_. One of these chemicals is _histamine_, which causes hypersensitivity of some body tissues. So the purpose of an antihistamine is to try to eliminate the annoying effects produced by histamine, and not by the allergen.

The first time an allergen enters the body, there is an incubation period during which antibodies are being formed, and no allergic symptoms occur. Subsequently, the allergen will trigger specific allergic reactions. These reactions are usually out of proportion to the actual need, but the body is effectively defending itself,

SUMMARY

1. _Respiration_ at a cellular level means the oxidation of food with the consequent release of energy. _Respiration_ at the gross anatomical level means _breathing_ (air containing oxygen enters the lungs, and air containing waste gases is expelled from the lungs).

2. Respiration is relatively simple for organisms like single-celled protozoa. They obtain oxygen by direct diffusion through their bodies, and they lose carbon dioxide wastes by diffusion also. But larger, more complex animals have a body surface area that is smaller in relation to their volume and less permeable, and for these animals a more effective system of gas exchange evolved.

3. _Gills_ are an efficient means of respiration in water. They are filaments that allow the blood to flow in a

direction opposite to the water flow, thus establishing an efficient _countercurrent flow_.

4. The more developed the organism, the more complex is the internal structure of the _lung_. Lungs have evolved from single sacs to highly folded and lobulated mammalian lungs, in which millions of tiny air sacs called _alveoli_ appear, each surrounded by a network of blood vessels.

5. Insects and some other land-dwelling arthropods have a system of tubes called _tracheae_, which branch out from surface openings called _spiracles_ to all parts of the body.

6. The respiratory system of mammals is made up of the lungs, the several passageways leading from the out-

and it has no way of knowing that the allergen is not indeed a serious threat. Ordinarily, two specific occurrences accompany allergic reactions: Certain body cells release chemicals, and smooth muscles contract. We do not know how such chemical substances as histamine, serotonin, and acetylcholine are released from cells, but we have learned that the location of the meeting between allergen and antibody determines the severity of the allergic reaction. If an antibody encounters an allergen in the bloodstream, the allergen is neutralized without any side effects. However, if antibody meets allergen in or on a cell, neutralization occurs, but the cell is disturbed also. It is this disturbance and release of toxic chemicals that produces the specific allergic reaction.

Histamine is the best-understood of the chemical substances released from cells. It dilates blood vessels and causes the contraction of smooth muscle, and it is probably involved in most allergic reactions to some degree. The most troublesome products of histamine are *wheals*, localized swellings that vary according to their location. Skin hives are external wheals. Wheals in the mucous membranes of the nose cause stuffiness and are usually accompanied by a runny nose. Such respiratory problems as bronchial asthma are caused by wheals in the air passages, and they can interfere with breathing. Some people have allergies that produce headaches, which are caused by wheals in the brain.

Hundreds of other types of allergies exist, and thousands of allergens have been identified. Some of the most common offenders are chemicals such as lacquers, animal danders (flaking skin) and hair, eggs, milk, wine, wheat flour, legumes, chocolate, tobacco smoke, coffee, nuts, citrus fruits, wool, and shellfish.

If you are allergic to strawberries, will your children inherit the allergy? Maybe, but it is more likely that they will inherit the susceptibility to allergy rather than an allergy to a specific substance.

Many people are allergic to pollen or cat hair. But recently a more dramatic allergy has come under suspicion. Some women may be infertile because they are allergic to their own eggs. These women may create antibodies that prevent sperm from entering the outer covering of the egg.

---

side to the lungs, and the muscular mechanism (*diaphragm*, abdominal, and rib muscles) that moves air in and out of the lungs.

7. The intake of air, *inspiration*, occurs when air rushes into the lungs to equalize a reduction of air pressure in the thoracic cavity. *Expiration* is the expulsion of air from the lungs.

8. The *respiratory center* in the medulla oblongata of the brain receives and coordinates information about the changing needs of the body for oxygen. A high carbon dioxide content of the blood causes an *increase* in breathing rate, and a low carbon dioxide content causes a *decrease*.

9. Through the process of diffusion, the blood that enters the lungs receives oxygen from the alveoli, and carbon dioxide and water diffuse from the blood into the alveoli and subsequently to the air.

10. *Hemoglobin*, the *respiratory pigment* of the blood, has such an affinity for oxygen that it binds about 98 percent of the available oxygen in the red blood cells. When oxygen combines with hemoglobin at high oxygen pressure, the resultant compound is *oxyhemoglobin*. Some toxic agents, such as carbon monoxide, bind to hemoglobin even more readily than oxygen does.

11. Carbon dioxide is transported by the blood in three ways: (1) About 67 percent of the carbon dioxide combines with water to form carbonic acid; (2) about 25 percent of the carbon dioxide reacts with hemoglobin and eventually is released in the lungs in exchange

for oxygen; and (3) the remaining 8 percent is dissolved in the blood as molecular carbon dioxide.

12. *Blood* is the term used for circulating fluids in general. Blood in the human body carries dissolved foods, transports carbon dioxide and oxygen, distributes hormones, defends against infection, helps maintain an even temperature, removes waste products, and performs several other functions.

13. The liquid portion of mammalian blood is *plasma* (water, dissolved solids including proteins, dissolved gases), and the solid portion is *formed elements* (red blood cells, white blood cells, platelets).

14. Blood *coagulates* at the site of a wound when platelets initiate the release of a series of enzymes and proteins. The ultimate product is a stringy protein, *fibrin*, which entangles the blood cells and forms a clot.

15. The *heart* is a double pump, each half consisting of an *atrium* and a *ventricle*. With the help of a series of one-way valves, blood enters the right atrium, passes to the right ventricle, then flows out through the pulmonary artery to the capillaries of the lung. It returns to the left atrium and then passes through the left ventricle and out again through the aorta. A complete heart cycle, or *beat*, consists of the ventricular contraction, *systole*, and the ventricular relaxation, *diastole*, when the ventricle receives blood from the contracting atrium.

16. The *strength* of heart contractions can be changed as body needs vary, and the *rate* and *rhythm* of cardiac muscle activity can be altered by several factors, including blood temperature, ionic concentration, and chemical activity. But the main factor is *nervous activity*, and the control center is located in the medulla. The heartbeat originates in a specialized section of muscle tissue called the sinoatrial node, or *pacemaker* of the heart. Accelerator and inhibitory nerves act in a feedback system.

17. Blood moves through a system of three types of vessels: *Arteries* carry blood away from the heart; *capillaries* are small, thin-walled vessels that permit diffusion between the blood and tissues; *veins* carry blood toward the heart.

18. The *lymphatic system* returns excess fluid and protein from the spaces between cells to the blood, and it filters and destroys harmful material.

19. When foreign substances (*antigens*) get into the bloodstream, they combine with specific *antibodies* and an inactive substance is formed, creating an *immunity*. An *allergy* may result when an antigen that does not normally evoke antibodies causes discomforting physiological symptoms.

RECOMMENDED READING

Helpful references for the entire physiology section, Chapters 18–24, are listed at the end of Chapter 18.

Baron, Samuel, and Howard M. Johnson, "Does Interferon Help Regulate Immunity?" *The Sciences*, April 1978.

Benditt, Earl P., "The Origin of Atherosclerosis." *Scientific American*, February 1977. (Offprint 1351)

Burke, Derek C., "The Status of Interferon." *Scientific American*, April 1977. (Offprint 1356)

Comroe, Julius H., Jr., "The Lung." *Scientific American*, February 1966. (Offprint 1034)

Cooper, Max D., and Alexander R. Lawton III, "The Development of the Immune System." *Scientific American*, November 1974. (Offprint 1306)

Edelman, Gerald M., "The Structure and Function of Antibodies." *Scientific American*, August 1970. (Offprint 1185)

Jones, Jack Colvard, "The Feeding Behavior of Mosquitoes." *Scientific American*, June 1978.

Kaj Jerne, Niels, "The Immune System." *Scientific American*, July 1973. (Offprint 1276)

Old, Lloyd J., "Cancer Immunology." *Scientific American*, May 1977. (Offprint 1358)

Perutz, M. F., "The Hemoglobin Molecule." *Scientific American*, November 1964. (Offprint 196)

Wessells, Norman K. (ed.), *Vertebrate Structures and Functions*, Readings from *Scientific American*. San Francisco: W. H. Freeman, 1974. (Paperback)

Wiggers, Carl J., "The Heart," *Scientific American*, May 1967. (Offprint 62)

Wood, J. Edwin, "The Venous System." *Scientific American*, January 1968. (Offprint 1093)

THE BIOLOGY
OF
POPULATIONS

PART SEVEN

25

Darwinian Evolution

Scanning electron micrograph of the surface of an egg of a silk moth, Antherea polyphemus, *showing the opening through which sperm can enter.*

EVOLUTION MEANS CHANGE OR, LITERALLY, "AN unrolling." In the biological sense, evolution specifically means *genetic* change, even though genetic changes almost always occur for the worse. But once in a while, once in a *great* while, a genetic change occurs that bestows a definite advantage. Such a change is the basis of evolution and the means by which entire new species may arise.

The concept of evolution pervades all biology and underlies all its major generalizations. We can start with the recognition of a demonstrable unity in all life on earth. Practically all organisms build the same amino acids into their proteins, use the same enzymes to carry on their energy-exchanging activities, construct their membrane systems in similar ways, and regulate their productivity by means of similar molecules, such as cyclic AMP in human livers and in slime molds. The more we learn about the chemistry and activities of organisms, the more striking the similarities become and the less important are the easily observed differences. Because of this underlying unity, biologists believe further that the present differences between organisms, bewildering in their diversity, have come about by divergence from earlier, simpler living things. These beliefs make possible the integration of isolated facts and give us confidence that general conclusions about evolution are valid.

Consider a single subcellular particle, a ribosome. If ribosomes are found to look and act in a special way in bacteria, that bit of information may be interesting to a microbiologist, but by itself it means relatively little. However, if ribosomes are found to look and act similarly in corn roots, human heart muscle, and tomato worm eggs, then it becomes safer to think that ribosomes are particles of general, fundamental importance to living things. Without an evolutionary connection between such apparently unlike organisms, a few scattered snippets of information would remain just that, interesting perhaps to a few specialists, but lacking in the broad significance that can make them interesting to every thinking human. As it is, we can look on all the earth's inhabitants as members of an enormously varied but ultimately related superfamily. As any satisfying scientific concept should do, the concept of evolution simplifies and unifies knowledge. That is important.

THE IDEA OF EVOLUTION

The earliest students of organisms were primarily interested in the human uses of plants and animals, and they tried to make schemes of classification purely for their own convenience. When they were making such schemes, similarities between kinds of organisms did become apparent, and those who looked closely noticed that the several sorts of birds known as hawks had features in common, or that the many varieties of beans, for all their little differences, were still beans. Such observations prompted speculation about the origin of species. In the Western world, at least, one usual explanation was a belief called "special creation," according to which all the living species were made at once by a Creator or God, and that once made, these species continued to reproduce without change.

The idea of special creation and the immutability of species was well established throughout the eighteenth century, although even then some doubts about it were beginning to be voiced. Carolus Linnaeus (1707–1778), considered the founder of useful taxonomy, expressed in his early works the idea of the immutability of species. However, in some of his later writing one can find hints that he was reconsidering this idea. The more species he observed and described, the more he saw *variation* from an ideal type. Variation suggested that species were not immutable—a direct contradiction of the idea of "special creation."

Before the concept of evolution was stated in its modern form by Charles Darwin in 1859, several theories of biological relationships were current. Darwin's own grandfather, Erasmus Darwin, had published *Zoonomia* in 1794, wherein he suggested that evolution probably takes place in plants and animals. But like many other scientists who had developed an interest in the idea of evolution, Erasmus was unable to describe a working mechanism of evolutionary change. Even Charles Darwin himself, for all his impact on scientific thought, was never able to explain exactly how evolution happens because he lacked an understanding of genetics. But what were the clues that helped Charles Darwin to convince most of the world that evolution *did* (and does) happen?

Geological Sources

Much of the basis for the growing interest in evolution came from geology. James Hutton, generally regarded as the father of modern geology, stated in 1785 that the landscape of the earth was created by a continuing process of natural change, not by sudden catastrophes, as propounded by contemporary scientists. Hutton's theory of slow and steady change was known as the *principle of uniformitarianism*, or simply the *uniformity of process*. Such a concept not only threatened biblical teachings but disputed the generally accepted *principle of catastrophism*.

The findings of geologists implied a long history of the earth. In order to harmonize these new ideas with biblical teachings, Baron Georges Cuvier and others proposed what came to be known as the "great compromise." According to Cuvier, the period from the Flood of Noah to modern times was 6000 years—the age of the earth and of humans—and this period was to be translated literally from the scriptures, even to the point of having 24-hour days during the Creation. Prior to the Flood, Cuvier proposed, was the supernatural time of "geology," when violent catastrophes occurred and all sorts of strange animals existed. Cuvier's "great compromise" was a welcome solution to the growing uncertainty about time, and his proposal attracted a large following.

All was calmly accepted until 1830, when Charles Lyell (Figure 25.1) published *Principles of Geology*, which supported Hutton's idea that the so-called geologic period was not a supernatural time. Lyell's book renewed worldwide interest in Hutton's notion that there was "no sign of a beginning or of an end." Lyell and Hutton both realized the dramatic and sudden effects of floods, earthquakes, volcanoes, and other geologic "catastrophes," but in terms of geo-

25.1

Sir Charles Lyell (1797–1875), the geologist whose interpretation of the earth's history influenced Darwin and helped prepare the minds of men for the idea of evolution. Besides advancing the idea that the earth had undergone changes, he gave Darwin a time scale that was great enough to make the theory of organic evolution tenable.

25.2

Jean Baptiste Lamarck (1744–1829), the French scientist whose ideas of evolution preceded those of Darwin. Although his explanation of evolution by inheritance of acquired characters was superseded by Darwin's "survival of the fittest," Lamarck made substantial contributions, especially in helping break the dogma of the perpetual constancy of species.

logic change they could see only an endless cycle of uniform processes. If Lyell and Hutton were correct, the earth had to be much older than the few thousand years claimed in current teachings. This was one of the critical issues that Darwin would have to confront. He would also have to confront the issue he never really solved: What *causes* evolution?

Lamarckian Evolution

One of the first scientists to attempt an explanation for the *causes* of evolution was Jean Baptiste Lamarck (Figure 25.2), who in the early 1800s proposed the following ideas:

1. As environmental conditions change, plants and animals develop new traits to cope with the change. If vegetation becomes scarce, for example, giraffes will be forced to stretch their necks higher and higher to reach tree leaves.

2. *The theory of use and disuse.* The more an individual uses a part of its body, the more developed that part becomes. Conversely, as an individual ne-

glects to use a body part, that part disintegrates and will eventually disappear. For instance, an athlete develops his muscles by constant exercise (*use*), whereas snakes used their legs less and less until those legs finally disappeared (*disuse*). (We know that the theory of use and disuse is fundamentally correct for an individual organism during its lifetime, but it is not correct from generation to generation.)

3. *The theory of the inheritance of acquired characteristics.* Individuals who have acquired or developed a characteristic as a result of use or disuse will pass that characteristic on to their progeny. In other words, the muscular athlete will foster offspring whose muscles are well developed. (There is no evidence to support this theory.)

Basically, Lamarck believed that animals strive to adapt to environmental changes, and in so doing, they acquire new characteristics; for example, the giraffe acquires a long neck by stretching for leaves. According to Lamarck, such an environmentally derived feature will be inherited by the offspring of the long-necked giraffe. Lamarck's explanation seemed reasonable in light of the information available at the time. But as appealing as his overall theory was, the theory of the inheritance of acquired characteristics is simply not correct.

Perhaps the greatest virtue of Lamarck's theory was that it could be tested by experiment. Numerous experiments over the last 200 years indicate that changes in body parts—shape, color, etc.—acquired by parents are *not* transmitted to their offspring. Only changes (mutations) in the DNA of the sex cells of the parents can be passed on to offspring. If the DNA of the sex cells is not affected, as it is not in the muscular athlete or the stretching giraffe, future generations will not be affected. Here is another way to show the inaccuracy of the theory of the inheritance of acquired characteristics: If you change the shape of your nose through plastic surgery, you are not guaranteeing a similar nose for your future offspring. Obviously, plastic surgery does not affect your sex cells, and only changes in your sex cells can affect your future offspring. Today biologists are able to explain that the thing that can be inherited is a *genotype*, not a *phenotype* as Lamarck unknowingly suggested.

Lamarck was a good scientist, and it is unfortunate that an original, creative, skillful scientist has been belittled because a small portion of his work turned out to be incorrect. Lamarck was wrong, but he had laid the groundwork for Charles Darwin.

25.3

Charles Darwin (1809–1882), whose name is inseparably linked with the theory of evolution. Science historian Erik Nordenskiold said of him: "If we measure him by his influence on the general cultural development of humanity, then the proximity of his grave to Newton's is fully justified."

CHARLES DARWIN AND THE VOYAGE OF THE BEAGLE

In 1831, when Charles Darwin left England on a five-year voyage around the world, he was only 22 years old (Figure 25.3). Darwin had signed on as ship's naturalist aboard the H.M.S. *Beagle*, a brig designed specifically for scientific research. The ship would chart the coastline of South America, make longitudinal and astronomical measurements, and collect samples of local wildlife (Figure 25.4).

Unlike Gregor Mendel, Darwin had no financial problems to complicate his life, and he was able to enjoy his formative years without too many worldly concerns. Young Charles had shown little aptitude for medicine, the law, or the clergy, but he had developed an avid interest in nature while studying theology at Cambridge for three years. Darwin's profession was finally chosen for him when Captain Robert Fitzroy invited him to become the *Beagle's*

naturalist. Fitzroy made it clear to Darwin that as part of his duties he would be expected to help refute the radical new ideas of geology by gathering as many specimens as he could to celebrate the wonders of nature. (Captain Fitzroy was quick to realize the irony of that situation when Darwin published his well-documented theory of evolution 28 years later. Another irony was that Darwin's former botany professor gave him a book to take along on the voyage, warning him not to take it seriously. The book was *Principles of Geology*, by Charles Lyell.)

Darwin did indeed find a glorious array of natural wonders and he meticulously accumulated a mass of information about them. But along with the many examples of nature, he also began to ask questions about the overwhelming diversity of living things. This was the question that began to nag at Darwin: What was the reason for so many different kinds of plants and animals? He had read the "heretical" book by Lyell, and he readily accepted the idea that the earth's landscape was gradually being altered through

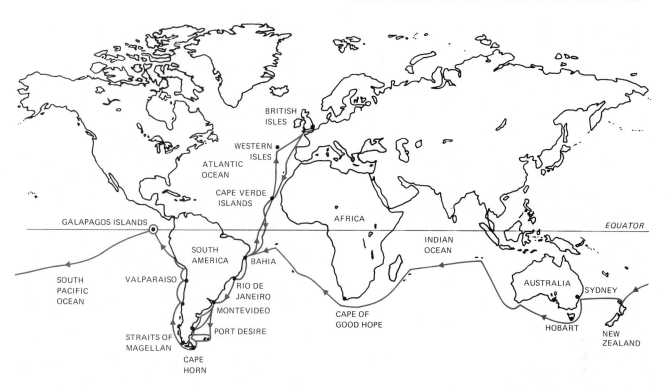

25.4

The route of Darwin's trip around the world on H.M.S. *Beagle*. On that voyage, described in his book, *The*

Voyage of the Beagle, Darwin accumulated the information that contributed heavily to the idea of organic

evolution, one of the few ideas that have had a deep and lasting effect on human thinking.

vast expanses of time. Was it also possible that plants and animals had been changed over long periods of time?

Darwin's observations during his trip aboard the *Beagle* were the basis for his later theories of evolution. Three items in particular seem to have impressed him: (1) Fossils found in South America appeared related to (but not the same as) the living animals of South America; in fact, Darwin realized later that these fossils were more like the living animals in the same country than like fossils of the same age in other countries. (2) Animals living in different climates of South America appeared related to (but not the same as) each other; here again, the similarities were greater among South American animals in hot and cool climates than between South American animals in a hot climate and animals of another country with a comparably hot climate. (3) Animals on islands appeared to be related to (but not the same as) animals of the closest mainland; in addition, some island animals seemed to show a relationship to other

species on the same island. These observations eventually suggested to Darwin the idea of the modification of species, an idea which was to become his life's work.

The voyage of the *Beagle* ended in 1836, and Darwin returned home with his head (and his notebooks) full of questions about the life he had seen. He began a systematic analysis of his collections. Darwin did not start his investigations with the intent of proving anything, certainly not to formulate a theory of evolution. Instead, he merely applied a lively curiosity to the natural world, and having observed thousands of plants, animals, and the earth, he sought an explanation that would integrate what he had learned. He wanted to know why and how, as well as what.

The Influence of Malthus

Shortly after Darwin began what he called his "systematic enquiry," he read *An Essay on the Principle of Population*, which had been written in 1798 by

25.5

Thomas Robert Malthus (1766–1834), the English economist whose tract on the possibilities of unlimited expansion of populations influenced Darwin's thinking in developing the theory of evolution. To this day, Malthusians, anti-Malthusians, and neo-Malthusians can become passionate in their beliefs concerning the thesis of this eighteenth-century theoretician.

25.6

Alfred Russel Wallace (1823–1913), the English explorer, naturalist, author, social reformer, and coauthor with Darwin of the principle of organic evolution. This modest, intrepid traveler made unbelievable journeys, frequently with no company but savages, collecting, describing, and observing, until he was forced to the conclusion that species are changeable.

Thomas Malthus, a political economist (Figure 25.5). Malthus had written that organisms possess the theoretical ability to breed at a geometric rate, thus eventually running out of food and space. This idea had been the cause of lively, often bitter debate ever since Malthus had first proposed it, but theoretically it is unassailable. If a single bacterium could divide and its progeny could divide every half hour without any hindrance, within three days the earth could not hold them all. This much is simple arithmetic. However, the anti-Malthusians pointed out that in reality such unlimited growth does not occur. Besides, observations of animals in the wild, as well as experiments with caged ones, have shown that many animals have a number of self-regulating devices, including failure to mate and infant abandonment. More recently, neo-Malthusians have looked at the human species, noted the exponential growth of the world population, and pointed out that here at least is a species that seems to be determined to crowd itself out of home and food.

Regardless of whether or not Malthus was completely realistic, the fact is that the reproductive potential is present in the form of innumerable gametes, and within the DNA of the gametes there is the potential for variation. It was the consideration of the potential for variation that made its impression on Darwin, and he noted in his autobiography: "In October 1838, that is fifteen months after I had begun my systematic enquiry, I happened to read for amusement Malthus on Population, and being well prepared to appreciate the struggle for existence which everywhere goes on, from long-continued observation in the habits of animals and plants, it at once struck me that under these circumstances favourable variations would tend to be preserved and unfavourable ones to be destroyed. The result of this would be the formation of a new species. Here, then, I had at last got a theory by which to work."

NATURAL SELECTION: THE KEY IS VARIATION AND SURVIVAL

Charles Darwin was a methodical man. Twenty-two years after the voyage of the *Beagle,* he was still working on his definitive study. Darwin, in fact, almost waited too long. In 1858, Alfred Russel Wallace also formulated a theory of evolution, based on his studies in Brazil and the East Indies (Figure 25.6). Wallace's interpretation was remarkably similar to Darwin's (both had been influenced by Malthus), and when Wallace sent the manuscript of his essay to

Darwin for his opinion, Darwin was astounded. Although Darwin's first instinct was to give Wallace full credit for the theory, the two men agreed to present their papers in the same issue of the *Journal of the Linnean Society.*

The next year, 1859, Darwin finally finished his book, *On the Origin of Species by Means of Natural Selection, or the Preservation of Favoured Races in the Struggle for Life;* the popular title is *The Origin of Species.* The book was so well thought out, so precisely documented and filled with examples, that it was almost futile to dispute the evidence. Darwin's theory contained the following points.

1. Organisms tend to produce more offspring than the environment can support.

2. Although organisms tend to reproduce at a geometric rate, the overall population does not increase at that same high rate. In fact, populations tend to remain fairly constant, probably because many individuals do not live long enough to reproduce.

3. Because so many individuals are introduced into a limited environment, they must compete for the available resources.

4. Individuals within any species vary considerably.

5. Individuals who inherit beneficial traits are better adapted to survive.

6. The organisms that survive transmit their favorable variations to their offspring.

25.7

A polar bear (*Thalarctos maritimus*), an example of an animal well suited to its environment. Its size, its feeding habits, its insulation, its metabolism, and its coloration all make polar bears able to live in frozen regions that would kill a less well-adapted animal.

Darwin suggested that new species may descend from the old ones. This theory contradicted the belief that once a species is created, it is incapable of change. The uproar was instantaneous, but there was no available evidence with which to mount a real counterattack. Darwin had shown that "descent with modification" had probably occurred, and no one would ever be able to think of living things in quite the same way again.

A crucial part of Darwin's theory stated that the best chance for survival came when offspring inherited those traits that best suited them to their environment. In a way, then, the environment was "selecting" those individuals who would survive. Darwin called the entire process *natural selection;* it is sometimes referred to as *the survival of the fittest.*

What Role Does the Environment Play?

There is no evolutionary rule that says evolution must proceed from the simple to the complex. If environmental conditions are such that a less complex mechanism is advantageous, then the simpler organisms will flourish. The main point is that organisms that are able to adapt to changing conditions will survive and produce offspring.

In Darwin's own words, "The more diversified the descendants of a species became in structure, constitution and habits, the better they would be able to seize on widely diversified places in nature." A plant that was able to get by with little water (a cactus, for example) would survive in the desert. Animals that were well insulated against the cold and had protective coloring besides (polar bears, for example) would survive in the arctic (Figure 25.7). The arctic environment did not create the polar bear. The polar bear already existed when the environment where it lived became colder and colder, or else the polar bear was forced to move northward for food when its resources dwindled or competition increased. The environment changed, and fortunately, some ancestor of the polar bear had the proper genetic makeup for survival. If the polar bear had been unable to exist in a frigid

THE COCKROACH: EVOLUTIONARY PERFECTIONIST

Nobody likes a cockroach, except perhaps the biologist, who recognizes a marvel of evolutionary perfection when he sees one. The cockroach, along with the shark, the opposum, the horseshoe crab, and several other organisms, is a truly prehistoric being, which no longer evolves in any major way because it is already so well suited to its environment.

Cockroaches flourished during the Carboniferous Period, about 300 million years ago, and fossil remains of cockroaches indicate that they have changed very little since then. How have cockroaches managed to survive for so long, and in what ways are they so perfectly evolved? Perhaps the most obvious answer to these questions is that the design of the cockroach is generalized, simple, and functional.

But besides their functional design, cockroaches are not without intelligence. In laboratory experiments roaches have demonstrated an ability to learn, and it is probably that ability that allows them to utilize fully their functional assets.

A cockroach either sees or senses a source of trouble and moves away from it in about 2/100 of a second. (Male sprinters competing in the 100-meter dash at the 1976 Olympic Games reacted to the starter's gun in about 12/100 of a second, about six times slower than the cockroach's reaction time.) The central nervous system of the cockroach, though relatively simple, is adapted for quick decision-making and movement. Two feelers, or *cerci*, at the base of its abdomen pick up deflections of air and relay the message of the interrupted environment to the nervous system. If the stimulus is sufficiently strong, the cockroach uses its six powerful legs to move quickly away from the intrusion.

Cockroaches do most of their prowling in the dark, when the danger of being discovered is minimized. When a cockroach does leave the safety of its nest, it is usually "searching" for a meal or a mate. When hunger is the stimulus, the cockroach uses its antennae to sense the presence of food, *any* food. Another reason the cockroach has survived so long is that it will eat just about anything, including soap and glue. In fact, the binding of the book you are holding has been put together with types of

climate, it probably would be extinct. Organisms that can adapt survive. Otherwise they die.

An important point should be reinforced here: *Individuals do not evolve; populations evolve.* The eminent geneticist G. Ledyard Stebbins has capsulized this issue admirably: "Individual animals, plants, or microorganisms develop, reproduce, and die; but they do not evolve. Evolution takes place in populations, and can be understood only by studying populations in their entirety, not just the individuals of which they are composed. Individuals retain the same genetic constitution throughout their lives; the genetic constitution of populations varies to a greater or lesser degree from one generation to another."

Recently, humans have created entirely new strains of some insects by using DDT as a pesticide and thereby introducing an environmental change. Theoretically, DDT is an effective killer, but invariably there were insects within a species that were naturally resistant to DDT, and these mutants survived the chemical onslaught that might have made their species extinct. These "superbugs," temporarily in the minority, multiplied rapidly because of the dwindling competition, and they passed on their nat-

glue, cardboard, and other cover materials that are intentionally made unattractive to mice and insects. Some older books make delightful homes for roaches, not to mention bookworms. Because they will eat anything, cockroaches can also live anywhere—another advantageous trait for survival.

The body of a cockroach is not only sturdy; it is flat enough to squeeze through minute openings to escape danger. This remarkable body can live without air for several hours and can withstand almost 100 times as much radiation as a human body can. It is no wonder that no insecticides are totally effective, and even the most successful repellent doesn't always work. Unfortunately, the cockroach's evolutionary process has not ceased altogether. An occasional mutant cockroach will survive even the most potent poison, and the unaffected strain will continue to live and reproduce, rapidly producing offspring that will inherit their parents' immunity.

Obviously, poison will not be the factor that finally eliminates the cockroach. If ever the cockroach is made extinct, it will probably be through permanent sterilization, preventing the birth of live and healthy offspring. Until then, the battle between humans and cockroaches continues, a draw at best.

ual resistance to their offspring. Before long, those insects that were resistant to DDT became more common than the sensitive ones. (DDT acted as the agent for selection; it did not create resistance in the insects.) Naturally, the DDT became less useful as an insecticide.

DARWIN'S THEORY UPDATED

Darwin was not aware of the importance of Gregor Mendel, even though their great theories were pre-sented only seven years apart. But remember, *most* people had never heard of Mendel until the twentieth century. As a result, Darwin was unable to explain natural selection in terms of genetics, and therefore he was incapable of answering one of the critical questions asked about his theory: What made evolution work?

If Darwin could have seen Mendelism rediscovered, if he could have witnessed the growth of cell biology, in short, if he had known about chromosomes, genes, and DNA, his theory might have started with the same ideas he actually expressed, but the information lacking in Darwin's time would have needed to be added to the original statements. Then Darwin's theory might have been more like this:

1.   Organisms tend to reproduce more offspring than the environment can support.

2.   Although organisms tend to reproduce at a geometric rate, the overall population does not increase at that same high rate. In fact, populations tend to remain fairly constant, because many individuals do not live long enough to reproduce, and because many animals have a number of self-regulating devices, including failure to mate, infant abandonment or infant destruction, and reduced fertility and litter sizes.

3.   Because so many individuals are introduced into a limited environment, they must compete for the available resources. Such a competitive struggle for existence is not limited to physical combat but includes such competitive activities as the selection of mates and obtaining nesting sites and available food.

4.   Individuals within any species vary considerably because genetic changes occasionally occur that modify the DNA sequence.

5.   Ordinarily, variations caused by gene mutations, chromosome mutations, or gene recombination are either harmful or useless. However, in the course of time, beneficial mutations may occur. Individuals who inherit beneficial mutations are better adapted to survive.

6.   In a changing environment, those organisms with favorable genetic variations survive. The surviving organisms then transmit their DNA to their offspring, and over long periods of time entirely new species may evolve. Organisms that have successful genetic variations not only live longer but produce more offspring who inherit the favorable variation.

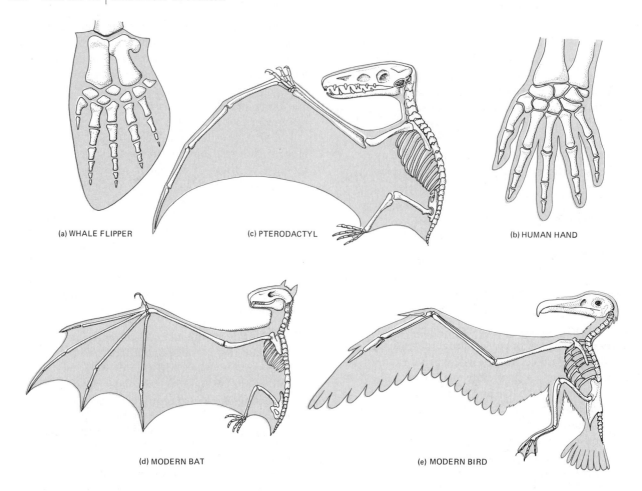

(a) WHALE FLIPPER (c) PTERODACTYL (b) HUMAN HAND

(d) MODERN BAT (e) MODERN BIRD

25.8

Homologous organs have structural and developmental similarity but functional diversity. Compare the bony structure of a human hand with that of a whale's flipper, or the wings of an extinct pterodactyl, a modern bat, and a modern bird. All are homologous. All have fundamentally the same plan of skeletal support, and all arise in similar fashion during embryonic life.

THE EVIDENCE FOR EVOLUTION

The facts that Darwin collected can be grouped into a small number of categories, each of which we will discuss briefly to introduce the kind of information that has convinced biologists of the validity of evolutionary theory. These categories are (1) comparative anatomy, especially of the vertebrate animals, (2) the embryological development of animals, (3) the variability of plants and animals under the special care of humans, (4) the distribution of organisms on the earth, and (5) the structure and distribution of fossil organisms.

Comparative Anatomy

Why, Darwin asked, is there such similarity in the structure of the forelimbs of such diverse animals as humans with hands, horses with hoofs, bats and birds with wings, and seals with flippers? He compared the bones of those organs and found many major ones in all instances. It seemed reasonable to Darwin that such similar forelimbs have similar origins, even though they may be used for different purposes now. When different species have similar body parts, those parts are called *homologous* (Figure 25.8). If organisms have parts that are used for similar purposes but are not similarly constructed, those parts are called *analogous* (Figure 25.9). Homologous parts have been derived from the same common ancestor, whereas analogous parts have not.

Considering further, Darwin asked why animals have structures they do not use. He noted the presence in humans of ear, tail, and skin muscles, of an appendix at the junction of the large and small intestines, and of a rudimentary "third eyelid," or nictitating membrane (Figure 25.10). None of these structures is used by humans, but all are functional in other vertebrate animals. A number of such structures are known, and they are called *vestigial* because they are believed to be lingering remnants of once useful parts. (In rabbits, for instance, the appendix serves as a temporary storage bin for the cellulose-rich diet of the rabbit. It thus allows extra time for the appropriate enzymes to break down the cellulose.) If evolution has occurred, vestigial organs are under-

25.9

Analogous organs have functional similarity but structural and developmental difference. (a) The proboscis of a butterfly and (b) that of an elephant look somewhat alike in spite of the difference in size, and they are used in essentially similar ways, at least in part, but in their embryological development and in their fundamental structure they are not in any way comparable.

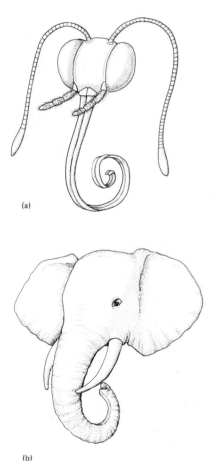

(a)

(b)

25.10

An organ that is functional in some species and vestigial in others: the nictitating membrane, or "third eyelid." Even though the membrane is consistently present in all vertebrates above fishes, it is frequently nothing more than a tiny fold of muscle. In the eyes of birds, however, it is a readily workable translucent membrane that can be flicked over the eyeball, cleaning it without requiring the bird to close its opaque "real" eyelids. In the human eye, it is barely visible and cannot be moved. The continued possession of a useless part by any animal is difficult to explain except in terms of evolution.

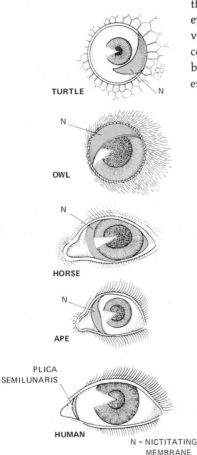

TURTLE

OWL

HORSE

APE

PLICA SEMILUNARIS

HUMAN

N = NICTITATING MEMBRANE

standable as structures that have almost but not quite disappeared. On the other hand, if evolution has not occurred, then vestigial structures seem beyond understanding.

Embryological Development of Animals

Darwin also considered the embryological development of animals (Figure 25.11). Ultimately tailless humans have a tail during the early stages of growth. Indeed, for a while a human embryo has a tail that is longer than the legs, and many a human infant is born with a tail, which is quickly and quietly snipped off by the attending doctor. All mammalian embryos have gill slits for a while, but they close eventually, leaving only slight evidence of their presence. In fishes, gill slits remain open and functional throughout life. If mammals evolved from a fishlike ancestor, then the possession of gill slits in the mammal embryo is understandable as a remnant from earlier times. Very young embryos of fishes, birds, and mammals are strikingly similar. They are in fact so similar that some biologists used to think that every individual passed through a sort of resumé of its evolutionary history—an idea expressed in the phrase "ontogeny recapitulates phylogeny." A superficial look at embryos can easily lead one to accept that idea, although detailed studies show that it is not tenable except in a vague, general sense.

Variations among Domestic Species

We do not usually see evolution happening. It is too slow to be readily seen in one or even a number of human generations, and it is consequently not obvious on casual observation. There are instances, however, that show that evolutionary change can occur within relatively short times, and these are notable in the plants and animals that have come under the influence of human cultivation. Consider, for example, the tough, weedy plant, *Brassica oleracea*, which has been subjected to artificial selection as a food plant for only a few centuries. It has given rise to a surprising diversity of garden vegetables: ordinary cabbage, red cabbage, Savoy cabbage, kale, kohlrabi, Brussels sprouts, cauliflower, and broccoli

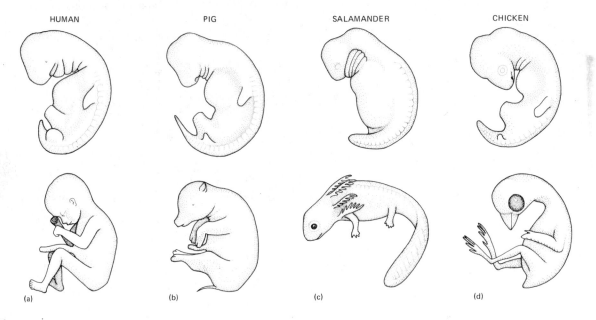

HUMAN   PIG   SALAMANDER   CHICKEN

(a)   (b)   (c)   (d)

25.11

Similarity among different species of animals during early embryo stages. Embryos of (a) a human, (b) a pig, (c) a salamander, and (d) a chicken are much alike at the stage when the head is beginning to develop, the gill arches are pronounced, the limbs have budded, and the tail is conspicuous. Only after specific peculiarities appear do the animals become readily recognizable. These similarities add evidence of the relatedness of such different vertebrates and support the theory of evolution.

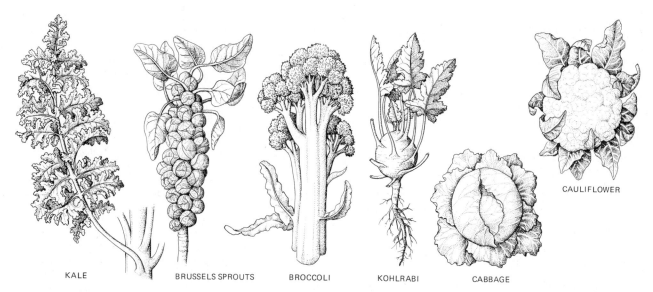

KALE BRUSSELS SPROUTS BROCCOLI KOHLRABI CABBAGE CAULIFLOWER

25.12

Varieties of a single species (*Brassica oleracea*). The original wild species, first cultivated in the eastern Mediterranean region, is something like modern collards, of which kale is a crinkly variety. Wherever man cultivated different varieties, the primitive types of cabbagelike plants were selected for different features and were adapted to different climates. Some varieties became hard-headed like modern cabbages; some grew swollen stems, as in kohlrabi; some made masses of flower buds, as in broccoli and cauliflower; and some made clusters of leaf buds, as in Brussels sprouts. The recorded appearance of new varieties adds more strength to the argument that species can change.

(Figure 25.12). These are all varieties of the same species, developed rapidly under the intense selection pressure of cultivation. Similarly, domesticated horses, sheep, dogs, cattle, and camels now exist in forms unknown in any wild state, having been artificially guided in their evolution by humans (Figure 25.13). Darwin called special attention to the many races of pigeons bred by pigeon fanciers, pointing out that evolution can indeed be observed within limited time.

The Geographic Distribution of Organisms

During the cruise on the *Beagle*, Darwin's curiosity was aroused by the distribution of species on the earth. He was particularly interested in island flora and fauna because these are usually special. The most famous of "Darwin's islands" are those in the Galapagos archipelago off the west coast of Ecuador in South America. There he studied many life forms but most particularly the tortoises and birds. He saw that the finches varied somewhat from island to island, but they were more like the finches of the South American mainland than they were like European

25.13

Variation within an animal species: dogs (*Canis familiaris*). Dogs have been domesticated and bred for special traits for such a long time that the original wild dog is not known. Hundreds of special breeds are developed and maintained all over the world, providing additional evidence of the plasticity of species and increasing the credibility of the theory of evolution.

DARWIN'S FINCHES

As part of his tour aboard the *Beagle,* Charles Darwin visited the Galapagos Islands, about 1000 kilometers (600 mi) off the west coast of Ecuador. These islands probably originated as active volcanoes more than one million years ago. It is believed that the Galapagos were never attached to the mainland, and if so, the descendants of the fauna that Darwin saw would have had to come by air or sea.

Because of the isolation of the islands, the assortment of wildlife was limited to a few kinds of birds, mammals, and reptiles, but there were 13 different species of finches. Darwin was not struck by the wide diversity of finches until he returned home to England, where he could leisurely examine his notebooks and specimens. He thought it unlikely that so many different species of the finch could have been created originally, especially since there were not very many other kinds of birds on the islands. It was more reasonable to assume that the 13 species had descended from a single species, and that they were modified through time to adapt to different environmental niches.

Darwin separated the types of finches into ground finches, tree finches, and a single species of warbler finches. The main difference in the birds was their beaks. Otherwise the birds resembled one another quite closely. Darwin saw that the beak shape was related to the bird's diet—vegetation, insects, seeds, cactus, or some combination of these. Not only were the beaks adapted to the kind of food the bird ate, but the beaks of the seed eaters varied in size, depending on the size of the preferred seeds.

The woodpecker finch has become the most famous of the group. Although its long beak resembles that of a woodpecker, the bird lacks the long tongue of a woodpecker, and therefore it cannot dig out insects from their hiding places within crevices. But the woodpecker finch carries a small twig or cactus spine in its beak, and it uses this tool to drive insects out into the open. If this bird lived on the South American mainland, it would have to compete with woodpeckers, and since it is less efficient than the woodpecker, it would probably lose out in an ongoing struggle for food. However, the woodpecker finch on the Galapagos Islands is unrivaled in its special environmental niche.

The study of the Galapagos finches proved to Darwin how species may descend from previous ones by way of geographical isolation. The relatively recent evolution of the finches gave Darwin the opportunity to see an unusually clear example of adaptation, and his experience with the finches may very well have been the most important factor in his formulation of the theory of natural selection.

finches (Figure 25.14). A reasonable explanation is that the island finches had come from the mainland, but they had been separated from the ancestral types long enough to have changed in form and habit. Other examples are available. Many caves harbor blind fishes and other blind animals, but such animals have useless eye sockets or remnants of eye stalks. The notable fact is that blind fishes in American caves are similar to the normal American fishes outside, and the blind fishes in European caves are similar to the normal European fishes outside. Once more, if we postulate an evolutionary development, the facts make sense; if we do not, they are baffling.

Fossil Organisms

Finally, the earth itself offered evidence. There were sea shells heaped in inland valleys, and layer upon layer of rocks contained fossilized remains of long-dead animals. Such fossils indicate that the world is not as everlastingly changeless as it may seem to an observer who has only one lifetime in which to study. Details of fossil study help establish some generalities, such as the fact that by and large the deeper the fossils are buried in rock, the simpler they are. In addition, many fossils are recognizably similar to animals and plants now living, but they are not identical. Fossils in the upper layers of rocks are more like

25.14

Divergence among species following geographical isolation and specialization of feeding habits. These finches from several of the Galapagos Islands are in many ways alike, but each species has its own separate home island and its own kind of chosen food. They have been separated from one another long enough to have become distinct species but not long enough to have lost their status as finches. They thus provide a living example of evolution as a working process. See also the essay at left.

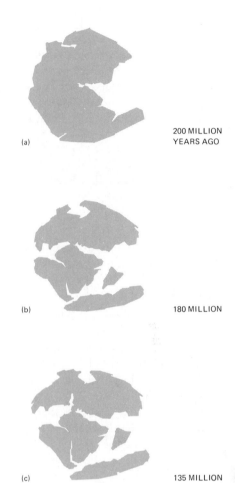

(a)   200 MILLION YEARS AGO

(b)   180 MILLION

(c)   135 MILLION

(d)   65 MILLION

(e)   PRESENT

25.15

Pangaea and the breakup of the continents. Many geologists think that some 200 million years ago, the part of the earth that was above sea level was united in one great land mass. At that time, all living things had access to each other by land, so that strict isolation was unlikely. When portions of Pangaea were separated from the main body, however, some populations were separated from the ancestral ones and thus had a chance to evolve in different ways. The series of pictures shows the progressive changes as they may have occurred at different times. (a) Pangaea, the original single great mass. (b) The separation of continents, 180 million years ago. (c) Further separation of continents, 135 million years ago. (d) By 65 million years ago, the face of the earth was approaching its present configuration. (e) The continents as they stand now. One can most easily understand the similarities and differences of species, genera, and families of organisms in different geographical regions by assuming that evolution has occurred.

25.16

A natural Linnean species, *Galeopsis tetrahit*, resynthesized by a plant breeder. Two species of hemp nettles, *Galeopsis pubescens* and *G. speciosa*, when hybridized, yielded a plant that was indistinguishable from *G. tetrahit* in all its structural details. Further, it was completely fertile with *G. tetrahit*. The chromosomes of the artificially produced *G. tetrahit* matched those of the wild species. If scientists can recreate a species in the few years that geneticists have been working, it is easy to see how new species could have arisen again and again in nature over the thousands of centuries during which organisms have had a chance to evolve.

modern living species than are fossils in deeper layers. Such facts puzzled the scientists of Darwin's time, as they had puzzled men for centuries. One fossilized log, for example, was included along with other treasures in an Etruscan tomb in Italy of about 500 B.C., and the tomb was not rediscovered until 1867. The Etruscans must have counted the log as something special, a treasure.

There are, it is true, problems with the study of fossils, or the science of *paleontology*. Sometimes sections of the earth have been flipped over like a pancake, leaving the deeper layers with their more ancient fossils uppermost, and burying the more

complex fossils deep in the earth. Then there is the perpetual difficulty of incomplete preservation. Few of the organisms that lived became fossilized, and those that were fossilized were usually imperfectly preserved, so that the real history of organic evolution is forever lost except in a fragmentary way. One may try to imagine how many of today's billions of humans will still be physically in existence as fossils 10 million years hence. Reading the fossil record is frequently compared to reading a book from which all but a few lines have been lost.

Darwin's information on paleontology, compared with the amount available today, was limited, but he knew enough to perceive the important principles. For one thing, geologists of the nineteenth century were showing that the earth was much older than anyone up to that time had suspected. If, as some historical calculations seemed to show, the earth was only some 6000 years old, and evolution was as slow as observations indicated, there simply was not time enough for plants and animals to have undergone drastic changes. By using calculations based on physical evidence, however, geologists began pushing the date of the origin of the earth further and further back. By Darwin's time, estimates had changed from thousands of years to millions, and it seemed likely that there might indeed have been time enough for evolution to work. Today's even more accurate dating techniques allow for billions instead of millions of years, so that we know there has been abundant time for evolution, slow as it is, to work its changes.

EXPERIMENTAL EVOLUTION

Darwin's hypothesis did not go unchallenged. People were reluctant enough to concede that the earth is not the center of the solar system, or that the solar system is not the center of the universe, but they did slowly come to accept those ideas. Darwinism, however, was the worst "heresy" of all. It was bad enough that the earth is not the center of the universe, but the idea that humans developed from some subhuman creature was hopelessly repugnant to many. Even today, well over a century since the publication of *The Origin of Species*, strong objections are expressed against the concept of organic evolution, in spite of the mass of evidence that biologists find convincing.

Two challenges have been made to proponents of evolution: Can you create a recognizable old species, or can you make a brand-new one? The answer to both challenges is yes, we can and have done so.

In 1930, Arne Müntzing crossed two species of weedy little mints, using pollen and ovules that were abnormal in their chromosome constituents, and the result was a plant well known as the hemp nettle, *Galeopsis tetrahit*. The new, artificially produced plant was indistinguishable from the wild type hemp nettle, it had the same chromosome configuration as the wild species, and it could fertilize or be fertilized by the wild species. In this instance, an old, successful species was made by direct human interference (Figure 25.16). Since then, many species of established plants have been similarly duplicated.

In the 1940s, G. Ledyard Stebbins produced a tetraploid grass from a diploid South African species, *Erharta erecta*. (A *tetraploid* plant has four sets of chromosomes, twice as many as a normal diploid.) The new tetraploid was a new species in that it was not interfertile with the plants from which it was derived, and it had different growth requirements. Stebbins planted the tetraploid in a number of places in California, and he reported 28 years later (in 1971) that it was surviving. In Stebbins' words, "The tetraploid did not occur anywhere until I produced and established it." We not only can observe evolution; we can make it happen.

SUMMARY

1. In the biological sense, *evolution* means *genetic change*. It is the unifying mechanism that provides for the continuation, diversification, and eventual improvement of all living things. Biologists believe that the present differences between organisms have come about by divergence from earlier living things.

2. *James Hutton's* theory of slow and steady geologic change, the *principle of uniformitarianism*, provided a basis for Charles Darwin's new theories of evolution.

3. An early scientist who attempted an explanation for the causes of evolution was *Jean Baptiste Lamarck*, who proposed the *theory of use and disuse* and the *theory of the inheritance of acquired characteristics*. Biologists have found that only the former theory is basically correct.

4. The population theories of *Thomas Malthus* provided a concept of a continuing struggle for existence that supplied a basis for Darwin's early speculations about evolution.

5. In 1859 *Charles Darwin* proposed his theory of *natural selection* in his book *The Origin of Species*. Two basic assumptions of the theory were that new species may arise from former ones, and that organisms that are best suited to adapt to environmental changes will survive.

6. The facts that Darwin collected to support his theory can be grouped into the following categories: *comparative anatomy, embryological development of animals, variability of plants and animals under the special care of humans, distribution of organisms on the earth*, and *structure and distribution of fossil organisms*.

7. Strong objections were expressed against Darwin's concept of organic evolution, and some criticism is still voiced today, in spite of the mass of evidence that biologists find convincing.

RECOMMENDED READING

The entire September 1978 issue of *Scientific American* was devoted to the subject of evolution.

Clarke, Bryan, "The Causes of Biological Diversity." *Scientific American*, August 1975. (Offprint 1326)

Darlington, C. D., "The Origins of Darwinism." *Scientific American*, May 1959.

Darwin, Charles, *The Autobiography of Charles Darwin* (edited by Nora Barlow). New York: W. W. Norton, 1958. (Paperback)

Darwin, Charles, *Facsimile of the First Edition of Charles Darwin's On the Origin of Species* (introduction by Ernst Mayr). Cambridge, Mass.: Harvard University Press, 1964.

Eisley, Loren C., "Charles Darwin." *Scientific American*, February 1956. (Offprint 108)

Kettlewell, H. B. D., "Darwin's Missing Evidence." *Scientific American*, March 1959. (Offprint 842)

Lack, David, "Darwin's Finches." *Scientific American*, April 1953. (Offprint 22)

Mayr, Ernst, *Animal Species and Evolution*. Cambridge, Mass.: Harvard University Press, 1963.

Mayr, Ernst, *Evolution and the Diversity of Life: Selected Essays*. Cambridge, Mass.: Belknap Press of Harvard University Press, 1976.

Mayr, Ernst, "Evolution." *Scientific American*, September 1978.

Moore, Ruth, *Evolution*, Revised Edition. New York: Time-Life Books, 1973. (Life Nature Library)

Moorehead, Alan, *Darwin and the Beagle*. New York: Harper & Row, 1969. (Available in paperback)

Wilson, J. Tuzo (ed.), *Continents Adrift and Continents Aground*, Readings from *Scientific American*. San Francisco: W. H. Freeman, 1976. (Paperback)

26

The Mechanisms
of Evolution

ONCE BIOLOGISTS HAD THE LOGIC OF THE IDEA of evolution brought to their attention, they began searching for the specific ways in which it could function. When *The Origin of Species* appeared in 1859, nothing was known of chromosomes or their action, or of any modern genetic principles, or of the biochemistry of DNA, RNA or proteins, or of the physiological effects of environmental forces. Only after an enormous amount of information was accumulated could an explanation of the mechanisms of evolution be developed. Much remains to be learned, of course, but the essential methods of evolution are reasonably well understood.

Darwin could see that variation within a species is common. Differences between individual people are accepted and used as everyday means of recognition. Pet dogs and cats are almost as easily seen as individuals. We do not usually pay close attention to nonhuman organisms, but it is still true that every pigeon, every starling, every plant of crab grass is unique. Although Darwin knew that variation occurs, he did not know the source of the variation as we do now.

VARIATIONS CAN BE INDUCED GENETICALLY OR ENVIRONMENTALLY

The differences between individuals are due in part to the organization of an organism's DNA (and that is the part that is important in evolution), and in part to the environmental forces that alter individual development (and that is of no direct importance in evolution). Each of these should be studied separately, but they work together simultaneously in a developing organism.

First, the special combination of nucleotide bases that makes up an individual's genetic potential is established when a sperm nucleus meets an egg nucleus. From the moment of fertilization, the biological capabilities of an individual are fixed, being determined by the molecular arrangement of the maternal and paternal DNA.

Second, the way in which that special DNA can express itself is influenced by the conditions surrounding the organism as it develops. For example, if a human zygote is genetically endowed with the capacity for growing straight black hair, then that is what the resulting person can expect, and he can pass only that capacity on to the next generation. However, variation in the appearance of the hair can be caused by alteration in the environment. Severe malnutrition or artificial bleaching can lighten the color, but the DNA in an individual's gonads remains unchanged by such environmental influences. Determining whether variations are genetically or environmentally caused is frequently difficult and sometimes impossible, especially because of the incompleteness of our knowledge. However, we do not need technical, biochemical expertise to see that a wheat plant will not produce apples, even though it is given optimal apple-growing conditions. What an organism *can* do is determined by its genes; what it *does* do is determined by its genetic ability operating in conjunction with the environmental forces that make possible the expression of that genetic ability.

Although some variation is environmentally induced, such variation cannot usually be transmitted to succeeding generations. There are some exceptions to that statement, however. For example, the one-celled green protist, Euglena, can be "bleached" by streptomycin or ultraviolet light. Not only will it remain colorless, but when it divides, the resulting daughter cells and subsequent generations of cells will be colorless, too. In this instance, it is the DNA of the chloroplast that is changed. Ordinarily, DNA is not affected by environmental forces, and it is mainly DNA that is passed to offspring.

Mutation and Recombinants as the Basis of Variation

Genetic variation is essential to evolutionary change. The next question is: What are the sources of genetic variation? The answer is: (1) *mutation*, or change in nucleotide sequences in DNA and (2) *genetic recombinations* of old nucleotide sequences. Some mutations are caused by such environmental factors as radiations or by chemicals, *mutagens*, in cells, but these must be recognized as quite different in action from environmental forces that act directly on some organ of an animal body. A severe radiation burn, for example, may cause loss of a limb, but that loss would not cause a DNA change in an animal's gametes, certainly not one that would result in limbless progeny. Rather than being caused by environmental factors, most mutations seem to be the result of mistakes in the copying of DNA sequences when replication occurs. Such mistakes are rare enough to be scarcely detectable in large, slow-growing animals, but they can be found readily in microorganisms in which billions of individuals can be screened. Rough estimates indicate that a mistake is made in bacteria about once in every million or so replications. In view of the numbers of DNA replications needed to carry a person to maturity or to form a whole population, it is clear that there are abundant opportunities for mutations to occur. These mutations have been called the raw material of evolution.

A new mutation cannot be of any evolutionary use unless it makes some change in the organism that possesses it, that is, unless it is phenotypically expressed. Because most mutations are recessive, they are not phenotypically expressed in the usual diploid organism, but they can be dispersed through a population by chromosome recombinations in sexual reproduction. Then if a new mutation does have a chance to express itself—that is, if it happens to occur in a homozygous condition—it is subject to the test of survival in its environment.

Even without new mutations, new phenotypes can be made by recombination. In breeding programs designed to produce new crop varieties, the actual number of new mutations must be relatively low, yet new varieties do appear after crosses between old varieties, and they appear quickly and in large numbers. When one mutant gene, already present in the gene pool, meets a second gene, they may both be expressed phenotypically, and when they are, you have essentially a new variety of organism.

Without knowing anything about mutations, chromosome recombinations, or other genetic events, Darwin could see that the numbers of gametes and the numbers of possible offspring were great enough

26.1
A mutation in sheep. The Ancon breed appeared spontaneously in New England. The owner, Seth Wright, bred it, but about 1870 the breed died out, only to occur a second time in Norway about 1920. This famous example illustrates how a single mutation can produce a striking result, which may or may not be advantageous. If short legs can occur suddenly in sheep, why not short tails in a Jurassic bird?

to allow for variation and consequent selective advantages. The mutations that confer some survival advantage on their possessors are the mutations that are most likely to be passed on to the next generation, even though the advantage may be slight. Change can pile on change, generation after generation, as long as each change helps an organism to survive and to beget new organisms. New mutations, plus new combinations of old and new genes, make new varieties, adding to the total variability of a species.

Great Changes or Small?

Most mutations cause slight changes, and therefore evolution has usually progressed in small moves. Occasionally, however, a single mutation can be responsible for a change of considerable magnitude There are known examples in which legs of animals are severely shortened, as in the famous Ancon sheep (Figure 26.1), or tails are eliminated, as in Manx cats. These examples confer no advantage, except to owners who like short-legged sheep or tailless cats, but it is conceivable that loss of a long, heavy tail by a flying reptile could be an enormous advantage, enabling the tailless one to outstrip his contemporaries in seeking food, escaping from hungry hunters, or overtaking a mate. A modern bird has a stumpy tail with a bunch of feathers on it. Archaeopteryx had a long, lizard-style tail with feathers down the sides (Figure 26.2). Because we have no intermediate fossils or transitional living birds, we are faced with an unanswered question: Did birds lose their tails by one striking mutation, or was there a step-by-step

reduction? Perhaps we shall never know how it happened, but we do know that both types have existed —that ancient birds had long tails and that modern birds have short ones.

ADAPTATION IS CRUCIAL IN A CHANGING ENVIRONMENT

In this unstable world, lands and seas and climates change. When an organism finds itself in a changed environment, it too must change or die. Whether its old home has altered or whether it was swept to a new one, it has only those two choices. In its old environment, the organism must have been well suited to its surroundings, or it would not have survived there; in the changed circumstances, it will probably be less well suited. The changing of an organism that brings the organism into better harmony with its environment is called *adaptation*.

Adaptation must not be thought of as a striving on the part of an organism, certainly not a conscious striving. Rather, it is a result of the combined effects of variation, necessarily already present or quickly available as mutation, and the selecting power of the environment. For example, plants in a population have differing capacities for producing cutin. Some individuals are heavily covered with a protective layer, and others are only thinly covered (Figure 26.3). If the climate becomes drier, as it did in the Sahara Desert, plants with thicker cutin will not dry as fast as those with thin cutin and may live to set a crop of seed. They have been "selected." Succeeding generations will also show variability, and steadily and inexorably those with the best protection against drying will be the only ones to live and reproduce. In this instance, only one feature, cuticular covering, has been pointed out, but in reality a plant would have to possess a whole mosaic of features that work together. It is the species, not the individual, that adapts.

Given the chancy nature of adaptation, one may wonder how any species ever succeeds in changing in the proper direction. Might not our cutinized plant have produced thinner coats? Indeed it might, but if so, it would have become extinct. Judging from available fossils, most of the species that have inhabited the earth are dead. Of the more than two million known species now living, plus many that are still unknown, few if any have been in existence for even 15 million years. In terms of the history of life, adaptation works part of the time, or we would not be here, but it has failed more often than it has succeeded. It is sufficient that some few species have changed fast enough and in the proper directions to keep life going.

Adaptation is a never-ending process. As organisms change to become suited to a changed environment, they then cause still further environmental changes, at which point they must change again. That changes the environment again, and the next round of changes has begun. Thus the story continues. Like a pair of antagonistic nations in an arms race, biological adaptation and environmental change keep advancing perpetually. Two examples illustrate the principle.

If, as is generally thought, the earth's original atmosphere contained little or no free oxygen, early organisms, like present-day obligate anaerobic bacteria, found oxygen toxic. But as photosynthetic organisms increased, they increased the proportion of oxygen in the air, and the nonphotosynthetic organism had to evolve a respiratory chemistry that would allow them to live in an oxygen-rich atmosphere. Meanwhile, green plants took advantage of available carbon dioxide and water, and they poured ever more oxygen into the air. Even now, too high a concentration of oxygen can be dangerous. We are adapted to a concentration of about 20 percent, and premature babies kept in high concentrations of oxygen to help them breathe may be blinded because of a disturbed blood supply to their eyes.

A second example concerns the interaction between plants and a dry, rocky terrain. As some plants adapted to a place where soil and water were scarce,

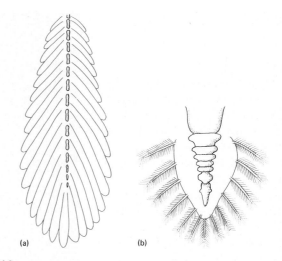

26.2
Tails of primitive and modern birds. (a) Archaeopteryx had a long tail with a number of separate slender bones and a row of feathers down each side. (b) Modern birds have short tails, with a few bones fused into a solid stub, and with the feathers produced so close together that they make a tuft.

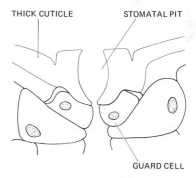

26.3
Adaptation for plant survival in a dry place. A section through the epidermis of an orchid leaf shows two protective features: a heavy coat of waterproof wax (cutin) and the stomatal opening at the bottom of a pit, where air currents and consequently water loss are reduced. Plants that live in arid regions generally have a thick coating of cutin and sunken stomata. × 1000.

26.4
Light and dark (melanic) forms of the British peppered moth, *Biston betularia*. The dark form is practically invisible against a dark background but conspicuous against a light background. Birds feed more readily on the moths they can see easily. Thus the conspicuous form becomes reduced in numbers, and the inconspicuous form increases. As a result, the dark ones survive in dirty regions and the light ones in cleaner regions.

26.5
Instant camouflage. Flounders and other flat, bottom-dwelling fish can change their pigment patterns. Placed against a checkerboard background, this sole is all but invisible, but it becomes apparent when placed against a plain, black background. If left against the black background, it can expand its dark pigment cells and within a few minutes make its entire skin practically black, reducing its visibility.

they acted on the soil to make it deeper and better able to hold what water was available. In so doing, they changed the environment in such a way that they had either to adapt to less dry conditions or to give way to other, better-adapted species. But stability is never achieved. There is instead a kind of seesawing, with continuing precarious balances enabling life to go on.

Any organism can be thought of as a collection of adaptations, but some special ones are so dramatic or so instructive that they are considered classic examples. As such they are known generally to biologists. One striking instance is the development of what is called *industrial melanism*, a darkening of color associated with the black deposits of soot on the landscape around coal-burning factories. Before about 1850, the peppered moth of England, *Biston betularia*, was light-colored, speckled with black. However, one practically black one was collected in 1848 (Figure 26.4). Over the next 50 years, the proportion of the dark form, called *melanic*, increased until melanics came to make up almost the entire population in those regions where heavy smoke darkened the tree trunks and killed the pale lichens that grew on them. In the cleaner areas of the south, the light forms continued to predominate. By releasing moths of both forms in industrial and nonindustrial counties, watching birds feeding, and counting how many of each form could be recaptured, H. B. D. Kettlewell of Oxford University showed that birds could easily find light-colored moths against a dark background. Conversely, dark-colored moths against dark backgrounds were difficult to find. He concluded that there was a strong se-

THE EVOLUTION
OF ANIMAL BEHAVIOR

The scientific study of behavioral evolution has been developed more slowly than that of structural, cellular, or biochemical evolution. The reasons are several. First, behavior is harder to reduce to quantitative data, and it is generally thought to be harder to deal with experimentally. Further, recognizable evolutionary trends are not usually apparent. Also, firm proof is lacking that there is a gene for any particular aspect of complex social behavior. Finally, the species with the most complex behavior, and the species most interesting to us, is the one on which behavioral experimentation is most severely restricted—the human species. The topic of behavioral evolution has been befogged by many books and articles based on speculation or intuitive thinking, or written by people who had some private cause to defend, but without convincing facts.

Nevertheless, in the last 25 years, studies on behavior with an evolutionary emphasis have become increasingly common, and by now a fair body of knowledge has accumulated. Some of the most informative studies have been made on insect societies, especially on ants, bees, wasps, and termites. Different species of wasps have different social patterns. Those presumed to be the most primitive are the solitary wasps, which raise their young without any formal social organization. Other species have small, rather loosely organized nests in which a number of individuals live together but show little or no specialization of duties. More advanced species have colonies in which certain individuals have definite roles in the life of the colony. The most advanced species are the vespine

wasps, with their rigid system of castes, in which queens, workers, and other types live rigorously ordered lives. In wasps, at least, there seems to be an evolutionary sequence of behavioral steps.

In other animal groups, by contrast, no patterns are discernible. Among the carnivores, for example, social habits appear to be random. African hunting dogs are highly organized, but raccoons and wildcats are solitary. Some carnivores have developed the habit of defending a home territory, and some have not. In some (hyenas) the females are dominant, and in others (wolves) it is the males. Generalizations that might be useful in deciding evolutionary trends cannot be made from such diverse types.

In 1975, Edward O. Wilson of Harvard University published *Sociobiology: The New Synthesis*, a book in which the author attempted to integrate ideas on genetics, evolution, and animal behavior, including human social behavior. The work was praised by some biologists as an important contribution to evolutionary thought, and it was damned by others as having reactionary political meaning and making unwarranted conjectures. The tone of the arguments was remindful of that of the debates following the original pronouncement of Darwin's evolutionary theory—colored by opinion, emotionally loaded, and frequently sounding as if the critics had not actually read the text they were criticizing. In time, new information and new insights will resolve the disagreements, and Wilson will be either vindicated or forgotten.

lective pressure against light-colored forms in industrial areas and against melanics in the cleaner regions. With stricter air-pollution standards, the selective pressure would once again favor the lighter moths.

An adaptation that helps camouflage an animal by providing it with a color that makes it inconspicuous is called *cryptic coloration*. The world abounds in cryptically colored creatures. Animals from fishes to antelopes are paler on the belly side, which is in shadow, than on the back, which is lighted. Striped

tigers are hard to see among the striped shadows of the grass in which they customarily hunt. Some animals, notably fishes and reptiles, can change color in a few minutes to match backgrounds. Some chameleons are so adept at switching from green to brown and back again that they are regarded by humans as the standard of inconstancy. Flounders can become practically invisible against a variety of backgrounds, even making a reasonably good attempt at duplicating a black-and-white checkerboard (Figure 26.5).

26.6

Mimicry in insects. (a) A caterpillar not only has the form and color of a dead twig but holds itself in a proper position and remains motionless.

(b) A katydid has wings so leaflike that they even have veins and torn edges. Until it moves, such an insect is not likely to be noticed by a hunter.

Another striking adaptation is *mimicry*, in which one species develops a resemblance to something different from its near relatives. The object copied may even be in a different kingdom. Among insects, famous examples are the stick insects that look like dead twigs and the green katydid, which, when resting, looks remarkably like a leaf, complete with midrib and leafstalk (Figure 26.6). In plants, the orchid, Ophrys, looks and especially smells so much like a female bee that pollinating males try to mate with the flower. More commonly, a harmless insect may resemble a poisonous or at least distasteful species. Insect-eating birds, having once tasted a repulsive type, will not attack a harmless mimic (Figure 26.7).

In contrast to cryptic coloration, *warning coloration* is conspicuous. A strikingly colored insect may be so noxious that preying birds learn to let it alone. It seems that each bird must learn for itself which bright-hued insects are not good, and therefore some of the insects are sacrificed. However, if a young bird kills only one or a few of the insects, it will from then on tend to avoid them. In the long run, the species profits by having a larger proportion of its members escape. That is worth the cost of a few losses, and selection pressure continues in favor of bright colors.

Adaptive coloration in birds is a compromise between being easy to find by a predator (that is bad) and easy to recognize by a possible mate (that is good). In birds that court, the female normally accepts only males of her own species. Since males are undiscriminating, and since interspecific hybrids are usually biological failures, it is important to a female to make certain that she chooses correctly, and strong coloration simplifies the choice. Female birds tend to be rather drab. Selection pressure seems to favor inconspicuous colors in females, thereby improving their chances of escaping predation, but conspicuous colors in males. The males thus have to take their chances with predators if they are to be successful breeders. The system functions; male birds are indeed generally the showy sex (Figure 26.8).

"SPECIES" IS DIFFICULT TO DEFINE

Until Darwinism disturbed their serenity, biologists felt comfortable about the meaning of a species. It was a group of individuals that resembled one another so closely that some careful student of the group could confidently assign them a name. There was no concern for mutability, little for variability,

26.7

A blue jay will eat a monarch butterfly if it has never tried one before, but the butterfly contains an emetic that causes the jay to vomit. Having learned that monarchs are nasty, the

jay will not attack a viceroy butterfly if he is offered one. The viceroy is harmless, but it looks so much like a monarch that the similarity is enough

to repel the jay. Both the monarch and the viceroy are protected, one by its own toxin and the other by its similarity to the noxious species.

and practically none for hybridization. The idea of evolution changed all that. Even though the very book that brought about the change had the word "species" in its title, exactly what a species is became a slippery concept, changing and eluding those who sought a simple firm definition. If a stand of flowers with red petals lives on one side of a mountain, and an almost identical stand, except that the flowers have pink petals, lives on the other side, are they the same species? One first tries to determine whether the two stands are interfertile. If they are, they may well be the same "species," that is, very closely related. Even if they are not, there is a new complication. Many apparently identical populations are loaded with genes that prevent fertilization, genes known as *incompatibility factors*.* Thus *taxonomy*, or the study of relationships as expressed by classification, changed from an observational to an experimental discipline. No longer satisfied with merely looking at their specimens, taxonomists tested them for interfertility and transferred them to different habitats in an

26.8

Adaptive coloration in the wood duck, *Aix sponsa*. The male is splashed with color: green, white, black, and brown. He is easy for a female to recognize. The female's brown and white flecked plumage is useful in making her hard to see when she is nesting and raising vulnerable ducklings.

---

* Flowers in the Phlox family have a series of incompatibility factors, which make self-pollination and even many cross-pollinations impossible.

26.9

The effects of environment on development. *Potentilla glandulosa* has several subspecies in California, including *Potentilla glandulosa glandulosa*, and *Potentilla glandulosa nevadensis*. When grown at different elevations, each subspecies changes. *P. g. glandulosa*, native at low elevations, shows reduced growth at 1400 meters (4600 ft) and fails at 3000 meters (10,000 ft). *P. g. nevadensis*, native at timberline (3000 meters), grows well at 1400 meters but does no better at low elevations than it does in the high mountains. Such responses to environmental change cannot be predicted but must be determined by trial.

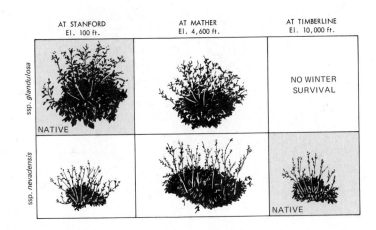

effort to discover which characters could be affected by environment and how much they could be changed (Figure 26.9).

The species is still the basic grouping of organisms for students of classification and of evolution, but there is no definition that meets the requirements of all the investigators. Many zoologists, who deal with sexually reproducing animals, define a species as a group of organisms capable of producing fertile offspring. That satisfactory working definition for most purposes is without meaning, however, when it is applied to an essentially asexual population of algae or fungi. One extreme group of taxonomists denies that there is any such thing as a species, claiming that the world is populated only by individuals, some of which more or less resemble others, and that a species exists only in the imaginations of classifiers. Meanwhile, for practical purposes, we think and talk and write about species, hoping to learn as much as we can about the earth's living things. The old aphorism concerning life itself comes back: It is easier to study something than to define it.

DEMES ARE GROUPS OF INDIVIDUALS ISOLATED FROM THEIR SPECIES

In nature, free passage of genetic material within a species may be restricted even when there is the biological possibility of such passage. The commonest restriction is distance, although in some animal groups, social practices may be effective as restrictions. Many animals range over a small territory and limit their mating to members of their immediate group. A group of individuals, partially or entirely isolated from the other members of the species, is a *deme*. City rats, for example, tend to form colonies, sometimes confined within a single vacant lot or even in a well-defined portion of a garbage dump. Mating, feeding, nesting, and raising young are carried on by the group as a tight unit. Members do not leave unless they are driven out by hunger or crowding, and strangers are not welcome. Several demes of mice may inhabit one barn, the deme in the cow stalls having little or no contact with the deme in the hay loft. On occasional chance meetings on neutral ground, the mice of one deme may ignore those of the other.

Still, demes have considerable fluidity. The areas they occupy are usually small and may be subject to change. When a vacant lot is cleared or built upon, the local rat population is scattered. If they are aggressive enough, the rats may try to work their way into other demes, carrying their genes with them. In nature, environmental instability is common. A deme of fishes in one pond may lose a few individuals to an adjacent pond during a flash flood. Two demes of strictly forest-inhabiting asters in separate patches of woods may be separated by fields in which they cannot grow. But occasional seeds may blow across the fields that form barriers, or the fields may be abandoned to grow up into brush and then into forest, at which time the demes can intermingle. Therefore they

HOW ANIMALS AVOID PREDATORS

Animals have ways to keep from being eaten. And if an individual animal stays alive, so may its species. Some animals, like leopards, kill before *they* are killed. Others, like gazelles, run and dodge to elude their enemies. And others, like elephants or rhinoceroses, are just too big or too formidable for most other predators to bring down. But in small and basically defenseless animals, more elaborate strategies have evolved. Probably the most common defense is to remain motionless. Many insects, for example, try to avoid being swallowed by looking as inconspicuous as possible. Some insects blend in so well with their surroundings that they are virtually invisible. But if they are discovered by a predator anyway, some insects may use a startling bluff, looking like a super-predator that scares away the intruder. In many cases, there is yet a third phase to the defense of the prey, and warning coloration or scare tactics may be accompanied by a noxious taste or a chemical repellent. If a noxious animal is killed anyway, it still benefits the species because the predator usually does not attack similar prey for a long time, if at all. Of course, no animal consciously knows what will deter a predator. Rather, an accidental change in appearance or behavior—a mutation— may prove to be favorable in the process of natural selection. The predator is the selector. If it is put off by the defense of its prey, then that prey will be "selected" to survive. If not, the species may soon become extinct.

Although this Australian leaf-tailed gecko is about 20 centimeters (8 in.) long, it can still remain safely camouflaged all day by resting motionless on a background that resembles its own coloring. In addition, the leaf-shaped tail helps to distract a predator away from the more vulnerable body. In the cover of night the lizard opens its eyes and creeps away in search of its own prey.

These insects survive by pretending to be something they are not. The specific device may help the insect blend in with its natural surroundings, or it may serve to bluff a predator into hunting for less "ferocious" prey, but one goal remains constant: Stay alive long enough to reproduce.

A larva of a swallowtail butterfly lies bent and motionless on a leaf, mimicking a bird dropping. This disguise may seem far-fetched, but apparently it works. Some small tropical frogs also mimic bird droppings. They tuck their legs under their bodies and remain motionless on leaves all day. At night they are free to move about.

The caterpillar of the Leucorhampa hawkmoth is sometimes called a snake caterpillar, with good reason. When disturbed, it moves away from its usual flat position on a twig and changes the shape of its rear end to look like a snake's head. The "eyes," hidden before, now appear menacingly. It is typical for an animal to present "eye spots" on parts of its body other than its head. A sphinx moth, for instance, has hidden eye spots on the expendable edges of its hind wings, and a hornworm tucks its head safely under its body when it displays its phony head and eyes further down on its body, where damage may not be serious if a predator strikes.

Looking remarkably like a battered leaf, this grasshopper hangs upside down from a twig, with its pointed head and thin antennae resembling a stem. Once the grasshopper moves, the adaptive concealment loses its effectiveness, but even if the grasshopper is threatened on the ground, it may be able to fool a predator into mistaking it for a fallen leaf as long as it remains absolutely still.

Insects are especially effective at using the evolutionary benefits of coloration because of their rapid reproductive cycles, but other animals, such as some frogs, also use color to warn or startle predators.

The "painted frog" of Panama uses its unusually dramatic colors to remind predators that its multicolored skin is highly poisonous.

Some animals have defense postures designed to avoid conflict completely, and they resort to chemical warfare only when other techniques do not work. This South American frog (*Physalaemus natteri*) tries to evade a predator by flashing two "eye spots" on its rear end. Perhaps the predator will think these monstrous "eyes" belong to one of its own predators, and it will either go away or remain startled long enough for the frog to escape.

These dashing little frogs live in Central and South America. They are popularly known as "arrow poison frogs" because glands in their skin secrete venom used by local Indians on arrowheads to paralyze prey.

Some animals can avoid conflict altogether if they detect a predator soon enough. A deer that spots a predator raises its tail to expose a white underside. Presumably this signal warns other deer that a predator is near, but it may also notify the predator that the deer has seen it early enough to escape. The signal effectively says: "I have seen you, and I can get away if you try to chase me." In most cases the predator does not bother to attack. Generally an animal will not expend energy in a fruitless pursuit.

do not diverge enough genetically to suffer reproductive isolation, and they remain within the species.

GENE POOLS

The *gene pool* is the total genetic material of all individuals in an interbreeding population (species). In a large breeding group, there may be thousands of rarely expressed recessive characteristics. Most of them are probably detrimental, some apparently neutral, and some beneficial. There is always the possibility that a new combination of already existing features may confer an advantage on the species. Or the advantage may go to humans rather than to the species itself if the species is a crop plant or an animal useful to us.

In developing new varieties of plants, for example, breeders take advantage of the genes that exist and try to bring them into combinations that give desired qualities (Figure 26.10). One variety of strawberries, for example, may have fine flavor but is susceptible to disease; a second may be disease resistant with little else to recommend it; a third has firm lasting qualities so that the fruit can be shipped. So long as all the necessary genes are present in the general pool, they can be brought together. Of course, some of the trials will produce berries that spoil easily, taste bad, and are prone to easy infection. But after enough crosses are made, the breeder should obtain the best combination of taste, durability, and resistance. Without a diversified gene pool to draw on, such results would be difficult if not impossible. It is therefore wise to maintain stocks of seeds and domestic animals, even though they may be unprepossessing, because no one knows what recessive genes are being perpetuated in the gene pools. Nor can anyone foresee what genetic reserves will be needed in the future. Who would have guessed a hundred years ago that durability during transport would be a necessary quality in a commercial strawberry? Is there in the gene pool of beans a gene that will confer drought resistance, so that submarginal land could be used to help feed hungry people?

THE HARDY-WEINBERG PRINCIPLE IS EFFECTIVE ONLY IF GENETIC STABILITY IS ASSUMED

If one is concerned with the genetics of individuals or at most of small breeding groups in controlled matings, the genotype of an individual can be found, provided that the phenotypes of ancestors or offspring

(a)

(b)

26.10
The results of selective breeding in plants. (a) Native wild strawberries are sweet and flavorful, but they spoil quickly after being picked and are discouragingly small. (b) Some varieties of cultivated strawberries have been bred for special qualities until they are scarcely recognizable as the original wild species.

are known. Dominance, recessiveness, and phenotypic ratios can be observed. Then, by the application of Mendelian principles, genotypes can be determined. With wild populations, however, and with human beings, controlled matings are seldom possible, and sufficiently detailed pedigrees are unobtainable. Consequently, when studying whole populations, we must use different methods.

In 1908, the English mathematcian G. H. Hardy and the German physician Wilhelm Weinberg independently developed a formula for analyzing the frequency of alleles in large populations. By using the

26.11
An albino deer. Most, perhaps all, animals occasionally produce albino individuals. Albino rats, mice, and rabbits are well known, but nonpigmented forms occur in robins, blackbirds, squirrels, fish, snakes, and even in humans and plants.

formula, now called the *Hardy-Weinberg principle*, we can calculate the approximate frequencies of different alleles in a population after measuring the frequency of phenotypes in a representative sample of the whole population.

Before we discuss the Hardy-Weinberg principle any further, we must make clear that it is effective only when the following strict conditions exist:

1. Natural selection does not occur.
2. The population does not gain or lose individuals through immigration or emigration.
3. No mutations occur.
4. Mating is random.
5. The population is large enough to be unaffected by random changes in the frequencies of genes.

Under actual conditions, such genetic stability is probably nonexistent, so why should we be concerned with the Hardy-Weinberg principle at all? As we shall see, the principle is useful as a practical approach. Consequently, it is employed by such people as medical advisors, who try to counsel prospective parents about the probability of certain gene frequencies in their future offspring.

The Hardy-Weinberg formula is usually expressed as

$$p^2 + 2pq + q^2 = 1,$$

where p is the frequency of a specific dominant allele, and q is the frequency of the corresponding recessive allele. In words, this formula states that "the frequency of the homozygous dominant genotype (p^2) plus the frequency of the heterozygous genotye ($2pq$) plus the frequency of the homozygous recessive genotype (q^2) equals all the genes in the population (1)."

If we can find the frequency of the expressed recessives, we can find the frequency of the recessive gene and of the dominant gene, and the number of heterozygous individuals in the populations. The heterozygotes are those in which the recessive gene is not expressed, but they are still "carriers" of the gene. A count of the number of people born with a cleft palate, a genetic recessive character, shows that about one out of every 2500 people shows the defect. By the Hardy-Weinberg formula we can calculate that one person out of 25 is heterozygous, or a carrier, for cleft palate.

Gene frequencies can be determined for such features as eye color, ear shape, blood groups, physical deformations, and predispositions to disease. Anthropologists can use data from such calculations in studying differences between human groups. For example, the hereditary recessive disease cystic fibrosis occurs among caucasians in the United States in 26 out of every 100,000 individuals. That means that one out of every 31 caucasians in the United States is a carrier. In Japan, by way of contrast, there is only one case of cystic fibrosis for every 100,000 Japanese. Using the Hardy-Weinberg formula, we can calcuate that one out of every 158 Japanese is a carrier.

Another recessive character is albinism, which occurs about once in every 20,000 births (Figure 26.11). The frequency of the albinism gene is the square root of 1/20,000 births. The Hardy-Weinberg formula shows that although the number of observed albinos is low, 1.4 percent of the population are carriers, or about one in every 71 individuals. The high number of recessive genes in populations is frequently surprising to people who are not familiar with the statistical laws of chance.

GENETIC DRIFT IS CAUSED BY THE ISOLATION OF SMALL GROUPS

Most breeding groups include large numbers, sometimes many thousands, of individuals. In a large population, even a rare mutation has a chance of being expressed occasionally. Not all breeding populations, however, are large. Small groups may be isolated by floods, landslides, or emigration, and a colony of perhaps as few as a hundred individuals may be restricted to interbreeding (Figure 26.12). With such a low number, a rare gene may exist in only one organism, and that organism may die before passing the gene on. For that population, then, the gene is permanently lost. In contrast, if there happen by chance to be two individuals with the gene, and it is expressed as a homozygous recessive, the once-rare gene may become more common in the isolated community than it was in the larger population from which it came. Genetic change resulting from the isolation of small groups is *genetic drift*. (See the essay on p. 534.)

SPECIATION IS THE PRODUCTION OF NEW SPECIES FROM OLD ONES

The development of new species, or *speciation*, can occur suddenly, or it can be a slow process extending over many generations. In either method, the result is a new population of organisms that cannot produce fertile offspring when mating with the descendants of the old ancestral line. If a population is maintaining itself and continuing to build up a load of new mutations, it may be expected that the species will become perhaps more variable, but that as long as breeding continues randomly and the environment remains stable, there will be no evolution in any particular direction. In biological history, however, populations have not remained static. We shall now see some of the ways in which new species are derived from old ones.

Polyploidy

An abrupt appearance of a new species can be brought about by a change in chromosome number. Animals have rarely used chromosome changes in the formation of new species, but plants have rather commonly done so. One way for chromosome alteration to bring about speciation is by the phenomenon of *polyploidy*, the replication of entire chromosome sets. Although plants usually form haploid gametes, in which there is one set of chromosomes, plants will occasionally produce an unreduced gamete with two sets, that is, a diploid gamete. Such gametes are rare, and the chance that one diploid gamete will fertilize a second diploid gamete is even rarer. With the huge number of gametes that are actually produced, how-

26.12

Genetic drift. (a) In a small woodlot, a population of mice consists of mostly gray individuals, but there are a few black ones. (b) A new highway through the woodlot effectively cuts the mouse population into two sepa-

rate groups. One group still has some black mice, but the other group happens to have only one individual carrying any black genes. (c) Before it has a chance to reproduce, the lone black mouse in the isolated group is

killed, leaving the little population all gray. When a population is small, a relatively minor chance event can alter the genetic makeup of the population.

(a)

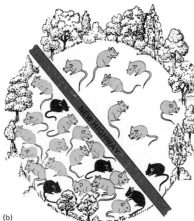
(b)

(c)

GENETIC DRIFT
IN A HUMAN POPULATION

Direct experimentation with human populations is impossible, but occasionally a group of people will voluntarily perform an act that turns out to be in effect a controlled experiment. One such "experiment" occurred when about 50 families of a religious sect migrated from Germany to Pennsylvania early in the eighteenth century. The German Baptist Brethren, usually known as Dunkers, settled in Franklin County, and some of them have remained there, although there has been considerable emigration to other parts of the United States. There has also been some immigration, because a small number of converts have come in. The feature of the Dunker practice that interests geneticists is their habit of marrying within the group, so that they are almost entirely separated genetically from other Americans, and of course, they have been long cut off from the original German gene pool. Another important factor was the size of the group, which numbered about 300 (at the time they were studied), with about 90 parents. The number is ideal for a study of genetic drift. With larger numbers, drift is not detectable; with smaller numbers, percentages of gene frequency are not reliable. The Dunker group fills the requirements admirably.

A team from Johns Hopkins University investigated several traits, chosen because they are apparently neutral as far as survival value is concerned. They would therefore not be subject to selection pressure, and differences of gene frequency between the Dunkers and either the German or American population would be random, as would be expected if genetic drift was operating. The traits were

blood groups (ABO, MN, and Rh), right- and left-handedness, hair on the middle joints of fingers, free or attached ear lobes, and the flexibility of the thumb joint ("hitchhiker's thumb"). As far as left-handedness and Rh blood groups are concerned, the Dunkers resembled the American population. In other ways, however, there were significant differences. The Dunkers had fewer individuals with hairy fingers, attached ear lobes, and hitchhiker's thumbs, and their blood types diverged as shown in the table. The figures represent percentages.

	U.S. POPULATION	GERMAN POPULATION	PENNSYLVANIA DUNKERS
Type A	40	45	60
Type AB plus B	15	15	5
Type M	30	30	44.5
Type MN	50	50	42
Type N	20	20	13.5

In about six or seven generations, the Type A and Type M individuals increased in frequency while Type N individuals were decreasing. H. Bentley Glass, who reported these results, said they show plainly that genetic drift can indeed occur when a small human population is reproductively separated from the general population. He suggested further that racial features may have arisen similarly, when human groups were small and isolated, and minor, nonselective traits became common in limited tribes, only to spread as populations increased in size.

---

ever, mating between pairs of diploids does happen. The resulting zygote, which has four chromosome sets, is a *tetraploid.* Tetraploid plants are frequently larger and more vigorous than diploids, and they are not interfertile with diploids. A new tetraploid is functionally a new species. Tetraploids can be artificially produced by treating plants with a toxic alkaloid, colchicine, derived from the autumn crocus, *Col-*

chicum autumnale (Figure 26.13). Such plants may be marketed as "tetra snaps" (tetraploid snapdragons) or "tetra tomatoes."

When a new species arises as the result of tetraploidy, or polyploidy of any kind, no new genetic material has been added. There are no mutations, no new nucleotide sequences; there is only a new combination of pre-existing genes. It is like making a new

26.13
The effect of doubling the chromosome number. The flowers on the left are standard diploid snapdragons (*Antirrhinum majus*). Those on the right are tetraploid as a result of crossing colchicine-induced diploid egg cells and diploid pollen. The tetraploid form not only is visibly larger but has larger cells.

building from an old one, merely rearranging the components.

Adaptive Radiation

The common way for speciation to occur is by slow steps, with geographic isolation providing the necessary conditions. An occasional bird or insect or seed may be blown or washed from its place of origin to new shores, perhaps far away, where it is isolated from the rest of the population, and where new environmental forces act on it. If a portion of a population is cut off from the main body of that population, it will most likely carry with it only a fraction of the entire genetic capabilities of that population (the *gene pool*). A small breeding population, sometimes consisting of only one pregnant female or a single seed, does not have access to the genetic potential (sources of genetic variability) that was available previously. Being thus separated, with a slightly different gene pool to draw on, and subject to somewhat different selection pressures, the isolated population can be on the way to evolving into a new species.

The results of such isolation are observed most clearly in islands far from a mainland, and indeed it was the peculiarity of island plants and animals that so stirred Darwin's curiosity. But organisms need not leave home to become isolated. Geological events may be responsible. Mountains are pushed up, rivers change courses, seas rise and fall, ice sheets advance and melt, hills erode, mountainsides slump in avalanches—all these change the continuity of land areas and of lakes and seas. Such changes leave some populations in places where they are cut off from larger populations, or conversely, other populations may be thrown in with populations from which they were previously separated. Events like these have been happening for billions of years. Countless mutations have occurred, and enormous numbers of plants and animals have been subjected to selection forces.

Of course, during such early stages of speciation, the possibility of successful interbreeding between individuals of divergent populations still exists, and while diverging, they do not yet constitute different species. There is not, in fact, any point in the history of an emerging species at which one could say: This generation is still the same species as the old, but the next generation is the new species.

Eventually, however, variation and selection contribute to the ever-widening separation of the two disparate populations until they have reached a state of reproductive incompatibility. The slow process of speciation by isolation is known as *adaptive radiation*, because organisms *radiate* out from one place and *adapt* to new environments. Adaptive radiation need not be exclusively geographical; it may be only the occupation of new niches.

Reproductive Isolation

The immediate causes for loss of interfertility between emerging species are many and various. Some simply prevent mating. An emerging species may have taken up a new habitat, say, a dry hillside, where it does not meet the other "species," which prefers moist bottom land. Or it may mature its gametes at a later season, so that fertilization is thrown out of timing. Since mating behavior in animals is frequently quite specific, the diverging populations may give different courtship signals and may not recog-

VIOLA CUCULLATA

VIOLA SAGITTATA

VIOLA LANGSDORFII

26.14

Sympatric and allopatric species. *Viola cucullata* and *V. sagittata* are sympatric species, both inhabiting the eastern United States. *V. langsdorfii*, which grows in Alaska, is by definition allopatric to both *V. cucullata* and *V. sagittata*. In general, sympatric species are not interfertile, but violets are notorious hybridizers, and hybrids between *V. cucullata* and *V. sagittata* are not rare.

nize each other as possible mates. Or structural changes in animal bodies may make copulation physically impossible. In any case, gametes may be incapable of fusion, or the female reproductive tract may not allow survival of sperm cells. Even successful fertilization of an egg does not ensure successful reproduction, because if the genes of the gametes are not functionally compatible, a number of disasters can follow. The embryo may abort, or the young animal may be (and frequently is) ill suited to the environment. Finally, even if the hybrid survives to maturity, it may not produce viable gametes. The mule is the hybrid offspring of mating a female horse with a male donkey. The mule is infertile, however, and it cannot reproduce itself. Because the mule is infertile, it does not constitute a species.

Sympatric and Allopatric Species

Although any one of the methods discussed in the paragraph above is enough to prevent the production of fertile offspring, usually several isolating mechanisms act at once. In nature, species that appear very similar in shape, behavior, color, etc., and occupy the same geographical region commonly have a number of differences that prevent interbreeding. Some species of violets, for example, live together in the north-

eastern United States (Figure 26.14). Such species are called *sympatric* (Latin, "together in the country"). Other species that are similar but that inhabit the prairies or the west coast are isolated both geographically and reproductively and are therefore called *allopatric* ("another country"). Even a species that appears uniform over a wide geographical range may have gradational changes in its gene pool. Consequently, in spite of the possibility for continuous gene flow through the whole population, the individuals in one locality can be so distant from individuals at the other end of the range that they are effectively isolated genetically. In such an instance, should we call the whole population a single species, or if not, where should we draw lines? One possible answer is that humans should not attempt to draw lines when nature has not done so.

CONVERGENT EVOLUTION: CHANGES TOWARD SIMILARITY

Much evolutionary discussion centers on change leading to differences, and indeed differences make the biological world interesting. Not rarely, however, changes can occur in such a way that organisms that are not closely related come to have certain similar characteristics. The process of changing toward similarity instead of difference is *convergent evolution*. Birds and bats have wings that are at least superficially similar, but birds have evolved from reptiles, and bats from a flightless mammalian ancestor. The eyes of a squid are remarkably like the eyes of vertebrates, although squids, which are molluscs, are enormously separated in evolutionary time from vertebrates. Probably the most striking example of convergent evolution is that of the Old World spurges and the New World cacti. Both, at least in some species, live in dry habitats, where a thick stem, reduced leaves, and a thick waterproof covering of wax are of great survival value. Except when they are flowering, some species of spurges are so cactuslike that one can scarcely distinguish between the two (Figure 26.15). They are not even in the same family. Dozens of such examples are known, all showing the shaping

force of the environment in directing the adaptive changes of which organisms are capable.

COEVOLUTION

The phrase "survival of the fittest" may indicate that in competition with members of its own species the powerful and pitiless are winners. That may be so sometimes. Surely a weak, timid lion is in trouble, and a diffident rat is a "social outcast." Nevertheless, genetic variations are usually tested against the total environment, which includes both living and nonliving factors. Any number of behavioral or structural features may confer an advantage. A rabbit that escapes the notice of a hunter by freezing into immobility survives as a result of a behavioral pattern. A green lizard on a green leaf escapes, too, because his color makes him inconspicuous. What an individual

(a)

(b)

(c)

(d)

26.15

An example of convergent evolution. (a) In the western hemisphere, most thick, spiny, succulent plants are cacti. (b) In the eastern hemisphere, plants that are vegetatively similar are spurges. A cactus and a spurge may look so much alike that telling them apart is difficult. Although they are in different families, each has evolved structurally to survive in arid climates. (c) The cactus flower, with its numerous petals and stamens, is unmistakable. The reproductive parts of these two families have remained distinctive, but the vegetative parts, in evolving, have *converged* toward great similarity. Cacti and spurges have many points of resemblance, but they are easy to separate when they are in flower. (d) The flower of the spurge (*Euphorbia mammilaris*) is obviously different from the flower of a cactus.

26.16

An example of coevolution. Bats and bat-pollinated flowers have in a number of instances developed structures that are mutually useful. The flower produces sugar-rich nectar and protein-rich pollen, both of which nourish the bats. In return, the flowers are fertilized and can set viable seed. On their side, bats have developed highly specialized tongues, which they can stick out, as shown in the photograph, to reach deep into a flower. By now, both bat and flower have become so completely dependent on each other that neither could survive without the other.

26.17

The phrase "dead as a dodo" refers to the fact that these flightless inhabitants of the island of Mauritius were all killed after the importation of pigs and rats to their island. They were also thoughtlessly slaughtered by people, and they have been extinct for more than two centuries.

has or does is obviously important. What individuals do in groups, however, is also important, and evolution is not all competition.

Cooperation is frequently conspicuous, not only within species, as in wolf packs and herds of buffalo, but between species. When two species undergo change in such a way that both can operate more effectively as a team, the process is called *coevolution.* Some elaborate examples of coevolution are found in flower-animal relations. Bat-pollinated flowers have odors attractive to bats and suitable physical structures that allow bats to get at the pollen. They even produce high-protein pollen, and the pollinating bats have specially adapted tongues (Figure 26.16). Similar coevolutionary trends can be seen in flowers pollinated by butterflies, with each partner adapting to the other.

Recently, Stanley A. Temple at the University of Wisconsin suggested a fascinating example of coevolution. According to Temple, it is not a coincidence that a large tree (*Calvaria major*) that still exists on the island of Mauritius in the Indian Ocean ceased to germinate at the same time that the dodo bird became extinct at the end of the seventeenth century (Figure 26.17). Apparently the dodo bird fed on the fruit of the Calvaria tree. The large pit inside the fruit was ground into small bits by the dodo's gizzard. (Most birds periodically swallow small stones, which are stored in the gizzard and are used to grind food into smaller, more digestible pieces.) In the course of coevolving with the dodo, the Calvaria's pit became harder and harder, protecting the seed within. Finally a point was reached where the seed could not be released naturally through the tough outer shell of the pit. And that is where the dodo comes in as a partner in coevolution. The dodo, keeping up with the hardening process of the pit, was still able to grind the pit in its gizzard and liberate the seed. As soon as the dodo was gone (probably because European settlers inadvertently hunted it into extinction), the Calvaria seeds were unable to germinate.

Although the Calvaria trees on Mauritius are slowly disappearing, those that are still there keep producing viable seeds within the pits of their fruit, but the seeds remain locked inside. Temple has fed some of these pits to turkeys, whose gizzards supposedly resemble those of the dodo birds. The 10 seeds that ultimately were freed by the grinding gizzards were planted, and three actually germinated.

There are several cases on record where animals became extinct when plants disappeared, but if Temple's hypothesis is correct, this will be the first time that a plant has been known to disappear because an animal became extinct.

SUMMARY

1. The differences between individuals are due in part to the organization of an organism's DNA (and that is the part that is important in evolution), and in part to the environmental forces that alter individual development (and that is of no direct importance in evolution).

2. *Mutations*, changes in nucleotide sequences in DNA or genetic recombinations of old nucleotide sequences, are the raw material of evolution. New mutations, plus new combinations of old and new genes, make new varieties, adding to the total variability of a species.

3. The changing of an organism that brings the organism into better harmony with its environment is called *adaptation*. Some striking examples of adaptation are *industrial melanism, cryptic coloration, mimicry*, and *warning coloration*.

4. The *species* is still the basic grouping of organisms, but there is no definition that meets the requirements of all the investigators.

5. A group of individuals, partially or entirely isolated from the other members of the species, is a *deme*.

6. The entire genetic stock of a population is a *gene pool*. In a large breeding group, there may be thousands of rarely expressed recessive characteristics. Most of them are probably detrimental, but some are apparently neutral, and some are beneficial.

7. The *Hardy-Weinberg formula*, in conjunction with the Mendelian principle of segregation, can be used in calculating the frequency of a gene in a large population. Genetic stability is assumed as a condition for the effectiveness of the Hardy-Weinberg formula.

8. Genetic change resulting from the isolation of small groups is *genetic drift*.

9. The development of a new species, *speciation*, can occur suddenly, or it can be a slow process extending over many generations. In either method, the result is a new population of organisms that cannot produce fertile offspring when mating with the descendants of the old ancestral line.

10. One way for chromosome alteration to bring about speciation is by the phenomenon of *polyploidy*, the replication of entire chromosome sets. When a new species arises as the result of polyploidy, no new genetic material has been added. There are no mutations, no new nucleotide sequences; there is only a new combination of pre-existing genes.

11. The slow process of speciation by isolation and increasing divergence is known as *adaptive radiation*, because organisms *radiate* out from one place and *adapt* to new environments.

12. *Reproductive isolation* occurs when the union of gametes is not accomplished. Although any one of many methods acting alone is enough to prevent the production of fertile offspring, usually several isolating mechanisms act at once.

13. Sometimes changes can occur so that organisms that are not closely related come to have certain similar characteristics. The process of changing toward similarity instead of difference is *convergent evolution*.

14. When two species undergo change in such a way that both can operate more efficiently as a team, the process is called *coevolution*.

RECOMMENDED READING

Ayala, Francisco J., "The Mechanisms of Evolution." *Scientific American*, September 1978.

Bishop, J. A., and Laurence M. Cook, "Moths, Melanism and Clean Air." *Scientific American*, January 1975. (Offprint 1314)

Bodmer, W. F., and L. L. Cavalli-Sforza, *Genetics, Evolution, and Man*. San Francisco: W. H. Freeman, 1976.

Calder, Wiliam A., III, "The Kiwi." *Scientific American*, July 1978.

Chelminski, Rudolph, "Polish Forest, a Time Machine." *Smithsonian*, May 1978.

Clarke, Bryan, "The Causes of Biological Diversity." *Scientific American*, August 1975. (Offprint 1326)

Duddington, C. L., *Evolution and Design in the Plant Kingdom*. New York: Thomas Y. Crowell, 1969.

Kurtèn, Björn, "Continental Drift and Evolution." *Scientific American*, March 1969. (Offprint 877)

Lerner, I. Michael, and William J. Libby, *Heredity, Evolution and Society*, Second Edition. San Francisco: W. H. Freeman, 1976.

Lewontin, Richard C., "Adaptation," *Scientific American*, September 1978.

Stebbins, G. Ledyard, *Processes of Organic Evolution*, Third Edition. Englewood Cliffs, N.J.: Prentice-Hall, 1977. (Paperback)

Wills, Christopher, "Genetic Load." *Scientific American*, March 1970. (Offprint 1172)

27

The Origin and Development of Life on Earth

FOR UNCOUNTED YEARS, THE HUMAN MIND HAS puzzled over the problem of where life came from and how the earth came to be. Every social group, no matter how simple or isolated, has a story of creation. The expressions range from the general and poetic, as in the Genesis account ("In the beginning God created heaven and earth") to the elaborate and fantastic, complete with monsters, superheroes, imaginary beasts of outlandish anatomy, hundreds of gods and goddesses, storms, earthly upheavals, floods, spirits, and magic. We can never know how life began; we were not there. But with experimentation we can begin to understand how it might have happened.

WAYS OF EXPLAINING THE ORIGIN OF LIFE

A number of explanations of the origin of life have been developed, all of them philosophically interesting, but each presenting its own problems. An ancient and widely followed belief is that of the Hebrew histories: God created life. Since such a belief is beyond experimental attack, it is outside the defined limits of biology as a science. *Special Creation* is a matter of faith and cannot be proved or disproved. A slightly less mystical idea is contained in *Vitalism*, according to which life has a force and quality of its own, something different and apart from the forces and mechanisms of the inanimate world. Like special creation, vitalism is not subject to experimentation, and consequently it will not be considered further in this text. A modification of vitalism is *Panspermia*, the idea that some kind of "seeds of life" are universally distributed everywhere, needing only suitable conditions for development. This idea approaches the notion of spontaneous generation, to be treated in some detail below.

The possibility that life came to the earth from outer space has been considered repeatedly over the years. One difficulty with this is that it explains nothing and pushes the problem away in time and space, making practical work impossible. Several bits of evidence, however, make it impossible to dismiss extraterrestrial life out of hand. Meteorites, investigated with precautions to exclude earthly contamination, have been shown to contain particles that resemble microscopic fossils and organic compounds. A meteorite that fell in Orgeuil, France, in 1864 has been intensively analyzed, and although the results have been criticized, they have so far not been flatly rejected. In addition, radio telescope data and spectroscopy of other planets and of space itself reveal the presence of a wide variety of organic compounds. The concentration of such carbon compounds as ethyl alcohol, for example, is spectacularly low, but their presence is undeniable. The ingredients necessary for the making of cells are out there.

The most promising efforts to find a plausible explanation for the origin of life came from a materialist philosophy. The hypothesis is that if the necessary energy, required chemical elements, and suitable physical conditions once permitted the synthesis of living stuff, we might be able to repeat the original event. Although no living cells have been manufactured with present techniques, some suggestive results have been obtained, and these will be the subject of much of this chapter.

Spontaneous Generation

For centuries it was believed that plants and animals could produce themselves without parents. That, superficially, is not unreasonable. Mold seems just to appear on bread; dead carcasses left exposed soon become crawling heaps of maggots; an old bottle of wine comes to contain a teeming mass of worms. One convincing experiment was made by a seventeenth-century Italian, Francesco Redi, who showed that meat exposed to air came to contain maggots

only if flies were allowed to reach it. Covered jars remained maggot-free. But the idea of spontaneous generation died hard. With the development of microscopes and the discovery of microorganisms, the problem became more complicated, because means of excluding bacteria, mold spores, and protozoa were not known. Preservation of food by the essentially modern technique of canning was practiced early in the nineteenth century, primarily as a means of feeding Napoleon's armies, but the process was found by trial and error, not by biological understanding of microorganisms.

No one understood how microorganisms got into things until Louis Pasteur in 1864 devised a simple means of overcoming all objections to earlier experiments. His flasks, filled with sterilized but easily fermentable substances, such as wine or urine, had openings to the air, but the necks were bent so as to trap foreign particles and prevent them from entering

27.1

Louis Pasteur (1822–1895), one of the great men of all time. A man of enormous patience, compassion, imagination, insight, Pasteur (pronounced Pa-STIR) has to his credit a list of accomplishments that no other scientist can equal: contributions to knowledge of molecular symmetry, fermentation, "spontaneous generation," the germ theory of disease, sterile surgery, the use of weakened pathogens in the treatment of disease.

27.2

Pasteur's demonstration that organisms of decay come only from pre-existing organisms of decay. This simple but ingenious procedure ultimately (but not immediately) stilled the voices of the proponents of spontaneous generation. It answered the objections to previous experiments, especially the objection that free air is necessary for spontaneous generation, and was so straightforward and convincing that it earned the highest praise scientists can give an experiment: It was elegant.

(a) NUTRIENT SOLUTION ADDED TO FLASK

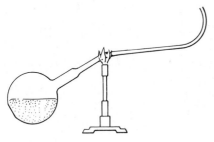

(b) NECK OF FLASK BENT INTO S-SHAPED CURVE USING HEAT

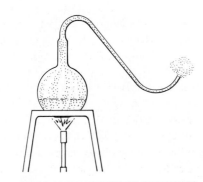

(c) SOLUTION BOILED VIGOROUSLY FOR SEVERAL MINUTES

DUST AND BACTERIA IN WATER DROPLETS

(d) SOLUTION IS COOLED SLOWLY AND REMAINS STERILE FOR MANY MONTHS

the flasks (Figure 27.2). His preparations remained clean indefinitely, showing that even bacteria are derived only from parent bacteria. Thus matters were settled as far as the life of cells goes, but the question of where the original cells came from remained as difficult as ever.

THE ORIGIN OF THE UNIVERSE AND THE BEGINNING OF LIFE

Wonder about the beginnings of life continued, but early in the twentieth century help was forthcoming from chemists, geologists, and paleontologists, who tried to look back in time and determine how the universe was born and how life began. Most scientists agree on what happened, even *when* it happened, but the unanswered question still remains: What *made* it happen?

The Origin of the Universe

The universe was born in a flash of light and energy 15–20 billion years ago.* Such a relatively precise date is possible because astronomers have been able to measure the speed at which galaxies (clusters of stars) are moving away from one another toward infinite space. If we go backward in time and reverse the direction of the moving galaxies, we can imagine

---

* It is probably impossible for most of us to imagine the reality of a number as large as 20 billion, but perhaps we can at least understand something of the enormousness of 1 billion. If you had a billion dollars and you spent a dollar a minute ($1440 a day), you would get rid of only a little more than half a million dollars a year. It would take about 2000 years to spend the entire billion. If you decided to invest your billion dollars at 5 percent interest, you would have to spend $95 a minute ($137,000 a day) just to spend the first year's *interest*. A billion years is a long time.

   To try to understand the magnitude of *20* billion, imagine this: If you bought a $10,000 sports car *every day*, you would have to live more than 5000 years to be able to spend 20 billion dollars.

27.3

A schematic view of the earth, with sections cut away to show the interior layers: the outermost *crust*, the *mantle*, and the two-part innermost *core*.

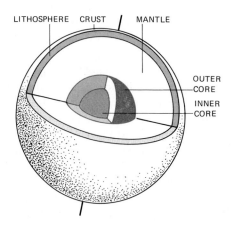

LITHOSPHERE CRUST MANTLE

OUTER CORE

INNER CORE

27.4

How the earth may have appeared during its formative period, with gases and vapor billowing from cracks in the ground. The picture is from a contemporary scene: the Nyamlagira crater in the Congo region of Africa.

that they were once closer together. Inevitably, at an even earlier time, the galaxies were intermingled into one huge mass. The temperature and pressure must have been enormous. When the pressure and temperature became too great to sustain the dense mass any longer, the entire mixture flew apart in a cosmic explosion of unimaginable power. At that moment, the universe was born. Astronomers have been able to detect, even now, traces of the radiant afterglow of that "big bang."

The Primordial Atmosphere and the Origin of the Earth

Out of that stupendous explosion came the primordial gases, the first "atmosphere," probably consisting mainly of helium and hydrogen. Clouds of primordial gas continued to swirl, and the atoms within became compressed. The temperature rose along with the increasing pressure, until finally, through billions of years, the hydrogen within the ball of gas exploded in a series of nuclear reactions. (All the elements originated from the nuclear reactions that ultimately converted hydrogen to all the more complex elements of the universe.) The cloud reached its maximum state of compression and became a star. Our sun is such a star, and it was probably formed along with the rest of our solar system five or six billion years ago.

The earth originated about 4.6 billion years ago as one of the cooling satellites that somehow formed around the central body of the sun, which even now continues to emit radiant energy. While the earth was

still in a somewhat gaseous state, the lighter elements, such as hydrogen, remained near the outer surface, and the heavier elements, such as nickel and iron, settled toward the center. Radioactive elements remained at the center of the earth, and as they disintegrated, colliding particles of matter increased the temperature until the rock at the interior of the earth began to melt, expanding and pushing hot lava through weak spots in the earth's crust. The primitive earth, first a seething mass of volcanic activity, gradually cooled enough to form an outer rim of rock, but at its center the earth remained a core of molten rock and metal (Figure 27.3).

Along with the eruption of molten rock, gases escaped; nitrogen, ammonia, hydrogen, water vapor, methane, hydrogen sulfide.* Some of the lighter gases, such as helium, had already escaped before the gases had become dense enough to succumb to the earth's gravity. Those gases were almost gone. As more and more gas escaped from the earth's core, the accumulation of vapor condensed, and the rain began (Figure 27.4). The oceans were forming, and

---

* These simple molecules of the earth's primitive atmosphere were the source of the elements of the present atmosphere. For instance, oxygen was present in the form of water vapor (H_2O), carbon was in the form of methane (CH_4), nitrogen was in the form of ammonia (NH_3), and obviously, hydrogen was abundant, both as a separate gas (H_2) and as a component of all the molecules mentioned above.

METHANOGENS: THE OLDEST FORM OF LIFE?

Microorganisms called _methanogens_ have been called a new form of life (older than and different from bacteria, plants, and animals), and in fact, they actually may be the oldest form of life yet known. Methanogens, which have been known to scientists since the 1950s, exist only in oxygen-free conditions (like sewers, underground hot springs, and the stomachs of sheep and cattle) that are remarkably similar to those of the original atmosphere. They live off carbon dioxide and hydrogen, which they convert to waste methane gas ("methanogen" means "methane producer"). Besides this fact, a new discovery shows that the RNA sequences of methanogens are not similar to those of bacteria, plants, or animals. Such evidence makes methanogens a likely key to the understanding of the earth's first life forms and subsequent evolution.

It is thought that the first cells (bacteria) appeared on earth about 3.5 billion years ago. If methanogens could exist in the primordial atmosphere (and apparently they could), and if a recent fossil embedded in a 3.4 billion-year-old rock from South Africa is indeed a methanogen, then possibly methanogens originated as much as four billion years ago, earlier than previous estimates of the origin of life.

But the main evidence for supposing that methanogens may be a form of life related to no other comes from an analysis of their cell walls. Biochemists in West Germany have shown that methanogens lack the peculiar cell envelope of ordinary bacteria. (Bacteria cell walls contain peptidoglycan, a combination of protein and carbohydrate; methanogens lack this component.)

Even if methanogens _are_ unrelated to other existing forms of life, they are not necessarily the "universal common ancestor" that scientists are always looking for. What the unrelatedness of methanogens probably does mean is that we have found one more organism in a long list of organisms whose living relatives seem to be missing.

the raw materials of life were there. But one immensely important component of the modern atmosphere was still lacking: free oxygen.

The Spontaneous Production of Organic Molecules

Imagine a moderately warm land, lashed by lightning, engulfed in an atmosphere without oxygen, and well supplied with water. The image of such a world has prompted many biologists to speculate on the ways life may have started. But it is not surprising that after Pasteur had so convincingly destroyed the theory of spontaneous generation, few scientists were eager to reopen the issue. Nevertheless, in the early 1920s, two scientists in particular, the Englishman J. B. S. Haldane and the Russian Alexander I. Oparin, independently speculated on how life might have arisen from nonliving matter (Figure 27.5). Either life was created instantaneously by a divine being, or it arose _gradually_ from materials that were once lifeless. The latter alternative was the only scientifically acceptable one. The important factor was _time_. Given enough time, there could be a gradual progression from the inorganic to the organic to the biological. Because of the work of men like Hutton, Lyell, and Darwin, scientists were finally able to imagine the earth, and even the universe, in terms of vast expanses of time. Nobel laureate George Wald has said, "What we regard as impossible on the basis of human experience is meaningless here. Given so much time, the 'impossible' becomes possible, the possible probable, and the probable virtually certain."

Haldane and Oparin discussed the chemical reactions that might occur where water, energy (in the form of heat, lightning, and ultraviolet light*), and a hydrogen-dominated atmosphere of methane (containing carbon), ammonia (containing nitrogen), and hydrogen were available. Haldane suggested that these factors would produce simple organic molecules, and he coined the phrase "hot dilute soup" (sometimes called the "primordial soup") to describe the teeming mixture of chemicals in the ancient seas, where life probably began. Oparin envisioned the production of organic molecules and went further to suggest that these primitive organic molecules would form into aggregates, or droplets, which he called _coacervates_ (Latin, "cluster together") (Figure 27.6). Some of the droplets, he argued, would be more stable than others; that is, they would "survive" longer,

---

* It is now believed that radioactivity was also an important energy source in the primitive earth.

27.5
Two speculators on the origin of life on earth: the Russian A. I. Oparin (left) and the Englishman J. B. S. Haldane. Their ideas have influenced present-day thinking on the problem of the origin of life. Oparin's book, *The Origin of Life on Earth*, was an original and convincing statement of how the first biological material might have developed.

have a chance to enlarge, and perhaps to divide. Such droplets, though not alive, could be thought of as prebiotic, and they might eventually lead to the formation of real biological cells.* Through millions of years of random combinations, molecules were formed that made possible the replication of entire organisms: amino acids and proteins, which provide the structural and functional compounds of cells, and nucleic acids, the only molecules capable of self-replication.

Theoretical suggestions, such as those of Haldane and Oparin, led to actual laboratory attempts to re-create a primitive climate. Would it be possible to actually generate the creation of amino acids from prebiotic conditions? In 1953, Stanley Miller, then a graduate student at the University of Chicago, experimented with a closed vessel containing a mixture

---

* Charles Darwin, more than half a century before, had suggested how life may have originated "in some warm little pond, with all sorts of ammonia and phosphoric salts, light, heat, electricity, etc., present." He went even further to suggest that "a protein compound was chemically formed ready to undergo still more complex changes." Darwin may have been more accurate than he thought about the possibility of life originating in "some warm little pond." In 1947, the British scientist J. D. Bernal proposed that primordial organic compounds in the ocean were still too dilute to produce the more complex molecules needed to initiate life. He suggested that the clay in lagoons and shallow pools near the oceans would be a perfect medium for the condensation and protection of organic compounds. Today many scientists have accepted the logic of Bernal's theory.

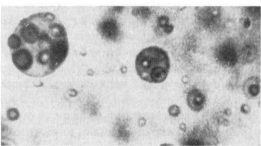

× 2000

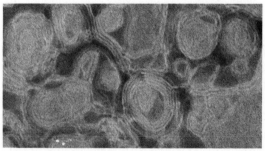

× 3200

27.6
Particles produced artificially from organic but nonliving materials. These bits, seen at high magnification, appear variously as microscopic droplets, balls within balls, or polygonal masses surrounded by membranelike layers. Objects resembling cells have been made by many ingenious techniques, but no indefinitely self-reproducing entities have so far been achieved.

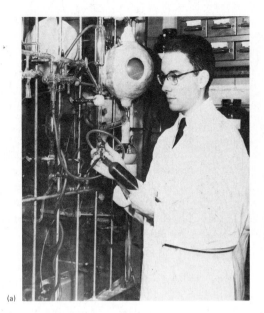

(a)

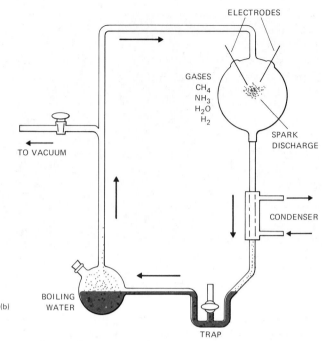

ELECTRODES

GASES
CH_4
NH_3
H_2O
H_2

SPARK
DISCHARGE

TO VACUUM

CONDENSER

BOILING
WATER

TRAP

(b)

27.7

The production of amino acids from methane, ammonia, hydrogen, and water, energized by electric sparks. (a) The young Dr. Stanley L. Miller, standing beside the setup with which he demonstrated his results. (b) Dia-gram of the apparatus Miller used to simulate a primordial atmosphere and to obtain compounds that are generally regarded as biological from nonbiological ones.

of gases that were supposed to have been present in prebiotic times (Figure 27.7). Miller introduced electric sparks into the vessel as a small substitute for lightning. After about a week of sparking, he found a number of organic molecules suspended in the water, including a variety of simple amino acids, from which proteins are made. This experiment showed that if ammonia, methane and water are supplied with energy in a *reducing* (hydrogen-rich) atmosphere, there will be produced many of those compounds that are characteristic of modern cells. Subsequent experiments have produced even more complex amino acids.

The Evolution of Cells

The first cells appeared in the form of bacteria in the primordial soup about 3.5 billion years ago. On a very small scale, photosynthesis probably started about 3 billion years ago, but at that time only in simple bacteria and blue-green algae. Cells were successful because they were efficient. Even the simplest cells, with their built-in chemical factories, speeded up the evolutionary process considerably, reproducing more rapidly than free-floating strands of DNA, and they provided the basis for the varied forms of all the multicellular life to come.

There are two ways in which the first cells may have arisen. According to one view, something resembling a membrane came first to give a cellular covering of cytoplasm. A contrasting view holds that the genetic material came first—the "naked gene"—and was later surrounded by cytoplasm. Which alternative is correct is not known, but *prokaryotic* cells must have come first. These were cells without nuclei, such as bacteria, but still containing the strand of DNA that makes replication possible. Prokaryotes, although simple and without chromosomes or mitosis, continue to flourish today. New daughter cells each receive a copy of the original DNA double helix as it unwinds and separates during cell division.

Then, perhaps a billion years after the appearance of the prokaryotic cells, the first cell nuclei appeared. These more complex cells, called *eukaryotic*, have not only nuclei but all the usual components and compartments of the familiar cells of higher organisms. The development of such cells permitted the regulation by DNA of many different kinds of cells. Once a cell appeared that we could now call a "plant cell," the stage was set for the revolution that brought some life out of the water and onto the land.

Photosynthesis: The Foundation for a World of Oxygen

Before the development of photosynthesis, life depended on fermentation, and the molecules that were being used to produce energy were not being replaced. It was simply a matter of time before life would die

LIFE BEYOND THE EARTH: EXOBIOLOGY

Since we know what we do of the universe and of the processes of evolution, it is reasonable to think that there is some kind of life on planets of other solar systems. The earth is the proper size and contains the proper ingredients for the life that has developed here. A planet that is too massive would have such a gravitational pull that mobility would be improbable; one that is too small would not have enough pull to retain the necessary gases. A habitable planet must be neither too hot nor too cold, and it must have an atmosphere that can be used. None of the planets in our own solar system seems to offer a hospitable environment, but there are some 10^{20} (100 million million million, or 100 billion billion) stars known to astronomers, each a sun in its own right, and each possibly with its planets. If we assume that only one star out of every billion carries a suitable planet where some kind of life might have evolved, that still leaves 10^{11} (100 billion) possible planets. (Some astronomers have estimated that the possibilities for life on other planets may be as high as 10^{18}, or 1 billion billion.)

"Some kind of life," then, appears to be most probable, but we can wonder if those other

evolutionary processes led to anything like the human species. The answer is "Probably not." Variability is the essence of evolution, and it is therefore highly unlikely that any two evolutionary streams would have led to identical or even similar types. Consider such recent islands as the Galapagos, with their scant million-year history, and see that life there has already diverged from that on the nearest mainland. It is reasonable, however, to expect the evolution of intelligence, that is, the ability to understand and perhaps to manipulate the environment, even to make efforts at communicating with us. Such thoughts have prompted the suggestion that we should try sending radio messages out into space and keeping radio watch for incoming messages. (Indeed, such experimentation is proceeding now.) The little green men that cartoonists like to draw may be smarter than we are, but if they could visit the earth, they would find some organisms here that might startle them. If you were a perceptive being from Beta Cygni, what would you think of such earth-born creatures as an octopus, a Venus fly-trap, a praying mantis, and a vampire bat?

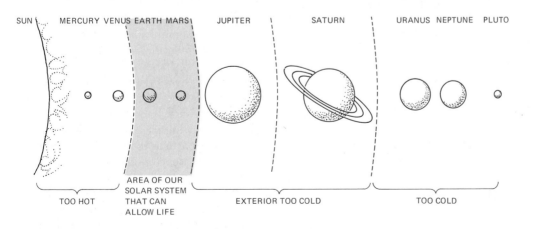

TABLE 27.1

Events in the Origin and Development of Life on Earth*

EVENTS IN THE ORIGIN OF LIFE	YEARS BEFORE PRESENT
Origin of the universe	15–20 billion
Origin of the sun and our solar system	5–6 billion
Origin of the earth	4.6 billion
Oldest rocks	3.8 billion
First cells (bacteria) appear	3.5 billion
Photosynthesis begins	3.3 billion
Multicellular organisms appear	1 billion
First animals with external skeletons	600 million
First fishes appear	400 million
First fishes migrate to land	350 million
First amphibians appear	325 million
First reptiles appear	300 million
Dinosaurs appear	175 million
First birds appear	150–200 million
First mammals appear	150 million
Dinosaurs disappear	75 million
First tree-dwelling mammals appear	60 million
First monkeys appear	35 million
First apes appear	15 million
Australopithecus (ape-man) appears	5 million
Homo erectus appears	3 million
Homo sapiens appears	200,000
Neanderthal appears	35,000–100,000
Cro-Magnon (modern *Homo sapiens*) appears	10,000–35,000

* Such a listing is necessarily approximate and temporary. Just a few years before this writing, for instance, the birth of the universe was marked at about 10 billion years ago; now the date is more firmly established at 15–20 billion years ago. Books go out of date even while they are still on press, as information about the earliest Homo fossils keeps moving the date further back in time. And that seems to be the pattern: The more we find out about the development of life, the further back in time we are pushed.

out and have to start all over again. But before the organic energy sources were used up, some cells developed chlorophyll, and photosynthesis evolved. (The waste product of fermentation, carbon dioxide, was essential to the process of photosynthesis.) When cells became able to convert the energy from the sun into chemical energy for their own use, they became independent, and as long as the sun shone on the land, they could evolve there as well as or even better than in the ocean. Before photosynthesis, the atmosphere was mostly nitrogen and carbon dioxide. Now oxygen began to be abundant, and a protective layer of oxygen (in the form of ozone) shielded the earth from harmful ultraviolet rays from the sun. Organisms did not *have* to leave the water, but for the first time, there was an option. Before photosynthesis, underwater organisms were protected from the harmful atmospheric radiation; now organisms were free to emerge and even migrate toward the dry land. Of course, organisms had to "learn" to adapt to free oxygen. That obstacle was overcome by the evolution of cellular respiration. The chemical chain of evolutionary events was ever more efficient, from fermentation to photosynthesis to cellular respiration.

WHY DOESN'T SPONTANEOUS GENERATION CONTINUE TODAY?

If living organisms *did* once develop from nonliving compounds, why hasn't the process continued? One answer to the question is that conditions have changed. The atmosphere is different from what it was, both in its components (which could be used in making amino acids) and in the radiation it allowed to reach the surface of the earth. Now the atmosphere has little free ammonia and a lot of oxygen, and that mix is not conducive to the synthesis of protein precursors. Besides, when the earliest organic molecules were formed and were followed by primitive cells, there was nothing to destroy them. If an occasional macromolecule now happens to be made spontaneously, any one of countless microorganisms is likely to be nearby, and the new molecule is digested or engulfed before it can do anything further. The air, the water, and the earth's present inhabitants all make it highly improbable, even impossible, for a living particle to start from zero and endure long enough to develop into something more complex.

In view of the variety of experimental conditions under which scientists have produced biological molecules, it seems likely that in the prebiotic world something approaching living entities may have been elaborated many times in many places and with many results. Whatever happened, we can be reasonably certain that if prebiotic bits accumulated, for a while they were safe from being eaten because there was nothing to eat them; they would not decay because there were no enzymes to act on them; and they would not be readily oxidized because there was little free oxygen. We have little hope of finding actual remains of the earliest prebiotic and biotic structures, and we may never know where or when life started on earth. Just as no clear distinction can be made between living and nonliving material, so there is no specific time to which we can point and say, "Before this, there was no life. After this, there is." But life did begin somewhere, sometime, and life forms have continued to increase in complexity, down to present-day organisms, including man, that creature who, as George Wald said, "turns back upon the process that created him and attempts to understand it."

GEOLOGICAL TIME

For some time, scientists have agreed that the universe began with a "big bang" about 15 billion years ago, and in the last few years estimates have gone back as far as 20 billion years. The earth was probably born about 4.6 billion years ago, and the first bacteria may have originated about a billion years later. Multicellular organisms most likely inhabited the earth about a billion years ago (Table 27.1). Once

organized creatures became established, they continued in uninterrupted progress until the present array evolved. Granted that remains of the earliest organisms are probably lost forever, succeeding ones did leave occasional fossils, and from them we have been able to attempt a reconstruction of the history of life on earth.

The Geological Eras

The major divisions of geologic time are called *eras*, and they are named for convenience simply by the estimated ages of fossils. A number of methods have been used to estimate the lengths of time involved, but the most recent and most accurate is one that uses the decay of radioactive elements, such as uranium (U^{238}), potassium (K^{40}), and, for some recent samples, carbon (C^{14}). If the amount of a radioactive element originally present can be known, as well as the rate at which it decays, then a measurement of the amount remaining in a sample will tell how old the sample is (Figure 27.8).

The oldest era is the *Precambrian*, which lasted from the beginning of rock formation (3.8 billion years ago) up to about 600 million years ago, or about three billion years—longer than all the other eras combined. The next was the *Paleozoic*, or the era of "old animals," which lasted about 350 million years. The third, the *Mesozoic*, or era of "middle animals," extended for about 160 million years up to the most recent, the *Cenozoic* (the era of "new animals"), in which we are living, and which dates from about 63 million years before the present (Table 27.2).

27.8

The principle used in radiocarbon dating. Knowing the amount of radioactive carbon in a modern biological sample and knowing the half-life of C^{14}, one can estimate the age of an unknown sample. A sample 5730 years old would have only half as much C^{14} as a modern sample, and a sample 11,460 years old would have

half that much, or one-fourth of the original amount. The figures are based on the supposition that there is a definite amount of carbon dioxide containing C^{14} in the air and that the proportion has remained constant through the years. When radioactive carbon dioxide is taken into a plant in photosynthesis, it may be locked in

when the plant dies, to remain perhaps as a piece of wood or to be eaten by an animal and incorporated into the animal's tissues. Thus biological samples indicate, by the amount of C^{14} they exhibit, how long they have been dead. (For a modification of this idea, however, see Figure 16.14.)

0 YEARS—
DEATH OF ANIMAL
OR PLANT

5730 YEARS—
1/2 LEFT

11,460 YEARS—
1/4 LEFT

17,190 YEARS—
1/8 LEFT

22,920 YEARS—
1/16 LEFT

60,000 YEARS—
ABOUT 1/1000 LEFT

TABLE 27.2
Geologic Time Scale and Important Biological Events

ERA	PERIOD	EPOCH	MILLIONS OF YEARS AGO	DURATION IN MILLIONS OF YEARS	PLANTS	ANIMALS
Cenozoic	Quaternary	Recent Pleistocene	2	2	Increase in herbaceous plants	Appearance of man
Cenozoic	Tertiary	Pliocene Miocene Oligocene Eocene Paleocene	65	63	Modern flowers	Birds, mammals; decline of reptiles
Mesozoic	Cretaceous		145	80	Increase of flowering plants	Increase of mammals; modern birds; dinosaurs dying
Mesozoic	Jurassic		190	45	First flowers; abundant conifers	Primitive birds and mammals; dinosaurs abundant
Mesozoic	Triassic		225	35	Conifers dominant	First dinosaurs; first mammals
Paleozoic	Permian		280	55	Decline of nonseed plants	Increase of reptiles and insects
Paleozoic	Carboniferous		345	65	Coal formation. Primitive conifers, many vascular nonseed plants	Early reptiles; insects; increase of amphibia
Paleozoic	Devonian		395	50	Development of vascular plants: club mosses, ferns	First amphibians; sharks; numerous invertebrates
Paleozoic	Silurian		430	35	Algae; first vascular plants	Insects; increase of fishes
Paleozoic	Ordovician		500	70	First land plants; aquatic algae	First fishes
Paleozoic	Cambrian		570	70	Algae	Marine invertebrates
Precambrian			3500	2930	Blue-green algae, bacteria	Protozoa

SUMMARY

1. A number of explanations of the origin of life on earth have been developed, such as *Special Creation, Vitalism, Panspermia,* and the introduction of organic compounds from outer space. The theory of *spontaneous generation* proposed that life could appear spontaneously from other life, but this notion was attacked as long ago as the seventeenth century.

2. The universe probably originated about 15–20 billion years ago as part of a cosmic explosion explained by the "big bang" theory. The first "atmosphere" probably consisted mainly of hydrogen and helium.

3. The sun was probably formed about five or six billion years ago, and the earth is thought to have formed

about 4.6 billion years ago. As the earth cooled, such gases as nitrogen, ammonia, hydrogen, water vapor, methane, and hydrogen sulfide escaped from the molten core. The oceans began to form but free oxygen was still lacking.

4. Given enough time, there could be a gradual progression from the inorganic to the organic to the biological. Scientists, such as J. B. S. Haldane and Alexander I. Oparin, discussed the chemical reactions that might occur in a reducing atmosphere where water, energy, ammonia, methane, and hydrogen were available. Haldane postulated the synthesis of numerous organic compounds, and Oparin went further to suggest that these primitive organic molecules would form prebiotic aggregates called *coacervates*, which might eventually lead to the formation of real biological cells.

5. Several scientists have succeeded in recreating a primitive climate. Stanley Miller demonstrated that simple amino acids could be produced if ammonia, methane, and water are supplied with electrical energy in a reducing atmosphere. Subsequent experiments have produced even more complex amino acids.

6. The first cells appeared in the form of bacteria about 3.5 billion years ago. *Prokaryotic* cells came first, followed by more complex, nucleated *eukaryotic* cells. According to one view of how first cells originated, something resembling a membrane came first to give a cellular covering of cytoplasm. A contrasting view holds that the genetic material came first—the "naked gene"—and was later surrounded by cytoplasm.

7. Before the organic energy sources for fermentation were used up, some cells developed chlorophyll, and photosynthesis evolved. Then some organisms adapted to free oxygen, using the process of cellular respiration.

8. Organisms have not continued to evolve from nonliving compounds because the primordial conditions have changed so drastically.

9. The major divisions of geologic time are called *eras*. From the oldest to the most recent, they are the *Precambrian*, *Paleozoic*, *Mesozoic* and *Cenozoic*.

RECOMMENDED READING

Dickerson, Richard E., "Chemical Evolution and the Origin of Life," *Scientific American*, September 1978.

Fox, S. W., and K. Dose, *Molecular Evolution and the Origin of Life*. San Francisco: W. H. Freeman, 1972.

Horowitz, Norman H., "The Search for Life on Mars." *Scientific American*, November 1977. (Offprint 389)

Layzer, David, "The Arrow of Time." *Scientific American*, December 1975.

Miller, Stanley L., "A Production of Amino Acids under Possible Primitive Earth Conditions." *Science*, May 15, 1953.

Miller, Stanley L., and Harold C. Urey, "Organic Compound Synthesis on the Primitive Earth." *Science*, July 31, 1959.

Oparin, Alexander I., *Life: Its Nature, Origin and Development*. New York: Academic Press, 1962.

Oparin, Alexander I., *The Origin of Life*. New York: Dover, 1953. (Paperback)

Orgel, L. E., *The Origins of Life*. New York: John Wiley, 1973.

Ponnamperuma, Cyril, *The Origins of Life*. New York: E. P. Dutton, 1972.

Wald, George, "The Origin of Life." *Scientific American*, August 1954. (Offprint 47)

28

Animal Behavior I: Perception and Communication

ANIMAL BEHAVIOR, LIKE EVERY OTHER BIOLOGI-cal activity, has developed through the action of natural selection. During the course of evolution, behavior from courtship to feeding to camouflage has been naturally selected when it has promoted the survival of the species. John Alcock (Arizona State University) describes the complexity of the study of animal behavior when he says that "the study of behavior is extremely broad-ranging, involving such diverse phenomena as the action of a gene, the structure of a nervous system, the relations of an animal to its environment, and evolutionary events lasting millions of years."

THE STUDY OF ANIMAL BEHAVIOR

In 1973, the Nobel Prize in Physiology or Medicine was awarded to three men who had studied and tried to explain the many facets of animal behavior. Konrad Lorenz, Nikolaas Tinbergen, and Karl von Frisch had already devoted about 50 years apiece to field studies of insects, birds, and animals of all sorts, but it was not until 1973 that their achievements could even be labeled in a sufficient manner for such a prestigious award to be presented. The study of animal behavior had acquired legitimacy, and with it a new name: *ethology*.

Lorenz, Tinbergen, and von Frisch made the study of ethology feasible by demonstrating methods and techniques that would soon become standard. Of course, animals were studied long before the twentieth century. Apparently Aristotle looked at *everything*—we have mentioned him throughout the book. He looked at animals, too, and recorded his observations in *Historia Animalium*, where he noted certain predictable patterns of animal behavior. Aristotle's observations of nature were not always complete or scientifically accurate. Nevertheless, Aristotle's inaccuracies were naive only in retrospect, and it is reasonable to imagine that new studies might have developed as a consequence of *Historia Animalium*. But for centuries none did. Instead, scientists and observers alike chose to think of animals in terms of human characteristics—a point of view known as *anthropomorphism* (Greek, "human form"). Anthropomorphism is nothing to be ashamed of. We practice it every time we think our pets are smiling at us. It just isn't scientifically valid.

During the 2000 years separating Aristotle and Charles Darwin, no explicit scientific hypotheses were set forth to indicate that animals were anything but simplified extensions of the human realm. Animals served purposes we noted and sometimes even assigned, but whatever we thought of animals, it was always in terms of our own visions of them. When we talk of bees, for example, it is common to refer to them as "busy," "industrious," or "loyal." In actuality, bees act according to strict physiological and genetic directions.

The world was not ready for Charles Darwin in 1859, when *The Origin of Species* presented the heretical notion that animals behave the way they do because of a continuing process of evolution. Living things, Darwin said, must be able to adapt to changing conditions. They are not created with fixed responses that were indelibly set eons ago. Darwin's more specialized publication, in 1872, of *Expression of the Emotions of Man and Animals* was the first scientific exploration of the causes of plant and animal behavior patterns.

Pioneers of animal behavior followed Darwin's revolutionary groundwork. Jean Henri Fabre in France observed insects in their natural habitats, recording them in terms of insect patterns rather than in the usual anthropomorphic fashion (Figure 28.1). Others were Sir John W. Lubbock and C. Lloyd Morgan in England and Jacques Loeb and Herbert S. Jennings in America. Sir Charles Sherrington and Edgar D. Adrian shared a 1932 Nobel Prize for their work with nerve and skeletal muscles. And of course,

Ivan Pavlov in Russia discovered the "conditioned reflex" during his experiments with dogs and won a Nobel Prize for his work. Classic works were published, and new work, inspired by the pioneers, continued and became refined, developing into a modern science with the complementary ideas of psychology, natural history, genetics, and evolution.

THE EVOLUTION OF ANIMAL BEHAVIOR

The study of animal behavior has revealed that as an animal's brain becomes more complex, its behavior patterns are less rigid. Insects seem to be programmed with an unshakable set of instructions, whereas "higher organisms" need not rely totally on inflexible behavior patterns.

A typical example of rigid insect behavior concerns dead ants and the specific odor they emit. Worker ants, recognizing the death odor, will cart the corpse to a garbage site away from the nest. However, if a live ant is experimentally dabbed with the

28.1

Jean Henri Fabre (1823–1915), whose detailed descriptions of insect behavior were different from anything written before. He was endlessly patient and meticulously careful; he kept minutely precise records of what he did; and he practically freed himself from the ancient human habit of interpreting animal behavior in anthropocentric terms.

substance that releases the "dead ant" odor, the ant will be promptly carried off to the garbage dump by the workers, even though the "dead" ant shows definite signs of life.

The Basic Conditions of Animal Behavior

In Chapter 19 we observed that complex behavior becomes possible when synaptic relationships and reflex arcs develop to the point where they provide more intricate pathways for a nerve impulse. And as the nervous system itself becomes more complex, it is ultimately the greater capacity of the brain that makes it possible for an animal to analyze and interpret stimuli in a selective way. When the response to any given stimulus becomes flexible, the organism is freed from *stereotyped behavior* or *fixed-action patterns*. Individuals who can adjust to a changing environment are better suited to find food or a mate and to avoid predators or unpleasant conditions. And of course, the improved senses are passed on to the

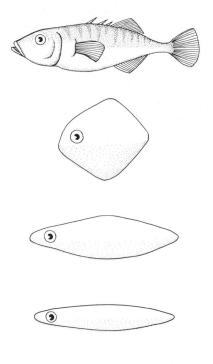

28.2

Some of Tinbergen's artificial fish, which he used to draw responses from live fish. His experimental animals, sticklebacks, were not interested in a model that to human eyes is a realistic copy (shown at the top), but their attention was caught by models with red bellies, regardless of how they were shaped. A male stickleback, seeing something with red on it, will either attack it as if it were an intruding male rival, or court it as if it were an acceptable female, depending on the time. A specific detail, whether of structure, color, or movement, that elicits a response is called a "releaser."

28.3

Sound as a sex attractant in mosquitoes. The silent tuning fork at left draws few male mosquitoes to the screening, but the humming fork at right has attracted a swarm. Males are not attracted to higher or lower frequencies than the one (or a close approximation of the one) emitted by the vibrating wings of female mosquitoes. The mating attractant is a tone of a particular pitch; the feeding attractant is temperature. Mosquitoes attack human ankles with greater frequency than they go for hands because ankles are thin-skinned and emit more detectable warmth than hands do.

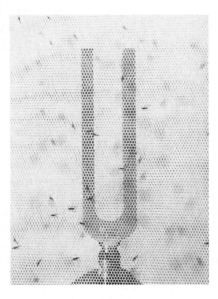
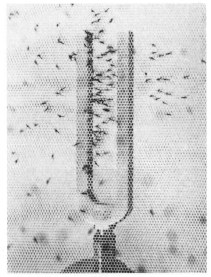

28.4
An attempt to compare the vision of a bee with that of a human being. The flower on the left was photographed in the usual way to make it look as nearly as possible like the flowers we see. The same flower shown on the right was photographed so as to accentuate the parts that reflect ultraviolet light. Since it can be experimentally demonstrated that bees are sensitive to ultraviolet (and human eyes are not), we can infer that the bees "see" the flower as it is on the right.

offspring, once more demonstrating the effectiveness of the natural process of selection.

Complex animals are able to rely on their own experience and specific needs when responding to stimuli. But even though the response of a complex animal may be more flexible than that of a simpler animal, all organisms possess the property of *irritability*, the basic ability to respond to stimuli. All animals must be able to sense and receive stimuli, conduct the information by way of a nerve impulse, and use an effector to initiate the proper response. In complex animals like humans, the once-simple sensory cells have evolved into sense organs, nerve cells have been elaborated into the brain, and effector cells have become glands and muscles. But the fundamental components of behavior remain the same for all living things.

THE RHYTHMS OF ANIMAL BEHAVIOR

Natural cyclical events, such as the rising and setting of the sun and moon, seasons, tides, light and dark, obviously affect the behavior of animals and, for that matter, of plants. The overall behavior of most organisms is certainly related to the great cycles and rhythms of nature, and it is instructive to investigate animal rhythms and the ways in which they depend on both genetically and environmentally imposed rhythms.

Biological Clocks
It has been known for hundreds of years that some plants have predictable, cyclical behavior. Sunflowers turn toward the rising and then toward the setting sun. More recently it has been found that bean leaves, which rise high in the morning and droop at night, can continue to rise and droop at the appropriate times for several days even after they have been kept in continuous darkness. Bean plants (and many other plants) must have some internally regulated timing mechanism, or "biological clocks." We now know that animals have such time-measuring "biological clocks," too, and we can observe definite cycles of behavior at specific times of the day or year, even if the visible environment is kept constant. One type of biological clock distinguishes the changing length of the days throughout the seasons. Many organisms respond to seasonal variations in day length by changing their metabolism, and such a response is called *photoperiodism*. Studies have shown that the photoperiodic mechanism is located in the brain, and, at least in insects, the eye is not involved. How does a biological clock work? Although we do not have a detailed understanding of the mechanism, it is likely that a biological clock is run by hormones, by oscillating responses to light and dark, or by a combination of these.

The term *circadian rhythm* has been applied to daily cycles of behavior (from the Latin, *circa*, "about," and *dies*, "day"). *Circannual* (for want of a more euphonious word) rhythms are cyclical behavior patterns on an annual basis. Whether circadian or circannual, the rhythms of a biological clock are genetically programmed, and they need not be learned. Biological clocks are usually complemented by a reaction to changes in heat or light. Environmental clues are valuable in providing a check on programmed

HOMING

One of the most fascinating aspects of animal orientation is *homing*. By homing, animals that leave their homes in search of food, for instance, are able to return without wandering aimlessly. Homing pigeons do not need landmarks to find their way; pigeons carried from their coops in covered containers are able to fly back without any trouble. It is not uncommon for pigeons to fly home in an almost direct route at about 80 km/hr (48 mph), covering almost 1000 kilometers (600 mi) a day. Homing pigeons and other airborne navigators use the sun or stars as a compass, but they probably also orient themselves according to the earth's magnetic field.

A sense of time is important to animals that must find their way over long distances, and animals that use the sun or stars to navigate have a time sense that makes allowances for the shifting positions of the celestial bodies. If an animal can make such a compensation, it presumably can determine its direction and distance away from its home base.

Honeybees and some other insects also have this ability to correlate time, distance, and direction according to the changing position of the sun. In fact, honeybees can navigate when the sun is obscured, using only patches of blue sky. Scientists have found that bees can detect polarized light.* By using the fixed relationship between the pattern of polarized light in the sky and the position of the sun, the bees can make the necessary adjustment to locate the sun. All skylight is polarized, but the honeybees make use of only the polarized light in the ultraviolet end of the spectrum, to which they are most sensitive. Recent evidence suggests that some birds may also use polarized light for navigation when the sun is hidden by clouds or mountains.

---

* We speak of light from the sun as being unpolarized, and waves of unpolarized light normally vibrate freely in all directions. But when unpolarized light from the sun is scattered by water droplets in the earth's atmosphere, the horizontal light waves are filtered out, and the light becomes "polarized." (Polarized light waves vibrate only up and down in a vertical plane.) The vibration direction of the polarized light that reaches the earth from a patch of sky is at right angles to the direction of the sun from that piece of sky. Bees seem to "know" this fact and adjust for the angle between the light and the sun.

---

rhythms because those rhythms must be reset occasionally when they deviate slightly from a 24-hour or 365-day pattern. Actually, no rhythm of a biological clock corresponds exactly to the length of a day or a year.

Migration

Some animals periodically leave their homes in favor of other locations for purposes of feeding or mating. This process is called *migration*. Ordinarily, migratory expeditions take place in the fall, and the animals return to their original homes in the spring. Even when environmental changes are not apparent, a biological clock provides the signal for action. Instead of waiting for an actual event like winter to occur, the animal is able to anticipate such an extreme change and prepare for it.

In a series of classic experiments, E. G. F. Sauer of the University of Freiburg in Germany investigated the annual migration of warblers to Africa. Each year, starting in August, the birds left their nests in Germany and continued their migrations until the end of October. The next spring the warblers returned to the exact sites they had left in the autumn, and even young warblers, who had never made the flight before, successfully navigated to Africa and back. Such competence is especially notable since warblers migrate singly, without the support of a familiar and experienced group. Warblers usually migrate at night, navigating by the stars, and they use a built-in time sense that coordinates with the seasonal position of the stars.

Birds have an impressive ability to migrate over long distances (Figure 28.5). A long-winged shearwater taken from its nest in Scotland returned from 4800 kilometers (2900 mi) across the Atlantic in 12½ days. Migrating birds feed off their own fat, which they build up in preparation for their flights. A typical migrating bird may store as much as 20 percent of its body weight in fat, and ruby-throated hummingbirds, which fly nonstop 800 kilometers (480 mi) across the Gulf of Mexico twice each year, store almost 50 percent of their weight as fat for fuel.

Some animals, such as the Cape buffalo of southern Africa, wild turkeys, white-tailed deer, and wolves, do not migrate every year. If the migratory impulse is primarily a search for better feeding grounds, then an animal apparently does not need to move from its home ground when food remains plentiful. But migration seems to be an inherited activity, and many organisms, including insects, fish, birds, and mammals, migrate when their environments become precarious (Figure 28.6). Whales may migrate thousands of kilometers during the summer in search of plankton, and fllimsy monarch butterflies travel from Canada to southern California and Mexico every fall. Locusts and shrimp are on the move ceaselessly, each shrimp packing a single grain of sand to

28.5

Map of the western hemisphere, showing migration routes of ten birds that travel hundreds or thousands of miles each year, going to new feeding grounds after the nesting season is over. The more spectacular journeys are those of the sooty shearwater (8), the knot (1), the golden plover (3), and Wilson's petrel (10). The key to the list: (1) knot, (2) red-eyed vireo, (3) golden plover, (4) bobolink, (5) gray kingbird, (6) sulfur-bellied flycatcher, (7) swallow, (8) sooty shearwater, (9) *Lessonia rufa*, and (10) Wilson's petrel.

(a)

(b)

28.6

Animal migrations. (a) Wildebeests move in great numbers over kilometers of African plains, seeking food as the seasons change. (b) Monarch butterflies fly south in late summer and fall, moving by the millions from the United States and Canada to spend the winter in Mexico. In their southern destination, they coat the forests with their bodies. Even while on their way, they can accumulate in clusters, as those shown here have done.

orient and balance itself. Antelopes and wildebeests migrate across the Serengeti Plains in Africa when the rains renew their favorite grazing land upcountry. Emperor penguins, instead of seeking a warmer climate during the winter, look for the coldest, most solitary spot they can find, where they can mate and raise new broods away from predator seals and sea leopards. Adelie penguins use the sun as a compass, and 300,000 or more of the birds may assemble at the same rookery every year (Figure 28.7).

Hibernation

Some animals cope with environmental problems by *hibernating* (Latin, "winter") instead of migrating.

Although animals usually hibernate during the winter to avoid the extreme cold and lack of food, some animals begin their hibernation periods as early as September 1. Hibernating animals use stored body fat at a very slow rate (compare the quick use of stored fat by migrating birds) and need little oxygen because of their lowered body metabolism—many endocrine glands stop working altogether. The ground squirrel, for instance, usually breathes between 100 and 200 times a minute. During hibernation the squirrel breathes only once or twice a minute, and its heart beats about five times a minute (Figure 28.8).

Two common misunderstandings about hibernation persist, encouraged by children's storybooks.

28.7

The end of a migration. The nesting ground of the Adelie penguins at Anvers Island in Antarctica is crowded with hundreds of thousands of birds who go there annually to nest and rear the young. The adults feed in the open water, visible in the distance, then climb onto the barren, stony ground, which they prefer for nesting, and feed the chicks. The main feeding areas are many kilometers from the breeding ground, but the penguins undertake mass movement so as to make the best of a difficult life, eating where food is available, and founding a rookery where reproduction is safe.

28.8

Awakening from hibernation. A ground squirrel passes from a state of lethargy, with most of its metabolic processes slowed almost to a zero rate, to an active condition as its temperature rises. (a) At 1.7°C, the squirrel is rolled into an inert ball, insensitive to most stimuli, barely breathing, and with minimal circulation. (b) At 3°C, it can move enough to stretch. (c) At 14.5°C, it can roll slowly over, but it is lethargic and unresponsive.

(d) At 17.5°C, it moves a little faster and more frequently, but it is far from being awake. (e) At 20°C it is no longer so completely relaxed. (f) At 26°C, it opens sleepy eyes and is alert enough to hold its tail in the characteristic erect position. (g) Finally, with the arrival of full warmth at 35°C, it achieves full alertness: eyes wide, muscles tense, standing up and ready to go.

(a)

(b)

(c)

(d)

(e)

(f)

(g)

From "Annual Biological Clocks" by Eric T. Pengelley and Sally J. Asmundson. Copyright © 1971 by Scientific American, Inc. All rights reserved.

(a)

(b)

(c)

First, bears do *not* hibernate. They sleep a great deal during the winter, and they consume much of their reserve body fat, but because their body temperatures are not lowered appreciably, we cannot consider bears as true hibernators. Second, hibernating animals do not usually remain asleep without interruption all winter, and some animals may actually leave their dens for short periods.

During periods of severe drought or heat, some animals, like the African lungfish, curl up in a ball in mucus-lined mud holes, slow down their body functions, and *estivate*, the summertime equivalent of hibernating.

INSTINCT AND LEARNING

Controversy accompanies any attempt to separate the concepts of instinct and learning. To further complicate the issue, there are no simple, clear-cut categories of learning, although we must attempt to separate certain basic kinds of learning patterns if we are to proceed at all. We will consider *learning* to be the modification of behavior as a result of changes in individual experience. Generally, such behavioral changes increase the organism's chances for survival, and hence they are advantageous to the species.

Instinct

The major strength of *instinctive*, or *innate*, behavior is that it is predictably correct and safe. The very nature of instinctive behavior makes it reliable, since it has survived the natural selection process and has proved to be advantageous to the organism. The reliability of instinctive behavior patterns is especially important when the organism has to make the correct choice the first time the action occurs. If an animal being pursued by a predator reacts improperly, it may not have the opportunity to react again.

Usually an animal has developed its behavior patterns by the time it needs to use them. Many animals have fully developed behavior patterns at birth because those patterns were genetically maturing during the embryo stage (Figure 28.9). Other behavior

28.9

Learning what to fear and what not to fear. When objects pass over an inexperienced chick, it drops down and stays still. Some animal behaviorists say that the action is innate, but others avoid use of the word "innate" because of the difficulties in making sure what is or is not innate. If an object passes over a chick again and again without hurting it, the chick ceases to crouch. However, if an unfamiliar object appears, the chick may assume the old crouching stance again. Apparently it learns to ignore the familiar and to be apprehensive—if a bird can be apprehensive—of the unknown. Thus a young goose (a) crouches at every moving thing at first. Later it becomes experienced and (b) ignores geese flying but still crouches (c) at the sight of a hawk.

28.10

Animals that take advantage of natural objects outside themselves to help them obtain food. Such actions can be called "tool using." The Galapagos woodpecker finch sticks a woody spine into an insect tunnel to extract insects. An Egyptian vulture lifts a rock to smash an ostrich egg. A chimpanzee gets termites out of their nest by poking a stem, which it has stripped of leaves, into the entrance hole. A sea otter floats on its back in the water, balancing a stone on its chest, and prepares to crack a clam against it.

patterns, such as the ability to mate, may develop along with the animal's growth and the maturation of the specific organs involved. They may also require the experience of watching and imitating adults.

Although instinctive behavior may be developed genetically, its application to a real situation may have to be learned. Chimpanzees in zoos may use thin sticks to poke at holes, with no obvious purpose. If the same chimpanzee had been born in the wild and remained there, it would have learned to apply its innate tool-using ability to probe for termites in holes in logs and termite nests (Figure 28.10). The tendency to perform behavior may appear spontaneously, but practice may perfect its execution.

Habituation

It is generally accepted that habituation is the least complex kind of learning. *Habituation* takes place when a stimulus no longer provides any expression of a survival advantage. In other words, the animal that has learned that a stimulus is meaningless will ignore it. By ignoring meaningless stimuli, an animal is better able to devote its full attention to stimuli that do have biological significance.

Conditioning

Another type of learning is *conditioning*, in which the animal is rewarded or punished according to its reaction to a given stimulus. Included in the general area of conditioned learning is the *trial-and-error* method, adopted by animals adjusting to unfamiliar stimuli in their environments. As with many types of learning, successes and failures, rewards and punishments are involved with conditioning and trial-and-error learn-

(a)

(b)

(c)

ing (Figure 28.11). Early in a learning task, 100 percent reinforcement gives optimal results, but once an animal is trained, it performs better and "forgets" less often if it is rewarded only occasionally.

Probably the most famous case of experimental conditioning was developed by the Russian physiologist Ivan Pavlov in the early 1900s. Using dogs in his experiments, Pavlov developed the idea of the conditioned reflex when he rang a bell just before he fed an experimental dog. The dog did not respond to the bell, but when the dog was actually fed, it salivated at the sight of the meat. Soon Pavlov was able to cause the dog to salivate at the sound of the bell, in anticipation of the meat to follow. This type of conditioning is called *classical conditioning.*

Another form of conditioning is *operant conditioning,* popularized in the 1930s by the American psychologist B. F. Skinner. The usual demonstration of operant conditioning involves placing an animal in a cage and providing a lever that, when pressed by the animal (typically a rat), will make food available through a small opening in the cage. As soon as the rat learns to make the association between the depression of the lever and the reward, it depresses the bar often. The difference between classical conditioning and operant conditioning is that the former depends on a *reflexive* response to a stimulus whereas the latter uses the promise of a positive reinforcement to trigger an arbitrary action, such as pressing a lever.

Latent Learning

This is probably a good time in the chapter to be reminded that animals do not need to be prodded to learn. *Latent learning* takes place when an animal ac-

(d)

(e)

28.11

An animal learns by punishment and reward. (a) A toad, accustomed to eating insects, is presented with a stingless robber fly. (b) The toad eats the fly. (c) A bumblebee, resembling the robber fly but equipped with a wicked sting, is dangled on a string before the toad. (d) As toads do, this one snaps at a moving insect, only to receive what is to a toad a serious dose of venom from the bumblebee. (e) The toad now backs off from a robber fly, having learned that anything resembling a bumblebee is best let alone.

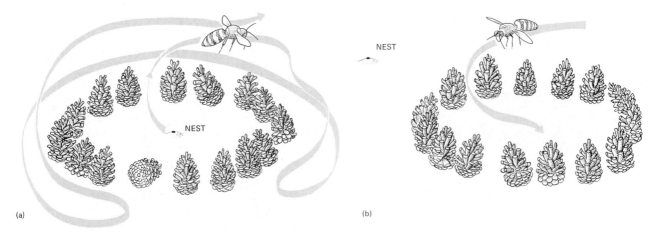

(a) (b)

28.12

Learning in a wasp. A female digger wasp, *Philanthus triangulum,* comes and goes from her sandy nest, finding her way by looking at landmarks. (a) Once, on emerging from her nest, she found that a ring of pine cones had been placed around it (by the experimenter, Niko Tinbergen). She made an exploratory flight around the new arrangement and then took off to forage. While she was gone, Tinbergen moved the cones away from the hole (b). When the wasp came back, she flew to the center of the ring to seek her nest. She had apparently "memorized" the landscape.

28.13

This photograph of Nobel Prize winner Konrad Lorenz and his "children" has achieved a modest fame as an example of what can be accomplished by *imprinting* of young animals. If newly hatched birds are presented with some object as soon as they can perceive anything, they will accept that object as a "parent," whether it actually is a parent, a mechanical toy, or Konrad Lorenz. Since the parent is in fact the first thing a young bird is most likely to meet, the system is effective in nature.

tively seeks to find out more about its environment. Such behavior may simply be called curiosity, but whatever the label, this exploratory learning needs no particular reinforcement or reward. Biologically, latent learning is important for an animal to prepare it for important changes in its environment. A rat that masters the idea of a maze or a digger wasp that can find its nest after wandering great distances in search of prey has an advantage over an animal that has not learned to explore and retrace its path or one that needs a reward to learn anything (Figure 28.12).

Imprinting

Rarely has a term been as completely associated with one person as "imprinting" is with the Austrian ethologist Konrad Lorenz. In the 1930s, Lorenz originated the term imprinting to show how newly hatched goslings permanently regard the first creature they can follow as their mother (Figure 28.13), but current research disputes the irreversibility of imprinting. Imprinting takes place during the first few hours of a bird's life, and it is no longer possible after that crucial early period has passed.

The sort of imprinting shown in Figure 28.13 is obviously artificially imposed upon animals. In a natural setting, the imprinting instinct will work perfectly well, with no sexual or other complications, because the first moving object the young animal sees is its mother, and it will follow her in a normal way.

We must constantly remind ourselves that students of animal behavior often use exceptional or unusual behavior in order to emphasize the natural and normal behavior of the animal that has never encountered an ethologist.

Insight Learning

When an animal is able to associate apparently unrelated experiences, it is demonstrating *insight*. Chimpanzees, of all the animals except humans, seem to have developed the highest degree of insight learning, although it can be observed in many of the higher animals. Animals capable of insight learning seem to practice a sort of mental trial-and-error process, analyzing the possibilities for the solution of a problem before actually setting out to tackle it. We have already mentioned that chimpanzees have a tool-using ability, which they can apply to locate and capture termites, for instance. But besides being able to attach sticks together to reach an object, or to pile boxes one on top of the other for the same purpose (Figure 28.14), two chimps may even cooperate, with minimal nonverbal communication, to accomplish other goals.

ANIMAL PERCEPTION AND COMMUNICATION

Of all the animals, only humans can speak, but all animals communicate. Lacking the ability to use words, animals have developed their own methods of communication, aided by acute senses of taste and smell, touch, hearing, and vision. In a murky underwater environment, where vision would be of little use, fish have well-developed chemical-sensing abilities. Bats, even more surrounded by darkness than fish are, use a sensitive and highly developed "radar" system to find their way. Birds can maintain a precise visual sense through vast expanses of air, where smells and sounds might be hopelessly jumbled. Whatever the medium, each organism has evolved its own best sensory method for survival and applied those methods toward communication within the species. Perception and communication are a matter of life and death, providing food, safety, and the propagation of the species.

Chemical Perception and Communication

The ability to respond to chemicals is probably one of the earliest senses developed by living organisms. Insects are especially reliant on chemical sensitivity, but most of the higher organisms, including humans, use a chemical sense to one degree or another. A female

28.14
Intelligent behavior in a chimpanzee. Meeting a situation he has never seen before, this hungry ape managed to get the inaccessible bananas by stacking boxes. Such action implies a certain amount of imagination.

28.15

Sources of pheromones. This diagram of a worker honeybee shows glands that secrete a number of substances useful in informing other bees about the bee's identity or the action that should be taken. The wax gland is used in comb construction and the poison gland in protection, but the mandibular glands, the sting chamber glands, and Nassanov's gland give out pheromones that function in communication within the hive.

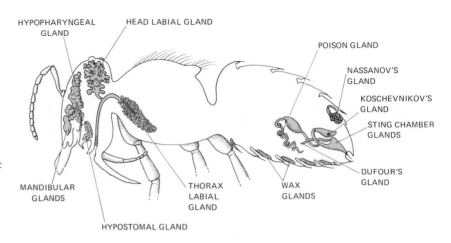

HYPOPHARYNGEAL GLAND

HEAD LABIAL GLAND

POISON GLAND

NASSANOV'S GLAND

KOSCHEVNIKOV'S GLAND

STING CHAMBER GLANDS

DUFOUR'S GLAND

MANDIBULAR GLANDS

THORAX LABIAL GLAND

WAX GLANDS

HYPOSTOMAL GLAND

ichneumon wasp is able to locate a moth larva several centimeters inside a piece of wood, where the larva has burrowed. After searching the surface of the wood with her antennae, the wasp drills a hole with her ovipositor through the wood into the larva. She then deposits an egg directly into the paralyzed larva. When the egg hatches, the newborn wasp will use the moth larva as its first food source. It is thought that the wasp *smells* the larva, even though the larva may have penetrated deep within a tree.

Foraging worker ants leave a scent trail for other ants when they have found food. The ants touch their extended stingers to the ground and secrete a substance that is usually detected only by members of the same species. The more satisfactory the food discovery, the more the ants will release their food scents onto the trail; in this way the trail becomes quantitatively stronger if the food is worth advertising. If another animal accidentally crosses the scent trail and disconnects it, the ants will become disoriented until they make a new connection with an ant on the other side of the break. (Army ants will be discussed in the next chapter.)

The scent trail of the ant is formed by a secretion that is effective in communicating information by odor to members of its *own* species. Such a species-specific chemical substance is called a *pheromone* (Greek, *phero*, "carry," and *hormon*, "arouse"). Pheromones have been referred to as social hormones (Figure 28.15). Water is the ideal medium for phero-

mones, and many species of fish communicate through odors. When one fish of a school is injured, for instance, it exudes an odor that serves to scatter the rest of the school away from the source of injury. Much interest has been shown in pheromones recently, especially because synthetic pheromones may eventually be successful in attracting and controlling insects that ruin agricultural crops. Certain species of insects seem to combine several pheromones to form more complicated messages. Ethologists are finding more and more evidence that pheromones usually work in combinations, sometimes with other stimuli.

The champion odor detector is probably the salmon. Shortly after they are hatched, salmon swim downstream to the open sea, where they grow to maturity. Two, three or even five years after a salmon has left its birthplace, it will unerringly retrace its path on a strenuous journey upstream to spawn in its original birthstream. It is believed that the salmon's incredible homing behavior is due to its ability to imprint the specific odor of the water where it was born, to remember it, and to isolate it from all the other odors in the water between the ocean and the original stream upriver.

Sound Perception and Communication

If humans could hear all the animal sounds in nature, they would be overwhelmed by the intense noise level. We all know that dogs hear sounds that we cannot hear, and yet, probably because it is biologically unimportant to our own survival, we do not customarily think of the enormous number of animal sounds we cannot hear. In fact, any given animal is unaware of much of what is going on around it. What is important is that the animal be able to perceive whatever it needs for its survival. Too much awareness of "reality" will merely reduce an animal's capacity to respond to the stimuli that are really important in its own natural world. Of course, this applies to all the senses, not just sound.

When we consider the subject of animal sounds, we usually think of dolphins or bats because of the sophisticated "sonar" and "radar" systems they have developed, and it is not common to consider rats or squirrels or moths in terms of auditory perception. In fact, however, rodents use ultrasonic frequencies to communicate with other members of their species, and moths have developed such sensitive hearing abilities that they can sometimes elude even the predator bat.

On a summer evening in 1956, Kenneth Roeder of Tufts University was giving an outdoor party at which one of his guests ran his finger around the damp edge of a glass, causing a whistling sound. Roeder noticed that at that moment the several moths flapping around a nearby light suddenly fell to the ground. The moths remained motionless for about a minute and then flew away, obviously alive and apparently unharmed. Roeder inferred a connection between the whistling sound of the glass and the swoon of the moths, and he and Asher Treat of City College of New York began an investigation.

Perhaps the findings of Roeder and Treat will be more significant if we first examine a related portion of the story: the precise auditory mechanism of the bat, a mechanism resembling radar. Bats are nocturnal, and they can navigate successfully and locate and capture prey in complete darkness. Donald R. Griffin and others have demonstrated that bats emit high-frequency sounds that are reflected back to them

(a)　　　　　　　　　　　　　× 6

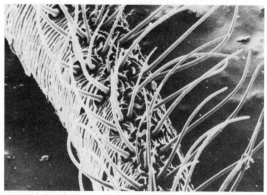

28.16

The sensitive smelling apparatus of a moth, *Samia cecropia*. (a) A male moth with feathery antennae, exposing to the air a great surface that is capable of catching stray molecules. (b) A portion of an antenna, showing the branches and slender sensory hairs to which such substances as pheromones from a female moth can stick. (c) A scanning electron micrograph of a male moth's sensory hairs. These are effective filters, which are likely to trap passing compounds of importance to the moth.

(b)　　　　　　　　　　　　　× 75

(c)　　　　　　　　　　　　　× 750

28.17

Specialized echo receptors. A bat's ears are constructed so as to take maximum advantage of sounds that the bat originally emitted and that are reflected off some object, whether it is an obstacle in the bat's path or an edible flying insect. The large, forward-pointing, concave ears funnel the faint echoes into the auditory canal. This lump-nosed bat is more spectacular than most bats, but even the little brown bats of the northern temperate regions have effective sound-gathering ears.

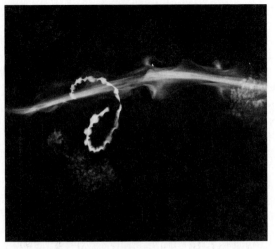

28.18

Record of a sound-guided kill. Photographed against a dark background, the long-exposure picture of a bat catching a moth shows the bat coming in from the right (the horizontal streak). The moth (the irregular central loop) tried to evade the hunter, but its flight ended at the point where the two streaks meet. The bat could pick up the echoes of its own squeaking from the moth's wings and body, and it could do so with such accuracy that it put its prey directly on target, despite the moth's wavering flight.

when the sound waves encounter objects of almost any size. (Bats can even locate tiny fruit flies in complete darkness.) This process of locating objects by the reflection of sound waves is called *echolocation*. By using echolocation, the bat can fly in a dark room filled with obstacles as unobtrusive as thin wires without bumping into anything. Besides being able to avoid obstacles, the bat is not distracted by the usual low-frequency sounds in the environment. However, if its ears are plugged, or if it is made mute by devocalization, or if high-frequency sounds are introduced, the bat will crash into obstacles and will stop flying until the interfering sounds are removed or its hearing and voice are restored. (Few humans can hear sound-wave frequencies higher than 20,000 Hz. Bats produce and hear sounds higher than 100,000 Hz.)

The echolocation system of a bat is so sensitive that it can detect differences in sound patterns of objects. Besides using echolocation for navigation, some bats can distinguish edible prey by echolocation. They can make the distinction successfully, even though the bat's outgoing signal is 2000 times stronger than the returning signal. Bats locate moths in the usual way, and moths can be further identified by the special alternating sound pattern produced by their up-and-down wing motion. If bats are such impressive navigators and hunters, do they ever miss? Indeed they do, and the experiments of Roeder and Treat will provide some interesting information, not only about sound perception and communication but about evolutionary adaptation.

If bats are interested in capturing moths, moths are equally interested in avoiding bats. Roeder and Treat discovered that Noctuid and other moths have developed simple ears near the thorax that allow them to detect the oncoming beeps of bats. When in danger, moths may collapse their wings and fall (as they did at Roeder's party), go into a "power dive," or begin a series of twists and turns like an airshow stunt pilot. Each tactic is an evasive one, designed to avoid the bat (Figure 28.18). It has also been suggested that the Arctiid moth may be capable of producing its own high-frequency sound to further distract predators.

The members of the cetacean order—dolphins, whales, and porpoises—possess an underwater "sonar" echolocation system equivalent to the bat's "radar." When Naval engineers T. G. Lang and H. A. P. Smith in Pasadena, California, conducted dolphin experiments in 1965, they noted discernible clicks, grunts, and squeaks, and six different kinds of whistles when the dolphins communicated with each other. Lang and Smith assigned meanings to some of the whistles and sequences of sounds, and other scientists have trained dolphins to react to simulations of dolphin sounds, but the language of the dolphins remains basically undeciphered.

Visual Perception and Communication

Many animals use visual displays as part of courtship rituals, and although such sexually motivated displays are not the only expression of visual communication in animals, they are certainly among the most interesting and appealing phenomena in nature.

The three-spined stickleback is a freshwater fish about the size of a goldfish, but unlike the goldfish, the stickleback has a drab gray-green color. During the mating season, however, the male stickleback abandons his protective dullness for colors befitting a

THE EARLY BIRD IS EARLIER THAN WE THOUGHT

On the basis of current research, it seems likely that chicks and their mothers use sounds to communicate with each other while the chick is still in the egg. What is more, the unhatched chick and its siblings, also in their shells, communicate with each other before the eggs are hatched. Think back to the concept of imprinting, and imagine how much easier it would be accomplished if mother and child where indeed able to establish a bond through prenatal communication. Even if we think of imprinting with Konrad Lorenz playing the role of mother goose, we can still accept the idea of prenatal communication since we know that Lorenz imitated bird sounds during imprinting and afterward. He may have imprinted his birds so successfully because he supplemented other factors of imprinting with appropriate bird sounds.

An embryo may be primed for the call of its mother by hearing its siblings. If an embryo has been denied any prenatal communication with its siblings, it will take longer than other birds to respond to the mother's calls. Apparently the chick is initially primed by hearing its own voice within the shell, and experimental embryos whose vocal cords have been cut do not recognize the sounds of their species until two days after hatching. The removal of the priming agent retarded but did not destroy the chick's receptivity.

One of the most interesting discoveries of embryonic behavior has been the phenomenon of synchronous hatching, which has been observed in bobwhite quails. Approximately 10 to 15 hours before hatching takes place, distinct clicking communications among embryos can be heard. It seems that the clicking sounds serve to synchronize hatching by slowing down the hatching impulse in the more developed embryos while speeding it up in embryos whose development has lagged.

Human prenatal communication cannot be tested precisely, although most doctors agree that the psychological state of a pregnant woman will affect the behavior of the fetus. Certainly after the seventh month of pregnancy the fetus will respond to external sounds, and at least one seven-month-old fetus is reported to have responded with vigorous movement when its mother attended concerts, with the most movement when Mozart was played.

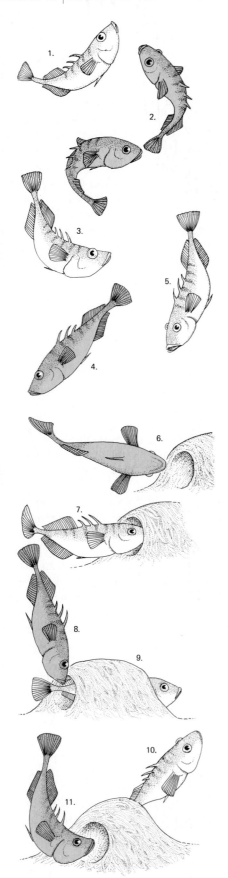

springtime suitor. Figure 28.19 illustrates the stickleback "zig-zag dance," which progresses in distinct stages from the establishment of a territory to courtship, mating, and caring for the newborn.

Mallard ducks provide another example of visual communication as a part of mating behavior. One of the 10 possible patterns is seen in the male duck as part of a full "ritualized" courtship sequence (Figure 28.20). The term *ritualization* was first used by Sir Julian Huxley in 1914 to describe display movements that have been modified through evolution to become stereotyped communication signals. Ritualized displays like those of the mallard duck communicate such specific messages that only members of the same species will respond.

Not everyone has seen a three-spined stickleback or a mallard duck, let alone their mating rituals. But the chances are excellent that we have all seen fireflies flashing in the dark. That, too, is a mating call. The mating signals of the firefly are simple and economical compared with the complex and ritualized displays of the mallard duck. And yet, each species of firefly produces a species-specific signal through differences in intensity, frequency, and duration of the flash (Figure 28.21).

28.19

Mating activities in fish. The three-spined stickleback illustrates stereotyped behavior in courtship and nesting. (1) At the appropriate season, a female approaches a male (gray) swimming upward. (2) The male answers by swimming toward her in a special side-to-side fashion called the "zig-zag dance." (3) The female comes closer to the male with her head turned upward. (4) The male leads the way to the nest. (5) The female follows him. (6) The male stops near the nest. (7) The female swims into it. (8) The male stands on his head and shakes with a motion known as the "tremble-thrust." (9) The female lays her eggs and (10) swims away. (11) The male enters the nest and deposits his sperm on the eggs.

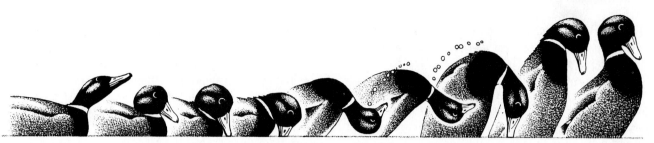

28.20

Mating display in a male mallard duck. Female ducks, dull of color, do the choosing in mating, but they normally accept only those males whose colors and movements are species-specific. A male mallard proves that he is a male mallard by having typical feather markings (green head, etc.) and by going through a set of exercises peculiar to mallards. In this series of pictures, a male mallard is shown in a part of the duck ballet that includes, among other actions, dipping the bill in the water, flipping a few drops, and raising the body high out of the water. This maneuver Tinbergen calls the "grunt-whistle" display. A female mallard, recognizing the signs, may accept the male duck and allow him to copulate.

28.21

Species-specific flash patterns in fireflies ("lightning bugs"). The males let the females know of their presence by showing themselves at the proper level—high in the trees, in low bushes, or near the ground—and emitting the proper sequence of flashes. Some give intermittent short bursts of flashes, others give longer sustained streaks of light, others give short zig-zag patterns, and still others make a still, steady glow. Thus the sexes find each other in the dark.

Because fireflies are probably distasteful to predators, they can practice an extremely conspicuous form of communication in a situation where it would normally be disadvantageous to reveal their nocturnal positions. As males fly over the grass, they emit regular flashes that advertise their positions to available females hidden in the grass below. A female emerges partially and flashes a response to a male. This sequence of calls and responses is repeated several times until the male finally approaches the female, and mating takes place. (For more information about biological luminescence in general, review the essay on page 32.)

Perception by Electricity and Nonvisible Radiant Energy

Adaptation to the environment is a crucial factor for survival, and some animals, like the African fish *Gymnarchus niloticus,* have evolved highly specialized sensory systems to cope with their environments.

The poorly sighted Gymnarchus perceives by using electricity. Weak electrical discharges are generated from highly modified muscle tissue in the fish's tail, and changes in the environment are detected when the resultant electric field from the tail to the sensory organs in the head is disturbed. Unlike electric eels and other powerful electric fish, Gymnarchus and other fish, such as the gymnotids and skates, use their electrical sense to locate objects and prey rather than to shock predators or paralyze prey.

Somewhat related to an electrical sense is the ability to locate objects through an awareness of radiant energy in the form of heat. Rattlesnakes and other pit vipers "see" their camouflaged victims by using 150,000 nerve cells in heat-sensitive organs on both sides of the reptile's head.

28.22

The round dance and the waggle dance. (a) When the food source is less than 75 meters (250 ft) from the hive, the scout bee will perform the *round dance,* a movement of alternating clockwise and counterclockwise circles. The richer the food supply, the more vigorous will be the dance. The direction of the nectar is not indicated, but the worker bees detect the scent of the flower on the scout bee, and a drop of nectar from the flower is also regurgitated by the scout and sampled by the worker bees. Besides these clues, the scout has deposited its own scent on the flower, and this odor is easily detected by the searching workers. (b) A scout bee whose food source is not in the immediate area of the hive will perform the *waggle dance,* a figure eight with a straight line between the loops. Usually the scout bee performs the waggle dance on the vertical surface of the dark hive. In this case, the straight run of the bee between loops will be translated into an angle related to the imaginary vertical line of gravity. The scout bee has transposed this same angle from its own orientation method of relating the nectar source to the position of the sun. As shown in the figure, both angles are the same. When the worker bees leave the hive they retranslate the gravity-related angle they have seen into the same angle in relation to the sun. A similar ability to transpose angles of orientation from the sun to gravity and vice versa is not uncommon among such insects as the ant and the dung beetle. If the scout bee performs the straight-run waggle upward, the worker bees will fly toward the sun; a downward waggle tells the bees to fly away from the sun. Information about the direction to the left or right of the sun is communicated in a similar manner. Distance is also indicated by the number of figure eights—the more performed, the closer the food source. Humming sounds made by the wings of the scout bee during the waggle dance probably provide redundant information about the food source. (c) Occasionally a scout bee will perform its waggle dance on a horizontal surface outside the hive. In this case, the straight portion of the dance will point directly to the food source.

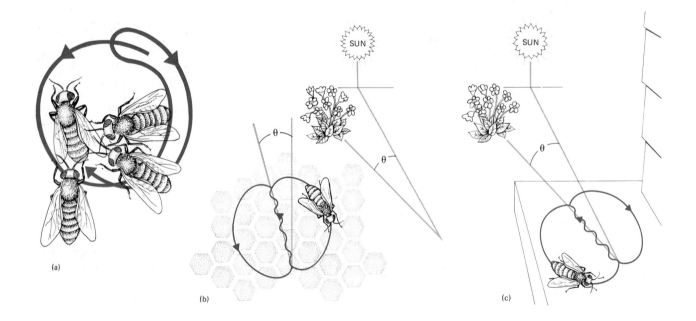

(a)

(b)

(c)

The Language of the Bees

The foraging worker honeybee leaves the hive in search of food. When she finds it, she returns to the hive and achieves a remarkable feat: She not only communicates to the other bees that food has been located, but informs them of its direction and distance from the hive. Therefore, without actually being led to the food, the other bees know exactly how to find it. Such precise information is relayed by either of two kinds of dances: the *round dance* or the *waggle dance* (Figure 28.22). Bees that have been raised outside the hive will respond accurately to the dances the first time they perceive them, indicating innate behavior.

Besides indicating the distance, direction, and quality of a food source, bees can also relay information about a new hive location when the present hive becomes overcrowded. Scout bees use the waggle dance to communicate with the waiting queen and other bees, and worker bees may investigate many potential sites. The quality of the sites is translated through the enthusiasm of the dance, until one site is finally chosen after a few days. Then all the inhabitants of the hive, as many as 40,000, will *swarm* to the new location.

Karl von Frisch deciphered what he called "the language of the bees" during a series of experiments that culminated in the 1950s. The awarding of the 1973 Nobel Prize to him was due primarily to his work with the communication system of honeybees. Recently, several biologists have presented arguments opposing von Frisch's interpretation of the language of the bees. They propose that the factors of communication are odor and sound, but further experiments tend to confirm the original findings of von Frisch, with the possibility that bees use pheromones to locate food sources in occasional instances when they do not use the dances.

SUMMARY

1. Animal behavior, like every other biological activity, is a product of evolution by natural selection, and it has been refined by selection so as to promote the survival of the species.

2. Although the response of a complex animal may be more flexible than that of a simpler animal, all organisms possess the property of *irritability*, the basic ability to respond to stimuli.

3. When an organism exhibits a metabolic change in response to seasonal variations in day length, the phenomenon is called *photoperiodism*. Daily cycles of behavior are known as *circadian rhythms*.

4. *Migration* occurs when animals periodically leave their homes in favor of other locations for purposes of feeding or mating. Migratory expeditions usually take place in the fall and spring. Some animals *hibernate* during the winter, using stored body fat at a slow rate and consuming very little oxygen because of a lowered body metabolism. *Estivation* is the summertime equivalent of hibernation.

5. *Learning* may be considered as the modification of behavior resulting from changes in individual experience. Some types of learning are *instinct, habituation, conditioning, latent learning, imprinting,* and *insight learning*.

6. Animals have developed their own special methods of *perception* and *communication*. In order to communicate, some animals (like social insects) use *pheromones*, species-specific chemical substances referred to as social hormones. Other animals (bats and dolphins, for example) use *echolocation*, a process of locating objects by the reflection of sound waves. Sticklebacks and mallard ducks use visual communication, including stereotyped, innate communication signals called *ritualization*. Perception by *electricity* and *nonvisible radiant energy* is possible for some animals.

7. Besides indicating the distance, direction, and quality of a food source, honeybee scouts can also relay other information to members of the hive by using the *round dance* or the *waggle dance*, part of the "language of the bees."

RECOMMENDED READING

One reading list for Chapters 28 and 29 will be found at the end of Chapter 29.

29

Animal Behavior II: Animal Societies

ORGANISMS LIVING TOGETHER IN ORGANIZED groups are said to live in _societies_. In a human society, the individual usually considers himself more important than the group, and relationships between members of the society are based on personal recognition. For instance, there is virtually no chance that one of your friends will mistake you for someone else or think that you are no different from any other friend. Humans at least _try_ to have personal relationships. In many nonhuman societies, the group is more important than the individual. In a society of army ants, for example, in spite of structural and functional differences between workers and soldiers, all members of a caste seem to be alike, and as far as we know, there are no individual relationships between members of an army ant society. In evolving a society that is like one giant organism, army ants have lost their individuality. Mammals, in contrast, have tended to evolve complex societies in which flexibility has been favored.

During the evolution of social groups, animals have evolved special ways of behaving that help groups to survive. In this chapter we will describe some of the activities by means of which animals stake out territories for themselves, establish hierarchies, recognize possible mates, and control their numbers. We will also consider several well-organized societies, among both mammals and invertebrates, which have been studied in detail by ethologists.

TERRITORIALITY AND AGGRESSION

A *territory* may be defined as a specific area guarded against intrusion by other members of the same species (Figure 29.1).

Establishing a Territory

Territories are usually established and defended for many reasons, although some reasons may appear to be more dominant than others. A hummingbird, for instance, seeks a territory mainly for food, colonies of bats use their territories primarily for shelter, and the Uganda kob selects a territory for mating.

Generally it is the male who establishes and defends a territory, and this is especially true during the mating season. Most vertebrates are territorial, but except for occasional occurrences in crickets, dragonflies, and fiddler crabs, territorial behavior among invertebrates is rare. Males tend to defend the center of their territories most vigorously (Figure 29.2), and as the male encounters intruders toward the edge of his territory he is just as likely to flee as to fight. The point where such ambivalent behavior may be observed marks the *boundary* of the territory.

Many animals declare their territories by placing scents in conspicuous locations around the area. A dog urinates on objects that define his concept of the boundaries of his territory. Antelopes gently deposit secretions from glands at the corners of their eyes on branches and other foliage. Male rabbits use chin and anal glands to mark territories directly or use their

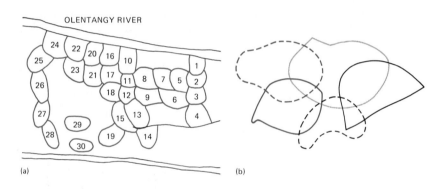

(a)   (b)

29.1

Territories and home ranges. (a) A territory is an area that an individual or group keeps exclusively for private use. Other members of the same species are challenged and usually driven off. The map shows the territories of a group of 30 male song sparrows. Adapted from Margaret Morse Nice. (b) Home ranges are less strictly defined than territories. Mice, grizzly bears, and gorillas have home ranges, over which they wander, tolerating some intrusion at the boundaries. They also go into adjacent home ranges themselves.

29.2

Defense posture. When the musk oxen of northern Canada are threatened, they form a circle, with the bulls facing outward and the cows and calves protected in the middle. None of their natural enemies are able to harm them when they are in such a position.

29.3

The private territory of a rabbit colony. The main nest is central, with tunnels radiating out from it, intersecting, and terminating at entrances on the outer rim. The boundaries, detectable by rabbits outside the colony, are marked by odoriferous secretions or by fecal pellets. Intruders that ignore the signs are driven off.

feces as territorial markings (Figure 29.3). Birds indicate their territories by singing, gibbons by howling, and crickets by chirping. Other animals signal their territorial claims in other ways, and inherent in any establishment of territory is a declaration of living rights (Figure 29.4). If a conflict arises anyway, especially among members of the same species, the animals will use familiar gestures of threat or appeasement. In either case, they avoid actual combat whenever possible.

Conflict Behavior

Threat gestures constitute an attempt to intimidate and drive off an intruder without actually fighting. One of the most studied threat displays is that of the herring gull, which will stand as upright as it can (most animals make themselves look as large as possible when they are threatening), point its head slightly downward, and display its wings outward without actually spreading them. In its *appeasement gesture,* however, the gull assumes a submissive position directly opposite to the threat position, lowering its body and shortening its neck, holding its wings in close, and pointing its bill upward. Figure 29.5 illustrates several other displays of threat and appeasement, always clearly opposite in character. Darwin called this tendency for signals of opposite meaning to have opposite form the *principle of antithesis.*

Courtship and Mating

Definite signals are used within a species to attract a sexual partner, to deter aggressiveness or threatening behavior, and to synchronize male and female actions

29.4

Undeclared territories. Many birds (such as these gulls on a roof) perch at recognized distances from each other, just beyond pecking range. When conditions become crowded, the high-ranking birds will defend their territories, pecking at the lower-ranking birds and forcing them to leave. Other animals, too, maintain spaces that are comfortable for them. Indeed, people usually establish temporary "territories" in public places, and intrusion on such territory is considered aggression.

so that copulation may take place. Conspicuous displays are important to bring the male and female together, but they are also a crucial factor in avoiding mating between members of different species, since such interspecific mating may produce offspring that are infertile or die soon after birth. Just as displays of sound by male birds both threaten males and attract females, so threat and appeasement gestures serve the dual purpose of avoiding fighting and initiating courtship and mating. In the previous chapter we discussed visual communication as it relates to the courtship and mating rituals, and now we will examine some further examples of displays and conflict behavior specifically related to sexual behavior.

Courtship often possesses the conflicting elements of approach and attack behavior because animals normally avoid close contact with other animals, even members of the same species. Therefore the body contact initiated during courtship and mating arouses ambivalent motivations of escape behavior and mating behavior at the same time. If the proper courtship displays have been signaled, the mating instinct will be stronger than the urge to escape, and mating will take place. But until both partners have demonstrated that no real danger exists, the presence of a potential mate intruding on a territory will arouse conflict possibilities as well as sexual ones. Such tendencies are vivid in animals like the female spider, which is just as likely to eat the male as to copulate with him. (See the description of the mating ritual of the praying mantis in the essay on p. 577.)

During courtship among birds it is not unusual for the male to offer food to the female. This action is

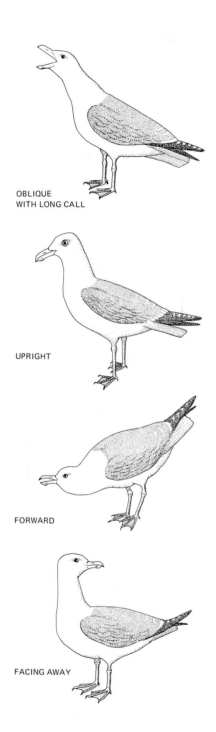

OBLIQUE
WITH LONG CALL

UPRIGHT

FORWARD

FACING AWAY

29.5

Body posture in herring gulls, which is understood by other gulls. The long call is a recognition signal of an intruder. The upright position is a threatening position. The head-ducking-forward position betokens appeasement, as does the gesture of facing away. George Schaller, a student of gorilla behavior, found that he could approach wild gorillas safely if he stopped walking forward and turned his head away at any sign of apprehension on the part of a gorilla. Gulls and gorillas interpret head turning as the opposite of a threat.

Fighting. Most encounters between animals of the same species stop short of physical combat. Posturing and bluffing are usually enough to send one potential combatant away, but many animals do fight, especially males fighting for territory or access to females, and in some instances injury is serious. A tag-eared, scar-faced, limping tom cat is a familiar example of a fighting male, but even small birds can on occasion tear into each other so violently that one is killed.

a likely evolution from the food-begging of young birds, a posture repeated by the female bird as an appeasement gesture toward the aggressive male. Some male insects also include a food offering in their courtship displays, even though such an offering is often merely symbolic (Figure 29.7). Mouth-to-mouth courtship feeding by bullfinches is almost certainly derived from similar gestures used by the parent bullfinches when they feed their chicks, and mouth-to-mouth courting among chimpanzees is also similar to one method of feeding the young. It is speculated that kissing in humans may have originated from the primate courtship practice of touching the lips and making smacking and sucking sounds.

29.7
Symbolic courtship. Some species of balloon flies practice what appears at first sight to be a meaningless ritual, as a male brings a silken balloon he has made to a female. After she accepts the present, they mate. Such action becomes understandable when one knows the activities of related species. In some species, a male brings the female a small captured insect for a food offering, possibly to keep her too busy to eat *him*. In other species, he attaches bits of silken fiber to his gift, making it more distinctive. In still other species, he encases the food in a full covering of silk. The final step is to make the wrapping so elaborate that the contents are omitted, and an empty balloon becomes as acceptable as a real present.

ANIMAL SOCIETIES

Animals in a society exhibit a high degree of cooperation, made possible by a species-specific system of communication. In the simplest case, the societal relationship of a mother and her child constitutes the primary unit of a social organization, but in a "true" society we look for a group of individuals working together. One aspect of this system of mutual benefit for the species genotype is *altruistic behavior*. In altruistic behavior, an individual sacrifices its own reproductive opportunity to help a genetically related

INNOVATIVE PREDATORY BEHAVIOR IN ANIMALS

The predatory behavior of animals is often bizarre, at least to humans. The scorpion, for instance, feeds on live bees in a surgical and spectacular fashion. The scorpion grabs its prey in its forward pincers, holds it steady, and paralyzes it with a slashing sting from its upcurled tail. The poison is already dripping on approach. Sometimes, if the bee is held in a particularly vulnerable position, the scorpion will merely stab it to death, conserving its venom. Now the most interesting sequence occurs. After the bee is paralyzed or freshly killed, the scorpion bites a tiny hole in the bee and injects digestive enzymes that liquefy everything but the shell. The scorpion then sucks out its meal, and the hollow shell of the bee remains intact, like an empty decorated Easter egg, lifeless but still beautiful.

Another ingenious predator is Mastophora, the Bolas spider. Its method of capturing its prey is so specialized that it might seem destined for certain failure. But the unusual prey-catching technique of the Bolas spider works. Instead of spinning a conventional web, the American Bolas spider spins a thread weighted at the end with an extra blob of silk. When an insect enters the striking area, the spider swings its line like a gaucho throwing his bola (hence the name, Bolas) and grabs its prey. The unorthodox method seems to be effective. The Bolas spider has relatives in Australia and Africa. The Australian cousin is called the Angler spider because of the way it casts its single line like a fisherman at its prospective dinner.

The anatomy of the praying mantis is perfectly designed for killing other insects. Furthermore, it is one of the few higher organisms that will attack, kill, and eat members of its own

species. It is not really unusual among insects for a female to kill and eat her mate, but the female praying mantis consumes her mate from the head down, *during* copulation. The male is as persistent as his counterpart. With head already eaten, what is left of him continues the strange intercourse until finally his sex organs are gone, too. (See photo above.)

In evolutionary terms, the seemingly inexplicable body control of the male praying mantis can be understood. In the process of natural selection, male mantises have been "favored" if their nervous systems make it possible for the lower abdomen to function without neural instructions from the head. Copulation, and therefore the survival of the species, can take place in spite of the cannibalistic female, which has been known to devour as many as seven willing mates in two days.

individual raise its offspring. Year-old scrub jays, for instance, sometimes give up their first reproductive opportunities in order to assist their mother with the care of the newest brood.

Animals do not make a conscious decision to be altruistic, or to concentrate on cooperative ventures that will include their genotype in the overall gene pool, but natural selection has favored those organisms that are programmed to propagate their genotype. An altruistic sibling might best perpetuate his genes through the successful reproduction of his brothers' and sisters' genes.

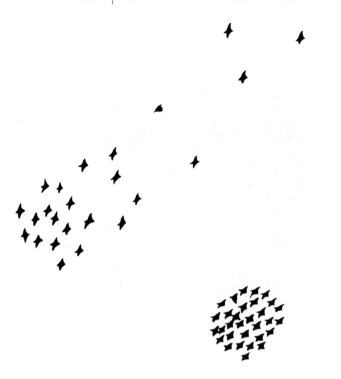

29.8
Alarm response in starlings. When threatened by a hawk, a loose flight of birds draws into a more compact flock. It may be more difficult for a hawk to single out a victim from among a dense flock, or diving through such a flock may be dangerous for the hawk, which can knock down one starling easily enough but may injure itself by hurtling into several. Whatever the reason, birds (and some fish) do habitually behave so.

Probably the most obvious advantage of groups is massed defense against predators (Figure 29.8), but there are many other advantages more subtle but equally important. Synchronous hatching gives the newborn birds the built-in protection of a group, and other advantages can be seen in predatory animals, such as lions and hyenas, that hunt in groups, and birds that flush out prey by flocking. Group grazing by prairie dogs exposes nutritious small plants, and emperor penguins conserve heat by huddling together while incubating their eggs. During the winter, bees increase the heat within their hives by constantly buzzing and moving their bodies to generate heat, sometimes producing an internal temperature 15 to 20 degrees higher than outside the hive. Insect societies provide countless examples of successful group living, and we will look at some of them now.

Social Insects

Most insects are among the most solitary of animals, partly because they do not live long enough to establish parent-child relationships. But termites, of Order Isoptera, may be the most social of all the insects.

Termites Termites cannot digest cellulose, a major ingredient of wood, which is nonetheless their main food source. They depend totally on symbiotic protozoa and bacteria living in their digestive tracts for the digestion of cellulose. Newborn termites acquire these microorganisms from their parents when they eat the elders' regurgitated food or fresh feces. Molting termites, which shed part of their digestive tracts along with their outer coverings, similarly depend on other termites to replenish their protozoa supply. Unlike most insects, then, the young termite must establish social contact with its parents and other termites if it is to survive.

The termite society is a caste system, as are the societies of other social insects. It consists of a primary, secondary, and tertiary queen, a king, and soldiers and workers. The term *caste* indicates a society regulated by a division of labor, one that is usually characterized by a rather limited and rigid system, where definite roles are determined by genetic, sexual, and nutritional factors. The blind workers groom and feed all the other members of the community, especially the primary queen, which secretes a presumably pleasant exudate and lays an egg every couple of seconds—between 40,000 and 80,000 each day (Figure 29.9). Despite the queen's amazing productivity, she receives none of the special old-age privileges visible in some vertebrate societies. When the queen shows signs of growing old, the workers starve her and then eat her. (African natives savor a termite cake made from termite paste and flour, and the queen is considered the greatest delicacy.)

Although termites are mostly blind, wingless, stingless, and relatively unprotected, they manage to exist in huge numbers. Part of the termite's durability is due to the functional architecture of its housing, which is complex, multi-leveled, air-conditioned, and custom-built to suit each environment. Termite dwellings in temperate zones are not usually visible, but in the tropics it is possible to see termite mounds more than five meters (16 ft) high (Figure 29.10).

The sterile workers are either male or female. Besides feeding all the other termites, they manage the daily operation and maintenance of the nest and carry off eggs as the queen produces them. Soldier termites have lobsterlike jaws half their total body size, with unreleasing pincers. They guard the nest, using their heads to plug holes until the workers have repaired the damage. Soldiers of some termite species have a specialized gland in the front of their heads, which releases a sticky chemical substance that incapacitates small enemies, but whatever the defense of the particular soldier termite, it is formidable. The king and queen exist only to produce a constant flow of nymphs. In an undisturbed society the nymphs develop into sterile workers, but there is a point in their development when their programs can be changed to produce soldiers or reproductive termites if a shortage has arisen among those castes. It is thought that such a drastic shift in development can be accomplished because of the absence of an adult-produced pheromone that is usually present and inhibits the nymphs from developing into anything but workers.

The winged reproductive males and females, called *alates*, are produced when seasonal conditions are favorable for them to leave the nests in swarms to seek mates and new nesting sites. Intraspecific swarming from several nearby nests appears to be synchronized, probably to ensure the survival of as

29.9
Termite society. A termite queen, with her great bloated abdomen, dwarfs the workers surrounding her. A helpless blob that must be fed and tended like a new baby, she is just an egg-laying machine with a head.

29.10

A termite mound in Africa. Such a mound is constructed, one grain of earth at a time, by thousands or even millions of individual workers in cooperation. They bind the soil with saliva and fecal matter, producing an architecturally elaborate structure that hardens into a product like concrete. No one knows how a termite, blind and all but mindless by human standards, can fit its bit of labor into a complex edifice that takes advantage of air currents for ventilation and may be five meters (16 ft) tall.

many termites as possible in an environment where predators are numerous. The wings of the alates are shed at the time of mating, the precise moment differing according to the species. As soon as mating occurs, the paired termites return to the darkness of a new nest, unable to survive in the light.

Honeybees As we know from our study of the language of the bees, the social organization of the honeybee is highly developed (Table 29.1). The honeybee society consists of one queen, sterile female workers, and male drones, totaling between 20,000 and 80,000 in each hive (Figure 29.11). Why do so many honeybees live together, and how is their society regulated to avoid conflict and chaos? It is clear that the overall benefit of the hive supersedes a bee's individual needs, and even the apparently rigid division of labor shown in Table 29.1 can be altered by means of the bees' remarkable communications system to cope with environmental changes. The dominant member of the honeybee society, the single female queen, does not take part in the routine activities of the hive. Her main function is to produce eggs, and she lays more than 1000 eggs each day of her life, which may last 5–10 years. The queen controls the population of the hive through the secretion from her mandibular glands of a pheromone popularly known as "queen substance." Actually, the queen secretes several pheromones, but it is a single compound, 9-ketodecenoic acid, that prevents the production of other queens and also inhibits the ovarian development of workers, so that no new eggs will be introduced into the hive. Thus the queen eliminates any rivals. The queen substance is licked off the queen by the workers. They quickly pass it on to all the other workers in a process of regurgitation known as *trophallaxis*, (from the Greek, *trophe*, "food," and *allaxis*, "exchange"), which we have already seen among termites. The exchange of food and pheromones through trophallaxis is essential for communication within insect societies. If the queen is removed from the hive, or if her pheromone production is otherwise decreased or halted, the workers will sense the loss in a matter of minutes and will become agitated.

Royal cells are not built as long as the mother queen is secreting "queen substance." But if the queen's production of 9-ketodecanoic acid decreases, as it usually does at the beginning of the spring reproductive season, the workers begin to construct large royal cells in preparation for a new queen. The queen lays one fertilized egg in each royal cell, and the workers feed a "royal jelly" to the larvae, enhancing the development of the larvae into queens. New queens, when one is needed, will develop into adults only 16 days after the eggs have been laid, compared with 21 days for a worker and 24 days for a drone. The old queen's decreased pheromone production signals another change in the behavior of the workers, who feed and groom her less and eventually expel her from the hive, which by now is overcrowded because of the new broods that have been produced during the peak spring nectar season. Soon afterward, the old queen and about half the workers will swarm to the site of a new hive, where they will form a new colony.

TABLE 29.1
Division of Labor in Honeybee Workers

1. *First three days.* The newborn worker leaves its cell, begins to clean cells.

2. After cells are cleaned, the queen deposits a new egg in each.

3. *Fourth day.* Worker collects honey and pollen from stores. Begins to feed the older larvae.

4. *Seventh day.* Begins feeding younger larvae, which receive an easily digested "milk" secreted by glands in the worker's head. Begins making short reconnaissance flights without gathering nectar or pollen.

5. *Tenth day.* Stops working with brood and takes on various household duties. Takes nectar from incoming foragers, deposits it in cells or feeds it to other bees, packs the pollen brought by foragers into the pollen cells, builds new cells using wax secreted by its wax gland, carries away dead bees and rubbish.

6. *Twentieth day.* Becomes a guard, inspects every bee arriving at the hive entrance. Twenty to thirty guards stand watch simultaneously, driving off intruders.

7. After a short period as guards, the workers become foragers, flying into the country and collecting nectar and pollen. They remain as foragers until they die about three weeks later.

8. Some foragers are scouts, looking for new food sources, and they communicate the position and quality of a food source by means of their waggle dance.

Adapted from N. Tinbergen, *Social Behavior in Animals*, Second Edition (London: Methuen; New York: John Wiley, 1965), pp. 101–102, from data originally formulated by Martin Lindauer.

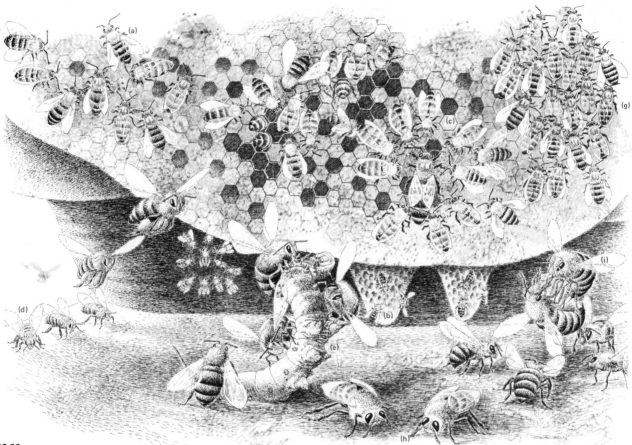

29.11

A beehive is a busy but well-regulated entity. Its seeming confusion is illusory. Many activities are going on, some all at once and others changing with the seasons, but each activity is precisely determined. Some of the main functions of a hive are shown here. (a) Most of the workers (sterile females) are engaged in bringing nectar and pollen from outside, furnishing the food for the colony and the material for building the waxy comb in which the young are raised. (b) Some "engineers" make wax from the pollen and construct the brood cells, objects that have long been a subject of admiration for their geometrical precision. (c) When workers, making up the great bulk of the population, emerge from their pupae and leave their cells, nurses prepare the empty cells for the next brood. (d) The duty of some workers is to ventilate the hive. They stand at the entrance and fan their wings to maintain the best combination of temperature and humidity for the health of the bees and the concentration of the honey. The evaporation of water from dilute nectar is like the boiling down of maple sap to make syrup. (e) Other workers defend the hive against accidental or threatening intruders, as shown by the attack against a wandering caterpillar. (f) One worker, having discovered a food source, is climbing the walls in a pattern that is understood by the other workers. They will take off to find the food source. (g) During cold weather, masses of bees clump together and vibrate their wings in an action something like shivering. They can generate enough heat to keep them from cold damage. (h) The males (drones) do nothing but wait for a chance to mate with a queen. Their lack of productivity is so well known that any lazy parasite on society is called a drone. The single queen in a hive is just an egg layer. She is fed and cared for by the workers, and in return she produces the thousands of eggs needed to keep the hive active. She can determine whether the eggs will be fertilized and yield diploid females, or will not be fertilized and yield haploid males. Whether the female will develop into a sterile worker or a fertile queen is determined by the nourishment provided to the growing larva. If queens are to be produced, special brood cells are constructed, and the larvae in them are fed with the queen-making diet. (i) When several queens emerge at once, they fight it out for supremacy until only one is left. She will leave the hive, fly high into the air, and there mate many times with accompanying drones. She stores enough sperm to last her lifetime. The whole intricate system works because each individual can react in response to what it sees, smells, tastes, and (probably) hears.

The old hive does not continue without a queen for long. Before leaving the hive, the queen also lays an unfertilized egg in each of the newly constructed drone cells. These unfertilized eggs develop into haploid male drones, which leave the hive and await the emergence from the hive of a new virgin queen. The young queen then disperses sufficient quantities of 9-ketodecenoic acid to attract drones waiting in a swarm downwind. The queen makes several nuptial flights a day, perhaps 15 flights in a week, until she has mated with several drones and accumulated enough sperm to fulfill the needs of a lifetime's production of fertilized eggs. These eggs will eventually become sterile female workers. During mating, the genitalia of the drone explode as he deposits his sperm in the queen's genital chamber, and he dies almost instantly. Drones that do not mate are driven out of the hive by the workers in the fall, when the food supply becomes low. After the new queen has completed her mating flights, she returns to the hive to join workers in an afterswarm, or she returns to the hive, kills the other emerging virgin queens, and begins a new colony within the hive.

Most species of bees are solitary. What are the advantages of living in a society as complex as the honeybees'? The greatest advantage is probably the increased efficiency of foraging, especially because flowers are a very spotty food source, and many honeybees working individually to acquire nectar would not be nearly as efficient as a highly communicative and cooperative society. Army ants, also highly social insects, benefit from societal living by mounting group attacks on their prey, an activity at which they excel.

Army ants Much research on army ants of the genus Eciton was done by Theodore C. Schneirla in Panama while he was associated with the American Museum of Natural History. Schneirla found that army ants, usually found in tropical areas, such as South America and Central America, are organized into nomadic colonies of 100,000 to 500,000 or more. They have the most complex animal society operating outside a fixed homesite. A colony of army ants consists of a single queen, a brood of young ants in varying stages of development, and sterile female workers. (Winged males appear during the mating season.) As with the honeybee, a division of labor exists among worker ants, but the division in the ants is formulated according to the worker's size instead of its age. The smallest workers feed the larvae; the medium-sized workers, making up the majority of the colony, spend most of their time raiding; and the largest workers, known as soldiers, are armed with large heads and powerful mandibles that make effective weapons against predators.

Army ants are raiders, typically living off other arthropods encountered on their systematic marches, which may advance at a rate of 35 meters (115 ft) an hour in an unbroken column 300 meters (1000 ft) long. Such an organized procession should be an unlikely activity for the near-blind ants, but they are guided by their own odor trail of pheromones, tipped onto the ground from the stingers of the lead ants and reinforced by the ants that follow (Figure 29.12). This *nomadic phase* of the ants' migratory activity lasts 17 days, during which they systematically move forward at dawn to conduct predatory raids, biting, stinging, and dismembering their prey (usually other

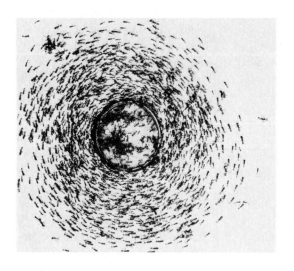

29.12

A whirlpool of army ants. When on the march, these ants follow a scent trail laid down by the leaders. In the experimental situation shown in the photograph, a circular trail was placed and ants started along it. If permitted to do so, they would probably keep going around until they starved, because they do not use their eyes or any individual initiative.

Army ants on the move. When a swarm of these carnivorous insects moves across the landscape, living creatures flee because the ants are too numerous, too effective in biting and tearing, and too inexorable to withstand. The mass advances like a living carpet, gliding along and conforming to every bump and depression on the surface of the ground, eating every unwary insect or larger animal they touch. Fortunately, they move at an ant's pace, and a person can walk slowly away from them, but the immensity of their numbers and their reputation for destructiveness make an observer watch his step.

insects). The raider workers doggedly lead the way, followed by soldiers and larvae-carrying workers. The huge queen, almost completely hidden by her retinue of tumbling workers, brings up the rear.

Unlike most species of ants, army ants do not build permanent underground nests. Each night after their raids are completed, army ants form temporary nests, or "bivouacs," by clustering together with their legs hooked to one another to form bristling, dangling cylinders that surround and protect the queen and brood. The nomadic stage ends abruptly, and the colony then halts its migration, beginning a *statary period* of 19 or 20 days. During the statary period infrequent raids take place, and very little prey is returned to the nest. The nomadic and statary phases coincide exactly with the queen ant's reproductive cycle.

The statary phase begins when the larvae from the last clutch of eggs have progressed to the pupal stage (Figure 29.14). After a week, the queen produces 20,000 to 30,000 new eggs, and the daily raids begin to increase. On the twentieth day a new generation of workers is born, and daily foraging raids resume at full speed. From these observations it can be concluded that the cyclical behavior pattern of the army ant is regulated not by external stimuli, such as the need for food, but rather by interactions between the adults and brood.

Like the social wasps and honeybees, army ants have relegated the male to a subordinate position in the society. Winged males are produced during the summer to prepare for their nuptial flights, which take place away from the nest. After mating, the males do not return to the nest, and without social contact with the colony they soon die.

Army ants carrying larvae. While the ants are traveling, the soldiers do the killing and supply the food, which is transported by workers to the members of the swarm. Meanwhile, nurses carry larvae along while the larvae continue to secrete a substance that stimulates the workers and soldiers. When the larval secretion diminishes and then ceases, the swarm stops and remains stationary until a new migratory phase is touched off by the appearance of a new generation of workers.

29.15

Baboons moving. This stylized representation of a traveling band has the quality of a prehistoric painting on a cave wall, showing the disposition of the different ages and sexes. The mature males, large and heavily maned, guard the fore and rear and sides of the troop, protecting the mothers with babies riding on their backs. Females in heat, indicated by their dark bottoms, are accompanied by mature males. The top-ranking male is in the middle. With such an organization, even the small and vulnerable are safe from the attacks of hunters, because the guardian males can be formidable antagonists (see Figure 29.17).

29.16

Peaceful coexistence. When a necessity must be shared, some species of animals mingle with little or no conflict. At this African water hole, zebras and antelope drink together. With water locally limited, it is to the advantage of all to drink without antagonism, and the animals have learned which ones are safe.

Social Organization in Vertebrates

Primates such as African baboons and Japanese macaques provide vivid examples of stratified vertebrate societies. As with other socially oriented animals, the behavior of baboons and macaques results from a combination of social and biological forces.

African baboons Baboon troops, usually containing about 60 members, are held together by a strict social order that permits individual relationships to function within the cohesive framework of the group. The troop functions and moves as a unit, in spite of inevitable tensions that accompany the competitive structure of the hierarchy and the unstable resources of the group. The central figure in a baboon troop, the dominant male, travels in the center of the troop, surrounded by females and juveniles (Figure 29.15). At the rim of the troop are the lower-ranking adult males, alert to join forces against predators that may attack from any direction. So well organized is the baboon defense system that perhaps only the lion is a real threat, and this ability to defend against predators is one of the major reasons for the evolutionary success of baboons. Indeed, baboons that leave the troop because of injury or for any other reason have little chance of surviving on their own.

Several troops of baboons will peacefully share a water hole where their territories overlap, and many other kinds of animals may be seen sharing the same water hole without frequent incidents of violence. Although individuals from adjacent baboon troops will rarely mingle, the groups are always aware of one another, and smaller troops will quietly make way for larger troops. Peaceful encounters among animals of different species are typical where a common source of food or water must be shared (Figure 29.16).

Within each baboon troop there is a firm dominance hierarchy ruled over by one or more high-ranking males. Dominant males protect lower-ranking members of the troop and exercise a general role of establishing and maintaining peace and order within the troop. Baboons usually avoid fights within the species, and low-ranking members are always willing to discontinue their disputes when the dominant male is in such a well-defined position of power that other members of the troop are naturally attracted to him and follow him whenever he decides to move (Figure

29.17). One of the benefits of his position is the sexual receptivity of females in heat (estrus), which approach the dominant male and make the formal gesture of presenting the swollen genitalia. When a female in estrus is accepted by the male, the two form a consort and remain together throughout the estrous period, grooming each other and copulating frequently. Usually the pair separates when the estrous period ends, but if the female has become pregnant, she may continue to follow the male closely.

Infant baboons are extremely attractive to all members of the troop. They will be fondled and groomed incessantly by males and females alike, even though they constantly cling to their mothers for one or two months (Figure 29.18). Apparently this early mother-child bond is a durable one; juvenile baboons retain their maternal attachments throughout their early years. As the infant develops, it spends more and more time with other young baboons in play groups, where it seems to practice adult gestures and begins to learn the social mannerisms of the adults. The importance of imitative play groups can be seen

29.17
Power display. When a male mandrill asserts his authority, he shows his teeth. An animal does not need to be very smart to recognize that such canine teeth as these can rip flesh to shreds. The mere threat of using them tends to be enough to subdue a potential rival. Struggles for supremacy are usually limited to a sort of tournament action, in which serious damage is rare.

29.18
A male baboon trying to get a reluctant baby to come to him. Males pay considerable attention to the young, fondling and grooming them, even though the very young ones tend to stay close to their mothers.

29.19
Behavior by imitation. The usefulness of having something to imitate is shown by baboons in successful and frustrated attempts to mate. In the left-hand pictures, a male mates with a female, using her legs as a platform in what is a normal procedure for baboons. The two right-hand pictures show the clumsy efforts (and failure) of a male in the presence of a receptive female. Having been raised in isolation and not having had the advantage of seeing others copulate, he has only an imperfect idea of how to behave. In troops, young baboons learn from their elders.

in experimental male juveniles that are unable to mount inexperienced females properly because they have never watched adult baboons copulating (Figure 29.19).

A complex communication system exists in most primate societies. The African baboon and Japanese macaque both have a repertory of vocal expressions ranging from a grunt to a shrill bark of warning. Friendly gestures may include lip-smacking and genital presentation. Scratching, shoulder-shrugging, and other bodily gestures are used to relate feelings of fear, escape, or uncertainty. Slapping the ground with the hands indicates threat or attack. An adult male will gently bite the nape of a female's neck to express sexual responsiveness. Besides tactile, visual, and auditory signals, baboons and macaques also communicate through a keen olfactory sense.

Japanese macaques Japanese macaque monkeys usually live in troops containing 50 to 150 members, substantially larger than a baboon troop but just as cohesive. A typically male-dominated social hierarchy exists, and a male's dominance rank is in direct proportion to the status of his mother. Each macaque participates in specific activities according to its sex, age, and rank, and it appears that these specialized activities are a major factor in the cohesiveness of the social system. As with baboons, the dominant male macaque has the highest-ranking position as leader and defender of the troop.

An example of the kind of behavioral flexibility that does not exist in the rigid societies of insects was observed in a confined macaque troop recently studied by G. Gray Eaton at the Oregon Regional Primate Research Center. One day after a snowstorm, a low-ranking male stopped eating a snowball and began to roll it in the snow until it was the size of a large beach ball. It is now common for this male and other macaques to roll large snowballs after every snowfall. Juvenile monkeys seem to delight in playing with the snowballs, and adults are content just to sit on them. Rather than having a definite and prescribed function, the snowballs seem merely to represent manifestations of the curiosity, adaptiveness, and tool-using skills of the monkeys. Such an ability to adapt their behavior to changes in their environment has certainly favored their survival.

Population Control in Animal Societies

Practically every animal species practices some form of instinctive population control to prevent starvation, crowding, and other typical problems of overpopulation. As soon as the population goes beyond the programmed "safe" level, animals will use birth-control methods as simple as continence (elephants in overpopulated situations copulate one-third as often as normal) to as drastic as cannibalism. During the 1950s John B. Calhoun of the National Institute of Mental Health placed 32 male and female Norway rats in a large enclosure with enough food and water to accommodate a rat population of 5000 after two years. Instead, the number of rats became stabilized at 150. In spite of the disease-free environment and adequate food supply, the rats became disrupted when their population reached the 150 level. Females were raped, and males became bisexual or sexually passive and fought and killed one another. Infants were neglected, and when they died, they were eaten by the adults. In all, only four percent of the young rats survived. In similar experiments with other animals, the results have been basically the same.

V. C. Wynne-Edwards of the University of Aberdeen has suggested that "man is almost alone in showing a long-term upward trend in numbers; most other animals maintain their population size at a fairly constant level." Wynne-Edwards hypothesizes that animal populations are held stable not by predators, starvation, and disease but by homeostatic regulatory systems within the animals themselves. Continuing his highly controversial hypothesis, Wynne-Edwards points out that many animals have synchronized community displays that serve as indicators of

daily population densities. These "population-pressure demonstrations" are most prominent prior to the mating season and at times of migrations. If the population is too high, the expendable members of the group (often determined by pecking orders) may be sacrificed. If the population is too low, new members will be accepted from among the waiting groups of lower-status animals. Such population displays may be seen among most birds, including starlings and blackbirds, and more vocal population displays are practiced by frogs, cicadas, and howler monkeys, to name a few.

Apparently animals have built-in regulatory mechanisms that direct their behavior when overpopulation threatens the survival of the species. Individuals acting in concert are the basis of a successful society, and in this chapter we have seen how animal behavior can be controlled for the benefit of the species. In the following chapters we begin to examine the human society and the parts we all play in shaping our relationships with the biosphere, the living world.

SUMMARY

1. Organisms living together in organized groups are said to live in *societies.*

2. A *territory* is a specific area guarded against intrusion by other members of the same species. Territories are most often established for purposes of food, shelter, or mating. Animals may declare their territorial claims in several ways, including scent markings and sounds; inherent in each method is a declaration of living rights.

3. *Threat gestures* are an attempt to intimidate and drive off an intruder without actually fighting, and *appeasement gestures* indicate a submissive position directly opposite to threat gestures.

4. Special conspecific signals are used to attract a sexual partner, to deter aggressiveness or threatening behavior, and to synchronize male and female actions so that copulation may take place. Sometimes courtship occurs as conflict behavior because animals normally try to avoid the closeness of other animals. Body contact initiated during courtship and mating arouses the ambivalent feelings of escape behavior and mating behavior at the same time.

5. Animals in a society exhibit a high degree of cooperation, made possible by a species-specific system of communication. One aspect of this system is *altruistic behavior,* in which an individual sacrifices its own reproductive opportunity to help a genetically related individual raise its offspring.

6. Most insects are solitary, but termites, army ants, and honeybees have rigid social organizations.

7. Some primates, such as African baboons and Japanese macaques, live in stratified societies. As with other socially oriented animals, the behavior of baboons and macaques results from a combination of social and biological forces.

8. Apparently animals have built-in regulatory mechanisms that direct their behavior when overpopulation or other factors, like food shortages, threaten the survival of the species.

RECOMMENDED READING

Alcock, John, *Animal Behavior: An Evolutionary Approach.* Sunderland, Mass.: Sinauer Associates, 1975.

Brower, Lincoln P., "Monarch Migration." *Natural History,* June/July 1977.

Dethier, V. G., and E. Stellar, *Animal Behavior,* Third Edition. Englewood Cliffs, N.J.: Prentice-Hall, 1970.

DeVore, Irven (ed.), *Primate Behavior: Field Studies of Monkeys and Apes*. New York: Holt, Rinehart and Winston, 1965.

Eibl-Eibesfeldt, Irenäus, *Ethology: The Biology of Behavior*, Second Edition (translated by Erich Klinghammer). New York: Holt, Rinehart and Winston, 1975.

Eisner, Thomas, and Edward O. Wilson (eds.), *Animal Behavior*. Readings from *Scientific American*. San Francisco: W. H. Freeman, 1975. (Paperback) (This collection contains many useful articles that would otherwise be recommended separately.)

Eisner, Thomas, and Edward O. Wilson (eds.), *The Insects*. Readings from *Scientific American*. San Francisco: W. H. Freeman, 1977. (Paperback)

Emlen, Stephen T., "The Stellar-Orientation System of a Migratory Bird." *Scientific American*, August 1975. (Offprint 1327)

Fabre, J. Henri, *The Insect World of J. Henri Fabre* (edited by Edwin Way Teale, translated by Alexander Teixeira de Mattos). New York: Dodd, Mead, 1966. (Also available in paperback)

Hailman, Jack P., "How an Instinct Is Learned." *Scientific American*, December 1969. (Offprint 1165)

Hölldobler, Berthold K., and Edward O. Wilson, "Weaver Ants." *Scientific American*, December 1977. (Offprint 1373)

Keeton, William T., "The Mystery of Pigeon Homing." *Scientific American*, December 1974. (Offprint 1311)

Lawick-Goodall, Jane van, *In the Shadow of Man*. New York: Dell, 1971.

Lore, Richard, and Kevin Flannelly, "Rat Societies." *Scientific American*, May 1977. (Offprint 577)

Lorenz, Konrad Z., *King Solomon's Ring: New Light on Animal Ways*. New York: Thomas Y. Crowell, 1952. (Also available in paperback)

Manning, Aubrey, *An Introduction to Animal Behavior*, Second Edition. Reading, Mass.: Addison-Wesley, 1972. (Paperback)

Menzel, Randolf, and Jochen Erber, "Learning and Memory in Bees." *Scientific American*, July 1978.

Morse, Roger A., "Environmental Control in the Beehive." *Scientific American*, April 1972. (Offprint 1247)

Olton, David S., "Spatial Memory." *Scientific American*, June 1977. (Offprint 578)

Pengelley, Eric T., and Sally J. Asmundson, "Annual Biological Clocks." *Scientific American*, April 1971. (Offprint 1219)

Sauer, E. G. F., "Celestial Navigation by Birds." *Scientific American*, August 1958. (Offprint 133)

Saunders, D. S., "The Biological Clock of Insects." *Scientific American*, February 1976. (Offprint 1335)

Schneider, Dietrich, "The Sex-Attractant Receptor of Moths." *Scientific American*, July 1974. (Offprint 1299)

Smith, John Maynard, "The Evolution of Behavior." *Scientific American*, September 1978.

Tinbergen, Niko, *Animal Behavior*. New York: Time-Life Books, 1965. (Life Nature Library)

Tinbergen, Niko, *The Study of Instinct*. New York: Oxford University Press, 1951. (Paperback)

Topoff, Howard, "Ants on the March." *Natural History*, December 1975.

Von Frisch, Karl, *Bees: Their Vision, Chemical Senses, and Language*, Revised Edition. Ithaca, N.Y.: Cornell University Press, 1971.

Von Frisch, Karl, "Dialects in the Language of the Bees." *Scientific American*, August 1962. (Offprint 130)

Wehner, Rüdiger, "Polarized-Light Navigation by Insects." *Scientific American*, July 1976. (Offprint 1342)

Wiley, R. Haven, Jr., "The Lek Mating System of the Sage Grouse." *Scientific American*, May 1978.

Williams, Timothy C., and Janet M. Williams, "An Oceanic Mass Migration of Land Birds." *Scientific American*, October 1978.

Wilson, Edward O., "Animal Communication." *Scientific American*, September 1972. (Offprint 1258)

Wilson, Edward O., "Pheromones." *Scientific American*, May 1963. (Offprint 157)

Wilson, Edward O., *Sociobiology: The New Synthesis*. Cambridge, Mass.: Belknap/Harvard University Press, 1975. (Available in paperback)

Wilson, Edward O., "Slavery in Ants." *Scientific American*, June 1975. (Offprint 1323)

Wilson, Edward O., *The Insect Societies*. Cambridge, Mass.: Harvard University Press, 1971. (Available in paperback)

30

The Individual and the Environment

ECOLOGY IS THE STUDY OF ORGANISMS AND their relation to all the forces that act upon them. These forces include general climate, local weather, the structure and composition of soil, the amount of contained materials in water, and the activities of organisms themselves. It is frequently necessary and always desirable to include the history of organisms and the history of the environment, because what has happened in the past helped determine what is happening now.

Ecology is not esthetics or nature study or self-preservation or concern for wildlife or antipollution measures, and yet all these may enter into ecological studies. Although we commonly hear it, the statement "That is bad for the ecology" is meaningless. Excess sulfur dioxide in the air may destroy lichens on rocks, and DDT in a river may poison salmon, and to be sure, that is bad for the lichens and the salmon, but it is not "bad for the ecology," that is, for the study of organisms with respect to their environment.

Farmers, hunters, fishermen, and sportsmen have been practicing ecologists ever since they began their pursuits, but a systematic study of environmental biology began only in the late nineteenth century. "Ecology" as a term was invented in 1886 by Ernst Haeckel, for use with animals only. Even then, productive investigations were hampered by a lack of information, since ecology is dependent on many other disciplines, especially physiology, genetics, taxonomy, geology, and of course physics, mathematics, and chemistry. Indeed, any ecologist working on a specific problem could say that the project is a neurological one, a biochemical one, or whatever. Conversely, almost any biological investigation is ecological, inasmuch as the activities of organisms occur only in some definite environment. Photosynthesis works only when light *in the environment* acts upon a chloroplast, which is provided with water, carbon dioxide, and inorganic nutrients *from the environment*. A DNA mutation in a bacterium is useless, either positively or negatively, unless the result of the mutation can be expressed phenotypically and tested *against the environment*. Viewed in this way, ecology is as much an attitude as a discipline, but the ecological way of thinking has influenced biology in all of its aspects. The very word "ecology," coming from the Greek word for house, refers to the fact that every organism exists in some dwelling place. The fact that ecology overlaps the other subdivisions of biology makes for some vagueness but at the same time provides an integrating force that helps the other subdivisions make sense.

Ecologists may study an organism mainly with regard to the reaction of individuals to their environment, or they may study the interactions of organisms in groups, paying attention to both the biological and the physical environment but thinking more of group activities than of individual organisms. The first is referred to as *autecology* (Greek, "one's own house"), and the second is *synecology* (Greek, "houses all together"). In the present chapter of this book, emphasis will be on autecology, although sharp separation of the two ways of studying ecology is obviously impossible. Chapter 31 will emphasize synecology.

ECOLOGY STARTS WITH THE CLASSIFICATION OF ORGANISMS

It is imperative in most biological work to know what organisms are being studied. For practical reasons alone, not to mention the human philosophical need to have satisfactory labels, ecologists must know what species they are dealing with, or else the collection of experimental, quantitative data is useless. One of the first things any ecological study requires is an inventory of the organisms being studied, and the inventory should be as exact as possible. It is not enough just to list "grass," for there are hundreds of

grasses, and they vary in their requirements and in their productive ability. Nor is it enough to write "frogs" or "the frog," as though all frogs were alike. For many years, serious scientific biologists sneered at ecology because it was descriptive and not experimental. Now that ecology has started to become experimental and quantitative, it has somewhat better standing in the academic world, but it still must rely heavily on the qualitative listing of organisms, that is, on their identification.

HABITATS ARE THE PLACES WHERE ORGANISMS LIVE

Ecologists, taxonomists, geneticists, and geographers use the term *habitat* (Latin, "living place") to mean a kind of local environment that has its own peculiarities (Figure 30.1). A habitat is a real space, whose features are determined by its climate, geological nature, past history, and present inhabitants. One example of a habitat is a ravine in the high hills of the eastern United States. There the climate, including temperatures, amount and distribution of rainfall, and time and intensity of illumination, are determined by the latitude and the configuration of the surrounding terrain, as well as by wind direction and distance from sources of atmospheric moisture. The geological nature of such a habitat is determined by the kind of bedrock that furnished the soil, the amount of erosion, the slope (which will affect water runoff and illumination), and the depth of the soil. What has happened in any given habitat can make great differences in any current situation. It may have been at one time under the sea (in the example under consideration this would have been many millions of years ago), it may or may not have been glaciated, or it may or may not have been cut for lumber in more or less recent times. Finally, the living inhabitants help determine the nature of the living space, because a place dominated by hemlock trees is different from a place dominated by locust trees. The plants and animals that live in a place are the ones that were able to get there. The plants grew from seeds blown or brought in, and the animals or their ancestors came by immigrating under their own power. The whole complex interplay of past and present forces extends back into geological history, but it exists now to make the habitat. Other examples of habitats, some natural and some made by human action, are a tropical sandy beach, a marine cliff in the Pacific northwest, an Indiana cornfield, a brackish Maryland estuary, a sugar maple

30.1
An example of a habitat. Here, on the edge of a pond, the light is bright, the soil is moist and heavily organic, and the plants are low-growing, mostly sedges and cattails. Animals may include redwing blackbirds, an occasional marsh hawk, garter snakes, land snails, and many insects.

grove in West Virginia, a Brazilian rain forest, and a dry hillside in southern Spain. Thousands of variations exist.

NICHES ARE DETERMINED BY THE ACTIONS OF A SPECIES

Within habitats, organisms survive by using *niches*. Not real spaces, niches are, rather, occupational specializations, determined by the physiological or behavioral actions of any given species. Consider, for instance, the *habitat* of a white cedar swamp in southern New Jersey. It experiences no great fluctuations in temperature, the underlying soil is mostly sandy, and there is abundant water and a high water table. The swamp has been little disturbed by geological events or by humans for many thousands of years. In that habitat the largest and most influential organisms are the white cedar trees, adapted physiologically to the special requirements of living in a swampy situation. There are high-bush blueberries, occupying their niche of being low shrubs, thriving in acid soil, and taking advantage of such sunlight as they can get. Swimming on the surface of the slow-moving water, adult Hydrophilid beetles fill a niche of scavenging organic debris, while over their heads fly insect-catching dragonflies filling their own niche. Water snakes fill theirs, living on frogs and other small animals. So each of the hundreds of species in the habitat survives by using a particular part of the habitat in such a way as to provide for itself shelter, a nutrient supply, and a chance to reproduce.

ECOLOGICAL FACTORS AFFECT ALL ORGANISMS

In studying the factors that influence the distribution and activities of organisms, ecologists have considered that the important factors—not listed in order of importance, because that would be useless as well as impossible—are the *climatic* factors, including light, water, etc., the *biotic* factors, including the influence of organisms on one another, and the *edaphic* factors. The last were originally intended to be factors determined by the soil, as the origin of the word (Greek, "soil") indicates. In practice now, any ecological factors that are neither climatic nor biotic are lumped into the edaphic category, as for example, radiation fallout or the action of residual pesticides.

CLIMATE AND MICROCLIMATE ARE IMPORTANT DETERMINANTS OF HABITATS

Climate includes all aspects of weather: rainfall, temperature, light, and all their variations. Climatic factors are the most immediately apparent forces determing the nature of any habitat, and they have an overriding influence on the dwellers therein. As it is broadly treated by weather-data collecting stations, climate is usually thought of as covering a rather wide geographical area, at least a number of square kilometers. Tables of data, compiled over the years, can give the monthly distribution of rainfall, maxima and minima of temperature for all the months, duration and intensity of insolation (sunlight energy), and

30.2

Examples of microclimates. In this stream through woods, there are hundreds of small variations in exposure to light and wind and water. A single rock may have areas that are different enough to encourage distinctly different populations. Part may be permanently submerged, part wet only during rain or unusually heavy water flow, part (on the south side) hot, and part (on the north side) cool. A spot under a dense bush is cooler, dimmer, and moister than a spot under a break in the overhead trees. Small plants, such as mosses, and many invertebrate animals, especially insects, have narrow, specific requirements, and they will live only in places that meet those requirements.

30.3

An ecosystem. This aerial view of a small pond shows a well-defined place that harbors a number of species, most of which live their entire lives here. Birds and other large animals may come and go, but many of the small animals and all the plants stay here, forming a complex system that maintains itself, changing little year after year.

wind speed and direction, as determined for, say, the city of Washington, D.C. But the data are averages, obtained from several arbitrarily placed, permanent recording stations, and those averages may be far from the actual weather conditions in the mall in front of the Lincoln Memorial or on the bank of the creek in Rock Creek Park. Small local differences in elevation, shading, direction of slope, proximity to hills or buildings or trees, or underground water can make considerable differences in *microclimate*. Microclimates, in turn, make microhabitats, which provide a diversity of niches for many plants and animals. Let us say, for example, that two rain gauges are placed under a large tree, one under the periphery of the longest branches and one close to the trunk. As measured by the gauges, the annual rainfall in the two places, perhaps only six meters (20 ft) apart, may differ as much as 50 percent. Ground-dwelling insects are sensitive to differences smaller than that, and as a result, species distribution of insects will be so strongly affected that one may find a whole different fauna in the soil of those two microhabitats (Figure 30.2).

ECOSYSTEMS ARE INDEPENDENT GROUPS OF ORGANISMS

Any partially or completely self-contained mass of organisms is an *ecosystem* (Figure 30.3). An ecosystem can be as simple and as exclusive as a culture of algae in a test tube, or as large and complex as the entire earth, which has been referred to, almost *ad nauseam*, as a space ship. In common usage, most ecosystems are somewhere in between those extremes, both in size and in complexity. Some common ecosystems are patches of woods, meadows, freshwater ponds, and brackish estuaries. In nature, none of these are closed. Organisms are more or less free to enter and to leave, and materials flow or are carried in and out. Like a living organism, an ecosystem is likely to be in a constant state of change, even while it is maintaining a degree of stability. Indeed, ecologists think of and commonly speak of ecosystems as superorganisms, each with some ability to achieve a kind of homeostasis comparable to the homeostasis that individual organisms maintain.

If all the ecosystems of the earth are considered as part of a still larger system, embracing all the earth's organisms, we arrive at the concept of the *biosphere*. The biosphere is the living portion of the earth, as distinct from the air, which is the *atmosphere*, and the crust, which is the *lithosphere*. Since these concepts are properly part of synecology, they may seem to belong in the next chapter, but some understanding of the ideas and jargon of ecology is necessary even to start. Besides it is important to emphasize, especially in ecological studies, that strict compartmentalization is neither possible nor desirable. Although each organism lives its own life and dies its own death, it never does anything completely alone. If it eats, it must get its food from some other organism. If it is photosynthetic, it depends on other organisms for their ability to recycle carbon dioxide and mineral nutrients. In a way, autecology is an artificial study because of this inescapable interdependence of organisms. However, many biologists find that progress in synecology requires that one first understand how individual organisms function *as individuals* before one can make useful generalizations concerning intraorganismic relations.

PRIMARY PRODUCTIVITY IS THE CONVERSION OF ENERGY TO A BIOLOGICALLY USEFUL FORM

The biosphere and all the ecosystems it comprises require energy, whose source is mainly the sun. The energy of sunlight is fed into the system by way of photosynthetic organisms. The capture of radiant energy and its transfer into usable food constitute the *primary productivity* of a system. Although the total energy gain in an ecosystem is the sum of the gains of all the individuals, it is of course up to each autotrophic organism to add to that total.

The rate at which photosynthesis can proceed is regulated by many factors, all working together:

1.   The intensity of illumination is affected by latitude, by the amount of particulate matter in the air (whether smoke, smog, biologically produced aerosols, or water droplets), by the direction of slope of a surface (a south-facing slope gets more direct sunlight than a north-facing one), by the overshadowing of higher plants, which varies from species to species and from season to season. For aquatic autotrophs, light is diminished by depth, since water itself has some absorbing power, by turbidity, and by the density of the autotrophs themselves.

2.   Carbon dioxide concentration is usually fairly constant, because winds keep the atmosphere well mixed, and water can dissolve as much as the atmosphere provides. However, plants could use a higher concentration than they actually have if they could get it. The amount of carbon dioxide is usually the limiting factor in photosynthesis in daylight in terrestrial plants, but in deserts, water may be the limiting factor.

3.   When the temperature is below freezing, water is limiting, and it may be so in places where there is water, but it is so full of dissolved salts that it is unavailable as usable water to living cells.

4.   Inorganic nutrients must be present in adequate supply. Some, such as nitrogen, potassium, phosphorus, and sulfur, are needed in relatively large amounts, and some in quite small amounts. Even if only a few molecules of a required element are needed but those few molecules are lacking, growth and photosynthesis are reduced or stopped. Some soils in Florida, for example, are deficient in boron, and some in Australia are deficient in zinc. In the sea, where there is enormous productivity, iron and phosphorus are usually in short supply. These elements are never really abundant in the sea, and the small amounts present are taken up by the marine autotrophs. If these autotrophs die and sink before they are eaten by some animal, they carry the scarce elements with them to the bottom. The elements are then out of circulation, but sometimes special water movements stir them up. The most famous of the ocean currents that brings about this *upwelling*, the Humboldt current along the west coast of South America, brings the deep, cold water, rich in mineral nutrients, to the illuminated surface. There productivity is so intense that the Peruvian coast is one of the richest biological

30.4

The biological effect of upwelling in
the ocean. In this view of the Peruvian
coast, the flocks of sea birds, especially cormorants, feed sumptuously
on the millions of fish in the surface
waters. The fish in turn eat smaller
animals, such as crustaceans, and the
small animals eat the *primary producers*, the photosynthetic algae. The
mass of plant material depends on the
flow of mineral nutrients, brought up
to the light by the Humboldt current,
a flow of cool water circulating around
the South Pacific Ocean and swinging
past the cold Antarctic region. This
stirring of the water brings a concentration of inorganic material to the
region and the result is a concentration
of biological activity unmatched in
any other place in the seas of the
earth.

30.5

A champion energy trap: a field of
sugar cane. An area like this can take
from the sun as much as 25,000 kilocalories per square meter per year. The
light is intense, the temperature high,
the water supply constant, and the
soil rich. Add all this to the natural
efficiency of a C-4 type photosynthesis, and the result is a highly productive system.

areas in the world, enabling thousands of tons of fish
and fish-eating birds to thrive (Figure 30.4).

Measurement of Productivity

In principle, it should be fairly simple to measure productivity by measuring increase in plant dry weight
or output of oxygen or loss of carbon dioxide, but in
practice, the accuracy of such methods is limited. Accuracy may be reduced by the respiratory activity of
the autotrophs themselves. In some samples, it may
be the respiratory activities of inevitably accompanying organisms. In many plants the accuracy-reducing
factor is the specially increased respiration caused
by light (photo-respiration) (Chapter 4).

When productivity of land plants is measured,
one common procedure is to weigh a measured area
of leaf at the beginning of a day and compare that
with the weight of a similar sample after a given time
in light. From such data, making corrections for respiration and transport of sugars away from the site
of photosynthesis, one can estimate the total productivity of an entire ecosystem. In aquatic systems, a
sample of water is illuminated, and the rate of oxygen
production is measured while a similar sample is kept
in the dark for measurement of oxygen consumption.
The actual productivity then is about equal to the
measured oxygen output of the illuminated sample
plus that which, it is assumed, is simultaneously being used in respiration. The amount of chlorophyll in
such a sample can be determined by measuring the
amount of light absorbed. Once these data are available, other samples can be studied merely by measuring light absorption, a simple matter with a spectrophotometer. Chlorophyll determinations are widely
used as indicators of the productivity of waters.

From data on productivity, it is possible to estimate the energy-trapping abilities of various ecosystems. The usual way of expressing the figures is in

Kilocalories per square meter per year (Kcal/m²/yr). As would be expected, deserts are the lowest in production, with a rough average of about 500 Kcal/m²/yr. Productivity in mountain forests, arid grasslands, and in the seas along continental shelves is variable but somewhat better, ranging between 500 and 3000. Moist grasslands or forests and ordinary croplands may trap between 3000 and 10,000 Kcal/m²/yr, but oceans beyond the continental shelves are relatively poor at around 1000. The truly heavy producers are such special marine ecosystems as brackish estuaries and coral reefs, or intensively cultivated croplands, whose totals may reach 10,000 to 20,000 Kcal/m²/yr, with exceptional places achieving up to 25,000. The most productive areas in the world are sugar cane fields (Figure 30.5), which have everything helping: dense plant cover, abundant water, full sun, adequate mineral supply, and a photosynthetic system that utilizes four-carbon acids (C-4 plants, described in Chapter 4). Theoretical values up to 50,000 Kcal/m²/yr are possible, but not on a large scale or for protracted times.

If we know the productivity figures for different kinds of territory and the areas covered by each of them, we can calculate about how much energy conversion occurs throughout the world in a year. The oceans cover about 360 million square kilometers (140 million square miles). With an average productivity of about 1000 Kcal/m²/yr, the total marine productivity is about 0.4 billion billion Kcal/yr. For many years it was thought that because the oceans cover so much of the earth, the greatest photosynthetic activity would be in the water, but more recent and careful measurements show that oceanic productivity is probably less than half the world's total (Table 30.1).

TABLE 30.1

Estimated Gross Primary Production (annual basis) of the Biosphere and Its Distribution Among Major Ecosystems

ECOSYSTEM	AREA, MILLIONS OF KM²	GROSS PRIMARY PRODUCTIVITY, KCAL/M²/YR	TOTAL GROSS PRODUCTION, 10^{16} KCAL/YR
Marine			
open ocean	326.0	1,000	32.6
coastal zones	34.0	2,000	6.8
upwelling zones	0.4	6,000	0.2
estuaries and reefs	2.0	20,000	4.0
Subtotal	362.4	—	43.6
Terrestrial			
deserts and tundras	40.0	200	0.8
grasslands and pastures	42.0	2,500	10.5
dry forests	9.4	2,500	2.4
boreal coniferous forests	10.0	3,000	3.0
cultivated lands with little or no energy subsidy	10.0	3,000	3.0
moist temperate forests	4.9	8,000	3.9
fuel-subsidized (mechanized) agriculture	4.0	12,000	4.8
wet tropical and subtropical (broadleaved evergreen) forests	14.7	20,000	29.0
Subtotal	135.0	—	57.0
Total for biosphere (round figures; not including ice caps)	500.0	2,000	100.0

From E. P. Odum, *Fundamentals of Ecology*, Third Edition (Philadelphia: W. B. Saunders, 1971), p. 51. Used with permission.

Limits to Productivity

Aside from the general need for photosynthesis to keep the biological world going, the human population has a special interest in productivity because that is the source of all food. With an ever-increasing number of people to be fed, even with many of those already alive undernourished, it is well to examine the reasons why we do not have an unlimited amount of food. The limitations can be in energy, materials, and space, and in some instances can derive from social prejudices.

The primary energy source for photosynthesis, sunlight, is present in abundance. Under ordinary growing conditions, plants manage to convert something less than one percent of the available sunlight energy into carbohydrate. The rest may be reflected, it may not reach a chloroplast, or it may be transmitted or dissipated as heat. So far, however, efforts to improve the efficiency of plants in natural light have been such as to require additional energy input from other sources. Closer planting, to make fuller ground cover, introduction of genetically improved strains of plants, and more intensive chemical fertilizing of soil, all require materials and machinery that use energy, usually petroleum. Nevertheless, one possible factor that, with better knowledge and skill, could be used more effectively is the energy of the sun.

Photosynthesis does not run just on energy, carbon dioxide, and water. It requires as well the mineral elements that necessarily go into the building of protoplasm. In nature, those elements are more or less uniformly distributed, but "more or less" is not good enough for intensive productivity. We have already stated that soils are sometimes deficient in certain elements and the sea is notably low in iron, phosphorus and potassium. Efforts can be made to provide artificial upwelling comparable to the natural upwelling of nutrients in the Peruvian Humboldt current, but that, too, takes energy. It also takes energy to process and distribute the nitrates, phosphates, and potassium in chemical fertilizers required by high-intensity agriculture. Answers to the problems of material supply are complex and still not available.

Space for growing things is limited. Until the present century, whenever land became unfit for cropping as a result of depletion or erosion, new land was opened. But most of the earth's usable land is already under cultivation and new land is rapidly running out. Even now, much inferior territory is being cultivated, and the more it is used, the poorer it becomes. Some optimists have looked to such unused tropical expanses as the Amazon basin. However, not only are these limited in extent, but tropical soils have less continuing durability than temperate soils do. Opinions and predictions vary, but a general belief is growing that with the ever-increasing demands of human population, agriculture will be harder and harder pressed to feed all the people. One hopeful group used to think that the vast expanses of the oceans could be made to provide practically unlimited amounts of food, but with more technically effective harvesting methods and more intensive fishing, the waters of the earth have been yielding less, not more. It may be within the scope of technology to use the available space to better advantage, but the necessary improvements are not yet with us.

Space for living is not necessarily connected to space for food production, but it is still pertinent to the discussion of space. Even if we make better use of agricultural and marine space by immensely improved procedures not now known, we will still need a certain amount of space. There is no agreement as to what constitutes minimal living space for a human being, but there is agreement that man does not live by bread alone. There is also agreement that if the human species continues to propagate at its present rate, even if sufficient food is somehow produced, individual lives will be restricted, dull, and depressing. There is a danger that human ecology will differ from the ecology of a termite heap only in that the individuals in the human heap will be conscious of their unhappiness.

One peculiarity of the utilization of food is that of human preferences. Said to be omnivorous, the human will eat almost anything, but not everybody will eat everything. Conventional Western diners are revolted at the thought of eating roasted beetles or parasitic grubs dug out of the skins of reindeer, or of drinking milk mixed with blood drawn from cattle veins. Japanese prisoners during World War II complained at being fed steak and potatoes when they wanted a "normal" diet of fish and rice. Attempts at using high-protein fish meal in undernourished parts of the world were met with resistance. People (and some dogs) will starve rather than eat an unaccustomed food. No one knows how this problem will be solved if and when we come to a time of worldwide starvation.

POLLUTION CAN ACT AS AN ECOLOGICAL FACTOR

Pollution does of course affect populations, but only through its effects on individuals. In a narrow sense, therefore, pollution is a fit subject for studies in autecology. Pollution is not new. When pre-Columbian

30.6
The polluters. Many industrial installations produce such quantities of particulate matter (soot and ash) and gases that they invite the use of such words as "spew" and "belch." Increasingly serious attempts are being made to change such scenes as this, because it is recognized that pollution from combustion is destructive to human beings, to wild and cultivated plants and animals, and even to inanimate objects.

Indians piled heaps of clam shells on the shores of the Chesapeake, or when ancient Chaldeans dumped their refuse outside their huts, that was pollution. What is new about pollution and what makes it suddenly ominous is its sheer enormity and the intrusion of new and dangerous components. Old kitchen refuse heaps may have been unsightly, but they harmed no one, and if they became too smelly, a tribe could move away easily enough. Now there's no moving away from the euphemistically named sanitary landfills, because there's no place to go. There is no escaping the wind-borne smog or the water-borne effluent from plastic factories (Figure 30.6).

Most pollution is like a weed: not intrinsically "bad" but out of place. Humans do not make mercury or sulfur or heat, but they do move mercury and sulfur and heat from a place where it does no biological harm to some other place, where it can and frequently does do harm. Shifts and concentrations of materials and energy can result in pollution, but modern industry has introduced still another threat in the form of synthetic chemicals. These are dangerous partly because the organisms of the earth have never evolved enzymes capable of digesting them and partly because organisms have no defenses against them. The sawdust and slabs from a sawmill are a temporary nuisance, which can be reduced within a few years to reusable water and carbon dioxide, but the polychlorinated biphenyls (PCBs) from plastics, poured into flowing streams, remain to do their biological damage for indefinite periods of time. Many products of modern chemical syntheses are neither attackable by enzymes—that is, they are not *biodegradable*—

nor readily oxidized or otherwise destroyed by such natural agents as air and water. This fact has been most fully documented for the chlorinated hydrocarbon pesticides such as dichloro-diphenyl trichloroethane (DDT) and its chemical relatives aldrin and dieldrin. The problem of nondegradable pesticides will be considered further in the discussion of food webs in Chapter 31.

The easiest of the pollutants to control is the waste from human biological activities. Techniques for reuse of sewage-laden water are available. Theoretically, water can be used over and over indefinitely. After all, rainwater is distilled and recondensed, and it is biologically and esthetically quite acceptable (provided that it does not carry down with it too much sulfur dioxide from the air as it descends). The organisms of this earth have been using recycled water for millions of years without visible harm. The difficulty is cost. The mere removal of solids from sewage, or the *primary treatment*, is relatively inexpensive. *Secondary treatment*, by which biological materials are mostly removed, costs two to three times as much. *Tertiary treatment*, depending on whether chemicals and nutrients are simply reduced or cleaned out enough to make the water drinkable, costs five to ten times as much as primary treatment. At present, the processes of filtration and chlorination can make water clear and free from living bacteria, but they have no effect on dissolved chemicals.

Air Pollution
Most of the serious air pollutants result from industrial activities and from heating and transport (Figure

30.7
The effects of sulfur dioxide, a common product of factory waste, on plants. Weeds are commonly thought of as tough, but they are alive and subject to the same kind of damage as are any other organisms. They can be used as indicators of pollution, as shown in these spotted leaves of blackberries, Rubus (left), and a common weed, lambsquarter, Chenopodium (right).

CANCER FROM THE AIR

Asbestos is one of the best-known cancer-producing agents. Made of fine fibers, it can float as dust in air and can be inhaled. Once asbestos is drawn into lungs, the sharp pointed particles, like microscopic needles, can penetrate cells, damaging interior lung surfaces. The fibers can also work completely through lung tissue and come to rest in the lining of the chest cavity, the mesothelium. In either place, asbestos can stimulate abnormal growth: lung cancer or cancer of the mesothelium (mesothelioma). Indeed, mesothelioma is not known to be caused by any agent except microscopic fibers. Even if asbestos does not produce cancer, it can cause asbestosis, a dangerous, even fatal scarring of lung tissue.

Cilia in the air passages can sweep some asbestos fibers back up from the lungs to the pharynx. When swallowed, they can cause cancers of the esophagus, stomach, large intestine, and rectum.

Asbestos is a special form of the mineral serpentine, frequently found as crocidolite in Africa or as chrysotile in the United States. Since asbestos occurs as soft fibers, it can be matted into feltlike sheets, woven into cloth, or added to such building materials as wall board and pipes. Like sand, it is a silicate and does not burn. That quality has been known for a long time. Marco Polo told his unbelieving thirteenth-century listeners about an oriental cloth that was not consumed by fire, and Benjamin Franklin used to amuse his guests by tossing a sheet of "paper" into flames and watching it lie there undamaged. He had made it of Pennsylvania asbestos. Because of the flame resistance and insulating properties of asbestos, millions of tons have been used in the building industry.

Since the early 1900s the public health menace of asbestos has been increasingly recognized. In Africa, Europe, and America, workers in asbestos-processing plants have been found to suffer from lung cancer eight to ten times more frequently than the rest of the population; in some places as many as 40 percent of the people in constant contact with asbestos were being killed by various diseases due to asbestos fibers. Even families of workers were stricken when they inhaled dust brought home on fiber-laden clothes. Men who sprayed asbestos onto walls to fireproof them released tons of dust into the air to be breathed in by everyone.

30.7). These include sulfur dioxide, which can be washed down in rain or inhaled to form sulfuric acid, oxides of nitrogen, which can be converted to nitrous acid, carbon monoxide, which combines with hemoglobin to the exclusion of oxygen, particulate matter such as carbon and siliceous dust, and lead from anti-knock additives in gasoline. (Lead is a *cumulative poison*, which is neither degraded nor excreted, but builds up concentration if more and more enters an organism.) An ironic aspect of smog pollution is the photochemical effect, which occurs when the same sunlight that powers photosynthesis also causes the conversion of hydrocarbons, mainly from automobile exhausts, into a variety of compounds that cause irritation of eyes and respiratory tracts.

One aspect of air pollution on organisms is the phenomenon of *synergism*. A single minor irritant may be scarcely noticed, and the same may be true of a second one, but the combined effects of the two are much greater than a simple addition. For example, if benzpyrene, a well-known carcinogen (cancer producer), is inhaled as an aerosol, it may or may not cause lung cancer, but if it is inhaled with sulfur dioxide, the resulting damage is worse than it would have been if either of the compounds had been inhaled separately. In view of the number of substances present in the air and the incomplete state of our knowledge concerning possible synergisms, it is clear that we are exposed to many dangers of which we are not even aware. (See the essay below.)

Radiation Pollution

Radiation polution from human activities has been a completely new problem since the end of World War

---

Industrial use of asbestos is now being controlled, and substitutes are being sought, but problems remain. One is that asbestos-bearing serpentine rocks are common in many parts of the world, and when they are used for highway paving, they can release deadly fibrous dust. In Montgomery County, Maryland, the concentration of air-borne asbestos (originating in a local quarry and spread on roads) was found in 1977 to be a thousand times higher than in regions where serpentine is not used for paving. One might ask if incidence of lung cancer is unusually high in that area. It is not, at least at present, because the quarrying and dispersal of asbestos-containing rock have not been carried on there for the 20 to 30 years it takes for the damage to become apparent. Now, however, roads, playgrounds, and parking lots bearing loose serpentine will have to be sealed to keep down dust.

A second problem appears when other tiny fibers are used. The hope that glass fiber could be used commercially instead of asbestos is weakened by the discovery that in the village of Karain in Turkey, 24 of the 55 people who died between 1970 and 1974 died of mesothelioma. (Strangely enough, Karain means "stom-achache" in Turkish.) The only possible cause found in Karain so far is a fibrous mineral called zeolite, which is present in drinking water. Zeolite is not asbestos, but it does have fibers of comparable size. If nonasbestos zeolite particles, about 3 micrometers thick and 15 micrometers long, can cause cancer, then perhaps fiberglass is also dangerous, and one more potential asbestos substitute becomes suspect.

Still another difficulty appears. Asbestos and cancer are clearly linked. Tobacco smoking and cancer are also linked. The insidious thing about asbestos, bad enough by itself, is that when it is coupled with tobacco smoke, the results are not simply added. The phenomenon of *synergism* emerges. Asbestos and tobacco smoke together are much more dangerous than either is alone. According to Irving Selikoff of the Mount Sinai School of Medicine in New York, "Asbestos workers who smoke have approximately eight times the lung cancer risk of other smokers and 90 times the risk of individuals who neither smoke nor work with asbestos." With respect to the children who live near the Maryland serpentine quarry, Selikoff has pointed out that they "have good reason never to smoke cigarettes."

II. Certainly there was radiation before we made and released radioactive materials. The cosmic radiation from the sun and the radiation of radioactive isotopes of elements in the earth were here, and they remain the principal source of radiation bombardment of organisms. Still, by-products from atom bomb explosions have been released in the air and are a cause for concern. These by-products, called *fallout*, descend like snow after an atomic blast. Because they can release enough energy to disrupt molecules, they are known as *ionizing*. Radioactive isotopes may emit *alpha* or *beta* particles, both of which have such poor penetrating power that simply wearing clothes during fallout is good protection. Other isotopes, such as carbon-14, emit *gamma* rays, which have great penetrating ability and are therefore capable of causing ionizations and potential damage deep within living tissues.

The durability of activity is as important as the kind of radiation. Some isotopes, say nitrogen-16, have a half-life of only 8 seconds. Others have long half-lives: carbon-14 at 5600 years, or thorium-232 at ten billion years. Strontium-90, with a half-life of 28 years, has received particular attention because it is readily taken into animal bones, and it may be passed through mother's milk to nursing babies. At present, even after massive statistical analysis of the Japanese survivors of the first atomic bombs and of victims of accidents, safe standards of acceptance for ionizing radiation have not been agreed upon. Lethal doses depend on the species under consideration, the usual measurement being a *rem* (roentgen equivalent man), a unit of delivery of energy. Some microorganisms, such as spore-forming bacteria and fungi, can survive doses larger than a million rem, but a dose of about 500 rem is considered lethal for a human being. In spite of differences of opinion as to the degree of danger from radioactivity, there is no disagreement that ionizing radiation is potentially destructive to organisms. In sufficient doses it can kill, and in lesser doses it can cause burns, cancer, and genetic change. Only the factual information can be provided by scientific study; the decisions on nuclear testing and on nuclear power production must be social and political decisions.

Noise Pollution

Dead silence is rarely met. On experiencing for the first time a heavily soundproofed room, most people feel a vague sensation of eeriness, for they have never before known what it is to have no stimulation in their ears. The industrial world is a noisy world, filled with the roar and clangor of machines, motors, sirens, planes, and amusement devices. Only in recent years has evidence been accumulating that high-intensity noise can be destructive. Aside from the obvious possible damage to ears, noise damage has been implicated in atherosclerosis, coronary heart disease, and increased blood pressure. David Lipscomb (University of Tennessee) demonstrated the destruction of cochleal cells (Chapter 20) in the ears of guinea pigs that had been subjected to noise levels equivalent to those encountered in loud rock music discotheques. The proprietor of one such establishment, on learning of the results, is reported to have said, "If I notice any appreciable number of guinea pigs coming to my discotheque, I will turn down the volume." As Melvin A. Benarde (Hahnemann Medical College) said, "The degree of annoyance and discomfort that people will endure is astonishing." It does appear, however, that noise-reduction programs are increasing, especially near airports, in the noisier factories, and in road-building activities.

Thermal Pollution

Another kind of nonmaterial pollution occurs when streams are used to cool large industrial activities. As a result, the streams become heated. The resulting *thermal pollution* can cause changes in the plant and animal life in an affected stream, since most species have temperature ranges that are fairly definite. Some species, the *eurythermal* types, are adapted to a wide range of temperatures. Others, the *stenothermal* types, can exist only within a narrow range with sharp limits. Thermal pollution will not affect eurythermal organisms, unless of course the temperatures exceed the limits; but stenothermal species may be eradicated by a change of only two or three degrees. The principle followed by many ecologists with regard to such environmental alterations is that any change is likely to bring about unforeseen and unforeseeable effects on organisms. For instance, if a nuclear power plant is planned with a river as a cooling system, everyone involved should try to know in advance how much heating of water will follow, what organisms inhabit the water, and what is likely to happen to them, basing predictions on models and on past experience. Recognition of thermal pollution, like that of noise pollution, is so recent that relatively little firm information is available, and the ecologically sensible guide is to proceed with caution. It is easier to prevent mistakes than to correct them.

THE PHYSIOLOGY OF THE INDIVIDUAL IN THE ENVIRONMENT

As pointed out repeatedly in this book, every organism is constantly maintaining a balance within itself and between itself and its environment. Much of what is academically classified as physiology and biochemistry could just as well be called ecology. If an animal has eyes, sees with those eyes, and behaves in some way as a result of the received images, he is responding to his external environment just as surely as he is modifying his internal environment. Whether an ecological or a physiological label is attached to the change is not important. What is important is to find out as much as we can about anything that excites our curiosity, and to integrate our findings with the findings of other people. The utility of the ecological view is that it helps in unifying a number of disciplines and in making sense of many disparate bits of information. Just as thinking in evolutionary terms has affected the course of biological history, thinking in evolutionary-ecological terms will affect its future.

SUMMARY

1. *Ecology* is the study of organisms and their relation to all the forces that act upon them. These forces include general climate, local weather, structure and composition of soil, amount of contained materials in water, and activities of the organisms themselves.

2. *Autecology* is the study of the reaction of individuals to their environment; *synecology* is the study of the interactions of organisms in groups.

3. A *habitat* is a real space, whose features are determined by its climate, geological nature, past history, and present inhabitants.

4. Within habitats, organisms survive by using *niches*, which are not real spaces but, rather, occupational specializations, determined by the physiological or behavioral actions of any given species.

5. The important factors that influence the distribution and activities of organisms are *climatic, biotic,* and *edaphic*.

6. *Climate,* usually thought of as covering a wide geographical area, at least a number of square kilometers, is determined by temperatures, wind speed and direction, and rainfall, and by the distribution of these through the yearly cycles. Small local differences in elevation, shading, direction of slope, proximity to hills or buildings or trees, or underground water can make considerable differences in *microclimate*.

7. Any partially or completely self-contained mass of organisms is an *ecosystem*. If all the ecosystems of the earth are considered as part of a still larger system, embracing all the earth's organisms, we arrive at the concept of the *biosphere*. The biosphere is the living portion of the earth, as distinct from the air (atmosphere) and the crust (lithosphere).

8. The biosphere and all the ecosystems that constitute it require energy, whose source is mainly the sun. The energy of sunlight is fed into the system by way of photosynthetic organisms. The capture of radiant energy and its transfer into usable food constitute the *primary productivity* of a system.

9. Most of the serious air pollutants (such as sulfur dioxide, oxides of nitrogen, carbon monoxide, and siliceous dust and other such particulate matter) result from industrial activities and from heating and transport. The phenomenon of *synergism* illustrates how one minor irritant alone may be scarcely noticed, and the same may be true of a second one, but when combined, the effects are much greater than a simple addition. *Radiation pollution, noise pollution,* and *thermal pollution* are other problems of our own making that affect our relationship to the environment.

RECOMMENDED READING

One reading list for Chapters 30 and 31 will be found at the end of Chapter 31.

31

Ecosystems, Communities, and Biomes

DATE	ASSUMED DENSITY PER SQUARE KILOMETER	TOTAL POPULATION (MILLIONS)
600,000 B.C.	.00425	.125
200,000 B.C.	.012	1
20,000 B.C.	.04	3.34
8,000 B.C.	.04	5.32
4,000 B.C.	1.0	86.5
1 A.D.	1.0	133
1650	3.7	545
1750	4.9	728
1800	6.2	906
1900	11.0	1610
1950	16.4	2400
2000	46.0	6270

THIS CHAPTER IS CONCERNED WITH INTERACtions between organisms rather than with the reactions of individual organisms to their environment. The subject matter here is *synecology* instead of *autecology*, which was emphasized in Chapter 30. Early students of biology, whether they called themselves professors, farmers, medical doctors, or hunters, first studied autecology, even though they did not think of what they were doing as such. They observed plants and animals, tried various ways of manipulating them, and experimented on them in an effort to learn how they lived. In so doing, they were founding the science of ecology, more specifically, autecology.

31.1

Growth of human population. After taking perhaps a million years to reach a count of one billion, the human species began expanding in the eighteenth century with unprecedented speed. In less than two centuries the population trebled, and it seems to be continuing the upward growth. Probably the greatest and most serious question the species has ever had to meet is: Where and how will the trend be changed?

31.2
Too many and too close. One of the results of having an enormous number of people squeezed into a limited space is the development of such slums as the "mud puddle" in San Juan, Puerto Rico.

It was part physiology, part anatomy, part meteorology, part geology, and part chemistry, but the general emphasis was on specific organisms and their relation to the world. The science of synecology arrived late, the first full-scale work being that of Kerner von Marilaun, who studied the Hungarian marshes of the Danube River basin in the 1860s, watched the vegetational changes that occurred spontaneously, and described in detail the process that we now call ecological succession.

Once born, ecology was slow to grow. As late as the 1950s, few people had even heard the word. A few academic courses were taught, some wildlife managers and some foresters worked on ecological principles, and a small number of books and a smaller number of scientific journals dealt with environmental problems, but there were no bumper stickers with little round symbols something like a Greek theta, Θ. When the senior author told his family in the 1930s that he was taking a course in ecology, he got a look as if he had brought home a dead snake, and they asked, "What is *that*?" And when he offered a course in ecology in the 1940s, only one student appeared.

But things changed, and they changed quickly, as they do when the annual-interest effect begins to show. (Annual interest on a dollar at 3 percent is three cents, but annual interest on a billion dollars at 3 percent is $30 million. The proportion is the same, but the absolute difference is enormous.) The effect is

forcefully shown in the population figures of human inhabitants of the earth. Although no accurate numbers can be obtained for early years, reasonable estimates indicate that the number of people living by about 6000 B.C. may have been around five million. At the height of the Roman Empire, that number may have grown to between 200 and 300 million, a 40- to 60-fold increase in some 6000 years. Much of the increase was due to agricultural improvements. Change came slowly, and by the year 1650, there were still only some 500 million people living (Figure 31.1). Then with the industrial revolution, the *rate of increase* began to increase. It took 200 years between 1650 and 1850 for the world population to double, reaching its first billion. It then took only 80 years to double again and only 45 years for the next doubling. It took something over a million years after the development of the first humans for the earth to have a billion living people, but it took 80 years to get its second billion, and only 30 for the third. World population in 1973 was estimated at 3.86 billion, and by 1980 it is expected to reach 5.5 billion according to low estimates or 7 billion according to high estimates. In the past, even high estimates have proved to be too low when actual times and figures were obtained.

Geometric increases cannot continue, no matter how you look at it. When we began adding some 75 million people a year to the world population, we suddenly began to see the effects. There are less space, more junk yards, more smog, more artificiality, and more starving people, mostly the result of the sheer number of human beings. It must be granted that, along with the great increase in population, there has been increased specialization and diversity of labor, sometimes resulting in more luxury for some fortunate countries. A twentieth-century American of moderate income enjoys such freedom of movement, physical comfort, and diversity of diet as would have been beyond the means of a Roman emperor. But for perceptive people and for unperceptive ones as well, whether they know it or not, the earth's potential is finite, and this fact has thrust itself harshly upon us. That is why ecology in the past 20 years has become more than a matter of academic concern even to businessmen and politicians.

ENERGY RELATIONS IN ECOSYSTEMS

Energy input for the biosphere, as primary productivity, was described in Chapters 4 and 30. The concern up to now has been with the mechanisms of energy trapping and with the buildup of food in plants, but the biosphere contains many nonphotosynthetic organisms: fungi, bacteria, and animals. These, in ecological terms, are the *consumers*, as contrasted with the photosynthetic *producers*. We can think of the biosphere as a single large organism, and we can treat it as such in discussing energy budgets, but the biosphere is so large and so full of variables that smaller units are easier to measure, and results are easier to understand. Specific ecosystems are more commonly used when energy flow is studied.

In an ecosystem, the main and for practical purposes the only energy input is sunlight. Most of the sun's energy, as much as 98 percent of the total, is transformed into heat, and it is therefore not part of an ecosystem's primary productivity. That energy is eventually lost by radiation back into space, but it serves more than one useful purpose: It maintains the temperature of soil and water; it causes differences in air temperature, which not only result in air currents but affect the evaporation and precipitation of water; and it keeps the green plants warm enough to be functional. To say that the heat energy is "lost to the system" is in keeping with the thermodynamics of the system, but the phrase must not be taken to mean that the heat energy is without usefulness. The small fraction of sunlight energy that is used in photosynthesis enters into the food production that maintains the system.

The energy trapped in food is released in respiration. Some of it is used by the same photosynthetic plants that trapped it, being drained off for use in the nongreen parts and for all living parts of plants during darkness. The rest of the energy is released as the green plants are eaten by herbivores (insects or other plant-eating animals) or as the plants die and are decomposed by fungi, bacteria, or *detritus eaters*. (Detritus is a term borrowed from geologists, who use it to mean particles of plant and animal material.) At last, all the energy that came into the system as light has dissipated from the system as heat, but meanwhile, as it flowed through the system, the energy was used to maintain the life of all the individuals within it.

A reasonably stable ecosystem achieves a state of *dynamic equilibrium*, which means that over a period of time the energy input will equal the outflow. Such a system has been compared to a water tank with a tap for inflow and an overflow drain. The rates of inflow and outflow are not constant, one sometimes exceeding the other, but in general the two rates keep an average level in the tank. In an ecosystem, input is faster than output during a summer day, but during darkness or in cold weather, when photosyn-

thesis slows down or stops while respiration continues, output is faster than input. As long as the dynamic equilibrium is maintained, the ecosystem remains stable.

If equilibrium is disturbed, the ecosystem changes. The energy input may exceed the outflow if a new species of more efficient autotrophs invades the system or a plant-eating insect species is eradicated. Then the total amount of organic material will increase and will continue to increase until a new equilibrium is reached. When that happens the system will be different from what it was. Another possibility is that primary productivity may drop, so that effective input is lower than respiratory outflow. A swamp may be drained, a fungus may almost wipe out the major producer, or a whole climate may be altered. As with cash flow, so with energy flow: Outgo cannot continue indefinitely to exceed income. In an ecosystem, the respiratory release of energy can continue only at the expense of stored food, and when that is reduced, the consumers will die, or at least be reduced in numbers, and the reduction will continue until a new equilibrium is achieved. Again, the new equilibrium will be at a new level, and the ecosystem will have changed.

The rate of energy flow is an indication of the activity of an ecosystem. A highly productive system, such as a warm tidal marsh, may have a primary productivity of 10,000 Kcal/m²/yr, and will support a large number of active consumers. At the same time, it will have a large amount of organic material, the *biomass*. In contrast, a desert may have a primary productivity of less than 500 Kcal/m²/yr, will support few consumers, and will have a small biomass. Each of these ecosystems can remain constant only so long as there is a balance between input and output. The Sahara Desert once supported an active human, plant, and animal population, but with a drastic change in rainfall, the photosynthetic energy available for that great ecosystem was cut back until now its total biological energy flow is one of the lowest in the world.

Trophic Levels

The primary producers make up only a part of the biomass of an ecosystem. They may be eaten by insects or grazing animals or seed-eating birds, which in turn may be eaten by carnivorous animals. Carnivorous animals themselves may then be eaten by still others. The distance of a consumer from the primary producer, in terms of the number of passages of food through organisms, determines the *trophic level* a

31.3

A simplified scheme of a *food*—that is, energy—*web*. Photosynthetic plants build up reserves great enough for their own needs, with enough extra to feed the herbivores, exemplified here by rabbits, mice, and a host of plant-eating insects. (The arrows indicate what is eaten by whom.) Intermediate feeders prey on some of the herbivores, with spiders, various amphibia, and insect-eating insects living one step away from the primary producers. Large carnivores, such as birds of prey and some mammals (a fox in this array), eat the smaller carnivores. In practice, the transfer of energy is more complex than the drawing suggests, because the number of kinds of organisms is reduced and the food sources are not all indicated.

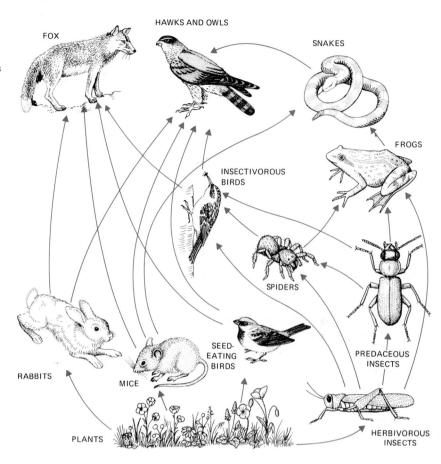

consumer occupies. A mouse that eats seeds is getting its energy directly from a photosynthetic plant. It is said to be a *first-level consumer*, with the plants as producers. A snake that eats mice is a *second-level consumer*, and a hawk that eats snakes is a *third-level consumer*. If a fox should eat a hawk, the fox would be a *fourth-level consumer*. These levels are not fixed, as can be seen on examination of human feeding habits. When people eat cereals, they are first-level consumers, but when they eat beef, which fed on grass, they are second-level consumers. Or a human may eat a fish, which lived on smaller fish, which fed on minute crustaceans, which fed on algae, and in those circumstances the human is a fourth-level consumer.

Feeding patterns in nature are further complicated by mixtures of possibilities. A plant may die and be decomposed by fungi and bacteria, which return the minerals and carbon immediately back to a form usable by plants again. A seed-eating bird may meet the same fate, or a parasitic fungus may destroy a plant-eating insect. The result is such an intricate series of possible events that a neat diagram of a *food chain* is too simple to be meaningful. A food chain, in the parlance of earlier ecologists, was the sequence of plant-herbivore-carnivore feeding. It is more realistic to speak of a *food web*, in which energy and materials may pass through any number of organisms before the energy is finally lost to the system in the form of heat (Figure 31.3).

31.4

An energy pyramid, based on productivity and consumption in an aquatic ecosystem in Silver Springs, Florida. The figures, in Kcal/m², show how little energy (one Kcal) is harvested by the main carnivores, such as the larger fish, in comparison with the amount (nearly 1000 Kcal) produced by the green plants.

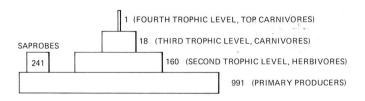

SAPROBES

241

1 (FOURTH TROPHIC LEVEL, TOP CARNIVORES)

18 (THIRD TROPHIC LEVEL, CARNIVORES)

160 (SECOND TROPHIC LEVEL, HERBIVORES)

991 (PRIMARY PRODUCERS)

Energy Relations in an Ecosystem

The only closed ecosystems are artificial ones, such as a balanced aquarium in which the materials are repeatedly recycled by plants and animals, and the only input is light energy. In nature, ecosystems are open to changes, with inflow and outflow of water, inwash and outwash of soil, wind blowing, immigration and emigration of organisms, and the carrying in and out of materials by animals. Still, for purposes of investigation and analysis, it is fair to treat a single ecosystem, such as a pond or a woodlot, as a unit for study.

Transfer of energy in ecosystems, which is fundamentally the same as it is in single organisms, conforms to the laws of thermodynamics. Recall that, these laws state (1) that energy is neither created nor destroyed, and (2) that with any energy conversion, some of the energy is lost to the system in the form of heat. In an ecosystem, all the entering energy in the form of sunlight can be accounted for either as chemical energy in food or as heat that simply warmed the place, and all the energy is eventually dissipated from the ecosystem back into space. That is in accordance with the first law.

The second law operates during every conversion of energy that occurs within the system, including the original photosynthetic buildup of food, the respiratory activity of the green plants themselves, and the movements and metabolism of every consumer at every trophic level. The energy loss has practical consequences. Some sunlight energy is lost to the system as heat even as it is being partially converted to chemical energy in food; more is lost in the general life processes of the green plant; still more is lost with every physical and biological action of any herbivore that eats a plant, and of any carnivore that eats the herbivore. At every trophic level, less and less of the energy originally stored photosynthetically is available as food.

One way to seek estimates of the action of food webs is to count individuals, but it is not instructive to compare the number of algae with the number of carp in a pond, or the number of grass plants with the number of hawks in a prairie. Another way is to determine the biomasses, in terms of weight, of the organisms at different trophic levels, but this, too, has disadvantages, because it is misleading to compare the mass of bacteria with the mass of redwings in a marsh. The amount of energy required for the lives of the various inhabitants of an ecosystem, however, can be calculated, and that figure gives a useful indication of what amounts are available at different trophic levels. Such information is important to people who are interested in the world's food supply. Figures can be used to construct what ecologists call _pyramids_. The base of a pyramid represents the energy in primary producers, and the upper levels represent the proportion of energy available in successive trophic levels.

The flow of energy in Silver Springs, Florida, based on data of H. T. Odum, can serve as an example (Figure 31.4). In this scheme, the main primary producers are freshwater eelgrass and algae; the saprobes are fungi, bacteria, and detritus eaters; the herbivores are insects, snails, and turtles; the third trophic level contains such carnivores as insects and small fish; and the so-called top carnivores are larger fish, such as bass. The amounts of energy, stated in Kcal/m², are relative. The striking conclusion that can be drawn from the figures is the small amount of energy at the top of the pyramid as compared with the amount captured by the primary producers. Note, too, the rapid diminution of energy flow at each succeeding trophic level.

It is also useful to compare two pyramids using the same organisms but different measures, this time weight (biomass) as well as energy (Figure 31.5). The primary producers are alfalfa, a leguminous hay crop, the herbivores are beef cattle, and the second-level carnivores are humans. The weights in Figure 31.5(a) are in kilograms, and the lengths of the bars in both pyramids are proportional to the amounts, except for the energy line in the third trophic level of (b), which, if drawn to scale, would be barely visible. Loss of biomass from one trophic level to the next is easy to see, but that loss is less than the energy loss, as is apparent when the biomass pyramid is compared with the energy pyramid.

Human Beings As Energy Consumers in the Biosphere

The great energy loss during passage to higher trophic levels is important in consideration of human populations and human food supply. Much of the energy that beef cattle obtain from plants is spent keeping warm and walking around. That energy is lost. Pigs are more efficient than cattle in producing protein, and chickens are still more efficient, especially when grown in batteries, as they usually are for commercial use. Instead of being allowed to run free, expending energy in movement and heat generation, chickens are confined in masses, warmed, fed a carefully regulated diet, and killed at the age when the greatest return from the energy input can be realized. The technique of raising millions of birds with maximum efficiency in the energy budget has made the price of poultry protein the lowest of all animal protein (at the cost, it might be argued, of the loss of old-fashioned chicken flavor). Still, a much greater human use of the energy from primary production can be made if the human being is the herbivore, occupying the second trophic level rather than the third. Most Americans are meat eaters from habit and personal choice, but the second law of thermodynamics urges us, at least for ecological efficiency, to be vegetarians.

A number of schemes have been devised in an effort to shorten the food chain for human food, thereby avoiding the energy losses incurred when people place themselves higher in the energy pyramid by eating animal protein. Some of these schemes are the cultivation of algae for food, the breeding of crop plants that are more nearly completely usable (we eat less than 30 percent of a wheat plant), the harvesting of yeasts grown on carbohydrates from hydrolyzed cellulose (of which millions of tons are wasted annually), even the growing of bacteria on petroleum

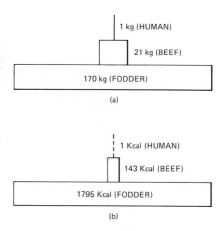

31.5
Pyramids illustrating the use of beef, fed on alfalfa, for human consumption. (a) A weight pyramid, showing that for every kilogram of human mass produced there must be 21 times as much beef taken in, representing about eight times as much again in plant material. (b) The same system, comparing energy rather than weight. Plants must produce enough to store 1795 Kcal to provide, by way of the intermediate beef, a single Kcal for people.

for their protein. By moving only one step farther from the primary producers, we might cultivate shellfish, such as oysters and mussels, in prepared beds or turn our attention to grazing animals that are now not generally eaten, such as some species of antelope. A number of other novel methods of producing more food and producing it more efficiently have been suggested, but so far they are prohibitively expensive, or they present technical problems, or they yield unaccustomed products that people would rather starve than eat.

The problem of feeding the people of the world becomes increasingly enormous. As long ago as 1967, an estimated 1.5 billion people were undernourished (simply had insufficient quantity of food) or malnourished (had food lacking in such essential components as protein and vitamins); and the numbers grow. Inadequate intake of calories results in *marasmus*, a disease in which children have spindly legs, wrinkled, old-looking faces, sluggish muscles, and no resistance to respiratory infections. Deficiency of pro-

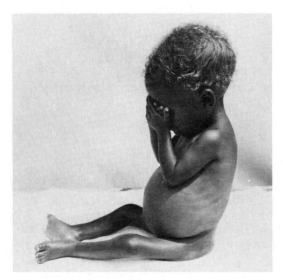

31.6
Victims of deficiency of protein:
kwashiorkor. This child in Kenya in
a time of famine was suffering from
extreme deficiency, and the results are
striking. Many children in regions
where the main nourishment is carbohydrate show milder but still weakening symptoms of kwashiorkor.

tein, regardless of how much food is eaten, produces
kwashiorkor (Figure 31.6). This malady, estimated
to afflict as many as 25 percent of African children,
is from a West African word meaning "sickness a
child has when another child is born." The shrunken
muscles, pot belly, skin sores, bleached and falling
hair, and intestinal upsets begin when a child is
weaned from its mother on the arrival of another
child or is put on an artificial diet because nursing is
becoming unfashionable. It can be cured, sometimes
with dramatic speed, by the inclusion in the child's
diet of sufficient protein containing the proper balance of amino acids.

In the United States, malnutrition is mainly a
matter of diet preferences, but it can and does occur,
mainly because of inadequate intake of vitamins. One
result of the recognition of American malnutrition
has been the spread of the nonsense phrase "empty
calories." A calorie is a calorie, whether it is included
in a sugar lump or a spoonful of the best-balanced
food in the world. True, one could obtain the daily
2500 Calories by eating pure starch, but one would
soon die of starvation because protein and vitamins

are as important as calories. One could also starve on
a strict diet of vitamin pills and protein concentrate.
For proper nutrition, both energy *and* specific kinds
of molecules are essential.

Concerned people turn to new sources for a food
supply. But the sea, once regarded as inexhaustible, is
yielding less rather than more. Any shopper who has
been paying for fish over the past few years knows
that without being told. Marginal land that is not
really good for agriculture is being forced into cultivation, frequently with poor results at best and disastrous results in some instances. Irrigation of deserts
has been moderately successful in some limited areas,
but it is not promising as a major means of feeding
masses of people, partly because it is both expensive
and difficult, partly because it causes salt accumulation. Richard H. Wagner (Pennsylvania State University) wrote: "Perhaps some deserts can be made to
bloom, but not in the foreseeable future." The Amazon River basin, which looks so lush and productive,
has relatively poor soil that does not stand up to the
continued, intensive cropping demanded by high-
yield agriculture. Much hope was generated by the
Green Revolution, which produced new high-yielding
varieties of grains, especially rice. It also produced
new combinations of amino acids, which promised to
overcome some of the nutritional deficiencies of improperly fed people. But it, too, has drawbacks. It
demands a high-energy input in tillage and fertilizer,
which the less-developed countries do not have and
cannot afford. And it depends on *monoculture*, the
use of a single crop with the built-in dangers of any
monoculture: susceptibility to unexpected disease.
The 1970 corn crop in the United States, for example,
was severely reduced by a fungus invasion, and the
classic example is the Irish potato, whose destruction
in Ireland in the late 1840s produced effects far beyond the boundaries of the country (Chapter 9).

With the birth rate continuing high, especially
in less-developed countries, and the death rate dropping because of better sanitation and medical care
everywhere, the population increases and increases.
Although food production is increasing, too, it is lagging behind, frightening those who study the facts.
People who look at population figures and world food
production figures and see trouble ahead are sometimes dismissed as "doom-sayers," but those doom-
sayers, who will not be here when the crunch comes,
have been proved too conservative already, both in
the magnitude of their numbers and in the projected
times. The pursuit of happiness is a vague abstraction
to someone who is hungry; happiness would be a full
belly. It behooves the students of today, who will

have to deal with tomorrow's already visible problems, to acquaint themselves with the reality of too many people and not enough food, for humans are too definitely a part of the biosphere to escape nature's regulations.

THE STUDY OF COMMUNITIES

A habitat was described in Chapter 30 as a place where organisms live. In any habitat, certain kinds of organisms live and make a _community_, whose characteristics are determined physically by the climatic and edaphic factors of the habitat and biologically by the species present, their sizes, numbers, distribution, and growth habits. A moist, temperate hillside in Germany, for example, could be climatically similar to one in Ohio, but since the two places have different histories, the plants and animals have evolved differently. Some of the genera of trees may be the same (but the species different), and some readily windborne organisms such as fungi, various microorganisms, and insects, may even be the same species. But the major inhabitants will be different. They will determine the makeup of the community, and it will be plants that are the primary determiners. They are large and stable, and they are the primary producers. The animals, both in their evolution and their migrations, will follow the plants that support them. For these reasons, the study of communities has concentrated on the plant life of regions and habitats, and indeed, it is sometimes called _plant sociology_, from the German word _Pflanzensoziologie_.

The makeup of communities may be studied simply in order to understand the workings of the biological world, that is, for theoretical or academic reasons. Practically, however, a knowledge of communities can be useful when that information can be applied to agricultural problems or when one is generating _impact statements_. In most developed countries and especially the United States, such major new developments as shopping centers, dams, housing developments, and industrial complexes cannot be legally started until an ecological "impact statement" has been prepared by a governmental or independent agency. Such a statement should include an inventory of existing conditions (soils, plants, animals, water, drainage, minerals, etc.) in a territory under consideration and areas outside it that might be affected.

One way in which studies of communities can be useful is in determining detrimental effects of human activities. It is reasonably well established that changes in the numbers and kinds of individuals in a community follow environmental changes. This principle can be applied, for example, to a stream that is receiving chemicals from a factory or washed silt from a construction site or heated water from a power plant. In an undisturbed stream, the number of different species is larger, though there may be fewer individuals, than in a disturbed one. Several equations are available for preparing mathematical statements that give a simple, quick indication of the degree of species diversity in a community. Such a statement, called an _index of diversity_, is obtained by taking samples or censuses of a community, listing the numbers and kinds of organisms, assigning them an "importance value," and calculating from the data the value of the index of diversity. Indices of diversity are useful to show changes of time or locality when questions of pollutants arise.

There is also an esthetic side—or perhaps it could be called a plaintive side—to the studies of natural communities. If the day comes when the inhabitable parts of the earth that are not paved are covered with tanks of high-protein alga cultures, some odd people may wish to know what used to live in the Pennsylvania mountains, the Louisiana swamps, or the Oregon forests. Perhaps only the artists will care, but it will be the scientists who can tell them.

The makeup of many communities has been examined, including those of lakes, ponds, running streams, estuaries, the open sea, coral reefs, mangrove swamps, temperate deciduous forests, montane forests, various grasslands, etc. Here we will describe two, chosen for variation and potential general interest: a temperate freshwater lake, and a temperate deciduous forest.

A Freshwater Lake Community

Freshwater lakes are said to contain lentic communities. The species present are determined by the climate, and their distribution is determined by the topography of the lake and by the season of the year. In the shallows around the margin of a lake, plants rooted in the bottom but with their leaves above the water level will be predominant. These include various rushes, sedges, cattails, and pickerelweed. In somewhat deeper water, water lilies, rooted in the muck, will have floating leaves on the surface, but still deeper, submerged plants (milfoil, waterweed) will be the main large plants. The floating, unattached fronds of tiny duckweeds frequently make an almost solid cover of some lakes. Many green and blue-green algae and diatoms, all photosynthetic, contribute to lake productivity. Certain species of animals are associated with the shallow-water plants, especially pond snails, water mites, small crustaceans, swimming bugs and beetles, and numerous insect nymphs. In the open water over deeper bottoms, the greatest number of organisms are members of the *plankton* (Greek, "wanderers"), small drifting plants and animals, such as algae, rotifers, water fleas, and beetles.

If a lake is deep enough, it can have water near the bottom that receives too little light to permit any overall photosynthetic gain. In this *profundal* zone, green plants cannot survive, but blood worms and other insect larvae, tube worms, and pea-shell clams can. Larger animals can move freely, but water snakes, turtles, frogs and their larvae (tadpoles), salamanders, toad tadpoles, and crayfish tend to stay in shallow water close to shore. The still larger "top consumers" are the fish, most commonly sunfish and bass.

Water movement due to currents (except wind-generated currents) is slight in lakes, but seasonal temperature changes can cause massive water movement (Figure 31.7). The movements occur because of the way temperature affects the density of water. Warm water is less dense than cooler water and therefore rises toward a lake surface. If it cools, it becomes denser and heavier, until it reaches a temperature of about 4°C (39°F), below which temperature it begins to be less dense and lighter again. In a lake, layers of water are formed as a result of temperature differences. An upper warm layer, the *epilimnion*, lies over an intermediate layer, the *thermocline*, whose temperature is between that of the epilimnion and the deep, cold *hypolimnion*. The hypolimnion remains static during summer, but when cold weather comes, the epilimnion chills, becomes dense, and sinks, stirring up the hypolimnion. This phenomenon is the *fall turnover*. If the surface water temperature drops below the critical 4° mark and becomes lighter again, it will remain on top and may even freeze. With the return of warm weather in spring, the surface water rises to 4°, becomes heavier, and sinks again, causing another mixing action with the hypolimnion. This is the *spring turnover*.

The seasonal turnovers do more than simply mix waters of different temperatures. Cold water can hold more dissolved gases than warm water, as anyone knows who has opened a warm bottle of carbonated drink. In summer, photosynthetic activity in the illuminated epilimnion provides oxygen in the upper layer, but the oxygen in the hypolimnion is depleted by the respiration of organisms and is not replaced. With the fall turnover, oxygenated water

from the upper layer sinks, with the result that during winter the oxygen supply in the hypolimnion is higher than in summer and, being colder in winter anyway, supports less respiration and less oxygen loss.

Lakes are sometimes described in terms of their biological productivity. Those with low productivity and a small biomass are called *oligotrophic* (Greek, "few foods"). They tend to be clear, look clean, have little odor, support few algae and water plants, and contain "game" fishes, such as trout. Oligotrophic lakes are geologically young, and they are preferred by people as sources of usable water, as well as for esthetic reasons. In contrast, *eutrophic* lakes have high productivity and a large biomass. They become turbid with algae, living and dead, and support such "coarse" fish as carp. Paradoxically, eutrophic lakes are less oxygenated than oligotrophic ones, although it might seem that an alga-laden lake should have more oxygen than a clear one. The reason lies in the activity of the algae themselves, which grow enormously at times in eutrophic lakes, forming "alga blooms." The algae photosynthesize rapidly and form oxygen bubbles, which float the algae to the surface, where the bubbles are liberated into the air. When the algae die, their remains sink and are decayed by bacteria, whose respiration lowers the oxygen content of the water. The total effect is to make a stagnant, foul-smelling, oxygen-deprived lake, in which only a few tough organisms can survive.

Eutrophication, a natural process, would occur without human help, but it would be slow, taking centuries instead of the few years required in artificial acceleration by the output of human activities. Exactly what causes speeded-up eutrophication is not known certainly, but the most likely guesses include the dumping of the phosphates contained in the detergents that were used by the millions of tons during the early 1970s, and the addition of nitrates. Nitrates enter lakes as runoff from croplands that have been fertilized and as outflow from sewage in thickly populated regions. Whatever the actual causes, it is evident that many lakes have rather quickly become eutrophic, and as a result they have lost their palatability as drinking water and have lost the quality and quantity of fish crops. In one of the better-known examples, the eutrophication of Lake Erie, the purity of the water has been much reduced in recent years, and the once-thriving white-fish industry has disappeared. Lake Washington in the Seattle region was moving toward eutrophication when measures were taken to stop the emptying of sewage into the lake, and the process has been reversed. This action shows that we can control pollution if we have a mind to.

A Temperate Forest Community

The original deciduous forests of the world are gone, cut for timber and cleared for farming. In the eastern United States a few small coves in the mountains, too steep and remote for lumbering, remain virgin, but the expanses of woods that do remain are second growth. In Europe the scene is even barer. Ireland and Scotland are all but treeless, and England is little better, though all were once wooded. On the continent, the woods are mostly planted, frequently with military-looking rows of trees chosen for maximum productivity and utility. In Italy, picturesque but scraggly pines hang over the Mediterranean cliffs, and dark cypresses accent the cemeteries. The Dalmatian coast is as nearly plantless as the Sahara Desert, cleaned of wood to build the sinking pilings of Venice. Fortunately for Americans, we do have many square miles of regrown woods, some undisturbed for a century or more, and it is one of such stands of trees that we shall describe as an example of a biological community.

In a patch of forest, such as still exists in the uplands of Maryland, the dominant organisms are black, white, and chestnut oak, mixed with shagbark and pignut hickories, a few beech trees, red and sugar maples, black gum and sweet gum, and tuliptrees (Figure 31.8). A few dozen species dominate the scene. Ecologists use the word "dominate" to mean that these are the organisms greatest in

31.7

A scheme of annual circulation of water in a temperate pond. During summer and winter, the upper layer, the *epilimnion,* and the lowest layer, the *hypolimnion,* become stabilized, but when temperatures change during spring and fall, changes in the densities of the layers cause mixing, known as the fall and spring turnovers. These turnovers are useful in redistributing oxygen and nutrients, and they keep a body of water from becoming stagnant.

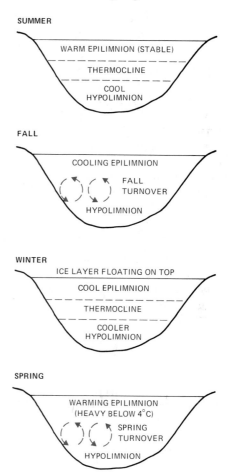

31.8

A mixed hardwood community. In the Blue Ridge of Virginia, the dominant trees are oaks, maples, tuliptrees, hickories, and a scattering of other deciduous trees, plus some evergreen hemlocks and pines. The understory consists of a middle layer of large bushes or small trees and the ground cover, slight as it is, of perennial herbs.

size, producing the most photosynthetic material, and determining the nature of the terrain, mainly (in this instance) by their shade. These are the tall trees, with their crowns forming the _canopy_ through which summer sunlight must filter. In the canopy, special animals find their niches. Squirrels gather acorns and hickory nuts, blue jays and sapsuckers feed in the branches, and during migrations a host of insect-eating birds stay mainly in the upper parts.

A lower level of woody plants makes an _understory_: Dogwoods, redbuds, several viburnums (haws and arrowwoods), some draped with greenbriers, lightly fill the space between the canopy and the ground. Where an occasional large tree has fallen, dead with age or blown down, the gap is filled with sumach bushes and covered with wild grapes. Along the ground, the plants visible at any one time are determined by the season, for the available light varies with the coming and going of the leaves above. In early spring, bloodroots, trilliums, and other flowers with an underground food storage send up blooms and leaves and make their year's food supply while there is still sun. Later the ferns that thrive in low light intensities unroll their fronds, and later still come woodland asters and wild lettuces. In a large enough area, animals as large as deer can survive. Wild turkeys, almost extinguished at one time, are increasing, and although they are still not common, they are no longer a rare curiosity. Birds feeding and nesting on the forest floor include ruffed grouse, oven birds, vireos and towhees. There are some carnivorous, herbivorous, and omnivorous animals: skunks, opossums, foxes, groundhogs, chipmunks, and racoons. Moles, voles, mice, and shrews live on the soil surface or in the soil.

The soil itself is alive with organisms (Chapter 30). The plant roots that make up the bulk of the living matter below the soil surface are easy to see, as are a few invertebrates (earthworms, centipedes, beetles, spiders, pill bugs), but most of the soil organisms are too small to see. Consequently they are ignored by the usual passerby. However, if efforts are made to find what is there, an enormous number of organisms can be taken from soil. If soil is slowly dried from above in a Berlese funnel, small insects can be made to drop in a trap below, or soil nematodes can be washed down (Figure 31.9). Many small creatures will find and inhabit the space under rocks and logs, or they can be provided with an artificial lure in the form of a board laid on the ground. Here small snails can live on decaying vegetation, beetles and their larvae eat smaller insects or plants, and hosts of ants, springtails, thrips, and other tiny

31.9

When small soil animals are too inconspicuous to pick out from among the soil particles, one can coax them to separate themselves by putting a soil sample in a Berlese funnel. The warmth from a lamp dries the soil from the top downward, and the mobile insects and worms work their way to the bottom, finally dropping into a receptacle. Thus freed from the soil, they can be identified and counted.

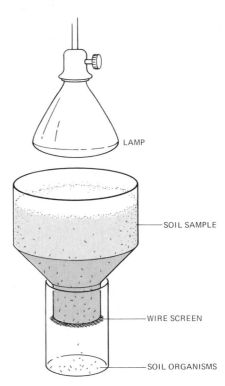

LAMP

SOIL SAMPLE

WIRE SCREEN

SOIL ORGANISMS

TABLE 31.1

Soil Components, By Percentage

Mineral material	93.0
Dead organic material	5.9
Living roots	0.7
Bacteria	0.3
Fungi and algae	0.3
Earthworms, insects, etc.	0.2
Total organic material	7.4

insects find niches. One can best detect the smallest organisms, the microfungi, bacteria, and protozoa, by culturing them in the laboratory. Though invisible in the woods, these last are present in great enough numbers to be important members of the community because of their ability to recycle the elements taken up by growing plants. In spite of the relatively slight biomass of microorganisms, their activity is great. The decomposing power of bacteria and fungi in soil keeps the minerals, carbon dioxide, and water moving through the community in a continuous cycle in an established, balanced ecosystem such as a forest. Table 31.1 shows, however, what a small part of the soil they occupy.

One important point concerning the community in an ecosystem must be emphasized: There is an integration that pervades the system. The primary producers must provide food for themselves and the consumers, the insects must pollinate, and the decomposers must recycle the elements. The final result is a self-perpetuating system, whose balance of change versus stability has caused ecologists to regard it as a kind of superorganism with a homeostasis of its own.

COMPETITION AND COOPERATION

Competition between species and between individuals in a species is one of the basic phenomena on which the idea of evolution is founded. Much has been made of competition as a force in the life of all organisms, and recognition of competition gave rise to the phrase "survival of the fittest." Both observationally and experimentally derived data attest to its reality. At the same time, another biological phenomenon has been working: *cooperation*. Examples of cooperation within a species are almost endless. Lions hunt in prides; colonies of penguins incubate communal eggs; insect societies have elaborately integrated systems of divided labor; musk oxen form organized rings of defense; wild dogs and wolves unite in hunting packs. Cooperation between species has also been going on so long that in some instances there has been time for progressive evolution, called *coevolution*, to alter the form or behavior, or both, of two cooperating species. In many such cooperative developments, coevolution has been carried so far that both species of the cooperating pair have become mutually dependent, each being helpless without the other. The yucca flower can be pollinated only by the yucca moth, and the yucca moth lives only by feeding its larvae on yucca ovules. The

fungi in roots, *mycorrhizas*, get nourishment from the roots, and the roots require the fungi for intake of minerals. The fungus component of a lichen derives food from the algal component, and the algal component is protected by the fungus. It is apparent that for survival and evolutionary change, cooperation has been a potent force, as Darwin himself made clear.

ECOLOGICAL SUCCESSION

For all its seeming stability, the biosphere changes, sometimes radically. Volcanoes erupt, changing the face of the earth and destroying life for miles around, as has happened many times within the period of human history. Land rises above the sea or sinks into it—or tilts, as England is now doing toward the east. Ice sheets descend, as did the great Wisconsin glacier, which last retreated into the Arctic only 18,000 years ago, leaving the whole northern United States and Canada scraped and sterilized. On a smaller scale, winds shift direction, leaving some areas without rain, as has been happening in northern Africa. Sometimes we are the cause, not nature. When an abandoned field is left to itself, it becomes an artificial "barren."

When an area is changed, it is open to colonization by plants and animals that are free to move into the unoccupied space. If the space has been completely cleared, it is called a _sere_. A sere may be a rocky area, scraped by a glacier and left high and dry and partly or entirely soilless; then it is a _xerosere_. A sere may be a pool of water, left in a depression after the retreat of a glacier; that is a _hydrosere_. Plants and animals coming into such biologically empty places will cause a series of population changes that constitute a _primary succession_. Plants and animals moving into a disturbed but not sterilized area, such as an abandoned field, do not have to start from zero, and the resulting changes after that constitute a _secondary succession_. In this chapter, we will follow the events in a xerosere, a hydrosere, both primary successions, and the secondary succession of an old field.

Primary Succession in a Xerosere
When a xerosere is opened for colonization, it offers an inhospitable environment. It has no soil, or very little, and it is devoid of organic matter. It experiences extremes of temperature, because rocks bake or freeze for lack of cover. Finally, it has no water-holding power. The first colonizers, appropriately

PLANTS THAT EAT ANIMALS

Some habitats offer special challenges to survival, perhaps by having too much or too little of some factor. One such habitat is a bog in which the essential element nitrogen is in short supply. Where such a shortage occurs, some organisms find means of survival anyway.

Plants, which are the primary food producers, are typically eaten by animals, but in places where nitrogen is limited, several genera and dozens of species of plants have evolved methods of reversing the old order. They can capture and digest their own meat. They manage the feat with traps like those used by human trappers: gums, pits, and springs.

The simplest traps are those of sticky plants. The glistening hairs on sundew leaves are tipped with adhesive droplets, to which wandering insects are effectively glued. When an insect is caught, nearby hairs bend in to meet it, secrete protein-digesting enzymes, and absorb nourishment for the plant. The shining leaves of butterworts can roll slowly up, enclosing insects captured by sticky hairs.

Bladderworts have underwater traps that suck in tiny swimming animals by allowing a rush of water into the bladders when a trigger is tripped.

The pitcher plants have leaves modified into cups that catch rain, and they are provided with downward pointing hairs that make it easy for an insect to go down but hard to climb up. Down in the leaf cups, insects drown and are digested. The pitcher plants in the eastern United States are Sarracenias; those in the west are Darlingtonias. The most spectacular cups are those of the oriental Nepenthes (meaning "without care"), named for the wine cup that Homer tells us Helen used when she spiked the wine to stupefy men. In addition to the insect trap, one species, *Nepenthes bicalcarata*, has at the cup opening a pair of downward-pointing spines. If a tarsier (a small primitive primate) tries to rob the plant of its prey, the spines prick it in the back of its neck. Leaves of pitcher plants usually have a number of dead insects floating in the contained water, but some species

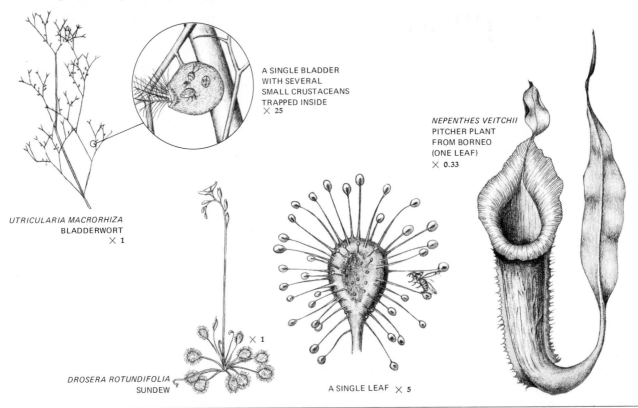

A SINGLE BLADDER
WITH SEVERAL
SMALL CRUSTACEANS
TRAPPED INSIDE
× 25

NEPENTHES VEITCHII
PITCHER PLANT
FROM BORNEO
(ONE LEAF)
× 0.33

UTRICULARIA MACRORHIZA
BLADDERWORT
× 1

DROSERA ROTUNDIFOLIA
SUNDEW

A SINGLE LEAF × 5

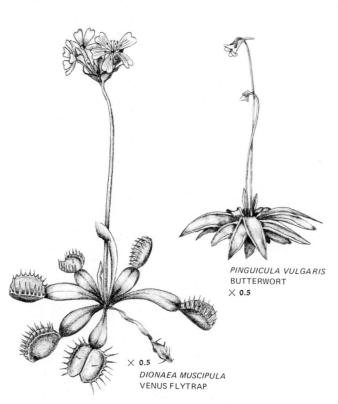

PINGUICULA VULGARIS
BUTTERWORT
× 0.5

× 0.5
DIONAEA MUSCIPULA
VENUS FLYTRAP

31.10
The beginning of vegetation on a dry, soilless area: a *xerosere*. Nothing can grow here unless it can withstand drying and extremes of temperature and can survive with minimal soil. In such a place the first arrivals, the *pioneers*, are lichens, seen here as pale splotches on the rock. In cracks, themselves enlarged by lichen action, rooted plants can become established, as shown by the small cacti in this picture. Lichens grow so slowly that they take years to arrive at this stage, but eventually, sometimes centuries later, they form a full cover of soil that supports a rich and varied flora with accompanying animals.

of ants, like people living in constant danger on the side of an active volcano, regularly inhabit the Nepenthes stems, hollowing out homes there.

The most astonishing of the insect catchers is the Venus flytrap, native to the coastal region of North and South Carolina. This famous plant has hinged leaves with toothed edges and sensitive hairs on the upper surfaces. When an insect touches a hair twice (only once is not enough), the blade neatly hinges shut like the jaws of a dredge. It takes about a second to cage a victim. The mechanism is incompletely understood, but it involves changes in turgor of the hinge cells. Some people walking in flytrap country for the first time walk with careful steps and wary eyes until they actually see the tiny plants, which are capable of holding ants but nothing much larger.

In all this, the most impressive fact is the multiplicity of ways in which plants have solved a single problem. The human species, for all its intelligence, has done no better in providing itself with protein.

called *pioneers*, must be photosynthetic and able to endure harsh living conditions. The organisms best adapted to such a life are lichens, of which small bits may be brought in by wind or water or migrating birds (Figure 31.10). They can cling to minute cracks in rocks, grow, and begin a living cover. Although lichens grow exceedingly slowly, expanding sometimes by millimeters per year, they do increase. As they do, they send microscopic rhizoids into crevices and secrete dissolving acids that break off rock particles (the beginning of soil). When they die, they leave organic matter to be added to the soil particles. After many years, perhaps hundreds, some of the larger rock crannies come to be filled with soil, which offers a starting place for the next invaders, usually mosses. Mosses are not as effective as lichens in

(a)

(b)

31.11

A pond drying up. (a) In this pond, open water is fringed with vegetation. A zone of floating plants is ringed about by plants rooted in the bottom of the pond. Outside them are other plants that tolerate soggy soil. Still farther from the water, woody plants help dry out the soil and even add to it by depositing organic material. (b) After many years, the encroaching plants continue to build up more soil and by transpiration make the area less marshy, until finally the whole pond is closed over. Given time, the site of the pond can become forested.

growing on bare rock, but they are better at colonizing small patches of soil, and as they do so, they carry on the work started by the lichens, creating more soil, adding more organic matter, and spreading. They increase at the expense of lichens, making the environment less favorable for the lichens and more favorable for the next wave of invaders. The typical third group coming in consists of *xerophytic* (Greek, "dry plant") grasses, which are drought-tolerant, as they must be to stand the sparse and infrequent watering they get. They are watered only when it rains, because a xerosere at this stage of succession still has little water-retaining capacity. The grasses take over where the mosses left off, continuing the work of making more soil and make it slightly deeper, richer in organic matter, and more widely spread. After the grasses come xerophytic shrubs, which keep the process going still further. In each of the successional stages, the current worker is outcompeting its predecessors and at the same time preparing the environment for a new type that will be better adapted to the change. When the new type comes, it will outgrow the current worker, and so it goes.

The question that arises is: Where does it end? It ends, as the phrase has it, when a *climax* is reached. A climax can be defined and recognized in two ways, one in terms of species stability and one in terms of productivity versus energy loss. In an early xerosere, which is unstable, xerophytic shrubs are replaced by xerophytic trees, such as pines. But pine seedlings, being rather intolerant of shade, do not grow as well

under old trees as do the seedlings of *mesophytic* (Greek, "middle plants") trees, such as maples. In that part of the world in which a succession like this can occur, that is, the northern United States, mesophytic trees can grow into a forest that will not be readily displaced by other incoming species. When a community becomes stabilized and self-perpetuating, it has reached a climax stage. Maple trees do not make their environment dangerous to themselves. The second way of determining when a climax has been reached is by comparing the buildup of organic matter with its breakdown. When the yearly amount of buildup exceeds the amount of breakdown, the community is unstable. It is like the so-called upward mobility of a family that is making more money than it is spending; it will not stand still. When productivity equals respiration, or when species stability has been attained, the community is at climax.

Succession is frequently described, as it has been here, mainly in terms of plants because they are more conspicuous and more readily identified as belonging to the community than animals. Animals may move in and out; they tend to follow the primary producers. Yet even so, certain species of animals are associated with species of plants at different stages of succession, and they of course form an active part of any community.

Primary Succession in a Hydrosere
As far as water is concerned, the conversion of a hydrosere is the opposite to that of a xerosere. The

31.12
Abandoned farm land returning to forest. When cultivation ceased, weedy plants sowed themselves, with broomsedge, a grass, covering most of the area. After a few years, seedlings of trees sprang up from seeds blown from the nearby woods. As the seedling trees matured, they in turn contributed more seeds. The process will continue until the old field is forested again as it was before the land was originally cleared. This *secondary succession*, with soil and plants already available, goes faster than a *primary succession*, as in a pond or on bare rock, and may reach completion within a century or two.

primary colonizers in a sterile pond must be photosynthetic and small enough to be transported. These are microscopic algae, which can grow, die, and sink to the bottom. They first form a film of organic matter and then build that up, layer on layer, year after year, until around the edges of the pond, the shallower parts come to have a mixture of detritus and small rock particles: *soil.* The next invaders are plants that can grow with their roots under water and their leaves in or above the surface (Figure 31.11). These are the cattails, pickerelweeds, and water plantains mentioned in the section on lake communities. These plants also help break up the parent rock and, when they die, add their remains to the organic material around the edge of the pond. As the muck at the bottom of the pond thickens, the pond itself becomes shallower, and the edge closes in on the open surface of the water. Around the definitely soil-filled margin will come sedges and *hydrophilic* (Greek, "water loving") grasses and, as the process continues, hydrophilic shrubs, such as buttonbushes, then alder trees. Meanwhile the pond gets shallower and soil in and around the pond gets deeper. As the pond shrinks, drier land is made available at a distance from it. Into the drier parts mesophytic tree seeds can come and can survive, eventually leading to the establishment of a climax mesophytic forest. In many ponds that are undergoing succession, clearly visible zones of vegetation show where one stage merges into the next. The changes are usually so slow that they are almost imperceptible, but by watching individual

places over long periods of time, ecologists have been able to show, even within one working lifetime, that these described stages are real.

Secondary Succession

Secondary succession has been documented by observations on land that was once used for farming (Figure 31.12). Such land differs from that in a sere in having soil already pulverized, some organic matter, and a store of ungerminated seeds. Another difference is that it is close to a supply of *propagules*, or reproductive organismal particles, such as seeds, spores, eggs, or viable bits of vegetative parts. The stages progress quickly, and they can be observed within a span of a few years.

After a field has been left untilled, the first invaders are annual weeds, especially ragweeds and evening primroses. They will be replaced by perennial weeds, whose underground rootstocks will persist for several years. These, particularly the goldenrods, are notably durable; they can be seen in old fields everywhere. After 10 to 15 years of weediness, shrubby plants begin to grow. These are the blackberries, poison ivy, sumach bushes, and black cherry seedlings, followed in many places by red cedar. In many an old field, the advancing tide of red cedars can be seen. There are small ones toward the middle, and the larger ones around the edges show plainly that they have been there longer. In the eastern United States, where this kind of succession has been most closely followed, the cedars will be slowly replaced by red maples, oaks, and the other trees that are part of the deciduous flora. It takes from 50 years to a century for the open field to begin to look once more like a typical deciduous forest community, a second century for it to achieve a measure of maturity, and as much as two more centuries for it to return to the impressive condition of its virgin state.

In all successions, the climax will depend on many forces, especially climatic ones. Only in a well-watered land can succession continue on to the mesophytic forest described in the previous paragraphs. If water is in short supply, a climax can be reached at some stage short of the mesophytic forest, and an area may be stabilized as grassland or xerophytic forest. These are *climatic climaxes*. Soils, too, can impose their restrictions, creating *edaphic climaxes*. Many beach communities, well rained on but sandy, are edaphic climaxes.

The study of successions raises questions beyond the vegetational and faunal changes. One may ask whether evolution is not a long series of successional changes and whether the human species is only one in a progression of organisms, each making way for the next and so altering his own habitat as to make it impossible for him to survive in it.

EFFECTS OF HUMAN ACTIVITY

In clearing land for cultivation, building cities, paving highways, constructing dams across streams, and herding domestic animals, humans have done more to alter the appearance of the earth than any other living thing has done since the beginning of time. Flying by day at an elevation of eight kilometers (5 mi), you can see the massive reduction of forests and the apparently endless geometry of cities and the net of roads between them. By night you can see the lights, stretching off into the distance until they are obscured by the smoke of a million chimneys. Consider the list. The air over the Sonoran Desert is hazy where once it was so clear that a mountain 50 kilometers (30 mi) away seemed just up the road. The air over New York City is so loaded with dissolved and particulate matter that several times a year most of the populace walk about with burning eyes and scratchy throats. The open ocean is awash with nylon nets that keep catching fish for years after the boats that lost the nets have returned to port. On remote tropical beaches, indestructible light bulbs and plastic bottles wash ashore. *Homo sapiens* is the prime defiler of the biosphere.

Some encouraging signs do begin to emerge. The very fact that there is an Environmental Protection Agency is an indication of awareness. In the past 20 years we have seen coming into being a number of wildlife protection societies and nature centers and wilderness clubs whose aim is to slow down the succession that leads to an unknown climax. Some of the planned dams have been scrapped, and some of the estuaries have been saved from real estate development. But still the pressure of population mounts and the demand for energy grows, so that "forward though I canna see I guess and fear."

BIOMES

A *biome* is a large, readily recognizable biogeographical unit whose plant life is determined largely by climate, and whose animal life is in turn determined by the kinds of plants that live there.

Classification of the earth's biomes includes about a dozen types. In this book we will not treat temperate rain forests, such as those of the Pacific Northwest (Figure 31.13); chaparral, such as the dry, scrubby vegetation of portions of the southwest United States and the countries around the Mediterranean Sea; tropical deciduous forests, such as occur in parts of South America, India, and Indochina; the tropical scrub forests found in Mexico (Figure 31.14), southern Africa, and Australia; or the mountain biomes of the high Rockies, the Andes, the Alps, and the Himalayas. We will list the main features of the oceans, the tundra, the northern coniferous forests, the temperate deciduous forests, the temperate grasslands, deserts, tropical rain forests, and tropical grasslands.

When such broad areas as biomes are described, much detail must be omitted. None of the biomes have a continuous, homogeneous vegetation cover. Thousands of local variations in climate, soil, elevation, and physical disturbances cause changes in the plants and animals, and when a biome is delimited and described, the weaknesses of generalities become apparent to anyone who travels through a region. Another difficulty with treating biomes as discrete entities is that they do not end abruptly. Where one meets another, there are usually gradual transition zones. The boundary between two different regions is an *ecotone*, and ecotones are likely to contain special plants and animals of their own. With these reservations, however, we can gain some idea of the distribution of organisms on earth and the forces that determine their survival.

31.13

A temperate rain forest in the Olympic National Forest. The temperature is never extreme, and water is abundant. The result is a special flora, marked by enormous conifers dripping with liverworts. The ground is soft and moist, deep in litter, and frequently covered with ferns.

31.14

Scrub forest. This scrub forest in the Yucatan Peninsula of Mexico scarcely deserves to be called a forest. The rainfall is scanty, the temperatures are generally high, and the soil is thin. During the season when rain does fall, the plants put out leaves and have a brief green period, but the rest of the year the landscape is gray. The small trees, hardly more than bushes, are dwarfed and widely spaced, but when they flower, occasional splashes of orange or yellow or red brighten the landscape.

The Ocean Biome

Traditionally, biomes are thought of as terrestrial, but the oceans can also be considered one great biome (Figure 31.15). Their very immensity demands attention; they cover some 70 percent of the earth's surface. In spite of their occasional great depth—down to 11,000 meters (36,000 ft) in the Marianas Trench—animals live in all parts, and they are theoretically free to go anywhere, since the oceans, unlike land masses, are continuous and always in motion. The endless mixing of the waters has produced a reliability of dissolved components with remarkably constant proportions of various salts. Where large rivers enter the sea, the salt water is diluted by fresh water, but even then it is the total concentration that is changed, not the relative amounts of each. Only in inland seas, not connected with the ocean, can departures from the standard sea-salt proportions be found. For these reasons we consider the oceans a biome, at the same time recognizing that the organisms in them are so numerous and so varied that we can give them only a superficial description here.

31.15

The edge of the sea. The immensity, the power, the mystery, and the variety of the oceans and their inhabitants have stirred the human imagination for thousands of years. Only in recent times, within the last couple of centuries, has scientific study of the oceans been pursued.

31.16
Tundra. The *tundra* is cold, windy, and relatively dry. Winters are long, dark, and frigid; summers are short, but summer days are long in the extreme latitudes. Soils thaw only down to a few centimeters. Because growing conditions are rigorous and the photosynthetic season is limited, plants are low, few, and sparse, and many are dwarfed down to miniature size. Animal life, too, is restricted to species that are specially adapted to the peculiarities of the environment, the only really abundant ones being insects.

The oceans present the sea dwellers with some variables that other dwellers never meet. The differences in pressure, from the surface to the deeps, are greater than anything terrestrial organisms ever have to endure. Some invertebrates in the deepest parts of the *benthic zone*, or *benthos*, sustain as much as 1000 atmospheres (about 1000 kg/cm² or 15,000 lb/in²). Also, inhabitants of the *littoral zone* along the edges of the oceans are affected by tides, which rise and fall, roughly twice a day. The tides change the water level from a few centimeters to as much as 16 meters (50 ft), thus inundating or exposing some organisms repeatedly. In addition, outflow of fresh water from land causes fluctuations in total salinity. But other factors make the sea a steadier place to live. The alkalinity of seawater is buffered at a fairly constant pH 8.2, and the temperature seldom goes above 27°C (81°F) or below 1°C (34°F) (seawater with 3.5 percent salt freezes at about −1.9°C). Compare that with a low of −68°C (−90°F) in Siberia and a high of 58°C (136°F) in the Lybian desert. As a place to live, the ocean is certainly different from land.

Plants in the sea are limited to the *euphotic zone* (Greek, "good light"), the upper levels into which light can penetrate. The larger plants are mostly seaweeds along shores (but see p. 162 for variations in the Sargasso Sea). However, microscopic algae, the *phytoplankton* (Greek, "plant wanderers") are distributed across the entire surface, and they are especially concentrated in colder waters and in upwellings such as the Humboldt current. Animals in oceans are as different from land animals as the two sets of

living conditions would lead one to expect. Whereas insects are the common terrestrial invertebrates, crustaceans are the marine ones, and whereas mammals and birds are the main terrestrial vertebrates, fish are the marine ones. Many of the familiar commercial fishes are taken from the euphotic zone: herring, mackerel, tuna, and sardines. In deeper water, the fish are less common, and they have evolved in ways that appear strange to people who are not accustomed to them. They have long thin tails, or huge eyes or practically none, or glowing spots, or gaping mouths, and the names given them reflect the feelings of the fishermen: gulpers, swallowers, hatchetfish, and viperfish. Some fish and echinoderms and sponges live even in the deepest parts of the sea that have been dredged, where it is perpetually dark and ice cold. In view of recent geological evidence of the activity of ocean bottoms, it is worth thinking that some of the benthic species have been there longer than the sea floor itself.

Tundra

Stretching all across northern America, Europe, and Asia is a flat, treeless expanse of territory, marked by long, cold winters and short, cool summers (Figure 31.16). The main plants are short grasses and sedges and lichens, with a mixture of dwarf woody willows. The surface thaws down to a depth of a few centimeters each summer, enough to allow the plants a meager living during only two months of growing weather. Below that, in some places hundreds of meters deep, is the *permafrost*, the permanently frozen ground. In spite of the sparse vegetation, animals abound, especially insects. Mosquitoes are traditionally the scourge of the arctic. A number of migratory birds nest in the tundra (curlews, for example); and arctic hares, foxes, bears, lemmings, musk oxen, reindeer, and caribou can exist on the available food. Isolated high-altitude plateaus on more southerly mountains have such tundralike plants and climates that they are frequently called tundra. The southern "tundras," however, have longer days than their arctic counterparts, and there is no permafrost, although the vegetation in both places is similar.

The Northern Coniferous Forests

South of the tundra and reaching like the tundra across northern Siberia and northern Canada, the northern coniferous forests, also called *taiga* (TIE-guh), are enormous stretches of spruce, pine, fir, and larch (Figure 31.17). Except for the larch, these are evergreen, and broad-leaved plants are scarce. The climate is cold, the growing season short, and the land poorly drained. The result is that it is a boggy territory providing ample breeding ground for mosquitoes. Because of the mainly evergreen plant cover, productivity is relatively high, and trees can grow well enough to provide the bulk of the world's timber. There is also food for such large animals as moose, which can wade through the wet, mossy bogs called *muskeag*, and for some of the world's most valuable fur-bearing animals: marten, mink, beaver, fisher. Other taiga dwellers are squirrels, lynx, snowshoe rabbits, and such birds as ptarmigans and pine siskins.

The Temperate Deciduous Forests

In contrast to the two previously listed biomes, which are circumpolar, the temperate deciduous forests are discontinuous. There is one large patch in eastern North America, one in western Europe, and one in eastern Asia, and there are some small island ones. Several aspects of the life of deciduous forests have been considered in connection with communities and with ecological succession, and that material need not be repeated. One should not assume, however, that these forests are of uniform biotic composition. Certainly oak trees are dominant in most of the regions, and members of the oak family (including beeches) are widespread. Even in the United States, however, differences can be found between the New England region with its maples, the north central states with their basswoods, the east central states with tulip-trees and gums, and the mixtures in the Appalachian Mountains. Scattered through the whole eastern

31.17
Northern coniferous forest, or taiga. *Taiga* occurs in regions, like those in south-central Canada, where the winters are long and cold but not truly arctic. The trees are mostly evergreen, and the soil is acid and wet. Sphagnum (peat) bogs, which are deep and soft, are locally called *muskeag*. Besides the ever-present mosquitoes, there are moose and many of the most sought-after fur bearing animals: marten, beaver, mink, and muskrat.

United States are stands of conifers, mixed with the deciduous trees. These are probably in edaphic climaxes in local areas, and they are not replaced by oaks because of elevation or soil peculiarities. It is notable that the extent of the temperate deciduous forests coincides with some of the most intense human activities: the American east coast, the European west coast, and China. This has meant that the forests, considered as a biome, are so radically changed from their original condition as to be generally unrecognizable.

Temperate Grasslands

Since the temperate grasslands furnish most of the world's grain and cattle, they are of critical importance to human economy. They occupy a large part of the inhabited earth: portions of the American middle west, through eastern Europe and all across

31.18

The Sonoran desert. The one factor deserts have in common is dryness; some may be cold and some hot. The deserts in the southwest United States are hot deserts, with temperatures of 50°C common during summer days. Many plants survive by doing their active growing during the rare seasons of rain. In American deserts, there are shrubby mints, legumes, lilies, and a wealth of cacti, some of which, like the saguaro, attain tree size. Along with the peculiar plants, animals adapted to the habitat can live effectively: snakes, lizards, rodents, wrens, doves, roadrunners, and quail.

31.19

Tropical grasslands. Where temperatures are moderately high and water supplies are intermediate—not wet enough to support forest but not dry enough to produce desert—*tropical grasslands* develop. Rain is usually seasonal, so that the turf grows luxuriantly during wet months and dries down to dormancy the rest of the year. Such territory supports a wealth of grazing animals, such as the herds of cattle on the Argentine pampas or the once-great assemblages of giraffe and antelope and other vegetarian animals of East Africa.

Asia north of the Himalayan Mountains, down the southeastern coast of South America, great stretches of eastern Africa, and parts of Australia. Grass is the vegetational climax in places where there is too little rainfall to allow forest development, that is, from 25 to 75 centimeters (10 to 30 in.) a year. Some grasslands are actually moist enough to support trees, but trees cannot establish themselves. The sod may be so thick with roots that tree seedlings cannot get started, or recurrent fires, which are not damaging to grass, kill any tree seedlings that do start. In such places, grass is the climax vegetation, but its dominance is determined by fire, not by climate. Such a climax is a *fire climax*. It is now thought that the eastern prairie grasslands of the United States were fire climaxes. We say "were" instead of "are" because such grasslands are gone except for a few "museum specimens" preserved for study and curiosity. Cattle, wheat, corn, and construction have taken the rest.

In the American grasslands, the eastern part was "high-grass" prairie, and the western part "short-grass" prairie. The high-grass prairie is now the corn belt, but before it was converted to agriculture, there was a stand of grass, two meters (6 ft) tall at the height of the season, with unbroken sod in soil whose fertility has been one of the main contributors to the country's wealth. The short-grass prairie to the west, which is now wheat and cattle country, gradually merges into bunch-grass territory, drier and with sparser plant cover. Where cattle are unprofitable, sheep can live, but moving westward from the grasslands, one meets ever less rainfall. The plants are thinner and less productive, and the country slowly becomes desertlike. From the beginning of history, grasslands have been the sources of great wealth, but they have also been the scenes of some of the world's most spectacular ecological disasters.

Deserts

In an area receiving less than about 25 centimeters (10 in.) of rainfall a year, the plant cover is so sparse that the region is called a *desert* (Figure 31.18). The great deserts of the world are the Asian Gobi Desert; the Australian desert, which takes up almost half the continent; and the Sahara, whose 8.3 million square kilometers (3.2 million mi²) rivals in size the entire United States (9.3 million km², or 3.6 million mi²). The Sonoran Desert and Death Valley in the United States, well known to us because they are near, are relatively small. One of the driest places in the world is the Atacama Desert in Chile: The first rain ever known there fell in 1971.

Deserts are not usually completely waterless, nor are they barren. Most have some plants, those adapted to survival in dry places, whether hot or cool. Some store water in juicy stems (cacti), some with long roots shed their leaves during the worst seasons (creosote bushes), and some are quick-growing annuals that can germinate, grow, and produce flowers and seeds during a short rainy season (poppies). Desert plants help themselves further by spacing themselves far enough apart to avoid competition for the scant moisture there is (see allelopathy in Chapter 17) and by using the intense heat and light efficiently (see C-4 photosynthesis in Chapter 4). Desert animals, too, have special adaptations for survival under stress (see the essay on p. 470). Some, like the kangaroo rat or its eastern hemisphere counterpart, the jerboa, live on dry food but conserve their metabolic water. Some, like the horned lizards ("horned toads"), go underground where there is some moisture, coming out to feed only in the cool of night. Others (desert rats) eat only juicy plants, and still others (camels) must drink but can wait a long time between drinks.

Tropical Grasslands

Tropical grasslands occur where cold weather does not impose limits on vegetation, but lack of water does (Figure 31.19). Some occur in east- central South America and in Australia, but the famous ones are in Africa south of the Sahara, where the grasslands are dotted with scrubby trees, making parklike *savannas*. Inhabited by the greatest concentration of large mammals on earth, the savannas are known for the aggregations of antelopes, lions, rhinoceroses, and elephants that make the east African plains a sort of Noah's Ark in the imaginations of humans the world over. The warm climate, the seasonal rainfall (100 to 150 cm of rain a year), and periodic fires make a combination of features that have produced a unique biome. The number of species is small, in contrast to the diversity in the rain forests, and all the grasslands species are adapted to long, hot, dry periods. Drought-enduring grasses dominate the vegetation, but some trees, rather widely separated, sur-

vive, especially the picturesque flat-topped acacias and the legendary baobab trees. Huge groups of consumers (grazers, carnivores, and birds) can be supported because the primary productivity is high, and each species has a niche it can fill effectively, getting the most out of what is there. The African grasslands are being watched with unequaled interest by ecologists, sportsmen, demographers, and plain "nature lovers" because they are being diminished. In many places they have already been destroyed by the growing demands of people for living space.

Tropical Rain Forests

In those few places where the seasons are always warm and wet, tropical rain forests grow: parts of Central America, the Congo lowland in Africa, the west coast of India, the west coast of Indochina, some of the south Pacific islands, and the greatest of all in the Amazon basin (Figure 31.20). A tropical rain

31.20

A tropical rain forest. The impressive features here are continuous warmth and moisture. The top of the forest canopy is brilliantly illuminated, but the lower levels are so heavily shaded that only selected plants can survive in the dim light. Here is a biome that has been called both an "experimental laboratory of evolution" and a "museum of retained species," because here the number of species is enormous, the variety beyond imagination.

forest is not a jungle. A jungle is full of tangled vegetation and may follow the destruction of a rain forest, but a virgin rain forest is open along the ground. In a true rain forest you can walk freely between the trunks of trees such as grow nowhere else: tall and straight and unbranched for perhaps 50 meters (165 ft), but with fluted or spiny or warted trunks that flare out at the base into serpentine roots like the flying buttresses of medieval cathedrals. The great crevices between such root buttresses, which are like rooms, harbor all manner of animals, especially colonies of bats. The forest floor is dark, moist, and littered with rotting fruits, leaves, and flowers. Even where an occasional shaft of sunlight pierces the dense canopy, water drips. The frequent rains do not come straight to the earth. They are caught by the upper leaves, which continue to let water fall for a while, so that it "rains" on the ground after the rain above has ceased. Massive, distorted vines climb to the tops of trees, and curtains of thin roots drop from such heights that you cannot see where they come from. Plants grow on plants. These are the *epiphytes*—orchids and bromeliads and ferns, stuck to trunks and branches, all vying for the little light that filters down. A rain forest may be dark, but it is never still or quiet. Besides the dripping of water, there are always dead branches snapping and falling, flocks of parrots screaming through the leaves, insects buzzing and stridulating, or the misnamed howler monkeys roaring like lions, so high in the branches that you need binocular glasses to see them. An occasional Morpho butterfly wavers silently past, flashing out a glint of electric blue when it passes through a spot of sunlight. Here evolution has figuratively "run wild." A Pennsylvanian who knows a couple of dozen common trees can recognize most of the trees around his home, but a Brazilian who knows that many knows nothing. You may see a tree, spectacularly massed with golden flowers, and not find another like it in the next hour of walking. Unlike the temperate woods, where black oaks grow by the millions in county after county, the rain forest is made up of hundreds of species, mixed and scattered in baffling profusion. If you have any enthusiasm at all for exotic nature, get to a rain forest before progress clears them out.

SUMMARY

1. It took over a million years after the appearance of the first humans for the earth to have a billion people, but it took only 80 years to get its second billion and only 30 for the third. But the earth's potential is finite, and ecology has become a matter of great concern.

2. In ecological terms, the photosynthetic organisms are *producers,* and the nonphotosynthetic organisms, such as fungi, bacteria, and animals, are *consumers.*

3. In an ecosystem the main energy input is sunlight. The small fraction of sunlight energy used in photosynthesis enters into the food production that maintains the ecosystem.

4. A reasonably stable ecosystem achieves a state of *dynamic equilibrium* when over a period of time the energy input equals the outflow. If equilibrium is disturbed, the ecosystem changes.

5. The rate of energy flow is an indication of the activity of an ecosystem. A highly productive system will support a large number of active consumers. It will also have a large amount of organic material, the *biomass.*

6. The distance of a consumer from the primary producer, in terms of passage of food through organisms, determines the *trophic level* a consumer occupies. In a *food web* energy and materials may pass through any number of organisms before the energy is finally lost to the system in the form of heat.

7. Transfer of energy in ecosystems, fundamentally the same as it is in single organisms, conforms to the laws of thermodynamics.

8. High-yield agricultural programs are possible, but they demand a high-energy input in tillage and fertilizer and depend on *monoculture,* the use of a single crop with the built-in dangers of unexpected disease.

9. In any habitat, certain kinds of organisms live and make a *community,* whose characteristics are determined physically by the climatic and edaphic factors of the habitat and biologically by the species present, their sizes, numbers, distribution, and growth habits.

10. In a lake, layers of water are formed as a result of temperature differences. An upper warm layer, the *epilimnion,* lies over an intermediate layer, the *thermocline,* whose temperature is between that of the epilimnion and the deep, cold *hypolimnion.* In many lakes there are two yearly changes of water, the spring and fall turnovers.

11. Lakes with low productivity and a small biomass are called *oligotrophic. Eutrophic* lakes have high pro-

ductivity and a large biomass. The total effect of eutrophication is a stagnant, foul-smelling, oxygen-deprived lake, in which relatively few species of organisms survive. Eutrophication can be speeded up by human action through the addition of waste phosphates and nitrates.

12. The feeding of animals and the decomposing power of bacteria and fungi keep the minerals, carbon dioxide, and water moving through the community in a continuous cycle in an established, balanced ecosystem such as a forest.

13. The final result of communal integration in an ecosystem is a self-perpetuating system, whose balance of change versus stability has caused ecologists to regard it as a kind of superorganism with a homeostasis of its own.

14. *Competition* between species and between individuals in a species is one of the basic phenomena on which the idea of evolution is founded. At the same time, *cooperation* has been working.

15. A region of the earth that has been completely cleared of organisms is a *sere*. A *xerosere* is a dry area, partly or entirely soilless; a *hydrosere* may be a pool of water left in a depression after the retreat of a glacier. Plants and animals coming into such biologically empty places will cause a series of population changes that constitute a *primary succession*. When plants and animals do not have to start from zero, the resulting changes constitute a *secondary succession*.

16. Humans have done more to alter the appearance of the earth than any other living thing has done since the beginning of time.

17. A *biome* is a large, readily recognizable biogeographical unit whose plant life is determined largely by climate, and whose animal life is in turn determined by the kinds of plants that live there. The boundary between two different biomes is an *ecotone*. Traditionally, biomes are thought of as terrestrial, but the oceans can also be considered one great biome.

18. The deepest part of the ocean is the *benthic zone*: the *littoral zone* is located along the edges of the oceans; plants in the sea are limited to the *euphotic zone*, the upper levels into which light can penetrate.

19. Some other biomes are the *tundra*, the *northern coniferous forests*, the *temperate deciduous forests*, the *temperate grasslands*, *deserts*, *tropical grasslands*, and *tropical rain forests*.

RECOMMENDED READING

Bernarde, M. A., *Our Precarious Habitat*, Second Edition. New York: W. W. Norton, 1973.

Benton, A. H., and W. E. Werner, Jr., *Field Biology and Ecology*, Third Edition. New York: McGraw-Hill, 1974.

The Biosphere. A Scientific American Book. San Francisco: W. H. Freeman, 1970. (Paperback)

Cohen, Bernard L., "The Disposal of Radioactive Wastes from Fission Reactors." *Scientific American*, June 1977. (Offprint 364)

Denison, William C., "Life in Tall Trees." *Scientific American*, June 1973. (Offprint 1274)

Ehrlich, P. R., and Anne H. Ehrlich, *Population, Resources, Environment*. San Francisco: W. H. Freeman, 1970.

Ehrlich, Paul R., John P. Holdren, and Richard W. Holm, *Man and the Ecosphere*, Readings from *Scientific American*. San Francisco: W. H. Freeman, 1971.

Frejka, Tomas, "The Prospects for a Stationary World Population." *Scientific American*, March 1973. (Offprint 683)

Gosz, James R., Richard T. Holmes, Gene E. Likens, and Herbert Bormann, "The Flow of Energy in a Forest Ecosystem." *Scientific American*, March 1978.

Harlan, Jack R., "The Plants and Animals that Nourish Man." *Scientific American*, September 1976.

Heslop-Harrison, Yolande, "Carnivorous Plants." *Scientific American*, February 1978.

Horn, Henry S., "Forest Succession." *Scientific American*, May 1975. (Offprint 1321)

May, Robert M., "The Evolution of Ecological Systems." *Scientific American*, September 1978.

Mayer, Jean, "The Dimensions of Human Hunger." *Scientific American*, September 1976.

McNaughton, S. J., and L. L. Wolf, *General Ecology*. New York: Holt, Rinehart, and Winston, 1973.

Newell, Reginald E., "The Global Circulation of Atmospheric Pollutants." *Scientific American*, January 1971. (Offprint 894)

Odum, E. P., *Fundamentals of Ecology*, Third Edition. Philadelphia: W. B. Saunders, 1971.

Richards, Paul W., "The Tropical Rain Forest." *Scientific American*, December 1973. (Offprint 1286)

Rousmaniere, John (ed.), "The Enduring Great Lakes." *Natural History*, August-September 1978. (A 51-page special report containing nine separate articles)

Siever, Raymond, "The Earth." *Scientific American*, September 1975.

Wagner, R. H., *Environment and Man*, Second Edition. New York: W. W. Norton, 1974.

Woodwell, George M., "The Carbon Dioxide Question." *Scientific American*, January 1978.

Epilogue: Some Biological Problems for the Future

ONE OF BIOLOGY'S ATTRACTIVE FEATURES IS THE amount of work remaining to be done. In spite of the uncounted man-years of labor that have been spent trying to find out why living things do what they do, the unknown remains enormous. We do not even know yet what questions to ask. However, we do know enough to guess some of the directions bio·logical investigation is likely to take in the near future, remembering at the same time that events have proved embarrassing to many a prophet.

Research already being done gives some indication of what may be the main topics of interest in the near future. We can therefore make some diffident conjectures about what we can reasonably expect biologists to do next. Here we will hint at some of the subjects of study that will probably take the energy of investigators during the lifetime of students reading this book. As always, some biologists will emphasize those kinds of research that promise practical results, and others will prefer investigations that seem to offer new knowledge simply for its own sake. Separating them into "applied" and "pure" lists is arbitrary. The most abstract theories have proved to be practical, sometimes surprisingly or even disastrously so, and straightforward functional technology has successfully stimulated thoughtful people to seek intellectually satisfying reasons for phenomena. As Derek Price of Yale University said, steam engines came before the statement of the laws of thermodynamics.

Biologists and their collaborators are already studying a number of problems and can be expected to keep on doing so for some time. In the minds of many people, including not only biologists but sociologists, theologians, and politicians, the most threatening development of the century is the increase in human population. Whether they like to think so or not, people are animals, with thousands of centuries of reproductive drive built into them, and given the opportunity, they reproduce. If life for the great majority of us is not to become unbearably miserable, our species must find a way to regulate our own numbers. That way will be partly biological (contraception) and partly social (persuasion). Physiological, chemical, and behavioral work on sexuality of animals and even of plants will bring new and possibly useful knowledge. The obvious urgency of the need almost guarantees that population problems will continue to occupy the attention of researchers for some time.

Closely connected to population is the difficulty of feeding even those who are already here. Many kinds of biologists must keep seeking cheaper, more efficient, tastier, and more nutritious food supplies. One continuing problem is finding space for edible plants and animals, producing them in adequate quantities, and still preserving the stock of genes that now exist in wild populations. You cannot breed selectively for high-protein corn or heat-tolerant animals unless you have sufficient variety in your living subject material. Even while agronomists, animal husbandmen, and microbiologists are pressing for more volume in food production, nutritionists are finding that our current knowledge of what food we really need is far from complete. The study of nutrition, for years thought to be a minor, unglamorous field of medicine, is becoming recognized as enormously important, and it will probably increase in interest.

Energy is popularly considered primarily an industrial rather than a biological commodity, but even with some hydroelectric and nuclear power plants, most of the energy used privately and industrially is of biological origin, as it has been for centuries. Fossil fuels are biological fuels, and as they run low and become increasingly expensive, the search is on for new, cheap supplies. Experimental efforts are already being made to utilize the abundant supply of cellulose, perhaps by fermentations, and to find more efficient producers in the form of quick-yielding, genetically selected forests. Another potential energy supply is wasted when refuse materials, especially paper products, are burned or allowed to rot. Retrieving energy microbially from garbage is possible but so far costs too much. The economics of waste use will change, and biologists should be ready to meet the demands of the future.

Human health is another concern of biologists. Even parasitic diseases are still important killers. Modern sanitation and drugs have reduced the suffering due to polio, diphtheria, typhoid fever, and many other diseases, but pneumonia, venereal diseases, malaria, and worm infections remain serious. Nevertheless, it is the noninfectious diseases—cardiovascular diseases and cancer—that are now the most conspicuous destroyers. Enormous efforts in both time and money have been and are being spent and will undoubtedly continue to be spent on research, but so far the main discovery, especially in cancer work, is that a so-called crash program is not enough when a problem demands basic biological information—information that we do not have. We can predict with some confidence that there will be no general, empirically discovered cure for cancers, but that when the fundamental activities responsible for cellular disruption become known, cures may turn out to be relatively easy. At any rate, today's students can be sure that into the foreseeable future there will be continuing emphasis on the search for cures of human and other animal diseases, and that (if past lessons mean anything) we will indeed keep finding ways to lessen our miseries.

Pollution has become a topic of popular concern because pollution is expensive, threatens public health, and offends people's sensibilities. Just as biologists have been the main ones to find and publicize the causes and effects of pollution, so they will have to share in and probably take a lead in reducing or at least controlling it. The enormous practical difficulties of regulating pollution will require the best efforts of trained people for many years to come.

One final practical aim of biology, in this incomplete list, is to understand what happens to an animal when it gets old. Serious biological study of aging is so new that even the term for the study, gerontology, is to be found only in newer dictionaries. Cancer cultures can be maintained indefinitely, and the growing regions of plants are permanently embryonic. These facts, indicating that cells are not necessarily doomed to a programmed death, cause some biologists to seek a sort of biochemical fountain of youth. This search, too, may continue for some time.

Turning now to the theoretical portions of biology that could keep researchers excited, several seem promising. One is the great field of developmental biology, which is concerned with all aspects of growth and differentiation. In spite of the dazzling recent advances in the understanding of the metabolic functioning of cells, the simple questions are not answered. How does an undifferentiated cell get started toward specialization? Why do cells *not* do most of the things they are capable of doing? What governs the gross architecture of a large complex organism? We do not know even how a plant cell builds a wall around itself, or why cells divide in effective directions, or how wandering cells find their way around in organisms, or how one cell divides mitotically while its nearest neighbor is dividing meiotically. The hints we have from experimental embryology indicate that developmental biology as a science is only beginning.

Another cellular study of great and increasing interest concerns membranes, as well as the physics and chemistry of surfaces. Some biologists have faith (and faith does have a place in science as well as in religion) that a fuller understanding of membranes and surfaces not only would lead us to solutions to many pesky biological problems but, more important, would lead us to new questions never before thought of.

Passing from the subcellular to the superorganismic, we come to a consideration of the investigation of the environment as an entity. Present information on the total effect of environment on organisms and of organisms on their environment is primitive, but at least we know enough to recognize our ignorance. That, say the philosophers, is the beginning of wisdom. Ecologists are passing from the first stages of descriptive science and are working toward experimental and quantitative studies, but they have made only a beginning. There is a long way to go.

One old-fashioned and unfashionable portion of biology is taxonomy, but there are and probably will continue to be individuals who believe that, as a part of knowledge, we should be aware of our fellow travelers, at least to the extent of knowing what they are. Anyone who seeks to learn the identity of any specific group of organisms soon finds undiscovered ones and is forced to the conclusion that many more remain unknown. Not only may we find more satisfactory systems of classification (some people seem to be genetically predetermined classifiers), but we need to know what is living where. Aside from a philosophical desire to find out, we must have taxonomists who can tell us what organisms we are using experimentally, which ones are possibly useful, and which ones can tell us what is happening as environments change. Taxonomy will be with us for a long time, though perhaps in unfamiliar form.

The last biological field we will mention here is psychobiology, which can be broadly understood to include neurobiology and all aspects of behavior. Like ecology, psychobiology has only recently started to become an experimental, scientific discipline. Also like ecology, it is so challenging in its complexity that its problems require great expenditure of time, energy, brains, and money. The last portion of the twentieth century may well see psychobiology as one of the most exciting and productive subfields of the larger discipline of biology.

Now that we have reviewed some of the past achievements in biology and tried to guess where the discipline is likely to go next, it is worth thinking what is to be gained from all this factual and speculative material. If you have given the text fair attention, you may have the impression that biology is a mass of facts integrated only by evolutionary theory. Such an impression is partly justifiable, but there is more to it than that. There are indeed two demonstrable bits of information that may be of lasting value, perhaps even of consolation or a source of strength in times of trouble. The first is the unity of all life, exemplified by the universality of the genetic code. Remember that an mRNA codon of GCU "means" alanine, whether the source of the codon is a sponge or a saint. It seems that all this world's organisms must be somehow related. The second bit, complementary to but opposed to the first, is the uniqueness of each individual's DNA. Unless you are one of a pair of identical twins, you have a sequence of base pairs unmatched on earth. For those who wonder, "Who am I?" there is an answer: "I am special. I am I, and there is not another like me in all creation."

Glossary

This glossary includes pronunciation, derivation, and definitions for technical terms, phrases, and abbreviations used in the text. Derivations from the French (Fr.), Latin (L.), and Greek (Gr.) are given as an aid to understanding, not as a scholastic exercise.

ABSCISIC ACID (ab-SIZ-ik) [L. cut off]. A plant growth-regulating substance that, among other things, regulates leaf fall and acts as an antagonist to auxin in such processes as cell elongation.

ACETYLCHOLINE (uh-SEET-ul-KO-leen). A chemical substance, released by synaptic vesicles in the nerve endings, that crosses the synaptic gap and excites the next nerve fiber.

ACID [L. sour]. A compound that yields hydrogen ions (protons) when added to water; turns blue litmus paper red; reacts with bases to form salts. (*See* BASE)

ACROMEGALY (ak-roh-MEG-uh-lee) [Gr. extremity + great]. A condition in which the skeleton is thickened and the face, hands, and feet continue to grow after skeletal development is complete because of excess growth-hormone secretion.

ACROSOME (AK-roh-sohm) [Gr. extremity + body]. A cuplike body at the anterior tip of sperm, containing several enzymes that aid the sperm in penetrating an egg.

ACTIN (AK-tin). A protein that makes up the thin filament in muscle. Actin and myosin, a larger protein, are the ultimate structural units of contractility.

ACTION POTENTIAL. The temporary reversal of relative charge on the inner and outer surfaces of a nerve or muscle cell membrane, passing along the membrane as a wave of excitement.

ACTIVE TRANSPORT. The movement, by a process requiring expenditure of energy, of molecules into or out of a cell, across a cell membrane.

ADAPTATION. A change in an organism that brings it into better harmony with its environment.

ADAPTIVE RADIATION. The process of speciation as isolated populations adapt to new environments.

ADDISON'S DISEASE. A disease caused by underactivity of the adrenal cortex. Symptoms include anemia, increased blood potassium, decreased blood sodium, and bronzing of the skin.

ADENINE (ADD-uh-neen). A purine, a nitrogenous base, which is a constituent of nucleic acids and of nucleotides such as ATP, ADP, AMP.

ADENOSINE TRIPHOSPHATE (ATP) (uh-DEN-oh-seen try-FOSS-fate). An organic compound containing adenine, ribose, and three phosphate groups; stores energy in chemical bonds and thus serves as an energy source for chemical reactions in living cells.

ADRENAL (uh-DREEN-ul) [L. upon + kidney]. The paired glands that rest on the kidneys; each is composed of inner medulla and outer cortex. Adrenal medulla chiefly secretes adrenalin. Adrenal cortex chiefly secretes aldosterone and cortisol.

ADVENTITIOUS ROOTS (ad-ven-TISH-us) [L. not properly belonging to]. A root produced from an organ other than the radicle or another root.

AEROBIC (air-ROH-bik) [Gr. air + life]. Processes or organisms that require free atmospheric oxygen.

AFFERENT (AF-uh-rent) [L. to bring to]. Bringing substances or impulses toward the heart or central nervous system; denoting directional action of arteries, nerves, etc.

AGGLUTINATICN (uh-GLOOT-in-AY-shun) [L. to glue to a thing]. The clumping of cells or particles in a fluid. For example, clumping of blood cells occurs when incompatible blood types are mixed.

ALDOSTERONE (al-DOSS-tuh-rone). An adrenal cortical hormone, which stimulates the reabsorption of sodium by the kidney and thereby reduces the amount of sodium in the urine.

ALLELE (uh-LEEL) [Gr. of one another]. Alternative form of a gene that occupies similar sites on homologous chromosomes.

ALLELOPATHY (al-uh-LAW-puh-thee) [Gr. other + suffer]. The phenomenon of keeping plant competitors at bay by the secretion of toxic chemical compounds.

ALLERGY (AL-er-gee). A hypersensitivity to some substance in the environment.

ALLOPATRIC SPECIES (al-oh-PAT-rik) [Gr. other, different + L. country]. Similar species that are isolated both geographically and reproductively.

ALL-OR-NONE LAW. The concept that an event happens to its fullest possibility or not at all; usually used to refer to the firing or lack of firing of a nerve impulse.

ALPHA PARTICLES (ALPHA "RAYS"). Helium nuclei emitted from radioactive isotopes of elements.

ALTERNATION OF GENERATIONS. A reproductive cycle in which the haploid gametophyte, or sexual phase, is followed by the diploid sporophyte, or asexual phase. Spores, produced from the sporophyte, give rise to new gametophytes.

ALTRUISM (AL-troo-iz-um) [Fr. others]. Willingness of an individual to sacrifice something or all of itself for the benefit of the species genotype; in humans, unselfish idealism.

ALVEOLUS (al-VEE-uh-lus) [L. hollow]. A tiny air sac in a lung, surrounded by a network of blood vessels; the site of gas exchange between the air and the capillaries.

AMINO ACID. An organic compound containing an amino group, —NH₂, and a carboxyl group, —COOH; the major components of proteins are amino acids.

AMNION (AM-nee-on) [Gr. lamb]. One of the special extra-embryonic membranes of reptiles, birds, and mammals; a fluid-filled sac in which the embryo floats.

AMYLASE (AM-uh-lace). An enzyme that hydrolyzes starch to sugar.

ANAEROBIC (an-uh-ROH-bik) [Gr. not + air + life]. Not requiring free atmospheric oxygen.

ANALOGOUS (an-AL-uh-gus) [Gr. conformable, proportionate]. Pertaining to parts of organisms used for similar functions but not similar in origin or in construction, such as wings of insects and birds. (See HOMOLOGOUS)

ANAPHASE (AN-uh-faze) [Gr. up, back, again + phase]. A stage in mitosis or meiosis in which the chromosomes move from the equator toward the poles of the spindle.

ANDROGENS (AN-druh-jenz) [Gr. man + to produce]. Sex hormones or other substances that have masculinizing effects, e.g., testosterone.

ANEMIA (uh-NEE-mee-uh) [Gr. no blood]. A condition resulting from a decreased production or increased destruction of erythrocytes in the human body; a lack of hemoglobin.

ANGIOSPERM (AN-jee-uh-sperm) [Gr. vessel + seed]. A flowering plant, whose ovules are enclosed within ovaries.

ANTAGONISTIC. In anatomy, opposing muscle pairs that permit movement in opposite directions.

ANTERIOR. Toward the front or head end.

ANTHERIDIUM (an-thuh-RID-ee-um). An ovoid, multicellular, male sex organ that produces sperm; found in mosses and ferns.

ANTHROPOMORPHISM (AN-thro-po-MORE-fiz-um) [Gr. man + form]. The practice of regarding animals in terms of human characteristics.

ANTIBIOTIC (an-tee-by-OTT-ik) [Gr. against + pertaining to life]. Substance formed by microorganisms that has the capacity to inhibit the growth of other microorganisms; used in the treatment of infectious diseases in humans and animals.

ANTIBODY [Gr. against + body]. A specific protein synthesized in response to the presence of a foreign material. (See ANTIGEN)

ANTICODON (an-tee-KOH-don) [L. against]. The sequence of three nucleotides in transfer RNA that pairs complementarily with three nucleotide codons of messenger RNA.

ANTIDIURETIC (an-tee-dy-yuh-RET-ik) HORMONE (ADH). A hormone, produced by the hypothalamus and released by the posterior lobe of the pituitary, that aids in the water absorption in the kidneys and stimulates the smooth muscles of the arteries.

ANTIGEN (AN-ti-jen) [Gr. against + to produce]. A foreign material that elicits the synthesis of a particular kind of molecule, the antibody, within an organism. (See ANTIBODY)

APHASIA (uh-FAY-zhuh) [Gr. without + speech]. The inability to understand written or spoken words, due to damage to the speech areas of the cerebrum.

APICAL BUD (AY-pi-kul). A mitotically active bud, at the tip of the stem of vascular plants, which adds new cells at the tip and thus adds to the height of the plant.

APPEASEMENT GESTURE. A submissive position directly opposite to a threatening posture.

APPENDICULAR SKELETON (ap-en-DIK-yu-ler). The structures that attach to the axial skeleton, such as arms and legs.

ARCHEGONIUM (ahr-ki-GO-nee-um) [Gr. the first of a race]. A multicellular female sex organ that produces eggs; found in mosses, ferns, and conifers.

ARTERY. A vessel that carries blood away from the heart.

ARTERIOSCLEROSIS (ahr-TEER-ee-oh-skloh-ROH-sis). [Gr. artery + hardness]. A disorder in which the arteries harden, with resultant higher-than-average blood pressure.

ARTICULATION (ahr-TIK-yu-LAY-shun) [L. joint]. Joint, the point where two separate bones meet or where a cartilage joins a bone; any place where two separate pieces of an organism are joined.

ASCUS (ASS-kus) [Gr. bag]. The spore-carrying cell common to the fungi in the Class Ascomycetes.

ASSOCIATION. The recombining of ions to form an unchanged molecule. (See DISSOCIATION)

ATHEROSCLEROSIS (ATH-uh-roh-skloh-ROH-sis). A deposition of fat, fibrin, cellular debris, and calcium on the inside of arteries.

ATOMIC NUMBER. The number that represents the number of protons in the nucleus of an atom of a chemical element.

ATOMIC WEIGHT. The average weight of an atom of an element relative to the weight of an atom of carbon 12.

ATP. (See ADENOSINE TRIPHOSPHATE)

ATRIUM (AY-tree-um) [Gr. hall]. A chamber of the heart, which receives blood from a vein and pumps it to the next chamber, the ventricle; formerly called an auricle.

AUTECOLOGY (ott-ee-KOL-ee-jee) [Gr. self + ecology]. The study of the reactions of individual organisms to their environment. (See SYNECOLOGY)

AUTONOMIC NERVOUS SYSTEM (ott-oh-NOM-ik). A portion of the vertebrate nervous system composed of nerve fibers that synapse in ganglia outside the central nervous system. It controls such "involuntary" activities as breathing and glandular secretion.

AUTOSOME (OTT-oh-sohm) [Gr. self + body]. A chromosome not directly associated with sex determination.

AUTOTROPH (OTT-oh-trohf) [Gr. self + to nourish]. An organism capable of synthesizing its food from inorganic sources. (See HETEROTROPH)

AUXIN (ox-in) [L. increase]. A plant growth-regulating substance, indoleacetic acid (IAA).

AXIAL SKELETON (AX-ee-ul). The skull, vertebrae, and ribs.

AXON (AX-on). A part of a neuron, a long process extending from the cell body. It usually carries impulses away from the cell body.

BACTERIOPHAGE (bak-TEE-ree-uh-fayj) [L. little rod + Gr. to eat]. A virus that infects bacteria.

BASE. A compound that accepts hydrogen ions when added to water; turns red litmus paper blue; reacts with acids to form salts. (See ACID)

BASIDIUM (buh-SID-ee-um) [L. a little pedestal]. A specialized reproductive cell in fungi of the Class Basidiomycetes in which meiosis occurs and which produces external spores.

BENTHIC ZONE (BEN-thik). The ocean deeps, below the region of light penetration.

BETA PARTICLES (BETA "RAYS") (BAY-tuh). Electrons emitted from radioactive isotopes of elements.

BILE. Organic alkaline substance, produced in the liver and stored and modified by the gall bladder, which emulsifies the fats in the small intestine and helps neutralize the acidic chyme.

BILIRUBIN (bil-ee-ROO-bin) [L. bile + red]. A bile pigment, converted from hemoglobin without iron, that is excreted in feces.

BIODEGRADABLE. Capable of being destroyed by enzymes from organisms.

BIOFEEDBACK. The conscious control of "automatic" body functions, such as blood pressure and rate of heart beat; any self-regulating biological system.

BIOLUMINESCENCE (by-oh-loo-mi-NESS-ens) [Gr. life + L. light]. The emission of light by living organisms or by enzyme systems prepared from living cells; the direct conversion of chemical energy to light energy.

BIOMASS. The total weight of organisms in an ecosystem.

BIOME (by-ohm). A major division of the earth's surface, largely determined by climate but characterized by the plants and animals inhabiting it.

BIOSPHERE. The living portion of the earth.

BIRAMOUS (by-RAY-mus) [L. two + branch]. Having two branches, such as the swimmerets of crustaceans.

BLASTOCYST (BLAS-toh-sist) [Gr. bud, sprout + bag]. The entire mass of trophoblast and embryo.

BLASTOMERE (BLAS-toh-meer) [Gr. germ + part]. A cell of the blastula in an embryo.

BLASTOPORE (BLAS-toh-pohr) [Gr. germ + a passage]. A small opening at one side of the double-walled ball of an embryo during the gastrula stage. It is the region where invagination occurred.

BLASTULA (BLAS-chuh-luh) [Gr. germ]. A hollow ball resulting from cellular cleavages early in embryonic development.

BOHR EFFECT [Christian Bohr, 19th-century Scandinavian physiologist]. High carbon dioxide level in the blood causes more oxygen to be released from hemoglobin.

BOLUS (BOH-luss) [L. morsel]. A moistened, softened mass of food being passed through the alimentary canal.

BOOK LUNGS. Thin-layered folds of tissue for gas exchange in organisms such as spiders.

BOWMAN'S CAPSULE [Sir William Bowman, 19th-century British physician]. In a kidney, one of the microscopic cup-shaped structures that filter water and other substances from blood.

BRAINSTEM. The stalk of the brain, relaying messages between the spinal cord and the brain. It is divided into the medulla, pons, and midbrain.

BRONCHUS (BRONG-kus) [Gr. windpipe]. One of the two posterior branches of the trachea serving as a passage for air to and from the lungs.

BUFFER. A compound, with dissociated ions, that resists change in pH.

BUNDLE SHEATH. A cellular covering around the veins of leaves.

CALORIE [L. heat]. The amount of energy required to raise the temperature of a gram of water one degree Celsius, specifically from 14.5° to 15.5°.

CALVIN CYCLE. The process by which carbon dioxide is reduced to carbohydrates during photosynthesis.

CAMBIUM (KAM-bee-um) [L. change]. A plant tissue, usually under the bark of trees, whose cells by repeated division increase the girth of the stem, adding new wood and bark.

CANCER [Gr. crab]. A malignant tumor characterized by abnormal, uncontrolled growth.

CANINES (KAY-nines) [L. dog]. Long sharp teeth adapted to tearing flesh.

CAPILLARY (KAP-uh-layr-ee) [L. hair]. A small, thin-walled vessel that permits exchange of nutrients, oxygen, and carbon dioxide between the blood and tissues; the connecting vessels between arteries and veins.

CAPILLARY WATER. Water that is present in the soil in liquid form but held firmly enough so that it does not drain away to the water table below. It is an important source of water to growing plants.

CARBOHYDRATES. Organic compounds composed of carbon, hydrogen, and oxygen, in the ratio of one carbon to two hydrogens to one oxygen, e.g., sugar, starch, cellulose.

CARBON MONOXIDE (CO). A common colorless gas, a product of internal combustion engines, dangerous to animals because it can displace oxygen in the red blood cells.

CARPELLATE (KAR-puh-lut) [Gr. fruit]. Ovulate; referring to the plant or plant part that bears the female gametophytes and eventually the embryos in flowering plants.

CASTE (KAST) [L. pure]. In insects, a stratum within a society rigidly parceled into different groups on the basis of the division of labor. Definite individual roles are determined by genetic, sexual, and nutritional factors.

CATALYST (KAT-uh-list) [Gr. a throwing down]. A chemical that speeds up a chemical reaction without permanently entering into the reaction itself.

CELL MEMBRANE. The outermost layer of the cell, consisting mainly of phospholipid and protein; about 100Å thick; selectively permeable; can initiate and control the active transport of substances.

CELLULOSE (SELL-yu-lohs). A polymer of glucose; the main component of the cell wall in land plants.

CENTRAL NERVOUS SYSTEM. The brain and spinal cord.

CENTRIOLE (SEN-tree-ohl) [L. center]. A self-replicating, cylindrical organelle located just outside the nucleus of animal and some plant cells; occurs in pairs, acting as focal points at opposite ends of the cell during nuclear division.

CENTROMERE (SEN-troh-meer) [Gr. center + part]. The region of spindle attachment on a chromosome, which ensures that a chromosome will proceed toward a pole in cell division.

CEPHALOTHORAX (SEF-uh-loh-THO-raks) [Gr. head + thorax]. The fused head and thorax of the chelicerates and crustaceans, such as spiders, crabs, and shrimp.

CEREBELLUM (ser-uh-BEL-um) [L. brain]. A part of the brain; the coordinating center for muscular movement.

CEREBRUM (se-REE-brum) [L. brain]. A part of the brain divided into two hemispheres, with each hemisphere divided further into lobes. Each lobe contains special functional areas, including areas for speech, vision, movement, learning, and memory.

CERVIX (SER-viks). The "neck" of the uterus, which opens into the vaginal canal.

CHELICERAE (kih-LISS-uh-ree) [Gr. claw + horn]. The first pair of appendages on the cephalothorax; the pinching claws in horseshoe crabs and the poison-injecting claws in spiders.

CHEMORECEPTOR (KEE-moh-ree-SEP-ter). Sensory cell or organ capable of perceiving chemical stimuli, such as those of smell or taste.

CHIASMA (kye-AZ-muh) [Gr. X]. Exchange of parts between two chromatids during prophase I of meiosis, when homologous chromosomes are paired.

CHITIN (KYE-tin) [Gr. tunic]. An insoluble, tough, horny polymer of acetyl glucose-amine, forming the cell wall of some fungi and the exoskeleton of the arthropods.

CHLOROPHYLLS (KLOR-oh-fills) [Gr. green + leaf]. The green pigments in plants that absorb light at the outset of photosynthesis.

CHLOROPLAST (KLOR-oh-plast) [Gr. green + formed]. A plastid that contains chlorophyll; photosynthesis is initiated when light strikes a chloroplast.

CHOLESTEROL (koh-LESS-tuh-role). A lipid that is an essential part of the human diet. Besides acting as a usual source of energy and assisting tissue maintenance, it is also needed for the production of bile acids and sex hormones.

CHOLINESTERASE (koh-lin-ESS-tuh-race). An enzyme that attacks and neutralizes the neuro-transmitter acetylcholine.

CHORION (KOHR-ee-on). An extraembryonic membrane that forms an outer membrane around the embryo; in mammals the chorion makes up most of the placenta.

CHORIONIC GONADOTROPIN (kohr-ee-ON-ik goh-nad-oh-TROH-pin). A female sex hormone, produced in the embryonic membranes and the placenta, which keeps the corpus luteum from disintegrating and stimulates steroid secretion from this structure.

CHROMATID (KRO-muh-tid). One of the two daughter strands of a newly duplicated chromosome.

CHROMATIN [Gr. color]. The readily stainable nuclear material of the chromosomes, composed of DNA and proteins.

CHROMATOPHORE (kro-MAT-uh-fore) [Gr. color + to bear]. A granule containing pigments in bacterial cells or in skin of some animals.

CHROMOPLAST (kro-moh-plast). A pigment containing plastid in a plant cell, usually red, orange, or yellow. Chloroplasts are green chromoplasts.

CHROMOSOME (KROHM-uh-sohm) [Gr. color + body]. The nucleoprotein structure that contains the hereditary units, the genes.

CHRYSALIS (KRISS-uh-liss) [Gr. gold, referring to the gold-colored pupa of a butterfly]. The pupal stage in butterflies in which the young is enclosed in a firm case.

CHYME (KIME). A soupy slurry of partially digested food.

CILIA (SILL-ee-uh) [L. eyelid]. Short bristlelike locomotor organelles on the free surface of cells.

CIRCADIAN RHYTHM (sir-KAY-dee-un) [L. about + day]. Daily cycle of behavior.

CLEAVAGE. The cell divisions, initiated after fertilization, that mark the beginning of embryo development.

CLIMAX. In ecology, a community of organisms that has achieved stability and can resist changes by invading species; in coitus, an orgasm.

CLITELLUM (kly-TELL-um) [L. a pack saddle]. A band around the body of earthworms, functioning during mating and providing a cocoon containing eggs.

CLITORIS (KLIT-uh-riss) [Gr. small hill]. Small erectile body just anterior to the vaginal opening; it is homologous to the male penis.

CLONE (KLOHN). A genetically uniform population of cells or organisms derived asexually from a single ancestor.

COACERVATES (koh-ASS-ur-vates) [L. cluster together]. Prebiotic aggregates formed by organic compounds, which may eventually have led to the formation of biological cells.

COCHLEA (KOCK-lee-uh) [L. snail]. The coiled tube in the inner ear, containing the sensitive cells that make sounds perceptible.

COCOON [Fr. shell]. The pupal stage in moths. The developing young is covered by an envelope, often of silk.

CODON (KOH-don). A sequence of three nucleotides in messenger RNA that codes for a single amino acid.

COENZYME (koh-EN-zime) [L. with + Gr. in + yeast]. A compound, such as a vitamin, that helps activate an enzyme.

COEVOLUTION (KOH-ev-oh-LU-shun). The evolution of two different species in ways that make them mutually dependent.

COITUS (KOH-i-tuss) [L. to come together]. The act of sexual intercourse.

COLEOPTILE (koh-lee-OP-tile) [Gr. sheath + feather]. The sheath that covers the first leaf in a grass seeding.

COLON (KOH-lun). The large intestine.

COMPANION CELL. A slender, nucleated cell in the phloem of vascular plants.

COMPENSATION POINT. A degree of light intensity at which the metabolism of a plant uses up the photosynthetic products as fast as they are produced.

COMPLETE METAMORPHOSIS (met-uh-MORE-fuh-sis) [Gr. after, beyond, over + shaping]. In insects, the change from egg to larva to pupa to adult.

COMPOUND. A substance whose molecules are composed of more than one kind of atom; every molecule of one compound contains the same elements joined in the same proportions by weight.

COMPOUND EYE. An arthropod eye that is composed of many tubular units, ommatidia, each with its own lens and light-sensitive receptor cells.

CONDENSATION. A reaction in which small, simple molecules are combined into larger, more complex ones, with the elimination of one or more molecules of water. (See HYDROLYSIS)

CONDITIONING. A type of learning in which an animal is rewarded or punished according to its reaction to a given stimulus.

CONDUCTIVITY. The ability of neurons to conduct a current along a nerve cell. The other basic property of neurons is excitability. (See EXCITABILITY)

CONE. (1) Light-sensitive cells in the retina of the eye; used in color vision. (2) Spore-bearing structures in nonflowering plants, such as pine trees.

CONIDIUM (kuh-NID-ee-um) [Gr. dust]. An asexual fungus spore.

CONJUGATION (kon-juh-GAY-shun) [L. together + yoke]. Sexual union of similar cells as in some algae, protozoa, and bacteria.

CONTRACTILE VACUOLE (kon-TRAK-tul VAK-yu-ohl). An intracellular organelle of water balance, which acts as a reservoir of excess water; common in protozoa. The excess water is accumulated and then emptied to the outside.

CONVERGENT EVOLUTION (kon-VER-jent) [L. together + to incline]. The process of changing toward similarity so that organisms that are not closely related come to be alike.

CORPUS CALLOSUM (KOHR-puss kuh-LOH-sum) [L. body + hard]. A tough bridge of nerve fibers that connects the two cerebral hemispheres and relays nerve impulses between them.

CORPUS LUTEUM (KOHR-puss LOO-tee-um) [L. body + yellowish]. A temporary endocrine gland in mammals that secretes the hormone progesterone; it is formed in an ovary after the discharge of an egg.

CORTEX (KOHR-teks) [L. bark]. (1) The outer portion of an organ, as in brains or kidneys. (2) The tissue in the outer bark of plants.

COTYLEDON (kot-uh-LEE-dun) [Gr. cup-shaped hollow]. Seed leaf of the embryo of a plant.

COVALENT BOND (koh-VAY-lent). A chemical bond between two atoms, formed by the sharing of one or more electrons.

CREATINE PHOSPHATE (KREE-uh-teen). A phosphate compound that is a source of energy in the contraction of muscle.

CRETINISM (KREET-in-iz-um) [Fr. Christian]. A disorder, characterized by irregular development of bones and muscles and mental retardation, resulting from underactivity of the thyroid gland during infancy. Often confused with Down's syndrome (mongolism).

CRISTAE (KRIS-tee) [L. cock's comb]. The inner membranes of mitochondria folded and doubled on themselves, forming incomplete partitions.

CROP. A storage segment in the digestive system of earthworms and birds, from which food is released at controlled intervals.

CROSSOVER. A process during prophase I of meiosis, whereby nonsister chromatids exchange parts; also the genetic results of such an exchange.

CRYPTIC COLORATION (KRIP-tik) [Gr. hidden, concealed]. An adaptation that helps camouflage an animal by providing it with a color that makes it inconspicuous.

CUSHING'S DISEASE. A condition usually caused by a cortical adrenal tumor that overproduces glucocorticoids. A tendency toward diabetes is caused by increased blood sugar.

CUTICLE (KYOOT-i-kul) [L. skin]. A waxy layer coating the outer wall of epidermal cells in plants.

CYCLIC AMP (CYCLIC 3', 5'-ADENOSINE MONOPHOSPHATE). A chemical substance that plays a major role in mediating cellular activity in animals, bacteria, slime molds, etc.

CYCLIC PHOSPHORYLATION (SIK-lik FOSS-for-i-LAY-shun). A process in some plants during which electrons are recycled and phosphorous is transferred; the only product is ATP. (*See* PHOSPHORYLATION)

CYTOCHROME (SIGH-toe-krohm). Protein-plus-pigment molecule containing iron; acts as a carrier of electrons from Krebs cycle during respiration in mitochondria and during photosynthesis in chloroplasts.

CYTOKININ (sigh-toe-KYE-nin) [Gr. cell + motion]. A plant growth-regulating substance, which stimulates cell divisions.

CYTOPLASM (SIGH-toe-plazm). Nonnuclear portion of a cell.

CYTOSINE (SIGH-toe-seen). A pyrimidine, a nitrogenous base that is a constituent of nucleic acids.

DEDIFFERENTIATION. Reversion of a specialized cell to an unspecialized condition.

DEGENERACY. The ability of more than one nucleotide triplet (codon) in DNA or mRNA to specify one amino acid.

DELETION. A chromosome change in which a segment of a chromosome, involving one or more genes, is lost.

DEME (DEEM). A group of individuals, partially or entirely isolated from the other members of the species.

DENATURATION (dee-nay-chu-RAY-shun). Alteration of three-dimensional structures and physical properties of a protein.

DENDRITE (DEN-dryte) [Gr. tree]. Short, threadlike process, extending out from the cell body of a neuron. Dendrites carry impulses toward the cell body.

DENDROCHRONOLOGY (DEN-droh-kruh-NOLL-uh-jee) [Gr. tree + time + study]. The technique of reconstructing past events, such as buildings or climatic changes, by the study of annual rings in wood.

DEOXYRIBONUCLEIC ACID (DNA) (dee-AHK-see-rye-boe-new-KLAY-ik). A double-stranded nucleic acid that is a constituent of chromosomes; contains genetic information coded in specific sequences of its nucleotides.

DETERMINATE CELLS. The cells of the early embryo that are so prearranged that they are destined to become some specific part of the future animal.

DETRITUS (deh-TRY-tuss) [L. rubbing away]. Microscopic particles in soil; in a biological sense, finely divided organic matter.

DIABETES MELLITUS (die-uh-BEE-teez MEL-it-us) [Gr. to pass through + honey]. An inherited metabolic disease caused by insufficient production of insulin. Cells are unable to use glucose, which is excreted in the urine.

DIALYSIS (die-AL-uh-sis). A technique of removing diffusible substances from a mixture of substances by allowing the diffusible components to pass through a selectively permeable membrane.

DIAPAUSE (DIE-uh-pawz) [Gr. pause]. A period of insect dormancy between the larval and mature stages.

DIAPHRAGM (DIE-uh-fram) [Gr. a muscular partition between the thoracic and abdominal cavities serving as part of the muscular mechanism for the respiratory system; any separating partition.

DIASTOLE (die-ASS-toe-lee) [Gr. expansion]. The time of ventricular relaxation, when the ventricle receives blood from the contracting atrium.

DICOTYLEDON (die-kot-uh-LEE-dun) [Gr. two + cotyledon]. A flowering plant whose seeds have two embryonic leaves, e.g., beans, walnuts, oaks.

DICTYOSOME (DIK-tee-uh-sohm). The Golgi body in cells.

DIFFERENTIATION. The process of developmental change from an unspecialized cell to a specialized cell.

DIFFUSION. The movement of molecules from a region of greater concentration toward a region of lesser concentration as a result of their kinetic energy.

DIGESTION. The process of breaking down large, complex, insoluble molecules of food into smaller, less complex, soluble molecules.

DIHYBRID CROSS (die-HY-brid) [Gr. two + L. mongrel]. A cross between individuals differing in two inheritable traits, or in which only two such different traits are considered by an experimenter.

DIKARYON (die-CARRY-on) [Gr. two + nut, kernel]. In fungi, a cell that contains two nuclei as a result of mating of two suitable types of mycelia.

DIPLOID (DIP-loyd) [Gr. double]. A cell having two full sets of chromosomes in the nucleus.

DISLOCATION. Such a severe strain of the muscles, tendons, and ligaments that a bone is pulled out of its socket and fails to return to its proper position.

DISSOCIATION. The separation of a molecule into ions. (*See* ASSOCIATION)

DOMINANT. Pertaining to a gene that expresses itself to the exclusion of its recessive allele.

DORSAL (DOHR-sul). Pertaining to a position toward the back surface. (*See* VENTRAL)

DOWN'S SYNDROME (MONGOLISM, TRISOMY 21). A genetic disorder in which an extra chromosome 21 is present. The afflicted person is mentally retarded.

DUCTUS ARTERIOSUS (DUCK-tus ahr-TEER-ee-OH-sis). The arterial duct that connects the pulmonary artery to the aorta in the fetal circulatory system of mammals.

DUODENUM (doo-oh-DEE-num). The portion of small intestine posterior to the stomach.

DUPLICATION [L. to double]. A chromosome change in which a segment of a chromosome is repeated.

DYNAMIC EQUILIBRIUM. In ecology, a condition that exists in a stable ecosystem when over a period of time the energy input equals the outflow.

ECDYSONE (EK-duh-sohn) [Gr. escape]. The insect molting hormone that favors the formation of the pupa.

ECG (ELECTROCARDIOGRAM) (uh-LEK-troh-KAR-dee-oh-gram) [Gr. electricity + heart + writing]. A record of the electrical activity of the heart.

ECHOLOCATION (EK-oh-loh-KAY-shun). A process of locating objects by the reflection of sound waves in such animals as bats and dolphins.

ECOLOGY (ee-KOL-uh-jee) [Gr. house + study]. The study of organisms in relation to all the forces that act upon them.

ECOSYSTEM (EE-koh-sis-tum). A space containing an interacting group of organisms, more or less self-contained, such as an enclosed field, a pond, or a small island.

ECOTONE (EE-kuh-tohn) [Gr. home + stretch]. A border region where two ecosystems meet, as at the edge of a forest.

ECTODERM (EK-toh-derm) [Gr. outer + skin]. The outer germ layer of an embryo, which gives rise to the skin, hair, fingernails, horns, hooves, brain with associated nervous tract, and eyes.

ECTOPIC PREGNANCY (ek-TOP-ik) [Gr. outside + place]. Abnormal pregnancy in which fertilization occurs outside the uterus and the zygote becomes implanted somewhere other than in the uterus.

EDAPHIC (ih-DAF-ik) [Gr. soil]. Pertaining to environmental forces that are neither biotic nor climatic, but are usually determined by the nature of the soil in which an organism lives.

EEG (ELECTROENCEPHALOGRAM) (uh-LEK-troh-en-SEF-uh-loh-gram) [Gr. electricity + the brain + writing]. A record of the electrical activity of the brain.

EFFERENT (EF-uh-rent) [L. to bring out]. Carrying substances or impulses away from heart or central nervous system; denoting directional action of arteries, nerves, etc.

EJACULATION (ee-JAK-yu-LAY-shun) [L. to throw out]. The forcible expulsion of the semen during male orgasm.

ELECTRON. A subatomic particle, the unit negative charge.

ELECTRON TRANSPORT SYSTEM. A group of enzymes closely associated with one another on the inner surface membrane of mitochondria. Electrons move from one of the enzymes to the next, resulting in alternatively reducing and oxidizing the enzyme and capturing energy by the synthesis of ATP from ADP and inorganic phosphate. (See CYTOCHROME)

ELEMENTS. Substances whose molecules consist of only one kind of atom.

EMBRYO SAC. The mature female gametophyte of the angiosperms, usually having eight nuclei.

EMULSIFICATION. The physical breakdown of fats.

END BULB OF KRAUSE [Wilhelm J. F. Krause, 19th-century German anatomist]. A specialized receptor in skin that is sensitive to cold.

ENDERGONIC (en-der-GAHN-ik) [Gr. within + work]. Activities or reactions that require energy.

ENDOCRINE GLAND (EN-doh-krin) [Gr. in + separate]. A ductless gland that secretes its product into the bloodstream without an emptying tube. (See EXOCRINE GLAND)

ENDODERM (EN-doh-derm) [Gr. inside + skin]. The inner germ layer of an animal embryo lining the archenteron. It gives rise to the lining of the digestive tract, liver, lungs, and pancreas.

ENDOMETRIUM (en-doh-MEE-tree-um) [Gr. inside + uterus]. The nutritious inner layer of the uterine wall.

ENDOPLASMIC RETICULUM (en-doe-PLAZ-mik reh-TIK-yu-lum) [Gr. within + L. net]. A labyrinthine complex of double membranes in the cytoplasm of a cell, acting as a system of internal channels through which various materials move.

ENDOSKELETON [Gr. within + a dried body]. Internal skeleton, which provides support and rigidity of form; typical of vertebrates.

ENDOSPERM. An amorphous mass of tissue, usually triploid, which functions as a nourishing tissue for growing embryos in the seeds of flowering plants.

ENDOTHERMIC [Gr. inside + heat]. Reactions that occur with absorption of heat or other energy; endergonic; the opposite of exothermic or exergonic.

ENERGY. The capacity to do work.

ENERGY-LEVEL SHELLS. The electrons in an atom are distributed in several different energy levels. When an electron changes from one shell to another, it either releases or absorbs energy.

ENZYME (EN-zime). A protein that speeds up a chemical reaction without permanently entering into the reaction itself; an organic catalyst.

EPICOTYL (EP-i-kot-ul) [Gr. upon + cup]. Stem above the attachment of seed leaf.

EPIDERMIS (ep-i-DER-mis). An outer covering. In plants it is the outermost layer or layers of leaf and of young stems and roots. In animals it is usually many cell layers thick.

EPIDIDYMIS (ep-i-DID-uh-mis) [Gr. upon + testes]. A coiled tube, lying on the testis, that stores sperm cells formed in the testis.

EPIGENESIS (ep-i-JEN-uh-sis) [Gr. upon + to be born]. The theory that development proceeds from the fertilized egg, and that the specialized body parts emerge during embryonic growth.

EPIGLOTTIS (ep-i-GLOT-is) [Gr. upon + tongue]. A valve that presses down and prevents food from entering the trachea or larynx during swallowing.

EPILIMNION (ep-i-LIM-nee-on) [Gr. upon + lake]. The upper layer of water in a lake.

EPINEPHRINE (ADRENALIN) (ep-i-NEFF-rin) [Gr. upon + kidney]. A neurohormone, secreted by the adrenal medulla, that produces a "fight or flight" condition, which permits the body to react quickly and strongly to emergencies.

EPIPHYTE (EP-i-fite) [Gr. upon + plant]. A plant that lives attached to another plant but does not parasitize it.

ERA. A major division of geologic time. From the oldest to the most recent, the eras are the Precambrian, Paleozoic, Mesozoic, and Cenozoic.

ERYTHROBLAST (ih-RITH-roh-blast) [Gr. red germ]. Cell in bone marrow, which is nucleated until it develops into a red blood cell.

ERYTHROCYTE (ih-RITH-roh-site) [Gr. red + cell]. A red blood cell.

ESOPHAGUS (ih-SOFF-uh-guss) [Gr. I shall carry + to eat]. A slender tube connecting the throat with the stomach.

ESTROGEN (ESS-troh-jen). One of the female sex hormones, secreted by ovarian follicles, which stimulates the thickening of the smooth muscle and the glandular lining of the inner walls of the uterus. It also promotes the development of secondary sex characteristics.

ETHOLOGY (ee-THOL-uh-jee) [Gr. character, habit + study]. The study of animal behavior.

ETHYLENE (ETH-uh-leen). A plant growth-regulating substance, C_2H_4, which hastens the ripening of fruits.

EUKARYOTE (you-CARRY-ote) [Gr. true + kernel]. An organism or cell with a membrane-bound nucleus; possesses mitochondria, plastids, flagella, and other organelles. (See PROKARYOTE)

EUPHOTIC ZONE (yu-FOTT-ik) [Gr. good + light]. The upper layers of water in seas, where light penetrates.

EURYTHERMAL (yur-i-THER-mul) [Gr. wide + heat]. Having the ability to endure a wide range of temperatures.

EUTROPHIC (yu-TROH-fik) [Gr. good + food]. Pertaining to a body of water containing an abundance of organisms, especially microscopic plants.

EVOLUTION [L. an unrolling]. Progressive change.

EXCITABILITY. The capacity of neurons to respond to stimuli; the other basic property of neurons is conductivity. (*See* CONDUCTIVITY)

EXCITATION-CONTRACTION COUPLING. The chemical process of muscle contraction, initiated by an electrical stimulus in the membrane of the muscle cell.

EXERGONIC (ek-ser-GON-ik) [L. out + Gr. work]. Pertaining to activities or reactions that liberate energy (*See* ENDERGONIC)

EXOBIOLOGY. The study of life beyond the earth.

EXOCRINE GLAND (ek-suh-krin) [Gr. out + separate]. A gland, such as a salivary or sweat gland, which exudes its secretion by a distinct duct or opening; contrasted with endocrine glands.

EXOSKELETON [L. outside + Gr. a dried body]. External coverings which provide protection, support, or prevention of dehydration; typical of invertebrates.

EXOTHERMIC (ek-soh-THER-mik) [L. outside + Gr. heat]. Pertaining to reactions that occur with liberation of heat or other form of energy; exergonic.

EXTENSION. A movement that straightens a joint. (*See* FLEXION)

FACILITATED TRANSPORT [L. easy]. The movement of molecules across a membrane by means of a postulated carrier that "escorts" a molecule across; no energy is required.

FALLOUT. Radioactive debris released by atomic explosions, emitting alpha, beta, and gamma rays for varying lengths of time, depending on the radioactive element involved.

FERMENTATION [L. leaven]. The partial decomposition of organic molecules in the absence of free oxygen. Such substances as alcohol and lactic acid may be produced.

FERTILIZATION [L. to bear, produce]. The fusion of a sperm and egg to form a zygote.

FETUS (FEE-tuss) [L. offspring]. The highly developed human embryo, from the third month of pregnancy until birth.

FISSION (FISH-un) [L. to cleave]. Asexual reproduction consisting of the division of a single cell and the separation of the two resultant cells, e.g., in protozoans, bacteria, some algae.

FLAGELLUM (fluh-JELL-um) [L. whip]. A long, whiplike locomotor organelle on the surface of a cell.

FLAME CELL. The excretory structure in flatworms, so called because of the wavering flagella that resemble a candle flame.

FLEXION (FLEK-shun). A movement that bends a joint. (*See* EXTENSION)

FLORIGEN (FLORE-uh-jen) [L. flower maker]. A postulated hormone responsible for the induction of flower buds.

FLUORESCENCE (flure-ESS-ens). Emission of light by a compound that is energized by photons or electrons.

FOLLICLE (FOLL-i-kull) [L. small bag]. Center of egg production in mammalian ovary. Each follicle houses an ovum that is in the process of being matured and released.

FOLLICLE-STIMULATING HORMONE (FSH) [L. small bag]. A female gonadotropic hormone that causes one or more immature follicles to develop.

FOOD WEB. The passage of food in an ecosystem from the primary producers through all the various consumers.

FOVEA (FOH-vee-uh) [L. small pit]. A spot at the center of the retina directly behind the lens where the concentration of the cones is greatest. This is where image formation is sharpest and color vision is most acute.

FUSIFORM INITIAL (FEW-zi-form) [L. spindle]. A long slender cell in the cambium of vascular plants that yields long slender xylem or phloem cells.

GAMETE (GAM-meet) [Gr. to marry]. A sexual reproductive cell; an egg or sperm.

GAMETOPHYTE GENERATION (guh-MEET-uh-fite). A gamete-producing haploid phase of plants having alternation of generations.

GAMMA RAYS. Electromagnetic rays emitted by radioactive isotopes of elements. Gamma rays have much greater penetrating power than alpha or beta rays.

GANGLION (GANG-glee-un) [Gr. knot]. A cluster of cell bodies of neurons located outside the central nervous system.

GASTRIN (GAS-trin) [Gr. stomach]. A hormone, secreted by certain mucosal cells of the stomach, that stimulates the production of hydrochloric acid when food enters the stomach.

GASTROVASCULAR CAVITY (gas-tro-VAS-kyu-lar) [Gr. stomach + L. a little vessel]. Saclike cavity in the body of coelenterates in which one opening serves both for intake of food and outflow of waste.

GASTRULATION (gas-truh-LAY-shun) [Gr. stomach]. The phase of embryonic development when there is infolding of the blastula to form the gastrula.

GENE (JEEN) [Gr. to produce]. A length of DNA that controls a specific cellular function, either by coding for a polypeptide or by regulating the action of other genes.

GENE POOL. The total genetic material of all individuals in an interbreeding population (species).

GENETIC CODE (jen-ET-ik). The complete set of triplet nucleotide symbols that indicate specific amino acids.

GENETIC DRIFT. The chance genetic changes resulting from isolation of small groups of organisms.

GENOTYPE (JEE-nuh-type) [Gr. to produce + type]. An individual's genetic makeup. (*See* PHENOTYPE)

GERMINATION [L. sprout]. The beginning of growth of a seed or spore.

GIANTISM. A disorder of the pituitary gland, caused by oversecretion of growth hormone during the period of skeletal development.

GIBBERELLIN (jib-ber-ELL-in). A plant growth-regulating substance that affects growth in height and starch digestion of seeds.

GILL. The respiratory organ of aquatic and some terrestrial organisms. Gills extract oxygen and release carbon dioxide.

GILL ARCH. A bar of tissue between gill slits.

GILL SLITS. Openings in the region of the pharynx leading from the internal alimentary canal to the outside of the animal; characteristic of chordate animals.

GIZZARD (GIZ-erd) [L. cooked entrails of poultry]. A thick-walled muscular compartment, formed in the digestive system of birds and earthworms, where food is ground.

GLIA (GLEE-uh) [Gr. glue]. Nonconducting cells of the nervous system that protect and nourish the neurons; they may also function to store memory traces.

GLOMERULUS (gluh-MARE-yu-luss) [L. a ball]. A ball of capillaries intimately associated with the Bowman's capsule in the kidney.

GLUCAGON (GLOO-kuh-gon). A hormone, secreted by alpha cells in the pancreatic islets of Langerhans; increases blood glucose concentration by stimulating the liver to convert glycogen to glucose.

GLYCOGENOLYSIS (gly-koh-jen-OL-i-sis) [Gr. sweet + a loosening]. The process whereby glycogen is reconverted into glucose.

GLYCOLYSIS (gly-KOL-i-sis) [Gr. sweet + solution]. The initial anaerobic breakdown of glucose to an intermediate compound occurring in the cytoplasm of a cell.

GLYCONEOGENESIS (gly-koh-nee-oh-jEN-i-sis) [Gr. sweet + new + origin]. The process whereby amino acids or fatty acids are converted into glycogen.

GOITER (HYPOTHYROIDISM) (GOY-ter). A condition characterized by an enlarged thyroid caused by insufficient iodine in the diet.

GOLGI BODY (GOAL-jee) [Camillo Golgi, 19th-century Italian histologist]. A collection of membranes associated with endoplasmic reticulum; functions in concentrating and delivering materials to be secreted.

GONORRHEA (gon-uh-REE-uh) [Gr. seed, reproductive organ + flow]. A venereal disease caused by a bacterium, *Neisseria gonorrheae*, usually contracted through sexual intercourse.

GRANA (GRAY-nuh) [L. grain]. Stacks of flattened, photosynthetic, membrane-bound vesicles in a chloroplast.

GRAVES' DISEASE. (*See* HYPERTHYROIDISM)

GUANINE (GWAH-neen). A purine, a nitrogenous base that is a constituent of nucleic acids.

GUSTATION [L. taste]. The sense of taste.

GYMNOSPERM (JIM-nuh-sperm) [Gr. naked + seed]. A type of seed plant whose seeds are not enclosed in an ovary but are borne on the surface of sporophylls.

HABITAT (HAB-i-tat) [L. living place]. The local physical dwelling place of an organism, such as a bare rock, a moist, cold, evergreen forest, or a tropical estuary.

HABITUATION (ha-bit-yu-AY-shun). A simple type of learning in which an animal eventually ignores a stimulus.

HAPLOID (HAP-loyd) [Gr. single]. Having only one complete set of chromosomes, that is, one genome in the nucleus of a cell.

HALF-LIFE. The amount of time required for half the atoms of a given radioactive sample to decay to a more stable form.

HARDWOOD. The wood of an angiosperm tree.

HARDY-WEINBERG PRINCIPLE. Mathematical formula used in calculating the approximate frequencies of different alleles in a population after measuring the frequency of phenotypes in a representative sample of the whole population.

HAVERSIAN CANALS (huh-VER-zhun) [Clopton Havers, an English physician]. Elongated tubes that carry blood vessels through bone.

HEMIZYGOUS (hem-ee-ZY-gus) [Gr. half + yoke]. Having only one allele, e.g., a sex-linked gene in XY males.

HEMOGLOBIN (HEE-moh-gloh-bin) [Gr. blood + L. globe]. The iron-containing, respiratory pigment in the red blood cells that transports oxygen and carbon dioxide throughout the body.

HEMOPHILIA (hee-moh-FILL-ee-uh) [Gr. blood + love]. The "bleeder's disease," in which blood does not clot normally.

HEPATIC PORTAL SYSTEM (huh-PAT-ik). A system of vessels in which the venous blood coming from the intestines passes through capillaries in the liver before moving on to the posterior vena cava.

HETEROGAMY (het-uh-RAH-guh-mee) [Gr. other + marriage]. A type of sexual reproduction resulting from the mating of two gametes that are of a different size or shape or both.

HETEROSPORY (het-uh-AHS-poh-ree) [Gr. other + seed]. Production of two kinds of spores: megaspores and microspores.

HETEROTHALLIC (het-uh-ruh-THAL-ik) [Gr. different + a sprout]. Requiring at least two mating types to complete a life cycle.

HETEROTROPH (HET-uh-ruh-trohf) [Gr. other + feeder]. An organism that must acquire its food because it cannot manufacture it from inorganic sources. (*See* AUTOTROPH)

HETEROZYGOUS (het-uh-ruh-zy-gus) [Gr. other + yoke]. Possessing two different alleles, at a given locus for a given characteristic on a pair of homologous chromosomes.

HIBERNATION [L. winter]. The dormant state of animals with concomitantly lowered body metabolism.

HISTOLOGY (hiss-TOL-uh-jee). The study of biological tissue.

HISTONE (HISS-tone). A basic protein associated with DNA in chromosomes.

HOMEOSTASIS (hoe-mee-oh-STAY-sis) [Gr. unchanging + standing]. The maintenance of a measure of physiological stability in spite of environmental changes.

HOMEOTHERMIC (hoe-mee-oh-THER-mik) [Gr. unchanging + heat]. "Warm-blooded"; having a body temperature that remains fairly constant, independent of external temperatures.

HOMOLOGOUS (huh-MOLL-uh-gus) [Gr. agreeing, corresponding]. Pertaining to similar body parts in different species, arising from common construction and development, such as bird wings and whale flippers. (*See* ANALOGOUS)

HOMOLOGOUS CHROMOSOMES [Gr. same]. Chromosomes occurring in matching pairs from each of two parents. The members of such a pair are called homologs.

HOMOSPORY (hoe-MAHSS-poh-ree) [Gr. same + seed]. Production of only one kind of spore. (*See* HETEROSPORY)

HOMOTHALLIC (hoe-muh-THAL-ik) [Gr. same + a sprout]. Pertaining to an organism that is self-fertile. (*See* HETEROTHALLIC)

HOMOZYGOUS [Gr. same + yoke]. Possessing two identical alleles at a given locus on homologous chromosomes.

HORMONE [Gr. to arouse]. A chemical compound secreted from an endocrine gland into the bloodstream and carried throughout the entire body. It affects a specific "target" organ or tissue, regulating and coordinating its activities.

HUMUS (HYU-mus). A hygroscopic, decay-resistant mix of protein and lignin in soil.

HYBRID (HY-brid) [L. mongrel]. A union of dissimilar biological entities, as in species hybrids, genetic hybrids, hybrid cells, or hybrid nucleic acids.

HYDROGEN BOND. A weak chemical bond formed by the attraction of a hydrogen bearing a slight positive charge to an oxygen or a nitrogen bearing a slight negative charge.

HYDROLYSIS (hy-DROL-uh-sis) [Gr. water + loosening]. The splitting apart of a compound into two molecules, with the addition of a water molecule.

HYDROPHILIC (hy-druh-FILL-ik) [Gr. water + loving]. Referring to organisms that live where water is abundant.

HYDROSERE (hy-druh-SEAR) [Gr. water + English "barren"]. A wet region cleared of organisms, open for invasion, such as a glacial lake.

HYMEN (HY-men). A fold of skin that partially blocks the vaginal entrance.

HYPERTENSION. High blood pressure.

HYPERTHYROIDISM. A condition caused by an oversecretion of TSH by the pituitary or by a thyroid tumor. Also called Graves' disease.

HYPHA (HY-fuh) [Gr. web]. A threadlike filament of a fungus.

HYPOGLYCEMIA (hy-poh-gly-SEEM-ee-uh) [Gr. under + sugar]. A condition of low blood sugar, caused by excessive secretions of insulin.

HYPOLIMNION (hy-poh-LIM-nee-on) [Gr. under + lake]. The lowest, coldest layer of water in a lake.

HYPOTHALAMUS (hy-poh-THAL-uh-mus) [Gr. under + inner chamber]. A portion of the brain, below the thalamus, that controls many physiological and endocrine activities.

HYPOTHYROIDISM. Underactivity of the thyroid gland. The resultant insufficiency of iodine usually causes a goiter, a swelling in the neck. (*See* GOITER)

IMAGO (ih-MAY-go) [L. image, likeness]. The adult stage of insects.

IMMUNOLOGY (im-yuh-NOL-uh-jee) [L. safe + Gr. study]. The science dealing with the process whereby organisms develop a chemical resistance, an antibody, to a foreign substance, its antigen, such as pollen, virus, bacteria, or feathers.

IMPLANTATION [L. into + to set]. The sinking of the blastocyst into the endometrium of the mammalian uterus.

IMPRINTING. The rapid fixing of social preferences. For example, newly hatched goslings will form an attachment to the first creature they can follow as their mother.

INCISOR (in-SIZE-er). The front teeth of mammals, adapted to chiseling action.

INCOMPLETE METAMORPHOSIS (met-uh-MORE-fuh-sis) [Gr. beyond, over + shaping]. The gradual transition of some insects, by several stages of development, from the nymph to the adult. The young are small copies of adults.

INDEPENDENT ASSORTMENT, LAW OF. A generalization of Mendel, which states that when one pair of alleles segregates during sexual reproduction, its manner of segregation is not affected by the manner of segregation of a second pair of alleles.

INDETERMINATE CELL (in-duh-TERM-uh-nut). A cell of the early embryo from which no preordained part of the future animal will necessarily differentiate.

INDUCTION [L. to lead in]. The influence of one type of tissue on the developmental pattern of another type of tissue.

INFLORESCENCE (in-fluh-RES-ens). A flower cluster.

INORGANIC. A chemical compound containing no carbon.

INSERTION. In anatomy, the point of attachment of a muscle to the bone that moves. (*See* ORIGIN)

INSIGHT LEARNING. A type of learning whereby an animal is able to associate apparently unrelated experiences.

INSTINCT. Innate learning, e.g., fixed action or stereotyped behavior; a controversial and poorly defined term, which is not generally acceptable to students of behavior.

INSULIN (IN-suh-lin). A hormone, secreted by beta cells in the islets of Langerhans portion of the pancreas; facilitates glucose transport across cell membranes. Insulin enhances the conversion of glucose to glycogen.

INTERNEURON (in-ter-NYU-ron) [L. between + Gr. nerve]. A nerve cell that lies entirely within the brain or spinal cord. It carries impulses from sensory neurons to motor neurons.

INTERPHASE (INT-er-faze) [L. between + Gr. phase]. The period between cell divisions, during which DNA duplicates.

INTERSTITIAL CELL (in-ter-STISH-ul). A cell among the seminiferous tubules in testes that secretes the male sex hormones, especially testosterone.

INVAGINATION (in-vaj-i-NAY-shun) [L. within + sheath]. The process of infolding, as in a blastula to form a double-layered cup.

INVERSION, CHROMOSOMAL. A type of chromosome change in which there is a reversal of a segment in a given chromosome.

ION (EYE-on). An electrically charged atom or group of atoms, the result of a gain or loss of electrons. Electrically charged groups of atoms are sometimes called complex ions.

IONIC BOND (eye-ON-ik). The chemical method of holding together two atoms by the transfer of one or more electrons from one atom to the other.

ISLETS OF LANGERHANS (EYE-lets; LAHNG-er-hanz) [Paul Langerhans, 19th-century German anatomist]. Cluster of pancreatic cells, containing two groups of cells: alpha and beta. The alpha cells secrete glucagon; the beta cells secrete insulin.

ISOGAMY (eye-SAHG-uh-mee) [Gr. equal + marriage]. Sexual reproduction resulting from the mating of two indistinguishable gametes.

ISOMER (EYE-soh-mer) [Gr. equal + part]. Compounds with identical components but different atomic arrangements in their molecules.

ISOTOPE (EYE-soh-tope) [Gr. equal + place]. An atom of an element with the same number of protons and electrons as other atoms of the same element but a different number of neutrons.

JAUNDICE [Fr. yellow]. A liver disease in which abnormally high accumulations of bile pigments cause the skin to turn yellow.

JUVENILE HORMONE. An insect hormone that prolongs the larval stage through several molts.

KARYOTYPE (CARRY-oh-type) [Gr. nucleus + type]. The size, shape, and appearance of a metaphase chromosome set of an individual or species.

KIDNEY. The organ that produces urine by filtration of water, salts, and nitrogenous waste from blood; the main water-regulating organ in land animals.

KILOCALORIE (KEE-loh-cal-uh-ree) [Gr. thousand + L. heat]. One thousand calories; also written as Kcal or Calorie.

KINESIS (kin-EE-sis) [Gr. movement]. A method of orientation in which a simple organism, such as a protozoan, is stimulated to move faster or slower, depending on the strength of the stimulus.

KLINEFELTER'S SYNDROME (KLYNE-fel-terz). A type of genetic disorder in which there is an extra X chromosome, giving an XXY karyotype. The afflicted individual is sterile and has underdeveloped male genitals.

KREBS CYCLE (TRICARBOXYLIC ACID CYCLE; CITRIC ACID CYCLE) (KREBZ) [Sir Hans Krebs, a British biochemist]. The cyclic series of reactions in mitochondria by which certain organic acids yield carbon dioxide and hydrogen.

KWASHIORKOR (kwash-ee-OHR-ker). A disease caused by insufficient protein in the diet, causing poor muscle development and skin difficulties.

LACTATION [L. milk]. The production of milk in mammals.

LARVA [L. ghost]. The young developmental stage in some insects, unlike the adult in appearance; a grub, maggot, or caterpillar.

LARYNX (LAR-inks) [Gr. upper part of windpipe]. The voicebox; a cartilagenous structure at the top of the trachea that contains the vocal chords.

LATENT LEARNING. An exploratory learning that takes place when an animal actively seeks to find out more about its environment; curiosity.

LATERAL BUD. A bud in the axil of a leaf. (*See* APICAL BUD)

LEAF PRIMORDIUM (pry-MORE-dee-um) [L. first + to begin to weave]. The incipient leaves, just below the apical growing point of a vascular plant.

LEARNING. In experimental psychology, the modification of behavior as a result of changes in individual experience.

LEUKEMIA (lyu-KEE-mee-uh) [Gr. white + blood]. A disease characterized by uncontrolled white blood cell production.

LEUKOCYTE (LYU-koh-site) [Gr. white + cell]. A white blood cell.

LEUKOPLAST (LYU-koh-plast). A plastid lacking pigment, containing fat or carbohydrate instead.

LIGAMENT. Tough sheet of fibrous tissue connecting two or more separate bones or cartilages.

LIPID (LIP-id). An organic compound insoluble in water but soluble in ethers and alcohols, e.g., fats, oils, steroids.

LITTORAL (LIT-uh-rul) [L. shore]. Referring to the region at the edge of a body of water, including both water and land, usually in oceans.

LIVER. The largest gland, which has many important functions, including among others the storage of foods, production of bile salts, conversion of stored glycogen to glucose, deamination of amino acids, and breakdown of hemoglobin.

LOOP OF HENLE (HEN-lee) [Friedrich Henle, 19th-century German anatomist]. A narrowly U-shaped portion of the nephron. It acts as a countercurrent multiplier.

LUCIFERIN (lyu-SIF-uh-rin) [L. light + to bear]. A substance, isolated from bioluminescent organisms, that emits light when acted upon by the enzyme luciferase.

LUTEINIZING HORMONE (LH) (LYU-tee-un-eye-zing) [L. yellowish]. One of the female gonadotropic hormones, which participates in egg maturation and in converting the old follicle, after ovulation, into the endocrine gland, corpus luteum.

LYMPH (LIMF). Clear circulating body fluid, which contains water, plasma, some proteins, electrolytes, and white blood cells.

LYMPHATIC SYSTEM (lim-FAT-ik). The "second circulatory" system made up of a network of thin-walled veins and capillaries that carry lymph fluid. It returns excess fluid and protein from the spaces between cells to the blood, and it filters and destroys harmful material.

LYSOGENIC (lye-soh-JEN-ik) [Gr. loosening]. The inactive state of a virus in a living bacterial cell.

LYSOSOME (LYE-soh-sohm) [Gr. loosening]. A cytoplasmic organelle rich in hydrolytic enzymes.

LYTIC (LIT-ik) [Gr. loosening]. Capable of causing dissolution; used to describe the stage in which an infecting phage destroys a bacterial host cell.

MACROPHAGE (MAK-roh-fayj) [Gr. large + to eat]. A scavenger cell, found in the liver, bone marrow, or spleen, that engulfs red blood cells.

MALPIGHIAN TUBULES (mal-PIG-ee-un) [Marcello Malpighi, 17th-century Italian anatomist]. The excretory system of insects and spiders.

MANDIBLE (MAN-duh-bul) [L. chew]. Jawlike appendage of the mouth in arthropods; the jawbone in vertebrate animals.

MARASMUS (muh-RAZ-mus). A disease caused by insufficient quantity of food, as contrasted with inadequate quality. It results in general weakness and lack of muscle.

MARSUPIAL (mar-SUPE-ee-ul) [L. pouch]. A mammal that bears young in poorly developed condition and then keeps them in an abdominal pouch. Each embryonic baby attaches itself to a nipple and nurses until it achieves a measure of independence.

MECHANORECEPTOR (MEK-uh-noh-ree-SEP-ter). Sensory cell or organ capable of perceiving changing mechanical stimuli, such as those of touch or hearing.

MEDULLA (muh-DULL-uh) [L. middle]. (1) The inner portion of an organ. (2) The posteriormost part of the brain, which controls such vital functions as heart rate, breathing rate, and blood pressure.

MEDUSA (muh-DEW-suh) [L. myth., *Medusa*, a woman with snakes for hair]. A free-swimming, reproductive, umbrella-shaped form in the life cycle of some coelenterates; a jellyfish.

MEGAGAMETOPHYTE (meg-uh-guh-MEET-uh-fite) [Gr. large + marriage + plant]. Haploid egg-producing plant; a female gametophyte.

MEGASPORE (MEG-uh-spohr) [Gr. large + seed]. A large spore that germinates to form a female gametophyte.

MEGASPORE MOTHER CELL. A diploid cell destined to undergo meiosis, producing four haploid megaspores.

MEIOSIS (my-OH-sis) [Gr. diminution]. The two nuclear divisions that result in the reduction of chromosome number from diploid to haploid.

MEISSNER'S CORPUSCLE (MICE-nerz KOR-pus-ul) [Georg Meissner, 19th-century German histologist]. A specialized receptor in the skin that is sensitive to direct touch.

MELANIN (MEL-uh-nin) [Gr. black]. A pigment in skin, eyes, or hair, giving brown to black color.

MELATONIN (mel-uh-TOE-nin). A chemical formed by the pineal gland. It affects skin pigmentation and seems to react to changes of light through the eye.

MENARCHE (MEN-ar-kee) [Gr. month + beginning]. The first menstrual period of a human female, which occurs during puberty.

MENINGES (meh-NIN-jeez) [Gr. membrane]. The three protective membranes enveloping the brain and spinal cord.

MENOPAUSE (MEN-oh-pawz) [Gr. month + cessation]. Permanent termination of the menstrual period in women.

MENSTRUAL FLOW (MEN-strew-al) [L. monthly]. The periodic sloughing off of the uterine wall along with cells and blood if no fertilization follows ovulation.

MESODERM (MEZ-oh-derm) [Gr. middle + skin]. The middle germ layer of an embryo, between the ectoderm and endoderm, giving rise to muscle, connective tissue, bone, circulatory system, lining of coelom, and urogenital system.

MESOPHYTIC (mez-oh-FIT-ik) [Gr. middle + plant]. Referring to plants that require moderate amounts of water.

MESOSOME (MEZ-oh-sohm) [Gr. middle + body]. In a bacterial cell, the infolding of plasma membrane into complex masses.

MESSENGER RNA (mRNA). A single-stranded nucleic acid that is transcribed from a DNA template; contains the "genetic message" to be translated into proteins.

METAMORPHOSIS. (*See* COMPLETE METAMORPHOSIS *and* INCOMPLETE METAMORPHOSIS.)

METAPHASE (MET-uh-faze) [Gr. after, beyond, over + to cause to appear]. A stage in nuclear division during which the shortened chromosomes are at the equatorial plane of the spindle.

MICROCLIMATE (MY-kroh-kly-met). The special conditions of light, temperature, and moisture that occur in a narrowly restricted area, as on one bank of a stream, under a bush, in a small woodland opening.

MICROGAMETOPHYTE (my-kroh-guh-MEET-uh-fite) [Gr. small + marriage + plant]. A haploid sperm-producing plant; a male gametophyte.

MICRON (MY-kron) [Gr. small]. A unit of measurement denoting one-millionth of a meter; one micrometer; one-thousandth of a millimeter.

MICRONUTRIENT [Gr. small]. One of the elements required in very small amounts by plants and by animals; a trace element, such as zinc, boron, chlorine, etc.

MICROPYLE (MY-kroh-pile) [Gr. small + gate]. An opening in the integuments of the ovules of seed plants through which a pollen tube can enter.

MICROSPORE [Gr. small + seed]. A spore that germinates to form a male gametophyte.

MICROTUBULE (my-kro-TOO-byul) [Gr. small]. A straight, thin cytoplasmic structure made of the protein tubulin; associated with movement and cytoplasmic structure.

MIGRATION. Seasonal movement of animals, usually in spring and fall, for feeding, mating, or escape from unfavorable weather.

MIMICRY (MIM-ik-ree) [Gr. to imitate]. The resemblance of an organism to some other animate or inanimate object, presumably increasing its chance of survival.

MITOCHONDRION (my-toe-KON-dree-on) [Gr. thread + granule]. A self-replicating cytoplasmic organelle, called the "powerhouse of the cell," the site of Krebs cycle activity and the cytochrome system. Its main work is the production of ATP for use in cellular respiration.

MITOSIS (my-TOE-sis) [Gr. thread + state, condition]. Nuclear division that results in the formation of two new daughter nuclei with identical chromosome complements.

MOLECULE (MOLL-i-kyul) [L. little mass]. The smallest unit into which a substance can be divided while still retaining the chemical properties of that substance.

MOLTING [L. to change]. The shedding and replacement of an old exoskeleton in crustaceans and other animals or of feathers in birds.

MONGOLISM (MON-guh-liz-um). A type of genetic disorder in which an extra chromosome 21 is present. The afflicted person is mentally retarded; Down's syndrome, trisomy 21.

MONOCOTYLEDON (MON-uh-kot-uh-LEE-dun) [Gr. single + cotyledon]. A flowering plant with one seed leaf in the embryo.

MONOHYBRID CROSS [Gr. one + mongrel]. A cross between individuals differing in only one heritable trait, or in which only one trait is considered by an experimenter.

MOTOR UNIT. A group of skeletal muscle fibers activated by the same nerve fiber.

MULTIPLE ALLELES (uh-LEELZ). Genes that are present in a population in more than two forms, although only two can be present in a single diploid organism, for example, the human ABO blood group.

MULTIPLE GENES. Two or more different pairs of alleles at different loci on chromosomes, capable of adding quantitatively to such traits as size and color.

MUTAGEN (MEW-tuh-jen) [L. to change + Gr. to produce]. An agent that causes mutation.

MUTATION (mew-TAY-shun) [L. change]. An inheritable change in a gene; a change in nucleotide sequences in a DNA molecule.

MYCELIUM (my-SEE-lee-um) [Gr. fungus]. Fungus tissue, composed of hyphae.

MYCORRHIZA (my-koh-RYE-zuh). [Gr. fungus + root]. Fungus inhabiting the roots of higher plants, feeding off the host, and aiding the host in the intake of minerals.

MYELIN (MY-uh-lin) [Gr. marrow]. A fatty protein material, wrapped around the axon of some nerve fibers, that aids in the conduction of electrical impulses.

MYOFIBRIL (my-oh-FYBE-ril [Gr. muscle + L. small fiber]. A subdivision of a muscle fiber.

MYOGLOBIN (my-oh-GLOW-bin) [Gr. muscle + ball, globe]. An iron-containing protein that acts mainly in muscle cells, quickening release of oxygen during increased activity.

MYOSIN (MY-uh-sin) [Gr. muscle]. A 100Å-thick protein that makes up the thick filaments of the muscle. Myosin and actin, a smaller protein, are the ultimate structural units of contractility.

MYXEDEMA (mix-uh-DEE-muh) [Gr. mucus + swelling]. A disorder, characterized by swollen facial features, dry skin, low basal metabolic rate, tiredness, and possible mental retardation, resulting from an underactive thyroid gland during adulthood.

NAD. (*See* NICOTINAMIDE ADENINE DINUCLEOTIDE)

NANOMETER (NAN-uh-mee-ter). One-billionth of a meter; used for measuring wavelengths of light.

NATURAL SELECTION. The process of the environment "selecting" for those individuals with traits that best suit them to their habitat; they are better able to pass on those traits to their offspring; the survival of the fittest.

NEMATOCYST (NEM-uh-tuh-sist) [Gr. thread + bladder]. A stinging cell on a tentacle of coelenterates, used for defense and capturing prey.

NEPHRIDIUM (nuh-FRID-ee-um) [Gr. kidney]. The excretory organ of the earthworm and other annelids.

NEPHRON (NEFF-ron) [Gr. kidney]. The functional unit of excretion of the kidney, separating water, minerals, and nitrogenous waste from the blood, and making urine.

NERVE [L. sinew]. A bundle of fibers enclosed in a connective tissue sheath through which stimuli are transmitted from the central nervous system to the peripheral nervous system or vice versa.

NERVE IMPULSE. An electrochemical impulse passing along a nerve fiber.

NEURON (NYU-ron) [Gr. nerve]. A nerve cell, the fundamental unit of the nervous system. It is made up of three parts: the cell body, the dendrites, and the axon.

NICHE (NITCH) [It. a hollow]. The physiological, behavioral, or nutritional specialty of an organism, permitting it to survive in a given habitat; determined by the physiological or behavioral actions of a given species.

NICOTINAMIDE ADENINE DINUCLEOTIDE (NAD). A coenzyme that functions as a hydrogen acceptor in cellular oxidation.

NODE [L. knot]. The part of a stem bearing a leaf; any differential structure along an otherwise uninterrupted line, e.g., a node of Ranvier in nerve fibers.

NONDISJUNCTION. Failure of homologous chromosomes to separate during meiosis.

NOTOCHORD (NOTE-uh-kord) [Gr. back + cord]. The rod-shaped body serving as internal skeleton in the embryos of all chordates and in the adults of some primitive ones; replaced by a vertebral column in adult vertebrates.

NUCLEIC ACIDS (new-KLAY-ik). Macromolecules that are polymers of nucleotides; DNA and RNAs.

NUCLEOLAR ORGANIZER (new-KLEE-uh-lar). A special chromosomal region where ribosomal RNA and other precursors of nucleoli are assembled.

NUCLEOLUS (new-KLEE-uh-lus). A nonmembrane-delimited assembly center in the nucleus, responsible for partial construction of ribosomes.

NUCLEOTIDE (NEW-klee-uh-tide). A molecule consisting of ribose or deoxyribose, a nitrogenous base, and a phosphate group; one of the units from which nucleic acids are synthesized.

NUCLEUS (NEW-klee-us) [L. kernel]. In cell biology, the control center of the cell containing the necessary information to direct the metabolism, replication, and heredity of cells.

NYMPH [L. young woman]. A young insect that appears somewhat like an adult but with juvenile features.

OBLIGATE. Allowing no alternative, as in obligate anaerobe, obligate aerobe, or obligate parasite.

OLFACTION [L. to smell]. The sense of smell.

OLIGOTROPHIC (oll-i-go-TROH-fik) [Gr. few + food]. Said of a body of water containing few organisms.

OMMATIDIUM (oh-muh-TID-ee-um) [Gr. small eye]. In arthropods, a single component of a compound eye, a slender cone with a light-sensitive receptor at the base.

OMNIVORE (OM-ni-vore) [L. all + eat]. An organism that can and does eat both plant and animal material.

OOGAMY (oh-OG-uh-mee) [Gr. egg + marriage]. Sexual reproduction with a large, nonmotile female gamete and a smaller, motile male gamete.

OOGENESIS (oh-oh-JEN-eh-sis) [Gr. egg + generation]. The process of maturation of eggs.

OPERATOR GENE. A gene that does not contribute directly to polypeptide synthesis but acts as a switch to "turn on" structural genes.

OPERCULUM (oh-PER-kyu-lum) [L. cover, lid]. (1) A tightly fitting, horny plate that covers the opening in the shells of some gastropods. (2) a bone-supported flap that covers the gill slits in fish. (3) The lidlike cover on the spore capsule of mosses.

ORGANELLE (or-guh-NELL) [Gr. bodily organ]. A specialized, subcellular structure, such as a mitochondrion, flagellum, lysosome, Golgi apparatus, etc.

ORGANIC. Pertaining to a chemical compound, usually complex, containing carbon; in the current popular sense, derived from living organisms.

ORIGIN. In anatomy, the point of attachment of the muscle to a bone that remains steady when the next bone moves. (See INSERTION)

OSMOSIS (oz-MOE-sis) [Gr. impulsion]. The simple movement of a solvent through a selectively permeable membrane.

OSMOTIC CONFORMER (oz-MOT-ik kon-FORM-er). Animal that is incapable of regulating the concentration of dissolved material in its body fluid.

OSSIFICATION (oss-uh-fi-KAY-shun) [L. bone]. A process of deposition of calcium phosphate and other minerals from the blood to form bones, replacing cartilage cells.

OSTIOLE (AHS-tee-ohl) [L. small opening]. An opening at the top of the central cavity of a sponge; any mouthlike opening.

OVARY (OH-vuh-ree) [L. egg]. An egg-producing organ.

OVIDUCT (OH-vuh-dukt) [L. egg]. A tube through which eggs pass from the ovary; in mammals, a Fallopian tube.

OVOVIVIPAROUS (oh-voh-vi-VIP-uh-rus) [L. egg + alive + to bring forth]. Retaining eggs within a mother's body until the young are partially developed. Embryos derive most of the nourishment from the yolk of the eggs, but the young are born alive.

OVULATION (awv-yu-LAY-shun) [L. little egg.] The release of a matured egg from a follicle of an ovary.

OVULE (AWV-yule) [L. little egg]. In seed plants, the structure that first forms the megasporangium and later the seed.

OVUM (OH-vum) [L. egg]. A female reproductive cell; an egg.

OXIDATION (awk-si-DAY-shun). A chemical reaction in which electrons are removed from the substance being oxidized. (See REDUCTION)

OXYGEN DEBT. The accumulation of lactic acid during heavy muscular activity, subsequently relieved by deep breathing.

OXYTOCIN (awk-si-TOE-sin). A female sex hormone secreted by the pituitary gland; it stimulates uterine contractions.

PACINIAN CORPUSCLE (puh-SIN-ee-un) [Filippo Pacini, 19th-century Italian anatomist]. A specialized skin receptor sensitive to firm pressure.

PALEONTOLOGY (PAIL-ee-on-TOL-uh-jee) [Gr. ancient + science]. The study of fossils.

PALISADE PARENCHYMA (puh-RENG-ki-muh) [Gr. in + pour]. Tissue of chloroplast-bearing, photosynthetic columnar cells under the upper epidermis of a leaf.

PANCREAS (PAN-kree-us). An elongated gland under the stomach, which secretes digestive enzymes into the small intestine. The islets of Langerhans in the pancreas secrete the hormones insulin and glucagon into the bloodstream.

PANSPERMIA (pan-SPER-mee-uh) [Gr. all + seed, semen]. The idea that "seeds of life" of some kind are universally distributed everywhere, needing only suitable conditions for development.

PARASYMPATHETIC SYSTEM (PAR-uh-sim-pa-THET-ik) [Gr. beyond + with + feeling]. A part of the autonomic nervous system that helps to regulate body functions. The action is mainly inhibitory.

PARATHYROIDS [Gr. near, beyond + shieldlike]. Four tiny glands, embedded in the thyroid gland, that secrete parathormone (PTH), which is responsible for increasing the calcium level and for decreasing the phosphate level in blood.

PARKINSON'S DISEASE. A condition caused by damage to the thalamus; characterized by involuntary tremors.

PARTHENOGENESIS (par-thin-oh-JEN-uh-sis) [Gr. virgin + production]. The development of an unfertilized egg, as in honeybees and wasps.

PASSIVE TRANSPORT. The movement of molecules across a membrane purely by diffusion.

PEDICELLARIA (ped-i-sell-AY-ree-uh). A pair of pincers that act as tiny pliers with crossed jaws, on the skin of echinoderms.

PEPSIN [Gr. digestion]. An enzyme, secreted by the stomach mucosa, that converts proteins to polypeptides.

PEPTIDE LINKAGE. A strong covalent bond between the carboxyl group of one amino acid and the amino group of another.

PERICYCLE (per-i-SY-kul) [Gr. around + circle]. In roots and stems, a layer of parenchymal cells between the phloem and the endodermis. It gives rise to lateral roots.

PERISTALSIS (per-i-STOLL-sis) [Gr. around + contraction]. A process that produces waves of muscular contractions and relaxations along the digestive tract. The alternating movements help to propel food along.

PERMAFROST (PER-muh-frost). The permanently frozen deeper layers of soil in the tundra regions of the northern hemisphere.

PERMEABILITY (per-mee-uh-BIL-i-tee) [L. through + to pass]. The ability to allow passage of substances, said of a membrane.

PETIOLE (PET-ee-ohl). A leaf stalk.

pH SCALE [p(otential of) H(ydrogen)]. A numerical scale from 0 to 14 that measures the concentration of hydrogen ions free in water. The smaller the number below 7, the more acidic is the solution; the higher the number above 7, the more basic the solution. Pure water is neutral, with a pH of 7.

PHAGE (FAYJ) [Gr. to eat]. A virus that infects bacteria; also bacteriophage.

PHAGOCYTE (FAG-oh-site) [Gr. to eat + cell]. A blood cell that has the ability to engulf and digest foreign material.

PHAGOCYTOSIS (fag-oh-sy-TOE-sis) [Gr. to eat + hollow vessel + state, condition]. Engulfing of bacteria or other particles by a cell.

PHARYNX (FAR-inks) [Gr. throat]. The part of the digestive tract from which the gill slits develop; a muscular ring in an alimentary tract, used by many animals for swallowing.

PHENOTYPE (FEE-nuh-type) [Gr. to show + type]. The appearance or discernible character of an individual, as determined by the genetic makeup and environmental influences. (*See* GENOTYPE)

PHEROMONE (FEH-roh-moan) [Gr. carry + arouse]. A compound, released into the environment by an individual, that is capable of causing a reaction in another individual, usually of the same species.

PHLOEM (FLOH-em) [Gr. bark]. The conducting tissue in vascular plants, usually in the bark, carrying dissolved nutrients.

PHOSPHOLIPIDS (fos-foh-LIP-idz). A group of fatty compounds that contain fatty acids, glycerol, and a negatively charged phosphate.

PHOSPHORYLATION (fos-for-i-LAY-shun) [Gr. light + to carry]. The addition of a phosphate group to an organic molecule.

PHOTOPERIODISM (fot-toh-PEER-ee-ud-iz-um). The ability of some organisms to respond metabolically to variations in day length.

PHOTORECEPTOR (foh-toh-ri-SEP-ter). Sensory cell or organ that is sensitive to light stimuli.

PHOTOSYNTHESIS [Gr. light + putting together]. The synthesis of carbohydrate from water and carbon dioxide, using the energy of light captured by chlorophyll.

PHYLLOTAXY (FILL-oh-tak-see) [Gr. leaf + arrangement]. The geometry of leaf arrangement.

PHYTOCHROME (FITE-uh-krohm) [Gr. plant + color]. A light-sensitive pigment involved in seed germination, flowering, morphogenesis, anthocyanin synthesis, etc.

PINEAL GLAND (PY-nee-ul) [L. relating to the pine]. A small round gland located in the midbrain; secretes melatonin; a light receptor in primitive vertebrates. In humans it may act as a "biological clock" that affects secretions of sex hormones.

PINOCYTOSIS (pin-uh-sy-TOE-sis) [Gr. to drink + cell + state, condition]. The engulfing of a droplet by a cell membrane.

PITUITARY (HYPOPHYSIS) (pi-TU-i-teh-ree) [L. secreting phlegm]. The so-called "master gland"; an endocrine gland that controls most of the other endocrine glands. The secretions of the anterior pituitary are controlled by hormones of the hypothalamus. Some of the regulatory hormones secreted by its two lobes are: TSH, STH, FSH, LH, ACTH, prolactin, vasopressin, and oxytocin.

PLACENTA (pluh-SEN-tuh) [L. flat cake]. In mammals, the tissues composed partly of uterine wall and partly of extraembryonic membranes, through which wastes and nutrients pass for a developing embryo. In plants, the place of attachment of ovules in a flower.

PLANKTON (PLANK-tun) [Gr. wanderers]. The microscopic, actively swimming plants and animals of lakes and seas.

PLANULA (PLAN-yu-luh) [L. flat]. Ciliated larva of coelenterates.

PLASMA (PLAZ-muh). The liquid portion of mammalian blood; composed of water, dissolved solids including proteins, and dissolved gases.

PLASTID (PLAS-tid) [Gr. formed + small]. A self-replicating cytoplasmic organelle in plants functioning in photosynthesis and/or nutrient storage, e.g., chloroplast, amyloplast.

PLATELETS (THROMBOCYTES). Minute disks that serve as starters in the process of blood clotting.

POIKILOTHERMIC (poy-KEE-luh-therm-ik) [Gr. varied + heat]. "Cold-blooded"; having a body temperature that varies with that of the environment.

POLAR BODY. In the maturation of animal eggs, the small, nonfunctional products of meiosis, in contrast with the large, functional egg cell.

POLAR MOLECULE. A molecule whose two ends bear different electrical charges.

POLIOMYELITIS (POH-lee-oh-my-uh-LY-tis). Infantile paralysis due to viral infection causing the destruction of the neurons leading to skeletal muscle.

POLLEN. The product of the anther in a flower, or of staminate cones in conifers. A pollen grain contains the sperm nuclei plus additional nonsexual nuclei.

POLLEN TUBE. A filament containing male sex nuclei, growing from a pollen grain and delivering the nuclei to the egg in the ovule.

POLLINATION. The movement of pollen from anther to stigma in a flowering plant, preceding fertilization.

POLYPEPTIDE (pol-i-PEP-tide). A chain of linked amino acids.

POLYPHYLETIC (pol-i-fy-LET-ik) [Gr. many + tribe]. Derived from different ancestors, said of groups of similar organisms that are believed not to have evolved from a common stock.

POLYPLOIDY (POL-i-ployd-ee) [Gr. many + folds]. The condition of more than two sets of chromosomes (genomes) in a single nucleus.

POLYSOME (POL-i-sohm). A string of five or more ribosomes linked together by a strand of mRNA during active protein synthesis; a polyribosome.

POSTERIOR. Situated in back of or in the back part of an organ, structure, or organism.

PREFORMATION. A hypothesis of development of new organisms which states that a fertilized egg contains a fully developed miniature animal which merely needs the nourishment of a female uterus for it to grow into a normal-sized baby.

PRIMARY TISSUE. Cells derived from the apical meristems in plants.

PRIMITIVE STREAK [L. first (in point of time)]. A strip of ectodermal thickening, developing on an embryo, marking the future longitudinal axis of the embryo.

PROBOSCIS (proh-BOSS-iss). Any elongated and usually tubular structure extending from the head of an animal; an elongated nose.

PROGESTERONE (proh-jes-tuh-rone) [L. before + to bear, carry]. One of the female sex hormones, produced by the corpus luteum. Together with estrogen, it regulates the menstrual cycle and maintains pregnancy.

PROGLOTTID (proh-GLOT-id) [L. before + the tongue]. A body segment of a tapeworm.

PROKARYOTE (pro-CARRY-ote) [L. before + Gr. nucleus]. A cell that has no membrane-delimited nucleus and no membrane-delimited organelles. Its chromosomes are "naked" DNA. (*See* EUKARYOTE)

PROLACTIN (proh-LAK-tin) [Gr. before + L. milk]. A female sex hormone, secreted by the pituitary gland, which allows mammary glands to secrete milk after childbirth.

PROPAGULE (PROP-uh-gyul) [L. plant cutting]. Any reproductive body of an organism, such as a fertilized egg, seed, spore, or living vegetative part.

PROPHASE [L. before + Gr. appearance]. The first named stage in nuclear division, during which chromosomes shorten and thicken, and spindle fibers form.

PROPRIOCEPTOR (proh-pree-oh-SEP-ter) [L. one's own]. Sensory cell or organ sensitive to changing movements and position of one's own body.

PROSTAGLANDIN (pros-tuh-GLAN-din). One of a group of fatty-acid hormones whose effects include contraction of uterine muscle, inhibition of progesterone, secretion by the corpus luteum, and lowering of blood pressure.

PROTEIN [Gr. first]. Macromolecule composed of chains of amino acids linked by peptide bonds.

PROTOPLAST (PROH-toh-plast). The content of any living cell; also a plant cell whose external coats (walls and lamellae) have been digested away, leaving the plasma membrane naked.

PSEUDOCOELOM (soo-doh-SEE-lum) [Gr. false + cavity]. A body cavity that is not lined inwardly and outwardly with mesoderm, but that has instead a body wall with an outer ectoderm over a muscular mesoderm.

PTYALIN (TIE-uh-lin). The amylase in saliva.

PUBERTY (PEW-burr-tee) [L. adult]. The period during which sexual maturation is achieved.

PULSE. The systolic pressure that can be felt in an artery near the surface of the body.

PUPA (PEW-puh) [L. doll]. The stage in the development of some insects between the larva and the adult.

PURINE (PURE-een). A nitrogenous base composed of a double ring of carbon and nitrogen atoms; a constituent of DNA, RNA, ATP, NAD, etc., e.g., adenine, guanine.

PYRIMIDINE (py-RIM-uh-deen). A nitrogenous base composed of a single ring of carbon and nitrogen atoms; a constituent of DNA and/or RNA, e.g., thymine, cytosine, uracil.

PYRUVIC ACID (py-ROO-vik) [L. pear]. An organic acid, $C_3H_4O_3$, which is an intermediate metabolic product in cellular respiration.

QUANTASOME. A unit of structure, containing chlorophyll, on the surface of thylakoid membranes in chloroplasts; thought to be photosynthetic units.

RADICLE [L. root]. The morphological lower tip of the embryo in a seed. The first root grows from its tip.

RADIOIMMUNOASSAY (RIA) (RAY-dee-oh-im-myu-noh-ASS-ay). A technique that measures hormone levels in the blood.

RADULA (RAJ-uh-luh) [L. scraper]. A mouth part in some molluscs, used as a rasping organ.

RAY. In plant anatomy, a strip of cells streaking across the grain of wood from the pith toward the bark.

RECEPTOR. A structure that is capable of perceiving changing stimuli. Receptors are capable of changing energy into nerve impulses.

RECESSIVE (ree-SESS-iv). Pertaining to a gene that is masked phenotypically in the presence of its dominant allele.

RECOMBINANT DNA (ree-KOM-buh-nent). A segment of DNA enzymatically removed from one chromosome, united with a strand of DNA from a second organism (frequently a virus), then inserted into a new host cell, sometimes into the host chromosome.

RED BLOOD CORPUSCLE (ERYTHROCYTE) (eh-RITH-ruh-site). One of the formed elements in the blood, responsible for the transport of oxygen and carbon dioxide.

REDUCTION. (1) Chemical: a reaction in which electrons are added to the substance being reduced. (See OXIDATION) (2) Biological: the change in chromosome number from diploid to haploid.

REFLEX. A simple behavior pattern consisting of an automatic response to an external stimulus.

REFLEX ARC [L. bent back]. A group of two or three neurons regulating a fairly simple body movement without conscious participation of the brain.

REGENERATION. Regrowth of lost or damaged tissue or part of an organism.

REGULATOR GENE. A gene that provides codes for synthesis of repressor proteins.

REM. A unit of measurement of the energy from radioactive elements; an acronym for "roentgen equivalent man."

REM SLEEP (for RAPID EYE MOVEMENT). A transition stage of sleep when dreams occur.

RENIN (REE-nin). A hormone from the kidney, which is important in the regulation of sodium balance and also plays a role in water balance by stimulating the thirst reflex in the brain.

RENNIN (REN-in). An enzyme, secreted by mucosa of the stomach, that curdles milk.

REPOLARIZATION. A condition that results when the outside membrane of a nerve fiber returns to a positive charge and the inside of the membrane becomes relatively negative again.

REPRESSOR. The protein that can combine with and repress action of an associated operator gene.

RESPIRATION (res-pi-RAY-shun) [L. again + breathe]. (1) Breathing. (2) The cellular processes by means of which energy for biological use is released from food.

RESTING POTENTIAL. The small voltage (1/10 volt) that normally exists across the surface membrane of many cells, especially muscle and nerve. This voltage is due to an imbalance in the distribution of ions. It charges rapidly if the cell is touched or stimulated. (See ACTION POTENTIAL)

RETICULAR FORMATION (rih-TIK-yu-lar) [L. net]. A net of nerve cells in the thalamus that filters all incoming stimuli, discarding unimportant ones and sending others to the proper decoding centers in the brain.

RETINA (RET-in-uh) [L. net]. The light-sensitive layer of the eye. It contains the rods and cones and a layer of sensory neurons.

RHODOPSIN (roh-DOP-sin) [Gr. rose + sight]. The light-sensitive pigment of rod cells in the eye, consisting of a colorless protein, opsin, and a colored carotenoid molecule, retinal (or retinene).

RIBONUCLEIC ACID (RNA) (rye-boe-new-KLAY-ik). A single-stranded nucleic acid containing the sugar ribose. It is transcribed from the DNA template; found in both nucleus and cytoplasm.

RIBOSOMAL RNA (rRNA) (rye-boh-SOHM-ul). The ribonucleic acid that is a constituent of ribosomes.

RIBOSOME (RYE-boh-sohm). A cytoplasmic structure composed of ribonucleoprotein; the site of protein synthesis.

RITUALIZATION. Display movements that have been modified through evolution to become stereotyped communication signals.

ROD. A photoreceptor in the retina of the eye, containing light-sensitive pigment (rhodopsin).

ROOT CAP. A thimble-shaped mass of loosely connected cells protecting the root tip.

ROUND DANCE. Part of the "language of the bees," which consists of a movement of alternating clockwise and counterclockwise circles to indicate the direction, distance, and quality of a food source, or to replay information about a new hive location.

RUFFINI ENDING (roo-FEEN-ee) [Angelo Ruffini, 18th-century Italian anatomist]. A specialized receptor in skin that is sensitive to warmth.

RUMEN (ROO-men). An enlargement of the esophagus in cows and other grazing animals. Partial digestion of cellulose occurs there.

SARCOMERE (SAR-koh-meer) [Gr. flesh + part]. A single muscle unit extending from one Z line to the next.

SARCOPLASM (SAR-koh-plaz-um) [Gr. flesh + something molded]. Viscous cytoplasm of a muscle fiber.

SARCOPLASMIC RETICULUM (sar-koh-PLAZ-mik reh-TIK-yu-lum) [Gr. flesh + molded + L. a network]. A complex network of fine tubules and sinuses in the sarcoplasm of a muscle fiber.

SATURATED FAT. Fat with fatty acids whose carbon atoms have the maximum number of hydrogen atoms attached to them. All the carbon atoms are hydrogenated with no double bonds between adjacent carbons.

SAVANNA (seh-VAN-uh). Grassland with scattered trees.

SCROTUM (SCROH-tum). An external sac between the male thighs containing testes and parts of spermatic cords.

SECOND-MESSENGER CONCEPT. A mechanism, proposed for the control of the endocrine system, stating that a hormone, the "first messenger," causes the conversion of ATP to cyclic AMP inside a target cell. The cyclic AMP, the "second messenger," then diffuses throughout the cell and causes the cell to respond with its specific endocrine function.

SECRETIN (sih-KREET-in). A hormone, secreted by the epithelial cells of the duodenum, that stimulates secretion of pancreatic enzymes. Secretin was the first hormone to be discovered (in 1903).

SEED. A mature ovule of a seed plant, containing an embryo.

SEGMENTATION [L. cut]. The growth pattern in many animals in which there is serial repetition of parts, notably in earthworms, abdomens of insects, and backbones of vertebrate animals.

SEGREGATION, LAW OF. One of the fundamental principles of Mendelian genetics, which states that contrasting characters in an individual separate when gametes are formed, each gamete receiving the gene for only one of each pair of alleles.

SEMEN (SEE-men) [L. to sow]. A secretion of the male reproductive organs consisting of sperm cells and fluids secreted by the seminal vesicles and prostate gland.

SEMICIRCULAR EAR CANALS. Structure in the inner ear, important in maintaining balance.

SEMINAL VESICLE (SEM-uh-nul VES-uh-kul). The portion of the male reproductive system that secretes nutritive fluids for sperm.

SEMINIFEROUS TUBULE (sem-uh-NIF-uh-rus) [L. sperm + bearing]. A coiled tube, contained inside testes, that produces the sperm cells.

SEPTUM. A partition, as between nostrils, between heart compartments, or between cells.

SERE (SEER). In ecology, a space that has been completely cleared of living organisms.

SEX-LINKED. Determined by a gene on a sex chromosome.

SIEVE CELL (SIV). A non-nucleated, elongated plant cell in the phloem, with perforations in its ends and sides through which materials, especially water and nutrients, can pass.

SIGN STIMULUS. In animal behavior, a major signal that initiates a given movement or reaction; releaser.

SINOATRIAL (SA) NODE (sy-noh-AY-tree-ul). The specialized section of muscle tissue where the heartbeat originates.

SINUSOID (SIGH-nuss-oyd). Relating to a cavity or hollow space. A form of terminal blood channel consisting of irregularly branching vessels.

SKELETAL MUSCLE (STRIATED MUSCLE, VOLUNTARY MUSCLE). A "voluntary" type of striated muscle under the control of the somatic nervous system.

SLIDING-FILAMENT THEORY. A theory stating that muscles are shortened when the thin actin and thick myosin filaments slide past each other.

SMALL INTESTINE. That part of the alimentary canal between the stomach and large intestine, the site of most digestion and absorption of nutrients.

SMOOTH MUSCLE (VISCERAL MUSCLE). Long, tapered cells, with one nucleus in the fiber of each cell, which line internal organs of the body (but not the heart). Smooth muscle contains no striations and moves slowly.

SOCIAL RELEASER. In animal behavior, a sign from members of the same species that initiates a given movement or reaction.

SOCIETY. Organisms living together in organized groups.

SOFTWOOD. The wood of a coniferous tree.

SOLUTE (SOL-yute). Substance dissolved in a medium.

SOLVENT. The medium in which substances are dissolved.

SOMA (SOH-muh). In general, a body. In neurobiology, that part of a nerve cell containing the nucleus.

SOMATIC NERVOUS SYSTEM (soh-MAT-ik). Nervous system composed of nerve fibers that lead from the brain and spinal cord to the skeletal muscles. It involves "voluntary" movements of the skeletal muscles.

SOMATOTROPIN (GH) (soh-MAT-oh-TROH-pin) [Gr. body + turn, change]. A hormone, secreted by the anterior lobe of pituitary, that regulates the growth of the skeleton.

SPECIATION (spee-shee-AY-shun) [L. sort, kind]. The development of new species.

SPECIES (SPEE-sheez) [L. kind]. (1) A group of individuals capable of producing fertile offspring. (2) A taxonomic unit considered satisfactory to a specialist in biological classification.

SPERMATID (SPUR-muh-tid). One of four haploid cells formed by the meiosis of spermatocytes during spermatogenesis. A spermatid develops directly into a sperm cell.

SPERMATOZOA (spur-muh-tuh-ZOH-uh) [Gr. seed + animal]. The sperm cells.

SPICULE (SPIK-yule) [L. a spike]. A crystalline, skeletal element, made of siliceous material, in the middle cell layers of sponges.

SPINAL CORD. A part of the central nervous system that passes through the protective vertebrae, from which 31 pairs of spinal nerves branch to each side of the body. It is the connecting link between the brain and most of the body.

SPINDLE. The aggregation of microtubules associated with chromosomes during nuclear division.

SPINNERET (spin-uh-RET). An appendage at the tip of the abdomen that is capable of extruding a silky filament, which is used for web-building by spiders.

SPIRACLE (SPEER-uh-kul) [L. to breathe]. A breathing hole along the sides of an insect abdomen, allowing for oxygen and carbon dioxide exchange.

SPONTANEOUS GENERATION. A theory that life can arise from nonliving matter.

SPORANGIUM (spuh-RAN-jee-um) [Gr. seed vessel]. Any plant part that produces spores.

SPORE [Gr. seed]. A reproductive cell in plants; in general, any nonsexual reproductive cell.

SPOROPHYLL (SPOHR-uh-fil) [Gr. seed + leaf]. A spore-bearing leaf.

SPOROPHYTE (SPOHR-uh-fite). The spore-producing, diploid phase of plants having alternation of generations.

STAMEN (STAY-men) [L. thread]. The pollen-producing organ in a flower, consisting of the pollen sac (anther) and usually a supporting filament.

STENOTHERMAL (sten-oh-THER-mul) [Gr. narrow + heat]. Having a limited tolerance for temperature extremes.

STERILE (STAIR-ul). (1) Incapable of reproduction. (2) Devoid of life (culture media, microbiological technique). (3) Sustaining little or no growth (soils).

STERNUM (STIR-num) [Gr. breastbone]. The breastbone.

STOLON (STOH-lun) [L. shoot]. A horizontal stem that grows along the ground and forms adventitious roots and shoots, as in a strawberry plant.

STOMA (STOW-muh) [Gr. mouth]. A microscopic opening between epidermal cells on leaf and stem surfaces, surrounded by a pair of guard cells.

STROMA (STROH-muh). [L. bed covering]. Nongrana regions of the chloroplasts. The sites of the "dark reactions" of photosynthesis; any relatively unorganized part or tissue.

STRUCTURAL GENE. A gene that produces mRNA and is therefore instrumental in the production of polypeptides. It functions in the synthesis of enzymes.

SUBSTRATE. Molecule on which an enzyme works.

SUCCESSION. In ecology, a series of species, each replacing the former, until a self-perpetuating community is established.

SYMBIOSIS (sim-bee-OH-sis) [Gr. with + life]. The close association of more than one species of organisms living together in some special physical or nutritional relationship.

SYMPATHETIC SYSTEM [Gr. with + feeling]. A part of the autonomic nervous system that controls metabolic activities. Its action is mainly stimulating.

SYMPATRIC SPECIES (sim-PAT-rik) [Gr. with, together + L. country]. Closely related species that occupy common geographical regions.

SYNAPSE (SIN-aps) [Gr. union]. The junction between the axon terminal of one neuron and the dendrite or cell body of the next neuron.

SYNAPSIS (si-NAP-sis) [Gr. union]. The pairing of homologous chromosomes during prophase I of meiosis.

SYNDROME (SIN-drohm). A group of symptoms and signs that, taken together, characterize a condition or disease.

SYNECOLOGY (sin-i-KOL-uh-jee) [Gr. together + ecology]. The study of interactions of organisms with their environment and with each other, with emphasis on populations and communities. (See AUTECOLOGY)

SYNERGISM (SIN-er-jiz-um). The enhanced effect of two or more factors working together. The final result is greater than it would be if either factor worked alone.

SYNOVIAL MEMBRANE (si-NOH-vee-ul). Special membrane covering bones at joints.

SYPHILIS (SIF-i-lis). A venereal disease, caused by the bacterium *Treponema pallidum*, contracted usually by sexual intercourse. It causes paralysis, insanity, and death.

SYSTOLE (SIS-toe-lee) [Gr. drawing together]. The ventricular contraction of the heart, by which the blood is forced out of the ventricle.

TAIGA (TIE-guh) [Rus.]. The cold, marshy, evergreen forest regions of the northern hemisphere.

TAXIS (TACK-sis) [Gr. arrangement]. A movement, usually by microorganisms, toward or away from a stimulus such as light, heat, or gravity.

TAXON (TACKS-on) [Gr. arrangement]. A category of classification of organisms.

TAXONOMY (tack-SON-uh-mee) [Gr. order + law]. The study of classification, especially with the aim of arranging organisms in ways that show their evolutionary relationships.

TELOPHASE (TELL-uh-faze) [Gr. end + phase]. The final stage in nuclear division, in which two daughter nuclei are formed; usually followed by cytoplasmic division.

TENDON [Gr. to stretch]. A fibrous band that connects a muscle to a bone.

TERRITORY. In animal behavior, a specific area guarded against intrusion by other members of the same species.

TEST [L. jug]. Any hard protective shell; especially the shells of sea urchins and certain protozoa.

TEST CROSS. A cross between an experimental organism and another with recessive phenotype to determine the genotype of the individual with dominant phenotype.

TESTES (TESS-teez) [L. jug]. Male gonads that produce sperm cells.

TESTOSTERONE (tess-TAHSS-tuh-rohn). The primary androgenic hormone that affects the production of spermatozoa, the development of sex organs, and the appearance of secondary sex characteristics.

TETRAD (TEH-trad) [Gr. four]. The complex of four chromatids during the pairing of homologous chromosomes during prophase I of meiosis; any quartet of objects, such as four microspores.

THALAMUS (THAL-uh-mus). A part of the brain, below the cerebrum, which acts as a monitor to all incoming and outgoing impulses of the central nervous system and modifies some of them.

THERMAL POLLUTION. A kind of nonmaterial pollution that occurs when streams are used to cool large industrial activities.

THERMOCLINE (THER-muh-kline) [Gr. warm + incline]. A middle layer of water in a lake, between the upper warm epilimnion and the lower cold hypolimnion.

THERMODYNAMICS (THER-moh-dy-NAM-iks) [Gr. heat + movement]. The study of energy relations in biological, chemical, and physical actions.

THERMORECEPTOR. Sensory cell or organ that is sensitive to changing temperature.

THRESHOLD. A minimum stimulus required to instigate a nerve impulse.

THYLAKOID (THIGH-luh-koyd) [Gr. sack, pouch]. A disklike, photosynthetic membrane-bound vesicle in a chloroplast.

THYMINE (THIGH-meen). A pyrimidine, a nitrogenous base that is a constituent of DNA.

THYMUS (THIGH-mus). A lymphoid gland, located under the breastbone, active during childhood. Its main function in humans may be the development of the immunological system.

THYROID. A gland in the neck that secretes the hormones thyroxine, which controls the rate of metabolism and growth, and thyrocalcitonin, which lowers the calcium level in the blood.

THYROXINE (thigh-ROCK-seen). A thyroid hormone, consisting of an amino acid containing iodine, which controls the rate of body metabolism and growth.

TISSUE. A mass of cells all similarly differentiated structurally and functionally.

TOTIPOTENCY (tote-i-POE-ten-see) [L. all + powerful]. The ability of a cell (or nucleus) to develop (or regulate the development) into any kind of differentiated cell; having complete genetic ability.

TRACHEA (TRAY-kee-uh) [Gr. rough]. In vertebrates, the windpipe. In plants, the open-ended water-conducting cell in wood. In insects and some other land-dwelling arthropods, an air tube in the body.

TRACHEID (TRAY-kee-ud) [Gr. rough]. A thickened and elongated water-conducting cell with closed tapered ends, occurring mainly in softwoods.

TRANSCRIPTION [L. across + to write]. Process whereby messenger RNA is synthesized from the DNA template.

TRANSDUCTION [L. to lead across]. (1) The ability of all receptor cells to change received stimuli into nerve impulses. (2) The genetic change in a bacterium, induced by introduction of foreign DNA via a virus carrier.

TRANSFER RNA (tRNA). A form of ribonucleic acid that serves as an adapter molecule in the synthesis of proteins. Each tRNA can combine with a particular amino acid and place it in its proper position in a growing polypeptide.

TRANSFORMATION. Phenotypic change in a bacterium caused by the passage and incorporation of DNA directly from one type of bacterium to another.

TRANSLATION. The process by which a protein molecule is synthesized from amino acids according to the specific nucleotide sequence of the mRNA molecule.

TRANSLOCATION [L. through + place]. (1) The transfer of a piece of one chromosome to another part of the same chromosome or to a different chromosome. (2) Movement of material from one part of an organism to another.

TROPHALLAXIS (troh-fuh-LAK-sis) [Gr. food + exchange]. The exchange of food and pheromones through regurgitation. In insect societies it is essential for communication.

TROPHIC LEVEL (TROH-fik). In ecology, the number of steps a consumer is removed from the primary producers. A vegetarian is a first-level consumer; a carnivore is higher.

TROPHOBLAST (TROH-fuh-blast) [Gr. food + sprout]. The growing mass of cells on one side of an embryo, which will give rise to the placenta and the extraembryonic membranes.

TROPISM (TROH-piz-um) [Gr. a turning]. A growth movement whose direction is determined by the direction from which the stimulus comes.

TUBULAR REABSORPTION. The process in the kidney that returns useful substances to the blood by active transport after those substances have been initially lost from the blood during glomerular filtration.

TUBULAR SECRETION. The process in the kidney that secretes waste products into the renal fluid as it passes through the nephric tubules.

TUNDRA (TUN-druh) [Rus.]. The cold, treeless plains of the northern hemisphere, with permanently frozen subsoil.

TURNER'S SYNDROME. A genetic disorder in which there is only one sex chromosome, the XO condition. The afflicted individual is female but has abnormal body structure.

TYMPANUM (TIM-puh-num) [Gr. drum]. The eardrum.

UMBILICAL CORD (um-BIL-i-kul) [L. the navel]. A cord that carries wastes from the embryo to the placenta, and oxygen and nutrients from the placenta to the embryo.

UNSATURATED FAT. Fat whose constituent fatty acids contain some carbon atoms joined by double bonds, instead of having two hydrogen atoms each.

URACIL (YUR-uh-sill). A pyrimidine, a nitrogenous base that is a constituent of RNA.

URETER (YUR-et-uh) [Gr. to urinate]. A tube carrying urine from the kidney to the urinary bladder.

URETHRA (yu-REE-thruh) [Gr. to urinate]. The canal that conveys urine from the bladder to the outside of the body. It also transports sperm outside the male body.

UTERUS (YU-tuh-rus) [L. womb]. The womb; the pear-shaped organ located just beyond the vagina, which houses, nourishes, and protects the developing fetus within its muscular walls.

VACUOLE (VAK-you-ole) [L. empty]. A membrane-bound vesicle, located in the cellular cytoplasm, containing mostly water with salts, proteins, crystals, and pigments.

VAGINA (vuh-JINE-uh). A muscle-lined tube that serves to receive sperm from the penis during sexual intercourse and to allow the fetus to pass down from the uterus during childbirth.

VASOPRESSIN (ADH) (vay-zoh-PRESS-in) [L. vessel + lower]. A hormone, produced by the hypothalamus and released by the posterior lobe of the pituitary, that aids water reabsorption in the kidneys and stimulates the smooth muscles of the arteries.

VEIN. A vessel that carries blood toward the heart.

VENEREAL (vuh-NEER-ee-ul) [L. myth. *Venus*, goddess of love]. Relating to or resulting from sexual intercourse. Examples of venereal diseases are syphilis and gonorrhea.

VENTRAL. Pertaining to a position toward the belly surface of an organ, structure, or organism. (*See* DORSAL)

VENTRICLE (VEN-tri-kul) [L. stomach]. A cavity in an organ, such as the heart chamber posterior to the atrium, or the cavities in the brain.

VESTIGIAL (ves-TIJ-i-ul) [L. trace]. Existing in a condition reduced from that of an earlier time. In a morphological sense, a part that has been diminished through evolution from a once useful condition; e.g., human tail bones, scale leaves of cacti.

VILLI (VILL-eye) [L. shaggy hair]. Fingerlike projections from a free surface of a membrane, as on the lining of the small intestine or the chorionic membrane of the placenta.

VITALISM [L. life]. An explanation of biological phenomena holding that life has a force and quality of its own, something different and apart from the forces and mechanisms of the inanimate world.

VITAMIN. An organic substance present in minute amounts in food, essential in supporting the work of enzymes, especially those involved in cellular respiration.

VIVIPAROUS (vi-VIP-uh-rus) [L. alive + to bring forth, produce]. Organisms that nourish the young within the uterus until they are born alive.

WAGGLE DANCE. Part of the "language of the bees," which consists of a movement of a figure eight with a straight line between the loops to indicate the direction, distance, and quality of a food source, or to relay information about a new hive location.

WARNING COLORATION. Conspicuous coloration that helps to warn predators of poisonous or other objectionable features.

WHITE BLOOD CORPUSCLE (LEUKOCYTE). One of the formed elements of the blood, serving as scavengers and immunizing agents.

XEROPHYTIC (zeer-oh-FIT-ik) [Gr. dry + plant]. Capable of surviving with relatively little water; used to describe such desert plants as cacti.

XEROSERE (ZEER-oh-seer) [Gr. dry + Eng. barren]. A dry region, cleared of organisms, open for invasion, such as a rocky outcrop.

XYLEM (ZYE-lem) [Gr. wood]. The wood tissue through which most of the water and minerals of a plant are conducted.

ZYGOSPORE (zy-goh-spohr) [Gr. yoke + pair + seed]. A thick-walled spore developing from the sexual fusion cells in some algae and fungi.

ZYGOTE (zy-goat) [Gr. yoked together]. The cell formed by the union of a sperm and an egg.

Metric Conversion Factors

ENGLISH TO METRIC   **METRIC TO ENGLISH**

Weight (Mass)

Ounces $\times$ 28.3495 = grams	Grams $\times$ .03527 = ounces
Pounds $\times$ .4536 = kilograms	Kilograms $\times$ 2.2046 = pounds
Tons $\times$ 1.1023 = tonnes (1000 kg)	Tonnes $\times$ .9072 = tons

Length

Inches $\times$ 25.4 = millimeters	Millimeters $\times$ .0394 = inches
Inches $\times$ 2.54 = centimeters	Centimeters $\times$ .3937 = inches
Feet $\times$ .3048 = meters	Meters $\times$ 3.2809 = feet
Yards $\times$ .9144 = meters	Meters $\times$ 1.0936 = yards
Miles $\times$ 1.6093 = kilometers	Kilometers $\times$ .621377 = miles

Area

Square inches $\times$ 6.4515 = square centimeters	Square centimeters $\times$ .155 = square inches
Square feet $\times$ .0929 = square meters	Square meters $\times$ 10.7641 = square feet
Square yards $\times$ .8361 = square meters	Square meters $\times$ 1.196 = square yards
Square miles $\times$ 2.59 = square kilometers	Square kilometers $\times$ .3861 = square miles
Acres $\times$ .405 = hectares	Hectares $\times$ 2.471 = acres
Square miles $\times$ 259.07 = hectares	Hectares $\times$ .00386 = square miles

Measure

Fluid ounces $\times$ 29.47 = milliliters	Milliliters $\times$ .0339 = fluid ounces
Quarts $\times$ .9433 = liters	Liters $\times$ 1.06 = quarts
Gallons $\times$ 3.774 = liters	Liters $\times$ .265 = gallons
Cubic feet $\times$ .02832 = cubic meters	Cubic meters $\times$ 35.3156 = cubic feet
Cubic yards $\times$ .7645 = cubic meters	Cubic meters $\times$ 1.308 = cubic yards

Pressure

lb/in.^{2} $\times$ .0703 = kg/cm^2	kg/cm^2 $\times$ 14.2231 = lb/in.^{2}

Temperature

When you know the Fahrenheit temperature:

$$°C = \frac{(°F - 32)}{1.8}$$

When you know the Celsius temperature:

$$°F = (1.8 \times °C) + 32$$

Photo Acknowledgments

Figure numbers precede the acknowledgments.

Prologue 2, Samuel Rushforth, Brigham Young University. 3(a), E. S. Beneke, Departments of Botany & Plant Pathology and Microbiology & Public Health, Michigan State University; 3(c), Runk/Schoenberger—Grant Heilman, Inc.

Chapter 1 Essay, p. 10, photo contributed by J. Herbert Taylor from an article originally published in *Genetics*, 1958. Essay, p. 26, T. Fujita, J. Tokunaga, and H. Inoue, *Atlas of Scanning Electron Microscopy in Medicine*, Igaku Shoin Ltd., Tokyo, 1971.

Chapter 3 3.1(b), The Bettmann Archive. 3.2(a), Don W. Fawcett; 3.2(b), Keith R. Porter, University of Colorado; 3.2(c), William Bloom and Don W. Fawcett, *A Textbook of Histology*, 10th ed., W. B. Saunders, Philadelphia, 1975 (micrograph courtesy of Dr. M. J. Karnovsky). 3.3(a), E. B. Small and D. S. Marszalek, courtesy of Gregory Antipa; 3.3(b), Paul E. Hargraves, University of Rhode Island. 3.5, Ralph W. G. Wyckoff, University of Arizona. 3.6, G. Cohen-Bazire. 3.7, courtesy of J. D. Robertson. 3.13, Peck-Sun Lin, New England Medical Center. 3.15, James G. Hirsch, Rockefeller University. 3.17, courtesy of W. A. Jensen, University of California, Berkeley. 3.18(a), Alexander Rich, M. I. T. 3.19(b), W. G. Whaley and H. H. Mollenhauer, Cell Research Institute, University of Texas at Austin. 3.20(c), H. Fernández-Morán, T. Oda, P. V. Blair, and D. E. Green. 3.21(a), Don W. Fawcett. 3.22(b), J. André. 3.23(b), Don W. Fawcett. Essay, p. 56 (top), Ellen Roter Dirksen, University of California, Los Angeles; (bottom right), A. D. Dingle, McMaster University.

Chapter 4 4.6(d,e), L. K. Shumway. 4.7(a), from S. B. Carpenter and N. D. Smith, 1975, "The Hidden World of Leaves." *Amer. For.* 81(5):28–30. Courtesy of Dr. S. B. Carpenter, Deparment of Forestry, University of Kentucky, Lexington.

Chapter 6 6.1, Rockefeller University Archives. 6.5(e,f), courtesy of Lee D. Simon, Waksman Institute of Microbiology, Rutgers University, New Brunswick, NJ. 6.7, The Bettmann Archive. 6.8, R. E. Franklin and R. Gosling, *Nature* 171(1953):740. 6.12, Philip Sharp, M. I. T.

Chapter 7 7.1, Peck-Sun Lin, New England Medical Center. 7.2, Eleanor Noback. 7.4(e), O. L. Miller, Jr., University of Virginia. Essay, p. 107, Hewson Swift and B. J. Stevens. 7.8(b), Alexander Rich, M. I. T.

Chapter 8 8.2(b), William T. Jackson, Dartmouth College. 8.9, J. B. Gurdon. Essay, p. 132, Martin Rotker/Taurus Photos.

Chapter 9 9.1(a), courtesy of Carl Zeiss, Inc., New York. 9.2, courtesy of A. K. Kleinschmidt. 9.3(b), H. Fernández-Morán, T. Oda, P. V. Blair, and D. E. Green. 9.4, courtesy of K. Maramorosch, Rutgers University. 9.5, courtesy of S.T.E.M. Laboratories, Kansas City, MO, and Fisher Scientific Company, Chicago. 9.6, courtesy of *Journal of Bacteriology* 132(3):950 (December 1977), Roger M Cole. 9.7, Phillipp Gerhardt, J. Stuart Pankratz, and René Scherrer in *Applied and Environmental Microbiology* 32(1976):438–439. 9.8, courtesy of the New York Public Library—Astor, Lenox, and Tilden Foundations. 9.9, Raymond C. Valentine, University of California, Davis. 9.10, William T. Hall, Ph.D., Electro-Nucleonics Laboratories, Inc. 9.13(d), courtesy of S.T.E.M. Laboratories, Kansas City, MO, and Fisher Scientific Company, Chicago. 9.16(b) and 9.17, Samuel R. Rushforth, Brigham Young University. 9.18, E. B. Small, courtesy of Gregory Antipa. 9.20(a) and 9.21(b), E. E. Butler and L. J. Petersen. 9.22(a), courtesy of R. G. Kessel and C. Y. Shih, *Scanning Electron Microscopy in Biology*, Springer-Verlag, Berlin, 1974. 9.23, Hal H. Harrison/Grant Heilman, Inc. 9.24(a), Grant Heilman. 9.25, J. M. Conrader/Photo Researchers, Inc. 9.26, Samuel Rushforth, Brigham Young University. 9.27, Jack Brotherson, Brigham Young University. 9.28(a,d), Jeremy Pickett-Heaps, University of Colorado; 9.28(b), Runk/Scheonberger—Grant Heilman, Inc.; 9.28(c), Andrew Staehelin, University of Colorado. 9.29, Professor W. Rauh, Institute of Systematic Botany, Heidelberg University. 9.30, Jack Dermid. 9.31, James

V. Allen, Brigham Young University. 9.33(a), Professor W. Rauh, Institute of Systematic Botany, Heidelberg University; 9.33(b) Hal H. Harrison/Grant Heilman, Inc.; 9.33(c) Jack Dermid. 9.35(b), T. E. Weier, C. R. Stocking, and M. G. Barbour, *Botany: An Introduction to Plant Biology*, 5th ed., © 1974, John Wiley & Sons, Inc. 9.36(a), U. S. Forest Service Photo; 9.36(b), Jack Dermid; 9.36(c), Verna R. Johnston/Photo Researchers, Inc. 9.37, courtesy C. L. Wilson (Wilson, Loomis, and Steeves, *Botany*, 5th ed., Holt, Rinehart and Winston). 9.38, Grant Heilman. 9.39, U. S. Forest Service Photo. 9.40, Edward Hacskaylo, U. S. Department of Agriculture. 9.41(a), U. S. Forest Service Photo; 9.41(b), Verna R. Johnston/Photo Researchers, Inc. 9.42, Grant Heilman.

Chapter 10 10.10, Grant Heilman. 10.11, 10.13–10.16, 10.19, 10.20, 10.22, Runk/Schoenberger—Grant Heilman, Inc. 10.23, Grant Heilman. 10.24, V. J. Okulitch. 10.25, Leonard Lee Rue III/Photo Researchers, Inc. 10.26(a), Runk/Schoenberger—Grant Heilman, Inc.; 10.26(b), Gordon Smith/Photo Researchers, Inc. 10.27(a,c), Runk/Schoenberger—Grant Heilman, Inc.; 10.27(b), Jack Dermid. 10.29, Hal H. Harrison/Grant Heilman, Inc. 10.31, Russ Kinne/Photo Researchers, Inc. 10.32(a,b), Hal H. Harrison/Grant Heilman, Inc.; 10.32(c,d,e), Grant Heilman. 10.33, Copyright © Sdeaurd C. Bisserot/Bruce Coleman, Inc. 10.34, Jerome Wexler/Photo Researchers, Inc. 10.35, S. Rannels/Grant Heilman, Inc. 10.37, Stephen Dalton/Photo Researchers, Inc. 10.38, Runk/Schoenberger—Grant Heilman, Inc. 10.39, John Hendry, Jr./Photo Researchers, Inc. 10.40, Runk/Schoenberger—Grant Heilman, Inc. 10.42, Russ Kinne/Photo Researchers, Inc. 10.43, Runk/Schoenberger—Grant Heilman, Inc. 10.45, Ron Church/Photo Researchers, Inc. 10.46, Thompson/Annan Photo Features. 10.48, courtesy of the American Museum of Natural History. 10.49(a), Hal H. Harrison/Grant Heilman, Inc.; 10.49(b), Karl H. Maslowski/Photo Researchers, Inc. 10.50(a), Hal H. Harrison/Grant Heilman, Inc.; 10.50(b), Dade W. Thornton/Photo Researchers, Inc. 10.52, Allan D. Cruickshank/Photo Researchers, Inc. 10.53, Grant Heilman. 10.54, Graham Pizzey/Bruce Coleman, Inc. 10.55, Gordon Smith/Photo Researchers, Inc. 10.56, Lynwood M. Chace. 10.57, Eric Hosking/Photo Researchers, Inc. 10.58, Scott Barry/Tom Stack & Associates. 10.59, Leonard Lee Rue III/Photo Researchers, Inc. 10.61, Richard Ellis, Ocean Research and Education Society. 10.62, Edgar Mönch/Photo Researchers, Inc. 10.63, Kristina Cannon-Bonventre/Anthro Photo. 10.65, courtesy of Dr. Dorothy Warburton, Columbia University, New York.

Chapter 11 11.1 The Bettmann Archive. 11.9, V. A. McKusick.

Chapter 12 12.1, 12.6, The Bettmann Archive. 12.13(a), A. M. Winchester. 12.19, Culver Pictures. 12.20, C. Stern, W. R. Centerwall, and S. S. Sarkar, *American Journal of Human Genetics*, vol. 16, 1964, University of Chicago. 12.22, courtesy of the President's Committee on Mental Retardation.

Chapter 13 13.2, Henry M. Reiswig. 13.9(a), Landrum B. Shettles. 13.10, Don W. Fawcett. 13.14, from *A Textbook of Histology*, 10th ed., by William Bloom and Don W. Fawcett, W. B. Saunders, 1975. 13.16, Landrum B. Shettles. 13.22, 13.23, E. S. Hafez, M.D. 13.24, Mia Tegner. 13.25(a) Mia Tegner and David Epel; 13.25(b), courtesy of Dr. Gerald Schatten, Department of Biological Science, University of Florida. 13.26, A. M. Winchester. Essay, p. 284, Clement L. Markert.

Chapter 14 14.1, Countway Library, from *Essai de Droptrique*, by Nicolaas Hartsoeker, Paris, 1694. 14.6, Marc Anthony DeGruccio. 14.13(a), reproduced with permission from *The Birth Atlas*, published and copyrighted by Maternity Center Association, New York; 14.13(b), © Marshall Henrichs. Essay, p. 302, Leonard McCombe, *Life* Magazine © 1971, Time, Inc.

Chapter 15 15.1, micrograph courtesy of F. D. Hess. 15.3, Copyright S. C. Bisserot/Bruce Coleman, Inc. 15.7, Dr. Charles C. Brinton, Jr., and Judith Carnahan. 15.30, photos courtesy of J. Heslop-Harrison, University of Wales, Welsh Plant Breeding Station. 15.35(b), Grant Heilman.

Chapter 16 16.7, Triarch, Inc. 16.8(a), from *Laboratory Manual for Plant Anatomy* by R. A. Popham; 16.8(b), micrograph courtesy of F. D. Hess. 16.9, 16.11, from *Laboratory Manual for Plant Anatomy* by R. A. Popham. 16.12, 16.15, 16.16, Triarch, Inc. 16.18, John Murphy, from *Joy of Nature*, copyright © 1977, Reader's Digest Association, Inc. 16.21, micrograph, courtesy of F. D. Hess. 16.23, 16.29, from *Laboratory Manual for Plant Anatomy* by R. A. Popham.

Chapter 17 17.2, James E. McMurtrey, Jr., Tobacco Laboratory, USDA. 17.5, M. H. Zimmermann. Essay, p. 368, P. V. Ammirato, from *Plant Physiology* 59:579–586, 1977. 17.7(b), Herbert Gehr, *Life* Magazine © 1948, Time, Inc. 17.17, courtesy of C. H. Muller, from *Bulletin of the Torrey Botanical Club*, v. 93(5), pp. 334–335, 1966.

Chapter 18 18.4, Photo Trends. 18.5, Clinical Pathological Conference, *American Journal of Medicine* 20:133, 1956. 18.7, Armed Forces Institute of Pathology, Neg. No. 62–10895. 18.8, Armed Forces Institute of Pathology Neg. No. 69–3573. 18.10, T. C. Fitzgerald/The Picture Cube. Essay, p. 394 (left), Photo #KNC 23425 in the John F. Kennedy Library; (right, top and bottom), The Bettmann Archive. 18.12, Copyright © 1965 CIBA Pharmaceutical Company, Division of CIBA-GEIGY Corporation, reproduced with permission of CIBA from the *Collection of Medical Illustrations* by Frank Netter, M.D. All rights reserved.

Chapter 19 19.3(a), from Cleveland P. Hickman, Sr., Cleveland P. Hickman, Jr., and Frances M. Hickman, *Integrated Principles of Zoology*, 5th ed., St. Louis, 1974, The C. V. Mosby Co. 19.7, from E. R. Lewis, T. E. Everhart, and Y. Y. Zeevi, *Science* 165: 1140–1143, 1969; copyright 1969 by the American Association for the Advancement of Science.

Chapter 20 20.9(a), Thomas Eisner, Cornell University. 20.12(a), E. R. Lewis, University of California; 20.12(b), E. R. Lewis, Y. Y. Zeevi, and F. S. Werblin, *Brain Research* 15:559–562, 1969. 20.15(a), Jen and Des Bartlett/Photo Researchers Inc.; 20.15(b), Arthur W. Ambler, NAS Photo/Photo Researchers Inc. 20.17(a), Pasquale P. Graziadei, Florida State University. 20.21, scanning electron micrographs by Pamela Henry and Robert E. Preston, courtesy of Joseph E. Haekins, Kresge Hearing Research Institute. 20.25, photos courtesy of Peter N. Witt, State of North Carolina, Mental Health Research.

Chapter 21 21.1, Andreas Feininger, *Life* Magazine © 1951, Time Inc., photo courtesy of Don Fawcett. 21.8, H. E. Huxley, *Journal of Molecular Biology* 37:507–520, 1968; copyright by Academic Press, Inc., London, Ltd. 21.10, © Lennart Nilsson, *Behold Man*, Little, Brown and Company, 1974.

Chapter 22 22.8, Dr. Jeanne Riddle, Henry Ford Hospital, Detroit. 22.9(b), Courtesy Keith R. Porter.

Chapter 23 23.8(a), © Leonard Lee Rue III/Animals, Animals; 23.8(b), Dr. E. R. Degginger.

Chapter 24 24.3(b), Dr. H.-R. Duncker, Zentrum für Anatomie und Cytobiologie der Justus-Liebig-Universität, Giessen. 24.6, from C. Kuhn III and E. H. Finke, *Journal of Ultrastructure Research* 38:161–173, 1972. 24.7, from Curtis Port and Irene Corvin, "Ciliated and Non-ciliated Epithelial Cells from Hampster Trachea," *Science* 177: cover, Sept. 22, 1972; copyright 1972 by the American Association for the Advancement of Science. 24.11, Pasquale P. Graziadei, Florida State University. 24.12, Patricia N. Farnsworth. 24.13, N. R. Rodman, Department of Pathology, West Virginia University. 24.18, The Bettmann Archive. Essay, p. 494, Warren Thayer. 24.21, Thomas Eisner, Cornell University. 24.25(a), Dr. Aaron Polliack; 24.25(b), American Optical Corporation.

Chapter 25 25.1, courtesy of the New York Public Library—Astor, Lenox, and Tilden Foundations. 25.2, The Bettmann Archive. 25.3, courtesy of the Burndy Library, Burndy Corporation. 25.5, 25.6, The Bettmann Archive. 25.7, Arthur Ambler, NAS Photo/Photo Researchers, Inc. 25.13, Copyright © Han Reinhard/Bruce Coleman, Inc. 25.14, Biological Sciences Curriculum Study, *Molecules to Man*, p. 58. Copyright © 1963 by the American Institute of Biological Sciences. Reprinted by permission of Houghton Mifflin Co.

Chapter 26 26.1, from W. Landauer and T. K. Chang, "The Ancon or Other Sheep," *Journal of Heredity* 40:105–112, 1949. 26.4, Dr. H. B. D. Kettlewell, Oxford University. 26.5, courtesy of Field Museum of Natural History, Chicago. 26.6, Edward S. Ross. 26.7, Lincoln P. Brower. 26.8, © A. M. Ambler, NAS Photo/Photo Researchers, Inc. 26.9, adapted from *Plant Systematics* by Otto T. Solbrig. 26.10(a), Grant Heilman; 26.10(b), John Colwell/Grant Heilman, Inc. 26.11, © Richard C. Finke, NAS Photo/Photo Researchers, Inc. 26.13, Bodger Seeds. 26.15(a,b), © Andrew Brilliant; 26.15(c,d), Grant Heilman. 26.16, Dale A. Zimmerman. 26.17, The Bettmann Archive.

Chapter 27 27.1, The Bettmann Archive. 27.4, photograph by Eliot Elisofon, Museum of African Art, Eliot Elisofon Archives. 27.5, Wide World. 27.6(a), courtesy of Dr. Cyril Ponnaperuma; 27.6(b), courtesy of Dr. Walther Stoekenius. 27.7(a), United Press International Photo.

Chapter 28 28.1 The Bettmann Archive. 28.3, Dr. Edwin R. Willis and Dr. Louis M. Roth. 28.4, © M. W. F. Tweedie, NAS Photo/Photo Researchers Inc. 28.6(a), © Norman Myers/Bruce Coleman Inc.; 28.6(b) Edward S. Ross. 28.7, Copyright © 1974 C. Haagner/Bruce Coleman Inc. 28.10(a), Copyright © Bruce Coleman. 28.10(b), Copyright © Jeff Foott/Bruce Coleman Inc.; 28.10(c), Baron Hugo van Lawick, © National Geographic Society; 28.10(d), Copyright © Norman Myers/Bruce Coleman Inc. 28.11, Lee Boltin with Jane and Lincoln P. Brower. 28.13, Nina Leen,

Life Magazine © 1964, Time Inc. 28.14, Lilo Hess. 28.16, © Dr. R. A. Steinbrecht. 28.17, Andreas Feininger, *Life* Magazine © 1952, Time Inc. 28.18, F. A. Webster. 28.21, courtesy of James Lloyd, University of Florida.

Chapter 29 29.2, Ted Grant/National Film Board of Canada, Photothèque. 29.4, Leonard Lee Rue III, NAS Photo/Photo Researchers, Inc. 29.6, Valerius Geist. Essay, p. 577, Edward S. Ross. 29.8, from *The Study of Instinct* by N. Tinbergen, Oxford University Press, London, 1951. 29.9, Edward S. Ross. 29.10, Copyright © Fritz Pölking/Bruce Coleman, Inc. 29.11, drawing by Bill Ogden. 29.12, courtesy of the American Museum of Natural History. 29.13, Copyright © Oxford Scientific Films/ Bruce Coleman, Inc. 29.14, Edward S. Ross. 29.15, from *Primate Behavior: Field Studies of Monkeys and Apes* by Hall and DeVore, Holt, Rinehart and Winston, New York and London, 1965. 29.16, Popperfoto, Paul Popper, Ltd. 29.17 © Frank Stevens, NAS Photo/Photo Researchers, Inc.; 29.18, Copyright © Norman Myers/Bruce Coleman, Inc.

Chapter 30 30.1, Hal Harrison/Grant Heilman, Inc. 30.2, Grant Heilman. 30.3, Laurence Lowry. 30.4, © Jen and Des Bartlett/Photo Researchers, Inc. 30.5, FAO photo by T. Mamphuis. 30.6, LeRoy Woodson/EPA Documerica. 30.7, Rutgers University, Plant Pathology Department, Cook College.

Chapter 31 31.2, Shostal Associates, Inc. 31.6, FAO photo by courtesy of the Government of Kenya; issued by FAO. 31.10, Grant Heilman. 31.15, Al Lowry/Photo Researchers, Inc.

PORTFOLIOS

I. CELLULAR DIFFERENTIATION 1 (top), Pasquale P. Graziadei, Florida State University; (bottom), scanning electron micrograph by Ch. Brucher/F. Spineli, CIBA/GEIGY Ltd., Basle, Switzerland. 2 (top), Dr. Aaron Polliack; (middle), Tsuneo Fujita, M.D., Niigata School of Medicine; (bottom left), courtesy of J. Gross, F. O. Schmitt, J. H. Highberger; (bottom right), J. Lawrence Katz, Rensselaer Polytechnic Institute, and Peter Frasca, Thomas Jefferson University. 3 (smooth muscle), Andrew P. Somlyo, M.D.; (heart muscle), Joachim R. Sommer, M.D.; (striated muscle), Andrew P. Somlyo, M.D.; (nervous tissue), W. Hild, *Zeitschr. f. Zellforsch.* 69:155, 1966. 4 (top), *Growth and Organization in Plants*, by Frederick C. Steward, Fig. 1–13; (middle), J. M. Tucker; (bottom left and middle), micrographs courtesy of F. D. Hess; (bottom right), Triarch Inc.

II. HOW A HUMAN EMBRYO DEVELOPS 1 (top), courtesy of Dr. Landrum B. Shettles; (four sequential photos), Carnegie Institution of Washington, Department of Embryology, Davis Division. 2 (top), courtesy of Dr. Roberts Rugh; (development of hand), Carnegie Institution. 3, 4 (top), courtesy of Dr. Roberts Rugh; 4 (bottom), from R. Rugh and L. Shettles, *From Conception to Birth: The Drama of Life's Beginnings*, New York, Harper and Row, 1971. 5–7, courtesy of Roberts Rugh. 8, drawing by Alexander Farquharson.

III. A SPECTRUM OF FLOWER COLOR 4 (hickory and maple), Al Bussewitz; (oak) George Sheng.

IV. WHAT ANIMALS SEE 1, Dr. E. R. Degginger. 2–3, Marshall Henrichs.

V. HOW ANIMALS MOVE 1, Animals, Animals/© Jerry Cooke. 2–3, Alan Rokach. 4 and 5 (top), © Frank Lane (A. Christiansen)/Bruce Coleman, Inc.; 4 and 5 (bottom), © Charles G. Summers, Jr./Bruce Coleman, Inc. 7, Animals, Animals/© Zig Leszczynski. 9, Isabelle Hunt Conant. 12 (left), © Russ Kinne/Photo Researchers, Inc.; (right), R. C. Snyder.

VI. HOW ANIMALS EAT 1 (cheetahs), Animals, Animals/© Fran Allan; (hyenas), © Norman Myers/Bruce Coleman, Inc. 2–3, © Norman Myers/ Bruce Coleman Inc. 4 (owl), Stephen Dalton/Natural History Photo Agency; (osprey), © Laura Riley/Bruce Coleman Inc. 5 and 6 (top), © Jen and Des Bartlett/Bruce Coleman Inc.; 6 (bottom), © Tom McHugh/Photo Researchers Inc. 7 (frog), Runk/Schoenberger—Grant Heilman, Inc.; (chameleon), © Alan Blank/Bruce Coleman, Inc.; (bee), © Treat Davidson, NAS Photo/Photo Researchers, Inc.; (okapi), © K. W. Fink/Bruce Coleman, Inc. 8 (tiger), © Frank W. Lane/Bruce Coleman, Inc.; (hippopotamus), © John Angell Grant/Bruce Coleman, Inc.; (rodent), © Russ Kinne/Photo Researchers, Inc.; (shark), Dr. E. R. Degginger.

VII. HOW ANIMALS AVOID PREDATORS 1, G. E. Schmida, F.P./ Bruce Coleman, Inc. 2 (larva), © Oxford Scientific Films/Bruce Coleman, Inc.; (grasshopper), © M. P. L. Fogden/Bruce Coleman, Inc.; (hawk moths), Carl W. Rettenmyer. 3 (top), © Alan Blank/Bruce Coleman, Inc.; (middle), Richard G. Zweifel, courtesy of American Museum of Natural History; (bottom), © Hans Reinhard/Bruce Coleman, Inc. 4, Mike Logan/Tom Stack & Associates.

PART OPENER PHOTO CREDITS 1, p. 6, E. Seling, Museum of Comparative Zoology, Harvard University. 2, p. 36, Dr. Aaron Polliack, from Polliack *et al.*, *Journal of Experimental Medicine* 138 (1973): 607. 3, p. 136, Dr. Barbara J. Panessa. 4, p. 224, Dr. J. Norman Grim, Northern Arizona University. Reprinted with permission from the June 1978 issue of *Bioscience* published by the American Institute of Biological Sciences. 5, p. 308, Dr. Barbara J. Panessa. 6, p. 382, scanning electron micrograph by Ch. Brucher/F. Spinelli, CIBA/GEIGY Ltd., Basle, Switzerland. 7, p. 504, G. D. Mazur and E. Seling, Museum of Comparative Zoology, Harvard University.

Index

Page numbers in italic type indicate illustrations. Roman numerals indicate coverage in portfolios, whose location is given in the table of contents.